D1475387

ASPARTAME

FOOD SCIENCE AND TECHNOLOGY

A Series of Monographs and Textbooks

1. Flavor Research: Principles and Techniques *R. Teranishi, I. Hornstein, P. Issenberg, and E. L. Wick (out of print)*
2. Principles of Enzymology for the Food Sciences *John R. Whitaker*
3. Low-Temperature Preservation of Foods and Living Matter *Owen R. Fennema, William D. Powrie, and Elmer H. Marth*
4. Principles of Food Science *edited by Owen R. Fennema*
   Part I: Food Chemistry *edited by Owen R. Fennema*
   Part II: Physical Methods of Food Preservation *Marcus Karel, Owen R. Fennema, and Daryl B. Lund*
5. Food Emulsions *edited by Stig Friberg*
6. Nutritional and Safety Aspects of Food Processing *edited by Steven R. Tannenbaum*
7. Flavor Research–Recent Advances *edited by Roy Teranishi, Robert A. Flath, and Hiroshi Sugisawa*
8. Computer-Aided Techniques in Food Technology *edited by Israel Saguy*
9. Handbook of Tropical Foods *edited by Harvey T. Chan*
10. Antimicrobials in Foods *edited by Alfred Larry Branen and P. Michael Davidson*
11. Food Constituents and Food Residues: Their Chromatographic Determination *edited by James F. Lawrence*
12. Aspartame: Physiology and Biochemistry *edited by Lewis D. Stegink and L. J. Filer, Jr.*

Other Volumes in Preparation

ASPARTAME
Physiology and Biochemistry

edited by
Lewis D. Stegink
L.J. Filer, Jr.
University of Iowa
College of Medicine
Iowa City, Iowa

MARCEL DEKKER, INC. New York and Basel

Library of Congress Cataloging in Publication Data

Main entry under title:
Aspartame: physiology and biochemistry.
(Food science; 11)
Includes index.
1. Aspartame–Metabolism. 2. Aspartame–Physiological effect. 3. Aspartame–Toxicology. 4. Aspartame–History. I. Stegink, Lewis D. II. Filer, L. J., Jr. III. Series: Food science (Marcel Dekker, Inc.); 11.
QP801.A84A84 1984 612'.01576 84-4297
ISBN 0-8247-7206-7

MARCEL DEKKER, INC.
270 Madison Avenue, New York, New York 10016

Current printing (last digit):
10 9 8 7 6 5 4 3 2 1

PRINTED IN THE UNITED STATES OF AMERICA

Preface

Aspartame may be the most thoroughly studied food additive ever approved by the U.S. Food and Drug Administration (FDA) in terms of research studies conducted prior to approval. Safety studies of aspartame were started in the late 1960s, when only the more classical toxicologic testing was required. These studies, however, extended into the 1970s, as more rigorous tests were required by both regulatory agencies and consumer groups. During these two decades, aspartame faced and survived a number of regulatory reviews (see Commissioner's summary in *Federal Register* 46, 38285, 1981). This book summarizes in part the research that resulted in aspartame's approval as a food additive as well as related topics regarding its function as a potential sweetening agent.

The sweet taste of aspartame was discovered serendipitously in 1965, a story described by Mazur in Chapter 1. Interestingly, this discovery opened up an entirely new area in the study of chemical structure and its relationship to taste. Indeed, as discussed by Schiffman, aspartame's taste was a sweet surprise, remarkably clean, very much like that of sucrose, and free of a metallic aftertaste.

Following the discovery of aspartame, Searle petitioned the FDA for approval to market it as a sweetening agent in certain foods (*Federal Register* 38, 5921, 1973). Searle's petition provided a large volume of data in support of claims for safety. Some of these studies are summarized by Molinary, Ishii, Bryan, Koestner, Cornell et al., Sturtevant, and Visek.

On July 26, 1974, the FDA approved Searle's petition and issued a regulation authorizing the use of aspartame in certain foods and for certain technologic purposes (*Federal Register* 39, 27317, 34520, 1974). As permitted by law, two parties formally objected to the regulation and requested a formal evidentiary hearing. The objections filed by Olney, and jointly by Turner and Label Inc., questioned the safety of the aspartic acid and phenylalanine moieties of aspartame. Later these parties waived their right to a formal evidentiary hearing conditioned upon the establishment of a Public Board of Inquiry (PBOI) composed of three qualified scientists from outside the FDA to review and evaluate the issue of safety. This novel approach represented a first as a means for resolving questions of food safety. Olney's concern was later expanded to include questions about a possible role of aspartame (or its diketopiperazine conversion product) in the induction of brain tumors in rats.

Before a PBOI could be convened, preliminary results from an FDA audit of the records of certain animal studies indicated the need for a comprehensive review of some aspartame research data. Thus the FDA formally stayed the regulation authorizing the marketing of aspartame (*Federal Register* 40, 56907, 1975). With the knowledge and approval of Searle, the aspartame data in 15 pivotal studies were thoroughly audited by either the FDA or the Universities Associated for Research and Education in Pathology, Inc. (UAREP). This massive undertaking required two years to complete. At the end of this review, both the FDA and UAREP agreed that the studies were authentic, and the FDA turned its attention to arranging the PBOI. During this interval, Searle carried out a large number of human and animal studies to further evaluate the safety of aspartame. Chapters by Ishii, Applebaum et al., Reynolds et al., Suomi, Stegink, Baker, Pitkin, Filer et al., and Koch and Wenz summarize these investigations.

In late January 1980, the Public Board of Inquiry heard three full days of testimony. Posthearing briefs and/or rebuttal statements were filed by all parties. On October 1, 1980 the PBOI issued its decision. The PBOI agreed with Searle and the FDA that the aspartate and phenylalanine content of aspartame did not constitute a risk when the additive was used according to the petition. However, the PBOI agreed with Olney that available data did not rule out the possibility that aspartame or its diketopiperazine caused brain tumors in rats. Accordingly, the Board withdrew approval of G. D. Searle & Co.'s food additive petition, and after vacating the stay on the aspartame regulation (21 CFR 172.804), revoked that regulation in its entirety.

Olney, Searle, and the FDA all filed detailed exceptions to those portions of the PBOI decision in which the PBOI disagreed with their respective positions. Turner also filed an exception, objecting to the scope of evidence considered by the PBOI. As a result, the FDA made a detailed reevaluation of the data considered by the PBOI, particularly the data relating to the brain tumor issue. Some of the brain tumor data reevaluated are reviewed in chapters by Ishii, Koestner, and Cornell et al. After considerable review of available data, a process that also included examination of new data (detailed in *Federal Register* 46, 38285, 1981), the Commissioner determined that aspartame had been shown to be safe for its proposed uses as a food additive and approved the food additive petition. Aspartame was later approved for use in beverages (*Federal Register* 48, 21378, 1983).

Just as this book is going to press, additional theoretical questions are being raised about the capacity of aspartame to affect brain neurotransmitter levels and behavior (*New England Journal of Medicine* 309, 429, 1983). Although these claims are not specifically addressed in this book, possible behavioral effects of aspartame are discussed in chapters by Butcher et al., Molinary, Reynolds et al., and Suomi. Fernstrom's chapter addresses the brain neurotransmitter issue.

The book is divided into five major sections: (1) the history and background of aspartame, (2) the metabolism of its component parts, (3) sensory and dietary aspects of the compound, (4) preclinical studies of aspartame and its diketopiperazine, and (5) clinical studies in man or nonhuman primates that address specific questions related to its food use.

In the first section Mazur describes its serendipitous discovery and Inglett discusses aspartame in relation to other sweeteners. Since aspartame is a dipeptide methyl ester (L-aspartyl-L-phenylalanine methyl ester), David Matthews was asked to prepare a chapter on the hydrolysis and absorption of peptides. Since orally ingested aspartame releases aspartate, phenylalanine, and methanol to portal blood, the metabolism of each of these components is discussed in chapters by Stegink, Harper, and Tephly and McMartin. The chapter by Tephly and McMartin also addresses the issue of potential methanol toxicity. Animal studies on the metabolism and distribution of radioactively labeled aspartame are described in chapters by Oppermann and by Matsuzawa and O'Hara.

The sensory and dietary aspects of aspartame are discussed in chapters by Schiffman and by Roak-Foltz and Leveille. The chapter by Schiffman reviews taste and taste perceptions of aspartame and other sweeteners. The review by Roak-Foltz and Leveille examines how projected intakes of aspartame might affect normal dietary intake of aspartate, phenylalanine, and methanol. Homler discusses specific problems that aspartame presents to the food scientist in new product formulation. Bowen discusses possible benefits of artificial sweeteners in preventing dental caries, and Porikos and Van Itallie discuss the value of aspartame in weight reduction.

The summary of preclinical studies includes a review of G. D. Searle & Co.'s chronic animal feeding studies by Molinary. Ishii reviews similar studies carried out in Japan. Bryan discusses the potential of sweetening agents to produce bladder cancer. Since dicarboxylic amino acids are known to cause brain damage in the infant rodent, Applebaum et al. discuss aspartate neurotoxicity in the rodent and Reynolds et al. describe neurotoxicity studies of aspartame in the infant nonhuman primate. The relative risk of the phenylalanine content of aspartame is discussed in chapters by Butcher and Vorhees, Reynolds et al., and Suomi. One major issue at the PBOI was the relationship of aspartame to brain tumors in rats fed large doses of the compound. This aspect of safety is discussed in chapters by Ishii, Koestner, and Cornell et al. This section is concluded by a chapter prepared by Sturtevant on possible neurohormonal effects of aspartame ingestion.

Specific issues relating to human consumption of aspartame are summarized in a series of chapters. Studies of chronic ingestion of aspartame carried out prior to the PBOI are summarized by Visek. Stegink discusses an extensive series of acute dosing studies in humans with aspartame carried out to resolve questions of safety relative to aspartate, phenylalanine and methanol content. Baker discusses the possible effects of aspartame ingestion on the amino acid composition of human milk, and Pitkin discusses aspartame ingestion during pregnancy. The question of whether the human infant metabolizes aspartame as effectively as adult subjects is resolved in a chapter by Filer et al., while Koch and Wenz discuss the use of aspartame by individuals homozygous or heterozygous for phenylketonuria.

The potential for interaction between aspartame and monosodium glutamate is reviewed by Stegink. The use of aspartame by diabetics is discussed by Horwitz. Finally, Fernstrom discusses possible interactions of aspartame with prescription drugs and considers the question of whether aspartame ingestion affects brain neurotransmitter concentrations. Food additives proposed for FDA approval face a more rigorous review today than compounds approved in past years. The extensive research program carried out to demonstrate aspartame safety may serve as a new standard for the study of food additives.

Lewis D. Stegink
L. J. Filer, Jr.

Contributors

Arnold E. Applebaum, Ph.D. Assistant Professor, Department of Anatomy, University of Iowa College of Medicine, Iowa City, Iowa

George L. Baker, M.D.* Professor, Department of Pediatrics, University of Iowa College of Medicine, Iowa City, Iowa

Anne F. Bauman Research Technologist, Department of Anatomy, and Research Assistant, Department of Surgery, University of Illinois at Chicago, Chicago, Illinois

William H. Bowen, B.D.S., Ph.D. Chairman, Department of Dental Research, University of Rochester School of Medicine and Dentistry, Rochester, New York

George T. Bryan, M.D., Ph.D. Professor, Department of Human Oncology and Associate Director for Laboratory Programs, Wisconsin Clinical Cancer Center, University of Wisconsin Center for Health Sciences, Madison, Wisconsin

Richard E. Butcher, Ph.D. Associate Director/Research, Western Behavioral Sciences Institute, La Jolla, California

Richard G. Cornell, Ph.D. Professor and Chairman, Department of Biostatistics, University of Michigan, Ann Arbor, Michigan

Present affiliations:
*Medical Director, Department of Medical Affairs, Mead Johnson and Company, Evansville, Indiana

Tahia T. Daabees, Ph.D.* Research Fellow, Department of Pediatrics and Biochemistry, University of Iowa College of Medicine, Iowa City, Iowa

John D. Fernstrom, Ph.D. Associate Professor, Departments of Psychiatry and Pharmacology, University of Pittsburgh School of Medicine, Western Psychiatric Institute and Clinic, Pittsburgh, Pennsylvania

L. J. Filer, Jr., M.D., Ph.D. Professor, Department of Pediatrics, University of Iowa College of Medicine, Iowa City, Iowa

Michael W. Finkelstein, M.S., D.D.S. Assistant Professor, Department of Oral Pathology and Diagnosis, University of Iowa College of Dentistry, Iowa City, Iowa

Alfred E. Harper Professor, Departments of Biochemistry and Nutritional Sciences, University of Wisconsin, Madison, Wisconsin

Barry E. Homler, Ph.D. Manager, Technical Services, Searle Food Resources, Inc., NutraSweet Group, G. D. Searle & Co., Skokie, Illinois

David L. Horwitz, M.D., Ph.D. Associate Professor, Department of Medicine, University of Illinois Health Science Center, Chicago, Illinois

George E. Inglett, Ph.D. Chief, Cereal Science and Foods Laboratory, Northern Regional Research Center, U. S. Department of Agriculture, Peoria, Illinois

Hiroyuka Ishii Life Science Laboratory, Central Research Laboratories, Ajinomoto Company, Inc., Yokohama, Japan

Richard Koch, M.D.† Professor of Clinical Pediatrics. Childrens Hospital of Los Angeles, and Professor, Department of Pediatrics, University of Southern California School of Medicine, Los Angeles, California

Adalbert Koestner, D.V.M., Ph.D. Professor and Chairman, Department of Pathology, Michigan State University, East Lansing, Michigan

Gilbert A. Leveille, Ph.D. Director, Nutrition and Health Sciences, General Foods Corporation, White Plains, New York

Kenneth E. McMartin, Ph.D. Assistant Professor, Department of Pharmacology, Section of Toxicology, Louisiana State University Medical Center, Shreveport, Louisiana

---

Present affiliations:
*Associate Professor, Department of Pharmacology, Faculty of Pharmacy, University of Alexandria, Alexandria, Egypt
†Head, Division of Medical Genetics, Children's Hospital of Los Angeles, Los Angeles, California

Yoshimasa Matsuzawa Life Science Laboratory, Department of Drug Metabolism and Biopharmacy, Central Research Laboratories, Ajinomoto Company, Inc., Yokohama, Japan

David M. Matthews, M.D., Ph.D. Professor, Department of Experimental Chemical Pathology, Vincent Square Laboratories, Westminster Hospital, London, England

Robert H. Mazur, Ph.D. Section Head, Sweetener Research Section, Department of Medicinal Chemistry, G.D. Searle & Co., Skokie, Illinois

Samuel V. Molinary, Ph.D. Scientific Consultant in Nutrition and Toxicology, Department of Scientific Regulatory Affairs, PepsiCo, Inc., Valhalla, New York

Sakkubai Naidu, M.B.B.S. Chief, Section of Pediatric Neurology. Department of Neurology, Loyola University Medical Center, Maywood, Illinois

Yuichi O'Hara* Life Science Laboratory, Central Research Laboratories, Ajinomoto Company, Inc., Yokohama, Japan

James A. Oppermann, Ph.D. Section Head, Department of Drug Metabolism, G. D. Searle & Co., Skokie, Illinois

Linda Parsons Department of Anatomy, University of Illinois College of Medicine, Chicago, Illinois

Roy M. Pitkin, M.D. Professor and Head, Department of Obstetrics and Gynecology, University of Iowa College of Medicine, Iowa City, Iowa

Katherine P. Porikos, Ph.D.† Research Associate, Obesity Research Center and Staff Associate, Department of Medicine, Columbia University College of Physicians and Surgeons, St. Lukes–Roosevelt Hospital Center, New York, New York

W. Ann Reynolds, Ph.D. Chancellor, California State University, Long Beach, California

Roberta Roak-Foltz, R.D. Research Specialist, Department of Nutrition and Health Sciences, General Foods Corporation, White Plains, New York

Paul G. Sanders Consulting Biostatistician, Department of Biostatistics, G. D. Searle & Co., Skokie, Illinois

Susan S. Schiffman, Ph.D. Professor, Department of Psychiatry, Duke University, Durham, North Carolina

---

Present affiliations:

*Section Manager, Product Safety and Assessment Department, Ajinomoto Company, Inc., Tokyo, Japan

†Adjunct Assistant Professor, Department of Psychiatry, University of Calgary, and Psychologist, Department of Psychiatry, Foothills Hospital, Calgary, Alberta, Canada

Lewis D. Stegink, Ph.D. Professor, Departments of Pediatrics and Biochemistry, University of Iowa College of Medicine, Iowa City, Iowa

Frank M. Sturtevant, Ph.D. Director, Office of Scientific Affairs, Research and Development Division, G. D. Searle & Co., Skokie, Illinois

Stephen J. Suomi, Ph.D.* Associate, Professor, Department of Psychology, University of Wisconsin, Madison, Wisconsin

Thomas R. Tephly, M.D., Ph.D. Professor, Department of Pharmacology, and Director, Toxicology Center, University of Iowa College of Medicine, Iowa City, Iowa

Theodore B. Van Itallie, M.D. Professor, Department of Medicine, Columbia University College of Physicians and Surgeons, St. Luke's—Roosevelt Hospital Center, New York, New York

Willard J. Visek, M.D., Ph.D. Professor, Division of Medicine, College of Medicine, and Department of Food Science, University of Illinois at Urbana-Champaign, Urbana, Illinois

Charles V. Vorhees, Ph.D. Associate Professor, Department of Pediatrics, Psychoteratology Laboratory, Institute for Developmental Research, Children's Hospital Research Foundation, University of Cincinnati College of Medicine, Cincinnati, Ohio

Elizabeth J. Wenz, R.D., M.S. Nutritionist, Division of Medical Genetics, Childrens Hospital of Los Angeles, Los Angeles, California

Robert A. Wolfe, Ph.D. Associate Professor, Department of Biostatistics, University of Michigan, Ann Arbor, Michigan

---

Present affiliations:
*Chief, Laboratory of Comparative Ethology, National Institute of Child Health and Human Development, National Institutes of Health, Bethesda, Maryland

Contents

HISTORY AND BACKGROUND

1

Discovery of Aspartame

Robert H. Mazur
G. D. Searle & Co., Skokie, Illinois

The discovery of the taste of L-aspartyl-L-phenylalanine methyl ester–aspartame (APM)–was accidental (1), as has been true for the discovery of previous novel chemical compounds having a sweet taste. Although there have been numerous attempts to rationalize structure-taste relationships of synthetic and natural sweeteners, the resulting theories and associated formulations have lacked predictive value (2-14); that is, one cannot plan the synthesis of a compound unrelated to known sweet chemicals with any reasonable probability of turning up a sweet substance. In addition, existing theories do not reliably predict the effects of structural changes on the potency and quality of taste within a series of closely related analogs of known sweeteners. It has been observed repeatedly that small structural changes can have dramatic effects on the taste of a compound. At the present time, therefore, it would be fair to say that new synthetic sweeteners are generally found by accident. The modern researcher is also handicapped because tasting one's compounds is no longer considered to be part of the characterization process.

In the early 1960s one of the projects in the Searle research laboratories was to find an inhibitor of the gastrointestinal secretory hormone gastrin as a possible treatment for ulcers. The C-terminal tetrapeptide of gastrin, Trp-Met-Asp-Phe-NH_2, was needed as a standard for a bioassay, and aspartame, Asp-Phe-OMe, was an intermediate in the synthesis. The following account of the actual discovery of aspartame by James M. Schlatter is taken from an affidavit.

In December 1965 I was working with Dr. Mazur on the synthesis of the C-terminal tetrapeptide of gastrin. We were making intermediates and trying to purify them. In particular, on an occasion in December, 1965, I was recrystallizing aspartylphenylalanine methyl ester (aspartame) which had been prepared by the hydrogenolysis of the protected dipeptide ester prepared and given to me by Dr. Mazur. I was heating the aspartame in a flask with methanol when the mixture bumped onto the outside of the flask. As a result, some of the powder got onto my fingers. At a slightly later stage, when licking my finger to pick up a piece of paper, I noticed a very strong, sweet taste. Initially, I thought that I must have still had some sugar on my hands from earlier in the day. However, I quickly realized this could not be so, since I had washed my hands in the meantime. I, therefore, traced the powder on my hands back to the container into which I had placed the crystallized aspartyl-phenylalanine methyl ester. I felt that this dipeptide ester was not likely to be toxic and I therefore tasted a little of it and found that it was the substance which I had previously tasted on my finger.

The sweetness of aspartame could not have been predicted from the tastes of the constituent amino acids; aspartic acid is tasteless to slightly sour, whereas phenylalanine is bitter (15). In addition, the L (natural) absolute configuration for both aspartic acid and phenylalanine was required for aspartame's sweetness. The other three diastereoisomers of aspartame—LD, DL, DD—were weakly bitter (1). This provides a good example of the stereochemical specificity possible in the binding of a substrate to a receptor. Also, the methyl ester was necessary; aspartylphenylalanine was tasteless!

The quality of the taste of aspartame is remarkably like sucrose, with no bitter aftertaste often associated with other sweeteners. Nor is there any menthol-like cooling effect or licorice flavor sometimes encountered with both natural and synthetic sweeteners. Tasters often cannot distinguish aspartame from sucrose and sometimes even prefer the former. The potency of aspartame is 150-200 times that of sucrose, depending on the particular application; this may be compared with 30 for cyclamate and 300 for saccharin, two well-known sweeteners.

In the following discussion only a very few selected examples are cited. As a result, it is not possible to evaluate properly structure-activity relationships of compounds related to aspartame. Those interested in a more detailed analysis will have to read the original papers.

Since 1965 many analogs of aspartame have been synthesized in laboratories all over the world. These may be classified into two broad groups: (a) amides of aspartic acid and (b) dipeptides of aspartic acid. Derivatives of these basic structures are, of course, subsumed and in the dipeptides, aspartic acid is always N-terminal. Only a few compounds fall outside of these classes, and they are especially interesting. The lower homolog, aminomalonylphenylalanine methyl ester (16), is sweet (230 times sucrose) (17), but the higher homolog, glutamylphenylalanine methyl ester (1), is tasteless. However, a derivative of glutamic acid, Ntrifluoroacetylglutamic acid α-p-cyanoanilide (18), is very sweet (3000 times sucrose). In aspartame the intact peptide bond is necessary, since N-methylation, changing the amide to an ester, or reversing the order of the carbonyl and amino groups all eliminate sweetness (19).

Simple α-amides of aspartic acid have not been especially sweet, for example, L-2-methylphenethylamide (50 times sucrose) (20) and L-1,4-dimethylpentylamide (100 times sucrose) (20). On the other hand, N-trifluoroacetylaspartic acid α-p-cyanoanilide (3000 times sucrose) (21) is quite potent. A number of complex amides based on D-alaninol, D-serinol, and D-valinol have been synthesized (22). The best were derivatives of aspartyl-D-alaninol, namely, the pivalate ester (240 times sucrose) and the cyclobutanecarboxylic ester (220 times sucrose).

In studies of aspartyl dipeptide esters, it was found that only a few amino acids could substitute for phenylalanine. Among L-amino acids, both the lower homolog, aspartylphenylglycine methyl ester (140 times sucrose) (23), and the higher homolog, aspartylhomophenylalanine methyl ester (100 times sucrose) (24), are less sweet than aspartame. Thus aspartame is the optimum member of this small series. Locking the aromatic ring into one conformation eliminates sweetness; aspartyl- Δ^{Z}- phenylalanine methyl ester (25) has no taste. Other L-amino acids replacing phenylalanine in aspartame with retention of sweetness are hexahydrophenylalanine (225 times sucrose) (26), tyrosine (50 times sucrose) (1), orthotyrosine methyl ether (100 times sucrose) (27), and methionine (100 times sucrose) (1), S-t-butylcysteine (900 times sucrose) (5), O-t-butylserine (65 times sucrose) (5), and 2-aminoisocaprylic acid (80 times sucrose) (26). Surprisingly, some aspartyl-D-amino acid esters and amides are sweet. These seem to be limited to D-amino acids with small alkyl or hydroxyalkyl side chains. Examples are aspartyl-D-alanine n-propyl ester (170 times sucrose) (26), aspartyl-D-alanine n-butylamide (125 times sucrose) (28), aspartyl-D-serine n-propyl ester (320 times sucrose) (29), aspartyl-D-threonine n-propyl ester (150 times sucrose) (29), and aspartyl-D-alanine t-butylcyclopropylcarbinylamide (1200 times sucrose) (30).

Compounds in which the phenylalanine methyl ester in aspartame was replaced by unsymmetrical diesters of aminomalonic acid were unusually sweet (31). Aspartylaminomalonic acid methyl, trans-2-methylcyclohexyl diester (7300 times sucrose), and the corresponding methyl, fenchyl diester (33,000 times sucrose) are examples. The latter compound is the sweetest substance ever reported, either natural or synthetic.

Table 1 lists the various analogs with their sweetness relative to sucrose. Since tasting was done under different conditions in a variety of laboratories, the potencies can only be used for approximate comparison of one compound with another. They are, however, a useful guide as to what is happening within a particular series.

The variety and generality of the above results suggest that a totally new type of biochemical nucleus has been found with the discovery of aspartame. That is,

Table 1 Analogs of Aspartame (Asp-Phe-OMe) and Their Sweetness Values Relative to Sucrose[a]

Asp-Phe-OMe	180
Homologs	
DL-Ama-Phe-OMe	230
Glu-Phe-OMe	0
Asp-Pgl-OMe	140
Asp-NHCH($CH_2CH_2C_6H_5$)CO_2Me	100
Anilides	
F_3CCO-Asp-NHC_6H_4CN(p)	3,000
F_3CCO-Glu-NHC_6H_4CN(p)	3,000
Simple amides	
Asp-NHCH(CH_3)$CH_2C_6H_5$	50
Asp-NHCH(CH_3)CH_2CH_2CHMe_2	100
Complex amides	
Asp-D-NHCH(CH_3)CH_2O_2CMe_3	240
Asp-D-NHCH(CH_3)$CH_2O_2CC_4H_7$(cyclo)	220
Dipeptides	
Asp-Tyr-OMe	50
Asp-Met-OMe	100
Asp-NHCH($CH_2C_6H_4$OMe(o))CO_2Me	100
Asp-Phe(6H)-OMe	225
Asp-NHCH(CH_2CH_2CHMe_2)CO_2Me	80
Asp-NHCH(CH_2OCMe_3)CO_2Me	65
Asp-NHCH(CH_2SCMe_3)CO_2Me	900
Asp-D-Ala-O$C_3H_7^n$	170
Asp-D-Ala-NH$C_4H_9^n$	125
Asp-D-Ala-NHCH(cyclo-C_3H_5)CMe_3	1,200
Asp-D-Ser-O$C_3H_7^n$	320
Asp-D-Thr-O$C_3H_7^n$	150
Asp-DL-NHCH(CO_2Me)CO_2– (Me) (t)	7,300
Asp-DL-NHCH(CO_2Me)CO_2	33,000
Miscellaneous	
Asp-Phe	0
Asp-Phe-OMe (DL,LD,DD)	0
Asp-MePhe-OMe	0
C_6H_5CH ‖ Asp-NHCCO_2Me	0
Asp-OCH($CH_2C_6H_5$)CO_2Me	0
H_2NCH(CO_2H)NHCOCH($CH_2C_6H_5$)CO_2Me	0
(DL)-H_2NCH(CO_2H)NH-(DL)-COCH($CH_2C_6H_5$)CO_2Me	

[a]Potencies refer to sucrose as 1. Nonsweet compounds are shown by a zero. Amino acids have the L configuration unless otherwise specified. Abbreviations are the conventional ones; in addition, Ama = aminomalonic acid, Pgl = phenylglycine, Phe(6H) = hexahydrophenylalanine, and MePhe = N-methylphenylalanine.

L-aspartic acid, while not at all sweet itself, when derivatized in a special way gives products which often have a sweet taste, sometimes even with astonishing potencies. Such a compound could be called a cryptophore or, in this case, a cryptoglucophore. The analogy is a chromophore, except that a chromophore is already colored, or at least has long-wavelength ultraviolet absorption. Additional groups merely enhance an already existing property. In the case of a cryptophore, further structural elaboration exposes or brings to light a property that was hidden in the original molecule. One could speculate that cryptophores might be discovered in other areas of biological activity, for example, those involving particular pharmacological properties.

REFERENCES

1. Mazur, R. H., Schlatter, J. M., and Goldkamp, A. H. (1969). Structure-taste relationships of some dipeptides. *J. Am. Chem. Soc.* 91, 2684.
2. Shallenberger, R. S., and Acree, T. E. (1967). Molecular theory of sweet taste. *Nature* 216, 480-482.
3. Kier, L. B. (1972). A molecular theory of sweet taste. *J. Pharm. Sci.* 61, 1394-1397.
4. Goodman, M., and Gilon, C. (1975). Peptide sweeteners: a model for peptide and taste receptor interactions. In *Peptides 1974* (Wolman, Y., ed.), Halstead Press, New York, pp. 271-278.
5. Brussell, L. B. P., Peer, H. G., and van der Heijden, A. (1975). Structure-taste relationship of some sweet-tasting dipeptide esters. *Z. Lebensm. Unters. Forsch.* 159, 337-343.
6. de Nardo, M., Runti, C., Ulian, F., and Vio, L. (1976). Rapporti fra constituzione chimica e scipore dolce. *Farmaco Ed. Sci.* 31, 906-916.
7. Lelj, F., Tancredi, T. Temussi, P. A., and Toniolo, C. (1976). Interaction of alpha-L-aspartyl-L-phenylalanine methyl ester with the receptor site of the sweet taste bud. *J. Am. Chem. Soc.* 98, 6669-6675.
8. van der Heijden, A. Brussel, L. B. P., and Peer, H. G. (1978). Chemoreaction of sweet-tasting dipeptide esters: A third binding site. *Food Chem.* 3, 207-211.
9. van der Heijden, A., Brussel, L. B. P., and Peer, H. G. (1979). Quantitative structure-activity relationships in sweet aspartyl dipeptide methyl esters. *Chem. Senses Flavor* 4, 141-152.
10. Crosby, G. A., DuBois, G. E., and Wingard, R. E. (1979). The design of synthetic sweeteners. In *Drug Design*, Vol. 8 (Ariens, E. J., ed.), Academic Press, New York, pp. 215-310.
11. Tinti, J. M., Durozard, D., and Nofre, C. (1980). Sweet taste receptor. Evidence of separate specific sites of carboxyl ion and nitrite/cyanide. *Naturwissenschaften* 67, 193-194.
12. Lelj, F., Tancredi, T., Temussi, P. A., and Toniolo, C. (1980). Interaction of alpha-L-aspartyl-D-phenylalanine methyl ester with the receptor site of the taste. *Farmaco Ed. Sci.* 35, 988-996.
13. Iwamura, H. (1980). Structure-taste relationship of perillartine and nitrocyanoaniline derivatives. *J. Med. Chem.* 23, 308-312.

14. Iwamura, H. (1981). Structure-sweetness relationship of L-aspartyl dipeptide analogues. A receptor site topology. *J. Med. Chem.* 24, 572-583.
15. Schiffman, S. S., and Englehard, H. H. (1976). Taste of dipeptides. *Physiol. Behav.* 17, 523-535.
16. Briggs, M. T., and Morley, J. S. (1972). Alpha-carboxylglycl dipeptide derivatives. British Patent 1,229,265 (December 13, 1972); *Chem. Abstr.* 78, 111760C, 7.
17. Fujino, M., Wakimasu, M., Tanaka, K., Aoki, H., and Nakajima, N. (1973). L-Aspartylaminomalonic acid diesters, new group of compounds with intense sweetness. *Naturwissenschaften* 60, 351.
18. Kawai, M., Nyfeler, R., Berman, J. M., and Goodman, M. (1982). Peptide sweeteners. 5. Side-chain homologues relating zwitterionic and trifluoroacetylated amino acid anilide and dipeptide sweeteners. *J. Med. Chem.* 25, 397-402.
19. MacDonald, S. A., Willson, C. G., Chorev, M., Vernacchia, F. S., and Goodman, M. (1980). Peptide sweeteners. 3. Effect of modifying the peptide bond on the sweet taste of L-aspartyl-L-phenylalanine methyl ester and its analogues. *J. Med. Chem.* 23, 413-420.
20. Mazur, R. H., Goldkamp, A. H., James, P. A., and Schlatter, J. M. (1970). Structure-taste relationships of aspartic acid amides. *J. Med. Chem.* 13, 1217-1221.
21. Lapidus, M., and Sweeney, M. (1973). L-4'-Cyano-3-(2,2,2-trifluoroacetamido) succinanilic acid and related synthetic sweetening agents. *J. Med. Chem.* 16, 163-166.
22. Miyoshi, M., Nunami, K., Sugano, H., and Fujii, T. (1978). Structure-taste relationship of novel alpha-L-aspartyl dipeptide sweeteners. *Bull. Chem. Soc. Jpn.* 51, 1433-1440.
23. Dahlmans, J. J., and Boesten, W. H. J. (1972). Sweet tasting asparaginyl amino acid esters containing them. Netherlands Appl. 7012899 (March 3, 1972); *Chem. Abstr.* 77, 33111e, p. 7.
24. Owens, W. H. (1982). Unpublished data.
25. King, S. W., and Stammer, C. H. (1981). Synthesis of L-aspartyl-delta-Z-phenylalanine methyl ester. Dehydroaspartame. *J. Org. Chem.* 46, 4780-4782.
26. Mazur, R. H., Reuter, J. A., Swiatek, K. A., and Schlatter, J. M. (1973). Synthetic sweeteners. 3. Aspartyl dipeptide esters from L- and D-alkylglycines. *J. Med. Chem.* 16, 1284-1286.
27. Kawai, M., Chorev, M., Marin-Rose, J., and Goodman, M. (1980). Peptide sweeteners. 4. Hydroxy and methoxy substitution of the aromatic ring in L-aspartyl-L-phenylalanine methyl ester structure-taste relationships. *J. Med. Chem.* 23, 420-424.
28. Sukehiro, M., Minematsu, H., and Noda, K. (1977). Structure-taste relations of aspartyl peptide sweeteners. I. Syntheses and properties of L-aspartyl-D-alanine amides. *Sci. Hum. Life* 11, 9-16.
29. Ariyoshi, Y., Yasuda, N., and Yamatani, T. (1974). The structure-taste relationships of the dipeptide esters composed of L-aspartic acid and beta-hydroxy amino acids. *Bull. Chem. Soc. Jpn.* 47, 326-330.

30. Brennan, T. M., and Hendrick, M. E. (1981). Branched amides of L-aspartyl-D-amino acid dipeptides and their compositions. European Patent Appl. 0034876 (September 2, 1981), p. 184.
31. Fujino, M., Wakimasu, M., Mano, M., Tanaka, K., Nakajima, N., and Aoki, H. (1976). Structure-taste relationships of L-aspartyl-aminomalonic acid diesters. *Chem. Pharm. Bull.* 24, 2112-2117.

2

Sweeteners: An Overall Perspective

George E. Inglett
Northern Regional Research Center, U.S. Department of Agriculture,
Peoria, Illinois

INTRODUCTION

A sweet taste is a quality of some chemical substances that the human race has always associated with pleasure. As a result, high value has been placed on materials exhibiting sweetness. Honey and fruits have been sought throughout history for their sweet properties; by the fourteenth century A.D., sugar was being refined. It was regarded, however, as a rare delicacy. Today we accept the presence of sugar as commonplace. Human cravings for sweetness are generally satisfied by beverages, breads, pastries, and confectionaries.

NATURAL SWEETENERS

Sucrose

In the twentieth century the vastly increased demand for sucrose was met commercially by extraction from cane (60%) and beet (40%) sources. Cane and sugar beets are the most agriculturally efficient crops for the production of sucrose. As a source of food calories, these crops produce more per acre than potatoes, corn, or wheat.

Sucrose's (Fig. 1) acceptability and palatability are unique. These qualities, along with its ready availability, low cost, simplicity of production, purity, and long history of use have assured its place in the diet.

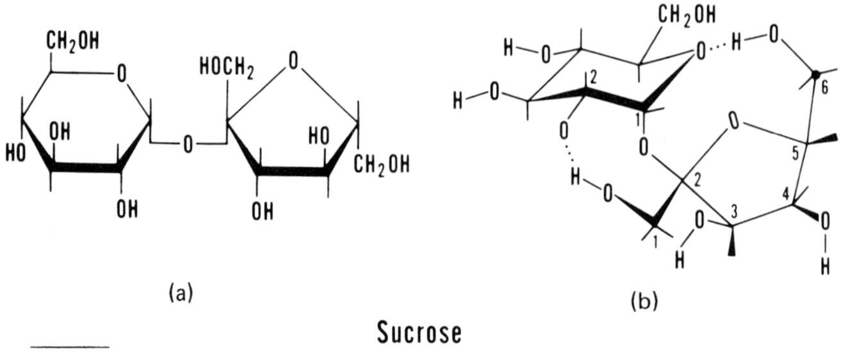

Sucrose

Figure 1 Sucrose.

The sweetness of fruits is attributed to the combination of sucrose and its component sugars, glucose and fructose. The high proportions of these sugars, in fruits such as grapes and sweet cherries are the reason for selecting them for yeast fermentations to produce wine (1).

Corn Sweeteners

During the past two decades the corn sweetener industry has been in an exciting and rapidly developing stage. These developments were largely due to advances in enzyme technology. First, enzyme technology using glucoamylase was applied to the starch hydrolysis process used to produce dextrose and dextrose syrups. Because of the higher starch concentrations that could be used, economical advantage resulted, since each batch required less steam in subsequent evaporation steps. Furthermore, higher yields and greater purity of the resulting dextrose were obtained (2).

A second advance in the application of enzyme technology to the corn sweetener industry involved the use of glucose isomerase in immobilized forms. Glucose isomerase catalyzes the conversion of glucose (dextrose) to fructose. This conversion allowed the dextrose industry to produce fructose for the first time. This was of importance, since fructose is much sweeter than glucose. Fructose was previously an expensive sugar derived from sucrose. The dextrose industry had long been handicapped by the low-level sweetness of their syrups. The use of glucose isomerase now made it possible to produce sweeter dextrose syrups due to the fructose introduced by glucose isomerization process. Isomerized dextrose syrups are called high-fructose corn syrups.

High-Fructose Corn Syrups

High fructose corn syrups, first produced commercially in the early 1970s, contained 42% fructose, 52% dextrose, and 6% higher saccharides in a syrup containing 71% solids.

The 42% fructose syrup is manufactured from starch. In the United States corn starch is used as the starting material. In this process an aqueous corn starch slurry is continuously liquified with α-amylase at elevated temperatures. The liquified hydrolysate produced has a dextrose equivalent in the range of 10-20. After pH and temperature adjustment, the hydrolysate is "saccharified" (hydrolyzed) with glucoamylase producing a solution with a dextrose content of 94-96%. The "saccharified" liquor is refined by filtering, to remove insolubles, and by subsequent carbon and ion exchange treatment. The refined dextrose solution is isomerized to yield fructose by passing it through reactors containing immobilized glucose isomerase. The resulting solution, with a fructose content of approximately 42%, is refined using carbon and ion exchange treatment before being concentrated to about 71% solids.

Second-generation high-fructose corn syrups contain fructose at 55-90% of the total solids. A 55% high-fructose corn syrup typically contains 55% fructose, 40% dextrose, and 5% higher saccharides in a syrup of 77% total solids. A 90% high-fructose corn syrup typically contains 90% fructose, 7% dextrose, and 3% higher saccharides in a syrup containing 80% total solids. These products have been available since 1976 and are prepared by fractionation of the 42% fructose syrups.

Polyhydric Alcohols

Sweet polyhydric alcohols include such substances as sorbitol, mannitol, maltitol, and xylitol. Although these materials impart only low levels of sweetness, individual polyhydric alcohols can be considered desirable for specific applications in foods. For example, xylitol is considered a potential aid to dental health by reducing dental caries.

INTENSE SWEETENERS OF NATURAL ORIGIN

Many people around the world use intensely sweet plant materials of natural origin to sweeten foods; these plant parts are often used for medicinal purposes. Intense sweetness is as desired by members of society today as it has been throughout the ages (3).

Phyllodulcin

A sweet tea, Amacha, is served at Hanamatsuri, the flower festival celebrating the birth of Buddha. Amacha is prepared from the dried leaves of *Hydrangea macrophylla* Seringe var. *Thunbergii* Makino. The sweet principle, phyllodulcin (Fig. 2), has been isolated, and its absolute configuration at the C_3 asymmetrical center determined by identification of D-malic acid from ozonized phyllodulcin.

Stevioside

The sweet herb of Paraguay (*yerba dulce*) has long been the source of an intense sweetener. Natives use the leaves of this small shrub, *Stevia rebaudiana* (Bert.)

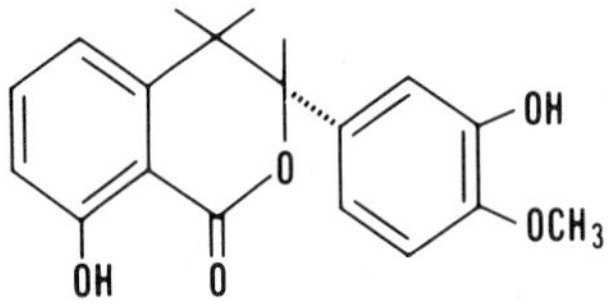

Figure 2 Phyllodulcin.

Hemsl., to sweeten their bitter drinks. The sweet, crystalline glycoside extracted from the leaves is named stevioside (Fig. 3).

In addition to stevioside, Japanese chemists have isolated other glucosides, including rebaudioside A (4) and dulcosides A and B (5). Rebaudioside A is obtained in a 1.4% yield from dried leaves and has superior taste properties. Stevioside is approved for food uses in Japan, where it is widely used as a sweetener.

Glycyrrhizin

Licorice, well known for centuries and widely used, is obtained from the roots of *Glycyrrhiza glabra*, a small shrub grown and hand harvested in Europe and Central Asia. The roots contain 6-14% glycyrrhizin.

Glycyrrhizic acid exists in licorice root as the calcium potassium salt in association with other constituents (6). Glycyrrhizic acid is a glycoside of the triterpene glycyrrhetic acid, which is condensed with O-β-D-glucuronosyl-(1' 2)-β-D-glucuronic acid. The absolute configuration of the aglycone glycyrrhetic acid was determined as the result of investigations too numerous to cite completely. Im-

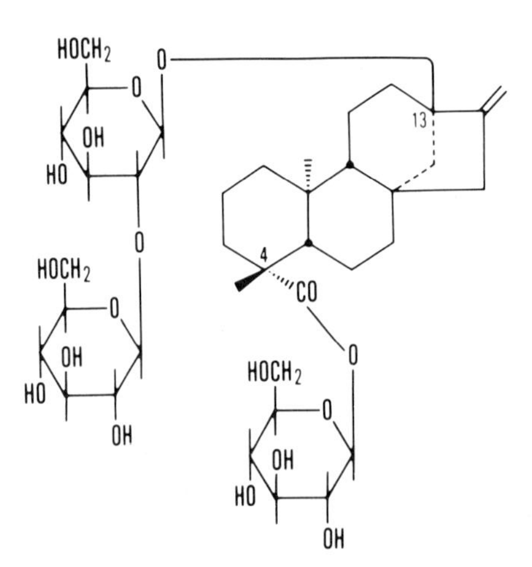

Figure 3 Stevioside.

portant contributions were made by Ruzicka et al. (7), Voss and Butter (8), Voss et al. (9), and Beaton and Spring (10). Although two isomers (18-*α* and 18-*β*) have been isolated, Beaton and Spring indicated that 18-*β*-glycyrrhetic is the only natural isomer that occurs in glycyrrhizin (Fig. 4).

Extracts containing glycyrrhizin are employed in the flavoring and sweetening of pipe, cigarette, and chewing tobaccos and in confectionery manufacture. Some segments of the flavor industry have long utilized these extracts in root beer, chocolate, vanilla, liqueurs, and other applications. Ammonium glycyrrhizin, the fully ammoniated salt of glycyrrhizic acid, is commercially available. Ammonium glycyrrhizin is the sweetest substance on the Food and Drug Administration list of natural generally recognized as safe flavors and is 50 times sweeter than sucrose.

Osladin

The sweet taste of rhizomes of the widely distributed fern *Polypodium vulgare* L. has attracted the interest of many chemists and pharmacists. Osladin comprises only 0.03% of the dry weight of the rhizomes; its chemical structure has been revealed as a bis-glycoside of a new type of steroidal saponin (11). The glycoside that results from replacement of the monosaccharide radical with hydrogen was isolated separately and named polypodosaponin. Its absolute configuration was determined by Jizba et al. (12). The disaccharide of osladin was shown to be neohesperidose, 2-O- -L-rhamnopyranosyl-B-D-glucopyranose.

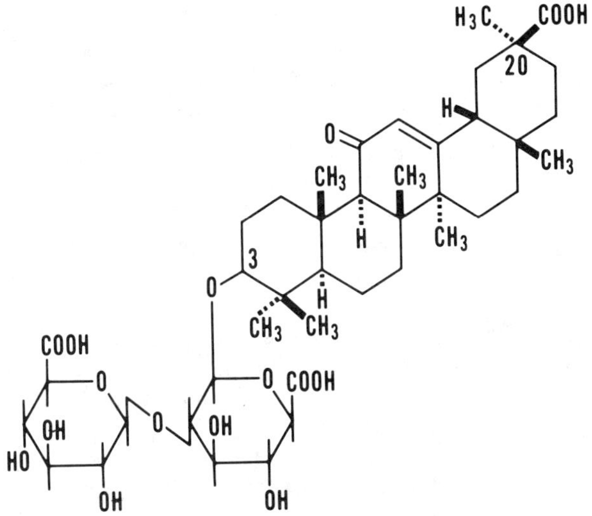

Figure 4 Glycyrrhizic acid.

Lo Han Fruit

Lo Han Kuo (Lo Han fruit), from *Momordica grosvenori* Swingle, is a dried fruit from southern China. The brownish gray pulp dries to a light fibrous mass. Swingle (13) reported that 1000 tons of the green fruits were delivered every year to the drying sheds at Kweilin (Kwangsi Province). The dried fruit is a valued folk medicine used for colds, sore throats, and minor stomach and intestinal troubles. Lee (14) found that the sweet principle could be extracted by water either from the fibrous pulps or from the thin rinds of Lo Han Kuo; 50% ethanol was also found to be a good extractant. Rinds afforded a more easily purified extract. Purified Lo Han sweetener is accompanied by a lingering taste described as licorice-like, somewhat similar to that of stevioside, glycyrrhizin, and the dihydrochalcones. Structural studies indicated the sweetener to be a triterpenoid glycoside with five or six glucose units (15). The purified sweetener has a more pleasant sweet taste than the impure material. The purest samples are about 400 times sweeter than sucrose.

PROTEIN SWEETENERS

Substances having sweet taste can vary immensely in their chemical structures. Sugars, sugar alcohols, and some amino acids are well recognized for their sweet taste, but only in the last decade has a new class of sweeteners from nature, the proteins, been found to be sweet. The intensity of their sweetness is especially surprising; all are more than 1000 times sweeter than sucrose.

Serendipity Berries

While studying various natural sweeteners previously mentioned, the author discovered the intense sweetness of red berries indigenous to tropical West Africa. The fruit was called the serendipity berry (Fig. 5). Its botanical name, *Dioscoreophyllum cumminsii* Diels, was established later (16-19).

Researchers at the Monell Chemical Senses Center (Philadelphia) and the Unilever Research Laboratorium (The Netherlands), working independently, confirmed the protein nature of the serendipity sweetener (20,21). The amino acid composition of this sweetener, also called monellin, was determined by van der Wel and Loeve (22) and Morris et al. (23). The most outstanding observation is the complete absence of histidine. Monellin is composed of two dissimilar polypeptide chains with known amino acid sequences that are noncovalently associated (24). Monellin is approximately 2500 times sweeter than sucrose on a weight basis. The molecular weight of the sweetener is 11,000.

Katemfe

Besides studies on miracle fruit and the serendipity berry, a large variety of plant materials were examined systematically by Inglett and May (18) for intensity and

Figure 5 Serendipity berries.

quality of sweetness. Another African fruit containing an intense sweetener is katemfe (Fig. 6), or the miraculous fruit of the Sudan. Botanically the plant is *Thaumatococcus danielli* (Benth.). The mucilaginous material around the aril at the base of the seeds is intensely sweet and causes other foods to taste sweet. Katemfe yields two sweet-tasting proteins, thaumatin I and II (25,26).

The purified sweetener is 1600 times sweeter than sucrose at 7% concentration. Thaumatin I contains 193 amino acids (25,26). Polyacrylamide gel electrophoresis in the presence of sodium dodecyl sulfate indicated that the protein is a single polypeptide chain with alanine as the N- terminal amino acid. Thaumatin I has been crystallized, and physical characteristics and diffraction data of the crystals obtained (27). A process for extraction of the thaumatins from the fruit was reported (28) and commercial interest in this sweetener is developing (29). Tate and Lyle Limited, in England, is marketing the sweetener under the name Talin. In addition to its sweetener value, it also have flavor potentiator utility. Large purchases of Talin have been made by companies in Japan, where it is used in food products.

Miracle Fruit

The study of the strange properties of the miracle fruit *Richardella dulcifiea* (Schum. and Thonn.) Baehni, formerly known as *Synsepalum dulcificum* (Schum.)

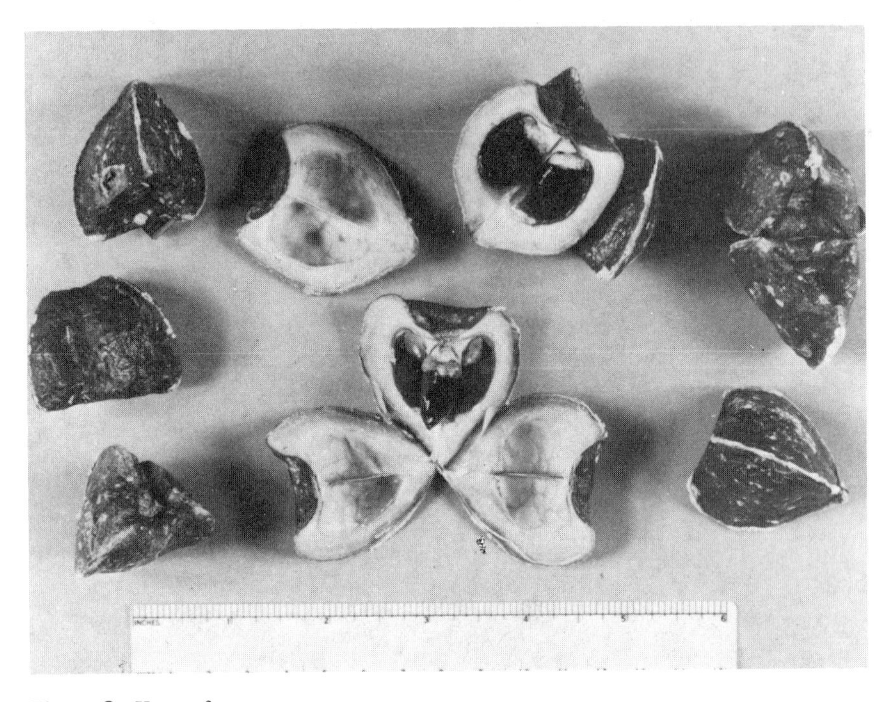

Figure 6 Katemfe.

Daniell, has added important new information on human perception of sweet taste. Although the miracle fruit's capacity to cause sour food to taste sweet has been known in the literature since 1725, scientific investigations of the fruit were not initiated until Inglett and his associates found experimental evidence that the active principle was macromolecular (30). This berry possesses a taste-modifying substance that causes sour foods such as lemons, limes, grapefruit, rhubarb, and strawberries to taste delightfully sweet. The berries are chewed by West Africans for their sweetening effect on some sour foods. The taste-modifying principle was independently isolated by two different research groups (31,32). Kurihara and Beidler (32) separated the active principle from the fruit's pulp with a carbonate buffer (pH 10.5). The destruction of the active principle by trypsin and pronase digestion suggested its proteinaceous character. The taste-modifying protein was also separated from the fruit's pulp with highly basic compounds, such as salmine and spermine (31).

The basic glycoprotein has a molecular weight between 42,000 and 44,000. The purified glycoprotein has no inherent taste. Sweetening of acid taste, observed at 5×10^{-8} M concentration of the glycoprotein solution, reached a maximum at 4×10^{-7} M and slowly declined over a period of 2 hr.

SYNTHETIC SWEETENERS

A variety of synthetic compounds have been prepared that have intense sweetness. Saccharin (Fig. 7) was discovered in 1879, which means that the history of synthetic sweeteners is slightly over 100 years old. Progress in the development of synthetic sweetener uses was especially active between 1950 and 1969, when saccharin and cyclamate (Fig. 8) were both approved for human use. The sweetness of saccharin-cyclamate blends was well perceived in soft drinks. As a result, these sweeteners were particularly attractive and widely used. Cyclamate production reached a peak rate of 21 million lb/year in the United States prior to being banned by the Food and Drug Administration in 1969. Some of the synthetic sweetener candidates being explored for filling intense sweetener needs in the U.S. market are as follows.

Aspartame, a Peptide-Based Sweetener

Like saccharin and cyclamate, the sweetness of L-aspartyl-L-phenylalanine methyl ester, also called aspartame (Fig. 9) (33,34), was discovered accidentally in the laboratories of G. D. Searle (Skokie, Illinois) in 1965. Many L-amino acids were substituted for aspartic acid and phenylalanine, and aspartic acid was shown to be required for sweetness. The taste of aspartame could not have been predicted from its constituent amino acids. L-Phenylalanine is bitter and L-aspartic acid is flat. When these amino acids are properly combined and the phenylalanine carboxyl group converted to a methyl ester, a product with a sucrose type of sweetness results. This sucrose-like sweetness allows it to blend well with other

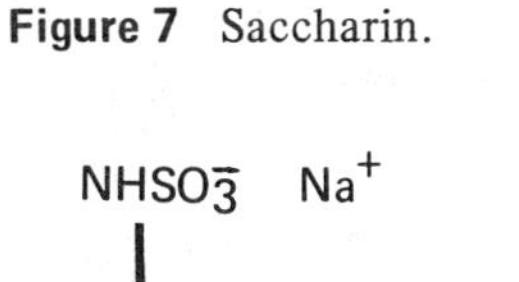

O
NH
SO_2

Figure 7 Saccharin.

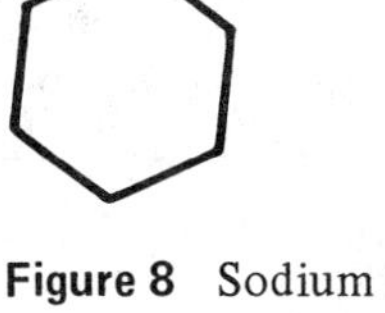

$NHSO_3^-$ Na^+

Figure 8 Sodium cyclamate.

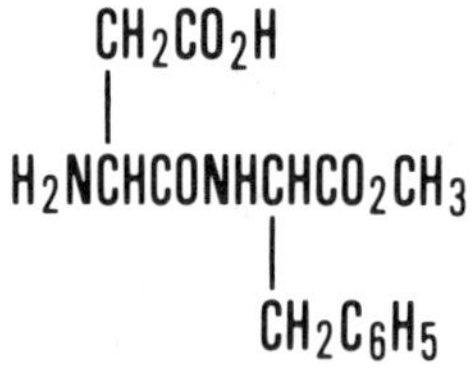

$$
\begin{array}{l}
\quad CH_2CO_2H \\
\quad | \\
H_2NCHCONHCHCO_2CH_3 \\
\qquad\qquad | \\
\qquad\qquad CH_2C_6H_5
\end{array}
$$

Figure 9 Aspartame.

food flavors. The sweetness of aspartame relative to sucrose is inversely related to the concentration of sucrose. At 3% sucrose, aspartame is 215 times the sweetness of sucrose, but it has only 133 times the sweetness potency at 10% sucrose concentration. This volume is concerned with the biochemistry and physiology of this sweetener.

Dihydrochalcone Sweeteners

Citrus peel contains flavonoids that can be converted by simple chemical modification to dihydrochalcones. Horowitz and Gentili discovered that neohesperidin isolated from the peel of the Seville orange gave an intensely sweet glycoside neo-

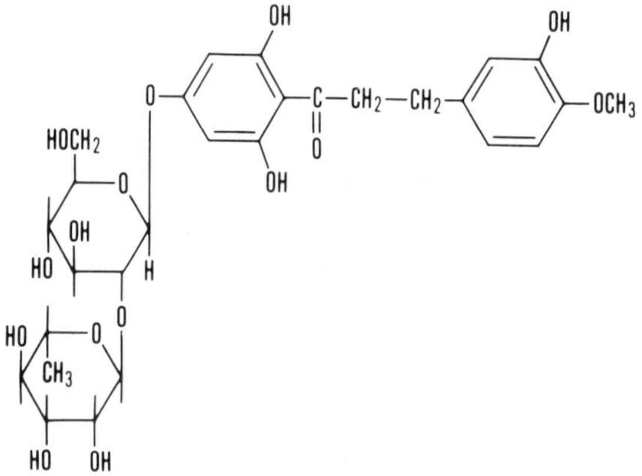

Figure 10 Neohesperidin dihydrochalcone.

hesperidin dihydrochalcone (Fig. 10) upon alkaline hydrogenation. Naringin from grapefruit peel also was found to give a sweet-tasting dihydrochalcone (35). Following these discoveries, a large number of analogs were prepared for structure-taste study. Some of the most interesting analogs, prepared by Krbechek et al. (36), involved substituting the methoxyl of the isovanillin group with ethoxyl and propoxyl groups.

Recent analogs prepared at Dynapol (Palo Alto, California) have shown that the structural elements of the dihydrochalcone responsible for inducing the sweet-taste response reside entirely on the aromatic nucleus. Nonglycoside dihydrochalcone derivatives were prepared having 4-O-carboxylalkyl and 4-O-sulfoalkyl substituents (37). They contained an intense sweetness that compared favorably with neohesperidin dihydrochalcone.

Acesulfame K

Acesulfame K (Fig. 11) is the potassium salt of 3,4-dihydro-6-methyl-1,2,3-oxathiazin-4-one-2,2-dioxide, which is a derivative of acetoacetic acid. This substance has some structural similarity to saccharin and represents one of the newest classes of intensely sweet substances. The sensory properties of acesulfame K are similar to those of saccharin. The sensory potency is roughly 130 times sucrose at a 4% sucrose concentration (38). Hoechst A.G. (Frankfurt, West Germany) is working on the applications of this sweetener. Hoechst reports that toxicological evaluations of acesulfame K have been completed successfully (39).

Other Synthetic Sweeteners

Perillartine is a naturally occurring aldoxine present in the oil of *Perilla namkemonsis* Deone. Aldoxin analogs have been synthesized, and some are claimed to be superior to perillartine (38). The relative sweetnesses of a wide variety of organic chemicals are summarized in Table 1 (38,40).

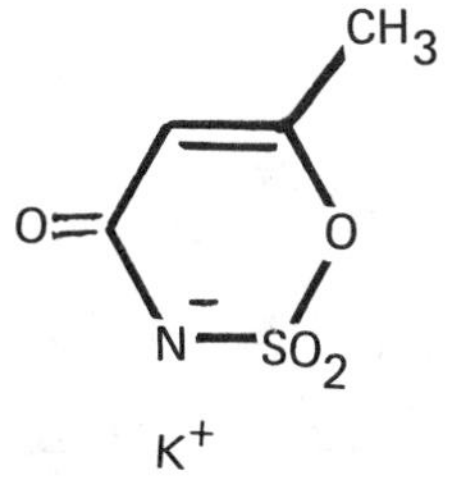

CH_3 O= O N^-—SO_2 K^+

Figure 11 Acesulfame K.

Table 1 Relative Sweetnesses of Various Sweeteners

Sweetener	Sweetness[a] (sucrose = 1)
Sucrose	1
Lactose	0.4
Maltose	0.5
Galactose	0.6
D-Glucose	0.7
D-Fructose	1.1
Invert sugar	0.7-0.9
D-Xylose	0.7
Sorbitol	0.5
Mannitol	0.7
Dulcitol	0.4
Glycerol	0.8
Glycine	0.7
Sodium 3-methylcyclopentylsulfamate	15
p-Anisylurea	18
Sodium cyclohexylsulfamate (cyclamate)	30-80
Chloroform	40
Glycyrrihizin	50
Acesulfame K	130
Aspartylphenylalanine methyl ester	100-200
5-Nitro-2-methoxyaniline	167
5-Methylsaccharin	200
p-Ethoxyphenylurea (dulcin)	70-350
6-Chlorosaccharin	100-350
n-Hexylchloromalonamide	300
Sodium saccharin	200-700
Stevioside	300
2-Amino-4-nitrotoluene	300
Naringin dihydrochalcone	300
p-Nitrosuccinanilide	350
Phyllodulcin	400
1-Bromo-5-nitroaniline	700
5-Nitro-2-ethoxyaniline	950
Perillaldehyde anti-aldoxime	2000
Neohesperidin dihydrochalcone	2000
Talin	2500
5-Nitropropoxyaniline (P-4000)	4000

[a]Many factors affect sweetness, and different methods have been used to determine sweetness ratios. The sweetness of sucrose, the usual standard, will change with age because of inversion. Sweet taste depends upon the concentration of the sweetener, the temperature, the pH, the type of medium used, and the sensitivity of the taster. The usual test methods are dilution to threshold sweetness in water and duplication of the sweetness of a 5 or 10% sucrose solution, although other techniques have also been employed. Where different sweetness values have been reported, the most commonly accepted ones have been cited in this table.

Source: Refs. 38 and 39.

REFERENCES

1. Bucke, C. (1979). Recent developments in production and use of glucose and fructose syrups. In *Developments in Sweeteners–1* (Hough, E. A. M., Parker, K. J., and Vlitos, A. J., eds.), Applied Science Publishers, London.
2. Inglett, G. E. (1963). Purification and recovery of fungal amylases. U.S. Patent 3,101,302.
3. Inglett, G. E. (1976). A history of sweeteners–Natural and synthetic. *J. Toxicol. Environ. Health* 2, 207-214.
4. Kobayashi, M., Horikawa, S., Degrandi, I. H., Ueno, J., and Mitsuhashi, H. (1977). Dulcosides A and B, new diterpene glycosides from *Stevia rebaudiana. Phytochemistry* 16, 1405-1408.
5. Kohda, H., Kasai, R., Yamasaki, K., Murakami, K., and Tanaka, O. (1976). New sweet diterpene glucosides from *Stevia rebaudiana. Phytochemistry* 15, 981-983.
6. Nieman, C. (1957). Licorice. Advances in food research. In *Advances in Food Research*, Vol. 7, Academic, New York.
7. Ruzicka, L., Jeger, O., and Ingold, W. (1943). New evidence for a different place of the carboxyl group in oleanolic acid. *Helv. Chim. Acta* 26, 2278-2282.
8. Voss, W., and Butter, G. (1937). The isomerism of glycyrrhetic acid. *Chem. Ber.* 70B, 1212-1218.
9. Voss, W., Klein, P., and Sauer, H. (1937). Glycyrrhizin. *Chem. Ber.* 70B, 122-132.
10. Beaton, J. M., and Spring, F. S. (1955). The configuration of the carboxyl group in glycyrrhetic acid. *J. Chem. Soc.* 3126-3129.
11. Jizba, J., Dolejs, L., Herout, V., and Sorm, F. (1971). The structure of osladin–The sweet principle of the rhizomes of *Polypodium vulgare* L. *Tetrahedron Lett.* 18, 1329-1332.
12. Jizba, J., Dolejs, L., Herout, V., Sorm, F., Fehlhaber, H., Snatzke, G., Tschesche, R., and Wulff, G. (1971). Polypodosaponin, a new type of saponin from *Polypodium vulgare* L. *Chem. Ber.* 104, 837-846.
13. Swingle, W. T. (1941). Lo Han Kuo, *Momordica grosvenori. J. Arnold Arbor. Harv. Univ.* 22, 198.
14. Lee, C. H. (1975). Intense sweetener from Lo Han Kuo (*Momoridca grosvenori*). *Experientia* 31(5), 533-534.
15. Lee, C. H. (1980). Personal communication, RJR Foods, Inc., Winston-Salem, N.C.
16. Inglett, G. E. (1974). Sweeteners, new challenges and concepts. In *Symposium: Sweeteners* (Inglett, G. E., ed.), Avi, Westport, Conn.
17. Inglett, G. E., and Findlay, J. C. (1967). Serendipity berry–Source of a new macromolecular sweeteners. Abstract paper 75A presented to the Division of Agriculture and Food Chemistry, 154th American Chemical Society Meeting, Chicago, Ill., September 10-15, 1967.
18. Inglett, G. E., and May, J. F. (1968). Tropical plants with unusual taste properties. *Econ. Bot.* 22, 326-331.
19. Inglett, G. E., and May, J. F. (1969). Serendipity berries (*Dioscoreophyllum cumminsii*)–Source of a new intense sweetener. *J. Food Res.* 34, 408-411.

20. Morris, J. A., and Cagan, R. H. (1972). Purification of monellin, the sweet principle of *Dioscoreophyllum cumminsii. Biochim. Biophys. Acta* 261, 114-122.
21. van der Wel, H. (1972). Isolation and characterization of the sweet principle from *Dioscoreophyllum cumminsii* (Stapf) Diels. *FEBS Lett.* 21, 88-90.
22. van der Wel, H., and Loeve, K. (1973). Characterization of the sweet-tasting protein from *Dioscoreophyllum cumminsii* (Stapf) Diels. *FEBS Lett.* 29, 181-184.
23. Morris, J. A., Martenson, R., Deibler, G., and Cagan, R. H. (1973). Characterization of monellin, a protein that tastes sweet. *Biol. Chem.* 248, 534-539.
24. Bohak, B., and Li, S.-L. (1976). The structure of monellin and its relation to the sweetness of the protein. *Biochim. Biophys. Acta* 427, 153-170.
25. van der Wel, H. (1974). Katemfe, serendipity berry, miracle fruit. In *Symposium: Sweeteners* (Inglett, G. E., ed.), Avi, Westport, Conn.
26. van der Wel, H., and Loeve, K. (1972). Isolation and characterization of thaumatin I and II the sweet-tasting proteins from *Thaumatococcus danielli. Eur. J. Biochem.* 31, 221-225.
27. van der Wel, H., Soest, Van T. C., and Royers, E. C. (1975). Crystallization and crystal data of thaumatin I, a sweet-tasting protein from *Thaumatococcus danielli* Benth. *FEBS Lett.* 56, 316-317.
28. Higginbotham, J. D. (1977). Extraction of a sweet substance from *Thaumatococcus danielli* fruit. U.S. Patent 4,011,206.
29. Higginbotham, J. D. (1979). Protein sweeteners. In *Developments in Sweeteners–1* (Hough, E. A. M., Parker, K. J., and Vlitos, A. J., eds.), Applied Science, London.
30. Inglett, G. E., Dowling, B., Albrecht, J. J., and Holgan, F. A. (1965). Taste-modifying properties of miracle fruit (*Synsepalum dulcificum*). *J. Agric. Food Chem.* 13, 284-289.
31. Brouwer, J. N., van der Wel, H., Francke, A., and Henning, G. J. (1968). Miraculin, the sweetness-inducing protein from miracle fruit. *Nature* 220, 373-374.
32. Kurihara, K., and Beidler, L. M. (1968). Taste-modifying protein from miracle fruit. *Science* 161, 1241-1243.
33. Mazur, R. H. (1974). Aspartic acid based sweeteners. In *Symposium: Sweeteners* (Inglett, G. E., ed.), Avi, Westport, Conn.
34. Mazur, R. H. (1984). The discovery of aspartame. In *Aspartame: Physiology and Biochemistry* (Stegink, L. D., and Filer, L. J., Jr., eds.), Marcel Dekker, New York, Chap. 1.
35. Horowitz, R. M., and Gentili, B. (1963). Dihydrochalcone sweetening agent. U.S. Patent 2,087,821.
36. Krbechek, L., Inglett, G., Holik, M., Dowling, B., Wagner, R., and Riter, R. (1968). Dihydrochalcones. Synthesis of potential sweetening agents. *J. Agric. Food Chem.* 16, 108-112.
37. DuBois, G. Z., Crosby, G. A., Stephenson, R. A., and Wingard, R. E., Jr. (1977). Dihydrochalcone sweeteners. Synthesis and sensory evaluation of sulfonate derivatives. *J. Agric. Food Chem.* 25, 763-772.

38. Crosby, G. A., and Wingard, R. E., Jr. (1979). A survey of less common sweeteners. In *Developments in Sweetners–1* (Hough, E. A. M., Parker, K. J., and Vlitos, A. J., eds.), Applied Science, London.
39. Gruettemann, K. H. (1981). Private communication, Hoechst A. G., Frankfurt, West Germany.
40. Beck, K. M. (1974). Synthetic sweeteners: Past, present, future. In *Symposium: Sweeteners* (Inglett, G. E., ed.), Avi, Westport, Conn.

METABOLISM

3

Absorption of Peptides, Amino Acids, and Their Methylated Derivatives

David M. Matthews
Vincent Square Laboratories, Westminster Hospital, London, England

INTRODUCTION

Pure proteins, though they contribute to the texture of foodstuffs, are said to be tasteless, but small peptides and amino acids are important in determining the flavor of various foods (1). They can be sour, bitter, savory, metallic-tasting, or sweet, though a number of peptides are without taste (2). The taste of a peptide is not necessarily related to its constituent amino acids, and may be totally different. Some years ago Mazur and colleagues (3) found that some small synthetic peptides were extremely sweet. In particular, aspartylphenylalanine O-methyl ester (aspartame) was found to be nearly 200 times as sweet as sucrose and a likely candidate for an innocuous synthetic sweetening agent. This discovery naturally made the question of the intestinal handling of peptides, methylated peptides, and amino acids a focus of great interest. In this chapter I shall outline present-day views on the intestinal handling of peptides and amino acids, paying particular attention to peptides and the treatment of methylated derivatives.

The history of the study of the digestion and absorption of protein digestion products is peculiarly fascinating and illustrates the folly of neglecting as irrelevant the literature of 50-100 years ago. This aspect of the subject, has, however, received extensive and fully documented coverage in recent years, (1,4-7). Until recently it was generally believed that proteins were completely hydrolyzed to free amino acids within the intestinal lumen before uptake by the absortive cells. This notion, the "classical hypothesis of protein absorption," remained virtually unshaken by publications during the 1950s and early 1960s, in particular the pioneering work of Newey and Smyth (see reviews in Refs. 4-8), suggesting the possibility of uptake of intact peptides. In 1968 independent publications from my laboratory and that of Adibi showed that di- and tripeptides could be absorbed more rapidly than the equivalent free amino acids in man and rat. At the time these findings, which are now so familiar as to be commonplace, seemed almost unbelievable, since they showed that the classic hypothesis of protein absorption must be incorrect, and strongly suggested mucosal uptake of intact peptides. It is no exaggeration to say that at the time some people doubted the mental stability of the heretical authors (9), but as the work was extended and confirmed by other laboratories, the initial skepticism began to subside. Acceptance of the idea that the intestinal transport of small peptides occurred was helped by demonstrations in 1972 and 1973 that di- and tripeptides were taken up by an active process, and by the realization that peptide transport took place elsewhere in the animal body (1,4,7,10). Furthermore, peptide transport had for some years been recognized as an essential process in the nutrition of certain microorganisms (1,4,7,10). The wide biological distribution of active transport of small peptides was emphasized when Sopanen demonstrated the occurrence of this process in germinating barley. The process has since been shown to occur in several other germinating seeds and probably occurs in the leaves of carnivorous plants (7).

The present accepted position by workers in the field of intestinal absorption is that the absorption of protein digestion products involves (a) uptake of intact di- and tripeptides by the absorptive cells followed by intracellular hydrolysis which may or may not be complete and (b) uptake of free amino acids liberated in the intestinal lumen by the peptidases of the brush border. It is not yet possible to state the relative quantitative importance of the two processes, though peptide uptake may well be the major mode of absorption of protein digestion products. Among the pieces of evidence supporting this position is the observation that patients with Hartnup disease, who have lost their ability to absorb most neutral amino acids, remain well nourished and are usually free from obvious intestinal disturbance. This apparently is due to their ability to absorb amino acids in peptide form.

The following is a brief summary of available information on the absorption of small peptides and of free amino acids. Original references will be given only when very recent or where a point is of particular interest. Where references are not given, they will be found in Refs. 1, 4-7, and 10-20. Standard three-letter abbreviations will be used for amino acid residues in peptides. Unless otherwise indicated, it will be assumed that amino acids which have stereoisomers are in the L form.

PROTEIN DIGESTION

The intralumenal phase of protein digestion is initiated by pepsin(s) (there are many isoenzymes) in the acid pH of the stomach and is continued in the small intestine by the proteolytic enzymes of the pancreas. Preliminary gastric digestion is relatively unimportant, and if it is absent, as in pernicious anemia, physiological disturbance is minimal. Absence of pancreatic proteases, however, leads to a severe defect in protein absorption. In man and animals most investigations suggest that 1-3 hr after ingestion of a protein meal, the lumenal contents consist of a mixture of peptides and amino acids in which peptides predominate. On the basis of a limited number of studies, the mean chain length of the peptides is estimated at three to six amino acid residues. Peptide hydrolysis is continued by brush border peptidases. The brush border may also contain adsorbed pancreatic proteases. Peptides of two and three amino acid residues and free amino acids are taken up by the absorptive cells, with peptide uptake probably the major mode of transport. The degree to which di- and tripeptides are hydrolyzed in the brush border is a function of their structure. Hydrolysis of absorbed peptides is continued by the peptidases present in the cytosol of the absorptive cells. These peptidases are quite distinct from those of the brush border.

Though it is generally believed that protein digestion products enter the portal blood largely as free amino acids, certain di- and tripeptides, which are exceptionally resistant to hydrolysis, enter the portal blood stream intact and may even appear in the urine. These include peptides of proline and hydroxyproline, found in gelatin, carnosine (β-Ala-His), which occurs in chicken meat, and peptides containing D-amino acid residues (which are not normally encountered on an appreciable scale in the human diet). In spite of textbook assertions that in general only free amino acids enter the portal blood, there is still uncertainty about the extent to which other peptides enter the blood after meals of different proteins (21). For example, Sleisenger et al. (22) provided indirect evidence compatible with substantial entry of intact peptides from a partial hydrolysate of casein (which was likely to contain hydrolysis-resistant phosphopeptides) into the blood of the guinea pig. Subsequent investigations (23) suggested that the proportion of dietary nitrogen entering the blood in peptide form was of the order of 10%. Gardner (21) suggested that the extent of entry of peptides into the circulation is mainly dependent on two factors: (a) the nature of the protein fed and (b) food processing, especially overheating, which can produce chemically altered hydrolysis-resistant peptides (17). Small peptides infused into the blood are readily taken up by liver, kidney, muscle, and other tissues and utilized (after hydrolysis) as effectively as free amino acids. It is remarkable that after a century of investigation we are uncertain about the extent to which peptides do enter the portal and peripheral blood. Early investigators were handicapped by inadequate analytical techniques; however, methods are now available that should be capable of solving this problem. Polypeptides of 8-10 or more amino acids do not appear to enter the portal blood on a substantial scale. Mixtures of such peptides have potent adverse biological effects and can produce hypotension and fatal "peptone shock," a fact known for more than 100 years. Much of the material present in the lumen of the small intestine is potentially lethal except in trace quantities. Only the barrier of the intestinal mucosa, with its numerous peptidases, protects us from such toxic compounds.

In man, the dog, and common laboratory animals, protein digestion and absorption are rapid processes, often being complete after a few hours. In some reptiles, capable of ingesting large or very large animals such as pigs, goats, or deer, the processes may take days or weeks, as befitting such Gargantuan repasts.

INTESTINAL ABSORPTION OF PEPTIDES

Active Absorption of Small Peptides

Though the work of Newey and Smyth and early work from my laboratory showed that peptide uptake by the absorptive cells of the small intestine was reduced by anoxia and metabolic inhibitors, this was not conclusive evidence for active uptake of peptides, that is, uptake dependent on metabolic energy. This is because the experiments, which were carried out in vitro, involved incubation times of 20 min or longer, and the intestine deteriorates over such periods. The only convincing experimental evidence of active transport is transport against an apparent electrochemical gradient. In 1971 Payne (10) told me that glycylsacrosine (Gly-Sar) was concentrated by the bacterium *Escherichia coli*. Gly-Sar may be regarded as Gly-Gly with the hydrogen of the peptide bond replaced by a methyl group, which makes the bond exceptionally resistant to hydrolysis. We immediately tried this dipeptide with hamster jejunum in vitro and found it to be concentrated against an apparent electrochemical gradient. The ability to concentrate the dipeptide was abolished by Na^+ replacement, anoxia, and metabolic inhibitors. Similar results were obtained with the tripeptide Gly-Sar-Sar, though the maximum concentration attained was not quite so great. We had high hopes for the tetrapeptide Gly-Sar-Sar-Sar, but this proved to be extremely poorly taken up. The results suggested that active intestinal uptake of peptides was confined to those of two and three amino acid residues. Anticipating the criticism that Gly-Sar and Gly-Sar-Sar were "unphysiological" peptides, we next tried carnosine (β-Ala-His), since this peptide occurs in the diet of carnivores and omnivores. Carnosine was also concentrated, providing evidence of active uptake similar to that given by Gly-Sar and Gly-Sar-Sar.

Effects of Na^+ Replacement on the Intestinal Transport of Small Peptides

Most experiments in vitro have shown peptide transport to be drastically reduced by Na^+ replacement. Rubino et al. (24), in a classic study of the uptake of Gly-Pro by rabbit ileum in vitro, showed that influx of this peptide was greatly reduced, though not completely abolished, by Na^+ replacement. The effect of Na^+ replacement by choline was to decrease V_{max} for Gly-Pro influx, whereas that on Gly influx was different, involving an increase in K_t. We have reported that Na^+ replacement abolished the ability of hamster jejunum to concentrate Gly-Sar, Gly-Sar-Sar, and carnosine; it also reduced the uptake of another hydrolysis-resistant peptide which was almost certainly actively taken up, β-Ala-Gly-Gly.

On the basis of these observations, we reported that active transport of small peptides was Na^+ dependent, and that peptide uptake, like uptake of amino acids, might be an Na^+-coupled transport process. Ward and Boyd (25), studying the effects of amino acids and peptides on the electrical properties of the brush border membrane of the small intestine of the mud puppy, noted that peptide-induced depolarizations were less markedly reduced in the absence of external Na^+ than the depolarizations caused by amino acids. They suggested that this might be because intracellular peptidase activity kept the intracellular concentration of the peptides very low, so that there was a maintained driving force for Na^+-coupled peptide entry even when extracellular Na^+ was substantially reduced; they did not commit themselves definitely to any one hypothesis.

Our first experiment on Na^+ replacement in peptide absorption studies in vivo, using high concentrations of Gly-Gly in rat ileum, showed that Na^+ replacement had no effect. This did not surprise us, since replacement of lumen Na^+ in vivo usually has little or no effect on the transport of neutral amino acids, a known Na^+-dependent process. However, Cheeseman and Parsons (26) subsequently reported that uptake of Gly-Leu (10 mM) by the small intestine of the frog *Rana pipiens* in vivo was unaffected by Na^+ replacement, whereas uptake of the equivalent mixture of free amino acids was inhibited. They suggested that peptide uptake might be Na^+ dependent only when concentrative. [This suggestion is incompatible with the observations Rubino et al. (24), made under conditions of influx.] It was further reported that *exit* from the absorptive cells of glycine and leucine taken up as Gly-Leu was reduced by removal of Na^+ from the vascular perfusate. As a result of this and of work with carnosine in frog intestine in vitro, it was suggested that Na^+ might have a role in peptide absorption, but at a later stage than that of peptide uptake. Himukai et al. (27) reported that influx into guinea pig small intestine of Gly-Gly and Gly-Leu was independent of the presence of Na^+.

Ganapathy and Leibach (28) have reported that uptake of Gly-Pro by intestinal brush border vesicles was unaffected by Na^+ replacement, that peptide transport across the brush border membrane was Na^+ independent, and that consequently "A long-standing question about the possible Na^+-dependence of transmembrane peptide transport has now been resolved." They did, however, concede that some form of linkage between metabolic energy and peptide transport, present in intact tissue, might be absent in the vesicle preparation.

At the moment all one can say is that there is no reasonable doubt that removal of Na^+ from preparations of relatively intact small intestine in vitro greatly reduces peptide uptake, but the exact details of the relationship between Na^+ and peptide transport are still obscure.

Effects of Peptides on the Electrical Potential Difference Across the Brush Border

The most recent and thorough investigation of the electrical effects of small peptides across the brush border has been reported by Ward and Boyd (25). They found that peptides, like amino acids, caused a reduction in the potential difference across the brush border, suggesting coupling of peptide flux across the border to that of another ionic species, probably Na^+. Their work also provided supporting evidence for a number of conclusions already reached about peptide uptake, such as its independence of uptake of amino acids.

Earlier results on the effects of peptides on the electrical properties of the small intestine have been reviewed by Matthews (4) and Matthews and Payne (7).

Effects of Peptides on the Intestinal Transport of Na^+ and Water

Individual peptides, like amino acids, stimulate absorption of Na^+ and water (see the review in Ref. 7). However, in 1980 Silk et al. (29) reported that a partial hydrolysate of fish protein, though extremely well absorbed in relation to the equivalent amino acid mixture, unexpectedly appeared to promote intestinal *secretion* of Na^+ and water. This contrasted with their results obtained with two partial hydrolysates of casein and one of lactalbumin, as well as the equivalent amino acid mixtures, which stimulated absorption of Na^+ and water from human jejunum. This effect of the fish protein hydrolysate, which they could not explain, would be a most undesirable feature if such a hydrolysate were used therapeutically. Thus the absorptive properties of partial hydrolysates probably depend on (a) the nature of the original protein and (b) the method of preparation of the hydrolysate.

Influence of Molecular Structure on the Uptake and Hydrolysis of Peptides by the Small Intestine

Because of the incomplete commercial availability of small peptides and their derivatives, and because no worker on intestinal peptide transport has yet had the full-time assistance of a chemist skilled in the synthesis of peptides and their derivatives, knowledge of the influence of structure on the transport and hydrolysis of small peptides is sketchy and incomplete; no systematic study has yet been made. The following account, which is based mainly on observations in rat and hamster, is condensed largely from Matthews and Payne (7).

Peptide Bond(s)

The presence of an α-peptide bond or bonds enables di- and tripeptides to make use of the specific peptide uptake system(s), while making them unacceptable to the uptake systems for free amino acids. A peptide link involving a β bond does not inevitably preclude transport, but a bond involving a γ linkage either reduces affinity for peptide transport or abolishes it, as in the case of glutathione (γ-Glu-Cys-Gly).

Methylation of the nitrogen of peptide bonds, as in Gly-Sar and Gly-Sar-Sar, leads to resistance (though not 100% resistance) to hydrolysis, but is compatible with active uptake. It probably does not cause much if any reduction in affinity for the uptake mechanism (Matthews and Burston, unpublished observations).

Terminal Groups

Substitution of either the amino terminal or carboxy terminal groups of small peptides reduces or abolishes affinity for transport. Methylation of the NH_2-terminal group, as in Sar-Gly, results in slow uptake and resistance to hydrolysis. Aspartame (Asp-PheOMe) appears to have some affinity for peptide transport, since it can inhibit uptake of Gly-Sar-Sar. However, it is a less powerful inhibitor than Asp-Phe, suggesting that either its affinity for the uptake mechanism(s) is less than that of the parent peptide or that it undergoes more brush border hydrolysis. Unfortunately, we have not investigated which of these possibilities applies (4,7,30). Recently, we have carried out further experiments (using rings of everted hamster jejunum in vitro) that show that Asp-PheOMe inhibits uptake of the hydrolysis-resistant dipeptide Gly-Sar. At infinitely high concentrations, Asp-PheOMe can totally inhibit mediated uptake of Gly-Sar (D. M. Matthews and D. Burston, unpublished data). This confirms that Asp-PheOMe can be taken up by the same pathway(s) available to small peptides of dietary origin and suggests that hydrolysis of this compound may be at least partly intracellular.

Amino Acid Side Chains

In the case of free amino acids, the charge on the amino acid side chain is extremely important, apparently dividing amino acids into three major "transport groups"—neutral, basic, and acidic amino acids—which seem to be taken up in large part by different mechanisms. If the results of a pilot investigation in our laboratory using Gly-Sar, Lys-Lys, and Glu-Glu (31) are representative, this is not true of dipeptides; all these peptides appeared to be taken up by the same system(s). In this case mammalian gut resembles *E. coli*, in which the peptide uptake systems are indifferent to the charge on the amino acid side chains.

In the case of amino acids, the kinetic characteristics of uptake are apparently related to the lipophilic properties of the side chain, being accompanied by an increase in apparent affinity for intestinal transport and a decrease in maximum transport velocity (12,13). Thus alanine, in which the single-hydrogen side chain of glycine is replaced by a methyl group, has a higher apparent affinity for transmural transport and a lower maximal transport velocity than glycine. These characteristics become progressively more marked in valine [side chain $(CH_3)_2CH-$] and leucine [$(CH_3)_2CHCH_2-$]. In this case, methyl and related groups have a clear-cut influence on the kinetic characteristics of transport. We are currently investigating whether the same is true of dipeptides, as has been suggested by several authors. The investigations so far are incomplete, though they have already yielded unexpected results (32). For example, we have found that Leu-Leu can completely inhibit the uptake of Gly-Sar in a competitive manner, but that neither Gly-Sar nor Gly-Gly can inhibit uptake of Leu-Leu (33). At this time it does not seem likely that the results will parallel those obtained with amino acids.

The three sections above show that the introduction of a methyl group or groups into the peptide molecule has different effects on intestinal transport and hydrolysis according to the site or sites of introduction, as common sense might suggest.

Stereochemical Specificity

The stereochemical specificity of intestinal peptide uptake is very marked, inclining strongly toward peptides of L-amino acids and glycine (which has no optically active isomers). The introduction of one D-amino acid residue into a dipeptide makes it very poorly taken up and strongly resistant to hydrolysis. With peptides containing more than one D-amino acid residue, these effects are intensified.

Maximum Size of Peptides Taken Up

With one exception, all reports indicate that large-scale intestinal uptake of peptides by active transport is confined to peptides of two and three amino acid residues. There has been a claim that Leu-Gly-Gly-Gly was absorbed intact by rat jejunum; however, we have been unable to confirm the large-scale uptake of tetrapeptides. However, I certainly would not deny the possibility of absorption of tetrapeptides, and perhaps even larger peptides, on a restricted scale.

Competition for Intestinal Uptake Between Peptides

A large number of experiments have demonstrated inhibition of uptake of one di- or tripeptide by another. In a substantial proportion of cases where inhibitory effects have been found, inhibition has been shown to be competitive in type. These data suggest that many di- and tripeptides, including the reference hydrolysis-resistant peptides Gly-Sar and Gly-Sar-Sar, share the same uptake system or systems. The question of whether there is one or more than one peptide uptake system is still open to debate. Over the years there have been claims that multiple peptide uptake systems exist in the small intestine. Initially, we were skeptical about these claims (7); however, our results are now compelling us to consider the possibility that there may be at least two peptide uptake systems. In a collaborative study with Fairclough, Silk, and others, we found that the effect of a high concentration of Gly-Gly on the uptake of peptides from a partial hydrolysate of casein was negligibly small. More recently, we found, using hamster jejunum in vitro, that the inhibitory effect of Val-Val on the uptake of Gly-Sar was greater than the apparent K_t of Val-Val would indicate, suggesting that there might be a second high-affinity component in the uptake of Val-Val which had passed undetected by kinetic analysis (32). Finally, we have had to consider the possibility of multiple uptake systems as a possible explanation of the failure of Gly-Sar and Gly-Gly to inhibit the uptake of Leu-Leu. In retrospect, the work of

Rubino et al. (24), indicating that there might be two systems (a "high-affinity" and a "low-affinity" system) for the uptake of Gly-Pro by rabbit ileum, supplies the best evidence suggesting the existence of more than one peptide uptake system, provided that one assumes that this peptide, as in two other species, undergoes no brush border hydrolysis and that nonmediated transport plays no role. The recent studies of Himukai and his colleagues (27) indicate that Gly-Gly is incapable of inhibiting the intestinal uptake of Gly-Leu. This suggests that the transport interactions among dipeptides may in fact be much less straightforward than previously supposed.

While it is accepted that *E. Coli* has at least two peptide uptake systems—a dipeptide and an oligopeptide system—our work with Gly-Sar and Gly-Sar-Sar suggests that the gut is not like this.

Independence of Intestinal Uptake of Peptides and Amino Acids

Though amino acids liberated from peptides at the brush border may inhibit the uptake of other free amino acids by the small intestine, intact peptides and free amino acids do not inhibit each other's uptake to any major extent, if at all. There is also no doubt that the active uptake of peptides and the active uptake of amino acids by animal small intestine are completely independent processes, as they are in other forms of life, including microorganisms and germinating seeds. Apart from the results of experiments on competition for transport, the independence of amino acid and peptide uptake is shown by the fact that it is possible to saturate mediated uptake of a free amino acid and show that additional uptake of this amino acid may be produced by adding a peptide containing the specific amino acid studied, or by replacing all of the free amino acid present by a high concentration of such a peptide.

It is interesting that the earliest and one of the best pieces of evidence for the independence of intestinal uptake of amino acids and peptides did not come from experiments on competition for transport in animals but, rather, from investigations of a genetic defect of amino acid transport in man—Hartnup disease. In this condition both intestinal absorption and renal reabsorption of most neutral amino acids, including essential ones, are defective. It had long been a mystery why such patients were usually well nourished and without obvious intestinal disturbance. In a crucial experiment, Milne, Navab, Asatoor, and colleagues showed that histidine was poorly absorbed in the free form, but was well absorbed from carnosine (β-Ala-His). Since it could have been argued that the peptide was being taken up by intact uptake sites for β-alanine, we devised a more conclusive experiment in which it was shown that an affected amino acid (phenylalanine) could be absorbed from the dipeptide Phe-Phe. Investigation of two further cases of Hartnup disease gave similar results, and investigators at St. Bartholomew's Hospital, London, obtained analogous results in the absorptive defect of cystinuria. In both disorders the defect in intestinal amino acid transport is one of entry into rather than exit from the absorptive cell. Finnish investigators have shown that there is a disorder, lysinuric protein intolerance, in which there is an exit defect (34,35). In this condition affected amino acids are absorbed no better from peptides than in the free form.

Two further lines of evidence corroborating the independence of amino acid and peptide uptake may be adduced. Firstly, in the rat and hamster the sites of maximal absorptive capacity for the two sets of compounds are different. Peptides are best taken up in the jejunum, and amino acids in the ileum. Secondly, in the same two animals the effects of dietary deprivation on peptide and amino acid absorption are different. In humans it has been reported that peptide absorption is more resistant to intestinal disease than the absorption of free amino acids.

Kinetics of Peptide Absorption

Since the publication in 1968 of the finding that dipeptides could be absorbed more rapidly than the equivalent amino acids, a great deal of work has been done comparing relative rates of absorption or influx into the absorptive cells of di- and tripeptides and their constituent amino acids. Most of this, together with work on the kinetic characteristics of peptide absorption, is summarized by Matthews and Payne (7). The experiments have been carried out in man and several different species of animals. In the majority of cases the relative rapidity of peptide absorption has been confirmed. An early and representative example is shown in Figure 1. When an equimolar mixture of free glycine and methionine is absorbed, absorption of glycine is depressed by the competitive effect of methionine, which has a much higher affinity for a common uptake mechanism. When the equivalent peptides Gly-Met or Met-Gly are absorbed, this depression of absorption of glycine is avoided. The superior absorption of the peptide is not, however, solely the result of avoidance of competitive inhibition. Methionine itself is not only more rapidly absorbed from the peptides than from the mixture, but also more rapidly absorbed than from free methionine alone.

In many instances absorption of peptides is more rapid than that of the equivalent free amino acids at high concentrations, although their rates of absorption are approximately equal at lower concentrations. In a few cases these differences in absorption rates have been observed only at low concentrations, and in some instances peptides have been reported to be absorbed more slowly than the equivalent amino acids. In much of the early work rates of absorption were studied at one concentration only. In 1980 we reported (36) a full kinetic investigation of the influx of lysine and Lys-Lys into hamster small intestine in vitro. At low concentrations (below 5 mM Lys-Lys) mediated uptake of lysine was *less* rapid from the peptide than from the equivalent amino acid, whereas at higher concentrations it was *more* rapid from the peptide than from the equivalent amino acid. This illustrates the danger of drawing conclusions from work carried

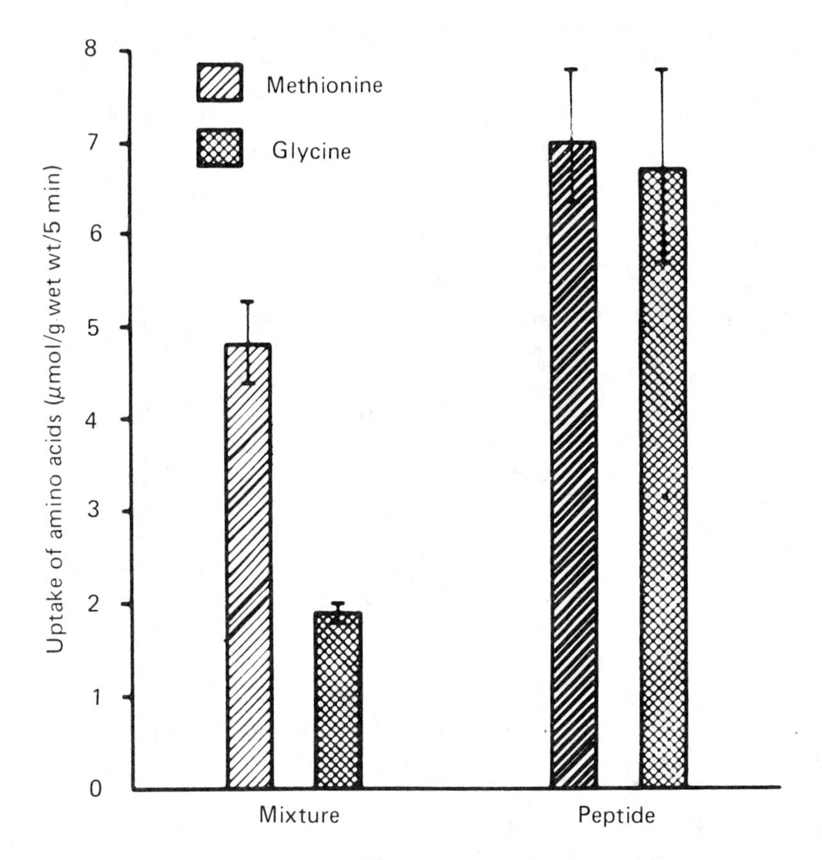

Figure 1 Absorption from rat small intestine of equivalent amounts of a mixture of methionine (5 mM) and glycine (5 mM) and the dipeptide Gly-Met (5mM). The slow absorption of glycine from the mixture is in large part due to inhibition of its absorption by methionine. When the peptide is absorbed, this competition is avoided, and both glycine and methionine are absorbed more rapidly than from a mixture of the free amino acids or from either amino acid (5 mM) alone. (From Ref. 16.)

out at single concentrations or over a restricted concentration range. More recently we found that uptake of valine from Val-Val was *slower* than uptake of valine from the equivalent free amino acid over a very wide concentration range (0.1-100 mM valine) (32). On the other hand, the reverse was true of leucine and Leu-Leu, with mediated uptake of leucine being faster from Leu-Leu than from free leucine at all concentrations from 0.1 to 100 mM (33).

Adibi has suggested that the absorption of amino acids from peptides is more rapid than the absorption of free amino acids because peptides have a higher V_{max}. This may be true in some cases, but our own experience suggests that it is not a valid generalization. The whole question of the kinetics of the intestinal uptake of peptides and the constituent amino acids requires systematic investigation. All that can be said at the moment is that (a) the kinetic characteristics of intestinal uptake of peptides and the equivalent amino acids are different and (b) for reasons which are not yet fully understood, amino acids are very frequently found to be absorbed more rapidly from peptides than from the free form(s).

Nearly all investigations of the kinetics of mediated uptake of peptides have suggested that this conforms to Michaelis-Menten kinetics, though under our conditions, using rings of everted hamster small intestine in vitro, we find a small nonsaturable component in uptake which may represent simple diffusion. If this is not corrected for [in addition to correcting for uptake of free amino acid(s) liberated in the brush border] plots for the relationship between uptake and concentration are biphasic. This could be interpreted as indicative of the presence of two uptake systems (32). Rubino et al. (24), whose work suggested the existence of two mediated systems for uptake of Gly-Pro, did not consider the possibility of a nonsaturable component in uptake. This makes their conclusions less firm than they might be. If there are multiple mediated uptake systems for peptides which cannot be resolved by ordinary kinetic analysis, then, unfortunately, we have a situation in which all our painstakingly extracted K_t and V_{max} values are false, intermediate between the "true" kinetic parameters of the uptake systems present. The study of the kinetics of uptake of intact peptides is in fact extremely difficult, particularly because in most cases corrections for the component representing possible diffusion and that representing uptake in the form of free amino acid(s) must both be carried out.

Absorption of Partial Hydrolysates of Proteins and the Possible Nutritional Significance of Peptide Absorption

It is obviously not feasible to undertake a study of the relative rates of absorption of all possible individual di- and tripeptides and their equivalent amino acids, since there are at least 400 (20^2) possible "normal" dipeptides and some 8000 (20^3) possible "normal" tripeptides.

Accordingly, we compared the rates of absorption in the rat of partial pancreatic hydrolysates of several proteins consisting largely of oligopeptides with those of the equivalent amino acid mixtures. The results were encouraging. In all cases total amino acid nitrogen was absorbed at a substantially greater rate from the partial hydrolysates than from the amino acid mixtures; in the case of a hydrolysate of lactalbumin the rate was about twice as rapid. Subsequently, I found that essentially similar experiments with "peptones" and amino acids, with similar results, had been performed before 1914.

Jejunal perfusion studies in man replicated the results found in the rat and showed an additional and very striking phenomenon. Percentage absorption rates of individual free amino acids from a mixture simulating casein vary greatly.

However, when a partial hydrolysate of casein containing both free amino acids and peptides was studied, these differences were largely "ironed out," owing to the fact that those amino acids absorbed slowly in the free form were much more rapidly absorbed in the peptide-bound form from the hydrolysates. These results and subsequent results obtained with hydrolysates of other proteins are summarized by Silk et al. (29) (Fig. 2). I have repeatedly stressed (1) that these results might be of utmost nutritional significance. When an amino acid mixture is absorbed, some amino acids are absorbed rapidly and others very slowly. This means that peak plasma concentrations occur at different times for different amino acids. This "temporal displacement" will lead to nonsimultaneous presentation of amino acids to the tissues, a state of affairs known to impair the efficiency of protein synthesis. This unnatural situation is avoided when a partial hydrolysate of protein is absorbed. This might account for the many reports that amino acid mixtures simulating proteins are nutritionally inferior to the proteins themselves. Though this hypothesis was originally mooted as early as 1971, it has, so far as I know, attracted little attention. However, nutritionists can no longer afford to ignore the possible implications of peptide absorption. Apart from the foregoing, the differences in the rates of peptide and amino acid

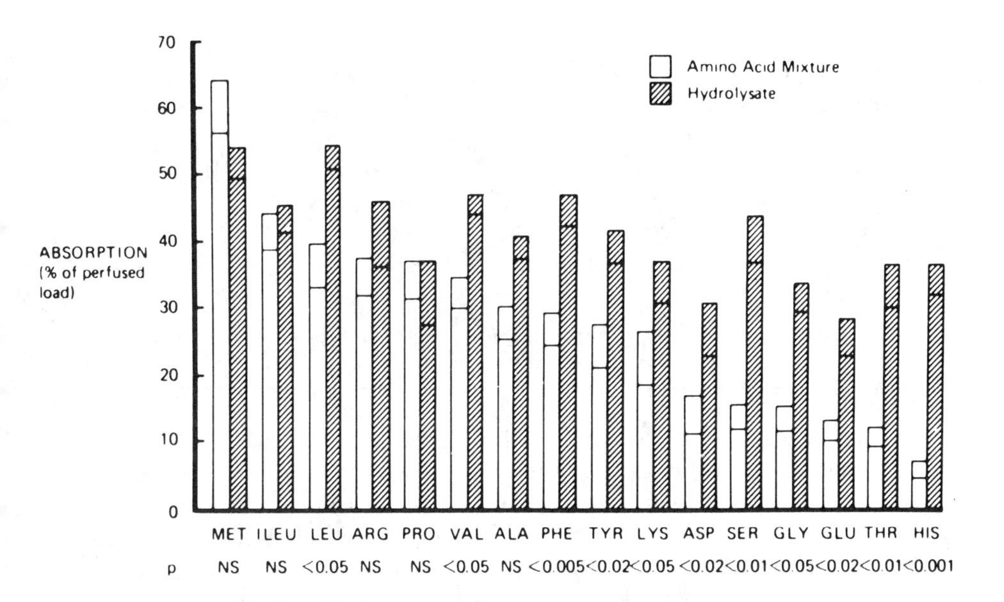

Figure 2 Absorption from human jejunum of (a) a mixture of free amino acids simulating lactalbumin and (b) a partial hydrolysate of lactalbumin containing largely peptides (open columns, free amino acid absorption; shaded columns absorption from partial hydrolysate). The horizontal lines across the columns represent the standard error of the mean. (From Ref. 29.)

absorption may have a bearing on the question of amino acid supplementation of inadequate diets. Should the supplement be in amino acid or peptide form? It may also be that the apparent requirement for an essential amino acid differs according to whether free or peptide-bound amino acid is given.

Most workers in the field consider that peptide uptake is likely to account for more absorption than that of free amino acids, and Adibi (personal communication) has suggested that it may be responsible for 90%. My own estimate would be more conservative, probably more than half.

Absorption of Biologically Active Peptides

Few biologically active peptides are absorbed on a large scale by the mechanism(s) responsible for the absorption of di- and tripeptides of dietary origin. This is because many are too large for mediated uptake, or if small enough structurally unsuitable. For example, the tripeptide thyroliberin (pyroGlu-His-ProNH_2) has both terminal groups "blocked," and though extremely resistant to hydrolysis, is only absorbed in significant amounts at high oral dose levels. Exceptions to these rules are carnosine (probably a neurotransmitter peptide and perhaps involved in wound healing) and the antibiotic cephalexin. In spite of the foregoing, it is probable that many biologically active peptides may be absorbed on a small scale by mechanisms which are poorly understood. Possible mechanisms have been outlined elsewhere (1,4,7). Even a peptide as large as insulin may be absorbed to some extent. Since many biologically active peptides are extremely potent, even small-scale absorption could lead to significant effects. The possibility of absorption of significant amounts of biologically active peptides of dietary origin, which might exert hormonal or kinin-like effects (1), or influence the central nervous system by enkephalin-like or neurotransmitter effects (21, 37,38), is exciting and calls for thorough investigation.

ABSORPTION OF METHYLATED AMINO ACIDS

Amino Acid Absorption

To understand the absorption of methyl derivatives of amino acids, a brief sketch of the mechanism of amino acid absorption is necessary. Useful references are Wiseman (11), Matthews and Laster (14), Lerner (18), and Johnson (20).

Amino acids are taken up from the intestinal lumen by several stereochemically specific active mechanisms which are almost entirely (or, in the case of dibasic amino acids, partially) Na^+ dependent. The structural requirements for transport are rather similar to those for peptides, except that a peptide bond is not acceptable to the transport systems for free amino acids. In man and experimental animals several uptake systems are recognized: (a) a system for most neutral amino acids; (b) a system for glycine, proline, and hydroxyproline; (c) a system for dicarboxylic amino acids; and (d) a system for dibasic amino acids and cystine. The existence of congenital methionine malabsorption ("blue diaper syndrome") suggests that there may be a distinct transport mechanism for this amino acid in man. Some amino acids are taken up by more than one system, for example, glycine is probably taken up by systems (a) and (b) and the dibasic amino acids may also be taken up, to some extent, by system (a) for neutral amino acids. Passage of amino acids out of the absorptive cells is by a mediated mechanism(s) that have been inadequately studied.

In general, amino acids are not extensively metabolized by the absorptive cells. However, Matthews and Wiseman (39) found that the dicarboxylic amino acids, glutamate and aspartate, were transaminated during passage through the intestinal wall, and it has since been shown that absorption of these amino acids may be represented, wholly or in part, by the appearance of alanine in the blood. One may speculate whether patients with the "Chinese restaurant syndrome," in which adverse effects apparently follow the ingestion of glutamic acid, may not be suffering from intestinal transaminase deficiency.

Absorption of Methyl Derivatives of Amino Acids

The effect of adding methyl groups to the side chain of amino acids, as in alanine, valine, leucine, and isoleucine, is to increase apparent affinity for transport. Replacement of the α-hydrogen by a methyl group, as in α-aminoisobutyric acid, reduces affinity for intestinal transport, transport at low concentrations being minimal (12,13). The transport system responsible for neutral amino acids prefers free terminal groups. It needs a free terminal carboxyl group, and if this is esterified with a methyl group, as in the methyl ester of L-histidine, active transport is abolished (14). N-Methylglycine (sarcosine) and N,N-dimethylglycine are transported by hamster small gut (through N-phenyl-glycine is not, possibly owing to the bulk of the substituent group). There is remarkably little systematic information in the literature on the effect of methylation on the intestinal absorption of amino acids. If one were forced to generalize, it might be said that the addition of methyl groups to the side chain of amino acids enhances affinity for transport. Insertion of methyl groups anywhere else in the molecule may be expected to reduce or abolish affinity for transport (11, 16, 18, 20).

ACKNOWLEDGMENTS

We are deeply grateful for financial support to the National Medical Research Fund, the Frank Odell Charity, Corporate Textiles Limited, and to many private individuals too numerous to mention. This support has temporarily averted closure of our department, enabling the work to be continued. We are indebted to Dr. M. L. G. Gardner for much helpful discussion and assistance with kinetic analysis.

REFERENCES

1. Matthews, D. M., and Payne, J. W. (eds.) (1975). *Peptide Transport in Protein Nutrition*. North Holland, Amsterdam.
2. Schiffman, S. S. (1984). Comparison of the taste properties of aspartame with other sweeteners. In *Aspartame: Physiology and Biochemistry* (Stegink, L. D., and Filer, L. J., Jr., eds.), Marcel Dekker, New York, Chap. 10.
3. Mazur, R. H., Schlatter, J. M., and Goldkamp, A. H. (1969). Structure-taste relationships of some dipeptides. *J. Am. Chem. Soc.* 91, 2684-2691.
4. Matthews, D. M. (1975). Intestinal absorption of peptides. *Physiol. Rev.* 55, 537-608.
5. Matthews, D. M. (1977). Memorial Lecture: Protein absorption–Then and now. *Gastroenterology* 73, 1267-1279.
6. Matthews, D. M., and Adibi, S. A. (1976). Peptide absorption. *Gastroenterology* 71, 151-161.
7. Matthews, D. M., and Payne, J. W. (1980). Transmembrane transport of small peptides. In *Current Topics in Membranes and Transport*, Vol. 14 (Bronner, F., and Kleinzeller, A., eds.), Academic, New York, pp. 331-425.
8. Barcroft, H., and Matthews, D. M. (1982). David Henry Smyth. *Biog. Mem. Fellows R. Soc.* 27, 525-561.
9. Milne, M. D. (1972). *In discussion following* Milne, M. D. Peptides in genetic errors of amino acid transport. In *CIBA Foundation Symposium: Peptide Transport in Bacteria and Mammalian Gut* (Elliott, K., and O'Connor, M., eds.), Associated Scientific Publishers, Amsterdam, p. 104.
10. Elliott, H., and O'Connor, M. (eds.) (1972). *Peptide Transport in Bacteria and Mammalian Gut*, Ciba Foundation Symposium, Associated Scientific Publishers, Amsterdam.
11. Wiseman, G. (1974). Absorption of protein digestion products. In *Biomembranes, Intestinal Absorption*, Vol. 4A (Smyth, D. H., ed.), Plenum, London, pp. 363-481.
12. Matthews, D. M., and Laster, L. (1965). The kinetics of intestinal active transport of five neutral amino acids. *Am. J. Physiol.* 208, 593-600.
13. Matthews, D. M., and Laster, L. (1965). Competition for intestional transport among five neutral amino acids. *Am. J. Physiol.* 208, 601-606.
14. Matthews, D. M., and Laster, L. (1965). Absorption of protein digestion products: A review. *Gut* 6, 411-426.
15. Matthews, D. M. (1971). Protein absorption. *J. Clin. Pathol. Suppl. R. Coll. Pathol.* 5, 29-40.
16. Matthews, D. M. (1971). Experimental approach in chemical pathology. *Br. Med. J.* 3, 659-664.
17. Elliott, H., and O'Connor, M. (eds.) (1977). *Peptide Transport and Hydrolysis*, Ciba Foundation Symposium, Associated Scientific Publishers, Amsterdam.
18. Lerner, J. (1978). A review of amino acid transport processes in animal cells and tissues. University of Maine Press, Orono, Maine.
19. Silk, D. B. A. (1981). Peptide transport. *Clin. Sci.* 60, 607-615.

20. Johnson, L. R. (ed.) (1981). *Physiology of the Gastrointestinal Tract*, Vol. 2, Raven Press, New York, pp. 1073-1122.
21. Gardner, M. L. G. (1982). Absorption of intact peptides. Studies on transport of protein digests and dipeptides across rat small intestine *in vitro*. *Q. J. Exp. Physiol.* 67, 629-637.
22. Sleisenger, M. H., Pelling, D., Burston, D., and Matthews, D. M. (1977). Amino acid concentrations in portal venous plasma during absorption from the small intestine of the guinea-pig of an amino acid mixture simulating casein and a partial enzymic hydrolysate of casein. *Clin. Sci. Mol. Med.* 52, 259-267.
23. Gardner, M. L. G., Lindblad, B. S., Burston, D., and Matthews, D. M. (1981). Entery of peptides into mesenteric blood during absorption of protein in the guinea-pig in-vivo: A reappraisal. *Gastroent. Clin. Biol.* 6, 96.
24. Rubino, A., Field, M., and Schwachman (1971). Intestinal transport of amino acid residues of dipeptides. I. Influx of the glycine residue of glycyl-L-proline across mucosal border. *J. Biol. Chem.* 246, 3542-3548.
25. Ward, M. R., and Boyd, C. A. R. (1982). Micro-electrode study of oligopeptide absorption by the small intestinal epithelium of *Necturus maculosus*. *J. Physiol.* 324, 411-428.
26. Cheeseman, C. I., and Parsons, D. S. (1976). The role of some small peptides in the transfer of amino nitrogen across the wall of vascularly perfused intestine. *J. Physiol.* 262, 459-476.
27. Himukai, M., Kano-Kameyama, A., and Hoshi, T. (1982). Mechanisms of inhibition of glycylglycine transport by glycyl-L-leucine and L-leucine in guinea pig small intestine. *Biochim. Biophys. Acta* 687, 170-178.
28. Ganapathy, V., and Leibach, F. H. (1982). Peptide transport in intestinal and renal brush border vesicles. *Life Sci.* 30, 2137-2146.
29. Silk, D. B. A., Fairclough, P. D., Clark, M. L., Hegarty, J. E., Marrs, T. C., Addison, J. M., Burston, D., Clegg, K. M., and Matthews, D. M. (1980). Use of a peptide rather than free amino acid nitrogen source in chemically defined 'elemental' diets. *J. Parenter. Enter. Nutr.* 4, 548-553.
30. Addison, J. M., Burston, D., Dalrymple, J. A., Matthews, D. M., Payne, J. W., Sleisenger, M. H., and Wilkinson, S. (1975). A common mechanism for transport of di- and tripeptides by hamster jejunum *in vitro*. *Clin. Sci. Mol. Med.* 49, 313-322.
31. Taylor, E., Burston, D., and Matthews, D. M. (1980). Influx of glycylsarcosine and L-lysyl-L-lysine into hamster jejunum *in vitro*. *Clin Sci.* 58, 221-225.
32. Burston, D., Wapnir, R. A., Taylor, E., and Matthews, D. M. (1982). Uptake of L-valyl-L-valine and glycylsarcosine by hamster jejunum *in vitro*. *Clin. Sci.* 62, 617-626.
33. Matthews, D. M., and Burston, D. (1983). Uptake of L-leucyl-L-leucine and glycylsarcosine by hamster jejunum in vitro. *Clin. Sci.* 65, 177-178.
34. Desjeux, J.-F., Rajantie, J., Simell, O., Dumontier, A.-M., and Perheentupa, J. (1980). Lysine fluxes across the jejunal epithelium in lysinuric protein intolerance. *J. Clin. Invest.* 65, 1382-1387.

35. Rajantie, J., Simell, O., and Perheentupa, J. (1980). Basolateral-membrane transport for lysine in lysinuric protein intolerance. *Lancet* 1, 1219-1221.
36. Burston, D., Taylor, E., and Matthews, D. M. (1980). Kinetics of uptake of lysine and lysyl-lysine by hamster jejunum in vitro. *Clin. Sci.* 59, 285-287.
37. Zioudrou, C., and Klee, W. A. (1979). Possible roles of peptides derived from food proteins in brain function. In *Nutrition and the Brain* (Wurtman, R. J., and Wurtman, J., eds.), Raven Press, New York, pp. 125-158.
38. Gardner, M. L. G., Lindblad, B. S., Burston, D., and Matthews, D. M. (1983). Trans-mucosal passage of intact peptides in the guinea-pig small intestine *in vivo*: A reappraisal. *Clin. Sci.* 64, 433-439.
39. Matthews, D. M., and Wiseman, G. (1953). Transamination by the small intestine of the rat. *J. Physiol.* 120, 55P.

4

Aspartate and Glutamate Metabolism

Lewis D. Stegink
University of Iowa College of Medicine, Iowa City, Iowa

INTRODUCTION

Aspartame (L-aspartyl-L-phenylalanine methyl ester) is approximately 40% by weight aspartic acid. Thus it is appropriate to review aspartate metabolism when considering aspartame. Because concerns have been raised about interactions between glutamate and aspartate when aspartame is added to foods already containing monosodium L-glutamate, this chapter will also discuss glutamate metabolism.

METABOLISM

The dicarboxylic amino acids, glutamate and aspartate, occupy unique positions in intermediary metabolism. Relatively high levels of these amino acids are found in the various body tissues. This is particularly true in mitochondria (1-3), where glutamate and aspartate play important roles in nitrogen and energy metabolism. As outlined in Figure 1, both aspartate and glutamate stand at major entry points to the tricarboxylic acid cycle, the functional energy-generating component of the cell.

Glutamate may enter its carbon skeleton into the common metabolic pathways in at least two ways. The first is by transamination, the mechanism commonly used to remove the α-amino group from most amino acids prior to utilization of the carbon chain. The carbon structure of glutamate (as α-ketoglutarate)

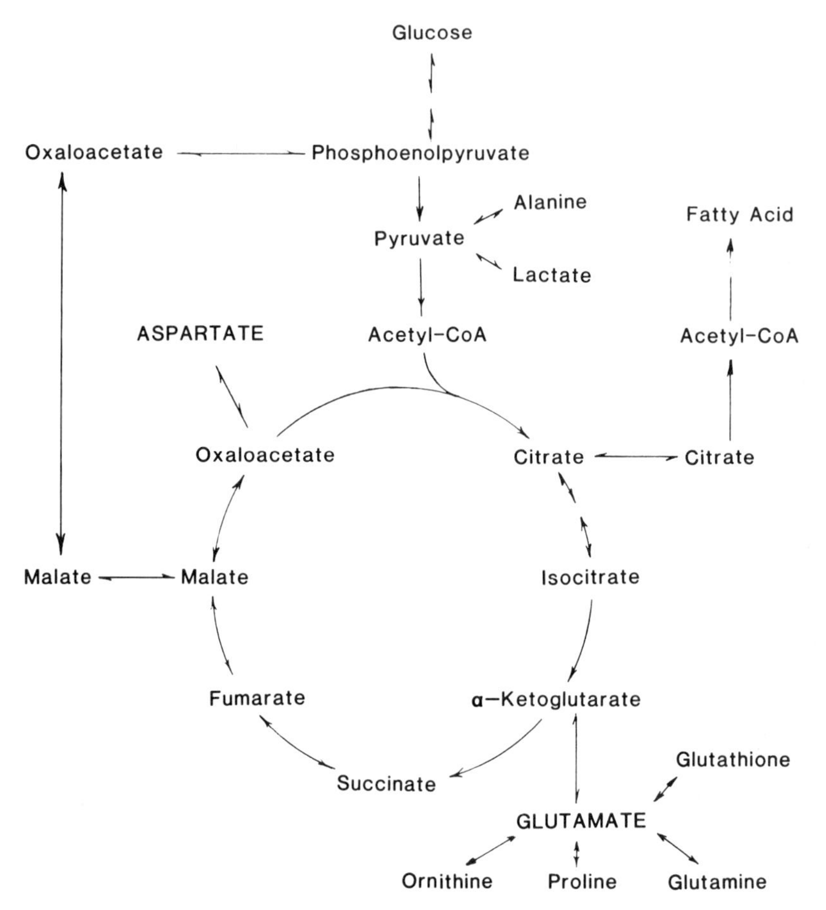

Figure 1 Available pathways of glutamate and aspartate metabolism.

enters metabolism at the level of the tricarboxylic acid cycle, where it is readily converted into carbon dioxide and energy (ATP) as required by cellular demands. Transamination reactions are readily reversible, having a K_{eq} of about unity; thus the flow of nitrogen between amino acids is shifted, depending upon exact cellular requirements. The transaminating enzymes are present in both the cytoplasmic and mitochondrial fractions of the cell, and activity levels of these enzymes are high in those tissues using large amounts of glutamate and aspartate.

Glutamate may also be converted into a α-ketoglutarate and ammonium ions by the action of the enzyme glutamate dehydrogenase. This reaction occurs in both cytoplasmic and mitochondrial fractions of the cell. In the mitochondrion, the NADH generated by this reaction is rapidly oxidized by the electron transport chain, pulling the reaction toward α-ketoglutarate formation. The α-ketoglutarate produced by this reaction also enters the tricarboxylic acid cycle, where it is oxidized to carbon dioxide and water with the generation of ATP. When sufficient energy is available to the cell, mitochondrial isocitrate dehydrogenase is inhibited by the increased levels of ATP and NADH. This results in citrate and malate accumulation. The citrate leaves the mitochondrion, where it is cleaved to oxaloacetate and acetyl-coenzyme A (acetyl-CoA) by the citrate cleavage enzyme. The acetyl-CoA produced can be converted to fatty acids and triglycerides, which are released from the liver to the circulation as the very low density lipoproteins. The oxaloacetate produced has several metabolic options. It can be converted to pyruvate by action of the malic enzyme, yielding NADPH required for fatty acid biosynthesis, or it can be converted to phosphoenolpyruvate by phosphoenolpyruvate carboxykinase for synthesis of glucose. The malate which accumulates also leaves the mitochondrion and is converted in the cytoplasm to phosphoenolpyruvate for oxidation to carbon dioxide, or for gluconeogenesis. Glutamate is also readily converted to glutamine, proline, γ-aminobutyrate, and ornithine, depending on the tissue involved.

Aspartate enters metabolism in a similar manner. Its major route of entry is via transamination with other α-ketoacids to produce oxaloacetate. Oxaloacetate released by the cytoplasmic transaminases does not enter mitochondria, but can be readily converted into phosphoenolpyruvate for either oxidation to carbon dioxide (via pyruvate and acetyl-CoA) or for gluconeogenesis. Aspartate entering mitochondria is transaminated to yield oxaloacetate, which is either oxidized in the tricarboxylic acid cycle or converted to citrate or malate, depending upon the mitochondrial energy level. Accumulated mitochondrial citrate and malate are metabolized as described previously.

Both glutamate and aspartate play important roles in the shuttle mechanisms required for the entry of reducing equivalents into mitochondria. The reduced pyridine nucleotides NADH and NADPH produced by the oxidation of substrates in the cytoplasm of the cell are unable to enter mitochondria. Thus the energy trapped in their bonds cannot be utilized by the mitochondrial electron transport chain to produce ATP. To solve this problem, mitochondria use oxidized and reduced substrates, which can pass through the inner mitochondrial membrane. A general system of this type is shown in Figure 2. Oxidized cytoplasmic substrate is reduced by NADH using a cytoplasmic dehydrogenase isozyme. The reduced substrate enters the mitochondrial matrix by an appropriate translocase system, with a counterion moving out of the matrix. Inside the mitochondrial matrix, the reduced substrate is oxidized by available NAD^+ in a reaction catalyzed by the mitochondrial dehydrogenase isozyme, yielding $NADH + H^+$ and the oxidized substrate. The oxidized substrate leaves the mitochondrial matrix via an appropriate translocase to the cytoplasm, where it can again be reduced, while a counterion moves back into the matrix. The $NADH + H^+$ produced within the mitochondria are oxidized by the electron transport chain to regenerate NAD^+ and produce ATP.

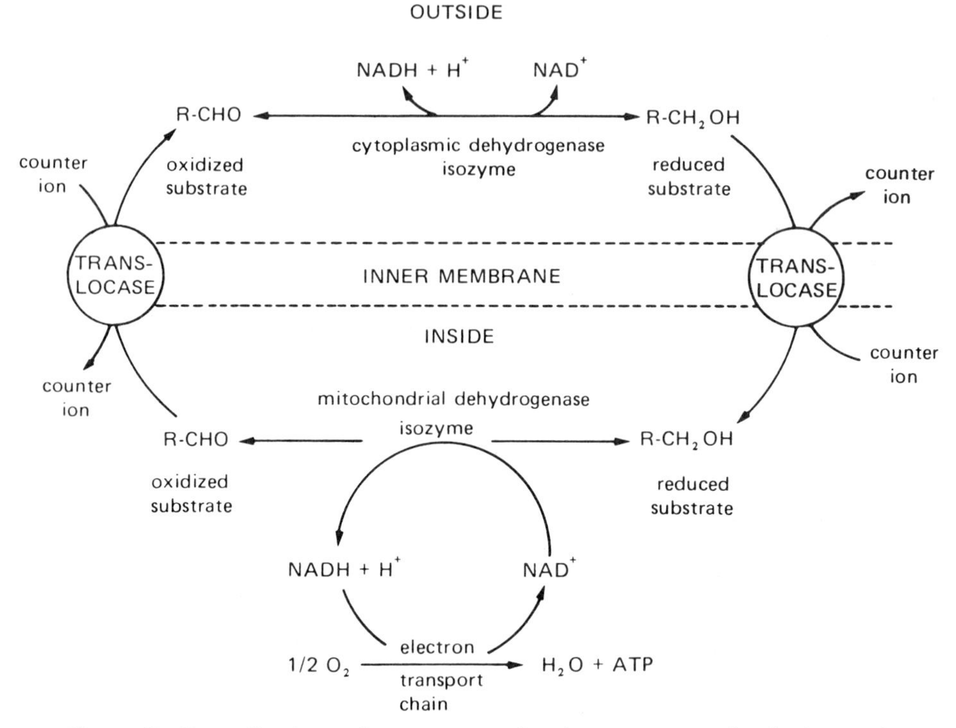

Figure 2 Generalized translocase system for the movement of reducing equivalents into the mitochondrial matrix. (From Ref. 7.)

Since no pair of oxidized and reduced substrates is known which functions singly in this manner in man, small integrated enzyme systems, sometimes referred to as "shuttle systems," are thought to serve this function. One such system is illustrated in Figure 3. In this specific example, aspartate, malate, glutamate, and α-ketoglutarate (abbreviated to αKG in Figure 3) cross the mitochondrial membrane on specific translocase systems. Cytoplasmic aspartate is converted to oxaloacetate and then to malate. The malate enters the mitochondrion, where it is oxidized to oxaloacetate with the production of NADH, accomplishing the goal of producing reducing equivalents inside the mitochondrion. The oxaloacetate is transaminated to yield aspartate, which leaves the mitochondrion to complete the cycle.

Glutamate and aspartate are also vital components of the urea cycle. As outlined in Figure 4, the nitrogen atoms found in urea originate from ammonium ions and aspartate. The ammonium ions are most readily formed by glutamate dehydrogenase. The α-ketoglutarate produced is readily recycled into glutamate by transamination with other amino acids, preventing glutamate depletion. Ammonium ion is also formed by the action of the L-amino acid oxidases. The

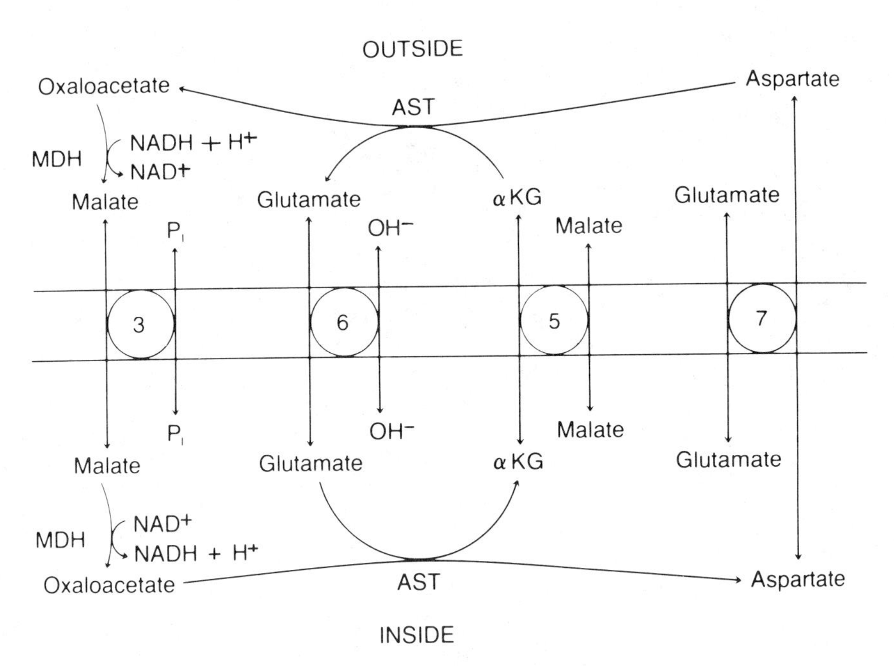

Figure 3 A "shuttle system" requiring the concerted action of four translocases. Note that this shuttle system depends on intra- and extramitochondrial forms of malate dehydrogenase (MDH) and aspartate aminotransferase (AST). Numbers identify specific translocase systems as described by Montgomery et al. (10). (From Ref. 10.)

ammonium ions produced by these reactions are converted to carbamyl phosphate, which condenses with ornithine to yield citrulline. Citrulline then condenses with aspartate to form argininosuccinate. The argininosuccinate is cleaved to yield arginine (which contains the nitrogen originally present in aspartate) and fumarate (which contains aspartate's carbon skeleton). The arginine is cleaved by arginase to yield urea (containing nitrogen found originally in ammonium ions and aspartate) and ornithine. The fumarate produced by cleavage of argininosuccinate is converted to malate and then oxaloacetate. Oxaloacetate is transaminated by appropriate amino acids to regenerate aspartate, which can again enter the cycle. These latter reactions prevent aspartate depletion.

Thus it is not surprising that the cell contains considerable quantities of free glutamate and aspartate. In particular, it is not surprising that these amino acids are the major amino acids found in the mitochondrion, where they may comprise 50-70% of the total free amino acids (1).

Considerable quantities of glutamate and aspartate are normally found in human brain and liver (4-6). Their concentration is particularly high in the brain, where they account for 25-30% of the total free amino acid pool (5,6).

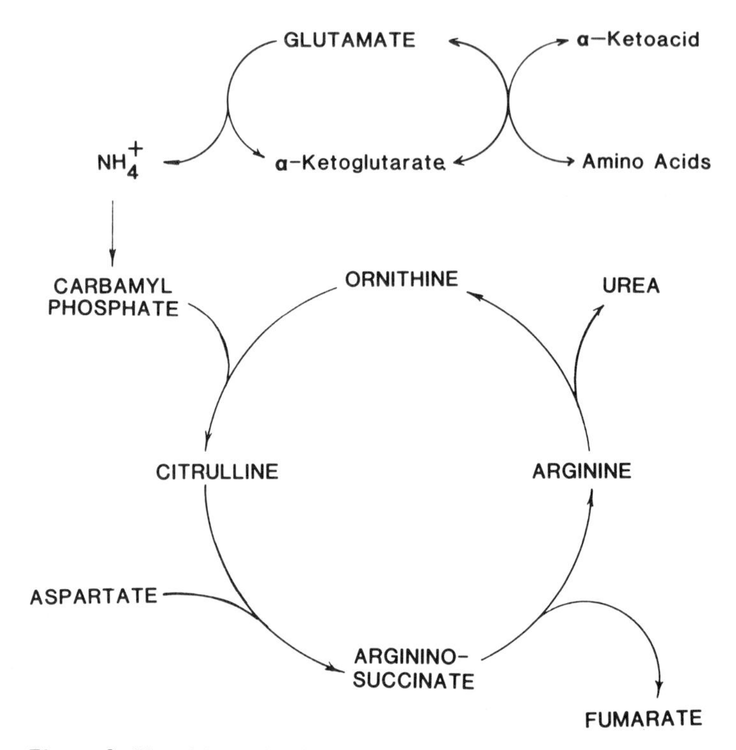

Figure 4 Urea biosynthesis, showing the roles of glutamate and aspartate during nitrogen transfer.

Human milk also contains large quantities of these amino acids (7-9). Free glutamate plus aspartate concentration is about 1.4 mM in human milk, and these amino acids comprise about 47% of the total free amino acids present in this fluid. In addition, large quantities of these amino acids are present in protein-bound form in human milk (11). Thus a 3.5-kg breast-fed infant may ingest between 394 to 788 mg of glutamate + aspartate daily when free and protein-bound sources of these amino acids are considered (8,9).

From this brief overview, two things should be obvious: (a) human cells normally handle large quantities of dicarboxylic amino acids, and (b) these amino acids are an essential part of normal cellular function.

DIGESTION AND ABSORPTION

Glutamate and aspartate account for approximately 20-25% of the total amino acid composition of dietary proteins, including those proteins found in milk (11). In addition, milk contains considerable quantities of these amino acids in free form (7-9). Thus adults and infants (formula or breast fed) normally obtain a considerable portion of their daily protein intake in the form of these amino acids.

Since glutamate and aspartate account for a significant percentage of the total amino acids present in dietary protein, it is important to consider how oral ingestion of these amino acids in protein-bound form affects plasma concentrations. Both glutamate and aspartate produce neurotoxic effects in the young of certain species (12-23). However, neurotoxicity is only noted when plasma levels of these amino acids are grossly elevated (21,22,24-26). Neurotoxicity has not been reported to occur from ingestion of protein-bound glutamate and aspartate present in meals, or when free glutamate is ingested with meals (27,28). Thus it is important to know how meal ingestion affects plasma glutamate and aspartate levels. In addition, data obtained from meal studies will be important when considering the metabolism of parenterally administered glutamate and aspartate.

The absorption of glutamate and aspartate in the intestinal lumen differs, depending on whether these amino acids are ingested in free or peptide-bound form. When ingested as free amino acids, they are absorbed from the intestinal lumen by active transport processes (29,30). When ingested in protein-bound form, the aspartate- and glutamate-containing peptides produced by proteolysis enter mucosal cells directly, where they are hydrolyzed to their constituent amino acids by specific intracellular peptidases (31-33). Thus the effect of dietary ingestion of these amino acids on plasma concentrations may differ, depending upon the form in which the amino acid is ingested.

Protein meal studies in man (34,35) strongly suggest that only protein-bound neutral and dibasic amino acids are quantitatively absorbed as free amino acids. The imino acids (proline and hydroxyproline), glycine, glutamate, and aspartate all appear to enter mucosal cells as constituents of small peptides. These small peptides are hydrolyzed within the mucosal cell to release their constituent free amino acids. Although large amounts of protein-bound glutamate and aspartate are absorbed in peptide-bound form, the gut also has specific transport sites for the free forms of these amino acids (29,30). Thus these amino acids are readily absorbed whether presented in free or peptide-bound form.

When amino acids are ingested in protein-bound form, proteolysis produces a relatively slow release of constituent amino acids and peptides for the absorptive process. Maximal plasma levels of most amino acids in peripheral blood are not reached until 2-6 hr after meal ingestion. In contrast, ingestion of an equivalent quantity of free amino acid produces a more rapid increase in plasma concentration. Thus the form in which amino acids are ingested affects plasma concentration.

We have studied the effects of a variety of meals given to normal volunteers upon plasma amino acid levels. These meals include a beverage meal providing no nutrients; Sustagen (Mead Johnson, Evansville, IN), a formula meal providing 0.4 g of protein per kilogram body weight (36); an egg-milk custard meal (37); and a

Table 1 Protein, Glutamate, and Aspartate Content of the Meals Studied

Meal	Protein (g/kg body weight)	Total glutamate[a] (mg/kg body weight)	Total aspartate[a] (mg/kg body weight)	Estimated glutamate[a] (mg/kg body weight)	Estimated aspartate[a] (mg/kg body weight)	Valine (mg/kg body weight)
Beverage	0	0	0	0	0	0
Sustagen	0.4	95	29	55	16	28
Custard	1.0	207	73	120	40	71
Hamburger-milk shake	1.0	162	82	94	46	56

[a]The total glutamate and aspartate values reported by Orr and Watt (39) were obtained after acid hydrolysis and thus include glutamine and asparagine. The approximate quantities of glutamate and aspartate present were calculated using the estimate of Jukes et al. (40), that 58% of the total glutamate and 55% of the total aspartate are actually present as these amino acids.

hamburger-milk shake meal that also provided protein at 1 g/kg body weight (37,38). The protein content and the estimated quantity of glutamate, aspartate, and valine provided by these meals are shown in Table 1.

When normal subjects ingest a beverage providing no protein, plasma valine levels remain unchanged or decrease slightly (Fig. 5). When normal subjects ingest Sustagen providing 0.4 g of protein and 28 mg of valine per kilogram body weight, plasma valine concentrations increase 30 min after meal ingestion and remain at an increased level over 6 hr (36). When subjects ingest larger quantities of protein (1 g of protein per kilogram body weight providing 56 or 71 mg of valine per kilogram body weight), changes in plasma valine concentration are greater (37,38). Plasma valine concentration increases slowly with time, reaching maximal values 3-6 hr after ingestion of the hamburger-milk shake meal. In general, the peak plasma concentration and the area under the plasma concentration-time curve increase in proportion to the quantity of valine ingested.

Plasma concentrations of glutamate and aspartate (Figs. 6 and 7) respond only slightly to the ingestion of large amounts of protein-bound glutamate and aspartate. Glutamate loads of 55-120 mg/kg body weight produce a slight increase (3-6 μmol/dl) in mean peak plasma glutamate levels above base line (Fig. 6) and have a minimal effect on the area under the plasma glutamate concentration-time curve. Similarly, aspartate loads of 40-46 mg/kg body weight increase

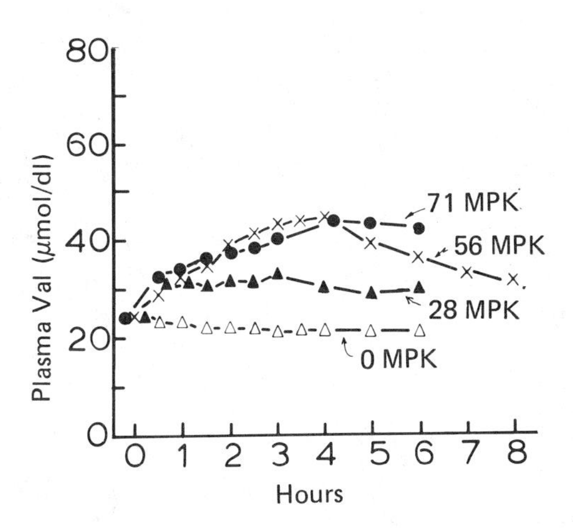

Figure 5 Mean plasma valine (Val) concentrations in normal adults (n = 6 per group) fed either a beverage meal providing no protein (△), Sustagen providing 0.4 g of protein and 28 mg of valine per kilogram body weight (MPK) (▲), a hamburger meal providing 1 g of protein and 56 mg of valine per kilogram body weight (X), or custard providing 1 g of protein and 71 mg of valine per kilogram body weight (●). The valine content of the meals was calculated from the data of Orr and Watt (39). (From Ref.41.)

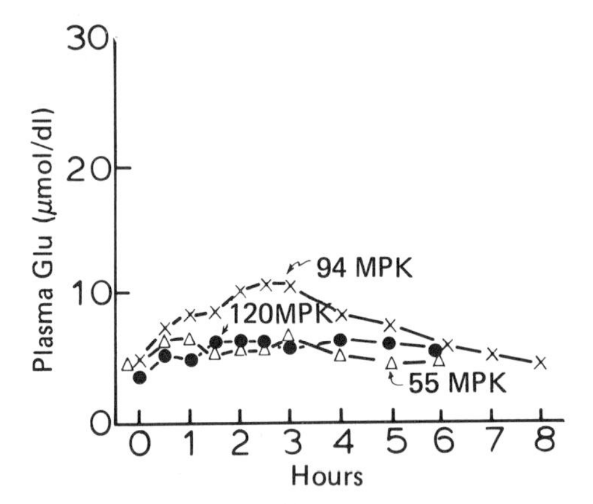

Figure 6 Mean plasma glutamate (Glu) concentrations in normal adults (n = 6 per group) fed either Sustagen providing 0.4 g of protein and 55 mg of glutamate per kilogram body weight (MPK) (△), a hamburger meal providing 1 g of protein and 94 mg of glutamate per kilogram body weight (X), or a custard providing 1 g of protein and 120 mg of glutamate per kilogram body weight (●). The glutamate content of the meals was calculated from the data of Orr and Watt (39) and corrected for glutamine using the values published by Jukes et al. (40) for the glutamate to glutamine distribution of a typical protein. (From Ref. 41.)

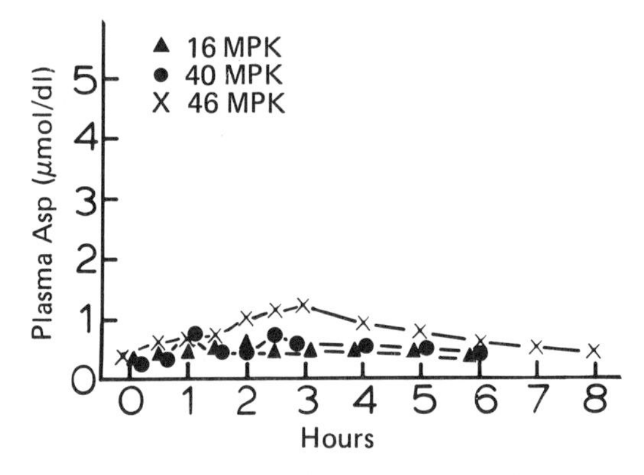

Figure 7 Mean plasma aspartate (Asp) concentrations in normal adults (n = 6 per group) fed either Sustagen providing 0.4 g of protein and 16 mg of aspartate per kilogram body weight (MPK) (▲), a hamburger meal providing 1 g of protein and 46 mg of aspartate per kilogram body weight (X), or a custard providing 1 g of protein and 40 mg of aspartate per kilogram body weight (●). The aspartate content of the meals was calculated from the data of Orr and Watt (39). The values were corrected for asparagine using the asparagine to aspartate distribution of a typical protein published by Jukes et al. (40). (From Ref. 41.)

mean peak plasma aspartate concentration by 0.53-0.81 μmol/dl over base line (Fig. 7). These data indicate that orally administered protein-bound glutamate and aspartate are metabolized more rapidly than valine and that ingestion of large amounts of these amino acids produce a slight change in the peripheral plasma concentration.

The slight increase in plasma concentration of aspartate and glutamate relative to the amount of each amino acid ingested suggests (a) a higher rate of catabolism for these amino acids than for indispensable amino acids like valine, (b) that some component of the meal may be affecting their metabolism and/or absorption, (c) that ingestion of these amino acids in protein-bound form produces a smaller plasma amino acid effect than ingestion as the free amino acid, or (d) some combination of the above.

In vivo studies of amino acid mixtures injected into loops of rat intestine suggest that the absorption of free aspartate and glutamate is much slower than that of other amino acids (42). Similar findings have been reported in humans by Adibi et al. (43) using an intubation technique to perfuse the jejunum with equimolar amounts of free amino acids. Glutamate and aspartate were the slowest absorbed. Silk et al. (44) reported that while free glutamate and aspartate were slowly absorbed, these amino acids were absorbed more rapidly from peptides.

Since these data may be somewhat misleading when extrapolated to meal feeding, we decided to evaluate factors affecting plasma glutamate and aspartate levels following meal ingestion. In an initial study, plasma glutamate levels in normal adult subjects administered 100 mg glutamate per kilogram body weight in water (45) were compared to plasma amino acid values noted in normal subjects ingesting a hamburger-milk shake meal that provided approximately the same quantity of glutamate in protein-bound form (38). The results of this study are shown in Figure 8. Subjects ingesting 100 mg/kg body weight free glutamate in water showed a rapid rise in plasma glutamate level, reaching a mean peak concentration of 50 μmol/dl. The mean area under the plasma glutamate concentration-time curve was 2220 units over the 8-hr time period. This contrasts with the mean peak plasma glutamate level of 10.8 μmol/dl and an 8-hr area under the curve of 1152 units when an equivalent quantity of protein-bound glutamate (94 mg/kg body weight) was ingested.

These data suggest that some meal component affected plasma glutamate concentration. To test this, we selected a formula meal (Sustagen) that provided smaller quantities of protein-bound glutamate (55 mg/kg body weight). Normal subjects were given Sustagen with and without added free glutamate (100 mg/kg body weight) to see whether meal ingestion affected plasma glutamate concentration when glutamate was ingested in free as well as in protein-bound form. Sustagen provided 0.4 g of protein, 1.12 g of carbohydrate, 0.06 g of fat, and 55 mg of protein-bound glutamate per kilogram body weight. Sustagen with added glutamate provided 55 mg of protein-bound glutamate and 100 mg of free glutamate per kilogram body weight. The effect of these feedings on plasma glutamate concentration is shown in Figure 9. For comparative purposes the response

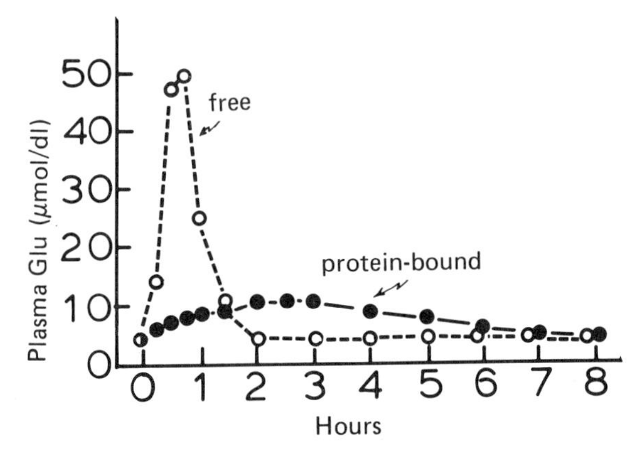

Figure 8 Mean plasma glutamate (Glu) concentrations in normal adults (n = 6 per group) ingesting either 100 mg of glutamate per kilogram body weight dissolved in 4.2 ml/kg body weight water (○), or a hamburger meal providing 1 g of protein, 94 mg of protein-bound glutamate, and 68 mg of protein-bound glutamine per kilogram body weight (●). (From Ref. 41.)

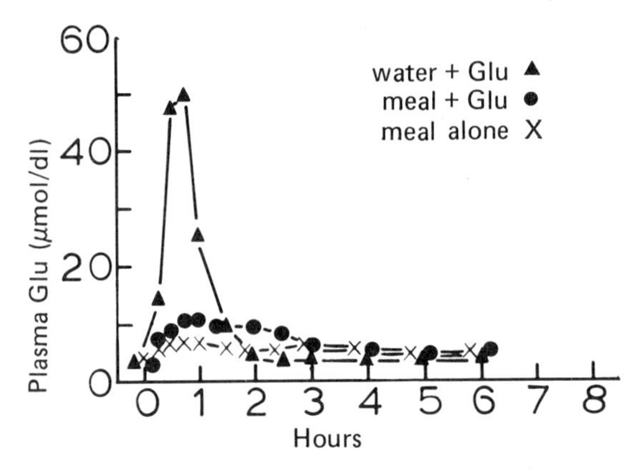

Figure 9 Mean plasma glutamate (Glu) concentrations in normal adults ingesting either Sustagen providing 0.4 g of protein and 55 mg of protein-bound glutamate (X), Sustagen with added free glutamate (100 mg/kg body weight (●), or water providing 100 mg of glutamate per kilogram body weight (▲). (From Ref. 41.)

to 100 mg of glutamate per kilogram body weight administered in water is also shown. Ingestion of glutamate in water increased plasma glutamate concentration to a mean peak value of 50 μmol/dl. When Sustagen was ingested, the mean peak plasma glutamate concentration was 6.7 μmol/dl. Free glutamate (100 mg/kg body weight) added to Sustagen increased the mean peak plasma glutamate concentration to 11.2 μmol/dl, significantly less ($p < 0.05$) than the value noted when glutamate was ingested with water. The area under the plasma glutamate concentration-time curve showed similar results. The mean 6-hr area under the curve value when 100 mg/kg body weight glutamate was ingested in water was significantly larger than the value noted after ingestion of Sustagen, or Sustagen with added glutamate.

These data suggest that glutamate is metabolized more rapidly when ingested with meals than with water; however, it was not clear from these studies whether the meal or one of its components modulated the absorption and/or metabolism of glutamate. A slower rate of absorption would permit greater catabolism of glutamate by the intestinal mucosa, resulting in a decreased release of glutamate to the portal blood. Alternatively, carbohydrate present in Sustagen could serve as a source of pyruvate, thereby facilitating the transamination of glutamate to α-ketoglutarate and increasing the rate of glutamate catabolism in the intestinal mucosa (Fig. 10).

Neame and Wiseman (46,47) have shown that the quantity of glutamate appearing in portal blood as glutamate or alanine varied with the glutamate load. A comparison of data from Parsons and Volman-Mitchell's study (48) with data from the study of Ramaswamy and Radhakrishnan (49) identifies a role of carbohydrate. In both studies glutamate was circulated in a Krebs-Ringer solution through the lumen of rat intestinal segments in vitro. The "sweat" secreted at the serosal side of the segment was collected at timed intervals, and the glutamate

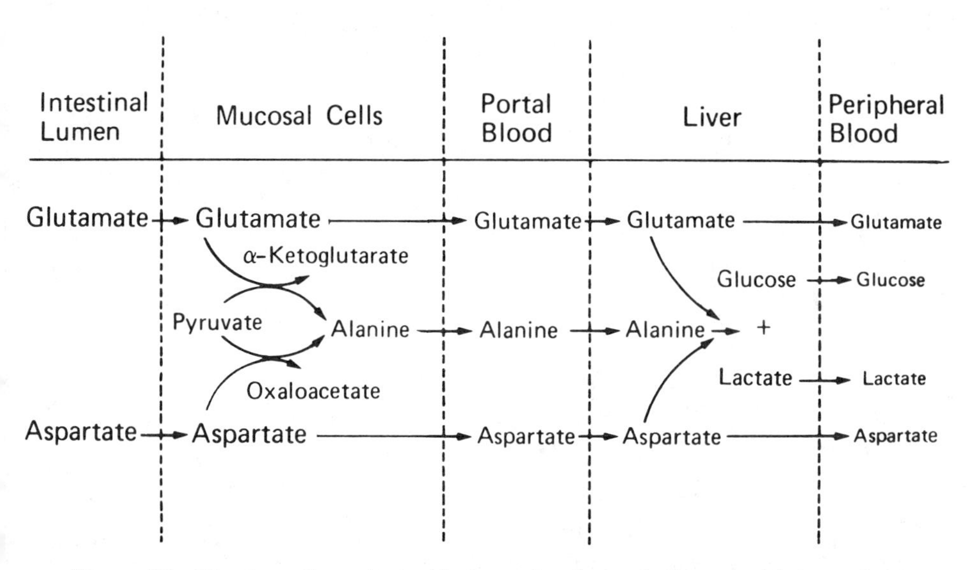

Figure 10 Dicarboxylic amino acid absorption from the intestinal lumen, showing mucosal cell transamination to yield alanine and hepatic conversion of those amino acids to glucose and lactate as factors modulating peripheral plasma levels. (From Ref. 7.)

α-ketoglutarate, and alanine content determined. Parsons and Volman-Mitchell (48) added 28 mM glucose to the perfusion solution and noted the presence of large amounts of alanine in serosal "sweat." Ramaswamy and Radhakrishnan (49) did not add glucose and found glutamate as the major amino acid in serosal "sweat." These studies suggest that the amount of alanine appearing at the serosal surface of the intestinal segment after glutamate administration increases greatly when glucose is present in the lumen.

To test the hypothesis that metabolizable carbohydrate was responsible for this effect, normal adults subjects were given 150 mg/kg body weight glutamate dissolved in either water or water containing sufficient carbohydrate as Polycose (Ross Laboratories, Columbus, Ohio) to provide 1.12 g/kg body weight (50). The response of normal adult subjects to ingestion of 150 mg of glutamate per kilogram body weight in either water or Polycose is illustrated in Figure 11. It is clear that the addition of carbohydrate had a striking effect on plasma glutamate concentration. Changes in blood glucose (not shown) indicate that gastric emptying has occurred. It seems likely that the rapid metabolism of glutamate after ingestion with meals reflects in part the carbohydrate content of the meal. Presumably carbohydrate is absorbed into the intestinal mucosa and converted to glucose and pyruvate, ultimately facilitating the transamination and metabolism of glutamate by mucosal cells.

However, it was possible that other meal components (protein?) affected mucosal cell metabolism, or that Polycose affected glutamate metabolism in

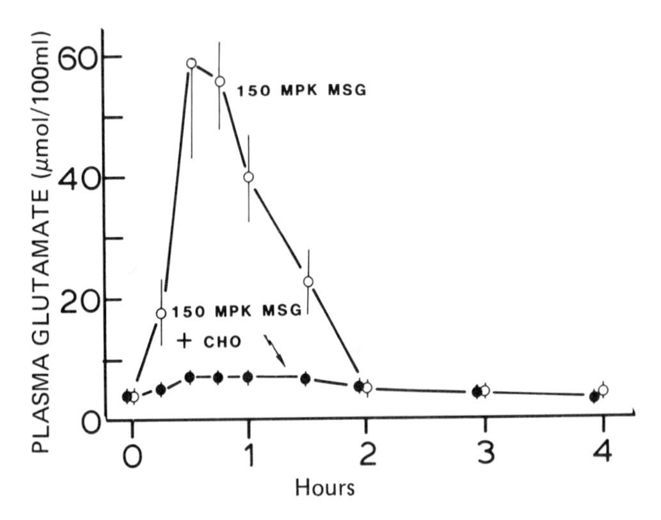

Figure 11 Mean (±SEM) plasma glutamate concentrations in normal adult subjects ingesting 150 mg/kg body weight (MPK) monosodium L-glutamate (MSG) with (●) and without (○) 1.1 g/kg body weight Polycose (CHO).

other ways, such as by blocking glutamate uptake. For example, Polycose addition may decrease the rate of glutamate absorption by increasing viscosity, thus altering the nature of the unstirred layer (51,52). To test this hypothesis, feeding tubes were placed in the jejunum of young pigs through gastrostomy openings. Portal vein catheters were chronically implanted through abdominal incisions. Following recovery from surgery the animals were administered 500 mg of glutamate per kilogram body weight dissolved in four different solutions. Glutamate was administered in (a) water (4.2 ml/kg body weight), (b) a water solution providing 1.12 g Polycose per kilogram body weight, (c) a water solution providing an amino acid mixture (0.4 g/kg body weight; Aminosyn, Abbott Laboratories, North Chicago, Illinois) not containing glutamate or aspartate, and (d) a water solution providing 1.12 g of a nonmetabolizable carbohydrate per kilogram body weight (β-cellobiose, Abbott Laboratories). The β-cellobiose was added to test whether increased viscosity resulting from Polycose addition affected the rate of glutamate absorption rather than facilitated its metabolism. Plasma glutamate concentrations in portal blood of these animals are shown in Figure 12. Mean ($\pm$SD) portal plasma glutamate concentration peaked at 96.8 $\pm$ 25.6 μmol/dl when glutamate was administered in water, at 114 $\pm$ 9.1 μmol/dl when glutamate was administered with cellobiose, and at 46.2 $\pm$ 7.9 μmol/dl when glutamate was administered with Polycose. The addition of amino acids (0.4 g/kg body weight) also had little effect on the mean peak plasma glutamate concentration of portal blood (85 $\pm$ 23 μmol/dl).

These data support the conclusion that only metabolizable carbohydrate affects glutamate metabolism in the intestinal mucosa. Other data from this study (53) demonstrate that carbohydrate ingestion with glutamate not only affects portal plasma glutamate concentrations, but also increases hepatic glutamate metabolism.

In considering these data, it is important to realize that the effect of metabolizable carbohydrate on glutamate metabolism may be greater in man than other species. We have compared the ability of Polycose to alter the peak plasma glutamate concentration resulting from a glutamate load in man, mice, and pigs. To carry out this study it was necessary to use a glutamate load administered in water that would produce a similar mean peak plasma glutamate response in all three species. A glutamate load was selected for each species that would produce a mean peak plasma glutamate response of 50-70 μmol/dl. Since humans metabolize glutamate less rapidly than mice or monkeys (54), a 100 mg/kg body weight dose was used in man and a 300 mg/kg body weight dose was used for mice and pigs. Although simultaneous administration of carbohydrate with glutamate had a dramatic effect on plasma glutamate concentrations in man, a much smaller effect was noted in mice and pigs (Fig. 13).

The precise mechanism by which metabolizable carbohydrate exerts its effect on plasma glutamate is not clear, and several different mechanisms may be

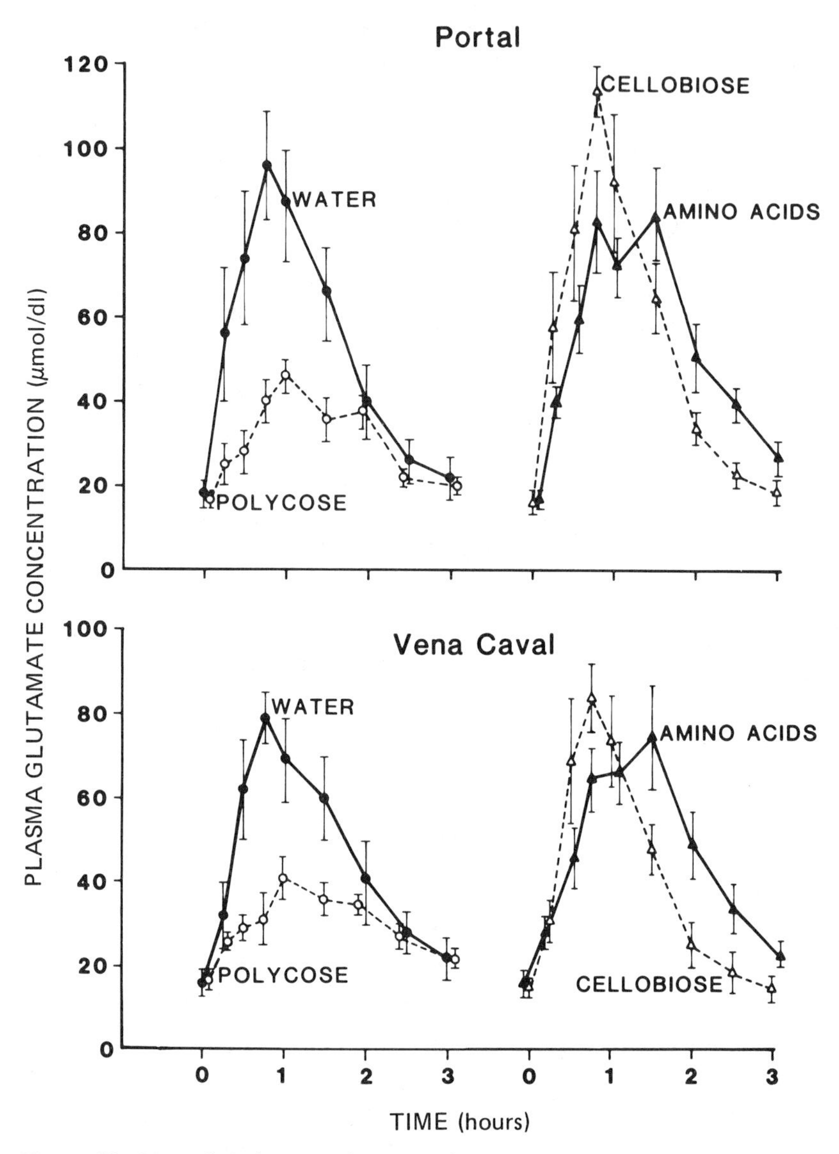

Figure 12 Mean (±SD) portal (top panel) and vena caval (bottom panel) plasma glutamate concentrations in young pigs administered 500 mg/kg body weight monosodium L-glutamate in water (●——●), with 1 g/kg body weight Polycose (○- - -○), with 1 g/kg body weight cellobiose (△- - -△), and with 0.4 g/kg body weight of an amino acid mixture (▲——▲). (From Ref. 53.)

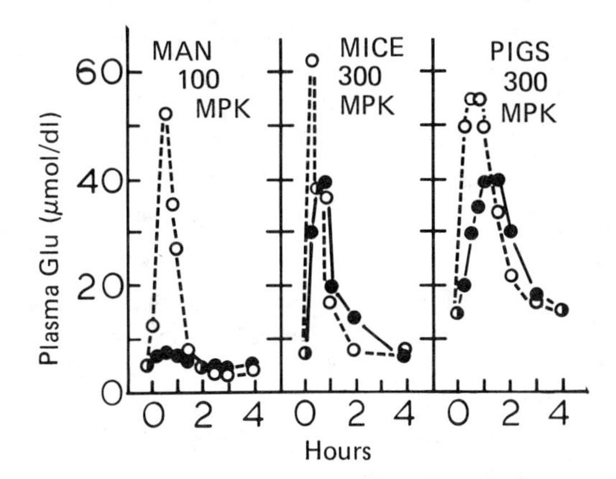

Figure 13 Mean plasma glutamate (Glu) concentrations in adult humans, infant mice, and young pigs administered glutamate dissolved in water (○), or in a water solution also providing 1.1 g of Polycose per kilogram body weight (●). Humans were administered glutamate at 100 mg/kg body weight, while mice and pigs were administered glutamate at 300 mg/kg body weight. (From Ref. 41.)

involved. Stegink (7) suggested earlier that carbohydrate might supply pyruvate, stimulating glutamate transamination in mucosal cells, decreasing the amount of glutamate released to the portal blood. This hypothesis is consistent with other data. During the absorption of free glutamate, considerable quantities of nitrogen originally ingested as glutamate appear in the portal blood as alanine (46-49,55, 56), indicating significant transamination. In vitro studies of glutamate uptake using isolated intestinal segments are also consistent with the hypothesis that carbohydrate increases the rate of mucosal glutamate metabolism (48,49).

Windmueller and Spaeth (57,58) reported that luminal glutamate, aspartate, and glutamine are major energy substrates for rat intestine. However, their data suggest that glucose metabolism provided relatively little of the intestinal energy requirement (57). In meal studies of rats fed a solution providing 70 mM glucose and a mixture of 20 amino acids (0.7-6.5 mM each), 39% of the total respired CO_2 was derived from luminal glutamine, glutamate, and aspartate, 38% was derived from arterial glutamine, while only 6% was derived from luminal glucose. They reported that only 3% of the luminal glucose transported across the intestine was metabolized (57). The low percentage of utilization of luminal glucose is difficult to reconcile with our pig data cited above. However, glucose utilization might have been higher had the glutamate load used by Windmueller and Spaeth been higher, requiring increased metabolism of glucose.

A carbohydrate-related effect on gastric emptying cannot explain the decreased plasma glutamate concentration noted when monosodium L-glutamate was administered with carbohydrate, since in our pig studies the feeding tube was positioned in the proximal jejunum.

It has been suggested that carbohydrate-induced insulin release might account for the effect of carbohydrate on portal plasma glutamate concentration. This seems unlikely. Although insulin increases uptake of the large neutral amino acids from plasma (59,60), plasma glutamate concentrations are not affected (61,62). Furthermore, the administration of certain amino acids stimulates insulin release (63-66). If increased insulin release had been the major factor affecting portal plasma glutamate concentration, the addition of amino acids to the monosodium L-glutamate solution should have also decreased portal plasma glutamate concentration. The concentrations of those amino acids implicated in insulin release were significantly increased during this arm of the study.

The carbohydrate effect on portal plasma glutamate concentration may also reflect increased blood flow to the intestine. Increased flow from the mesenteric artery to the portal vein would lower portal plasma glutamate concentrations if the following rates were not affected: glutamate absorption from the intestinal lumen, glutamate catabolism in mucosal cells, and glutamate transfer from mucosal cells to the portal blood. This suggestion could not be evaluated in our pig study (53), since blood flow measurements were not carried out. However, this mechanism seems unlikely to account for all of the carbohydrate effect on portal plasma glutamate concentration. Blood flow to the intestine would have to increase two- to three-fold to account for the observed differences noted in either peak portal plasma glutamate concentration or the area under the curve. While some differences in blood flow may occur owing to simultaneous ingestion of carbohydrate, it is difficult to imagine a change of this magnitude in view of the relatively constant response noted in the other three arms of the study.

However, a carbohydrate-related increase in blood flow cannot be eliminated as a partial cause of the decreased portal plasma glutamate concentrations noted. For example, the portal plasma alanine response noted in our pig study could be interpreted as consistent with a carbohydrate-induced change in blood flow. We expected the addition of metabolizable carbohydrate to increase portal plasma alanine concentration by increasing glutamate transamination in mucosal cells. During the absorptive process of both free (46-49,55,56) and peptide-bound (67-73) glutamate, a considerable quantity of nitrogen appears in portal blood as alanine. This was also true in our pig study, as indicated by the large increase in portal plasma alanine concentration after monosodium L-glutamate administration in water. However, portal plasma alanine concentration was not significantly increased by the addition of metabolizable carbohydrate. This lack of response would be consistent with an increase in glutamate transamination coupled with an increased blood flow to the intestine during loading with monosodium L-glutamate and carbohydrate. If a carbohydrate-induced increase in blood flow occurred without increasing mucosal glutamate metabolism, portal plasma alanine concentrations should have been significantly decreased when carbohydrate was administered. Thus both factors would be required to account for the observed data.

The increased portal plasma aspartate concentrations noted after glutamate administration arise from mucosal cell glutamate metabolism. If the rate of aspartate synthesis was constant in all four arms of the study, an increased portal blood flow, sufficient to cause the observed decrease in portal plasma glutamate concentration, should also produce a similar decrease in portal plasma aspartate concentration. Although the mean peak portal plasma aspartate concentration did not differ significantly between tests, values of the area under the curve were decreased when metabolizable carbohydrate was present. These data suggest that carbohydrate-induced changes in blood flow were a partial factor decreasing portal plasma glutamate concentration.

GLUTAMATE AND ASPARTATE METABOLISM IN PARENTERAL FEEDING

Since considerable amounts of enterally administered glutamate and aspartate are metabolized by the intestinal mucosa, it is possible that parenterally administered glutamate and aspartate might be metabolized less well, leading to significant elevation of plasma glutamate and aspartate concentrations.

Although protein hydrolysates containing glutamate and aspartate were used during the development of parenteral nutrition (74-78), most amino acid solutions in current use within the United States do not contain these amino acids. One reason for omitting glutamate and aspartate from parenteral solutions is based on their potential neurotoxicity as reported in some animal species.

The administration of large amounts of glutamate and/or aspartate to neonatal rodents has been shown to produce elevated plasma glutamate and aspartate concentrations and hypothalamic neuronal necrosis (12-23). However, the capacity of these amino acids to produce neuronal necrosis in the nonhuman primate infant is questionable. While Olney and colleagues reported neuronal necrosis in infant nonhuman primates given large doses of glutamate (79,80), four other research groups were unable to produce lesions (81-91) even when plasma glutamate plus aspartate concentrations were grossly elevated (86,90). These groups generally had no difficulty in producing the glutamate-induced lesion in infant rodents. Aspartate- or aspartame-induced lesions have not been reported in primates. In fact, Reynolds et al. (88,89) were unable to detect lesions in infant nonhuman primates given large doses of aspartame plus monosodium L-glutamate, although these doses produced grossly elevated plasma glutamate and aspartate levels.

The controversy over nonhuman primate sensitivity to dicarboxylic amino acids led us to explore additional ways of evaluating the potential of these compounds to produce neurotoxicity. Under certain circumstances, even the highly sensitive mouse tolerates substantial doses of glutamate and aspartate without developing neuronal necrosis. For example, our data (22,24) and those of Takasaki and colleagues (25,26) indicate that plasma glutamate plus aspartate concentrations must reach at least 60-100 μmol/dl (5-10 times normal) in infant mice before neuronal necrosis is noted. Thus dicarboxylic amino acid toxicity requires two factors: a sensitive animal species and grossly elevated plasma levels of glutamate and/or aspartate.

One way of evaluating potential risk to human infants ingesting nutritional products containing glutamate and aspartate is to measure plasma and erythrocyte amino acid concentrations when such preparations are given parenterally or enterally. If plasma concentrations of glutamate and aspartate are not increased above normal values, little risk of dicarboxylic amino acid-induced neuronal necrosis would be expected.

We first examined this question in formula-fed infants. Plasma amino acid concentrations were determined in term and premature infants fed either conventional formulas containing little free glutamate and aspartate, or formulas based on protein hydrolysates that contained large quantities of free glutamate and aspartate. Plasma glutamate and aspartate levels of formula-fed infants were considered to establish normal ranges for these amino acids for both term and premature infants. Glutamate and aspartate levels were not significantly affected when the formulas provided large amounts of free glutamate and aspartate (37, 92,93).

During parenteral feeding, the gut and liver are bypassed to some degree, since amino acids are administered directly into the peripheral circulation. Thus, while enterally fed glutamate and aspartate may not elevate plasma concentrations, parenteral administration of these amino acids might increase plasma levels.

This question was first examined in normal, healthy adult volunteers given amino acid solutions (94). Subjects were infused with one of two parenteral solutions (25% dextrose-5% protein as a hydrolysate or a mixture of amino acids). One group received nitrogen in the form of a casein hydrolysate preparation containing glutamate and aspartate (Amigen, Baxter-Travenol, Morton Grove, Illinois). The other group was given a mixture of crystalline amino acids (Aminosyn, Abbott Laboratories, North Chicago, Illinois) free of glutamate and aspartate. The mean daily intake and plasma amino acid concentrations of glutamate and aspartate are shown in Table 2. No significant differences in plasma glutamate or aspartate concentrations were noted between the two groups, despite the wide

Table 2 Plasma Amino Acid Levels in Normal Adults Given Parenteral Solutions With and Without Glutamate and Aspartate

	Amount infused (mg/kg body weight per day)		Plasma level (μmol/dl)[a]	
Amino acid	Aminosyn	Amigen	Aminosyn	Amigen
Glutamate	0	113	4 ± 2	5 ± 2
Aspartate	0	27	0.8 ± 0.4	0.9 ± 0.3

[a]Data expressed as mean ± SD.

difference in glutamate and aspartate intake. This indicated that infusion of glutamate and aspartate at a level of 113 mg of glutamate and 27 mg of aspartate per kilogram body weight per day had no significant effect on the plasma concentrations of these amino acids.

Next we determined whether parenterally administered glutamate and aspartate were metabolized as efficiently by infants as adults. Although Olney (95) postulated that human infants would metabolize glutamate and aspartate less efficiently than adults, we had reported 10 years earlier plasma amino acid values for a series of young infants given 22.5% dextrose-2.5% protein hydrolysate-based solutions parenterally (96). The nitrogen source for these infusions was either enzymatically digested casein or fibrin, each providing large amounts of glutamate and aspartate. The glutamate and aspartate intakes of these infants and their plasma glutamate and aspartate concentrations are shown in Table 3. Plasma glutamate and aspartate concentrations during infusion of both preparations remained within the values seen in orally fed infants. These data indicate that intravenously fed infants metabolize glutamate and aspartate efficiently at these levels of infusion.

The low birth weight infant fed parenterally might be representative of a population with decreased ability to metabolize dicarboxylic amino acids. Such infants are biochemically immature and are further compromised by a feeding technique that bypasses the gut. Filer et al. (37) reported plasma amino acid concentrations in two low birth weight infants (1.4 and 1.5 kg) fed parenterally with a casein hydrolysate-based preparation providing 165-201 mg of glutamate and 39-48 mg of aspartate per kilogram body weight per day. Mean plasma glutamate levels in these infants ranged from 3.6 to 3.9 μmol/dl, while plasma aspartate concentrations ranged from 1.6 to 1.7 μmol/dl (Table 4). Both sets of values are well within the range of plasma concentrations seen in orally fed low birth

Table 3 Mean (±SD) Plasma Aspartate and Glutamate Concentration in Infants Infused with Protein Hydrolysate Based Solutions

	Plasma concentration (μmol/dl)		
Amino acid	Casein hydrolysate[a]	Fibrin hydrolysate[b]	Normal, orally fed[c]
Glutamate	6.12 ± 2.64	4.37 ± 1.12	10.7 ± 3.6
Aspartate	1.06 ± 0.54	0.42 ± 0.21	2.60 ± 1.40

[a]Regimen provided 173 mg of glutamate and 40 mg of aspartate per kilogram body weight per day.
[b]Regimen provided 32 mg of glutamate and 53 mg of aspartate per kilogram body weight per day.
[c]Data of Filer et al. (37).

Table 4 Plasma Glutamate and Aspartate Concentrations in Low Birth Weight Infants on Total Parenteral Nutrition

		Intake (mg/kg body weight)		Plasma concentrations (μmol/dl)	
Patient	Weight (kg)	Glutamate	Aspartate	Glutamate	Aspartate
1	1.4	165	39	3.6	1.6
2	1.5	201	48	3.9	1.6
Normal, orally fed infants[a]				10.7 ± 3.6	2.6 ± 1.4

[a]Data of Filer et al. (37); mean ± SD.

weight infants. Thus even premature infants metabolize parenterally administered glutamate and aspartate.

Recently, a solution of amino acids containing glutamate and aspartate became available for study. We have used that solution to further test the effect of parenterally administered glutamate and aspartate on plasma amino acids in a group of low birth weight infants.

Eight infants with birth weights between 1.05 and 2.68 kg and gestational ages between 28 and 41 weeks were studied in a crossover design (97). All infants were fed parenterally with dextrose-amino acid solutions and lipid emulsion. Mean daily intakes (per kilogram body weight) were 80 kcal of energy, 2 g of lipid, 15 g of dextrose, and 2 g of amino acids. All infants received a similar intake of minerals and vitamins.

In regimen I, Neopham (Cutter Laboratories, Berkeley, California) was used as the amino acid source; Travasol (Baxter-Travenol, Morton Grove, Illinois) provided the amino acids in regimen II. Regimen I provided a mean intake of 226 mg of glutamate and 130 mg of aspartate per kilogram body weight per day. Regimen II provided no glutamate or aspartate. Subjects were infused with each regimen for successive 3-day periods. Initial base-line observations were obtained while these infants were receiving intravenous dextrose alone.

Mean (±SD) plasma glutamate and aspartate concentrations are shown in Table 5. The mean plasma glutamate concentration was slightly but significantly higher (8.69 ± 2.85 μmol/dl) during infusion of regimen I, containing dicarboxylic amino acids, than during infusion of regimen II (6.65 ± 1.98 μmol/dl), which did not contain glutamate and aspartate. However, the plasma glutamate levels observed were all within two standard deviations of the mean observed for orally fed low birth weight infants (10.7 ± 3.6 μmol/dl; mean ± SD) (37). No significant differences in plasma aspartate concentrations were noted between infusion regimens ($p < 0.05$, paired t test).

Table 5 Plasma Aspartate and Glutamate Concentrations in Low Birth Weight Infants on Total Parenteral Nutrition

	Intake (mg/kg body body weight per day)		Plasma concentrations (μmol/dl)	
Solution	Aspartate	Glutamate	Aspartate	Glutamate
Glucose	0	0	1.92 ± 0.43[a]	4.98 ± 3.41[a]
Neopham	130	226	3.44 ± 0.97	8.69 ± 2.85
Travasol	0	0	2.74 ± 1.68	6.65 ± 1.98
Normal, orally fed infants[b]			2.60 ± 1.4	10. 7 ± 3.6

[a]Data shown as mean ± SD.
[b]Data of Filer et al. (37).

Table 6 Erythrocyte Glutamate and Aspartate Concentrations in Low Birth Weight Infants on Total Parenteral Nutrition

	Erythrocyte concentration (μmol/100 g)	
Solution	Glutamate	Aspartate
Dextrose	43.4 ± 13.2[a]	11.9 ± 6.08
Neopham	60.8 ± 15.0	18.2 ± 8.54
Travasol	52.6 ± 14.4	16.1 ± 7.53

[a] Data expressed as mean ± SD.

Erythrocyte concentrations of glutamate and aspartate are given in Table 6. No significant differences in erythrocyte aspartate or glutamate concentrations were noted between infusion regimens. Thus both plasma and erythrocyte data indicate that low birth weight infants adequately metabolize the infused glutamate (226 mg/kg body weight per day) and aspartate (130 mg/kg body weight per day). Such observations do not support Olney's hypothesis (95) that human infants metabolize dicarboxylic amino acids less well than adults.

REFERENCES

1. King, M. J., and Diwan, J. J. (1972). Transport of glutamate and aspartate across membranes of rat liver mitochondria. *Arch. Biochem. Biophys.* 152, 670-676.
2. Kovacevic, Z., and McGivan, J. D. (1983). Mitochondrial metabolism of glutamine and glutamate and its physiological significance. *Physiol. Rev.* 63, 547-605.

3. Duszynski, J., Mueller, G., and LaNaoue, K. (1978). Microcompartmentation of aspartate in liver mitochondria. *J. Biol. Chem.* 253,6149-6157.
4. Ryan, W. L. and Carver, M. J. (1966). Free animo acids of human foetal and adult liver. *Nature* 212, 292-293.
5. Perry, T. L., Berry, K., Hansen, S., Diamond, S., and Mok, C. (1971). Regional distribution of amino acids in human brain obtained at autopsy. *J. Neurochem.* 18, 513-519.
6. Perry, T. L., Hansen, S., Berry, K., Mok, C. and Lesk, D. (1971). Free amino acids and related compounds in biopsies of human brain. *J. Neurochem.* 18, 521-528.
7. Stegink, L. D. (1976). Absorption, utilization and safety of aspartic acid. *J. Toxicol. Enviorn. Health 2*, 215-242.
8. Baker, G. L., Filer, L. J., Jr., and Stegink, L. D. (1979). Factors influencing dicarboxylic amino acid content of human milk. In *Glutamic Acid: Advances in Biochemistry and Physiology* (Filer, L. J., Jr., Garattini, S., Kare, M. R., Reynolds, W. A., and Wurtman, R. J., eds.), Raven Press, New York, pp. 111-123.
9. Baker, G. L. (1984). Aspartame ingestion during lactation. In *Aspartame: Physiology and Biochemistry* (Stegink, L. D., and Filer, L. J., Jr., eds.) Marcel Dekker, New York, Chap. 28.
10. Montgomery, R., Dryer, R. L., Conway, T. W., and Spector, A. A. (1983). *Biochemistry. A Case-Oriented Approach*, Mosby, St. Louis, p. 288.
11. Macy, I. G., Kelly, H. J., and Sloan, R. E. (1953). *The Composition of Milks*, National Academy of Sciences, National Research Council, Publication 254, Government Printing Office, Washington, D.C.
12. Olney, J. W. (1969). Brain lesions, obesity, and other disturbances in mice treated with monosodium glutamate. *Science* 164, 719-721.
13. Lemkey-Johnston, N., and Reynolds, W. A. (1974). Nature and extent of brain lesions in mice related to ingestion of monosodium glutamate. A light and electron microscope study. *J. Neuropathol. Exp. Neurol.* 33, 74-97.
14. Olney, J. W., Ho, O. L., Rhee, V., and de Gubareff, T. (1973). Neurotoxic effects of glutamate. *N. Engl. J. Med.* 289, 1374-1375.
15. Tafelski, T. J., and Lamperti, A. A. (1977). Effects of a single injection of monosodium glutamate on the reproductive neuroendocrine axis of the female hamster. *Biol. Reprod.* 17, 404-411.
16. Lamperti, A., and Blaha, G. (1976). The effects of neonatally-administered monosodium glutamate on the reproductive system of adult hamsters. *Biol. Reprod.* 14, 362-369.
17. Olney, J. W., and Ho, O.-L. (1970). Brain damage in infant mice following oral intake of glutamate, aspartate or cysteine. *Nature* 227, 609-611.
18. Okaniwa, A., Hori, M., Masuda, M., Takeshita, M., Hayashi, N., Wada, I., Doi, K., and Ohara, Y. (1979). Histopathological study on effects of potassium aspartate on the hypothalamus of rats. *J. Toxicol. Sci.* 4, 31-46.
19. Olney, J., W., Labruyere, J., and de Gubareff, T. (1980). Brain damage in mice from voluntary ingestion of glutamate and aspartate. *Neurobehav. Toxicol.* 2, 125-129.
20. Olney, J. W., Ho, O.-L., and Rhee, V. (1973). Brain damaging potential of protein hydrolysates. *N. Engl. J. Med.* 289, 391-395.

21. Applebaum, A. E., Finkelstein, M. W., Daabees, T. T., and Stegink, L. D. (1984). Aspartate-induced neurotoxicity in infant mice. In *Aspartame: Physiology and Biochemistry* (Stegink, L. D., and Filer, L. J., Jr., eds.), Marcel Dekker, New York, Chap. 17.
22. Finkelstein, M. W., Daabees, T., T., Stegink, L. D., and Applebaum, A. E. (1983). Correlation of aspartate dose, plasma dicarboxylic amino acid concentration, and neuronal necrosis in infant mice. *Toxicology* 29, 109-119.
23. Takasaki, Y. (1978). Studies on brain lesions by administration of monosodium L-glutamate to mice. I. Brain lesions in infant mice caused by administration of monosodium L-glutamate. *Toxicology* 9, 293-305.
24. Stegink, L. D., Shepherd, J. A., Brummel, M. C., and Murray, L. M. (1974). Toxicity of protein hydrolysate solutions: Correlation of glutamate dose and neuronal necrosis to plasma amino acid levels in young mice. *Toxicology* 2, 285-299.
25. Takasaki, T., Matsuzawa, Y., Iwata, S., O'Hara, Y., Yonetani, S., and Ichimura, M. (1979). Toxicological studies of monosodium-L-glutamate in rodents: Relationship between routes of administration and neurotoxicity. In *Glutamic Acid: Advances in Biochemistry and Physiology* (Filer, L. J., Jr., Garattini, S., Kare, M. R., Reynolds, W. A., and Wurtman, R. J., eds.), Raven Press, New York, pp. 255-275.
26. O'Hara, Y., and Takasaki, Y. (1979). Relationship between plasma glutamate levels and hypothalamic lesions in rodents. *Toxicol. Lett.* 4, 499-505.
27. Takasaki, Y. (1978). Studies on brain lesions by administration of monosodium L-glutamate to mice. II. Absence of brain damage following administration of monosodium L-glutamate in the diet. *Toxicology* 9, 307-318.
28. Anantharaman, K. (1979). *In utero* and dietary administration of monosodium L-glutamate to mice: Reproductive performance and development in a multigeneration study. In *Glutamic Acid: Advances in Biochemistry and Physiology* (Filer, L. J., Jr., Garattini, S., Kare, M. R., Reynolds, W. A., and Wurtman, R. J., eds.), Raven Press, New York, pp. 231-253.
29. Ramaswamy, K., and Radhakrishnan, A. N. (1966). Patterns of intestinal uptake and transport of amino acids in the rat. *Indian J. Biochem.* 3, 138-143.
30. Schultz, S. G., Yu-Tu, L., Alvarez, O. O., and Curran, P. E. (1970). Dicarboxylic amino acid influx across brush border of rabbit ileum. *J. Gen. Physiol.* 56, 621-639.
31. Matthews, D. M. (1984). Absorption of peptides, amino acids and their methylated derivatives. In *Aspartame: Physiology and Biochemistry* (Stegink, L. D., and Filer, L. J., Jr., eds.), Marcel Dekker, New York, Chap. 3.
32. Matthews, D. M. (1975). Intestinal absorption of peptides. *Physiol. Rev.* 55, 537-608.
33. Matthews, D. M., and Payne, J. W. (1980). Transmembrane transport of small peptides. In *Current Topics in Membranes and Transport,* Vol. 14 (Bronner, F., and Kleinzeller, A., eds.), Academic, New York, pp. 331-425.
34. Nixon, S. E., and Mawer, G. E. (1970). The digestion and absorption of protein in man. I. The site of absorption. *Br. J. Nutr.* 24, 227-240.

35. Nixon, S. E., and Mawer, G. E. (1970). The digestion and absorption of protein in man. 2. The form in which digested protein is absorbed. *Br. J. Nutr.* 24, 241-258.
36. Stegink, L. D., Baker, G. L., and Filer, L. J., Jr. (1983). Modulating effect of Sustagen on plasma glutamate concentrations in humans ingesting monosodium L-glutamate. *Am. J. Clin. Nutr.* 37, 194-200.
37. Filer, L. J., Jr., Baker, G. L., and Stegink, L. D. (1979). Metabolism of free glutamate in clinical products fed infants. In *Glutamic Acid: Advances in Biochemistry and Physiology* (Filer, L. J., Jr., Garattini, S., Kare, M. R., Reynolds, W. A., and Wurtman, R. J., eds.), Raven Press, New York, pp. 353-362.
38. Stegink, L. D., Baker, G. L., and Filer, L. J., Jr. (1982). Plasma and erythrocyte amino acid levels in normal adult subjects fed a high protein meal with and without added monosodium glutamate. *J. Nutr.* 112, 1953-1960.
39. Orr, M. L., and Watt, B. K. (1957). *Amino Acid Content of Foods,* Home Economics Research Report No. 4, Household Economics Research Division, Institute of Home Economics, Agricultural Research Service, USDA, Superintendent of Documents, U.S. Government Printing Office, Washington, D.C.
40. Jukes, T. H., Holmquist, R., and Moise, H. (1975). Amino acid composition of proteins. Selection against the genetic code. *Science* 189, 50-51.
41. Stegink, L. D., Bell, E. F., Daabees, T. T., Andersen, D. W. Zike, W. L., and Filer, L. J., Jr. (1983). Factors influencing utilization of glycine, glutamate and aspartate in clinical products. In *Amino Acids, Metabolism and Medical Applications* (Blackburn, G. L., Grant, J. P., and Young, V. R., eds.) John Wright-PSG, Boston, pp. 123-146.
42. Gitler, C. (1964). Protein digestion and absorption in nonruminants. In *Mammalian Protein Metabolism,* Vol. 1 (Munro, H. N., and Allison, J. B., eds.), Academic, New York, pp. 35-69.
43. Adibi, S. A., Gray, S. J., and Menden, E. (1967). The kinetics of amino acid absorption and alteration of plasma composition of free amino acids after intestinal perfusion of amino acid mixtures. *Am. J. Clin. Nutr.* 20, 24-33.
44. Silk, D. B. A., Marrs, T. C., Addison, J. M., Burston, D., Clark, M. L., and Matthews, D. M. (1973). Absorption of amino acids from an amino acid mixture simulating casein and a tryptic hydrolysate of casein. *Clin. Sci. Mol. Med.* 45, 715-719.
45. Stegink, L. D., Filer, L. J., Jr., Baker, G. L., Mueller, S. M., and Wu-Rideout, M. Y.-C. (1979). Factors affecting plasma glutamate levels in normal adult subjects. In *Glutamic Acid: Advances in Biochemistry and Physiology* (Filer, L. J., Jr., Garattini, S., Kare, M. R., Reynolds, W. A., and Wurtman, R. J., eds.), Raven Press, New York, pp. 333-351.
46. Neame, K. D., and Wiseman, G. (1957). The transamination of glutamic and aspartic acids during absorption by the small intestine of the dog *in vivo. J. Physiol.* 135, 442-450.
47. Neame, K. D., and Wiseman, G. (1958). The alanine and oxo acid concentrations in mesenteric blood during the absorption of glutamic acid by the small intestine of the dog, cat, and rabbit *in vivo. J. Physiol.* 140, 148-155.

48. Parsons, D. S., and Volman-Mitchell, H. (1974). The transamination of glutamate and aspartate during absorption *in vitro* by small intestine of chicken, guinea pig and rat. *J. Physiol.* 239, 677-694.
49. Ramaswamy, K., and Radhakrishnan, A. N. (1970). Labeling patterns using C^{14}-labelled glutamic acid, aspartic acid and alanine in transport studies with everted sacs of rat intestine. *Indian J. Biochem.* 7, 50-54.
50. Stegink, L. D., Filer, L. J., Jr., and Baker, G. L. (1983). Effect of carbohydrate on plasma and erythrocyte glutamate levels in humans ingesting large doses of monosodium L-glutamate in water. *Am. J. Clin. Nutr.* 37, 961-968.
51. DeSimone, J. A. (1983). Diffusion barrier in the small intestine. *Science* 220, 221-222.
52. Smithson, K. W., Millar, D. B., Jacobs, L. R., and Gray, G. M. (1981). Intestinal diffusion barrier: Unstirred water layer or membrane surface mucous coat? *Science* 214, 1241-1244.
53. Daabees, T. T., Andersen, D. W., Zike, W. L., Filer, L. J., Jr., and Stegink, L. D. (1984). Effect of meal components on peripheral and portal plasma gluamate levels in young pigs administered large doses of monosodium L-glutamate. *Metabolism* 33, 58-67.
54. Stegink, L. D., Reynolds, W. A., Filer, L. J., Jr., Baker, G. L., Daabees, T. T., and Pitkin, R. M. (1979). Comparative metabolism of glutamate in the mouse, monkey and man. In *Glutamic Acid: Advances in Biochemistry and Physiology* (Filer, L. J., Jr., Garattini, S., Kare, M. R., Reynolds, W. A., and Wurtman, R. J., eds.), Raven Press, New York, pp. 85-102.
55. Matthews, D. M., and Wiseman, G. (1953). Transamination by the small intestine of the rat. *J. Physiol.* 120, 55P.
56. Peraino, C., and Harper, A. E. (1962). Concentrations of free amino acids in blood plasma of rats force-fed L-glutamic acid, L-glutamine or L-alanine. *Arch. Biochem. Biophys.* 97, 442-448.
57. Windmueller, H. G., and Spaeth, A. E. (1980). Respiratory fuels and nitrogen metabolism *in vivo* in small intestine of fed rats. *J. Biol. Chem.* 255, 107-112.
58. Windmueller, H. G. (1980). Enterohepatic aspects of glutamine metabolism. In *Glutamine: Metabolism, Enzymology, and Regulation* (Mora, J., and Palacios, R., eds.), Academic, New York, pp. 235-257.
59. Pozefsky, T., Felig, P., Tobin, J. D., Soeldner, J. S., and Cahill, G. F. Jr. (1969). Amino acid balance across tissue of the forearm in post absorptive man. Effects of insulin at two dose levels. *J. Clin. Invest.* 48, 2273-2282.
60. Felig, P., and Wahren, J. (1971). Influence of endogenous insulin on splanchnic glucose and amino acid metabolism in man. *J. Clin. Invest.* 50, 1702-1711.
61. Aoki, T. T., Brennan, M. F., Muller, W. A., Moore, F. D., and Cahill, G. F., Jr. (1972). Effect of insulin on muscle glutamate uptake. Whole blood versus plasma glutamate analysis. *J. Clin. Invest.* 51, 2889-2894.
62. Weber, F. L., Veach, G. L., and Friedman, D. W. (1981). Effects of insulin and glucagon on the uptake of amino acids from arterial blood by canine ileum. *Dig. Dis. Sci.* 26, 113-118.
63. Floyd, J. C., Jr., Fajans, S. S., Conn, J. W., Thiffault, C., Knopf, R. F.,

and Guntsche, E. (1968), Secretion of insulin induced by amino acids and glucose in diabetes mellitus. *J. Clin. Endocrinol. Metab.* 28, 266-276.
64. Cremer, G. M., Molnar, G. D., Taylor, W. F., Moxnes, K. E., Service, F. J., Gatewood, L. C., Ackerman, E., and Rosevear, J. W. (1971). Studies of diabetic instability. II. Tests of insulinogenic reserve with infusions of arginine, glucagon, epinephrine and saline. *Metabolism* 20, 1083-1098.
65. Floyd, J. C., Jr., Fajans, S. S., Pek, S., Thiffault, C. A., Knopf, R. F., and Conn, J. W. (1970). Synergistic effect of certain amino acid pairs upon insulin secretion in man. *Diabetes* 19, 102-108.
66. Floyd, J. C., Jr., Fajans, S. S., Pek, S., Thiffault, C. A., Knopf, R. F., and Conn, J. W. (1970). Synergistic effect of certain amino acids and glucose upon insulin secretion in man. *Diabetes* 19, 109-115.
67. Christensen, H. N. (1949). Conjugated amino acids in portal plasma of dogs after protein feeding. *Biochem. J.* 44, 333-335.
68. Dent, C. E., and Schilling, J. A. (1949). Studies on the absorption of proteins: The amino acid pattern in the portal blood. *Biochem. J.* 44, 318-333.
69. Elwyn, D. H., Launder, W. J., Parikh, H. C., and Wise, E. M., Jr. (1972). Roles of plasma and erythrocytes in interorgan transport of amino acids in dogs. *Am. J. Physiol.* 222, 1333-1342.
70. Elwyn, D. H., Parikh, H. C., & Shoemaker, W. C. (1968). Amino acid movements between gut, liver, and periphery in unanesthetized dogs. *Am. J. Physiol.* 215, 1260-1275.
71. Peraino, C., & Harper, A. E. (1961). Effect of diet on blood amino acid concentrations. *Fed. Proc.* 20, 245.
72. Pion, R., Fauconneau, G., and Rerat, A. (1964). Variation de la composition en acides amines du sang porte au cours de la digestion chez le porc. *Ann. Biol. Anim. Biochim. Biophys.* 4, 383-401.
73. Wolff, J. E., Bergman, E. N., and Williams, H. H. (1972). Net metabolism of plasma amino acids by liver and portal-drained viscera of fed sheep. *Am. J. Physiol.* 223, 438-446.
74. Cox, W. M., Jr., Mueller, A. J., Elman, R., Albanese, A. A., Kemmerer, K. S., Bargon, R. W., and Holt, L. E., Jr. (1947). Nitrogen retention studies on rats, dogs and man: The effect of adding methionine to an enzymic casein hydrolysate. *J. Nutr.* 33, 437-457.
75. Filler, R. M., Eraklis, A. J., Rubin, V. G., and Das, J. B. (1969). Long-term total parenteral nutrition in infants. *N. Engl. J. Med.* 281, 589-595.
76. Shohl, A. T., Butler, A. M., Blackfan, K. D., and McLachlan, E. (1939). Nitrogen metabolism during the oral and parenteral administration of the amino acids of hydrolyzed casein. *J. Pediatr.* 15, 469-475.
77. Wilmore, D. W., and Dudrick, S. J. (1968). Growth and development of an infant receiving all nutrients exclusively by vein. *J. Am. Med. Assoc.* 203, 860-864.
78. Wilmore, D. W., Groff, D. B., Bishop, H. C., and Dudrick, S. J. (1969). Total parenteral nutrition in infants with catastrophic gastrointestinal anomalies. *J. Pediatr. Surg.* 4, 181-189.
79. Olney, J. W., and Sharpe, L. G. (1969). Brain lesions in an infant rhesus monkey treated with monosodium glutamate. *Science* 166, 386-388.

80. Olney, J. W., Sharpe, L. G., and Feigin, R. D. (1972). Glutamate-induced brain damage in infant primates. *J. Neuropathol. Exp. Neurol.* 31, 464-488.
81. Reynolds, W. A., Lemkey-Johnston, N., Filer, L. J., Jr., and Pitkin, R. M. (1971). Monosodium glutamate: Absence of hypothalamic lesions after ingestion by newborn primates. *Science* 172, 1342-1344.
82. Abraham, R., Dougherty, W., Golberg, L., and Coulston, F. (1971). The response of the hypothalamus to high doses of monosodium glutamate in mice and monkeys. Cytochemistry and ultrastructural study of lysosomal changes. *Exp. Mol. Pathol.* 15, 43-60.
83. Newman, A. J., Heywood, R., Palmer, A. K., Barry, D. H., Edwards, F. P., and Worden, A. N. (1973). The administration of monosodium L-glutamate to neonatal and pregnant rhesus monkeys. *Toxicology* 1, 197-204.
84. Wen, C., Hayes, K. C., and Gershoff, S. M. (1973). Effects of dietary supplementation of monosodium glutamate on infant monkeys, weanling rats and suckling mice. *Am. J. Clin. Nutr.* 26, 803-813.
85. Abraham, R., Swart, J., Golberg, L., and Coulston, F. (1975). Electron microscopic observations of hypothalami in neonatal rhesus monkeys (*Macaca mulatta*) after administration of monosodium L-glutamate. *Exp. Mol. Pathol.* 23, 203-213.
86.. Stegink, L. D., Reynolds, W. A., Filer, L. J., Jr., Pitkin, R. M., Boaz, D. P., and Brummel, M. C. (1975). Monosodium glutamate metabolism in the neonatal monkey. *Am. J. Physiol.* 229, 246-250.
87. Heywood, R., and James, R. W. (1979). An attempt to induce neurotoxicity in an infant rhesus monkey with monosodium glutamate. *Toxicol. Lett.* 4, 285-286.
88. Reynolds, W. A., Butler, V., and Lemkey-Johnston, N. (1976). Hypothalamic morphology following ingestion of aspartame or MSG in the neonatal rodent and primate: A preliminary report. *J. Toxicol. Environ. Health* 2, 471-480.
89. Reynolds, W. A., Lemkey-Johnston, N., and Stegink, L. D. (1979). Morphology of the fetal monkey hypothalamus after in utero exposure to monosodium glutamate. In *Glutamic Acid: Advances in Biochemistry and Physioloy* (Filer, L. J., Jr., Garattini, S., Kare, M. R. Reynolds, W. A., and Wurtman, R. J., eds.), Raven Press, New York, pp. 217-229.
90. Reynolds, W. A., Stegink, L. D., Filer, L. J., Jr., and Renn, E. (1980). Aspartame administration to the infant monkey: Hypothalamic morphology and plasma amino acid levels. *Anat. Rec.* 198, 73-85.
91. Heywood, R., and Worden, A. N. (1979). Glutamate toxicity in laboratory animals. In *Glutamic Acid: Advances in Biochemistry and Physiology* (Filer, L. J., Jr., Garattini, S., Kare, M. R., Reynolds, W. A., and Wurtman, R. J., eds.), Raven Press, New York, pp. 203-215.
92. Filer, L. J., Jr., Baker, G. L., and Stegink, L. D. (1977). Plasma aminograms in infants and adults fed an identical high protein meal. *Fed. Proc.* 36, 1181.
93. Filer, L. J., Jr., Stegink, L. D., and Chandramouli, B. (1977). Effect of diet on plasma aminograms of low-birth-weight infants. *Am. J. Clin. Nutr.* 30, 1036-1043.

94. Stegink, L. D. (1977). Peptides in parenteral nutrition. In *Clinical Nutrition Update*—Amino Acids (Greene, H. L., Holliday, M. A., and Munro, H. N., eds.), American Medical Association, Chicago, IL., pp. 192-198.
95. Olney, J. W. (1981). Excitatory neurotoxins as food additives: An evaluation of risk. *Neurotoxicology*. 2, 163-192.
96. Stegink, L. D., and Baker, G. L. (1971). Infusion of protein hydrolysates in the newborn infant. Plasma amino acid concentrations. *J. Pediatr.* 78, 595-602.
97. Bell, E. F., Filer, L. J., Jr., Pon Wong, A., and Stegink, L. D. (1983). Effects of a parenteral nutrition regimen containing dicarboxylic amino acids on plasma, erythrocyte, and urinary amino acid concentrations of young infants. *Am. J. Clin. Nutr.* 37, 99-107.

5
Phenylalanine Metabolism

Alfred E. Harper
University of Wisconsin, Madison, Wisconsin

INTRODUCTION

Phenylalanine, which makes up over half of the aspartame molecule, is an essential component of body proteins and an important precursor of many aromatic compounds required for normal body function. As phenylalanine (Phe) cannot be synthesized by mammals, it is nutritionally indispensable for them. They must therefore obtain it from foods. The nutritional requirement for Phe, however, depends upon the amount of tyrosine (Tyr) present in the diet.

In 1913 Embden and Baldes (1) discovered that phenylalanine could be converted to Tyr by mammalian liver. Subsequently this conversion was shown to be the initial step in the degradation of Phe. Thus, beyond this step, the catabolic pathway of Phe is that of Tyr. Also, Tyr is the immediate precursor of many of the aromatic compounds formed from Phe. Thus both the nutritional roles and the metabolism of Phe and Tyr are intimately linked. These two amino acids must therefore be considered together.

NUTRITIONAL ESSENTIALITY OF PHENYLALANINE: AROMATIC AMINO ACID REQUIREMENTS AND INTAKES

The nutritional essentiality of aromatic amino acids for mammals had been recognized prior to 1930. The specificity of the aromatic amino acid requirement could not be established, however, until McCoy et al. (2) had identified threonine,

---

Supported in part by grant AM 10748 from the National Institutes of Health.

the last of the nutritionally essential amino acids to be discovered, and were then able to devise nutritionally adequate diets in which mixtures of amino acids could be substituted for proteins. Subsequently, by feeding rats diets from which each amino acid in turn was deleted, Rose and his associates (3) were able to establish that Phe was indispensable for this species. They found that if enough Phe was included in the diet, Tyr was not required. A few years later, they demonstrated that for human adults also, Phe was indispensable and Tyr was dispensable (4). Similar findings have been reported for all animal species studied.

After all of the nutritionally indispensable amino acids had been identified, quantitative requirements for each of them could be established by observing growth rates of the young or the nitrogen balance of adults consuming diets in which the amount of one indispensable amino acid was increased incrementally while the others were provided in amounts that were more than adequate. Using this procedure, the quantitative requirement of the growing rat for Phe was found to be 0.9% of the diet when Tyr was deleted, but only 0.7% when tyrosine was present (5). The aromatic amino acid requirements of the rat have been reinvestigated several times since then with more nutritionally adequate diets. These studies indicate that the total aromatic amino acid requirement is 0.7% of the diet, of which 45% can be provided by Tyr (6).

The Phe requirement of human adults was estimated by Rose and his associates (4) to be about 16 mg/kg body weight per day, but subsequent estimates suggest that the average requirement is only about 12 mg/kg per day (7,8). Close to 75% of the human adult requirement for Phe can be met by Tyr (9). The requirement of human infants (2-6 months of age) for aromatic amino acids has been estimated from studies of infants fed amino acid diets and from the amounts of aromatic amino acids provided by minimum intakes of human milk proteins that support satisfactory growth rates to be 141 mg/kg body weight per day (7,8). Based on the proportions of these two amino acids in the quantity of human milk proteins that will meet the protein requirement of the infant, the aromatic amino acid requirement can be met by 65 mg/kg of Phe and 76 mg/kg of Tyr.

From comparison of the adult and infant requirements, it is evident that the need for amino acids decreases sharply with increasing age (7,8). The rate of change in the Phe requirement during the first 2 years of life is shown in Figure 1. This is from a study by Snyderman and associates (10) in which they determined the amount of Phe needed to maintain satisfactory growth and normal blood Phe concentrations in children with phenylketonuria (PKU), who cannot degrade Phe. The diet used in this study contained Tyr, so the values obtained are specifically for the Phe requirements of infants from birth to 24 months of age. The values for 2- to 6-month-old infants correspond quite well with the average value estimated from milk intakes (7,8). As the PKU infants used in the study lacked the major pathway for Phe degradation, this figure also illustrates how rapidly amino acid needs for protein synthesis decline as growth rate falls.

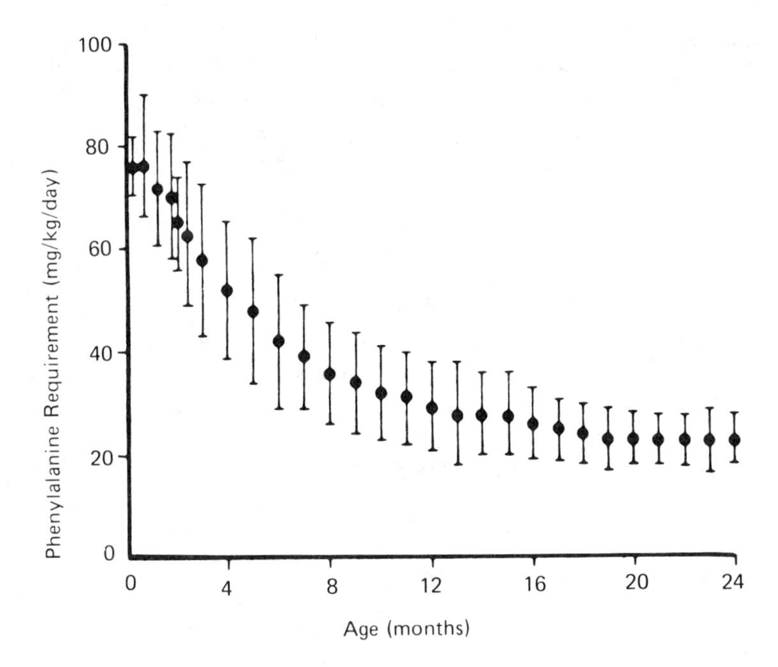

Figure 1 Change in phenylalanine requirement with increasing age from birth to 2 years, based on a phenylalanine intake that supports satisfactory rate of growth and normal blood phenylalanine concentration of phenylketonuric infants. Points represent mean ± SD for 50 infants. (From Ref. 10.)

The adult requirement for aromatic amino acids (Phe + Tyr), that is, the maintenance requirement after growth has ceased, is only about 1/10 of that for early rapid growth of the infant.

The daily consumption of aromatic amino acids by adult males consuming the usual U.S. intake of about 100 g of protein from a mixed diet composed largely of foods from the major food groups is about 8.2 g/day, 4.6 g of Phe and 3.6 g of Tyr (11). This intake represents between 9 and 10 times the estimated requirement. In several nitrogen balance studies in which protein was provided almost entirely from cereal grains, the quantities of aromatic amino acids ingested still exceeded the requirement by three- to fourfold (7). Hence the probability of a dietary deficiency of Phe occurring in the adult is negligible. Even in infants and young children, if caloric need is met, the probability of an aromatic amino acid deficiency occurring is remote. A specific Phe deficiency is unlikely in any event. as poor-quality diets of the type associated with development of malnutrition are usually much lower, relative to the requirements, in their content of amino acids such as lysine, tryptophan, threonine, and methionine than they are in the aromatic amino acids.

With average intakes of protein and amino acids for all age groups in the U.S. exceeding considerably the average requirements (12), efficiency of the body's mechanisms for disposal of surpluses of the aromatic amino acids is of greater practical interest and concern than is the question of how to meet requirements for them.

METABOLISM OF PHENYLALANINE

Overview

The major catabolic pathway for phenylalanine is outlined in Figure 2.

Conversion of Phe to Tyr was found to occur in mammalian liver long before Tyr had been shown to be nutritionally dispensable (1). Direct evidence that Tyr was formed from Phe in mammalian tissues was obtained in 1940 by Moss and Schoenheimer (13), who isolated deuterated Tyr from tissue proteins of rats that had been fed deuterated Phe. This conversion, which subsequently proved to be the first step in the major pathway for catabolism of Phe, consists of an irreversible hydroxylation. Thus the pathway for degradation of Phe, apart from the initial reaction, is primarily that for degradation of Tyr. Subsequent isotopic

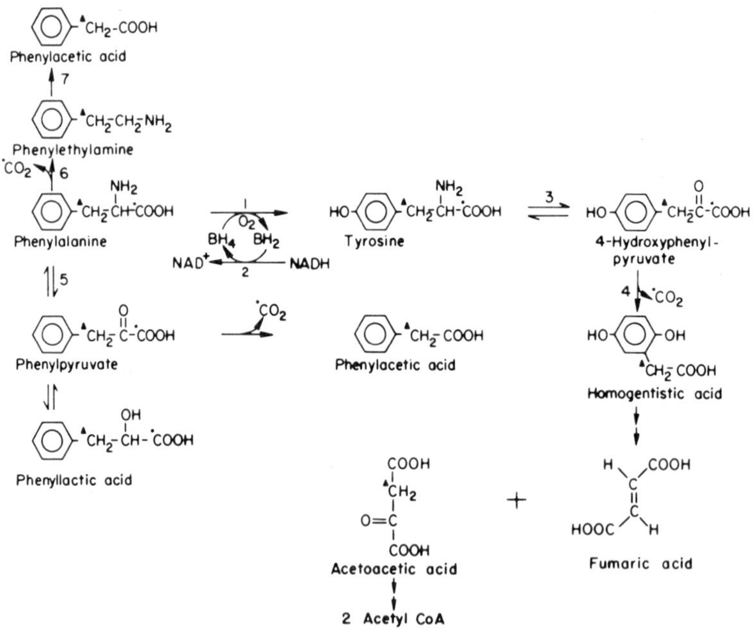

Figure 2 Pathway of phenylalanine catabolism (BH_4, tetrahydrobiopterin; BH_2, dihydrobiopterin).

(14-17) and enzymatic studies (18,19) had revealed by the mid-1950s that Tyr arising from hydroxylation of Phe was converted, through transamination with α-ketoglutarate, to p-hydroxyphenylpyruvate, which was then oxidized, with the loss of the carboxyl carbon as CO_2, to homogentisic acid. This intermediate was further oxidized to maleylacetoacetate, which on isomerization was converted to fumarylacetoacetate. Hydrolysis of the latter yielded fumarate, a potential glucose precursor, and acetoacetate. Phenylalanine is thus both glucogenic and ketogenic.

In addition to the pathway by which it is converted to Tyr, Phe can also undergo transamination with pyruvate (Fig. 2) and to a lesser extent with α-ketoglutarate to yield phenylpyruvate (20), which may be reduced to phenyllactate (21) or undergo oxidative decarboxylation to phenylacetate. Phenylalanine can also be converted by direct carboxylation via the reaction catalyzed by aromatic L-amino acid decarboxylase to phenylethylamine, which can then be oxidized to phenylacetate (22). These are minor pathways in normal animals and human subjects (23), as is evident from the low urinary excretion of the end products which are not further metabolized by mammals. Only if phenylalanine hydroxylase is inhibited (24), defective (25), or overloaded (26) do these products appear in the urine in quantity, in most animals conjugated with glycine, but in humans with glutamine (27,28).

Genetic Defects of Phenylalanine Metabolism

Discovery of the human genetic disease of phenylalanine metabolism, phenylketonuria (PKU) (phenylpyruvic oligophrenia) by Folling in 1934 (25), and identification of the defect as a deficiency in the Phe-oxidizing system by Jervis (29), stimulated interest in the pathways of Phe metabolism and provided direct evidence that conversion of Phe to Tyr was an essential step in the major pathway for degradation of excess Phe. Phenylalanine accumulates in the blood and tissues of infants with PKU, blood Tyr concentration is often low, and Phe is excreted in the urine together with large quantities (1-2 g/day) of phenylpyruvic acid and phenyllactic acid, products of the phenylalanine transamination pathway which are ordinarily excreted in only trace amounts. Other aromatic compounds, including o-hydroxyphenylacetate and phenylacetylglutamine, are also excreted. The amounts of all of these products excreted depends upon the amount of protein consumed (25).

The common form of the disease results from the hereditary absence, or a defect in the synthesis, of the enzyme, phenylalanine hydroxylase (Fig. 2) owing to the homozygous state of a single autosomal recessive gene that is carried by about 1 in 60 individuals, the exact proportion varying somewhat from population to population. The incidence of the disease in the United States is about 1 in 15,000 newborn infants. The disease results in delayed development and electroencephalographic abnormalities and leads to mental retardation early in life;

pigmentation is often sparse, so that untreated children with PKU usually have fair complexion, blue eyes, and blond hair. A proportion of them also suffer from seizures and eczema. In most hospitals newborn infants are tested for this defect, which can be detected by the occurrence of blood Phe concentrations of 20 mg (120 μmol)/dl or more (30).

The specific cause of the mental retardation is not known, but rigid control of Phe intake immediately after birth and for the first few years of life prevents elevation of blood Phe concentrations to hazardous levels and the accumulation of the end products of the transamination pathway. It also largely prevents the mental deterioration (31). Thus the defects presumably result from some impairment caused by the accumulation of phenylalanine or its metabolic products or, when food intake is depressed, to the lack of tyrosine or its metabolic products, or to some combination of these (25,30). The biochemical bases for the defects are not known, but among the possible causes of impairment are the following: High blood Phe can inhibit uptake of other large neutral amino acids into brain (32) and, in animals, is associated with depressed brain serotonin concentration (33); high concentrations of phenylpyruvate will inhibit mitochondrial pyruvate utilization (34); depletion of tyrosine can limit the supply of precursors of essential aromatic compounds such as dopamine and norepinephrine.

Persons who are heterozygotes for this autosomal, recessive trait do not have overt signs of PKU. They usually have slightly elevated plasma Phe concentrations, but, after a loading dose of Phe (100 mg/kg body weight), their blood Phe concentration will rise well above the value of 12-14 mg/dl observed for normal controls and the time required for clearance of Phe will be prolonged (35). During the consumption of a mixed diet, the plasma Phe concentrations of such individuals typically do not rise above 9 mg/dl. Also, hyperphenylalaninemia of between 6 and 20 mg/dl occurs in as many as 20 infants in 15,000, but in most of these plasma Phe concentrations will fall to within the normal range within a few weeks, and in most of the remainder within a few months. In about 5% of them, moderate elevation, usually not over 8 mg/dl, will persist when they consume the usual mixed diets (30). In some in whom liver biopsies have been done, liver phenylalanine hydroxylase is about 10% of normal, well above values for individuals with PKU. They do not develop mental retardation, even without restriction of dietary Phe. As Phe concentrations in maternal plasma below 10 mg/dl are not considered to be a hazard to the health of the offspring, the moderate elevation of plasma Phe concentration observed in this type of hyperphenylalanimia has not been found to affect development of the fetus (35). For the homozygous PKU female who becomes pregnant, dietary Phe restriction is recommended, as otherwise blood Phe concentration is likely to be elevated to levels which will pose a hazard for the developing fetus (36).

There is also a rare form of PKU in which the genetic defect is not in Phe hydroxylase, but in the enzyme dihydropteridine reductase (Fig. 2) that is required for regeneration of the tetrahydrobiopterin cofactor needed for the hydroxylation reaction (30). Infants with this defect do not respond to a low phenylalanine intake, but may to more elaborate therapy involving treatment with the cofactor (37).

Phenylalanine Hydroxylase (Phenylalanine 4-Monoxygenase)

This enzyme system, which is responsible for the conversion of Phe to Tyr, has been extensively studied. It was thought initially to be confined solely to liver (19), but small amounts have been detected in kidney and pancreas (38). Early work had indicated that the hydroxylation reaction was NADH dependent and that the enzyme system consisted of at least two protein components (19,39). The system has been characterized, as shown schematically in Figure 3, mainly by Kaufman and his associates (40,41). They have established that the complete hydroxylation system consists of two enzymes, an iron-containing phenylalanine hydroxylase and dihydropteridine reductase, and requires two cofactors, tetrahydrobiopterin and a reduced pyridine nucleotide. It consumes equimolar amounts of phenylalanine, oxygen, and reduced pyridine nucleotide, and produces an equivalent amount of tyrosine. Dihydrofolate reductase is an ancillary component required for the initial conversion of the inactive 7,8-dihydrobiopterin to 5,6,7,8-tetrahydrobiopterin, the active form of the cofactor. Molecular

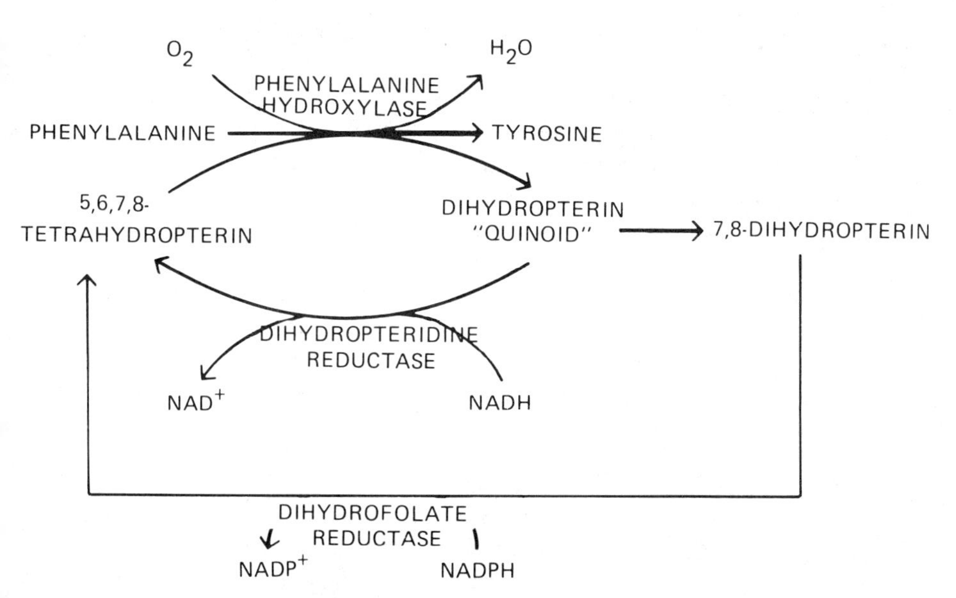

Figure 3 Enzyme system for hydroxylation of phenylalanine.

oxygen is the source of the oxygen in the tyrosine hydroxyl group (42), and NADH is more active than NADPH in the dihydropteridine reductase reaction (43).

Tryptophan will also serve as a substrate for phenylalanine hydroxylase of liver, but the rate of hydroxylation is much slower than with phenylalanine (44); also, tryptophan hydroxylase of brain will hydroxylate phenylalanine to a limited extent (45). Purification of these two enzymes has made it possible to demonstrate that they are distinctly different proteins, although they have some properties in common, including dependence on tetrahydrobiopterin as a cofactor (46,47).

As Phe hydroxylation is the pathway for synthesis of Tyr, the initial step in Phe catabolism, and the site of one of the most common human genetic defects of amino acid metabolism, information about the way in which this enzyme system is regulated is of considerable interest. Many factors are known to influence the activity of the system in vitro, but the significance of most of these in vivo has not been established. With a Michaelis constant of 0.15-0.2 mM for Phe, the rate of hydroxylation should increase in response to an increase in substrate concentration over the range of 0.2-0.3 mM up to 2.0 mM or more, as observed in liver of rats that have ingested a large Phe load (26). The rapid clearance by normal human subjects of a dose of Phe of 100 mg/kg (35) and the rise of only twofold in blood Phe concentration of rats when 5% Phe was added to their diet (48) would be compatible with this type of substrate control.

There have been many reports that a high dietary intake of Phe depresses liver phenylalanine hydroxylase activity in both the rat and the hamster (33). Woods and McCormick (49) found the depression to be proportional to the Phe content of the diet, but could not account for it on the basis of inhibition of the enzyme due to accumulation of Phe metabolites (50). The concluded that excess Phe suppressed enzyme synthesis. In most of these studies, however, the elevation of plasma Phe concentration was accompanied by elevated, rather than depressed, plasma Tyr concentration (33), and in two of them (6,48) plasma Phe concentration was not elevated much more when Tyr was included in the diet with Phe. From these observations it would appear that the physiological significance of this effect in vivo is questionable. Even when phenylalanine hydroxylase activity was suppressed in animals consuming a high Phe diet, Phe metabolism seemed to be limited by the rate of conversion of Phe to Tyr only when the Phe load was unusually large; and by removal of Tyr, only when a large amount of Tyr was included in the high Phe diet.

There have also been reports that phenylalanine hydroxylase is allosterically activated by Phe in vitro (38,51). The enzyme, which is a tetramer having a subunit size of 50,000 daltons, appears to bind one Phe per monomer (52). The purified enzyme is inactive in the absence of Phe, but is 50% activated by 0.06 mM Phe at pH 7.4. Activation should thus occur over the range of Phe concentrations observed physiologically, and this mechanism would be expected to function in conjunction with the control by substrate concentration that occurs independently of activation, to increase the rate of Phe catabolism still further in response to an increase in Phe intake. Regulation of Phe hydroxylation in this way could account for the rapid rate of clearance of Phe from the blood of normal human subjects administered a load of Phe of 100 mg/kg body weight (35). When Phe hydroxylase is preincubated with 0.1 to 1.0 mM Phe, the enzyme is activated 20-fold or more. It is also activated by preincubation with tryptophan, methionine and a few other amino acids; thus after ingestion of a high protein meal or a load of Phe, greatly increased Phe hydroxylase activity would be expected (52a).

A variety of other factors have been shown to influence the activity of the Phe-hydroxylating system in vitro. Among these are limited proteolysis of the enzyme, alkylation of SH groups, interaction with lysolecithin, state of phosphorylation of the enzyme, an inhibitory substance (possibly Mn^{2+}) in the liver cytosol (41), and the amount of Fe^{3+} bound to the enzyme (53). The importance of several of these factors in vivo is unclear. There is evidence, however, from observations on activation of the Phe hydroxylation system by glucagon treatment of rats, that a cAMP-dependent phosphorylation-dephosphorylation mechanism is involved in regulation of Phe hydroxylation in vivo (41). Also, from studies of the response of the hydroxylation system to iron, control of activity through modification of iron binding may also provide a mechanism for regulation of Phe hydroxylation in vivo (53).

Measurements of the relative rates of release of deuterium from deuterated Phe and Tyr (54) indicate that phenylalanine hydroxylation is ordinarily the rate-limiting step for Phe degradation. Besides the potential for control of Phe hydroxylase activity discussed above, the possibility that regulation of dihydropteridine reductase may also be involved in control of the rate of hydroxylation of Phe was suggested from observations on effects of glucagon on this enzyme and on Phe oxidation in vivo (55). Evidence from studies of substrate (52), iron (53), and phosphorylation (41) activation of the hydroxylation system all indicate that the system is not fully activated in vivo. As the major need for Phe is for protein synthesis, low activity of the Phe hydroxylation system, which represents the initial step in Phe degradation, could serve as a mechanism for conserving Phe when intake of this essential amino acid is low. A control system that increased the rate of Phe oxidation when Phe intake is high would contribute to rapid removal of excess Phe by increasing the supply of substrate for Tyr-α-ketoglutarate aminotransferase, the second enzyme in the major degradative pathway.

Tyrosine Oxidation

Regulation of tyrosine-α-ketoglutarate aminotransferase has been investigated intensively (56,57). It appears to be rate limiting for oxidation of tyrosine to CO_2 in vivo (58), but, as it can be induced rapidly by tyrosine, dietary protein, cortisol, insulin, and glucagon, whereas p-hydroxyphenylpyruvate dioxygenase and homogentisate oxygenase, the next two enzymes in the reaction sequence (Fig. 2), are relatively resistant to induction (58), control may be exerted at different steps in the degradative pathway, depending on the nutritional and hormonal state of the organism. Although Phe degradation would not ordinarily appear to be limited by reactions in the tyrosine catabolic pathway (54), this may depend on the amount of Phe consumed and on the amount of Tyr that must be degraded at the same time (6,48). Plasma Tyr concentration of rats fed increasing increments of Phe increased when Tyr was also included in the diet (6). Also plasma Phe rose from 0.02 to 0.3 mM when 7% of Phe was included in a standard laboratory diet, but plasma Tyr increased from 0.02 to 1.7 mM (59).

Phenylalanine-Pyruvate Aminotransferase

Phenylketonuric subjects (25), rats given a loading dose of Phe (27,55), and rats that have been injected with p-chlorophenylalanine to inhibit phenylalanine hydroxylase (60) excrete products of the phenylalanine transamination pathway, phenylpyruvate and its reduction product phenyllactate (Fig. 2). Unless phenylalanine hydroxylase activity is suppressed or overloaded, little Phe would appear to be metabolized via transamination, as only small amounts of the products of this pathway are ordinarily present in urine. As these products are not degraded by animals beyond phenylacetate (23), the amounts of these compounds excreted in urine should be a reliable indicator of the amount of Phe channeled into this pathway.

The specificity of the enzyme designated as phenylalanine-pyruvate aminotransferase has been investigated (61,62). It is apparently identical with isozyme 1 of histidine-pyruvate aminotransferase found in rat liver cytosol. Histidine-pyruvate aminotransferase isozyme 1 of liver mitochondria is identical with serine-pyruvate aminotransferase, but is much more active with Phe than with either histidine or serine as substrate. Tyrosine and tryptophan will also serve as substrates for it. The enzyme in the basal state has only one-half to one-third the activity of phenylalanine hydroxylase, but can be induced to 10-fold the basal level by treatment of rats with glucagon (55). Isozyme 2 is specific for histidine transamination and occurs in many other organs besides liver. It may be identical with glutamine aminotransferase. Isozyme 1 is found only in liver and would appear to be a relatively nonspecific phenylalanine-pyruvate aminotransferase. There is also some transamination of Phe with α-ketoglutarate (25), but this activity is even lower than that of the Phe-pyruvate aminotransferase and appears not to be induced by glucagon (55).

PHYSIOLOGICAL ASPECTS OF PHENYLALANINE METABOLISM

Oxidation of Phenylalanine In Vivo in the Rat

The fate of orally administered tracer doses of [U-^{14}C]Phe or [U-^{14}C]Tyr was investigated by Dalgliesh and Tabechian (63) in rats that had been without food overnight. As they point out, the results from these studies, with small doses (1 mg) of the amino acids and measurements of limited accuracy, do not justify quantitative conclusions, but should permit valid comparisons between Phe and Tyr. The proteins of gut, liver, and muscle were labeled rapidly after administration of either Phe or Tyr, with about half of the Phe and one-third of the Tyr radioactivity being incorporated into tissue proteins within 1 hr. The free amino acid pool of liver was labeled very little by either amino acid, probably because of rapid incorporation and degradation of these amino acids by this organ. The free pool of muscle was labeled highly, about 20% of the radioactivity of Phe or Tyr recovered between 10 and 15 min being in this fraction. By 1 hr there was little label in the free amino acid pool of muscle, indicating very rapid turnover of the free amino acids. Depletion of radioactivity from the free pools was accompanied by a steady release of labeled CO_2 in expired air. Within 4 hr 20% of the radioactivity from Phe and 40% of that from Tyr had been expired. The authors interpreted this as evidence that hydroxylation of Phe was rate limiting for Phe oxidation. This is in agreement with the more recent observations of Milstien and Kaufman (54), but as the specific radioactivities of the liver free pools could not be measured, it cannot be assumed that the quantitative relationship implied by the results provides an accurate measure of the relative rates of oxidation of these two amino acids.

In another study of the effect of Phe intake on oxidation of Phe and Tyr in vivo, Godin and Dolan (59) administered tracer (1 mg) doses of either DL-[1-^{14}C] Phe, DL-[3-^{14}C]Phe, or DL-[2-^{14}C]Tyr to rats consuming diets that contained inadequate (0.1%), adequate, or excessive (5-7%) amounts of Phe. Less than 10% of the administered radioactivity from [3-^{14}C]Phe was recovered in expired CO_2 during the 24 hr after administration when Phe intake was inadequate or adequate, indicating that, as with many other amino acids (64), Phe is conserved well when intake is inadequate to marginally adequate. When Phe intake was excessive, 26% of the radioactivity from [3-^{14}C]Phe and 42% of that from [1-^{14}C]Phe was recovered in expired CO_2 (59), indicating that the rate of oxidation of Phe increased greatly in response to the increased load. A similar response was observed for tyrosine oxidation. As blood Phe concentration increased by 15-fold in rats consuming the high-Phe diet, considerable dilution of radioactivity must have occurred; therefore these figures are undoubtedly underestimates of the extent of oxidation, as they are not based on specific radioactivity measurements of the liver amino acids. Despite their limitations, the results indicate that whether the increased rate of Phe oxidation was due to increased substrate concentration or to activation of phenylalanine hydroxylase by substrate, phosphorylation, or some other mechanism, the rate of Phe hydroxylation increased rapidly to remove surplus Phe, although not enough with this very large load to prevent some 10% of the administered radioactivity from spilling over into the urine.

It has been generally accepted that the hydroxylation pathway (Fig. 1) is the major route for phenylalanine degradation in mammals (65) and that transamination to phenylpyruvate or decarboxylation to phenylethylamine are minor pathways. David et al. (66) suggested, however, from studies of tyrosine degradation that decarboxylation of the aromatic amino acids by aromatic L-amino acid decarboxylase may be a major route for degradation of Phe and Tyr when these amino acids are consumed in excess. This report served as a stimulus for investigation of the relative importance of hydroxylation, transamination, and direct decarboxylation in the catabolism of a large load of Phe in vivo in the rat (23).

Initially in this study by Haley and Harper (23) the fate of phenylacetate, the major end product of the decarboxylation pathway of Phe catabolism, was examined to determine whether it might be degraded further in vivo. When rats were administered a loading dose of [1-^{14}C]phenylacetate, 95% of the radioactivity was excreted in the urine as phenylacetate or phenylacetylglycine within 48 hr, and less than 1% of the radioactivity was recovered in expired CO_2. Thus urinary phenylacetate should provide a measure of the extent of direct decarboxylation of Phe, although it might give an overestimate if some phenylpyruvate arising from the transamination pathway were also decarboxylated.

Subsequently, the fate of L-[1-^{14}C]Phe or L-[3-^{14}C]Phe, administered orally or intraperitoneally with a loading dose (60 or 130 mg) of L-Phe, was investigated in rats that had been kept without food for at least 24 hr. Some of the results of this study (23) are summarized in Figure 4. Urine contained 3.3% of the radioactivity from [1-^{14}C]Phe and 5.6% of that from [3-^{14}C]Phe. Only nine-carbon

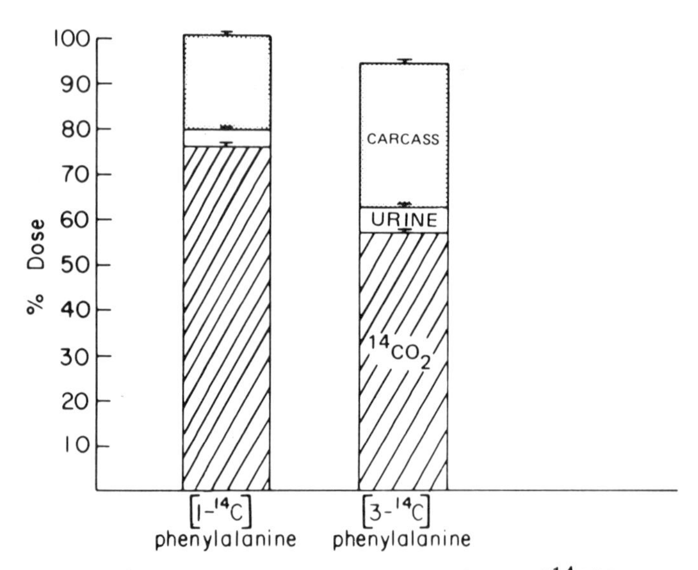

Figure 4 Distribution of radioactivity from L-[^{14}C]Phe in expired CO_2, urine, and carcass of rats 24 hr after they were administered a loading dose (130 mg/100 g body weight) of Phe (crosshatched area, CO_2; open area, urine; stippled area, carcass). (From Ref. 23.)

compounds would be labeled after administration of [1-^{14}C]Phe, whereas compounds arising subsequent to decarboxylation of Phe, such as phenylacetate, would also be labeled after administration of [3-^{14}C]Phe. From chemical analysis of urine and carcass samples for end products of the transamination and decarboxylation pathways, it was concluded that only 6% of the Phe load was metabolized via these routes, and, from measurement of phenylacetate excretion specifically, that, at the maximum, 2% of the Phe load was metabolized via direct decarboxylation. A total of 2% of the Phe load was excreted as Phe and Tyr and unidentified metabolites, and 17% remained in carcass. When the label was in the 3-position, more radioactivity was retained in the carcass than when the label was in the 1-position, mainly in fatty acids and small water-soluble molecules, and less was expired as CO_2, as Godin and Dolan (59) had observed. From this information and from measurements of expired $^{14}CO_2$ after administration of [1-^{14}C]Phe, it would appear that 75% of a Phe load, representing 1.5 times the daily requirement of the rat, was degraded via the Phe hydroxylase pathway within 10 hr.

Fellman et al. (67) have reported that direct decarboxylation is a trivial pathway of tyrosine metabolism in liver preparations even with high tyrosine concentrations and, according to a note in that paper, David and associates have also concluded, contrary to their earlier report, that direct decarboxylation of tyrosine accounts for less than 1% of the dietary tyrosine degraded, even when tyrosine intake is high. It is thus evident that metabolism of Phe and Tyr occurs mainly via the Phe hydroxylation-Tyr transamination pathway, and that major control of the metabolisms of Phe and Tyr, respectively, must be exerted at these two steps.

As indicated earlier, a significant fraction of the metabolism of Phe evidently occurs via the alternate pathways only when Phe hydroxylase is defective, inhibited, or greatly overloaded. The capacity of the rat for Phe oxidation when the Phe dose (130 mg/100 g) is large enough to saturate Phe hydroxylase can be estimated roughly from the in vivo oxidation studies to be about 100 μmol/100 g rat per hour (23). This value agrees quite well with that calculated from the results of another similar study in which rats were administered a much larger (640 mg/100 g) dose of Phe (55).

The Transamination Pathway In Vivo

As Phe hydroxylase can be activated by a phosphorylation mechanism, presumably mediated at least in part by cAMP and glucagon, and as Phe-pyruvate aminotransferase is induced by glucagon, the question arises as to whether Phe metabolism may be altered either qualitatively or quantitatively by treatments that alter the activities of key enzymes in these pathways of metabolism. In studies by Fuller and associates (60) and Brand and Harper (68) on rats, glucagon treatments that induced Phe-pyruvate aminotransferase were not found to increase the fraction of a Phe load that was converted to phenylpyruvate. Brand and Harper (55) did observe, however, that glucagon treatment increased by close to 50% the capacity of rats to oxidize a large load of Phe, despite the fact that Tyr accumulated in their blood and tissues. Interestingly, this response was not associated with evidence of increased activity of Phe hydroxylase, but was accompanied by a twofold increase in the activity of dihydropteridine reductase. It should be noted that Phe hydroxylase was assayed in this study with the synthetic dimethyltetrahydropterin cofactor, which, according to more recent studies (41), provides a measure of the fully activated enzyme. As the enzyme is not fully activated in tissues, this assay probably overestimated the Phe hydroxylase activity of control animals. In a subsequent study (26) in which the enzyme was assayed with the naturally occurring tetrahydropterin cofactor, Phe hydroxylase in liver from chronically glucagon-treated rats was found to be elevated.

Relationships between activities of Phe-degrading enzymes, oxidation of Phe in vivo, and excretion of metabolites of the alternative Phe degradation pathways in rats administered a loading dose of Phe were investigated further by Haley and Harper (26). As in the previous study (55), chronic glucagon treatment increased the rate of oxidation of [1-^{14}C]Phe, but, in this study, the increased rate of Phe oxidation was associated with an increase in Phe hydroxylase activity. Similar responses were not observed after a single injection of glucagon, possibly because a single injection results in only a transitory increase in Phe hydroxylase (69).

Chronic glucagon treatment also increased the activity of Phe-pyruvate aminotransferase by 10-fold and increased the quantity of Phe degradation products, mainly phenylpyruvate, excreted in the urine by about 4-fold (26). This result differs from that obtained in previous studies (55,60) in which urinary phenylpyruvate excretion was not increased by glucagon treatment, despite a comparable increase in Phe-pyruvate aminotransferase activity. This discrepancy has raised a question about the importance of Phe-pyruvate aminotransferase in Phe metabolism in vivo (68). Fellman et al. (70) have suggested on the basis of studies of phenylketonuria and tyrosyluria that aromatic amino acid transamination may occur in extrahepatic sites.

In relation to this, it is perhaps significant that in the earlier study from our laboratory (55) the dose of Phe administered was much greater (640 mg/100 g) than that (127 mg/100 g) used subsequently (26) and that the amount of phenylpyruvate excreted by the glucagon-treated rats in the former study was over twice that observed in the latter. Thus considerable phenylpyruvate was being formed and excreted by the glucagon-treated rats, even though the level did not exceed that for the controls. The major difference between these two studies was in the quantity of phenylpyruvate excreted by control animals that were not treated with glucagon. With the lower Phe load very little phenylpyruvate was excreted by controls (26), but with the larger load (55) controls excreted somewhat more phenylpyruvate and much more Phe than the glucagon-treated group, indicating that their capacity for Phe catabolism was greatly exceeded. The results for both

Phe oxidation and urinary phenylpyruvate excretion obtained with rats administered the smaller Phe load fit well with predictions based on the pattern of activities of Phe-degrading enzymes; but if excretion of phenylpyruvate is solely a function of Phe-pyruvate aminotransferase activity, much less phenylpyruvate excretion by untreated rats administered the large load of Phe would have been anticipated. These results suggest, as do the observations of Fellman et al. (70), that when the Phe load is great and the activity of Phe-pyruvate aminotransferase is low, Phe may be used as a substrate in other transamination reactions. Shih et al. (24) have observed that when rats in which Phe hydroxylase has been inhibited by p-chlorophenylalanine are given a large dose of Phe, glucagon treatment, which induces Phe-Tyr aminotransferase, results in a substantial increase of phenylpyruvate excretion.

TOLERANCE FOR PHENYLALANINE: BLOOD POOLS OF PHENYLALANINE AND TYROSINE

There have been many reports of effects of feeding rats and other animals large amounts of Phe, or both Phe and Tyr together, in efforts to produce biochemical and behavioral changes resembling those seen in PKU patients. Much of this information has been reviewed (33). The animal studies indicate that the growth rate of rats is depressed by large intakes of Phe, especially when protein intake is low, but rats fed from weaning on adequate diets containing as much as 7% of Phe do not exhibit specific signs of toxicity such as are observed in animals fed high-tyrosine diets. Growth is depressed moderately in rats fed a diet containing 5% L-Phe—this would be comparable to an intake of 25 g of L-Phe per day in addition to that in the dietary protein for an adult human. There was a close inverse correlation between plasma Phe concentration and weight gain (48).

In a number of the animal studies elevated plasma Phe concentration was associated with urinary excretion of phenylpyruvate, evidence that, when Phe intake was in the range of 5% of the diet, the capacity of the Phe hydroxylation system was exceeded. Also, a high Phe intake was found to depress serotonin concentration in several tissues, including brain (71). The studies of animal models of PKU also indicated that a high Phe intake would induce behavioral changes in both rats and monkeys, but these were not claimed to mimic those due to PKU. A major problem with this type of animal model is the ability of the animal to degrade Phe and to adapt over time to a high Phe intake. The use of an inhibitor of Phe hydroxylase, such as p-chlorophenylalanine, in conjunction with a high Phe intake provides the potential for a better model (60,72).

Of more interest in relation to normal metabolism is the effect of changing protein intake on blood and tissue pools of Phe and Tyr. In a study (73) in which young rats that had been consuming a nutritionally adequate diet were fed for 4 hr diets ranging in casein content from 0 to 55%, plasma Phe concentrations, measured at the end of the feeding period, increased from 9 to 14 μmol/dl as casein content was increased from 15 to 55%. Blood Tyr concentrations increased from 21 to 59 μmol/dl. Brain Phe concentration (5 μmol/100 g) was not affected by increasing protein intake, but brain Tyr concentration increased from 15 to 23 μmol/100 g. Considering that the 15-55% casein diets represent a range of 0.8-3% of the diet as Phe, the relatively small responses in blood and brain Phe concentrations indicate that Phe is degraded efficiently by the normal animal.

When rats were allowed to adapt to diets differing in protein content from 5 to 75% of casein for a 2-week period, plasma Phe concentration was not significantly altered from the value of 9 μmol/dl by altering protein content of the diet over the range of 15-75% casein. Plasma Tyr concentration, however, decreased with increasing protein intake, from 27 μmol/dl (15% casein) to 18 μmol/dl (75% casein). The difference between the responses of the adapted animals and the responses of those studied after a single meal to increasing protein intake presumably reflects the increased capacity of rats adapted to a high protein intake to degrade amino acids (73). Neither brain Phe nor Tyr concentration increased in response to increasing protein intake; in fact, both tended to decline steadily in the adapted rats as the protein content of the diet increased.

Aoki et al. (74) measured blood amino acid concentrations in human subjects at hourly intervals for 4 hr after they had consumed 200 g of broiled sirloin which might be expected to contain 1.6 g of Phe. Plasma Phe concentration rose from 3.0 μmol/dl to a peak after 3 hr of 3.9 μmol/dl. Plasma Tyr concentration rose from 3.5 to 5.1 μmol/dl during the same time period. Thus ingestion of this amount of meat in a single meal increased the plasma pools of Phe and Tyr only 30 and 50%, respectively.

Phenylalanine tolerance tests can be used to estimate the phenylalanine-degrading capacity of human subjects. The results presented in Figure 5 show the pattern of clearance of Phe from the blood of a normal control and three hyperphenylalaninemic subjects who consumed a Phe load of 100 mg/kg body weight (35). This load, which would represent over eight times the daily requirement for Phe in a single dose, was completely cleared from blood of the normal subject in between 4 and 8 hr. In the hyperphenylalaninemic subjects the time for clearance was greatly prolonged, presumably because of the inherently low Phe hydroxylase activity of such subjects (75). The hyperphenylalaninemic subjects showed smaller and less persistent rises in blood Phe concentration when they consumed the equivalent of 180 mg/kg body weight per day of Phe from food proteins. From these observations it is evident that the phenylalanine hydroxylation system of normal human subjects can be temporarily overloaded if they consume a large load of Phe as a single bolus, but the capacity of the system is such that an amount equivalent to eight times the daily requirement of Phe is cleared rapidly from the blood.

Stegink and associates (76) have used the tolerance test to study the effects of large doses of aspartame on blood amino acid concentrations. In one such study normal adults were given oral doses of 100, 150, 200 mg of aspartame per kilogram body weight in a single dose. These doses provided approximately 56,

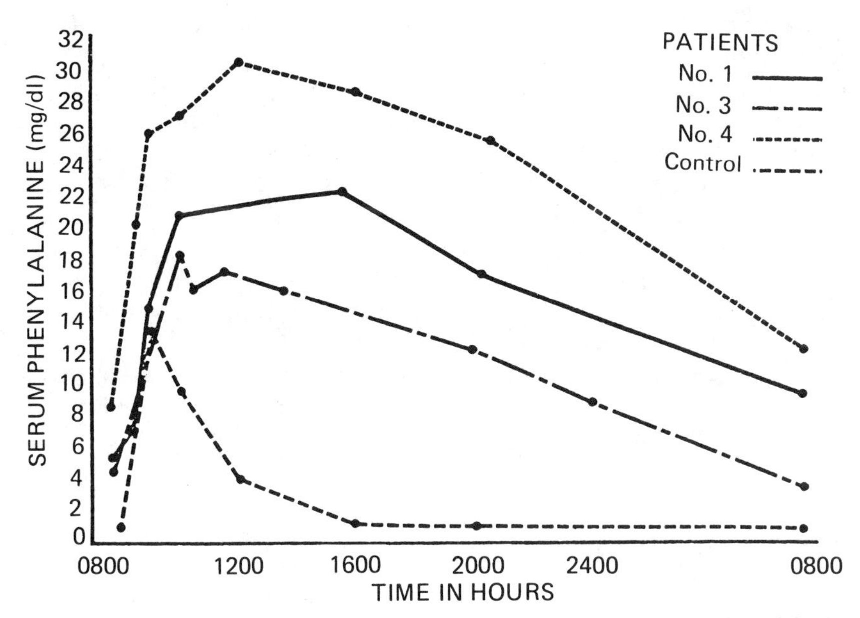

Figure 5 Serum Phe concentrations of normal control and hyperphenylalaninemic subjects given a single dose of 100 mg/kg body weight of Phe. Regular meals were ingested after 12:00. (From Ref. 35.)

84, and 112 mg/kg body weight of Phe, or about 5-10 times the requirement of Phe in a single dose. Peak plasma Phe concentrations occurred between 1 and 2 hr after ingestion of the aspartame load. The increases in plasma Phe concentrations were from 5-6 μmol/dl to 20, 35, and 49 μmol/dl, respectively, with the increasing Phe intakes. These values were well below the value of 120 μmol/dl used as an indicator of PKU. Also, plasma Phe concentration declined steadily, so that even with the highest dose the plasma Phe value was below 15 μmol/dl within 6 hr. Plasma Tyr concentrations increased from 5-6 μmol/dl to peak values of 10-14 μmol/dl by 3 hr. The much greater rise in plasma Phe than Tyr concentration further supports the conclusion that in normal subjects Phe hydroxylation is the limiting reaction in Phe degradation.

In PKU heterozygotes who consumed 100 mg of aspartame per kilogram (56 mg/kg Phe) in a single dose (about five times the daily requirement), plasma Phe concentration rose to a peak value of 42 μmol/dl within 2-3 hr and declined to about 12 μmol/dl by 6 hr (77), indicating that heterozygotes for PKU catabolize Phe at about half the rate of normal individuals, but can still clear a loading dose of 56 mg Phe per kilogram body weight, in dipeptide form, close to 85% of the usual daily intake of Phe, efficiently within a period of about 6 hr. As would be expected when Phe hydroxylase is limiting, Tyr concentration in these subjects rose less after the aspartame load than it did in normal subjects.

In human subjects in the postabsorptive state infused with tracer amounts of Phe and Tyr labelled with stable isotopes, Clarke and Bier (77a) found Phe and Tyr turnover rates to be 36 and 40 μmol, respectively, per kg of body weight/hr and conversion of Phe to Tyr to be 6 μmol/kg of body weight/hr. These values are lower than those observed in the loading studies indicating, as would be predicted from observations on Phe hydroxylase in vitro, that the enzyme is activated when Phe concentration is elevated and that this contributes to rapid clearance of a large load of Phe.

The evidence from a variety of studies of both animals and human subjects who have ingested large amounts of protein, a dipeptide containing Phe or Phe itself, indicate that Phe is cleared efficiently by normal subjects and even by PKU heterozygotes who have half or less the normal level of phenylalanine hydroxylase. Only when this enzyme is inhibited in animals or deficient, as in PKU patients, and the ability to convert Phe to Tyr is severely impaired does Phe accumulate to hazardous levels in body fluids.

CONDITIONAL INDISPENSABILITY OF TYROSINE

As was indicated in the discussion of the nutritional essentiality of Phe, Tyr is synthesized by normal animals and human subjects and is therefore nutritionally dispensable. Under conditions in which Phe hydroxylase activity is deficient, as in PKU patients, Tyr becomes an essential nutrient and must be provided in the diet.

Chipponi et al. (78) have discussed the concept of conditionally indispensable nutrients, citing a number of examples in which a disease or genetic defect results in impairment of the ability to synthesize substances required for normal body function, with the result that these substances must be obtained from food. They cite an example of a group of cirrhotic patients, maintained parenterally, in whom negative nitrogen balance was accompanied by severe declines in plasma cystine (Cys) and Tyr concentrations. Upon provision of these two amino acids as an oral supplement, the patients went into positive nitrogen balance, and plasma Cys and Tyr concentrations returned to normal. The ability of these patients to synthesize Cys and Tyr was presumably impaired as the result of liver disease and both amino acids became conditionally indispensable.

Snyderman (10) cites the premature infant as another example of tyrosine dependency. In such infants depletion of tyrosine from the diet results in classic manifestations of amino acid deficiency: reduced weight gain and nitrogen retention, and low plasma concentration of tyrosine. These changes are reversed when tyrosine is included in the diet. Phenylalanine hydroxylase is one of many amino acid-degrading enzymes that increase sharply in activity at the time of birth. Evidently in premature infants, and in some full-term infants, delayed development of this enzyme results in tyrosine becoming conditionally indispensable for a period of time.

There have also been a number of observations indicating that severe kwashiorkor results in a fall in plasma Tyr and an elevated plasma Phe/Tyr ratio and increased exertion of phenylpyruvate, presumably as a result of depletion of Phe hydroxylate in such infants and the low food intake characteristic of this disease (79,80). The extent to which such infants develop Tyr dependency is not clear, but it is evident that their ability to synthesize Tyr from Phe is impaired. Antener et al. (81) observed that young mothers with protein energy malnutrition who were given a Phe tolerance test (100 mg/kg body weight) had higher blood Phe and lower blood Tyr concentrations than healthy controls, evidence of Phe hydroxylase deficiency. Also, in some of the patients, high excretion of p-hydroxyphenlypyruvate after a tyrosine load indicated the probability of p-hydroxyphenylpyruvate oxidase deficiency as well.

PHENYLALANINE AND TYROSINE IN THE SYNTHESIS OF BIOLOGICALLY ACTIVE COMPOUNDS

Tyrosine is the immediate precursor of (a) the catecholamine neurotransmitters dopamine and norepinephrine in brain; (b) the hormones epinephrine and norepinephrine in the adrenal medulla, and thyroxine and triiodothyronine in the thyroid gland; (c) the pigment melanin in skin and various other organs; and (d) the electron transport carrier ubiquinone (coenzyme Q). As the amount of Phe that is converted to Tyr (Fig. 2) depends upon the amount consumed, the extent to which Phe serves as the precursor of these substances will depend upon the amount of Phe consumed in excess of the needs for protein synthesis and the proportions of Phe and Tyr in the diet.

Catecholamine Synthesis in Brain

The sequence of reactions by which the catecholamines dopamine and norepinephrine are synthesized in brain is outlined in Figure 6. The initial reaction, hydroxylation of Tyr by Tyr hydroxylase to form 3,4-dihydroxyphenylalanine

TYROSINE — TYROSINE HYDROXYLASE → DIHYDROXYPHENYLALANINE (DOPA)

DOPA DECARBOXYLASE (Aromatic amino acid) decarboxylase

DOPAMINE — DOPAMINE-BETA-HYDROXYLASE → NOREPINEPHRINE

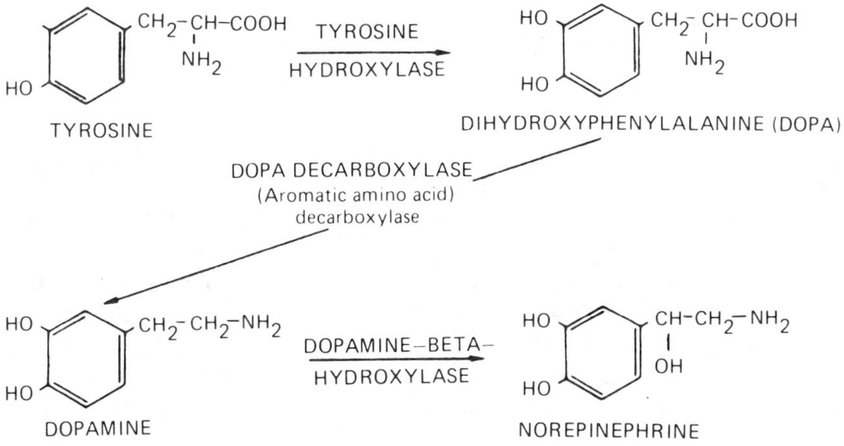

Figure 6 Pathway of catecholamine synthesis.

(DOPA), is considered to be the rate-limiting step in the pathway (82,83). This hydroxylation requires the same tetrahydrobiopterin cofactor as the phenylalanine hydroxylase reaction. Decarboxylation of DOPA by aromatic L-amino acid decarboxylase yields dopamine, and in dopaminergic neurons the pathway proceeds no further. In noradrenergic neurons dopamine is hydroxylated by dopamine β-hydroxylase to form norepinephrine. In certain nerve tracts and in the adrenal medulla, norepinephrine is converted to epinephrine by N-methylation, which requires S-adenosylmethionine as the methyl donor and the enzyme phenylethanolamine-N-methyltransferase.

As the K_m of tyrosine hydroxylase for tyrosine, as measured with the natural tetrahydrobiopterin cofactor, is 4-20 μM (84), and as normal brain Tyr concentration is about 75 μM (83), it has been considered unlikely that catecholamine synthesis would be sensitive to variations in blood and tissue tyrosine concentrations. Failure of injections of Tyr in the rat to alter brain catecholamine concentrations supported this view (85). The major controls are considered to be feedback inhibition of Tyr hydroxylase by norepinephrine and dopamine and modulation of the activity of the enzyme in response to changes in the rate of neuronal firing (83,86). Wurtman and associates (87,88), however, observed an increase in brain Tyr content and a small but significant increase in DOPA concentration following the administration of 50 mg of Tyr per kilogram body weight to rats that had been treated with a decarboxylase inhibitor. Injection of tryptophan or leucine, which compete with Tyr for uptake into brain, depressed both Tyr and DOPA concentrations in brain. The authors concluded that, in addition to factors that control the activity of Tyr hydroxylase, catecholamine synthesis probably depends also on the availability of the precursor amino acid tyrosine.

Gibson and Wurtman (89,90) have investigated the possibility that catecholamine synthesis may be influenced by protein intake. They found that brain Tyr concentration was increased in proportion to increasing protein intake and, although brain DOPA concentration was also elevated in the rats treated with a decarboxylase inhibitor, the relationship between changes in brain Tyr and DOPA concentrations was not strictly proportional. Subsequently, when they studied the effect of protein intake on the accumulation of a metabolic end product of the norepinephrine synthetic pathway, they found that differences in brain Tyr content were not associated with increased accumulation of the norepinephrine metabolite, except when the animals were in a cold environment, which increases the rate of firing of noradrenergic neurons. They interpreted this to mean that catecholamine synthesis in brain may be affected by diet-induced increases in Tyr supply only when catecholaminergic neurons are firing rapidly.

In a series of experiments in vivo, Wurtman and associates (91,92) have shown that administration of tyrosine increases catecholamine synthesis in spontaneously hypertensive rats and in other conditions in which catecholaminergic neurotransmission is stimulated. Their earlier studies (87) suggest that dietary conditions that lower brain Tyr concentration depress catecholamine synthesis, but the subsequent studies suggest that increases in brain Tyr content following consumption of a high protein meal may have little effect upon catecholamine synthesis and release except when catecholaminergic neurons are highly activated (93). As blood Tyr concentration is responsive to Tyr intake, these effects would be expected to depend on both the absolute intakes of Phe and Tyr and the relative proportions of these two amino acids in the diet.

Hormone Synthesis

Epinephrine and norepinephrine, besides being neurotransmitters, are hormones produced by the adrenal medulla which are released in response to neural stimuli. They have a wide variety of effects, mediated in large measure by stimulation of adenylate cyclase and production of cAMP in target tissues. Epinephrine promotes glycogenolysis in liver and muscle; both of these catecholamines stimulate the release of nonsterified fatty acids from adipose tissue; both also affect blood pressure, epinephrine by increasing heart and pulse rate and cardiac output, norepinephrine by increasing peripheral resistance (94,95).

These hormones are synthesized from Tyr, and hence also from Phe, as outlined in the preceding section on neurotransmitters (Fig. 6). There is no known deficit of adrenal medullary function or of control of the production of these hormones through alteration in precursor supply.

Thyroid hormones, the most active forms of which are triiodothyronine and thyroxine, are also produced from Tyr, and therefore indirectly from Phe. Tyrosine side chains in thyroglobulin in the thyroid gland are iodinated to form mono- and diiodotyrosine residues. Some of these residues become linked by ether linkages to form triiodothyronine and thyroxine, which are released subsequently from thyroglobulin by the action of proteases (94,95).

Formation of thyroid hormones is influenced by the amount of iodine available for their formation, but is not known to be influenced by the supply of their amino acid precursors.

Ubiquinone

The benzoquinone moiety of ubiquinone, or coenzyme Q, a component of the electron transport system, is also synthesized from tyrosine and, therefore, also from phenylalanine. Investigations of the incorporation of [U-^{14}C]Phe into lipid fractions of liver preparations by Olson and associates (96) indicated that p-hydroxyphenyl derivatives of phenylalanine and tyrosine were incorporated into the aromatic nucleus of the ubiquinone molecule. From studies of compounds that, on addition to the incubation medium, diluted the radioactivity in the final product, and from studies of the effectiveness of various aromatic precursors (97), they concluded that p-hydroxybenzoate was the immediate precursor and that this was derived from tyrosine.

There is still some question about the nature of all of the intermediate products, but Olson (96) has suggested from the accumulated evidence that a highly likely sequence is as follows:

Phe
↓
Tyr
↓
p-hydroxyphenylpyruvate
↓
p-hydroxyphenyllactate
↓
p-hydroxycinnamate
↓
p-hydroxybenzoate
↓decarboxylation
↓hydroxylation
↓methylation
↓isoprenyl alkylation
ubiquinone-9

This represents a unique pathway for Phe, and Tyr as p-hydroxyphenyllactate, which is formed from p-hydroxyphenylpyruvate, the product of tyrosine transamination, is excreted in large amounts by patients with tyrosinemia, a disease in which p-hydroxyphenylpyruvate oxidase activity (Fig. 2) is very low (98). This and other similar observations have led to the conclusion, discussed earlier, that this metabolite was not further degraded by mammals; however, the amount that would be required for ubiquinone synthesis would represent a small fraction of usual Phe and Tyr intakes. This would also make it unlikely that the dietary supply of the aromatic amino acids would influence ubiquinone synthesis, as the amounts of the precursor amino acids released from tissue protein turnover undoubtedly provide adequate amounts of substrate for effective functioning of this synthetic pathway.

Melanin

Tyrosine is a precursor of the dark pigment melanin formed in melanocytes in the basal layer of the epidermis. The copper-containing enzyme tyrosinase required for its formation is found in animals only in the organelles of these melanin-producing cells. This enzyme catalyzes the conversion of tyrosine to DOPA (dihydroxyphenylalanine; Fig. 5) and then to dopaquinone, which undergoes a series of oxidation reactions and interconversions to produce indole-5,6-quinone and 5,6-dihydroxyindole, which condense to form polymers responsible for the color of skin and hair (94,95). The light skin and hair color of many PKU patients is sometimes attributed to inadequate intakes of tyrosine in the absence of the ability to convert Phe to Tyr.

PLASMA AMINO ACID CONCENTRATIONS AND BRAIN FUNCTION

Interest in relationships between plasma and brain concentrations and patterns of free amino acids and brain function has been stimulated by the following observations: In PKU and other genetic diseases of amino acid metabolism, accumulation of one or more amino acids in blood is associated with impariment of mental function (25); dietary disproportions of amino acids which result in distortions of plasma and brain amino acid patterns can lead to altered feeding behavior (99); and alterations in plasma amino acid concentrations and patterns can alter uptake into brain of amino acids that are precursors of neurotransmitters and, thereby, the brain concentrations of the neurotransmitters themselves (100).

One of the consequences of the high blood Phe concentrations produced in animals during efforts to develop a laboratory model for the study of PKU was depressed brain serotonin concentration. This could be prevented by providing the animals with additional tryptophan (33). Yuwiler and associates (71) suggested that these observations could be accounted for by high plasma Phe depressing the uptake into brain of the serotonin precursor, tryptophan.

Experiments on amino acid transport in vitro established that, as in other tissues (101), there are specific transport systems in brain, which overlap to a limited extent, for acidic, small neutral, large neutral, and basic amino acids (102, 103). Phenylalanine, tyrosine, the branched-chain amino acids (leucine, isoleucine, and valine), tryptophan, histidine, and methionine are all transported mainly by the large neutral amino acid system and hence will compete with each other for entry into brain. A high blood concentration of Phe had been found in early studies of animal models of PKU to depress tryptophan uptake (33). Fernstrom and Wurtman (6,100) demonstrated that brain tryptophan uptake and brain serotonin concentration were directly proportional to the ratio of the plasma concentration of tryptophan to the sum of the concentrations of the large neutral amino acids that compete with tryptophan for transport. Peng et al. (104) observed that in rats with high plasma Phe concentration after being fed a high Phe diet, brain Phe and Tyr concentrations were greatly elevated, whereas brain concentrations of leucine, isoleucine, valine, methionine, and histidine were all depressed, despite the fact that blood plasma concentrations of this group of amino acids were not. Wurtman et al. (87) have shown that brain catecholamine formation can be influenced by brain Tyr concentration, which in turn is directly proportional to the ratio of plasma Tyr concentrations to the plasma concentrations of the other large neutral amino acids that compete with it for transport (105). These reports are representative of a large number (93,106) indicating that brain amino acid and neurotransmitter pools can be altered by modifications in dietary protein content or amino acid balance or by metabolic defects of amino acid metabolism that lead to imbalances in blood amino acid patterns.

The consequences of these changes in brain amino acid pools for brain function and their significance in relation to behavior have been explored to only a limited extent. A large intraperitoneal dose of Phe (1 g/kg body weight) in the suckling rat, which resulted in depletion of brain tryptophan, was found to depress brain protein synthesis. Provision of tryptophan with the injected Phe prevented the depression (107). Large doses of other amino acids will also depress protein synthesis in brain (108), and Munro et al. (109) have shown that administration of precursors of dopamine and serotonin will cause disaggregation of brain polysomes and depress incorporation of amino acids into brain proteins. This effect apparently depends on the formation of the neurotransmitters, as it did not occur when an inhibitor of decarboxylation, which prevents the formation of dopamine and serotonin, was administered with the precursors. Inhibitors of decarboxylation did not, however, prevent the depression of protein synthesis caused by a large dose of phenylalanine (110), indicating that this effect of Phe does not depend on phenylethylamine formation.

Associations between depletion of brain pools of individual amino acids and feeding behavior have been extensively documented (99,106,111). Depressed food intake and preference for a diet that will produce a balanced plasma amino acid pattern over one that will not have been observed consistently in rats as responses to dietary imbalances of amino acids. These effects on feeding behavior with imbalances involving several different amino acids do not appear to be associated with depletion of any specific amino acid or with depletion of neurotransmitters (106). On the other hand, Anderson (112) has proposed that control of protein and energy intakes of rats are mediated by the changes in brain serotonin and catecholamine content, respectively, that occur in response to dietary influences on the ratios of the plasma concentrations of tryptophan and large neutral amino acids (Trp/LN), and tyrosine and phenylalanine (Tyr/Phe), respectively. The hypothesis was based on evidence of an inverse relationship between the plasma Trp/LN ratio and long-term protein intake and of a direct relationship between the plasma Phe/Tyr ratio and long-term caloric intake in rats allowed to select between two diets differing in protein content. Subsequent studies (73,111, 113,114) have not provided confirmation of this hypothesis, as differences in the protein intakes of rats selecting between diets differing in protein content have not been associated with consistent changes in either serotoninergic or catecholaminergic neurotransmitters.

Despite the convincing evidence that serotoninergic (115) and catecholaminergic (116) neuronal systems are involved in feeding behavior, and that brain concentrations of neurotransmitters can be influenced by plasma concentrations of their amino acid precursors and amino acids that compete with them for transport into brain, it appears that the changes observed in Tyr and tryptophan concentrations in brain are reflections of the type of diet consumed rather than the source of signals for control of subsequent feeding behavior.

Observations indicating that under some conditions the supply of precursors of catecholaminergic neurotransmitters may influence behavior were cited in the discussion of neurotransmitter synthesis. In connection with this, a plasma amino acid imbalance of metabolic origin is observed in patients with liver failure (117). Plasma concentrations of Phe, Tyr, methionine, and free tryptophan are elevated, and concentrations of the branched-chain amino acids which compete with them for uptake into brain are depressed (118). Also, patients with liver failure consuming the amount of protein needed for their nutritional support often develop encephalopathy. On the basis of these observations, Fischer and Baldessarini (119) proposed that the deterioration of mental function observed in such patients was associated with increased uptake of neurotransmitter precursors into brain and increased production of neurotransmitters derived mainly from phenylalanine and tryptophan. Subsequent investigations of this subject have been reviewed recently by Bower and Fischer (120) and Bernardini and Fischer (121). Animals with portacaval shunts were used to simulate hepatic failure. These animals developed hepatic encephalopathy in association with elevated cerebrospinal fluid concentrations of Phe, Tyr, Trp, and the false neurotransmitters octopamine and phenylethanolamine. On the assumption that increased brain uptake of aromatic amino acids and tryptophan contributed to the accumulation of the false neurotransmitters and serotonin, the animals were treated with a solution of glucose and branched-chain amino acids, which would be expected to compete with the aromatic amino acids and tryptophan for uptake into brain. This treatment restored the brain amino acid and neurotransmitter patterns to normal and resulted in recovery of the animals from the encephalopathy.

Subsequently, Fischer and associates (122) treated a series of patients who had severe hepatic failure and encephalopathy with an amino acid solution that was rich in branched-chain amino acids. They observed that the mental function of over half of the patients improved; that if the patients were able to tolerate 70-80 g of amino acids, the plasma amino acid pattern returned toward normal; and that mortality was lower than anticipated in patients with acute alcoholic hepatitis. These results and others in which beneficial responses have been observed after administration of elevated amounts of branched-chain amino acids suggest that altered brain uptake of Phe, Trp, and possibly Tyr contribute to development of hepatic encephalopathy. More stringent tests are being undertaken to establish under carefully controlled conditions the efficacy of this approach to the management of liver failure (122).

The observations on both animals and those with patients (120) support the view that increased brain uptake of aromatic amino acids can play a role in behavioral changes by leading to imbalances in the pattern of synthesis of brain neurotransmitters. They also lend support to the idea that competition among plasma neutral amino acids for uptake into brain can lead to behavioral changes as the result of effects of precursor amino acid supply on neurotransmitter synthesis. Interestingly, as observed by Wurtman and associates (93), behavioral effects from alterations in neurotransmitter precursor supply have been observed primarily when neuronal activity is abnormal.

In conjunction with these observations on relationships between blood plasma amino acid concentrations and patterns and brain uptake of amino acids that are precursors of neurotransmitters, it is of interest to note that Berry et al. (123) have investigated the possibility that modification of the plasma concentrations of amino acids that compete with Phe for transport may suppress brain uptake of Phe by PKU patients. They noted that a supplement of leucine, isoleucine, and valine, administered orally, significantly reduced the concentration of Phe in the cerebrospinal fluid of PKU patients.

REFERENCES

1. Embden, G., and Baldes, K. (1913). Uber de Abbau des Phenylalanins im tierischen Organismus. *Biochem. Z.* 55, 301-322.
2. McCoy, R. H., Meyer, C. E., and Rose, W. C. (1935-1936). Feeding experiments with mixtures of highly purified amino acids. VIII. Isolation and identification of a new essential amino acid. *J. Biol. Chem.* 122, 283-302.
3. Rose, W. C. (1937). The nutritive significance of the amino acids and certain related compounds. *Science* 86, 298-300.
4. Rose, W. C. (1957). The amino acid requirements of adult man. *Nutr. Abstr. Rev.* 27, 631-647.
5. Rose, W. C., and Womack, M. (1946). The utilization of the optical isomers of phenylalanine, and the phenylalanine requirement for growth. *J. Biol. Chem.* 166, 103-110.
6. Stockland, W. L., Lai, Y. F., Meade, R. J., Sowers, J. E., and Oestemer, G. (1971). L-Phenylalanine and L-tyrosine requirements of the growing rat. *J. Nutr.* 101, 177-184.
7. Williams, H. H., Harper, A. E., Hegsted, D. M., Arroyave, G., and Holt, L. E., Jr. (1974). Nitrogen and amino acid requirements. In *Improvement in Protein Nutriture* (Harper, A. E., and Hegsted, D. M., eds.), National Academy of Sciences, Washington, D.C., pp. 23-63.
8. World Health Organization (1973). Energy and protein requirements. Report of a Joint FAO/WHO Ad Hoc Expert Committee. *WHO Tech. Rep. Ser.* 522.
9. Rose, W. C., and Wixom, R. L. (1955). The amino acid requirements of man. XIV. The sparing effect of tyrosine on the phenylalanine requirement. *J. Biol. Chem.* 217, 95-101.
10. Snyderman, S. E. (1984). Human amino acid nutrition. In *Genetic Factors in Nutrition* (Velazquez, A., and Bourges, H., eds.), in press.
11. Food and Nutrition Board (1959). *Evaluation of Protein Nutrition*, National Academy of Sciences, Publication 711, Washington, D.C. pp. 35-42.
12. Phipard, E. F. (1974). Protein and amino acids in diets. In *Improvement of Protein Nurtriture* (Harper, A. E., and Hegsted, D. M., eds.), National Academy of Sciences, Washington, D.C.
13. Moss, A. R., and Schoenheimer, R. J. (1940). The conversion of phenylalanine to tyrosine in normal rats. *J. Biol. Chem.* 135, 415-429.

14. Schepartz, B., and Gurin, S. (1949). The intermediary metabolism of phenylalanine labeled with radio active carbon. *J. Biol. Chem.* 180, 663-673.
15. Lerner, A. B. (1949). On the metabolism of phenylalanine and tyrosine. *J. Biol. Chem.* 181, 281-294.
16. Weinhouse, S., and Millington, R. H. (1949). Ketone body formation from tyrosine. *J. Biol. Chem.* 181, 645-653.
17. Dische, R., and Rittenberg, D. (1954). The metabolism of phenylalanine-4-^{14}C. *J. Biol. Chem.* 211, 199-212.
18. Ravdin, R. G., and Crandall, D. I. (1951). The enzymatic conversion of homogentisic acid to 4-fumarylacetoacetic acid. *J. Biol. Chem.* 189, 137-149.
19. Udenfriend, S., and Cooper, J. (1952). The enzymatic conversion of phenylalanine to tyrosine. *J. Biol. Chem.* 194, 503-511.
20. Lin, E. C. C., Pitt, B. M., Civen, M., and Knox, W. E. (1958). The assay of aromatic amino acid transaminations and keto acid oxidation by the enol-borate-tautomerase method. *J. Biol. Chem.* 233, 668-673.
21. Knox, W. E., and Pitt, B. M. (1957). Enzymatic catalysis of the keto-enol tautomerization of phenylpyruvic acids. *J. Biol. Chem.* 225, 675-688.
22. Porter, C. C. (1973). Inhibitors of aromatic acid decarboxylase–Their biochemistry. *Adv. Neurol.* 2, 37-57.
23. Haley, C. J., and Harper, A. E. (1978). The importance of transamination and decarboxylation in phenylalanine metabolism in vivo in the rat. *Arch. Biochem. Biophys.* 189, 524-530.
24. Shih, J. C., Lien, C.-C., and Malcolm, R. D. (1982). An in vivo function of glucagon-induced phenylalanine: Pyruvate transaminase in p-chlorophenylalanine-treated rats. *Arch. Biochem. Biophys.* 215, 66-71.
25. Knox, W. E. (1972). Phenylketonuria. In *Metabolic Basis of Inherited Disease* (Stanbury, J., Wyngaarden, J., and Fredrickson, D., eds.), McGraw-Hill, New York, pp. 266-295.
26. Haley, C. J., and Harper, A. E. (1982). Glucagon stimulation of phenylalanine metabolism. The effects of acute and chronic glucagon treatment. *Metabolism* 31, 524-532.
27. Goldstein, F. B. (1961). Biochemical studies on phenylketonuria. I. Experimental hyperphenylalanemia in the rat. *J. Biol. Chem.* 236, 2656-2661.
28. Woolfe, L. I. (1951). Excretion of conjugated phenylacetic acid in phenylketonuria. *Biochem. J.* 49, ix-x.
29. Jervis, G. A. (1953). Phenylpyruvic oligophrenia: Deficiency of phenylalanine-oxidizing system. *Proc. Soc. Exp. Biol. Med.* 82, 514-515.
30. Holzman, N. A. Batshaw, M. L., and Valle, D. L. (1980). Genetic aspects of human nutrition. In *Modern Nutrition in Health and Disease* (Goodhart, R. S., and Shils, M. E., eds.), Lea and Febiger, Philadelphia, pp. 1202-1207.
31. Koch, R., Friedman, E. G., Williamson, M. L., and Azen, C. G. (1982). Preliminary report of the effects of diet discontinuation in phenylketonuria. *J. Inher. Metab. Dis. Suppl.* 1, 63-64.
32. Pratt, O. E. (1982). Transport inhibition in the pathogenesis of phenylketonuria and other inherited metabolic diseases. *J. Inher. Metab. Dis. Suppl.* 2, 75-81.

33. Harper, A. E., Benevenga, N. J., and Wohlhueter, R. M. (1970). Effects of ingestion of disproportionate amounts of amino acids. *Physiol. Rev.* 50, 439-448.
34. Halestrap, A. P. (1974). Inhibition of mitochondrial pyruvate transport by phenylpyruvate and α-ketoisocaproate. *Biochim. Biophys. Acta* 367, 102-108.
35. Koch, R., and Blaskovics, M. (1982). Four cases of hyperphenylalaninemia: Studies during pregnancy and of the offspring produced. *J. Inher. Metab. Dis.* 5, 11-15.
36. Lenke, R. R., and Levy, H. L. (1980). Maternal phenylketonuria and hyperphenylalaninemia: An international survey of the outcome of untreated and treated pregnancies. *N. Engl. J. Med.* 303, 1202-1208.
37. Kaufman, S., Kapatos, G., McInnes, R. R., Schulman, J. D., and Rizzo, W. B. (1982). Use of tetrahydropterins in the treatment of hyperphenylalaninemia due to defective synthesis of tetrahydrobiopterin: Evidence that peripherally administered tetrahydropterins enter the brain. *Pediatrics* 70, 376-380.
38. Tourian, A., Goddard, J., and Puck, T. T. (1969). Phenylalanine hydroxylase activity in mammalian cells. *J. Cell. Physiol.* 73, 159-170.
39. Kaufman, S. (1957). The enzymatic conversion of phenylalanine to tyrosine. *J. Biol. Chem.* 226, 511-524.
40. Kaufman, S. (1971). The phenylalanine hydroxylating system from mammalian liver. *Adv. Enzymol.* 35, 245-319.
41. Hesegawa, H., and Kaufman, S. (1982). Spontaneous activation of phenylalanine hydroxylase in rat liver extracts. *J. Biol. Chem.* 257, 3084-3089.
42. Kaufman, S., Bridgers, W. E., Eisenberg, F., and Friedman, S. (1962). The source of oxygen in the phenylalanine hydroxylase and the dopamine-β-hydroxylase catalyzed reactions. *Biochem. Biophys. Res. Commun.* 9, 497-502.
43. Nielsen, K. H., Simonsen, V., and Lind, K. E. (1969). Dihydropteridine reductase. A method for the measurement of activity and investigations of the specificity for NADH and NADPH. *Eur. J. Biochem.* 9, 497-502.
44. Renson, J., Weissbach, H., and Udenfriend, S. (1962). Hydroxylation of tryptophan by phenylalanine hydroxylase. *J. Biol. Chem.* 237, 2261-2264.
45. Bagchi, S. P., and Zarycki, E. P. (1971). Occurrence of phenylalanine hydroxylation in the pineal gland *in vivo*: Possible role of tyrosine and tryptophan hydroxylases. *Res. Commun. Chem. Pathol. Pharmacol.* 2, 370-381.
46. Nakamura, S., Ichiyama, A., and Hayaishi, O. (1965). Purification and properties of tryptophan hydroxylase in brain. *Fed. Proc.* 24, 604.
47. Gal, E. M. (1972). Molecular basis of inhibition of monooxygenases by p-halophenylalanines. *Adv. Biochem. Pharmacol.* 6, 149-163.
48. Kerr, G. R., and Waisman, H. A. (1967). Dietary induction of hyperphenylalaninemia in the rat. *J. Nutr.* 92, 10-18.
49. Woods, M. N., and McCormick, D. B. (1964). Effects of dietary phenylalanine on activity of phenylalanine hydroxylase from rat liver. *Proc. Soc. Biol. Med.* 116, 427-430.
50. McCormick, D. B., Young, S. K., and Woods, M. N. (1965). Effects of acid catabolites on activity in vitro of phenylalanine hydroxylase from rat liver.

Proc. Soc. Exp. Biol. Med. 188, 131-133.
51. Shiman, R., and Gray, D. W. (1980). Substrate activation of phenylalanine hydroxylase. *J. Biol. Chem.* 255, 4793-4800.
52. Shiman, R. (1980). Relationship between substrate activation site and catalytic site of phenylalanine hydroxylase. *J. Biol. Chem.* 255, 10029-10032.
52a. Kaufman, S., and Mason, K. (1982). Specificity of amino acids as activators and substrate for phenylalanine hydroxylase. *J. Biol. Chem.* 257: 14667-14678.
53. Shiman, R., and Jeffereson, L. S. (1982). Iron-dependent regulation of rat liver phenylalanine hydroxylase activity in vivo, in vitro and in perfused liver. *J. Biol. Chem.* 257, 839-844.
54. Milstien, S., and Kaufman, S. (1975). Studies on the phenylalanine hydroxylase system in vivo. *J. Biol. Chem.* 250, 4782-4785.
55. Brand, L. M., and Harper, A. E. (1974). Effect of glucagon on phenylalanine metabolism and phenylalanine-degrading enzymes in the rat. *Biochem. J.* 142, 231-245.
56. Pitot, H. C., and Yatvin, M. B. (1973). Interrelationships of mammalian hormones and enzyme levels in vivo. *Physiol. Rev.* 53, 228-326.
57. Freedland, R. A., and Szepesi, B. (1971). Control of enzyme activity. In *Enzyme Synthesis and Degradation in Mammalian Systems* (Rechcigl, M., ed.), Karger, Basel, pp. 103-140.
58. Ip, C. C. Y., and Harper, A. E. (1973). Effects of dietary protein content and glucagon administration on tyrosine metabolism and tyrosine toxicity in the rat. *J. Nutr.* 103, 1594-1607.
59. Godin, C., and Dolan, G. (1966). Metabolism of radioactive phenylalanine in rats with different dietary intakes of phenylalanine. *J. Nutr.* 90, 284-290.
60. Fuller, R. W., Snoddy, H. D., Wolen, R. L., Coburn, S. P., and Sirlin, E. M. (1971). Effect of glucagon and p-chlorophenylalanine on hepatic enzymes that metabolize phenylalanine. *Adv. Enzyme Regul.* 10, 153-167.
61. Noguchi, T., Okuno, E., and Kido, R. (1976). Identity of isozyme I of histidine-pyruvate aminotransferase with serine-pyruvate aminotransferase. *Biochem. J.* 159, 607-613.
62. Shih, J. C., Chiu, R. H., and Chan, Y. L. (1976). Histidine:pyruvate transamination and phenylalanine:pyruvate transamination may be catalyzed by the same enzyme. *Biochem. Biophys. Res. Commun.* 68, 1348-1355.
63. Dalgliesh, C. E., and Tabechian, H. (1956). Comparison of the metabolism of uniformly ^{14}C-labelled L-phenylalanine, L-tyrosine and L-tryptophan in the rat. *Biochem. J.* 62, 625-633.
64. Kang-Lee, Y.-A., and Harper, A. E. (1978). Threonine metabolism in vivo: Effect of threonine intake and prior induction of threonine dehydratase in rats. *J. Nutr.* 108, 163-175.
65. Meister, A. (1965). *Biochemistry of the Amino Acids*, Academic, New York, p. 895.
66. David, J.-C., Dairman, W., and Udenfriend, S. (1974). On the importance of decarboxylation in the metabolism of phenylalanine, tyrosine and tryptophan. *Arch. Biochem. Biophys.* 160, 561-568.
67. Fellman, J. H., Roth, E. S., and Fujita, T. S. (1976). Decarboxylation to tyramine is not a major route of tyrosine metabolism in animals. *Arch. Bio-*

chem. Biophys. 174, 562-567.

68. Brand, L. M., and Harper, A. E. W. (1974). Studies on the functional significance of rat liver phenylalanine-pyruvate aminotransferase. *Proc. Soc. Exp. Biol. Med.* 174, 211-215.
69. Donlon, J., and Kaufman, S. (1978). Glucagon stimulation of rat hepatic phenylalanine hydroxylase through phosphorylation in vivo. *J. Biol. Chem.* 253, 6657-6659.
70. Fellman, J. H., Buist, N. R. M., Kennaway, N. S., and Swanson, R. E. (1972). The source of aromatic ketoacids in tyrosinemia and phenylketonuria. *Clin. Chim. Acta* 39, 243-246.
71. Yuwiler, A., Geller, E., and Slater, G. G. (1965). On the mechanism of brain serotonin depletion in experimental phenylketonuria. *J. Biol. Chem.* 240, 1170-1174.
72. Anderson, A. E., and Guroff, G. (1972). Enduring behavioral changes in rats with experimental phenylketonuria. *Proc. Nat. Acad. Sci. U.S.A.* 69, 863-867.
73. Peters, R. A. (1983). *Biochemical and Neurochemical Studies on the Control of Protein Intake by the Rat*, Ph.D. thesis, University of Wisconsin, Madison, Wisconsin.
74. Aoki, T. T., Brennan, M. F., Muller, W. A., Soeldner, J. S., Alpert, J. S., Saltz, S. B., Kaufman, R. L., Tan, M. H., and Cahill, G. F., Jr. (1976). Amino acid levels across normal forearm muscle and splanchnic bed after a protein meal. *Am. J. Clin. Nutr.* 29, 340-350.
75. Justice, P., O'Flynn, M. E., and Hsia, D. Y. Y. (1967). Phenylalanine hydroxylase activity in hyperphenylalanemia. *Lancet* 1, 928-929.
76. Stegink, L. D., Filer, L. J., Jr., and Baker, G. L. (1981). Plasma and erythrocyte concentrations of free amino acids in adult humans administered abuse doses of aspartame. *J. Toxicol. Environ. Health* 7, 291-305.
77. Stegink, L. D., Filer, L. J., Jr., Baker, G. L., and McDonnell, J. E. (1980). Effect of an abuse dose of aspartame upon plasma and erythrocyte levels of amino acids in phenylketonuric heterozygous and normal adults. *J. Nutr.* 110, 2216-2224.

77a. Clarke, J. T. R., and Bier, D. M. (1982). The conversion of phenylalanine to tyrosine in man. Direct measurement by continuous intravenous tracer infusions of L-(ring$^2H_5$) phenylalanine and L-(1-^{13}C) tyrosine in the postabsorptive state. *Metabolism* 31:999-1005.

78. Chipponi, J. X., Bleier, J. C., Santi, M. T., and Rudman, D. (1982). Deficiencies of essential and conditionally essential nutrients. *Am. J. Clin. Nutr.* 35, 1112-1116.
79. Whitehead, R. G., and Milburn, T. R. (1962). Metabolites of phenylalanine in the urine of children with kwashiorkor. *Nature* 196, 580-581.
80. Edozien, J. C., and Obasi, M. E. (1965). Protein and amino acid metabolism is kwashiorkor. *Clin. Sci.* 29, 1-24.
81. Antener, I., Verwilghen, A. M., Van Geert, C., and Mauron, J. (1981). Biochemical study of malnutrition. 5. Metabolism of phenylalanine and tyrosine. *Int. J. Vitam. Nutr. Res.* 51, 297-306.
82. Nagatsu, T., Levitt, M., and Udenfriend, S. (1964). Tyrosine hydroxylase:

The initial step in norepinephrine biosynthesis. *J. Biol. Chem.* 239, 2910-2917.

83. Fuller, R. W., and Steinberg, M. (1976). Regulation of enzymes that synthesize neurotransmitter monamines. *Adv. Enzyme Reg.* 14, 347-390.
84. Kaufman, S. (1974). Properties of the pterin-dependent aromatic amino acid hydroxylases. In *Aromatic Amino Acids in the Brain* (Wolstenholme, G. E. W., and Fitzsimmons, D. W., eds.), Elsevier, Amsterdam, pp. 85-108.
85. Dairman, W. (1972). Catecholamine concentrations and the activity of tyrosine hydroxylase after an increase in the concentration of tyrosine in tissues. *Br. J. Pharmacol.* 44, 307-310.
86. Joh, T. H., Park, D. H., and Reis, D. J. (1978). Direct phosphorylation of brain tyrosine hydroxylase by cyclic AMP-dependent protein kinase: Mechanism of enzyme activation. *Proc. Nat. Acad. Sci. U.S.A.* 75, 4744-4748.
87. Wurtman, R. J., Larin, F., Mostafapour, S., and Fernstrom, J. D. (1974). Brain catechol synthesis: Control by brain tyrosine concentration. *Science* 185, 183-184.
88. Wurtman, R. J., and Fernstrom, J. D. (1975). Control of brain monamine synthesis by precursor availability and nutritional state. *Am. J. Clin. Nutr.* 28, 638-647.
89. Gibson, C. J., and Wurtman, R. J. (1977). Physiological control of brain catechol synthesis by brain tyrosine concentration. *Biochem. Pharmacol.* 26, 1137-1142.
90. Gibson, C. J., and Wurtman, R. J. (1978). Physiological control of brain catecholamine synthesis by brain tyrosine concentration. *Life Sci.* 22, 1399-1406.
91. Sved, A. F., Fernstrom, J. D., and Wurtman, R. J. (1979). Tyrosine administration reduces blood pressure and enhances brain norepinephrine release in spontaneously hypertensive rats. *Proc. Nat. Acad. Sci. U.S.A.* 76, 3511-3514.
92. Oishi, T., and Wurtman, R. J. (1982). Effect of tyrosine on brain catecholamine turnover in reserpine-treated rats. *J. Neural Transm.* 53, 101-108.
93. Wurtman, R. J., Hefti, F., and Melamed, E. (1981). Precursor control of neurotransmitter synthesis. *Pharmacol. Rev.* 32, 315-335.
94. Metzler, D. E. (1977). *Biochemistry. The Chemical Reactions of Living Cells,* Academic, New York, pp. 805-890.
95. White, A., Handler, P., Smith, E. L., Hill, R. L., and Lehman, I. R. (1978). *Principles of Biochemistry*, McGraw-Hill, New York, pp. 727-755.
96. Olson, R. E. (1965). Anabolism of the coenzyme Q family and their biological activities. *Fed. Proc.* 24, 85-92.
97. Trumpower, B. L., Houser, R. M., and Olson, R. E. (1974). Studies on ubiquinone. Demonstration of the total biosynthesis of ubiquinone-9 in rat liver mitochondria. *J. Biol. Chem.* 249, 3041-3048.
98. La Du, B. N., and Gjessing, L. R. (1972). Tyrosinosis and tyrosinemia. In *The Metabolic Basis of Inherited Disease* (Stanbury, J. B., Wyngaarden, J. B., and Frederickson, D. S., eds.), McGraw-Hill, New York, pp. 296-307.
99. Peng, Y., Tews, J. K., and Harper, A. E. (1972) Amino acid imbalance, protein intake, and changes in rat brain and plasma amino acids. *Am. J. Physiol.* 222, 314-321.

100. Fernstrom, J. D., and Wurtman, R. J. (1972). Brain serotonin content: Physiological regulation by plasma neutral amino acids. *Science* 178, 414-416.
101. Christensen, H. N. (1962). *Biological Transport*, Benjamin, New York.
102. Blasberg, R., and Lajtha, A. (1965). Substrate specificity of steady-state amino acid transport in mouse brain slices. *Arch. Biochem. Biophys.* 112, 361-377.
103. Neame, K. D. (1968). A comparison of the transport systems for amino acids in brain, intestine, kidney and tumor. *Prog. Brain Res.* 29, 185-199.
104. Peng, Y., Gubin, J., Harper, A. E., Vavich, M. G., and Kemmerer, A. R. (1973). Food intake regulation: Amino acid toxicity and changes in rat brain and plasma amino acids. *J. Nutr.* 103, 608-617.
105. Fernstrom, J. D., and Faller, D. V. (1978). Neutral amino acids in the brain: Changes in response to food ingestion. *J. Neurochem.* 30, 1531-1538.
106. Harper, A. E., and Peters, J. C. (1981). Amino acid signals and their integration with muscle metabolism. In *The Body Weight Regulatory System: Normal and Disturbed Mechanisms* (Cioffi, L. A., James, W. P. T., and Van Italie, T. B., eds.), Raven Press, New York, pp. 33-38.
107. Siegel, F. L., Aoki, K., and Colwell, R. E. (1971). Polyribosome disaggregation and cell-free protein synthesis in preparations from cerebral cortex of hyperphenylalaninemic rats. *J. Neurochem.* 18, 537-547.
108. Roberts, S. (1974). Effects of amino acid imbalance on amino acids utilization, protein synthesis and polyribosome function in cerebral cortex. In *Aromatic Amino Acids in the Brain*, CIBA Foundation Symposium 22, Elsevier, Amsterdam, pp. 299-318.
109. Munro, H. N., Roel, L. E., and Wurtman, R. J. (1973). Inhibition of protein synthesis by doses of L-dopa that disaggregate brain polyribosomes. *J. Neural Transm.* 34, 321-323.
110. Weiss, B. F., Roel, L. E., Munro, H. N., and Wurtman, R. J. (1974). L-Dopa, polysomal aggregation and cerebral synthesis of protein. In *Aromatic Amino Acids in the Brain*, CIBA Foundation Symposium 22, Elsevier, Amsterdam, pp. 325-332.
111. Peters, J. C., and Harper, A. E. (1981). Protein and energy consumption, plasma amino acid ratios, and brain neurotransmitter concentrations. *Physiol. Behav.* 27, 287-298.
112. Anderson, G. H. (1979). Control of protein and energy intake: Role of plasma amino acids and brain neurotransmitters. *Can. J. Physiol. Pharmacol.* 57, 1043-1057.
113. Romsos, D. R., Chee, K. M., and Bergen, W. G. (1982). Protein intake regulation in adult (ob/ob) and lean mice: Effects of non-protein energy source and of supplemental tryptophan. *J. Nutr.* 112, 505-513.
114. Reeves, P. G., and O'Dell, B. L. (1981). Short-term zinc deficiency in the rat and self-selection of dietary protein level. *J. Nutr.* 111, 375-383.
115. Blundell, J. E. (1977). Is there a role for serotonin (5-hydroxytryptamine) in feeding? *Int. J. Obesity* 1, 15-42.

116. Leibowitz, W. S. (1980). Control of feeding and drinking. In *Handbook of the Hypothalamus*, Vol. 3A (Morgane, P., and Panksepp, J., eds.), Marcel Dekker, New York, pp. 299-437.
117. Iber, F. L., Rosen, H., Levenson, S. M., and Chalmers, T. C. (1957). The plasma amino acids in patients with liver failure. *J. Lab. Clin. Med.* 50, 417-422.
118. Fischer, J. E., Yoshimura, N., Aguirre, A., James, J. H., Cummings, M. G., Abel, R. M., and Deindorfer, F. (1974). Plasma amino acids in patients with hepatic encephalopathy: Effects of amino acid infusions. *Am. J. Surg.* 127, 40-47.
119. Fischer, J. E., and Baldessarini, R. J. (1971). False neurotransmitters and hepatic failure. *Lancet* 2, 75-79.
120. Bower, R. H., and Fischer, J. E. (1983). Nutritional management of hepatic encephalopathy. *Adv. Nutr. Res.* 5, 1-11.
121. Bernardini, P., and Fischer, J. E. (1982). Amino acid imbalance and hepatic encephalopathy. *Annu. Rev. Nutr.* 2, 419-454.
122. Fisher, J. E., Rosen, H. M., Ebeid, A. M., James, J. H., Keane, J. M., and Soeters, P. B. (1976). The effect of normalization of plasma amino acids on hepatic encephalopathy in man. *Surgery* 80, 77-91.
123. Berry, H. K., Botinger, M. K., Hunt, M. M., Phillips, P. J., and Guilfoile, M. B. (1982). Reduction of cerebrospinal fluid phenylalanine after oral administration of valine, isoleucine and leucine. *Pediatr. Res.* 16, 751-755.

6

Methanol Metabolism and Toxicity

Thomas R. Tephly
University of Iowa College of Medicine, Iowa City, Iowa

Kenneth E. McMartin
Louisiana State University Medical Center, Shreveport, Louisiana

INTRODUCTION

Methanol is commonly used in industry for organic synthetic procedures or as a solvent. As a result, it is accessible to the general public in a variety of products such as antifreeze, fuels (Sterno), duplicating machine fluids, and in gasoline as a fuel extender. Methanol and other alcohols have been employed as sources of energy or fuel for many years, particularly in times of war. Methanol's use as an automobile fuel, as well as other proposed uses for energy production, will increase human methanol contact from a limited laboratory or industrial exposure to a general environmental exposure. Although methanol theoretically represents a "clean" substance capable of oxidation to water and carbon dioxide, in humans biochemical reactions produce metabolites that are clearly toxic.

A consideration of the toxicity of methanol, especially in species which demonstrate signs and symptoms, seems appropriate for several reasons. First, humans are sensitive to methanol poisoning, and limits of tolerance must be considered. Second, nutritional factors may play an important role (e.g., folate deficiency) in determining susceptibility. Our current understanding of the mechanisms involved in methanol toxicity is described.

CHARACTERISTICS OF POISONING IN MAN

The toxicity of methanol in humans has been appreciated since the early part of the twentieth century. In 1855 MacFarlan (1) proposed that a mixture of 1 part of impure methanol ("wood naphtha") to 9 parts ethanol would be a cheap substitute for the use of ethanol in dissolving resin or in chemical synthesis. This "methylated spirit" did not affect the eyes as the wood naphtha was known to do and soon became widely used in industry. Because of the toxicity of the vapors, wood naphtha itself was not used a great deal in industry, and since it was a foul-smelling and vile-tasting liquid, it was not consumed. After 1896, when methanol purification processes improved, the use of methanol increased dramatically, both as an industrial solvent and as an inexpensive substitute for ethanol. Concurrent with this increased use, reports of blindness and death followed. By 1904 Buller and Wood had collected 235 cases of blindness or death connected with exposure of the victims to methanol (2,3), including 10 cases involving inhalation or absorption of methanol through the skin. Although the dissemination of knowledge concerning methanol toxicity to industrial users decreased the number of industrial poisonings, cases of severe methanol poisoning continued to be observed due to the ingestion of methanol as a substitute for ethanol. Epidemics were not uncommon, especially in areas where extreme poverty existed, where prohibition was the rule, or where war was waged.

Since the initial report by Buller and Wood (2), numerous accounts of individual toxic responses to methanol and epidemics of methanol poisoning have appeared in the literature (4-6). The pattern of signs and symptoms of methanol poisoning is clear. A central nervous system depression is observed similar to that produced by ethanol, but to a much lesser degree. The inebriating effects of methanol have been described as disappointing by numerous users (7). The initial depressant period is followed by an asymptomatic latent period which occurs about 8-24 hr after ingestion of the alcohol and during which patients describe no overt symptoms or signs. Then headache, dizziness, weakness, and nausea are reported, followed in more severe cases by intense vomiting and excruciating abdominal and muscular pain. The abdominal pain has been described by Roe (8) to be so intense as to cause the patients to throw themselves out of bed. Patients may be disoriented and may display severe mental disintegration. With more severe cases, classic respiratory difficulties of metabolic acidosis are noticed, that is, Kussmaul breathing (severe dyspnea characterized by both marked increases in depth and rate of respiration). Coincident with the onset of the respiratory problems, patients complain of visual defects ranging from blurred vision to complete loss of vision. Prior to death, coma deepens, respiration becomes shallower, and convulsions may occur. Death is attributed to respiratory failure, and patients are usually blind prior to death (8). Often a patient may not die and may be left partially or totally blind.

Visual impairment in methanol poisoning is a characteristic feature. Upon ophthalmoscopic examination the optic fundus displays distinctive characteristics which appear in a defined sequence with various intensities (9). Initially, when symptoms of visual disturbance are reported, hyperemia of the optic disk is observed. This may disappear within 1-7 days and may be succeeded by peripapillary edema characterized by a whitish, striated edema which blurs the margin of the optic disk and extends over the adjacent retina. The optic nerve head appears swollen, as are the retinal veins. Generally, patients who exhibit marked degrees of edema suffer permanent visual loss. Optic disk edema may persist for 10-60 days, and if the damage is sufficiently severe, pallor of the optic disk, indicating optic nerve atrophy, will eventually ensue. This is seen as a contraction of the retinal vessels, a lesion that has been reported by Buller and Wood (2) and by Benton and Calhoun (9).

Harrop and Benedict (10) first reported metabolic acidosis in a patient in whom the plasma carbon dioxide binding power was 36.4 vol % of CO_2 (equivalent to 15.8 mEq/liter bicarbonate), and, in this case, the titratable organic acids in the urine were increased markedly. These findings have been confirmed by other workers (11,12). However, because metabolic acidosis was generally not seen following methanol administration to lower animals, acidosis was not accepted as a major feature of methanol poisoning until Chew et al. (13) and Roe (8) showed the beneficial effect of alkali therapy in methanol poisoning. The administration of sodium bicarbonate intravenously provided rapid relief from dyspnea, abdominal pain, and visual disabilities, with a rapid return of normal metabolic and mental function. Indeed it has subsequently been shown many times that methanol intoxication leads to marked depletion of bicarbonate, with plasma bicarbonate levels reaching as low as 4.0 mEq/liter, with a blood pH as low as 7.04 (14). Recently we have observed patients with blood pH values as low as 6.9 (5). The plasma CO_2 combining power of four moribund patients examined by Bennett et al. (15) was 0, and urinary pH had declined to 4.5. The concurrent decrease in blood pH and blood pCO_2 indicates that the acidosis produced by methanol is an uncompensated metabolic acidosis (16), as suspected by earlier workers such as Roe (8). In some studies the degree of metabolic acidosis closely parallels the severity of the ocular symptoms (8,9).

One would expect that the identification of the characteristic signs and symptoms of methanol poisoning should be confirmed by appropriate laboratory tests which would indicate the presence of methanol. Whereas the methods employed for the measurement of blood methanol have been accurate and useful, one must remember that there is no correlation between the blood methanol and the methanol toxicity syndrome. The most severe toxicity occurs many hours following the peak blood level or tissue level of methanol, and the identification of methanol in blood or tissues, while important, does not necessarily provide an accurate indication of the toxicity.

Other observations have led to confusion concerning the methanol poisoning syndrome. For example, there are large individual differences in the duration of the latent period, and there may be great variations in the amount of methanol needed to produce toxicity in individuals. According to Bennett et al. (15), as little as 15 ml of 40% methanol produced death, whereas as much as 500 ml did not induce permanent damage in other patients. Symptoms of methanol poisoning appeared within a few hours or were delayed up to 72 hr. The severity of the disease was not related to the length of the latent period or to the amount of methanol consumed, observations which have led some authors (2,15) to suggest that, within the general population, different susceptibilities to methanol exist.

The apparent variability in sensitivity of humans to methanol ingestion may have several causes. The variation could be due to the inability to obtain exact information from patients who were relatively disoriented. Roe (8) proposed another theory, that the variance in reaction to methanol could be explained by the different amounts of ethanol consumed with the methanol. In his studies those patients who had consumed ethanol either before or after methanol had a longer latent period before the appearance of poisoning than those who had ingested only methanol. Furthermore, the ethanol consumers were more apt to demonstrate no toxic effects of methanol. In uncomplicated methanol poisoning (where ethanol was not a factor), those patients who had ingested more methanol generally presented with symptoms sooner and of much greater severity than those who had consumed less methanol. Death occurred more quickly in those who drank more methanol. Although the amount of methanol ingested was not the only factor determining the degree of acidosis, individual predisposition was thought not to play a major role (8).

Recent information suggests that another explanation for the individual variation may account for the variable sensitivities observed among humans who have ingested methanol. Formate is an intermediate produced during the oxidation of methanol to carbon dioxide and water, and is thought to be responsible for many of methanol's toxic effects. Susceptibility to methanol poisoning may depend on the activity of folic acid-requiring metabolic reactions which are involved in formate metabolism. Nutritional differences among individuals, such as folic acid deficiency, may play an important part in the ability of an individual to metabolize formate. Different degrees of nutritional deficiency may be observed in debilitated and inebriated persons who have not had an adequate diet. In monkeys we observed variability in the metabolism of methanol to formate and carbon dioxide when the animals were studied at different times. Some laboratories have been unable to duplicate results obtained by others (17). This failure may not be due to differences in experimental design or differences in the procedures of those individual laboratories. Instead, it is possible that animals maintained on the best nutritional regimens may be less susceptible to methanol poisoning, owing to a better hepatic capacity to metabolize methanol and formate to carbon dioxide. This will be discussed further under the role of folic acid in formate metabolism.

CHARACTERISTICS OF POISONING IN ANIMALS

Nonprimates

One of the obstacles that has retarded our understanding of the mechanism by which methanol produces its toxicity in humans has been the difficulty in extrapolating results obtained from experiments with certain common laboratory animals to humans. A fundamental difference exists between the characteristics of methanol poisoning in humans and those in nonprimate animals (18,19). Metabolic acidosis and ocular toxicity, the usual symptoms of methanol poisoning, are not observed in lower species. The effect of methanol in nonprimate animals is manifested almost exclusively as a central nervous system depression such as that observed with other aliphatic alcohols. The species differences in susceptibility to methanol poisoning have not been recognized by some authors and have contributed to certain misleading statements when results from experimental animals have been applied to human methanol poisoning. For instance, despite clinical evidence of the effectiveness of ethanol therapy in methanol poisoning, Gilger et al. (20) once recommended that the use of ethanol therapy be discontinued, since it significantly increased the toxicity of methanol in mice.

In 1902 Hunt (21) observed that the signs of acute poisoning with methanol in rabbits were similar to those observed with ethanol poisoning and that ethanol was more toxic than methanol. The lethal dose for a single oral dose of methanol in dogs was about 8-9 g/kg (18,22), 7-9 g/kg for rabbits (18,21), and 10 g/kg for mice and rats (18,20). Gilger and Potts (18) reported that in most nonprimate laboratory animals, methanol intoxication was observed as ataxia, a loss of righting reflex, and other symptoms consistent with central nervous system depression.

Despite the ingestion of lethal doses of methanol, nonprimate species generally do not develop significant metabolic acidosis. Haskell (22) concluded from studies of 14 dogs poisoned with methanol that, although acidosis was sometimes observed, the severity of intoxication was not correlated with the degree of acidosis. Roe (23) found no decrease in the alkali reserve or any signs of acidosis in rats or rabbits that had received toxic oral doses of methanol. Gilger and Potts (18) found little effect on the CO_2 combining capacity of blood after oral administration of very high doses of methanol in rats, rabbits, and dogs.

No impairment of vision has been observed in methanol-poisoned nonprimate animals. Although the production of clinical visual impairment with methanol has been reported in nonprimates (21,24,25), Gilger and Potts (18) pointed out that these claims were often based on four common sources of confusion not related to the typical visual disturbances seen in humans: (a) interpretation of ataxic manifestation following methanol treatment as blindness, (b) nonspecific visual impairment following exposure keratitis which results from the eyelids remaining open for extended periods of time during methanol-induced coma, (c) alteration of pupillary size and reaction due to the anesthetic action of methanol in nonprimates, and (d) interpretation of the lack of response of comatose animals to visual stimuli as blindness. Furthermore, in well-conducted studies of ocular effects of methanol in rabbits (18,23,26), chickens, dogs (18,26,27,28), and rats (18,23), there was neither evidence of visual disturbances nor any changes in the appearance of the fundus.

No consistent histopathology has been demonstrated in nonprimate species. Roe (23) showed no histological changes in the retina, particularly those lesions that have been reported on human autopsy specimens. Although Fink (28) did not observe any ophthalmoscopic or clinical evidence of ocular dysfunction in rabbits and dogs poisoned with methanol, he reported retinal ganglion cell degeneration and occasional edema within the optic nerve. Cooper and Kini (29) pointed out that any histological changes in the retina of nonprimates were probably the result of narcotic effects of high doses of methanol.

Monkeys

Although the differences between the effects of methanol in humans and in laboratory animals were known prior to 1955 (19), there had been few investigations of the toxicity of methanol in the nonhuman primate. In a study of the ocular effects of methanol in three rhesus monkeys as well as in various nonprimates, Birch-Hirschfeld (26,27) reported clinical and ophthalmoscopic evidence of ocular damage in only one animal, a monkey. Tyson and Schoenberg (25) observed ocular changes in one monkey poisoned with methanol; however, these results were considered by Gilger and Potts (18) as an artifact of their experimental methods. Scott et al. (30) reported histopathological evidence of retinal ganglion cell degeneration in monkeys, but they did not discuss their ophthalmoscopic findings.

In 1955 Gilger and Potts (18) reported the first of a series of studies which centered on the toxicity of methanol in the rhesus monkey. The monkey was much more sensitive to methanol than were other laboratory animals. They reported a minimum lethal dose of 3 g/kg body weight for the monkey. Clinically, the signs observed in the monkey were similar to those noted in humans. There was a slight initial central nervous system depression followed by a latent period, a progressive weakness, coma, and death usually in about 20-30 hr. All four monkeys given a lethal dose became severely acidotic (plasma bicarbonate less than 6.5 mEq/liter) within 24 hr. Two of the monkeys showed signs typical of methanol amblyopia observed in humans including dilated, unresponsive pupils and changes of the retina on ophthalmoscopic examination. One monkey showed evidence of optic disk hyperemia and retinal edema.

Potts (31) examined the efficacy of alkali treatment following the administration of lethal doses (6 g/kg) of methanol to the monkey. If adequate bicarbonate was initially given to reverse the metabolic acidosis, four of six animals survived. However, despite correction of the acidosis, two animals eventually died and both of these monkeys showed severe optic disk and retinal edema. Data from one representative monkey indicated that the excretion of organic acids in the urine was markedly elevated, but only a minor part of the acid excretion was accounted for as formic acid. Since only a slight excretion of formic acid was found in this monkey, formaldehyde was suggested to be the toxic agent in the methanol poisoning syndrome in monkeys.

Gilger et al. (32) evaluated the efficacy of ethanol therapy in methanol poisoning in the monkey. Five monkeys were given oral doses of methanol (4-6 g/kg) together with small repeated doses of ethanol (initially 0.75 g/kg than 0.5 g/kg every 4 hr for about 2½ days). All five survived; however, four animals died when administered methanol only. No monkey displayed visual impairment during or after the methanol plus ethanol exposure; when given methanol alone, two demonstrated eye changes. Four of five monkeys did not develop acidosis when administered both ethanol and methanol; the first monkey was given ethanol for only 1½ days and became moderately acidotic when the ethanol was stopped. Monkeys administered only methanol became severely acidotic prior to death.

Although ethanol could prevent methanol poisoning in the monkey, ethanol therapy was known to be ineffective in some clinical cases when administered late in the course of the poisoning (15). When ethanol therapy was delayed as much as 12 hr, the lethality of methanol could be fully prevented (33), but when ethanol was administered 16 hr or more after methanol, monkeys died. In all monkeys, ethanol reversed or slowed the development of acidosis. In four of five monkeys in which ethanol treatment was delayed no eye changes were observed; the fifth showed minor peripapillary edema which lasted until death. Ethanol therapy decreased the rate of disappearance of methanol from the blood, which is further evidence that ethanol inhibits the metabolism of methanol in vivo. This study showed that there is a period beyond which ethanol is not effective as a therapy for methanol poisoning in monkeys. The explanation for this lack of effect beyond a given time relates to damage caused by a methanol metabolite. Once the metabolite concentration is high enough to induce irreversible damage, inhibition of the metabolism of methanol to the metabolite producing the damage cannot alter the course of the toxicity.

Although the studies by Potts, Gilger, and co-workers appeared to establish that the rhesus monkey is a model for methanol poisoning in man, their results could not be reproduced by Cooper and Felig (17). They administered methanol orally to rhesus monkeys, reporting a failure to produce the typical symptoms observed in man or in the monkey by Gilger and Potts (18). Cooper and Felig (17) observed inebriation, narcosis, coma, and death within 24 hr (usually without a latent period). However, no obvious visual impairment was observed. The minimal lethal dose was 7 g/kg (17), with 16 animals surviving on 6 g/kg or less. Acidosis (plasma bicarbonate value of 4.9 mEq/liter) was reported in only one of three cases, and in the other two monkeys the results were equivocal.

Results obtained by McMartin et al. (34) and Clay et al. (35) agreed with the earlier studies done by Potts and his co-workers (18,31-33,36). The administration of methanol (3 g/kg) to monkeys produced a syndrome similar to that described for humans. An initial slight central nervous system depression was followed by a latent period of 12-16 hr, during which time the monkeys displayed no obvious signs of toxicity. This was followed by a progressive deterioration in their condition characterized by anorexia, vomiting, weakness, hyperpnea, and tachypnea. Then they went into coma with shallow and infrequent respiration, followed by death due to respiratory failure 20-30 hr after methanol administration. When an attenuated but prolonged syndrome was produced by the administration of an initial 2 g/kg body weight of methanol with subsequent supplemental doses of methanol, a profound ocular toxicity was observed approximately 40-60 hr after the initial dosage (37-39).

These studies of methanol poisoning in the monkey have allowed for the examination of several features which would not be apparent in human cases. First the latent period appears to represent a period of compensated metabolic acidosis when an increase in formic acid levels and a decrease in plasma bicarbonate levels (Fig. 1) follow in an inverse relationship. When the compensatory mechanisms become exhausted and acidity increases further, the blood pH begins to drop (uncompensated metabolic acidosis). In monkeys the ocular toxicity takes time to become obvious, and when high doses were used (3 g/kg), animals died rapidly, probably owing to metabolic acidosis (34). Clinically evident ocular toxicity developed in methanol-poisoned monkeys after only 40 hr of exposure

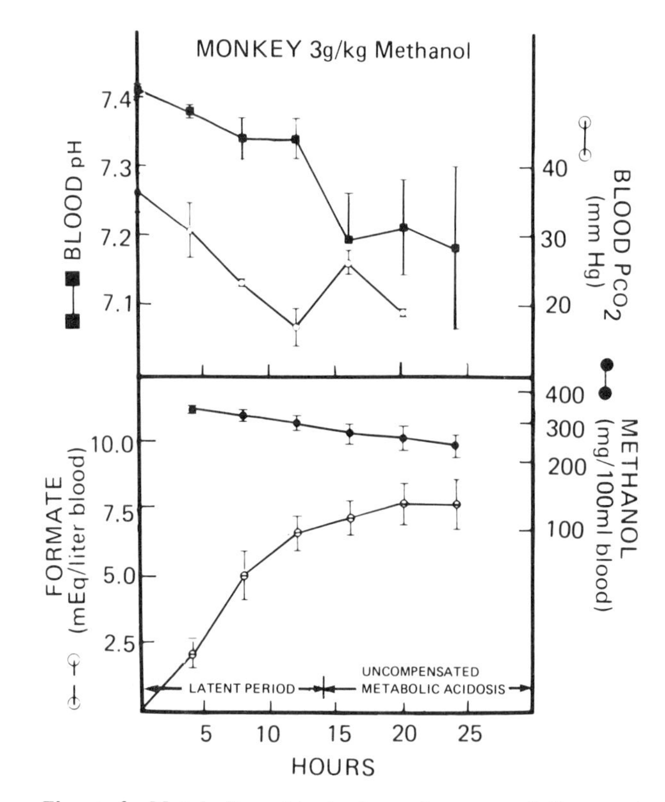

Figure 1 Metabolic acidosis, formate accumulation, and methanol metabolism in the monkey. Each point represents the mean ± SEM of at least three animals. (From T. R. Tephly, *Fed. Proc.* 36,*5*: 1627-1628, 1977.)

to methanol and its metabolites (39). More subtle changes in the ocular system probably occur earlier, but these changes were not assessed by objective methods.

Potts (31) found that only a small fraction of the increased urinary excretion of organic anions which occurred during metabolic acidosis was present as formate. McMartin et al. (34) demonstrated that the depletion of bicarbonate in methanol-poisoned monkeys occurred coincident with an accumulation of formate in the blood. Clay et al. (35) also showed that the increase in blood formate levels in monkeys poisoned with methanol was completely accounted for by a decrease in bicarbonate. Thus formate accumulation appears to be responsible for the generation of metabolic acidosis in the monkey, at least during the early phases of metabolic acidosis.

Potts suggested (31) that formaldehyde may be responsible for the development of methanol toxicity. Studies by McMartin et al. (40) showed that in methanol-poisoned monkeys with profound metabolic acidosis, no formaldehyde could be detected in either tissues or body fluids. The lack of accumulation of formaldehyde should be expected because, when formaldehyde was injected intravenously into monkeys, it disappeared from the blood with an extremely short half-life (1 min). Blood formate levels rose accordingly (40). The failure to detect formaldehyde after methanol administration does not necessarily preclude a role for formaldehyde in methanol toxicity, since formaldehyde may be formed in situ and still interfere with normal cellular functions (41) in a "hit and run" fashion. However, direct evidence for formaldehyde involvement is lacking, and formate accumulation is important, at least in the metabolic acidosis seen in the monkey after methanol administration.

In humans the accumulation of formic acid following methanol ingestion is marked (5), ranging from 11 to 26 mEq/liter. Acidosis with blood pH values of 6.9 and plasma bicarbonate concentrations of about 3 mEq/liter has been a common observation. Decreases in bicarbonate coincide well with the increase in formate concentration, and, under many conditions, formate accumulation is a major factor in the acidosis observed in methanol-poisoned humans. Hemodialysis, as an integral part of the treatment of methanol toxicity has been shown to rapidly decrease the levels of formate as well as methanol (5).

The accumulation of formate in monkeys and humans, but not in rats, after methanol exposure indicates that either methanol oxidation to formate occurs at a faster rate in monkeys and humans, or that formate oxidation to carbon dioxide proceeds at a slower rate in monkeys and humans. Watkins et al. (42) and Clay et al. (35) observed that the rate of methanol oxidation to carbon dioxide is approximately the same in the rat and monkey. Studies conducted in our laboratory have determined that methanol disappearance from blood occurs at the same rate in both rats and monkeys. At a dose of 1 g/kg methanol, methanol disappeared at a rate of 3.7 mg/dl per hour in rats and at 3.9 mg/dl per hour in monkeys. Thus the species difference in blood formate accumulation seen between rats and monkeys cannot be explained by different rates of renal clearance in these species.

Furthermore, differences in the renal disposition of formate do not appear to account for the species difference in blood formate levels. Clay et al. (35) found, in rats, that during a 24-hr period after the administration of methanol (6 g/kg), the urine collected from these animals contained only 0.2% of the administered dose as formate. In contrast, urine collected from a monkey for 23 hr after methanol administration (4 g/kg) contained more than 2% of the dose as formate. Thus the renal excretion of formate appears to be directly related to the systemic formate concentrations, and differences in excretion probably do not account for the observed species differences in blood formate accumulation.

The accumulation of formate in monkeys after methanol administration is probably related to the fact that monkeys metabolize formate to CO_2 slower than rats. McMartin et al. (43) administered various doses of $[^{14}C]$formate (ranging from 0.5 to 15 mmol/kg) to rats and monkeys. They noted that the rate of formate oxidation to CO_2 was dose dependent in both species. More importantly, however, the rate of formate metabolism in the monkey was at least 50% slower than that found in rats at all doses of formate administered. The results of these studies confirmed those of Clay et al. (35), who determined the half-life of formate elimination in both rats and monkeys. In the rat doses of sodium formate up to 100 mg/kg were eliminated with a half-life of 12 min, whereas in the monkey doses of 50 and 72 mg/kg yielded half-lives of 31 min. These values for the monkey are in agreement with those reported by McMartin et al. (43). Thus the slower rate of formate metabolism to CO_2 in the monkey explains why formate accumulates and produces acidosis in the methanol-poisoned monkey but not in rats.

METHANOL METABOLISM

Absorption and Distribution of Methanol

The absorption and distribution of methanol have been characterized by numerous workers (44,45) and are similar to the absorption and distribution of ethanol. Methanol, like ethanol, is rapidly absorbed from the gut and distributes uniformly to body water (45). Methanol may also be absorbed by inhalation, and this property has resulted in several cases of poisoning in humans (2,3). Concentrations of methanol over 300 ppm in the air are considered hazardous (46). Percutaneous absorption of methanol occurs and has been shown to lead to poisoning in children (47).

No differences exist in capabilities for absorption of methanol between various animal species, and blood levels are entirely predictable based on the concept that methanol distributes uniformly to body water. This is not true for the metabolites of methanol. Thus a direct comparison of the toxicity of certain metabolites of methanol to that of methanol itself is difficult when these substances are administered systemically. In other words, metabolites generated in situ within a particular cell locus or within a given organ may have a different toxic potential than those same substances administered parenterally or enterally.

Methanol Elimination

The elimination of methanol from the blood appears to be very slow in all species, especially when compared to ethanol. Where studies of methanol disappearance from the blood have been reported, some reports indicate zero-order disappearance from blood, and some studies indicate first-order kinetics. Bildsten (48) and Widmark and Bildsten (49) reported that, at a dose of 0.8 g of methanol per kilogram body weight, rabbits eliminated an average of 0.76 μg of methanol per gram of animal per minute. Bernhard and Goldberg (50) reported an average elimination rate of 0.67 μg/g per minute when 0.56-2.10 g of methanol per kilogram body weight was administered to rabbits. The elimination curves in those studies were linear and it was noted that 4-6 hr after administration there was a temporary increase in the blood methanol concentration. On the other hand, Koivusalo (51) found that when methanol was administered to rabbits at doses of 0.2-3.4 g/kg body weight, the elimination rate of methanol from the blood was dependent on the blood concentration of methanol. Haggard and Greenberg (44) also concluded that the rate of elimination of methanol from the blood is dose dependent and proportional to the concentration of methanol in the blood. We recently found that the rate of elimination of methanol from the blood is linear with time in monkeys given 1 g/kg of methanol where peak blood levels are about 110 mg/dl in blood. In contrast, first-order decay curves are obtained in monkeys administered 3 g of methanol per kilogram body weight (peak blood methanol levels over 300 mg/dl). A half-life of about 24 hr has been reported (34) for the higher dose. We also (34) observed that, when inhibitors of methanol oxidation are administered to monkeys receiving 3 g of methanol per kilogram body weight, the half-life of methanol disappearance from the blood increases from 24 hr to about 49 hr. Thus at doses of 3 g of methanol per kilogram body weight, first-order disappearance from the blood is established in monkeys.

Metabolism of Methanol

Two enzymes are important in the oxidation of methanol to formaldehyde, alcohol dehydrogenase, and catalase (Fig. 2). The existence of relatively selective inhibitors for each enzyme has made it possible to test their importance in methanol oxidation in animals. It had been known for many years that the metabolism of methanol was blocked by the administration of ethanol and that methanol toxicity was attenuated by ethanol. Roe (8) suggested that humans who had taken ethanol simultaneously with methanol had less severe toxicity than when methanol was ingested alone. The assumption had existed for years that alcohol dehydrogenase was the major enzyme involved in methanol oxidation. Studies on alcohol dehydrogenase by Lutwak-Mann (52) showed that a partially

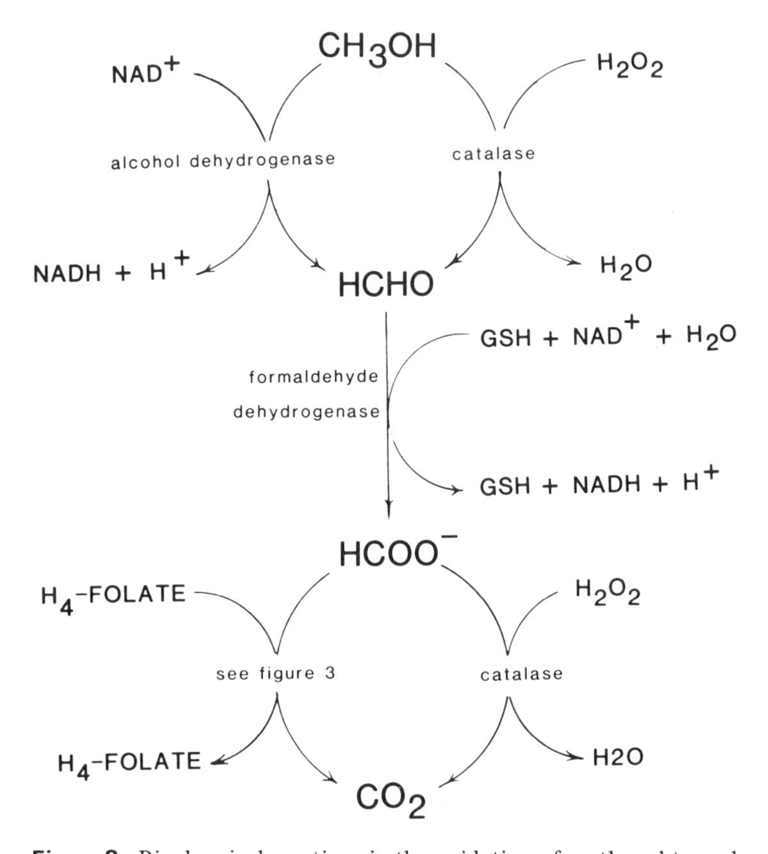

Figure 2 Biochemical reactions in the oxidation of methanol to carbon dioxide.

purified preparation of horse liver alcohol dehydrogenase oxidized methanol, although at a slower rate than ethanol. However, when crystalline horse liver alcohol dehydrogenase was prepared, it appeared to be incapable of catalyzing the oxidation of methanol (53,54), an observation that directed the attention of investigators to the catalase-peroxidative system as a mediator of the metabolism of methanol.

Interest returned to alcohol dehydrogenase and its role in methanol oxidation when Kini and Cooper (55) showed that, at high substrate concentrations, methanol was metabolized by crystalline horse liver alcohol dehydrogenase. Kini and Cooper (55) also showed that it was possible to copurify ethanol and methanol dehydrogenase activities from monkey liver. Their results conclusively demonstrated that alcohol dehydrogenase from monkey liver was capable of catalyzing methanol oxidation in vitro. Makar and Tephly (56) repeated these studies and showed that monkey liver alcohol dehydrogenase catalyzes methanol oxidation in vitro and that this activity is inhibited by the alcohol dehydrogenase inhibitors pyrazole and 4-methylpyrazole. They reported that the Michaelis constant for methanol was about six times higher than that observed for ethanol, results similar to those found by Kini and Cooper, who had reported a K_m of 17 mM for methanol and 2.7 mM for ethanol. Makar and Tephly (56) observed K_m values of 20 mM for methanol and 3.2 mM for ethanol with the monkey liver enzyme. Pyrazole and 4-methylpyrazole were found to be competitive inhibitors when methanol and ethanol were utilized as substrates for monkey liver alcohol dehydrogenase. 4-Methylpyrazole yielded a K_i value of 9 μM which was about one-fourth that observed for pyrazole. Makar and Tephly (56) also showed that 4-methylpyrazole had no inhibitory properties toward catalase activity of rat liver homogenates in vitro or in vivo. Pyrazole, on the other hand, inhibits hepatic catalase activity when injected in vivo (57). Other studies have shown that purified hepatic alcohol dehydrogenase from rats (58) and humans (59,60) catalyze methanol oxidation. Although the Michaelis constant of methanol for alcohol dehydrogenase appears to be relatively high (10-100 mM), concentrations of this magnitude (20-30 mM) can be achieved in vivo after drinking a sizable quantity of either methanol or ethanol.

The inhibition of methanol oxidation by ethanol does not necessarily mean that the alcohol dehydrogenase system functions for methanol oxidation in a given animal species. Catalase can mediate the oxidation of a variety of alcohols to their corresponding aldehydes in the presence of a hydrogen peroxide-generating source (61). A study performed by Keilin and Hartree (62), using purified catalase and various peroxide-generating systems, showed that methanol and ethanol were metabolized at similar rates. Both rates were more rapid than those obtained with alcohols possessing higher molecular weights. Thus methanol and ethanol had about equivalent reactivities with the catalase peroxidative systems, whereas propanol and butanol appeared to display lower substrate reactivity. In fact, Keilin and Hartree suggested (62) that the physiological function of catalase might be involved with the metabolism of certain alcohols. Previously, it had been presumed that the exclusive function of catalase in the living organism was to decompose hydrogen peroxide. An important understanding of how alcohols such as ethanol or methanol might react with catalase in the presence of hydrogen peroxide was provided by Chance (63). He showed that substrates for catalase peroxide (complex I) react with substrates such as methanol and ethanol and promote the decomposition of the catalase peroxide complex, the rate of which was dependent upon the rate of reactivity with the substrate and the catalase-hydrogen peroxide complex. Chance postulated that catalase could conceivably account for most of the metabolism of methanol in the animal organism in vivo (63).

Definitive studies on methanol oxidation in vivo began with the use of selective inhibitors. Heim et al. (64) discovered that the herbicide 3-amino-1,2,4-triazole could inhibit hepatic and renal catalase activity in rats when injected intraperitoneally. This provided a means to test the direct participation of hepatic catalase in the oxidation of methanol in vivo. Aminotriazole has since been a very useful and effective substance for studying the role of hepatic or renal catalase in the oxidation of agents in vivo. It does not inhibit erythrocyte catalase activity, nor does it affect liver cytochrome c content, blood hemoglobin levels, or urobilinogen excretion (64).

Nelson et al. (65) showed that aminotriazole had no effect on ethanol elimination in the dog, although hepatic catalase activity was markedly reduced. Mannering and Parks (66) showed that aminotriazole inhibited hepatic catalase activity in rats in vivo and that, in livers from rats whose hepatic catalase activity had been reduced by 90%, a marked inhibition of methanol oxidation to formaldehyde was observed in vitro. When crystalline beef liver catalase was added to reaction mixtures employing homogenates of rat liver obtained from aminotriazole-treated animals, methanol-oxidizing capacity was restored to control values. These results indicated that hepatic catalase activity was important for methanol oxidation in vitro and, furthermore, that the rate-limiting step in the process was likely to be the capacity of the liver to generate hydrogen peroxide (66). Thus, when peroxide-generating systems were added to hepatic homogenates in addition to crystalline liver catalase, a marked stimulation of activity beyond control values was observed. Mannering and Parks (66) also employed aminotriazole in order to determine whether catalase participated in the metabolism of methanol by rats in vivo. However, they found that aminotriazole had no effect on the rate of disappearance of methanol from the blood of rats. This apparent discrepancy was later explained (67) on the basis that considerable amounts of methanol are eliminated via excretory routes, as well as by metabolism, at the high doses of methanol which were employed in their studies (3 g/kg). When [^{14}C]methanol oxidation was studied by measuring $^{14}CO_2$ formation in vivo in rats, aminotriazole treatment markedly inhibited the oxidation of methanol to CO_2 (67).

Further evidence for a catalase-peroxidative system functioning in the metabolism of methanol in rats was provided in studies where ethanol and 1-butanol were employed as alternate substrate inhibitors of methanol oxidation. Ethanol and methanol have about equal reactivities with catalase peroxide complex I, while ethanol is 6-10 times more reactive than methanol with alcohol dehydrogenase (68). Thus if catalase was functioning in the oxidation of methanol by the rat, one would have expected a 50% inhibition of methanol oxidation by ethanol, and, if alcohol dehydrogenase were functionary, a 90% inhibition would have been expected. Tephly et al. (67) showed that when equimolar doses of ethanol and methanol were injected into rats, a 50% decrease in the rate of methanol oxidation occurred. When 1-butanol, which has only a slight reactivity with the catalase-hydrogen peroxide complex I, was injected, only a very slight inhibitory effect on methanol oxidation in the rat occurred. These results are consistent with the concept that the catalase-peroxidative system is the major catalyst of methanol oxidation in rats. Similar conclusions have been reached in isolated perfused rat liver experiments (69).

Although the role of a catalase-peroxidative system for the metabolism of methanol in rats was clear, different results were obtained with monkeys. Makar et al. (70) showed that pretreatment of monkeys with 1 or 3 g/kg body weight of aminotriazole 1 hr prior to methanol injection did not inhibit the rate of methanol metabolism, although hepatic catalase activity in livers from monkeys was reduced to 10% of control values. Studies were also performed using substrate inhibitors. When equimolar doses of ethanol and methanol were injected in monkeys, an 80% inhibition of the rate of methanol oxidation was observed (70). When 1-butanol, which produced only a slight effect on methanol oxidation in the rat, was injected into monkeys along with [^{14}C]methanol, a 90% inhibition of methanol oxidation was observed. Butanol is a highly reactive substrate for alcohol dehydrogenase, and, if alcohol dehydrogenase were functioning, one would have expected a 90% inhibition of methanol oxidation by 1-butanol. These results support the concept that the catalase-peroxidative system is not functional in methanol oxidation in the primate and that the metabolism of methanol in the monkey is dependent on the activity of alcohol dehydrogenase.

Other evidence for a major role of alcohol dehydrogenase in methanol oxidation in the monkey was provided by Watkins et al. (42), who showed that pyrazole markedly inhibited methanol oxidation in the rhesus monkey. Although pyrazole rapidly inhibited methanol metabolism in vivo, there was a possibility that inhibition of hepatic catalase activity by a pyrazole metabolite could be responsible for the inhibition of methanol oxidation in the monkey. Thus 4-methylpyrazole, a more potent inhibitor of alcohol dehydrogenase activity than pyrazole and one which does not inhibit hepatic catalase activity (56), was tested in the monkey (34). 4-Methylpyrazole was found to be a potent inhibitor of methanol oxidation with little or no effect on hepatic catalase activity.

Thus a major role of alcohol dehydrogenase in the metabolism of methanol in vivo in the monkey has been established. McMartin et al. (34) also showed that 4-methylpyrazole prevents the development of methanol poisoning in the monkey.

The question of why the peroxidative system does not function in the monkey has been examined. It should be recalled that Mannering and Parks (66) showed that when a peroxide-generating system was added to rat hepatic homogenates, peroxide generation appeared to be a rate-limiting factor. When a glucose and glucose oxidase preparation was added, marked stimulation of methanol oxidation occurred. When catalase activity had been reduced markedly, such as from aminotriazole-treated rats, glucose and glucose oxidase addition did not stimulate methanol oxidation (66). Goodman and Tephly (71) suggested that the monkeys may not metabolize methanol through a catalase-dependent system owing to decreased activity levels of peroxide-generating enzymes. Since peroxide-generating systems appear to be rate limiting for methanol oxidation via a catalase-dependent system in the rat, these workers proposed that this system should be rate limiting in the monkey, perhaps to an even greater degree (71) than noted in the rat. It is well known that urate oxidase activity is essentially absent in human liver, and

Goodman and Tephly (71) have shown that urate oxidase activity was also very low in monkey liver. Furthermore, glycolate oxidase activity, xanthine oxidase activity, and other peroxide-generating enzyme activities are also very low in monkey and human liver (72). This could account for why methanol oxidation in the monkey via a catalase-peroxidative system is difficult to demonstrate. Makar and Mannering (58) also suggested that the catalase distribution in the cell may be a consideration.

A third possible mechanism by which methanol could be oxidized to formaldehyde has been suggested by Rietbrock et al. (73) and Teschke et al. (74). This system, the hepatic microsomal mixed-function oxidase system, employs the hepatic endoplasmic reticulum, NADPH, and molecular oxygen.

METABOLISM OF FORMALDEHYDE

Formic acid was considered as the toxic agent in the acidosis seen in methanol poisoning until Van Slyke and Palmer (11) discredited the toxic role of formate. They failed to account for the increased organic acid excretion observed in methanol toxicity as due to formate. Potts (31) also failed to account for the organic acids excreted in the urine as due to formate following methanol poisoning in monkeys. Thus formaldehyde became a candidate as a causative agent in the toxicity of methanol poisoning (41,75,76), even though no one had demonstrated the presence of elevated formaldehyde levels in body fluids or tissues following methanol administration. Keeser (77) appeared to demonstrate the presence of formaldehyde in the cerebrospinal fluid, vitreous humor, and peritoneal fluid of rabbits which had been administered methanol. However, these studies were rather incomplete, lacked appropriate controls, and the method employed to measure formaldehyde lacked sensitivity and specificity. No formaldehyde could be detected in blood, urine, or tissues obtained from methanol-intoxicated animals in studies performed by Koivusalo (51) and Scott et al. (30) or from methanol-poisoned humans (8,78).

There are several ways by which formaldehyde can be disposed of in biological systems. First, formaldehyde has a high degree of reactivity with proteins and other endogenous compounds containing active hydrogen atoms (79). Formaldehyde can combine with any number of functional groups found in proteins or nucleic acids. Thus it may immediately form adducts with cellular constituents, leading to the formation of stable intermediates.

Strittmatter and Ball (80) isolated a formaldehyde-specific, NAD-dependent formaldehyde dehydrogenase from beef liver in 1955 and pointed out that this enzyme required reduced glutathione (GSH). This enzyme, which appears to be quite specific for formaldehyde, is often isolated with glutathionine thiolase (81, 82). In the reactions catalyzed by this enzyme (Fig. 2), formaldehyde combines with reduced GSH to form S-formyl glutathione, and in the presence of the thiolase, the product hydrolyzes to form formic acid and reduced glutathione. Reduced glutathione is therefore a key agent in the generation of formate from formaldehyde. The first reaction appears to be freely reversible, but the second reaction is not, a feature which explains the apparent irreversibility of the two-step reaction as described by Strittmatter and Ball (80). Formaldehyde dehydrogenase activity is present in rat liver, human brain, and a number of other species and tissues such as retina (83). These tissues have not been examined adequately for the presence of S-formyl glutathione hydrolase. The specific activity of this enzyme in crude preparations appears to be quite high, and its presence would be expected in other tissues (81).

Formaldehyde oxidation can also occur in liver mitochondria through an aldehyde dehydrogenase activity (or activities) which is likely to be similar to the aldehyde dehydrogenases of mitochondria that have been described previously (84-86). Aldehyde dehydrogenase activity of mitochondria appears to be very high and is capable of reacting nonspecifically with many aldehyde substrates. Thus it is likely that formaldehyde-oxidizing capabilities of liver are extremely high, either through the formaldehyde dehydrogenase-S-formyl glutathione hydrolase system or through aldehyde dehydrogenase activities in mitochondria or cytosol. Goodman and Tephly (87) have shown that the formaldehyde dehydrogenase activity of human liver is, in fact, higher than that of rat liver. Thus one cannot explain, at this time, the fact that methanol poisoning is uniquely present in humans or monkeys on the basis of an inability to metabolize formaldehyde, since the conversion of formaldehyde to formate can apparently proceed as readily in humans as it does in rats.

Formaldehyde can be metabolized through the tetrahydrofolic acid-dependent one-carbon pool which is capable of utilizing one-carbon units at various oxidation levels and transferring these one-carbon moieties to various endogenous acceptors. Apparently, free formaldehyde enters these reactions by combining with tetrahydrofolate nonenzymatically (88) or through the formaldehyde-activating enzyme to form N^5,N^{10}-methylenetetrahydrofolate. This enzyme has been demonstrated in pigeon liver by Osborn et al. (89) and has been found to be present in a number of mammalian tissues (90).

The metabolism of formaldehyde has been studied by Malorny et al. (91) in dogs and cats in vivo. These investigators administered formaldehyde intravenously and orally to dogs and showed that there was a rapid appearance of formic acid in blood plasma and the presence of only negligible levels of formaldehyde in blood. Experiments in vitro with human blood showed that formaldehyde was oxidized to formic acid (92,93). Rietbrock (94) showed that in dogs, cats, rabbits, guinea pigs, and rats the infusion of formaldehyde resulted in a rapid disappearance of formaldehyde from the blood with a half-life of approximately 1 min. Malorny et al. (91) found that when equimolar amounts of formaldehyde, formic acid, or sodium formate were infused in dogs, the peak concentrations of formic acid in the plasma were equivalent in all three cases, indicating that formaldehyde was rapidly metabolized to formic acid.

Although it is possible that formaldehyde may be responsible for certain of the toxic findings in methanol poisoning, it would be unlikely that it could account for the metabolic acidosis, since formate appears to be the major factor in the metabolic acidosis seen in monkeys and humans poisoned with methanol. It is also unlikely that formaldehyde can be generated in the liver and delivered to the optic nerve in an intact state. Therefore either formaldehyde forms a product with some endogenous acceptor which is responsible for the ocular toxicity, or formaldehyde is generated in situ in the eye, where it may exert an effect on the ocular system. Although these possibilities cannot be ruled out at this time, the responsibility of formaldehyde for the ocular toxicity of methanol is unlikely, since formate itself can produce ocular toxicity in the monkey (95). In studies where blindness in monkeys was produced from formate, no formaldehyde could be detected in body fluids or tissues (95). In any case, more studies need to be performed on the fate of formaldehyde in the organism in order to disregard it as a toxic agent in the methanol poisoning syndrome in man.

FORMATE METABOLISM

Nonprimates

The ability of animal tissues to oxidize formate into CO_2 was first reported by Batelli (96) and Battelli and Stern (97), who observed that tissues obtained from a variety of animals, such as the horse, cow, sheep, dog, and rabbit, were capable of oxidizing formate into CO_2 in the presence of hydrogen peroxide (Fig. 2). More that 40 years later, Chance (63) studied the kinetics of the catalase-hydrogen peroxide system with different substrates and showed that formate reacts with the hydrogen peroxide catalase complex (complex I).

In subsequent years, a number of in vitro investigations strongly indicated a key role of the catalase-hydrogen peroxide system in the oxidation of formate. Some of the experimental results leading to this conclusion are the following:

1. There is a good correlation between the formate-oxidizing ability and the catalase activity in liver preparations of different species (98), in different tissues within one animal species (99), and in the subcellular compartments from tissue preparations (100).
2. Administration of aminotriazole to guinea pigs greatly lowered the formate-oxidizing ability of liver fractions in vitro (98).
3. Certain types of neoplasms in rats (101), mice (102), and humans (103) lead to a marked lowering of both catalase activity and the formate-oxidizing ability in vitro.
4. Folate-deficient rats possess a marked impairment in formate-oxidizing ability (104) and lowered hepatic catalase levels.
5. Decreased formate oxidation in vitro results from decreased hydrogen peroxide generation caused by factors such as a decreased hepatic xanthine oxidase activity, vitamin B_6 deficiency (105), or thyrotoxicosis (100). On the

other hand, factors that increase hydrogen peroxide generation stimulate formate oxidation. This can be accomplished by supplementing liver preparations with hypoxanthine, a known substrate of xanthine oxidase.

Another path of formate oxidation to CO_2 is the folate biochemical pathway (88,90,106,107). Formate enters into the folate pool by combining with tetrahydrofolate (THF) to form 10-formyl-THF, a reaction catalyzed by 10-formyl-THF synthetase, an enzyme widely distributed among mammalian tissues (108). Kutzbach and Stokstad (109) showed that 10-formyl-THF oxidoreductase catalyzes the oxidation of the formyl group directly to CO_2. Thus there is a two-step conversion of formate to CO_2.

Rietbrock et al. (73) suggested that exogenously administered formate, or formate arising from methanol metabolism in vivo, is oxidized via the folate-dependent pathway. They found an inverse correlation between plasma levels of folate in different animal species and the half-life of exogenously administered formate. They also reported that dogs accumulated formic acid to a small extent (2 mEq/liter) in their blood following methanol administration. Pretreatment of dogs with folic acid prior to methanol produced a lower blood formate level, whereas methotrexate (an inhibitor of dihydrofolate reductase) had the opposite effect (110).

Palese and Tephly (111) measured $^{14}CO_2$ formation following [^{14}C] formate administration to rats and showed that folate deficiency resulted in a greatly diminished rate of formate oxidation. In contrast, administration of aminotriazole, the potent catalase inhibitor, did not inhibit the rate of formate oxidation to CO_2. Administration of ethanol in molar ratio of 22:1 (ethanol:formate) did not alter the rate of formate oxidation in the rat. However, in folate-deficient rats, the catalase-hydrogen peroxide system may serve as an alternate pathway, since, in folate-deficient rats aminotriazole or ethanol administration did result in some inhibition of the rate of formate oxidation (111).

The knowledge that formate is being metabolized in vivo via a folate-dependent system has been utilized to advantage in order to produce metabolic acidosis in rats after methanol treatment. Rats, made folate deficient, oxidize formate at a markedly slowed rate (111,112), and administration of methanol (4 g/kg) to folate-deficient rats leads to high formate levels and severe metabolic acidosis (113). Blood formate levels reached values as high as 18 mEq/liter in these animals. This value is higher than blood formate levels noted in methanol-poisoned monkeys (34).

Monkeys

In monkeys the folate-dependent pathway is also the major route of formate oxidation to CO_2. Makar et al. (70) showed that aminotriazole had no effect on methanol oxidation to CO_2 in the monkey, and McMartin et al. (114) demonstrated that neither the rate of formate oxidation nor the half-life of formate in the blood was altered by aminotriazole. However, the rate of formate metabolism in folate-deficient monkeys was approximately 50% lower than that observed in control monkeys. Formate oxidation was stimulated in monkeys by the administration of either folic acid (114) or 5-formyl-THF (115).

McMartin et al. (114) also showed that the sensitivity of monkeys to methanol was related to folate, since folate-deficient monkeys became especially sensitive to the toxicity of methanol relative to the amounts of formate produced. Thus, when 0.5 g/kg of methanol was given to either folate-deficient or control monkeys, the level of blood formate in the folate-deficient animals was more than two times greater than that observed in the control animals.

Noker and Tephly (115) then showed that methanol toxicity can be modified considerably in monkeys by the administration of folate derivatives. These workers followed the course of methanol toxicity in monkeys administered [^{14}C] methanol (2 g/kg) or [^{14}C]methanol with repetitive doses of 5-formyl-THF. In monkeys treated with 5-formyl-THF (2 mg/kg at 0, 4, 8, 12, and 18 hr after methanol), blood formate levels were significantly decreased (by at least 50%) from those observed in the untreated animals. Similar results were obtained when sodium folate was employed instead of 5-formyl-THF. In both treated and untreated monkeys, the elimination of methanol from blood followed zero-order kinetics and proceeded at a rate of 7.9 mg/dl per hour in the 5-formyl-THF-treated animals, and at 7.1 mg/dl per hour in the untreated animals. Therefore the clearance of methanol was not altered by folate administration. In addition, the distribution and route of metabolism of [^{14}C]methanol did not appear to be changed by 5-formyl-THF treatment, since the total amount of ^{14}C label recovered in urine as either expired [^{14}C]methanol or $^{14}CO_2$ was the same for both treated and untreated monkeys. However, the rate of methanol oxidation to CO_2 was significantly increased in those animals treated with 5-formyl-THF, and folate treatment was effective in reducing blood formate levels by increasing the rate of formate metabolism to CO_2. Blood pH and blood bicarbonate levels remained within the normal range in animals treated with 5-formyl-THF, in contrast to the marked bicarbonate depletion, high blood formate levels, and metabolic acidosis observed in animals not given 5-formyl-THF.

Noker and Tephly (115,116) have also shown that 5-formyl-THF (when given in repetitive doses) is effective in reversing methanol toxicity in the monkey once it has developed. The accumulation of blood formate in monkeys could be markedly altered by 5-formyl-THF, even when administered after toxicity became apparent. A rapid decline in blood formate levels was observed in methanol-poisoned animals several hours after the initiation of 5-formyl-THF treatment. In monkeys not given 5-formyl-THF, formate levels continued to climb. The decline in formate concentrations in monkeys treated with folate was coupled to an increase in the rate of CO_2 formation from methanol.

The results demonstrate that the severity of methanol toxicity in monkeys is correlated with accumulation of formate in the blood and that this can be significantly modified by procedures which provide the monkey with more folate. These results suggest that there is a reciprocal relationship between the formate oxidation rate and the hepatic folate level of the animal. They suggest the possible use of folates for the treatment of human methanol toxicity.

Regulation of Formate Oxidation Through Regulation of Tetrahydrofolate

Since the folate biochemical pathway is primarily involved in the metabolism of formate, the regulation of the rate of formate metabolism is governed by the regulation of the hepatic tetrahydrofolate concentrations in liver. This concept has been advanced recently by studies which have explored the role of 5-methyl-THF:homocysteine transmethylase (methionine synthetase). This cytosolic enzyme is reponsible for the methylation of homocysteine to form methionine as well as for the conversion of 5-methyl-THF to THF (Fig. 3). It requires methyl-

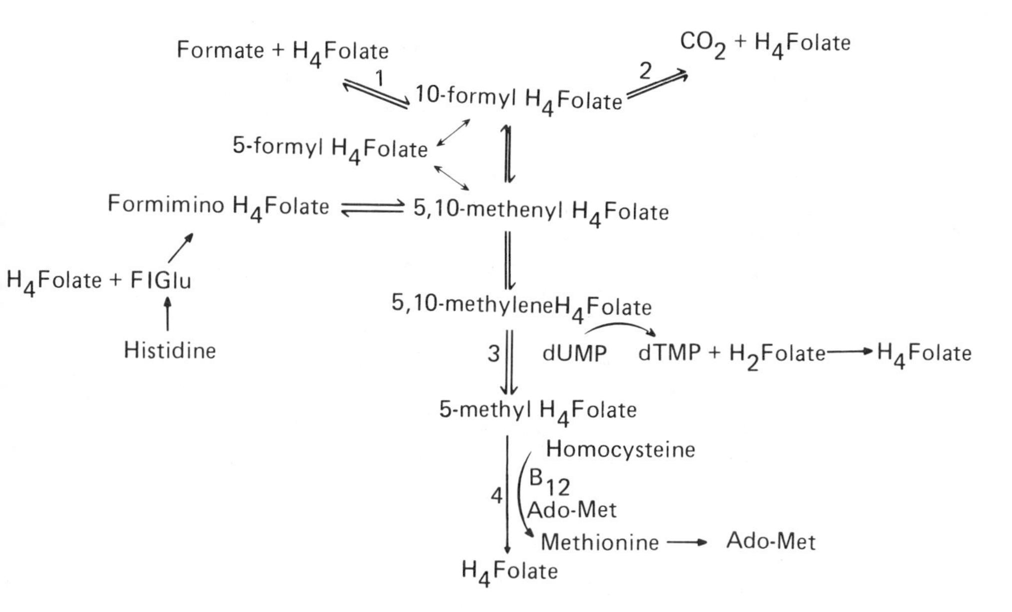

Figure 3 Pathway of folate-dependent formate metabolism (H_2 folate, dihydrofolate; H_4 folate, tetrahydrofolate; B_{12}, vitamin B_{12}; Ado-Met, S-adenosylmethionine). Reaction 1 is catalyzed by formyl-tetrahydrofolate synthetase and requires activation of formate by ATP. Reaction 2 is catalyzed by formyl-tetrahydrofolate dehydrogenase and utilizes NADP+. Reaction 3 is catalyzed by methylene-tetrahydrofolate reductase and is thought to be irreversible. Reaction 4 is catalyzed by methyl-tetrahydrofolate homocysteine methyltransferase (methionine synthetase) and is dependent upon vitamin B_{12} and catalytic amounts of adenosylmethionine, a reducing system.

cobalamin and S-adenosylmethionine for maximal activity. As far as we know, methionine synthetase is the only methlycobalamin-dependent enzyme in the mammalian organism. The anesthetic gas nitrous oxide has been reported to react with transition methyl complexes, such as the cobalt-ligand complex in vitamin B_{12}, and oxidizes the coenzyme from the active cob(I)alamin form to the inactive cob(III)alamin form (117). Deacon et al. (118) have shown the inhibition of hepatic and brain methionine synthetase activity in vivo by nitrous oxide, and Eells et al. (119,120) demonstrated that, following nitrous oxide treatment of rats, there was a significant decrease in hepatic levels of tetrahydrofolate forms and an increase in hepatic 5-methyl-THF. Rats treated with nitrous oxide also exhibited a marked decrease in the rate of formate oxidation to carbon dioxide. When methanol (4 g/kg) was administered to rats which were exposed to nitrous oxide:oxygen (50:50) for 2 hr, there was a marked metabolic acidosis in these animals, with accumulation of blood formate, a decrease in blood pH to 7.2, and a depletion of blood bicarbonate. This metabolic acidosis produced after the administration of methanol to rats had not been demonstrated previously, except where rats were made folate deficient (104). Hepatic methionine synthetase activity was reduced to 10% of control levels in animals treated with $N_2O:O_2$ (50:50), a finding which accounts for the depletion of hepatic tetrahydrofolate. Recently, Eells et al. (120) demonstrated an excellent correlation between the rate of formate oxidation in rats with hepatic tetrahydrofolate levels. Since S-adenosylmethionine levels are also dependent upon hepatic methionine levels, one would expect alteration of S-adenosylmethionine concentrations in liver. S-Adenosylmethionine levels are depleted by the treatment of rats with nitrous oxide, and a good correlation between tetrahydrofolate levels and S-adenosylmethionine was also recorded (120).

Methionine administration to rats which have been treated with nitrous oxide leads to a reversal of the depletion of tetrahydrofolate levels in liver and a reversal of the inhibition of formate oxidation produced by nitrous oxide (120). However, the mechanism by which methionine is capable of reversing the depletion of tetrahydrofolate brought on by nitrous oxide treatment is still unexplained; that is, although nitrous oxide inhibits methionine synthetase activity and depletes tetrahydrofolate levels, methionine administration does not reverse the inhibition of methionine synthetase activity, although it restores tetrahydrofolate in liver. Therefore methionine cannot be exerting its effect by a direct action on methionine synthetase activity. It is possible that methionine exerts its effect through the elevation of S-adenosylmethionine concentrations in liver. Following methionine treatment, there is a marked elevation of S-adenosylmethionine levels in rat liver (120) and S-adenosylmethionine acts as an inhibitor of 5,10-methylene-THF reductase (121). More work is needed in order to determine the mechanism by which methionine exerts its reversal of the nitrous oxide depletion of hepatic tetrahydrofolate.

Recent studies in our laboratory have shown that treatment of monkeys with a nitrous oxide:oxygen (50:50) mixture leads to marked sensitization of the monkey to methanol toxicity. Following a dose of 1 g/kg of methanol (a dose which produces only a slight increase in blood formate in monkeys), there was a marked accumulation of formate (4 mEq/liter) 12 hr after methanol. These values are greater than blood formate levels observed when 2 g/kg of methanol were given to air-breathing monkeys.

A great deal more work is needed in order to understand which step of the many enzymatic reactions in the folate biochemical pathway regulates the regeneration of tetrahydrofolate in monkeys. However, it is important to realize that primates are at some risk with respect to their folate regulation; and it would appear to be important for future work to determine that step or process which is deficient and which places the primate at a distinct liability when it comes to the disposition of one-carbon moieties.

REFERENCES

1. MacFarlan, J. F. (1855). On methylated spirit, and some of its preparations. *Pharm. J. Trans.* 15, 310-315.
2. Buller, F., and Wood, C. A. (1904). Poisoning by wood alcohol: Cases of death and blindness from Columbian spirits and other methylated preparations. *J. Am. Med. Assoc.* 43, 1117, 1132, 1289-1296.
3. Wood, C. A., and Buller, F. (1904). Poisoning by wood alcohol: Cases of death and blindness from Columbian spirits and other methylated preparations. *J. Am. Med. Assoc.* 43, 972-977.
4. Gonda, A., Gault, H., Churchill, D., and Hollomby, D. (1978). Hemodialysis for methanol intoxication. *Am. J. Med.* 64, 749-757.
5. McMartin, K. E., Ambre, J. J., and Tephly, R. T. (1980). Methanol poisoning in humans: Role for formic acid accumulation in the metabolic acidosis *Am. J. Med.* 68, 414-418.
6. Naraqi, S., Dethlefs, R. F., Slobodniuk, R. A., and Sairere, J. S. (1979). An outbreak of acute methyl alcohol intoxication. *Aust. N.Z. J. Med.* 9, 65-68.
7. Kobro, M. (1946). Methanol poisoning. *Acta Pharmacol.* 2, 95-108.
8. Roe, O. (1948). The ganglion cells of the retina in cases of methanol poisoning in human beings and experimental animals. *Acta Ophthalmol.* 26, 169-182.
9. Benton, C. D., and Calhoun, F. P. (1953). The ocular effects of methyl alcohol poisoning: Report of a catastrophe involving 320 persons. *Am. J. Ophthalmol.* 36, 1677-1685.
10. Harrop, G. A., and Benedict, E. M. (1920). Acute methyl alcohol poisoning associated with acidosis. *J. Am. Med. Assoc.* 74, 25-27.
11. Van Slyke, D. D., and Palmer, W. W. (1920). Studies of acidosis XVI. The titration of organic acids in urine. *J. Biol. Chem.* 41, 567-585.
12. Ziegler, S. L. (1921). The ocular menace of wood alcohol poisoning. *J. Am. Med. Assoc.* 77, 1160-1166.

13. Chew, W. B., Berger, E. H., Brines, O. A., and Capron, M. J. (1946). Alkali treatment of methyl alcohol poisoning. *J. Am. Med. Assoc.* 130, 61-64.
14. Kane, R. L., Talbert, W., Harlan, J., Sizemore, G., and Cataland, S. (1968). A methanol poisoning outbreak in Kentucky. *Arch. Environ. Health* 17, 119-129.
15. Bennett, I. L., Jr., Cary, F. H., Mitchell, G. L., Jr., and Cooper, M. N. (1953). Acute methyl alcohol poisoning: A review based on experiences in an outbreak of 323 cases. *Medicine* 32, 431-463.
16. Keyvan-Larijarni, H., and Tannenberg, A. M. (1974). Methanol intoxication. Comparison of peritoneal dialysis and hemodialysis treatment. *Arch. Intern. Med.* 143, 293-296.
17. Cooper, J. R., and Felig, P. (1961). The biochemistry of methanol poisoning II. Metabolic acidosis in the monkey. *Toxicol. Appl. Pharmacol.* 3, 202-209.
18. Gilger, A. P., and Potts, A. M. (1955). Studies on the visual toxicity of methanol V. The role of acidosis in experimental methanol poisoning. *Am. J. Ophthalmol.* 39, 63-86.
19. Roe, O. (1955). The metabolism and toxicity of methanol. *Pharmacol. Rev.* 7, 399-412.
20. Gilger, A. P., Potts, A. M., and Johnson, L. V. (1952). Studies on the visual toxicity of methanol II. The effect of parenterally administered substances on the systemic toxicity of methyl alcohol. *Am. J. Ophthalmol.* 35, 113-126.
21. Hunt, R. (1902). The toxicity of methyl alcohol. *Johns Hopkins Hosp. Bull.* 13, 213-225.
22. Haskell, C. C., Hileman, S. P., and Gardner, W. R. (1921). The significance of the acidosis of methyl alcohol poisoning. *Arch. Intern. Med.* 27, 71-82.
23. Roe, O. (1948). The ganglion cells of the retina in cases of methanol poisoning in human beings and experimental animals. *Acta Ophthalmol.* 26, 169-182.
24. Holden, W. A. (1899). The pathology of the amblyopia following profuse hemorrhage and of that following the ingestion of methyl alcohol, with remarks on the pathogenesis of optic-nerve atrophy in general. *Arch. Ophthalmol.* 28, 125-134.
25. Tyson, H. H., and Schoenberg, M. J. (1914). Experimental researches in methyl alcohol inhalation. *J. Am. Med. Assoc.* 63, 915-922.
26. Birch-Hirschfeld, A. (1901). Experimentelle Untersuchugen uber die Pathogenese der Methylalkoholamblyopie. *Arch. Ophthalmol.* 52, 358-383.
27. Birch-Hirschfeld, A. (1902). Weiterer Beitrag zur Pathogenese der Alkoholamblyopie. *Arch. Ophthalmol.* 54, 68-98.
28. Fink, W. H. (1943). The ocular pathology of methyl-alcohol poisoning. *Am. J. Ophthalmol.* 26, 694, 802-815.
29. Cooper, J. R., and Kini, M. M. (1962). Biochemical aspects of methanol poisoning. *Biochem. Pharmacol.* 11, 405-416.
30. Scott, E., Helz, M. K., and McCord, C. P. (1933). The histopathology of methyl alcohol poisoning. *Am. J. Clin. Pathol.* 3, 311-319.
31. Potts, A. M. (1955). The visual toxicity of methanol VI. The clinical aspects of experimental methanol poisoning treated with base. *Am. J. Ophthalmol.* 39, 86-92.

32. Gilger, A. P., Potts, A. M., and Farkas, I. S. (1956). Studies on the visual toxicity of methanol IX. The effect of ethanol on methanol poisoning in the rhesus monkey. *Am. J. Ophthalmol.* 42, 244-252.
33. Gilger, A. P., Farkas, I. S., and Potts, A. M. (1959). Studies on the visual toxicity of methanol X. Further observations on the ethanol therapy of acute methanol poisoning in monkeys. *Am. J. Ophthalmol.* 48, 153-161.
34. McMartin, K. E., Maker, A. B., Martin-Amat, G., Palese, M., and Tephly, T. R. (1975). Methanol poisoning I. The role of formic acid in the development of metabolic acidosis in the monkey and the reversal by 4-methylpyrazole. *Biochem. Med.* 13, 319-333.
35. Clay, K. L., Murphy, R. C., and Watkins, W. D. (1975). Experimental methanol toxicity in the primate: analysis of metabolic acidosis, *Toxicol. Appl. Pharmacol.* 34, 49-61.
36. Potts, A. M., Praglin, J., Farkas, I., Orbison, L., and Chickering, D. (1955). Studies on the visual toxicity of methanol VIII. Additional observations on methanol poisoning in the primate test object. *Am. J. Ophthalmol.* 40, 76-83.
37. Baumbach, G. L., Cancilla, P. A., Martin-Amat, G., Tephly, T. R., McMartin, K. E., Makar, A. B., Hayreh, M. S., and Hayreh, S. S. (1977). Methyl alcohol poisoning IV. Alterations of the morphological findings of the retina and optic nerve. *Arch. Ophthalmol.* 95, 1859-1865.
38. Hayreh, M. S., Hayreh, S. S., Baumbach, G. L., Cancilla, P., Martin-Amat, G., Tephly, T. R., McMartin, K. E., and Makar, A. B. (1977). Methyl alcohol poisoning III. Ocular toxicity. *Arch. Ophthalmol.* 95, 1851-1858.
39. Martin-Amat, G., Tephly, T. R., McMartin, K. E., Makar, A. B., Hayreh, M., Heyreh, S., Baumbach, G., and Cancilla, P. (1977). Methyl alcohol poisoning II. Development of a model for ocular toxicity in methyl alcohol poisoning using the rhesus monkey. *Arch. Ophthalmol.* 95, 1847-1850.
40. McMartin, K. E., Martin-Amat, G., Noker, P. E., and Tephly, T. R. (1979). Lack of a role for formaldehyde in methanol poisoning in the monkey. *Biochem. Pharmacol.* 28, 645-649.
41. Koivusalo, M. (1970). Methanol. In *International Encyclopedia of Pharmacology and Therapeutics*, Vol. 2 (Tremolieres, J., ed.), Pergamon, New York, Section 20, pp. 465-505.
42. Watkins, W. D., Goodman, J. I., and Tephly, T. R. (1970). Inhibition of methanol and ethanol oxidation by pyrazole in the rat and monkey *in vivo*. *Mol. Pharmacol.* 6, 567-572.
43. McMartin, K. E., Martin-Amat, G., Maker, A. B., and Tephly, T. R. (1977). Methanol poisoning: Role of formate metabolism in the monkey. In *Alcohol and Aldehyde Metabolizing Systems*, Vol. 2 (Thurman, R. G., Williamson, J. R., Drott, H., and Chance, B., eds.), Academic, New York, pp. 429-439.
44. Haggard, H. W., and Greenberg, L. A. (1939). Studies in the absorption distribution and elimination of alcohol IV. The elimination of methyl alcohol. *J. Pharmacol. Exp. Therap.* 66, 479-496.
45. Yant, W. P., and Schrenck, H. H. (1937). Distribution of methanol in dogs after inhalation and administration by stomach tube and subcutaneously. *J. Ind. Hyg. Toxicol.* 19, 337-345.

46. Leaf, G., and Zatman, L. J. (1952). A study of the conditions under which methanol may exert a toxic hazard in industry. *Br. J. Med.* 9, 19-31.
47. Gimenez, E. R., Vallegjo, N. E., Roy, E., Lis, M., Izurieta, E. M., Rossi, S., and Capuccio, M. (1968). Percutaneous alcohol intoxication. *Clin. Toxicol.* 1, 39-48.
48. Bildsten, N. V. (1924). Mikrobestimmung von Methylalkohol im Blute. *Biochem. Z.* 146, 361-369.
49. Widmark, E. M. P., and Bildsten, N. V. (1924). Die Elimination des Methylalkohols and die Bedingungen fur die Akkumulation desselben. *Biochem. Z.* 148, 325-335.
50. Bernard, C. G., and Goldberg, L. (1934). Uber die Einwirkung der durch Kohlensaure gesteigerten atmung auf die Ausscheidung des Methylalkohols beim Kaninchen. *Skand. Arch. Physiol.* 67, 117-128.
51. Koivusalo, M. (1956). Studies on the metabolism of methanol and formaldehyde in the animal organism. *Acta Physiol. Scand. Suppl.* 131,
52. Lutwak-Mann, C. (1938). Alcohol dehydrogenase of animal tissues. *Biochem. J.* 32, 1364-1374.
53. Bonnichsen, R. K., and Wassen, A. M. (1948). Crystalline alcohol dehydrogenase from horse liver. *Arch. Biochem.* 18, 361-363.
54. Theorell, H., and Bonnichsen, R. (1951). Studies on liver alcohol dehydrogenase I. Equilibria and initial reaction velocities. *Acta Chem. Scand.* 5, 1105-1126.
55. Kini, M. M., and Cooper, J. R. (1961). Biochemistry of methanol poisoning III. The enzymatic pathway for the conversion of methanol to formaldehyde. *Biochem. Pharmacol.* 8, 207-215.
56. Makar, A. B., and Tephly, T. R. (1975). Inhibition of monkey liver alcohol dehydrogenase by 4-methylpyrazole. *Biochem. Med.* 13, 334-342.
57. Lieber, C. S., Rubin, E., De Carli, L. M., Misra, P., and Gang, H. (1970). Effects of pyrazole on hepatic function and structure. *Lab. Invest.* 22, 615-621.
58. Makar, A. B., and Mannering, G. J. (1968). Role of the intracellular distribution of hepatic catalase in the peroxidative oxidation of methanol. *Mol. Pharmacol.* 4, 484-491.
59. Blair, A. H., and Vallee, B. L. (1966). Some catalytic properties of human liver alcohol dehydrogenase. *Biochemistry* 5, 2026-2034.
60. von Wartburg, J.-P., Bethune, J. L., and Vallee, B. L. (1964). Human liver-alcohol dehydrogenase. Kinetic and physicochemical properties. *Biochemistry* 3, 1775-1782.
61. Keilen, D., and Hartree, E. F. (1936). Coupled oxidation of alcohol. *Proc. R. Soc. London Ser. B.* 119, 141-159.
62. Keilen, D., and Hartree, E. F. (1945). Properties of catalase. Catalysis of coupled oxidation in alcohols. *Biochem. J.* 39, 293-301.
63. Chance, B. (1947). An intermediate compound in the catalase hydrogen peroxide reaction. *Acta Chem. Scand.* 1, 236-267.
64. Heim, W. G., Appleman, D., and Pyfrom, H. T. (1956). Effects of 3-amino-1,2,4-triazole (AT) on catalase and other compounds. *Am. J. Physiol.* 186, 19-23.

65. Nelson, G. H., Kinard, F. W., Aull, J. C., and Hay, M. G. (1957). Effect of aminotriazole on alcohol metabolism and hepatic enzyme activities in several species. *Q. J. Stud. Alcohol* 18, 343-348.
66. Mannering, G. J., and Parks, R. E., Jr. (1957). Inhibition of methanol metabolism with 3-amino-1,2,4-triazole. *Science* 126, 1241-1242.
67. Tephly, T. R., Parks, R. E., Jr., and Mannering, G. J. (1964). Methanol metabolism in the rat. *J. Pharmacol. Exp. Ther.* 143, 292-300.
68. Sund, H., and Theorell, H. (1963). Alcohol dehydrogenases. In *The Enzymes*, 2nd ed., Vol. 7 (Boyer, P. D., Lardy, H., and Myrback, eds.), Academic, New York, pp. 25-83.
69. Van Harken, D. R., Tephly, T. R., and Mannering, G. J. (1956). Methanol metabolism in the isolated perfused rat liver. *J. Pharmacol. Exp. Ther.* 149, 36-42.
70. Makar, A. B., Tephly, T. R., and Mannering, G. J. (1968). Methanol metabolism in the monkey. *Mol. Pharmacol.* 4, 471-483.
71. Goodman, J. I., and Tephly, T. R. (1968). The role of hepatic microbody and soluble oxidases in the peroxidation of methanol in the rat and monkey. *Mol. Pharmacol.* 4, 492-501.
72. Goodman, J. I., and Tephly, T. R. (1970). Peroxidative oxidation of methanol in human liver: The role of hepatic microbody and soluble oxidases. *Res. Commun. Chem. Pathol. Pharmacol.* 1, 441-450.
73. Rietbrock, N., Stieren, B., and Malorny, G. (1966). Beeinflussung der Methanolstoffwechsels durch Folsaure. *Klin. Wochenschr.* 44, 1318-1319.
74. Teschke, R., Hasumura, Y., and Lieber, C. S. (1975). Hepatic microsomal alcohol-oxidizing system: Affinity for methanol, ethanol, propanol and butanol. *J. Biol. Chem.* 250, 7397-7404.
75. Cooper, J. R., and Marchesi, V. T. (1959). The possible biochemical lesion in blindness due to methanol poisoning. *Biochem. Pharmacol.* 2, 313-315.
76. Potts, A. M., and Johnson, L. V. (1952). Studies on the visual toxicity of methanol I. The effect of methanol and its degradation products on retinal metabolism. *Am. J. Ophthalmol.* 35, 107-113.
77. Keeser, E. (1931). Uber die Ursache der Giftigkeit des Methyl Alkohols. *Dtsch. Med. Wochenscr.* 57, 398-399.
78. Alha, A. R., Raekallio, J., and Mukula, A.-L. (1958). Detection of methanol poisoning: With special consideration of the estimation of formic acid in solid viscera, blood and urine; investigation of 11 fatal cases. *Ann. Med. Exp. Biol. Fenn.* 36, 444-451.
79. French, D., and Edsall, J. T. (1945). The reactions of formaldehyde with amino acids and proteins. *Adv. Protein Chem.* 2, 277-335.
80. Strittmatter, P., and Ball, E. G. (1955). Formaldehyde dehydrogenase, a glutathionine-dependent enzyme system. *J. Biol. Chem.* 213, 445-461.
81. Uotila, L., and Koivusalo, M. (1974). Formaldehyde dehydrogenase from human liver: Purification, properties, and evidence for the formation of glutathione thiol esters by the enzyme. *J. Biol. Chem.* 249, 7653-7663.
82. Uotila, L., and Koivusalo, M. (1974). Purification and properties of S-formylglutathione hydrolase from human liver. *J. Biol. Chem.* 249, 7664-7672.

83. Kinoshita, J. H., and Masurat, T. (1958). Effect of glutathione on formaldehyde oxidation in the retina. *Am. J. Ophthalmol.* 46, 42-46.
84. Cinti, D. L., Keyes, S. R., Lemelin, M. A., Denk, H., and Schenkman, J. B. (1976). Biochemical properties of rat liver mitochondrial aldehyde dehydrogenase with respect to oxidation of formaldehyde. *J. Biol. Chem.* 251, 1571-1577.
85. Koivula, T., and Koivusalo, M. (1975). Different forms of rat liver aldehyde dehydrogenase and their subcellular distribution. *Biochim. Biophys. Acta* 397, 9-23.
86. Siew, C., Deitrich, R. A., and Erwin, V. G. (1976). Location and characteristics of rat liver mitochondrial aldehyde dehydrogenases. *Arch. Biochem. Biophys.* 176, 638-649.
87. Goodman, J. I., and Tephly, T. R. (1971). A comparison of rat and human liver formaldehyde dehydrogenase. *Biochim. Biophys. Acta* 252, 489-505.
88. Blakley, R. L. (1969). *The Biochemistry of Folic Acid and Related Pteridines*, Wiley, New York.
89. Osborn, M. J., Hatefi, Y., Kay, L. D., and Huennekens, F. M. (1957). Evidence for enzymic deacylation of N^{10}-formyl tetrahydrofolic acid. *Biochim. Biophys. Acta* 26, 208-210.
90. Huennekens, F. M., and Osborn, M. J. (1959). Folic acid coenzymes and one-carbon metabolism, *Adv. Enzymol.* 21, 370-446.
91. Malorny, G., Rietbrock, N., and Schneider, M. (1965). Die Oxydation des Formaldehyde zu Ameisensaure im Blut, ein Beitrag zum Stoffwechsel des Formaldehyde. *Naunyn-Schmiedebergs Arch. Exp. Pathol. Pharmakol.* 250, 419-436.
92. Matthies, H. (1957). Untersuchungen uber eine Aldehydedehydrogenase in kernlosen Erythrocyten. *Biochem. Z.* 329, 421-527.
93. Mattheis, H. (1958). Vergleichende untersuchungen uber die Aldehydedehydrogenase kernloser Erythrocyten. *Biochem. Z.* 330, 169-173.
94. Rietbrock, N. (1965). Formaldehydroxydation bei der Ratte. *Naunyn-Schmiedebergs Arch. Exp. Pathol. Pharmakol.* 251, 189-201.
95. Martin-Amat, G., McMartin, K. E., Hayreh, S. S., Hayreh, M. S., and Tephly, T. R. (1978). Methanol poisoning: Ocular toxicity produced by formate. *Toxicol. Appl. Pharmacol.* 45, 201-208.
96. Battelli, F. (1908). Oxydation de l'acide formique par les extraits des tissus animaux en presence de peroxyde d'hydrogene. *Compt. Rend. Soc. Biol.* 138, 651-665.
97. Battelli, F., and Stern, L. (1908). Uber die Peroxydasen der Tiergewebe. *Biochem. Z.* 13, 44-51.
98. Aebi, H., Fiei, E., Knab, R., and Siegenthaler, P. (1957). Utersuchungen uber die Formiatoxydation in der Leber. *Helv. Physiol. Acta* 15, 150-167.
99. Nakada, H. I., and Weinhouse, S. (1953). Studies of glycine oxidation in rat tissues. *Arch. Biochem.* 42, 257-270.
100. Venkataraman, S., and Sreenivasan, A. (1966). Formate oxidation in rat liver. *Enzymologia* 30, 91-96.
101. Stein, A. M., and Mehl, J. W. (1955). Reduction of total formic acid oxi-

dase and liver catalase in leukemic and tumor bearing mice. *Fed. Proc.* 14, 286.
102. Aebi, H., and Portwich, F. (1959). Formiatoxydation und Katalase-aktivitat bei Ratten mit Walker Sarkom. *Helv. Physiol. Acta* 17, 189-201.
103. Portwich, F., and Aebi, H. (1960). Erfassung der Peroxydbildung tierischer gewebe Mittels peroxydatischer Umsetzungen. *Helv. Physiol. Acta* 18, 1-16.
104. Friedmann, B., Nakada, H. I., and Weinhouse, S. (1954). A study of the oxidation of formic acid in the folic acid-deficient rat. *J. Biol. Chem.* 210, 413-421.
105. Schulman, M. P., and Rickert, D. A. (1959). The oxidation of glycine and formate to CO_2 by rat liver homogenates. *J. Biol. Chem.* 234, 1781-1783.
106. Krebs, H. A., Hems, R., and Tyler, B. (1976). The regulation of folate and methionine metabolism. *Biochem. J.* 158, 341-353.
107. Stokstad, E. L. R., and Koch, J. (1967). Folic acid metabolism. *Physiol. Rev.* 47, 83-116.
108. Whitely, H. R. (1960). The distribution of the formate-activating enzyme and other enzymes involving tetrahydrofolic acid in animal tissues. *Comp. Biochem. Physiol.* 1, 222-247.
109. Kutzbach, C., and Stokstad, E. L. R. (1968). Partial purification of a 10-formyltetrahydrofolate: NADP oxidoreductase from mammalian liver. *Biochem. Biophys. Res. Commun.* 30, 111-117.
110. Rietbrock, N., Stieren, B., and Malorny, G. (1966). Beeinflussung der Methanol-Stoffwechsels durch folsaure. *Klin. Wochenschr.* 44, 1318-1319.
111. Palese, M., and Tephly, T. R. (1975). Metabolism of formate in the rat. *J. Toxicol. Environ. Health* 1, 13-24.
112. Plaut, G. W. E., Betheil, J. J., and Lardy, H. A. (1950). The relationship of folic acid to formate metabolism in the rat. *J. Biol. Chem.* 184, 795-805.
113. Makar, A. B., and Tephly, T. R. (1976). Methanol poisoning in the folate-deficient rat. *Nature* 261, 715-716.
114. McMartin, K. E., Martin-Amat, G., Makar, A. B., and Tephly, T. R. (1977). Methanol poisoning V. Role of formate metabolism in the monkey. *J. Pharmacol. Exp. Ther.* 201, 564-572.
115. Noker, P. E., and Tephly, T. R. (1980). The role of folates in methanol toxicity. In *Alcohol and Aldehyde Metabolizing Systems*, Vol. 4 (Thurman, R. G., Williamson, J. R., Drott, H., and Chance, B., eds.), Plenum, New York, pp. 305-315.
116. Noker, P. E., Eells, J. T., and Tephly, T. R. (1980). Methanol toxicity: Treatment with folic acid and 5-formyltetrahydrofolic acid. Alcoholism. *Clin. Exp. Res.* 4, 378-383.
117. Banks, R. G. S., Henderson, R. J., and Pratt, J. M. (1968). Reactions of gases in solution. Part III. Some reactions of nitrous oxide with transition-metal complexes. *J. Chem. Soc. A*, 2886-2898.
118. Deacon, R., Lumb, M., Perry, J., Chanarin, I., Minty, B., Halsey, M., and Nunn, J. (1980). Inactivation of methionine synthetase by nitrous-oxide. *Eur. J. Biochem.* 104, 419-422.
119. Eells, J. T., Makar, A. B., Noker, P. E., and Tephly, T. R. (1981). Meth-

anol poisoning and formate oxidation in nitrous oxide-treated rats. *J. Pharmacol. Exp. Ther.* 217, 57-61.

120. Eells, J. T., Black, K. A., Makar, A. B., Tedford, C. E., and Tephly, T. R. (1982). The regulation of one-carbon oxidation in the rat by nitrous oxide and methionine. *Arch. Biochem. Biophys.* 219, 316-326.
121. Kutzbach, C., and Stokstad, E. L. R. (1967). Feedback inhibition of methylenetetrahydrofolate reductase in rat liver by S-adenosylmethionine. *Biochim. Biophys. Acta* 139, 217-220.

7

Aspartame Metabolism in Animals

James A. Oppermann
G. D. Searle & Co., Skokie, Illinois

INTRODUCTION

Aspartame is the methyl ester of the dipeptide l-aspartyl-1-phenylalanine. In considering the metabolism of this compound, it was expected that extensive degradation of the compound would occur in the gastrointestinal tract during the absorption process. Initially it was anticipated that the methyl group would be cleaved by intestinal esterases, such as chymotrypsin (1), thus liberating the naturally occurring dipeptide aspartylphenylalanine. Peptide hydrolases located in the microvillus membrane lining the small intestine would then metabolize the dipeptide to its constituent amino acids, which would then enter the systemic circulation. To test this hypothesis, three separate radiolabeled compounds were synthesized (2), each with the label in a different portion of the molecule, (aspartyl, phenylalanyl, or methyl ester portions). The disposition of each of these compounds was then compared to that of the individual ^{14}C-labeled amino acids or [^{14}C] methanol (3) in various animal species. Some of these studies reported here have been published in detail elsewhere (4-7).

DISPOSITION OF [^{14}C-METHYL] ASPARTAME

The excretion of radioactivity in the urine, feces, and expired air following the oral administration of 20 mg/kg body weight [^{14}C-methyl] aspartame and an equimolar dose of [^{14}C] methanol to rats is given in Table 1. No major differences

Table 1 Excretion of Radioactivity Following the Oral Administration of 20 mg/kg Body Weight [^{14}C-Methyl] Aspartame and an Equimolar Dose of [^{14}C]Methanol to Male Rats (N = 6)

	Percentage of administered dose (± SEM)[a]		
	Urine	Feces	Expired air (CO_2)
[^{14}C-Methyl] aspartame	1.82 ± 0.17	0.06 ± 0.01	58.8 ± 2.5
[^{14}C]Methanol	4.65 ± 0.75	1.09 ± 0.37	63.9 ± 2.7

[a]Urine and feces were collected for 3 days; expired air was collected for 8 hr.

in the recoveries of radioactivity were found between these two compounds. The major fraction of the ^{14}C was excreted in the expired air, accounting for approximately 60% of the dose. Only small quantities of ^{14}C were excreted in the urine and feces. These data suggest that the methyl group of aspartame was removed as a result of metabolism and that its disposition was similar to that of exogenously administered methanol.

A similar comparative study was conducted in the rhesus monkey. Figure 1 gives the cumulative excretion of $^{14}CO_2$ in expired air after the oral administration of equimolar quantities of [^{14}C-methyl] aspartame (20 mg/kg body weight) and [^{14}C]methanol. In this study 73% of the label was recovered in expired air during the first 8 hr after methanol administration, and 67% of the dose was excreted after aspartame administration. The initial rate of $^{14}CO_2$ excretion was slower after labeled aspartame administration than after methanol. Comparison of the appearance of ^{14}C in the plasma (Fig. 2) revealed a similar time course, that is, a slower rate of appearance of label in the plasma after aspartame administration. Nevertheless, total plasma concentrations of ^{14}C and its decline were quite similar after administration of these two compounds. The initial rate differences have been attributed to the time required for aspartame to leave the stomach and for its methyl group to be hydrolyzed by intestinal esterases. In addition, methanol can be absorbed from the stomach. Thus it would be anticipated that its absorption would be slightly more rapid.

It is doubtful that a significant degree of metabolism of aspartame occurs in the stomach. In vitro incubation of aspartame in simulated gastric fluid (pH 1.0) containing the enzyme pepsin for 15 min at 37°C resulted in no observable degradation as delineated by thin-layer chromatography. In addition, administration of [^{14}C-methyl] aspartame to rats in which the pylorus region of the stomach was ligated resulted in no measurable excretion of $^{14}CO_2$ and nearly complete recovery of the label from the stomach 4 hr after compound administration. Administration of [^{14}C]methanol under these conditions results in a significant excretion of radioactivity in expired air.

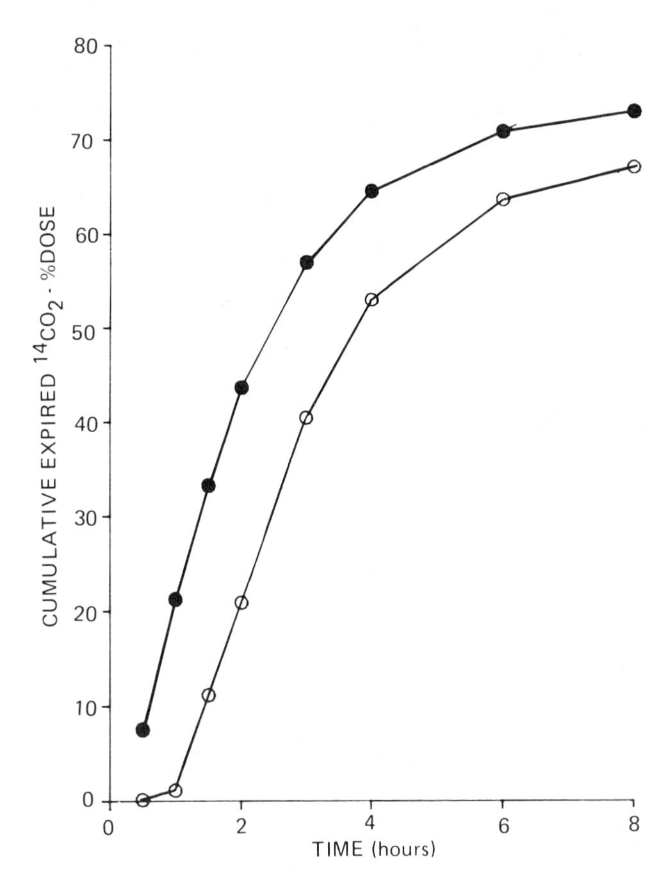

Figure 1 The mean cumulative excretion of radioactivity in expired air after the oral administration of 20 mg of [^{14}C-methyl] aspartame (○) per kilogram body weight or an equimolar dose of [^{14}C] methanol (●) to rhesus monkeys. The results are the average of four (aspartame) or five (methanol) experiments. (From J. Oppermann, E. Muldoon, and R. E. Ranney, Metabolism of aspartame in monkeys. *J. Nutr.* 103: 1454-1459, 1973.)

Thus the initial hydrolytic reaction to release methanol probably occurs in the small intestine:

$$\text{Aspartame} \xrightarrow[\text{esterases}]{\text{intestinal}} \text{methanol + aspartylphenylalanine}$$

The major enzyme catalyzing this reaction is probably chymotrypsin, which has potent esterase and proteolytic activity (8).

The rapid appearance of $^{14}CO_2$ in the expired air after methanol or [^{14}C-methyl] aspartame administration suggests oxidation of the liberated methanol

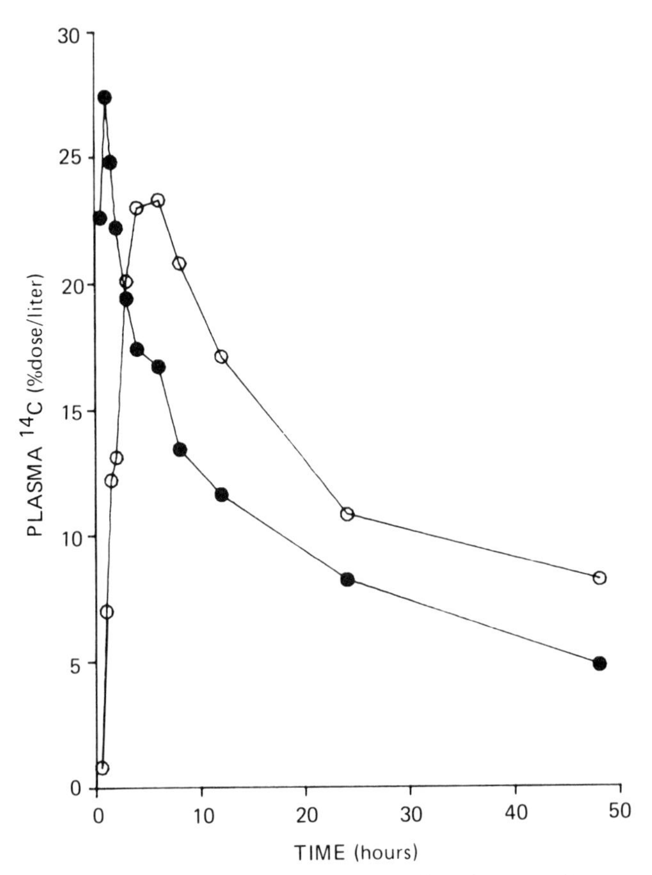

Figure 2 The mean plasma concentrations (percent dose per liter) of total radioactivity after the oral administration of 10 mg of [^{14}C-methyl] aspartame (○) per kilogram body weight or an equimolar dose of [^{14}C] methanol (●) to rhesus monkeys. The results are the average of four (aspartame) or five (methanol) experiments. (From J. Oppermann, E. Muldoon, and R. E. Ranney, Metabolism of aspartame in monkeys, *J. Nutr.*, Vol. 103: 1454-1459,1973).

as soon as the compound enters the liver. The first reaction is the oxidation of methanol to formaldehyde. Two enzymatic systems have been reported to catalyze this reaction. In the rat (9) it appears that this reaction is mediated through a catalase-peroxidase system according to the following scheme:

$$2H_2O + O_2 \xrightarrow{\text{catalase}} 2H_2O_2$$

$$CH_3OH + H_2O_2 \xrightarrow{\text{peroxidase}} HCHO + 2H_2O$$

However, in the monkey, and presumably in man, another enzyme system appears to predominate (10,11). In these species the transformation of methanol to formaldehyde is mediated by the enzyme alcohol dehydrogenase according to the following scheme:

$$CH_3OH + NAD^+ \xrightarrow{\text{alcohol dehydrogenase}} HCHO + NADH + H^+$$

The next reaction in the formation of CO_2 from methanol is the oxidation of formaldehyde to formic acid. A specific dehydrogenase–formate dehydrogenase (FDH)–has been described (12) which is glutathione (GSH) and NAD dependent. The sequence of events is as follows:

$$HCHO + GSH \rightarrow HO{-}CH_2{-}SG$$

$$HO{-}CH_2{-}SG + FDH \rightarrow (HO{-}CH_2{-}SG)FDH$$

$$(HO{-}CH_2{-}SG)FDH + NAD^+ \rightarrow (HCO{-}SG)FDH + NADH + H^+$$

$$(HCO{-}SG)FDH + H_2O \rightarrow FDH + GSH + HCOOH$$

This enzyme system occurs both in rat and human liver (11). The activity of this enzyme in human liver has been shown to be several fold higher than in rat liver.

The last step is the conversion of formate to CO_2. This is carried out by the catalase-peroxidase system, which, in the rat, is also responsible for the initial oxidation of methanol. For formate the oxidative reactions are (13)

$$2H_2O + O_2 \xrightarrow{\text{catalase}} 2H_2O_2$$

$$HCOOH + H_2O_2 \xrightarrow{\text{peroxidase}} CO_2 + 2H_2O$$

In rats this route of formate oxidation was confirmed in vivo by Nakade and Weinhouse (14).

In the primate formate oxidation occurs by a folate-dependent pathway. A more detailed description of the metabolism of methanol and its toxicity is given in the chapter by Tephly and McMartin (15).

DISPOSITION OF [^{14}C-ASPARTYL] ASPARTAME

The excretion of radioactivity in urine, feces, and expired air after the oral administration to rats of 20 mg/kg body weight [^{14}C-aspartyl] aspartame and an equimolar dose of [^{14}C] aspartic acid is given in Table 2. Each carbon atom of the aspartic acid moieties was labeled. The majority of radioactivity after administration of either compound was excreted in expired air as $^{14}CO_2$, 68% after aspartame and 59% after aspartic acid. Approximately 3-4% of the radioactive dose was recovered in the urine, and less than 2% was recovered in feces. The excretion profile was similar after administration of either compound, thereby suggesting that aspartic acid was released as a result of aspartame metabolism.

A similar study was conducted in rhesus monkeys. Four rhesus monkeys were fed equimolar doses of aspartame or aspartic acid for 5 days. On the sixth day each monkey was given an aqueous solution of either [^{14}C-aspartyl] aspartame or [^{14}C] aspartic acid by oral intubation. Figure 3 gives the rate of excretion of

Table 2 Recovery of Radioactivity Following the Oral Administration of 20 mg/kg Body Weight [^{14}C-Aspartyl] Aspartame and an Equimolar Dose of [^{14}C] Aspartic Acid to Male Rats (N = 6)

	Percentage of administered dose (± SEM)[a]			
	Urine	Feces	Expired air	Carcass
[^{14}C-Aspartyl] aspartame	3.71 ± 0.41	1.13 ± 0.56	68.3 ± 1.4	9.33 ± 0.54
[^{14}C] Aspartic acid	3.47 ± 0.53	0.26 ± 0.03	58.5 ± 8.7	8.00 ± 0.45

[a]Urine and feces were collected for 3 days; expired air was collected for 8 hr.

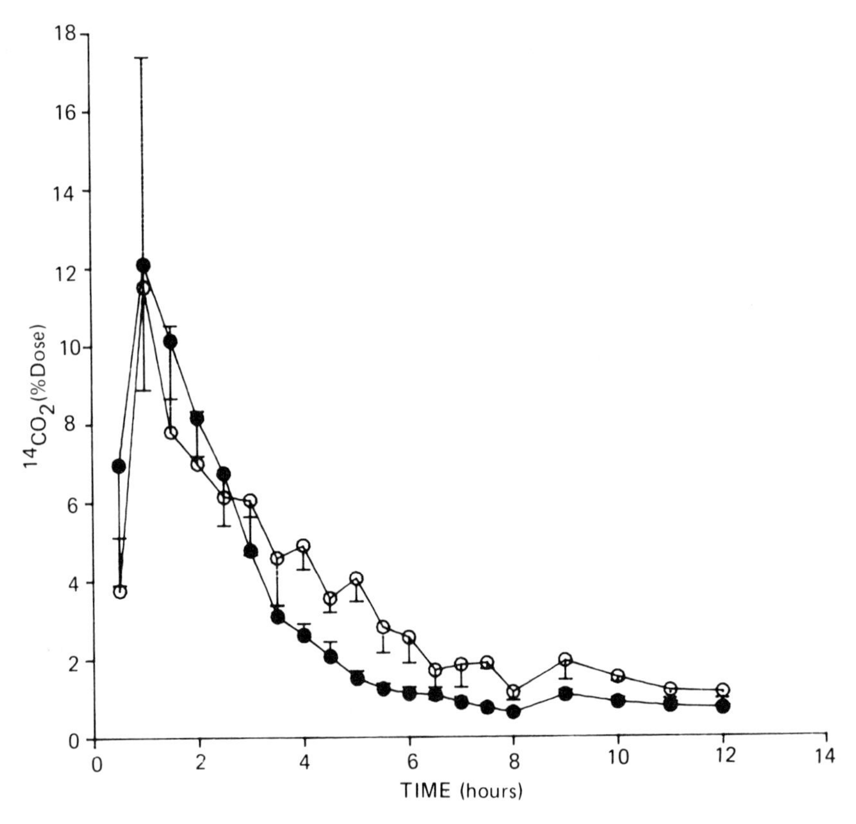

Figure 3 The mean (N = 4) excretion of radioactivity in expired air after the oral administration of 20 mg [^{14}C-aspartyl] aspartame (○) per kilogram body weight or an equimolar dose of [^{14}C] aspartic acid (●) to rhesus monkeys. The values are expressed as the percent of dose excreted per 0.5 hr interval. The vertical bars represent the standard errors of the mean.

^{14}C in expired air after administration of these two compounds. The excretion profile was nearly identical for these two compounds. This suggests that aspartic acid was rapidly liberated from aspartame and thus entered into the same metabolic pathways as exogenously administered aspartic acid. After 12 hr the total recovery of radioactivity in expired air was 77.0 ± 3.9% (± SEM) after [^{14}C-aspartyl] aspartame administration and 67.3 ± 1.9% after [^{14}C] aspartic acid administration.

Figure 4 gives the mean plasma ^{14}C concentrations after the administration of each of these labeled compounds. A large degree of variation occurred between animals after administration of either compound. The range of maximum plasma ^{14}C concentrations was between 52.5 and 120.3 (values expressed as percent dose per liter) after aspartame administration and 71.5 and 167 (percent dose per liter) after aspartic acid administration. The time required to reach maximum plasma ^{14}C levels varied between 20 min and 5 hr. Thus the plasma ^{14}C curves

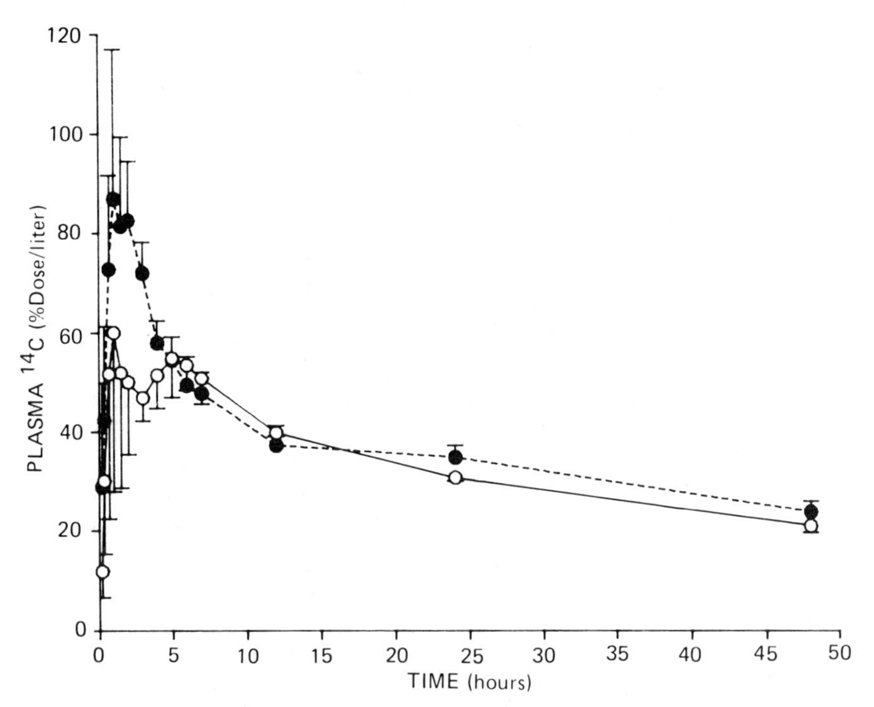

Figure 4 The mean (N = 4) plasma concentrations of radioactivity after the oral administration of 20 mg of [^{14}C-aspartyl] aspartame (○) per kilogram of body weight or an equimolar dose of [^{14}C] aspartic acid (●) to rhesus monkeys. The values are expressed as the percent of administered dose per liter of plasma. The vertical bars represent the standard errors of the mean. (From J. Oppermann, E. Muldoon, and R. E. Ranney, Metabolism of aspartame in monkeys, *J. Nutr.* Vol. 103: 1454-1459, 1973.)

appear to be different at early times after compound administration. However, at later times, both the absolute plasma concentrations and their rate of decline were similar.

The excretion of label into the urine and feces was similar after administration of either compound. Urinary excretion accounted for 2.12 ± 0.31% and 3.80 ± 0.38% after [^{14}C-aspartyl] aspartame and [^{14}C] aspartic acid dosing, respectively. Fecal excretion was 1.63 ± 0.36% after [^{14}C-aspartyl] aspartame and 1.57 ± 0.32% after [^{14}C] aspartic acid.

A number of metabolic reactions are responsible for the conversion of aspartate to CO_2. Neame and Wiseman (16) have shown that when aspartic acid was presented to intestinal mucosal cells, alanine was found to be the major amino acid appearing in the mesenteric blood. This occurs as a result of two enzymic reactions, the sum total of which has been estimated to account for 85% of the loss of aspartate in rat in vitro preparations (17). The two reactions are

$$\text{Aspartate} \xrightarrow{\text{aspartic acid-4-decarboxylase}} \text{alanine} + CO_2$$

$$\text{Aspartate} + \text{pyruvate} \xrightarrow{\text{alanine-2-oxoacid aminotransferase}} \text{alanine} + \text{oxaloacetate}$$

The alanine formed in these reactions enters the tricarboxylic acid cycle via pyruvate and acetyl coenzyme A (CoA) through participation of the following enzyme systems:

$$\text{Alanine} + \text{2-oxoglutarate} \rightarrow \text{pyruvate} + \text{glutamate}$$

$$\text{Pyruvate} + \text{CoA} + \text{NAD}^+ \rightarrow \text{acetyl CoA} + CO_2 + \text{NADH} + H^+$$

The oxaloacetate produced by the transamination is oxidized via the tricarboxylic acid cycle. Thus the 70% excretion of label as $^{14}CO_2$ observed after [^{14}C-aspartyl] aspartame or [^{14}C] aspartic acid administration probably results from these reactions.

Approximately 25% of the label was not excreted in the expired air, urine, or feces, but appeared to be incorporated into normal body constituents. The long terminal plasma half-life of radioactivity (Fig. 4) is evidence for incorporation of aspartate's carbon atoms into such compounds. Aspartic acid itself, of course, is a normal constituent of proteins and much of the label is probably found in body protein. In addition, aspartate can be converted to uridine monophosphate in a five-step synthesis, and to methyl aspartate in a one-step reaction. All of these pathways would lead to incorporation of the label into endogenous compounds.

Thus studies in rats and monkeys with [^{14}C] aspartyl-labeled aspartame indicate that aspartic acid is rapidly and completely liberated from aspartame as a result of metabolism and that the released aspartic acid participates in a variety of metabolic reactions, the predominance of which leads to oxidation to [^{14}C] carbon dioxide.

DISPOSITION OF [^{14}C-PHENYLALANYL] ASPARTAME

The disposition of aspartame labeled with ^{14}C in the phenylalanine portion of the molecule was studied in the rat, dog, and monkey. Each carbon atom in the phenylalanine moiety was labeled. Table 3 summarizes the recoveries of radioactivity following administration of 20 mg/kg body weight [^{14}C-phenylalanyl] aspartame or an equimolar dose of [^{14}C]phenylalanine to male rats. In this study the recovery of label was significantly lower in all matrices after aspartame administration. The cause of the low recoveries of label with aspartame dosing is not known; however, these results were not reflective of findings in the rhesus monkey. It is important to note that only a minor fraction of the label was excreted in the expired air, in contrast to results obtained with aspartyl- and methyl-labeled compounds. With the phenylalanine-labeled compound, or with phenylalanine itself, the majority of label remained within the body for the first 3 days following drug administration.

Table 3 Average Excretion of Radioactivity Following the Oral Administration of 20 mg/kg Body Weight [^{14}C-Phenylalanyl] Aspartame and an Equimolar Dose of [^{14}C]Phenylalanine to Male Rats (N = 6)

	Percentage of administered dose (± SEM)[a]		
	Urine	Feces	Expired Air
[^{14}C-Phenylalanyl] aspartame	1.62 ± 0.57	0.30 ± 0.08	10.4 ± 1.3
[^{14}C]Phenylalanine	3.87 ± 0.32	0.44 ± 0.09	17.0 ± 1.6

[a]Urine and feces were collected for 3 days; expired was collected for 8 hr.
Source: Ref. 5.

Table 4 Recovery of Radioactivity Following the Oral Administration of 20 mg/kg Body Weight [^{14}C-Phenylalanyl] Aspartame or an Equimolar Dose of [^{14}C] Phenylalanine to Female Beagle Dogs (N = 4)

	Percentage of administered dose (± SEM)[a]		
	Urine	Feces	Expired air
[^{14}C-Phenylalanyl] aspartame	2.63 ± 0.30	6.17 ± 0.77	17.31 ± 4.20
[^{14}C]Phenylalanine	5.32 ± 0.43	5.38 ± 0.92	23.77 ± 2.26

[a]Urine and feces were collected for 4 days; expired air was collected for 8 hr.

A similar study to that described above was conducted in female beagle dogs. The dogs were pretreated for 5 days with unlabeled aspartame (20 mg/kg body weight) or phenylalanine prior to the oral administration of the labeled compounds. Table 4 gives the recoveries of label following administration of these two compounds. There was slightly less ^{14}C excreted in the urine of dogs given aspartame than in phenylalanine-treated animals; fecal excretions, however, were similar.

The total excretion of label in expired air was slightly less after [^{14}C-phenylalanyl] aspartame administration. Figure 5 illustrates that major differences in

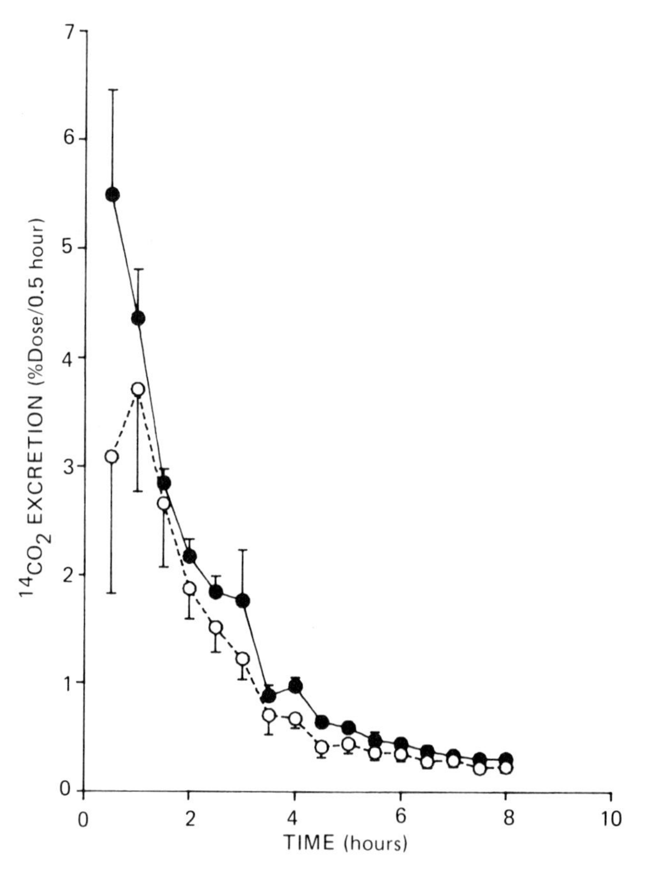

Figure 5 The mean (N = 4) excretion of radioactivity in expired air after the oral administration of 20 mg of [^{14}C-phenylalanyl] aspartame (○) per kilogram body weight or an equimolar dose of [^{14}C] phenylalanine (●) to female beagle dogs. The values are expressed as the percent of dose excreted per 0.5-hr intervals. The vertical bars represent the standard errors of the mean.

CO_2 excretion occurred within the first 2 hr; thereafter the rates of excretion were similar. Similarly, the appearance of radioactivity in plasma (Fig. 6) was slightly lower after administration of aspartame, but thereafter slightly higher concentrations were found up to 15 days following dosing. These results taken as a whole suggest that phenylalanine was liberated from aspartame. The slightly lower rate of ^{14}C appearance in expired air and in plasma at early times after aspartame administration is taken as an indicator of the time needed for hydrolysis of the methyl moiety and then hydrolysis of the resulting dipeptide before the phenylalanine moiety of aspartame could enter the normal metabolic pathways for this amino acid.

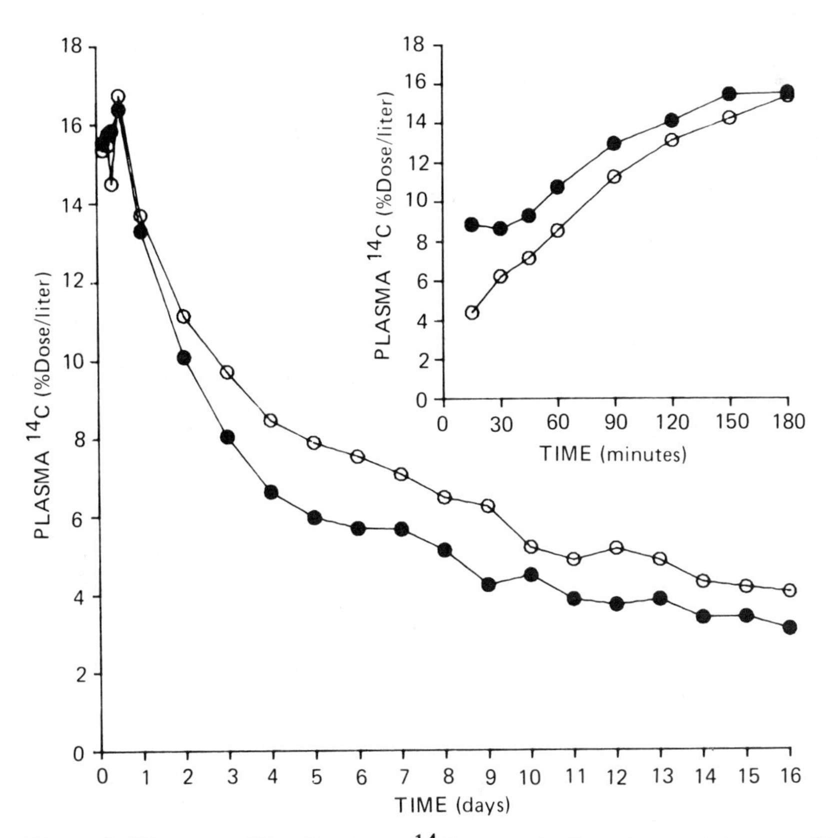

Figure 6 The mean (N = 4) plasma ^{14}C concentrations (percent dose per liter) following the oral administration of 20 mg of [^{14}C-phenylalanyl] aspartame (○) per kilogram body weight or an equimolar dose of [^{14}C] phenylalanine (●) to female beagle dogs. The insert gives the plasma concentration-time curves during the first 3 hr after compound administration.

The nature of the plasma radioactivity present after [^{14}C]aspartame administration was investigated by passing the plasma through ultrafiltration membrane cones (18). These membranes retain molecules with molecular weights greater than 50,000 and have essentially no retention for molecules with molecular weights less than 5000. Figure 7 gives the distribution of radioactivity into the low and high molecular weight fractions at various times after the administration of [^{14}C-phenylalanyl]aspartame. Within the first hour the majority of radioactivity was present in the low molecular fraction. At later times the plasma radioactivity was associated with high molecular weight (>50,000) compounds. In vitro incubation of [^{14}C]phenylalanine or [^{14}C-phenylalanyl]aspartame with dog plasma resulted in less than 3% of the added radioactivity being associated

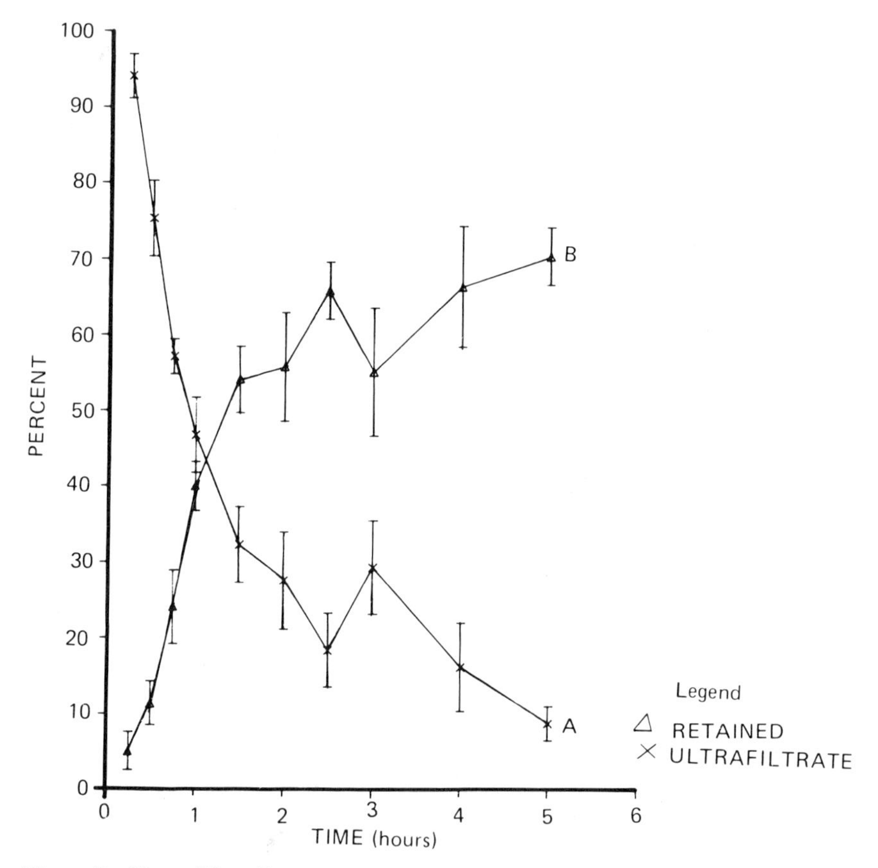

Figure 7 Mean (N = 4) percentages of total plasma radioactivity in the ultrafiltrate (A) and retained fractions (B) at various times after the oral administration of [^{14}C-phenylalanyl]aspartame to beagle dogs. The vertical bars represent the standard errors of the mean.

with the higher molecular weight fraction. It therefore appears that the ^{14}C in the high molecular weight fraction of plasma occurred as a result of incorporation into plasma proteins of the [^{14}C]phenylalanine (or ^{14}C-labeled nonessential amino acids derived from phenylalanine) which were released as a result of metabolism of the aspartame molecule.

The rapid appearance of $^{14}CO_2$ after the oral administration of [^{14}C-phenylalanyl]aspartame indicates that aspartame was rapidly metabolized. Figure 8 shows a thin-layer radiochromatogram of the plasma low molecular weight fraction 30 min after the oral administration of [^{14}C-phenylalanyl]aspartame to a dog. At this early time interval the radioactivity was associated primarily with phenylalanine and its metabolite tyrosine. No intact aspartame was evident. The 15-, 45-, and 60-min plasma samples gave essentially the same results. This was direct evidence that aspartame was completely hydrolyzed prior to the appearance of ^{14}C in the systemic circulation.

The disposition of [^{14}C-phenylalanyl]aspartame was also compared to [^{14}C] phenylalanine in rhesus monkeys. Table 5 summarizes the recoveries of radio-

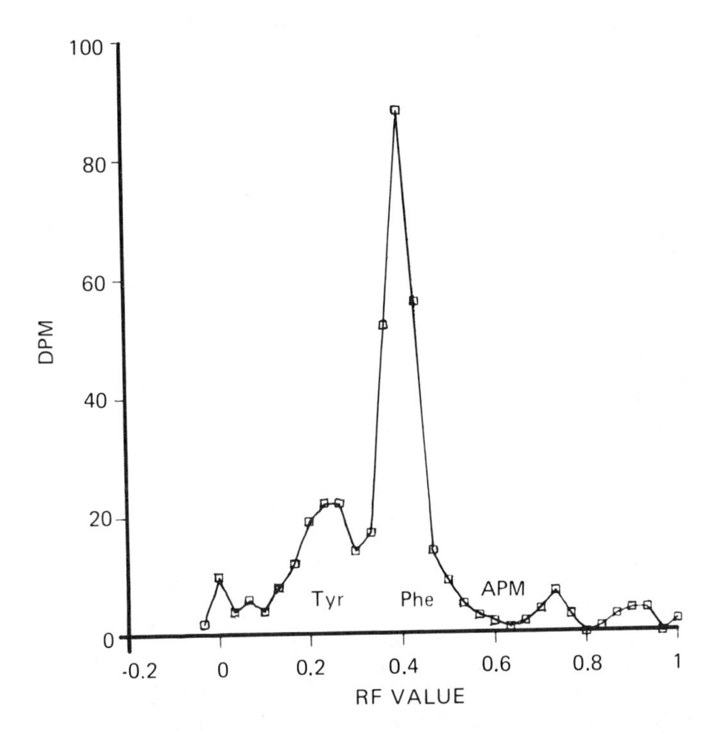

Figure 8 Thin-layer radiochromatogram of plasma sample from a dog given [^{14}C-phenylalanyl]aspartame 30 min previously. The Rf values of authentic aspartame, phenylalanine, and tyrosine are shown. The ordinate gives the number of disintegrations per minute.

Table 5 Average Excretion of Radioactivity Following the Oral Administration of [^{14}C-Phenylalanine]Aspartame (20 mg/kg Body Weight) or an Equimolar Dose of [^{14}C]Phenylalanine to Female Rhesus Monkeys

	Percentage of administered dose		
	Urine	Feces	Expired air
[^{14}C-Phenylalanyl] aspartame	2.8	1.6	17.5
[^{14}C]Phenylalanine	3.1	4.7	18.0

Source: J. Oppermann, E. Muldoon, and R. E. Ranney, Metabolism of aspartame in monkeys, *J. of Nutrition*, vol. 103: 1454-1459, 1973.

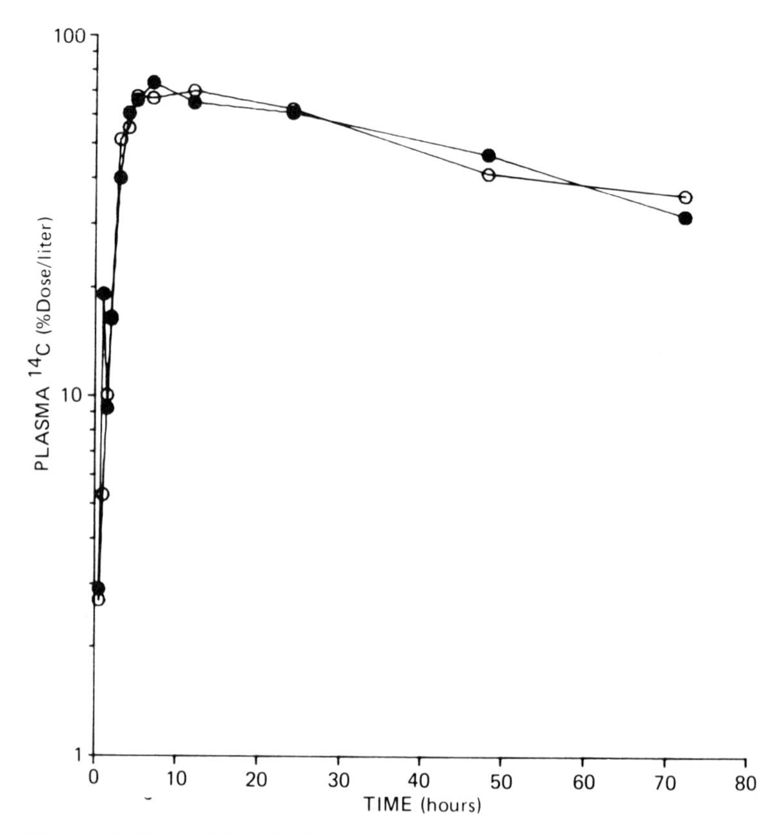

Figure 9 Mean (N = 4) plasma concentrations of total radioactivity at various times after the oral administration of 20 mg of [^{14}C-phenylalanyl] aspartame (○) per kilogram body weight or an equimolar dose of [^{14}C]phenylalanine (●) to female rhesus monkeys. The values are expressed as the percent of dose per liter. (From J. Oppermann, E. Muldoon, and R. E. Ranney, Metabolism of aspartame in monkeys, *J. Nutr.*, vol. 103: 1454-1459, 1973.)

activity in the various routes of excretion in this species. Except for a slightly greater fecal excretion after phenylalanine administration, the total recoveries were nearly identical. Likewise, the plasma ^{14}C concentration-time curves (Fig. 9) were essentially the same after administration of either compound.

In summary, comparative studies with [^{14}C-phenylalanyl] aspartame and [^{14}C] phenylalanine have shown in the rat, dog, and monkey that aspartame is metabolized to release free phenylalanine. Studies in the dog have shown this metabolism to be presystemic, since even at 15 min after compound administration no intact aspartame could be demonstrated in the plasma. However, both phenylalanine and tyrosine were present in the plasma at early time points, whereas at later times the plasma radioactivity was primarily associated with large molecular weight compounds, presumably plasma proteins.

DISPOSITION OF [^{14}C] PHENYLALANINE METHYL ESTER

As indicated above, the presence of the aspartame molecule itself has not been demonstrated in the systemic circulation after its oral administration to experimental animals. Although it has been assumed that the initial reaction in aspartame metabolism was hydrolysis of the methyl ester, producing aspartylphenylalanine and methanol, it is also possible that the initial metabolic attack may occur at the peptide bond, thus liberating aspartic acid and phenylalanine methyl ester. Burton et al. (19) examined the potential for absorption of phenylalanine methyl ester by administering [^{14}C] phenylalanine methyl ester hydrochloride ([^{14}C] PM) either intraduodenally or intragastrically to rhesus monkeys containing indwelling portal vein cannulae.

In these studies, α-[^{14}C-U-phe] phenylalanine methyl ester hydrochloride was prepared at a high specific activity (1 mCi/mg) (20). This was required since rapid metabolism of this compound was anticipated and it was of interest to determine if very small quantities of [^{14}C] PM survived the absorption process and entered the blood intact.

Two female rhesus monkeys were prepared with indwelling portal vein cannulae. One monkey (9 days after surgery) was administered 20 ml of an aqueous solution containing 20 mg and 20 mCi of [^{14}C] PM intragastrically through a nasogastric tube. The other monkey received a similar dose administered directly into the duodenum after reopening the abdominal wound (under methoxyfluorane anesthesia) made during the portal vein cannulation surgery (11 days earlier). Blood samples were then taken from both the portal vein and saphenous vein (peripheral blood) at various times after dosing. After removing an aliquot for total ^{14}C determinations, blood was transferred immediately to a centrifuge tube which contained $AgNO_3$ (to inhibit esterase activity) and unlabeled carrier PM. Each sample was then extracted with a methanol-acetonitrile mixture and the extract analyzed for total radioactivity ("extractable ^{14}C") and for unchanged [^{14}C] PM by thin-layer radiochromatography.

The recovery of [^{14}C]PM from control blood samples by this procedure was approximately 70%. The identity of [^{14}C]PM in selected blood samples was confirmed by high-performance liquid chromatography.

The concentrations of total ^{14}C, extractable ^{14}C, and [^{14}C]PM in portal and peripheral venous blood are shown in Figures 10 and 11. In the monkey dosed intragastrically, the total ^{14}C in portal blood increased to a maximum at 30 min, declined to a minimum by 1 hr, and then increased during the next 4 hr. In the monkey dosed intraduodenally a similar time course was observed, except that absorption was more rapid. Peak portal plasma concentrations occurred at the earliest sampling point (10 min). In both monkeys, the total ^{14}C in peripheral venous blood was initially much lower than that noted in portal blood, but reached similar levels 1 hr after administration. The time required for ^{14}C in peripheral blood to reach an equivalent level to portal blood presumably represents the major part of the absorption phase. During the first 40 min after dose administra-

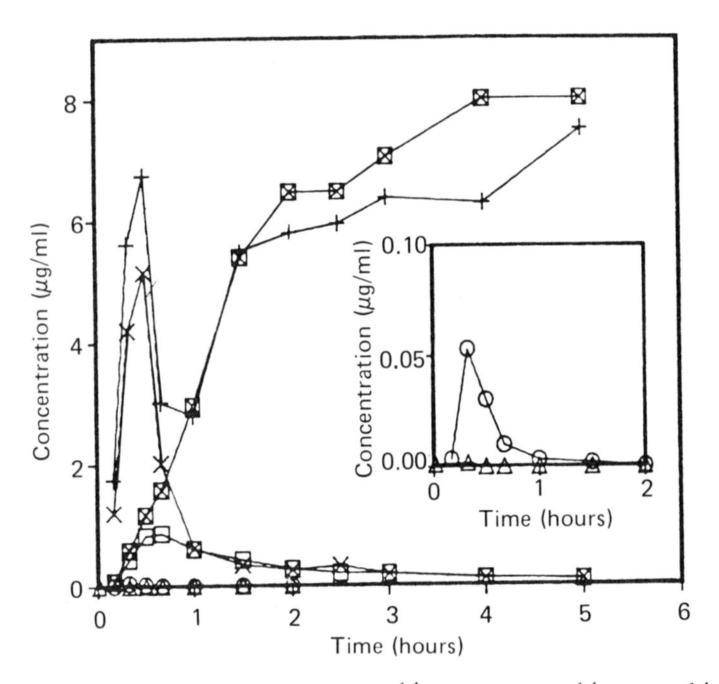

Figure 10 Concentration of total ^{14}C, extractable ^{14}C, and [^{14}C]PM in blood after intragastric administration of 20 mg of [^{14}C]PM to monkey 266. Curves represent the portal blood concentration of total ^{14}C (+), extractable ^{14}C (x), and [^{14}C]PM (o), and the peripheral blood concentration of total ^{14}C (⊠), extractable ^{14}C (□), and [^{14}C]PM (△). The inset shows the concentration of [^{14}C] PM in portal (o) and peripheral (△) blood, with the ordinate expanded 40-fold for clarity. (From Ref. 14.)

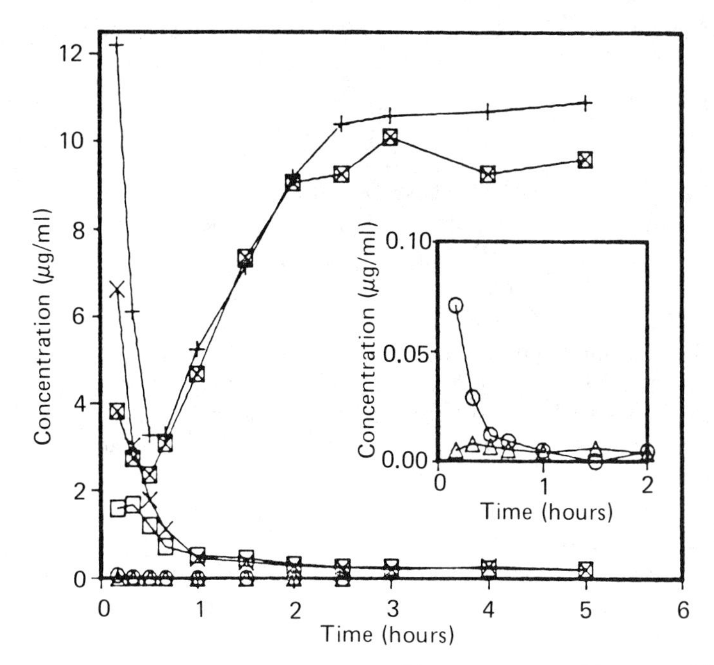

Figure 11 Concentration of total ^{14}C, extractable ^{14}C, and [^{14}C]PM in blood after intraduodenal administration of 20 mg of [^{14}C]PM to monkey 291. Curves represent the portal blood circulation of total ^{14}C (+), extractable ^{14}C (x), and [^{14}C]PM (o), and the peripheral blood concentration of total ^{14}C (⊠), extractable ^{14}C (□), and [^{14}C]PM (△). The inset shows the concentration of [^{14}C]PM in portal (o) and peripheral (△) blood, with the ordinate expanded 40-fold for clarity. (From Ref. 14.)

tion, the concentrations of extractable radioactivity from portal blood were similar to total ^{14}C concentrations. Thereafter the concentration of extractable radioactivity rapidly declined reaching levels of less than 5% of total ^{14}C by 2 hr. During the first hour after compound administration most of the extractable radioactivity was phenylalanine. At later times the percentage of extractable ^{14}C identifiable as phenylalanine declined.

Very small amounts of [^{14}C]PM were detected in portal and peripheral blood samples during the first 1-2 hr after dosing. This is shown in the inserts in Figures 10 and 11. At times later than 2 hr no [^{14}C]PM was detectable (<0.001 μg/ml) in blood samples. During the first 2 hr after dosing, [^{14}C]PM was present in portal blood at 0.2% of the total ^{14}C level (based on the respective areas under the blood concentration-time curves) and less than 0.1% of peripheral total ^{14}C.

These in vivo results demonstrate the rapid and extensive metabolism of [^{14}C] PM during the absorption process. The amount of unmetabolized PM entering the portal blood represented at most only 0.2% of the total compound absorbed. The amount of [^{14}C]PM entering the systemic circulation following passage through the liver was even smaller and no PM (<0.001 μg/ml) was detectable in either the portal or peripheral circulation at times later than 2 hr after dosing.

The fraction of PM absorbed unchanged is dependent on the relative rates of transport versus ester hydrolysis. Previous studies with amino acid methyl esters have indicated they are neither substrates nor inhibitors of the intestinal amino acid active transport systems (21-23); thus their absorption is assumed to be a passive diffusion process. On the other hand, PM is known to be a good substrate for esterases (24-26) and for amidases such as chymotrypsin (8). These enzymes are found both extracellularly and intracellularly. Thus PM may be hydrolyzed both within the intestinal lumen prior to absorption and within intestinal mucosa cells prior to entering the hepatic portal system. In this study greater than 99.7% of a 20-mg dose of [^{14}C]PM was hydrolyzed prior to reaching the hepatic portal system in the monkey.

In conclusion, studies conducted in a number of animal species indicate that aspartame is rapidly and extensively metabolized to its constituent amino acids and methanol. This hydrolysis occurs completely during the absorption process, primarily during its passage from the intestinal lumen into the portal circulation.

REFERENCES

1. Neurath, H., and Schwartz, G. W. (1950). The mode of action of the crystalline pancreatic proteolytic enzymes. *Chem. Rev.* 46, 69-153.
2. These three ^{14}C-labeled compounds were synthesized by J. M. Schlatter and D. J. Zitzewitz of the Research and Development division of G. D. Searle and Co.
3. These compounds were obtained from Amersham/Searle Corporation.
4. Oppermann, J. A., Muldoon, E., and Ranney, R. E. (1973). Effect of aspartame on phenylalanine metabolism in the monkey. *J. Nutr.* 103, 1460-1466.
5. Ranney, R. E., Oppermann, J. A., Muldoon, E., and McMahon, F. G. (1976). Comparative metabolism of aspartame in experimental animals and humans. *J. Toxicol. Environ. Health* 2, 441-451.
6. Ranney, R. E., and Oppermann, J. A. (1979). A review of the metabolism of the aspartyl moiety of aspartame in experimental animals and man. *J. Environ. Pathol. Toxicol.* 2, 979-985.
7. Oppermann, J. A., and Ranney, R. E. (1979). The metabolism of aspartame in infant and adult mice. *J. Environ. Pathol. Toxicol.* 2, 987-998.
8. Purdie, J. E., and Benoiton, N. L. (1970). The interaction of α-chymotrypsin with phenylalanine derivatives containing a free α-amino group. *Can. J. Biochem.* 48, 1058-1065.
9. Tephly, T. R., Parks, R. E., and Mannering, G. J. (1964). Methanol metabolism in the rat. *J. Pharmacol. Exp. Ther.* 143, 292-300.

10. Makar, A. B., Tephly, T. R., and Mannering, G. J. (1968). Methanol metabolism in the monkey. *Mol. Pharmacol.* 4, 471-483.
11. Goodman, J. I., and Tephly, T. R. (1970). Peroxidative oxidation of methanol in human liver: The role of hepatic microbody and soluble oxidases. *Res. Commun. Chem. Pathol. Pharmacol.* 1, 441-450.
12. Strittmatter, P., and Ball, E. G. (1955). Formaldehyde dehydrogenase, a glutathione dependent enzyme system. *J. Biol. Chem.* 213, 445-461.
13. Abeles, R. H., and Lee, H. A., Jr. (1960). The dismutation of formaldehyde by liver alcohol dehydrogenase. *J. Biol. Chem.* 235, 1499-1503.
14. Nakade, H. I., and Weinhouse, S. (1953). Studies of glycine oxidations in rat tissues. *Arch. Biochem. Biophys.* 42, 257-270.
15. Tephly, T. R., and McMartin, K. E. (1984). Methanol metabolism and toxicity. In *Aspartame: Physiology and Biochemistry* (Stegink, L. D., and Filer, L. J., Jr., eds.), Marcel Dekker, New York, Chap. 6.
16. Neame, K. D., and Wiseman, G. (1957). The transamination of glutamic and aspartic acids during absorption by the small intestine of the dog in vivo. *J. Physiol.* 135, 442-450.
17. Parsons, D. S., and Volman-Mitchell, H. (1974). The transamination of glutamate and aspartate during absorption in vitro by small intestine of chicken, guinea pig and rat. *J. Physiol.* 239, 677-694.
18. Amicon Centriflo CF50A membrane cones, Amicon Corporation, Lexington, Mass.
19. Burton, E. G., DalMonte, P., Spears, C., Frank, P., and Oppermann, J. A. (1984). Absorption of phenylalanine methyl ester by the rhesus monkey and species differences in its metabolism in blood, plasma and intestinal mucosa. *J. Nutr.* (in press).
20. The [^{14}C]phenylalanine methyl ester was prepared by Don Zitzewitz of the Drug Metabolism Department of G. D. Searle and Co.
21. Spencer, R. P., Weinstein, J., Sussman, A., Bow, T. M., and Markulis, M. A. (1962). Effect of structural analogues on intestinal accumulation of glycine. *Am. J. Physiol.* 203, 634-636.
22. Lin, E. C. C., Hagihiri, H., and Wilson, T. H. (1962). Specificity of the transport system for neutral amino acids in the hamster intestine. *Am. J. Physiol.* 202, 919-925.
23. Lerner, J. (1978). *A Review of Amino Acid Transport Processes in Animal Cells and Tissues*, University of Maine Press, Orono, Maine, pp. 1-8.
24. Goldberg, M. I., and Fruton, J. S. (1969). Beef liver esterase as a catalyst of acyl transfer to amino acid esters. *Biochemistry* 8, 86-97.
25. Stoops, J. K., Horgan, D. J., Runnegar, M. T. C., de Jersey, J., Webb, E. C., and Zerner, B. (1969). Carboxylesterases (EC 3.1.1). Kinetic studies on carboxylesterases. *Biochemistry* 8, 2026-2033.
26. Heymann, E. (1980). Carboxylesterases and amidases. In *Enzymatic Basis of Detoxication*, Vol. 2 (Jakoby, W. ed.), Academic, New York, pp. 291-323.

8

Tissue Distribution of Orally Administered Isotopically Labeled Aspartame in the Rat

Yoshimasa Matsuzawa and Yuichi O'Hara
Life Science Laboratory, Central Research Laboratories,
Ajinomoto Company, Inc., Yokohama, Japan

INTRODUCTION

Aspartylphenylalanine, L-aspartic acid, and L-phenylalanine are natural constituents of diet. As such, they and aspartame (the methyl ester of aspartylphenylalanine) should be absorbed, distributed, metabolized, and excreted as though they were dietary components (1). To test this hypothesis, studies on the whole-body autoradiography of adult rats given radiolabeled aspartame have been compared to the whole-body autoradiography of rats given radiolabeled aspartic acid or phenylalanine.

Although whole-body autoradiograms do not provide data that can be expressed quantitatively, it is an effective method for the study of the absorption, distribution, metabolism, and excretion of chemical compounds.

EXPERIMENTAL METHODS

Radioactive Compounds

^{14}C-Labeled aspartame, [U-^{14}C-Phe] aspartame and [U-^{14}C-Asp] aspartame, was provided by Dr. D. J. Zitzewitz of the G. D. Searle Laboratories. [U-^{14}C] Aspartic acid and [U-^{14}C] phenylalanine were obtained from New England Nuclear. The specific activities were adjusted to 0.5 Ci/mol for [U-^{14}C-Phe] aspartame, [U-^{14}C-Asp] aspartame, and [U-^{14}C] aspartic acid and to 0.25 Ci/mol for [U-^{14}C] phenylalanine by diluting with the nonlabeled compounds.

Animals

Male Sprague-Dawley rats (Charles River Japan Co., Inc.), weighing about 250 g, were used. They were housed in an air-conditioned room with a 12 hr light-12 hr dark schedule. A single dose of each compound equivalent to 20 mg/kg body weight, and containing 8 μCi of radioactivity, was administered by stomach tube.

Whole Body Autoradiography

Animals were sacrificed by ether anesthesia at 0.5, 2, 6, and 24 hr and 7 days after dosing. Each animal was placed in a thin rubber bag and frozen solid in acetone cooled with dry ice (solid carbon dioxide). Sagittal whole-body sections (20 μm) were prepared in a Bright Cryostat at -20°C according to the technique described by Ullberg (2). After freeze drying, the sections were apposed to x-ray film (LKB Co., Inc. Sweden) under slight pressure at room temperature for 4 weeks. The film was then developed.

RESULTS

Phenylalanine Moiety

The pattern of distribution of [U-^{14}C-Phe] aspartame following its oral administration was very similar to that of [U-^{14}C] phenylalanine after 0.5, 2, 6, and 24 hr and 7 days (Figs. 1-10).

Thirty minutes after administration of these compounds, very high levels of radioactivity were observed in the lumen of the stomach and upper small bowel. Significant uptake of radioactivity was observed in the pancreas, gastrointestinal mucosa, hair follicles, salivary gland, and liver. Radioactivity was observed in the kidney, adrenal gland, bone marrow, spleen and eye. Some radiolabel was localized to the brain, spinal cord, heart, thymus, lung, and testis.

Two hours after administration of these compounds, radioactivity within the stomach decreased, with the highest activity localized to the lumen of the small intestine. In addition to those organs that contained radiolabel at 30 min, a high level of radioactivity was seen in the nasal mucosa, salivary, and sublingual glands.

Six hours following administration, radioactivity within the gastrointestinal tract moved from the small intestine to the cecum and the feces within the rectum, while radioactivity of the whole body decreased. At this time its distribution was similar to that of earlier phases.

Twenty-four hours after administration of radiolabeled compounds, the radioactivity within the gastrointestinal tract had vanished. Although the total radioactivity level in the body had decreased, its general distribution remained unchanged. Little radioactivity was observed in the whole body 7 days after administration of these compounds.

Aspartic Acid Moiety

The distribution of radiolabel following oral administration of [U-^{14}C-Asp] aspartame and [U-^{14}C] aspartic acid (Figs. 11-18) was similar to that of [U-^{14}C-Phe] aspartame and [U-^{14}C] phenylalanine. However, the turnover of the aspartic acid moiety, as measured by disappearance of the radiolabel, was faster than that of the phenylalanine moiety.

DISCUSSION

The whole-body autoradiogram of rats given ^{14}C-labeled aspartame was similar to that of rats given the individual ^{14}C-labeled amino acids of which aspartame is comprised.

Radioactivity within the gastrointestinal tract decreased rapidly, as the compounds were distributed to the whole body, especially to those organs where protein turnover is high. This signifies that the amino acid moieties of aspartame are utilized for protein synthesis. There was no evidence of aspartame accumulation from this single-dose study. The results of this study support other metabolic studies of aspartame indicating rapid metabolism (3).

ACKNOWLEDGMENTS

The authors are grateful to Dr. D. J. Zitzewitz, G. D. Searle Laboratories, for synthesis of radioactive aspartame, and thank Professors L. J. Filer, Jr. and L. D. Stegink for helpful advice and suggestions.

REFERENCES

1. Ranney, R. E., Oppermann, J. A., Muldoon, E., and McMahon, F. G. (1976). Comparative metabolism of aspartame in experimental animals and humans. *J. Toxicol. Environ. Health* 2, 441-451.
2. Ullberg, S. (1954). Studies on the distribution and fate of ^{35}S-labeled benzylpenicillin in the body. *Acta Radiol. Suppl.* 118, 1-110.
3. Oppermann, J. A. (1984). Aspartame metabolism in animals. In *Aspartame: Physiology and Biochemistry* (Stegink, L. D., and Filer, L. J., Jr., eds.), Marcel Dekker, New York, Chap. 7.

Figures 1-18 follow

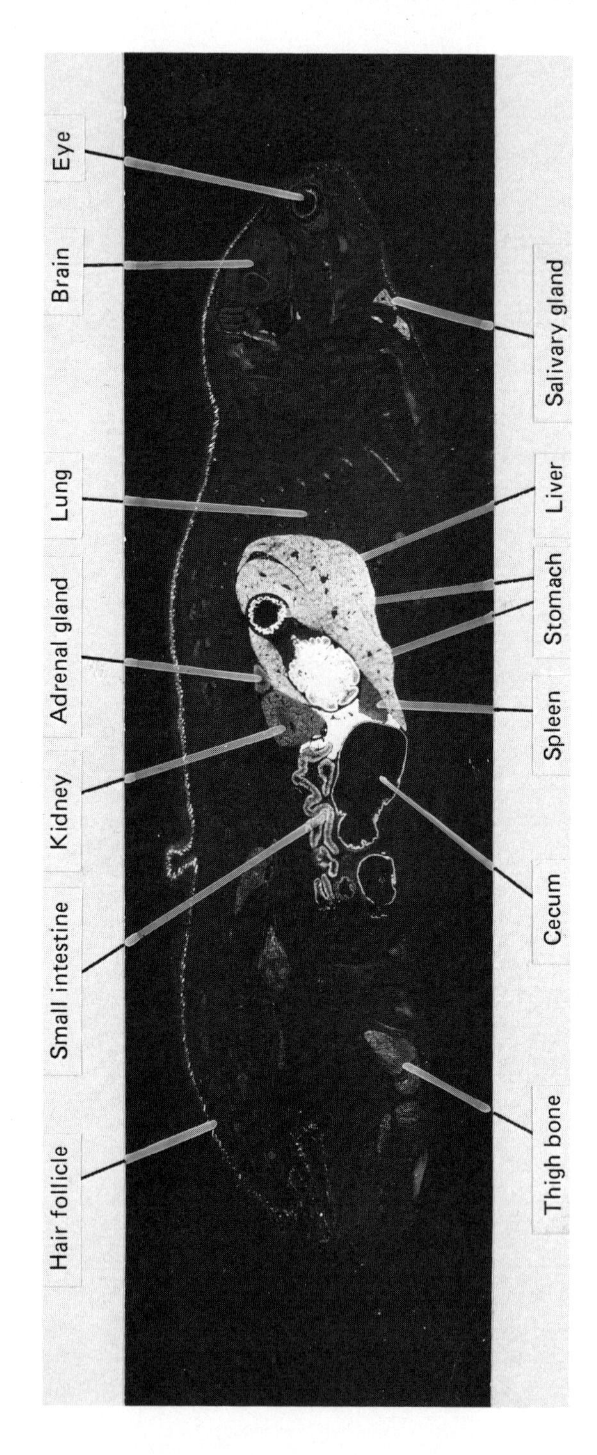
Eye
Brain
Salivary gland
Lung
Liver
Stomach
Adrenal gland
Spleen
Kidney
Cecum
Small intestine
Thigh bone
Hair follicle

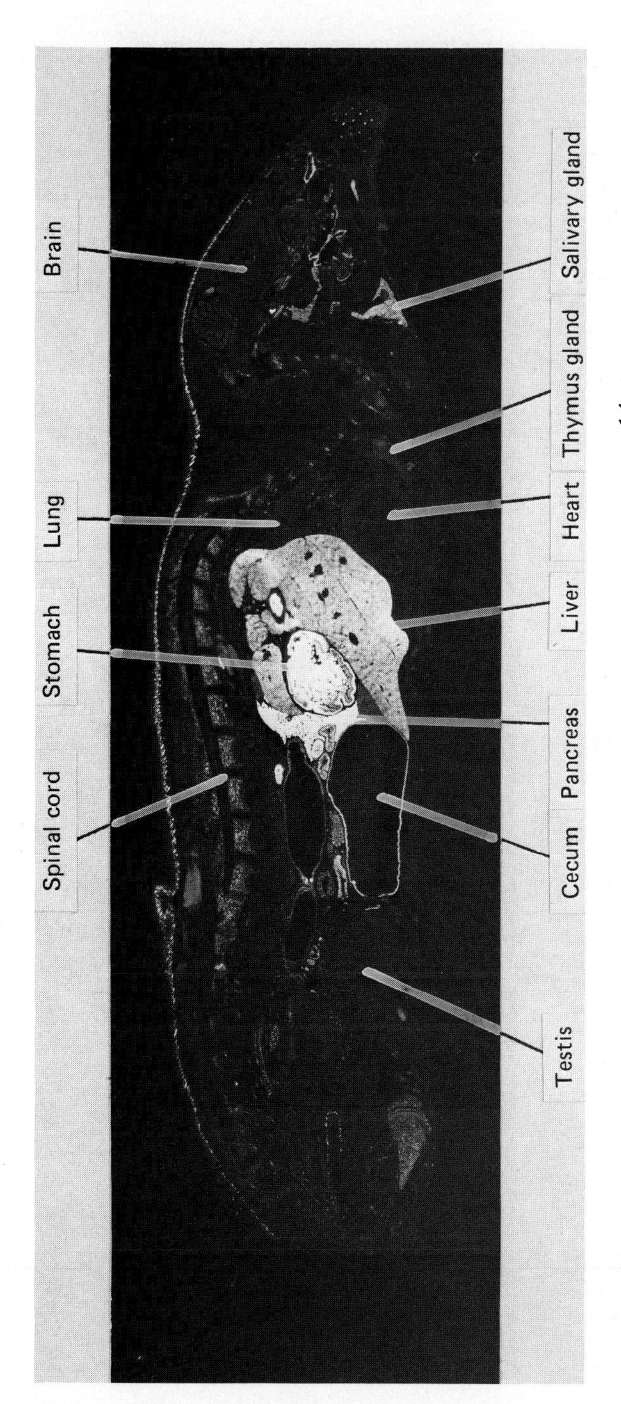

Figure 1 Whole-body autoradiogram of a male rat 30 min after oral administration of [U-^{14}C-Phe] aspartame.

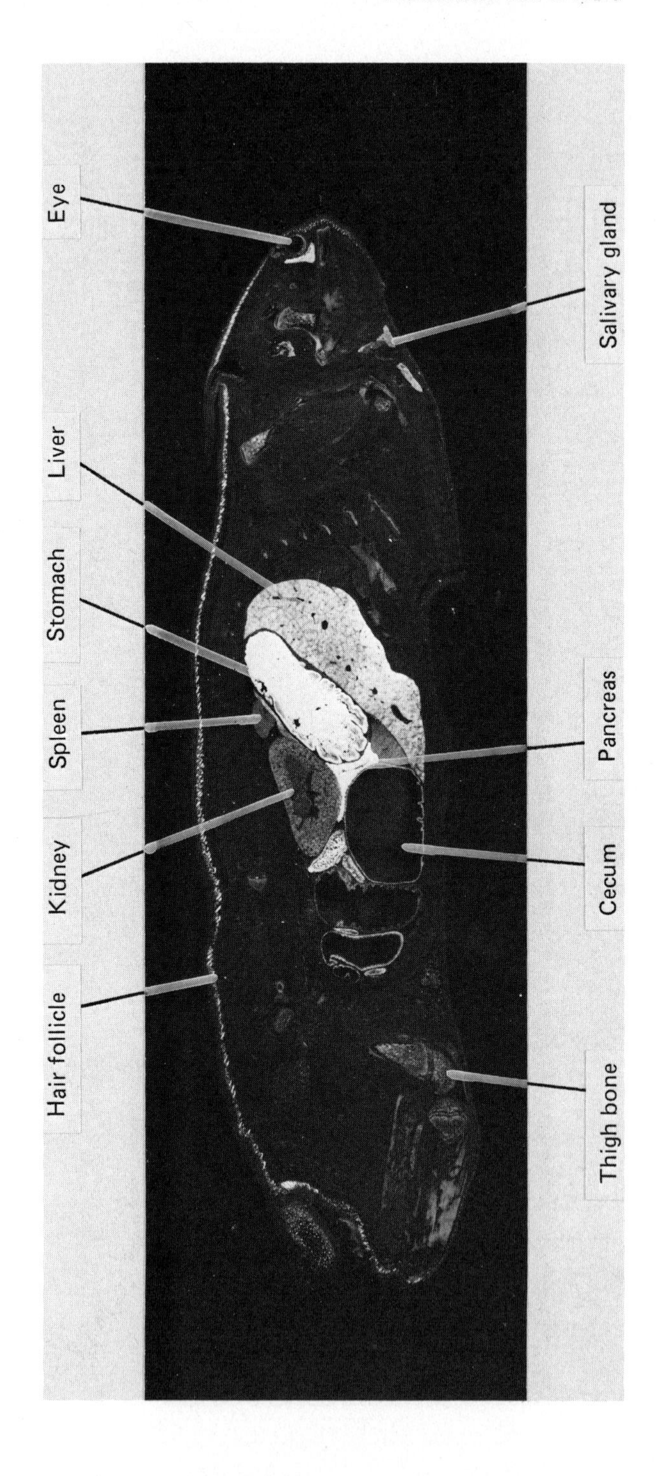
Eye
Liver
Stomach
Spleen
Kidney
Hair follicle
Salivary gland
Pancreas
Cecum
Thigh bone

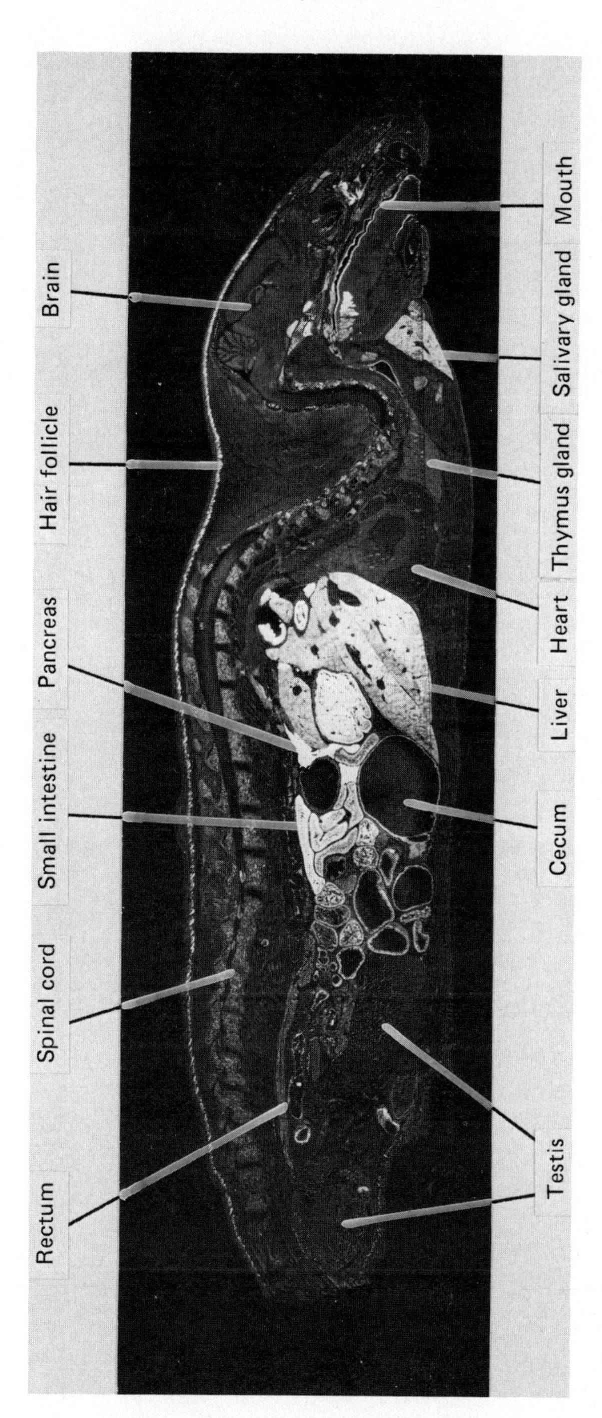

Figure 2 Whole-body autoradiogram of a male rat 30 min after oral administration of [U-^{14}C] phenylalanine.

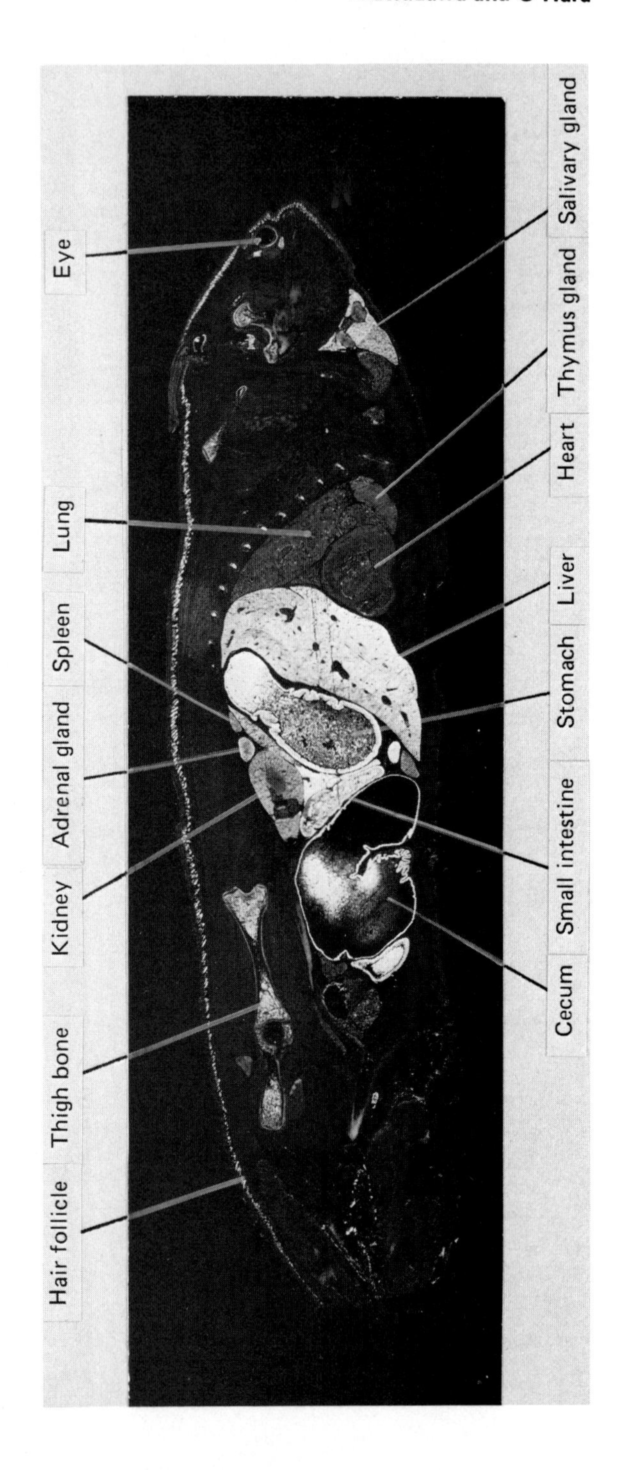
Eye
Lung
Spleen
Adrenal gland
Kidney
Thigh bone
Hair follicle
Salivary gland
Thymus gland
Heart
Liver
Stomach
Small intestine
Cecum

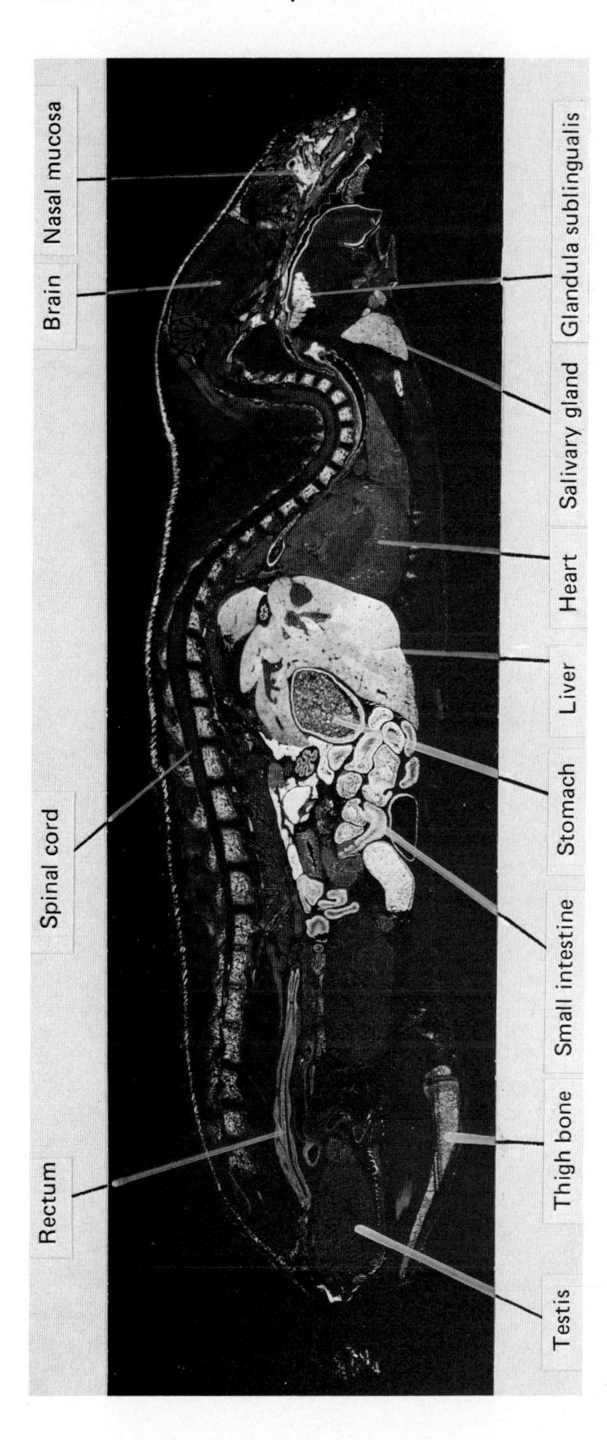

Figure 3 Whole-body autoradiogram of a male rat 2 hr after oral administration of [U-^{14}C-Phe] aspartame.

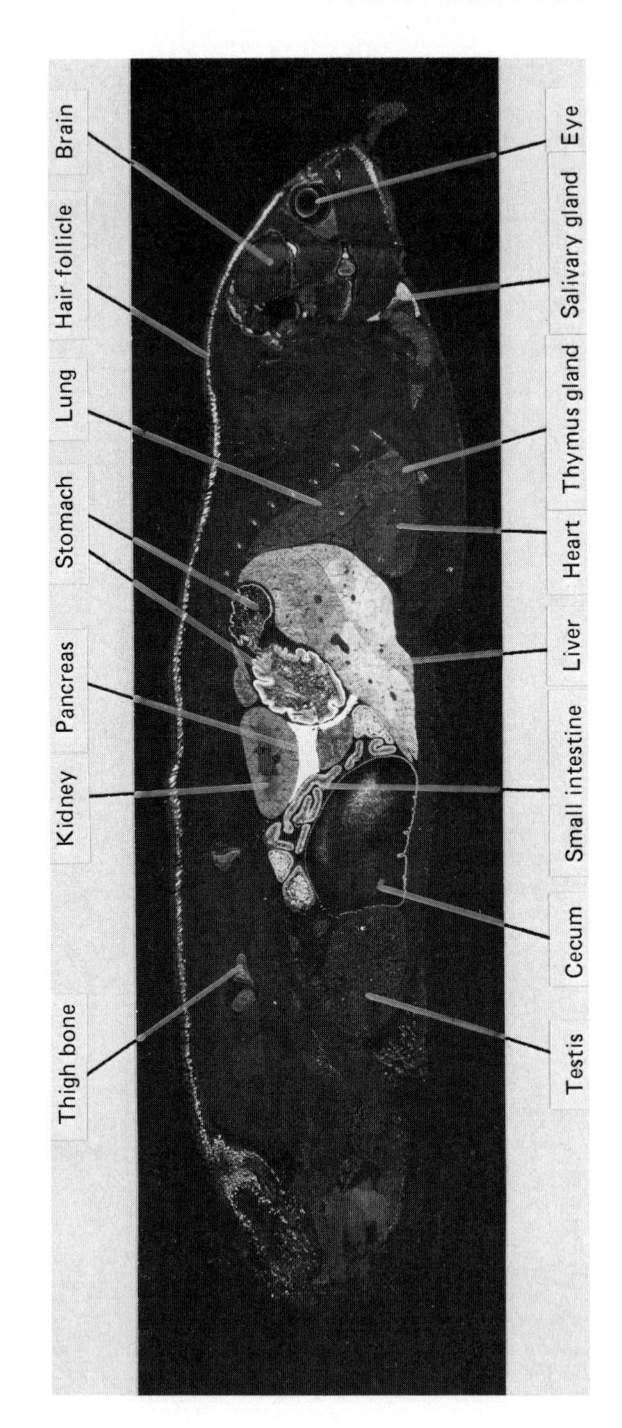
Brain
Hair follicle
Lung
Stomach
Pancreas
Kidney
Thigh bone
Eye
Salivary gland
Thymus gland
Heart
Liver
Small intestine
Cecum
Testis

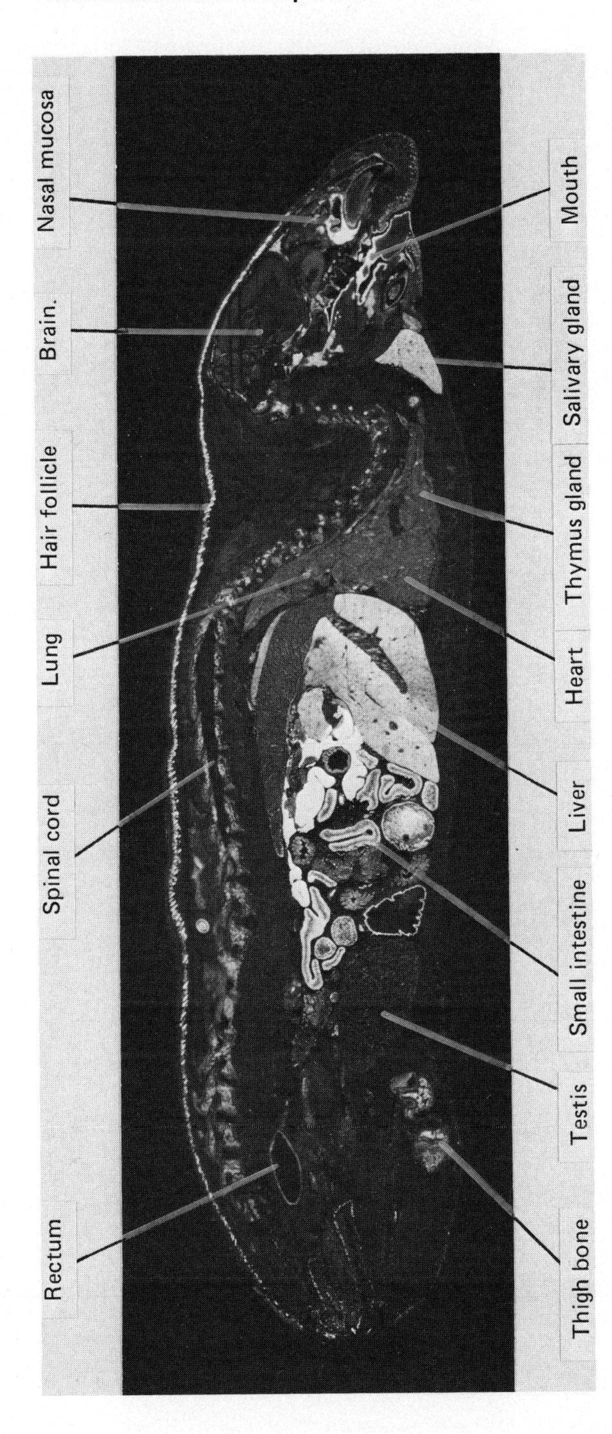

Figure 4 Whole-body autoradiogram of a male rat 2 hr after oral administration of [U-^{14}C]phenylalanine.

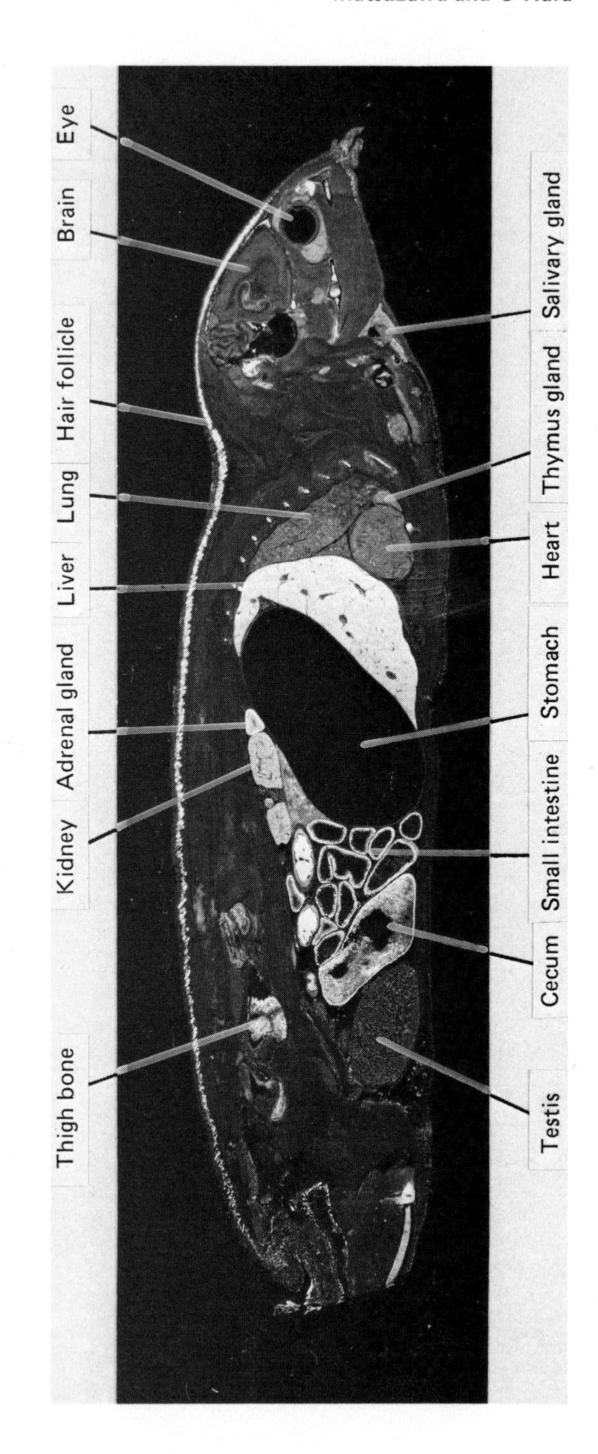
Eye
Brain
Hair follicle
Lung
Liver
Adrenal gland
Kidney
Thigh bone
Salivary gland
Thymus gland
Heart
Stomach
Small intestine
Cecum
Testis

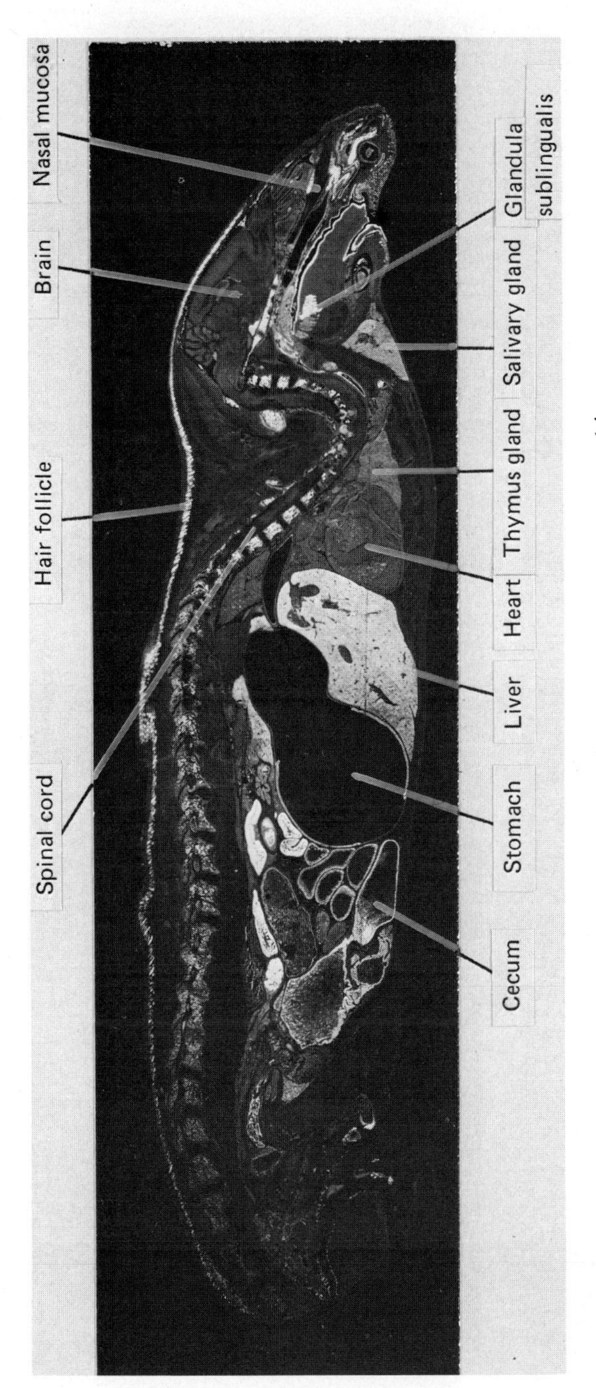

Figure 5 Whole-body autoradiogram of a male rat 6 hr after oral administration of [U-^{14}C-Phe] aspartame.

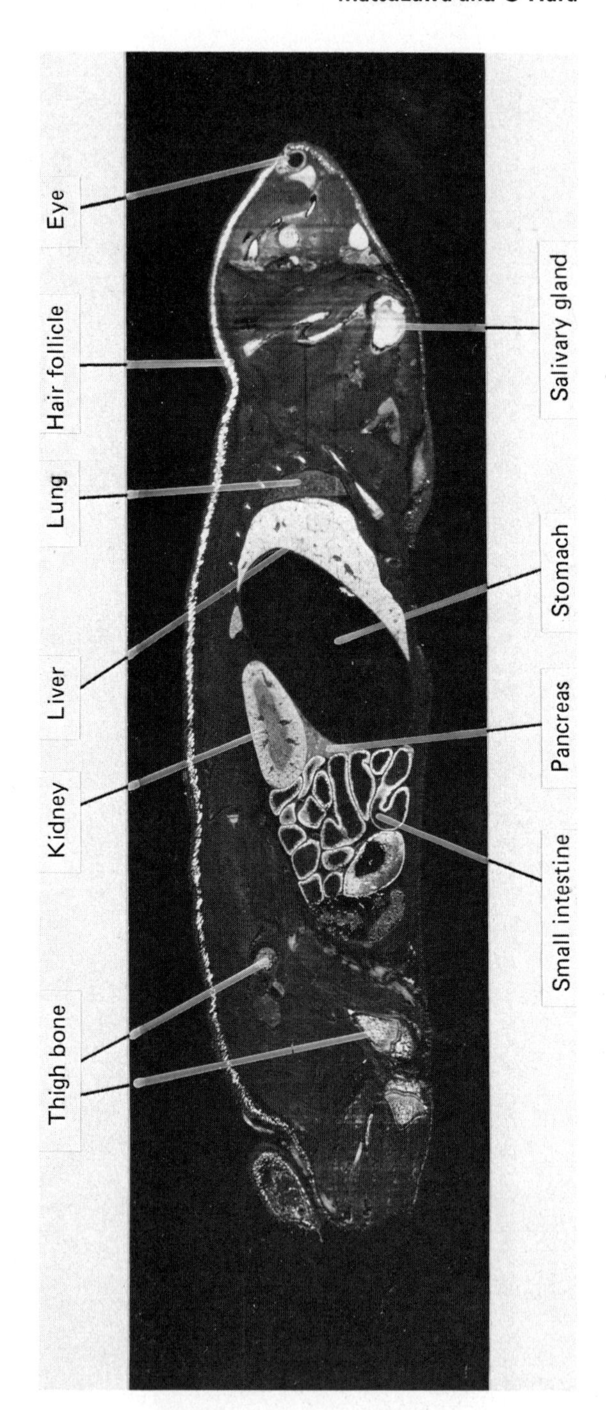
Eye
Hair follicle
Lung
Liver
Kidney
Thigh bone
Salivary gland
Stomach
Pancreas
Small intestine

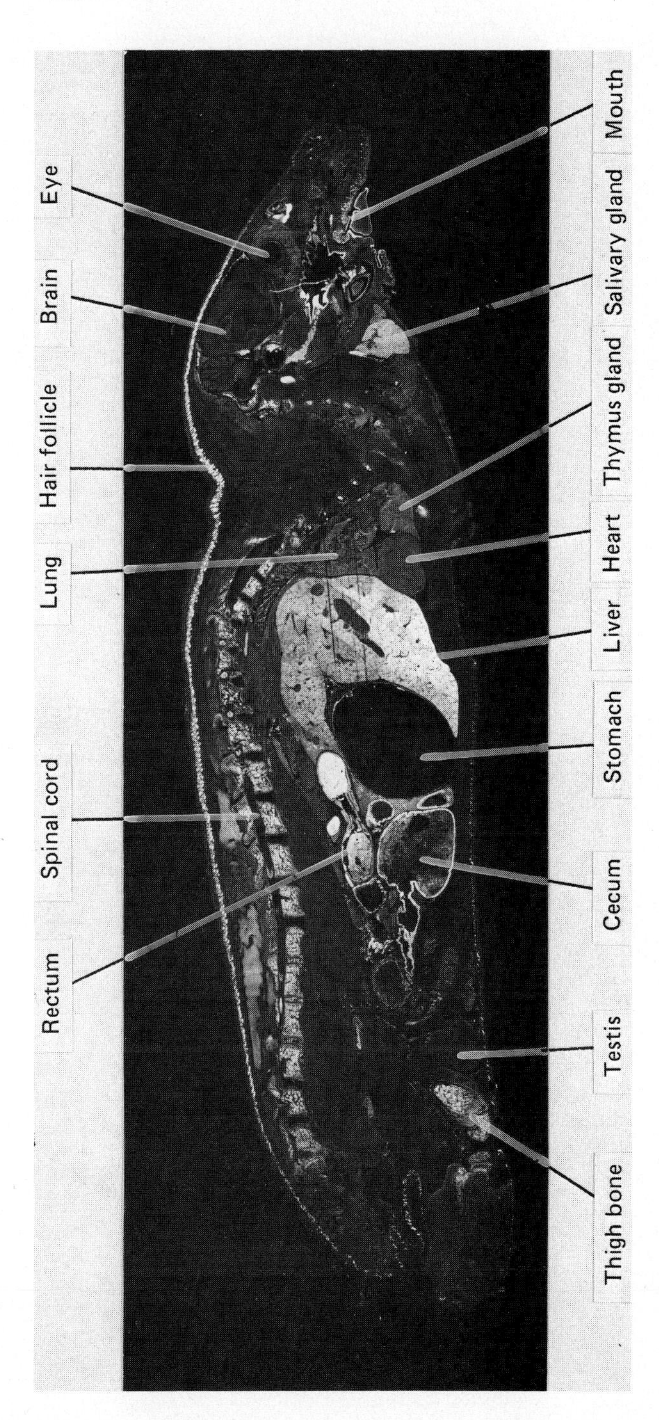

Figure 6 Whole-body autoradiogram of a male rat 6 hr after oral administration of [U-^{14}C]phenylalanine.

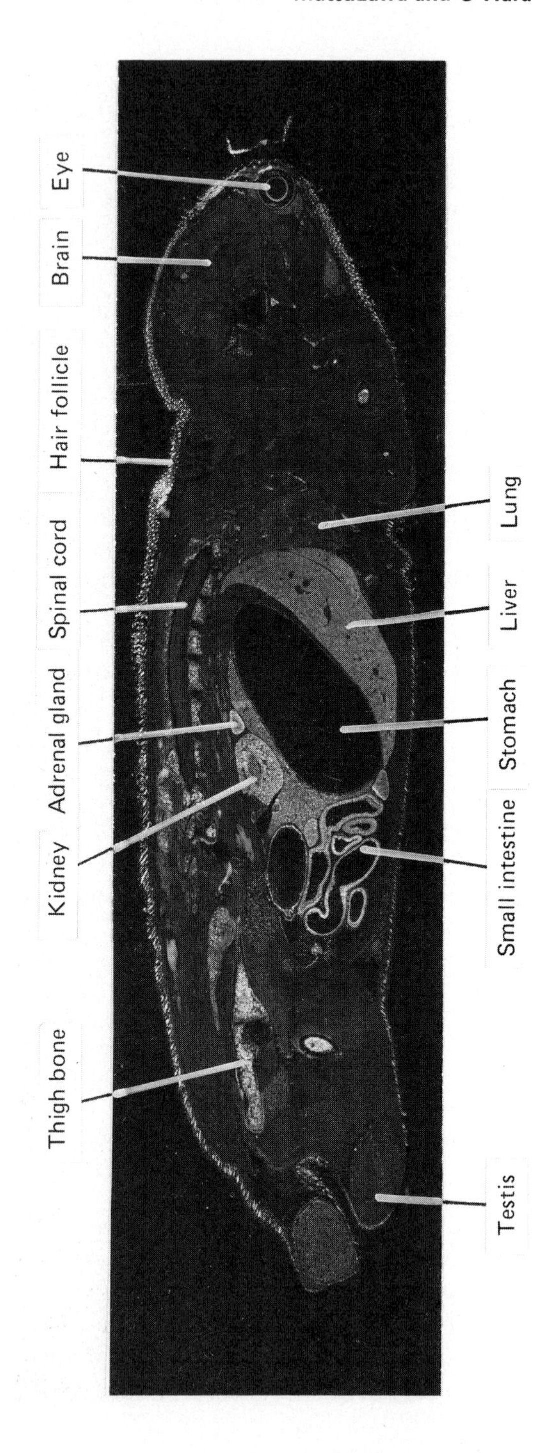
Eye
Brain
Hair follicle
Spinal cord
Adrenal gland
Kidney
Thigh bone
Lung
Liver
Stomach
Small intestine
Testis

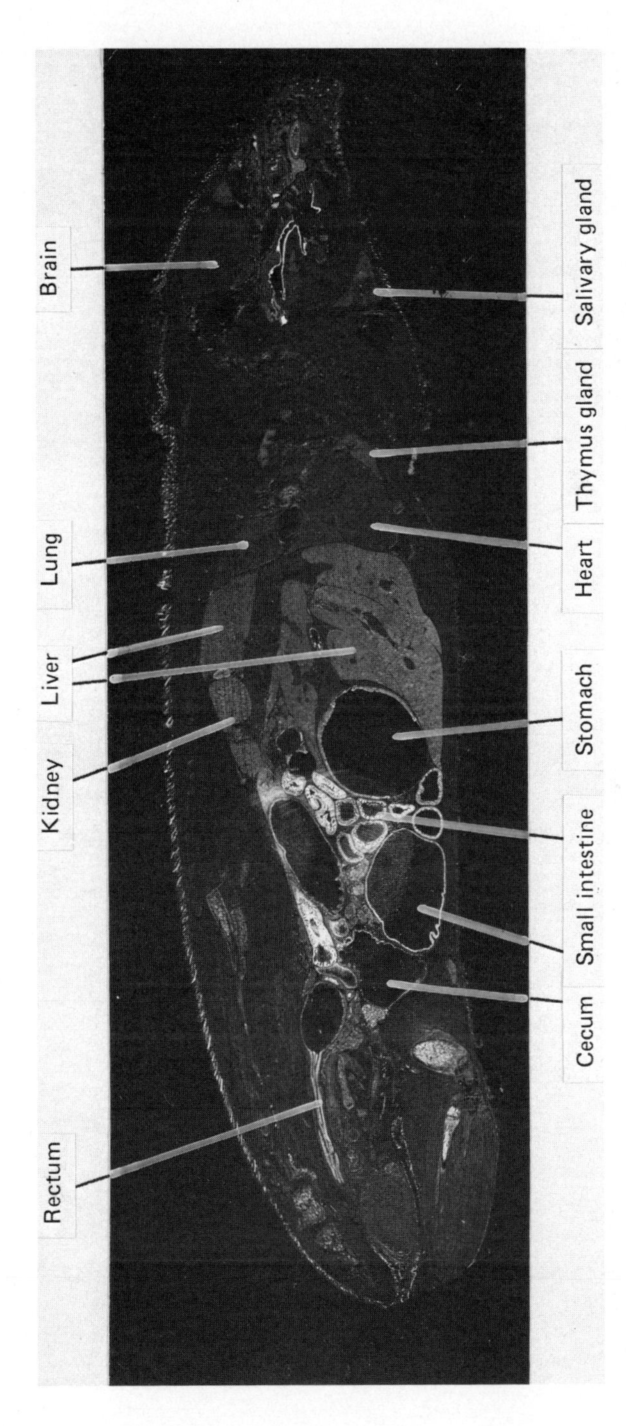

Figure 7 Whole-body autoradiogram of a male rat 24 hr after oral administration of [U-^{14}C-Phe] aspartame.

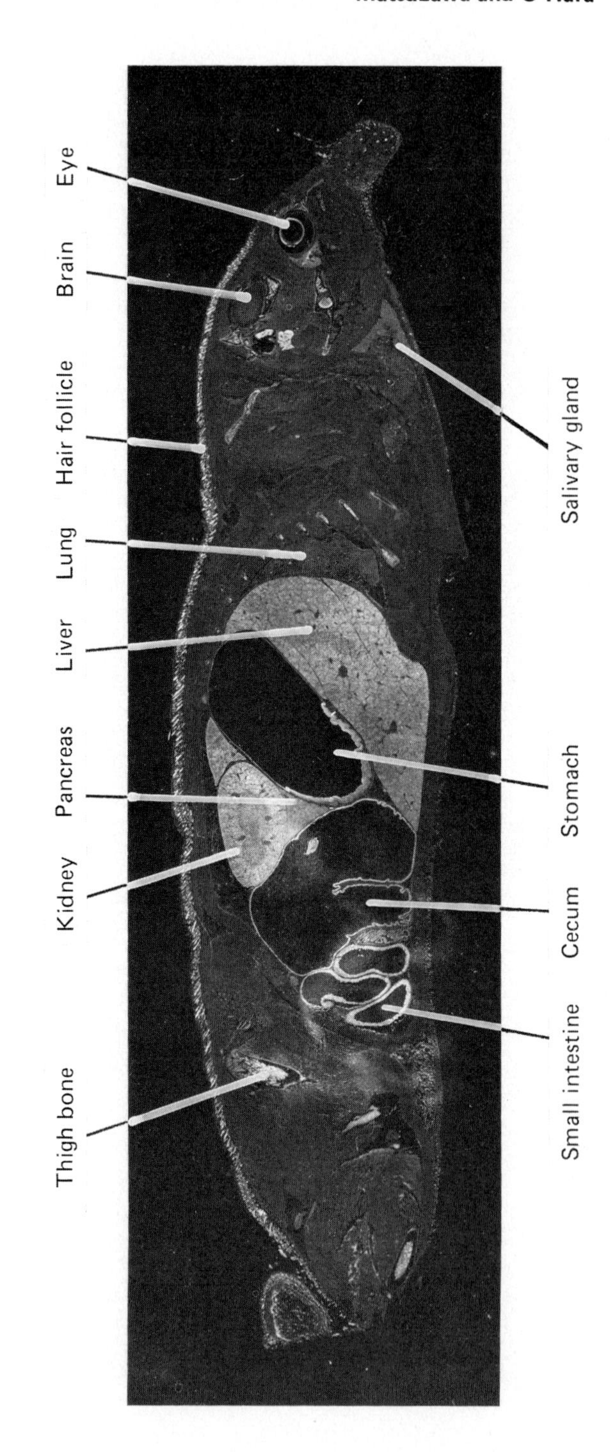
Eye
Brain
Hair follicle
Lung
Liver
Pancreas
Kidney
Thigh bone
Salivary gland
Stomach
Cecum
Small intestine

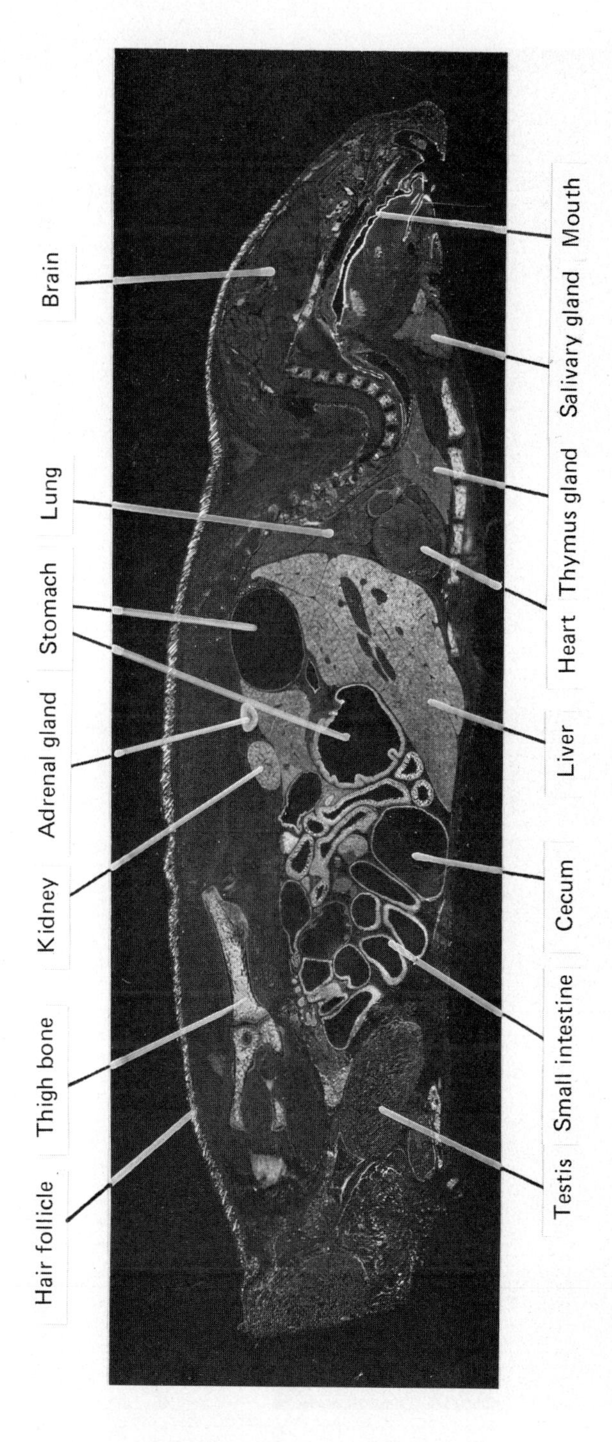

Figure 8 Whole-body autoradiogram of a male rat 24 hr after oral administration of [U-^{14}C] phenylalanine.

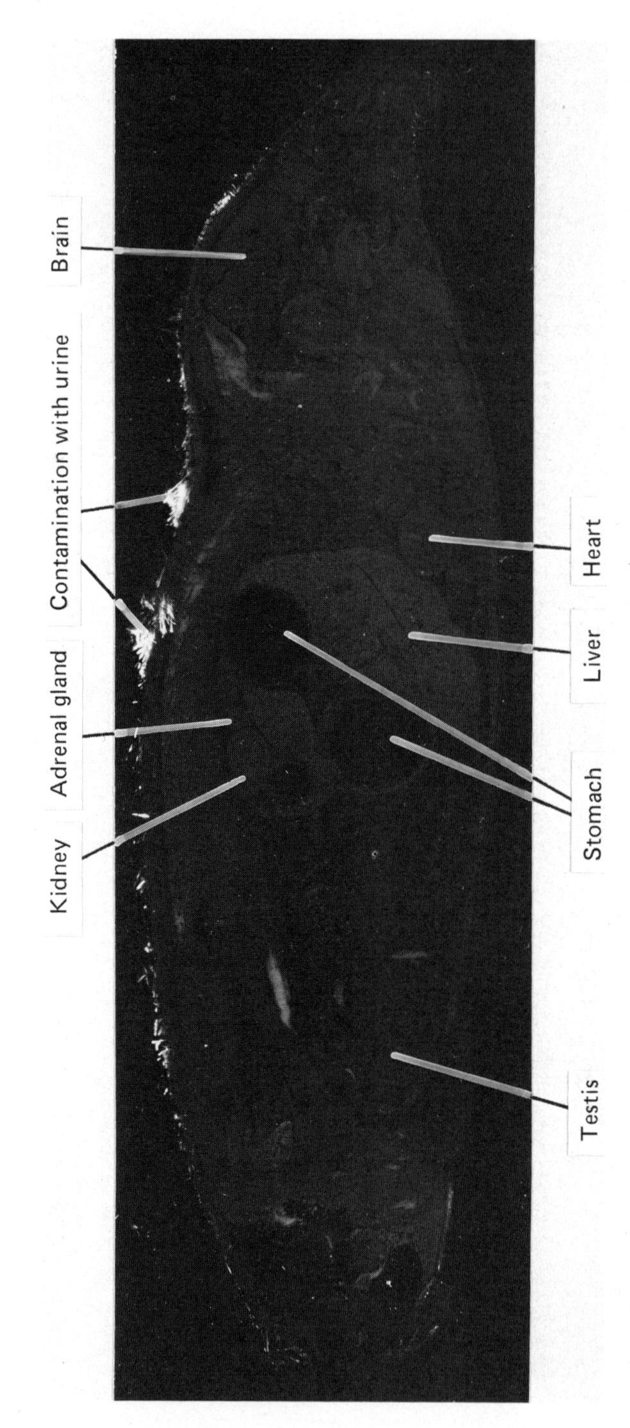

Figure 9 Whole-body autoradiogram of a male rat 7 days after oral administration of [U-^{14}C-Phe] aspartame.

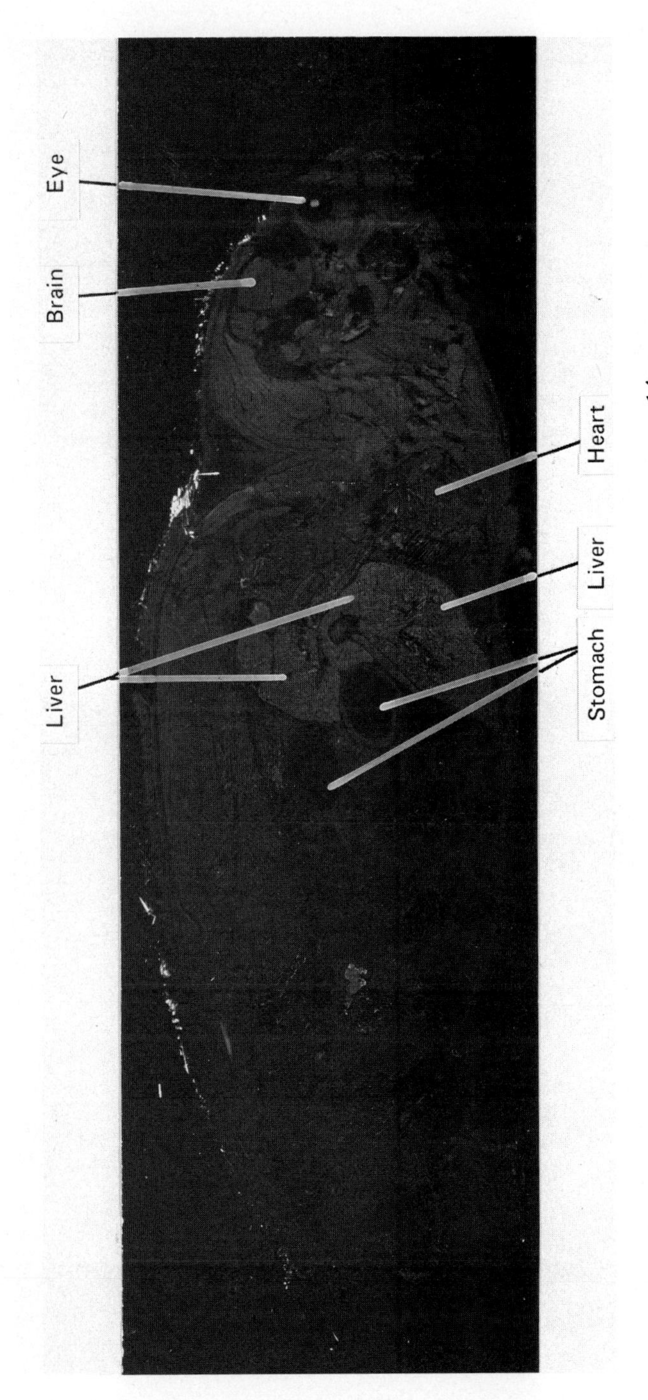

Figure 10 Whole-body autoradiogram of a male rat 7 days after oral administration of [U-^{14}C] phenylalanine.

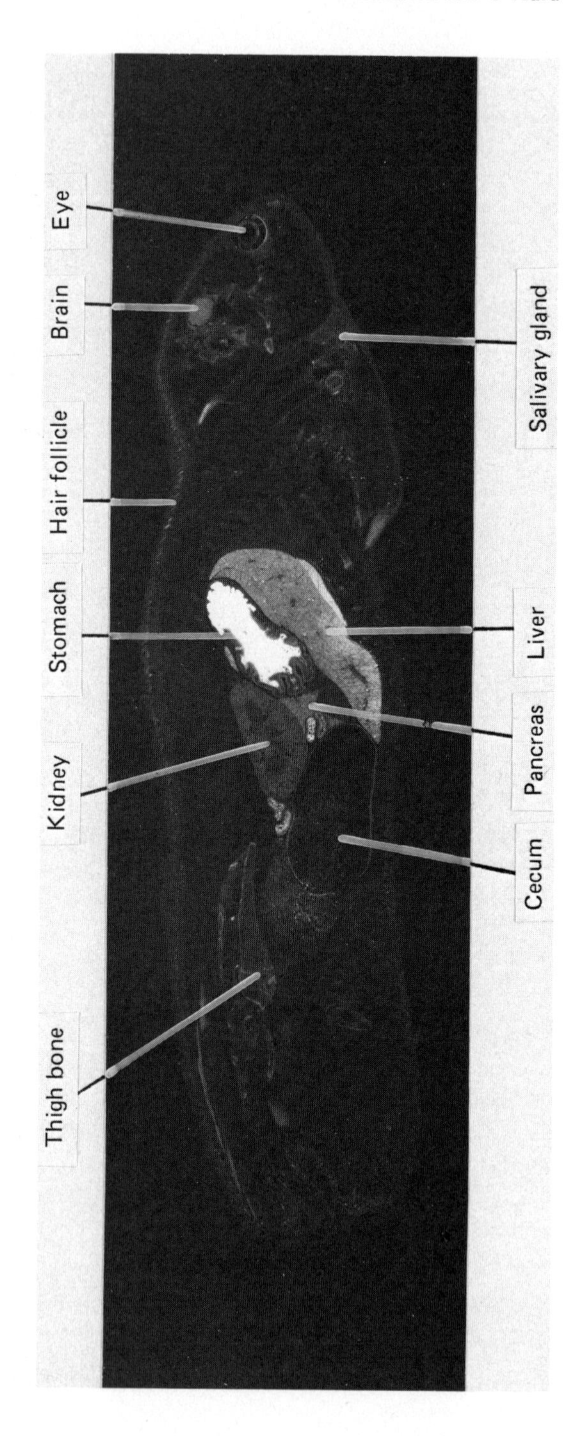
Eye
Brain
Hair follicle
Stomach
Kidney
Thigh bone
Salivary gland
Liver
Pancreas
Cecum

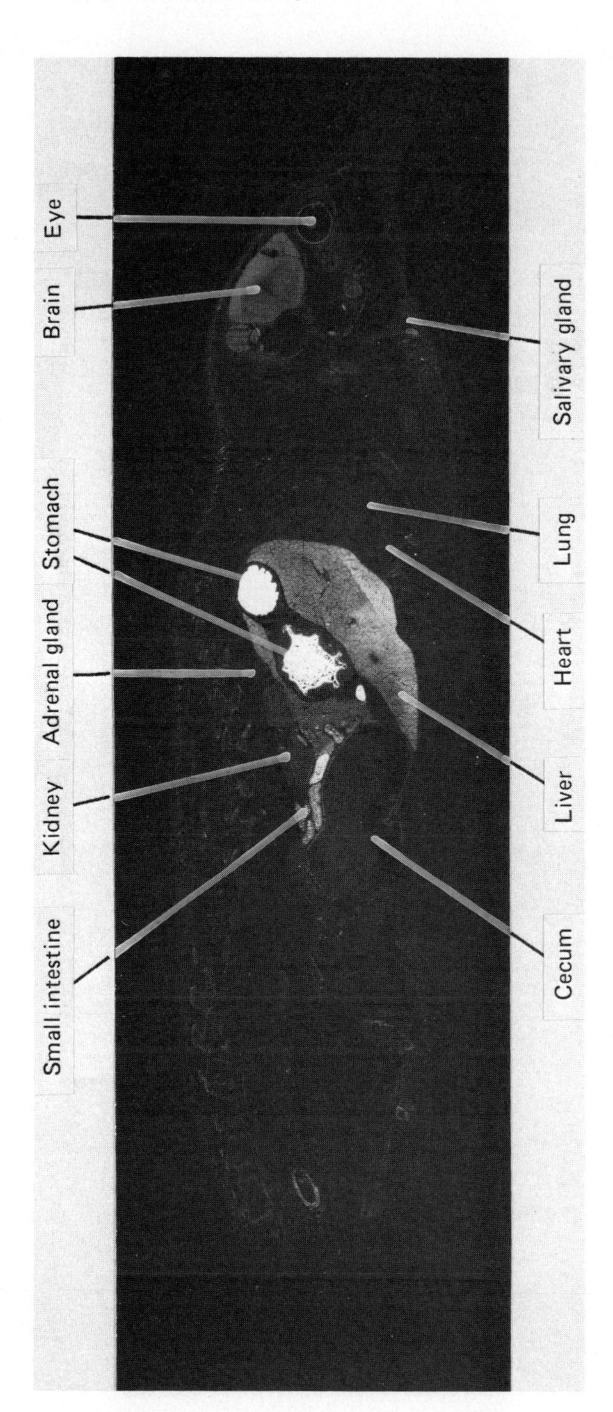

Figure 11 Whole-body autoradiogram of a male rat 30 min after oral administration of [U-^{14}C-Asp] aspartame.

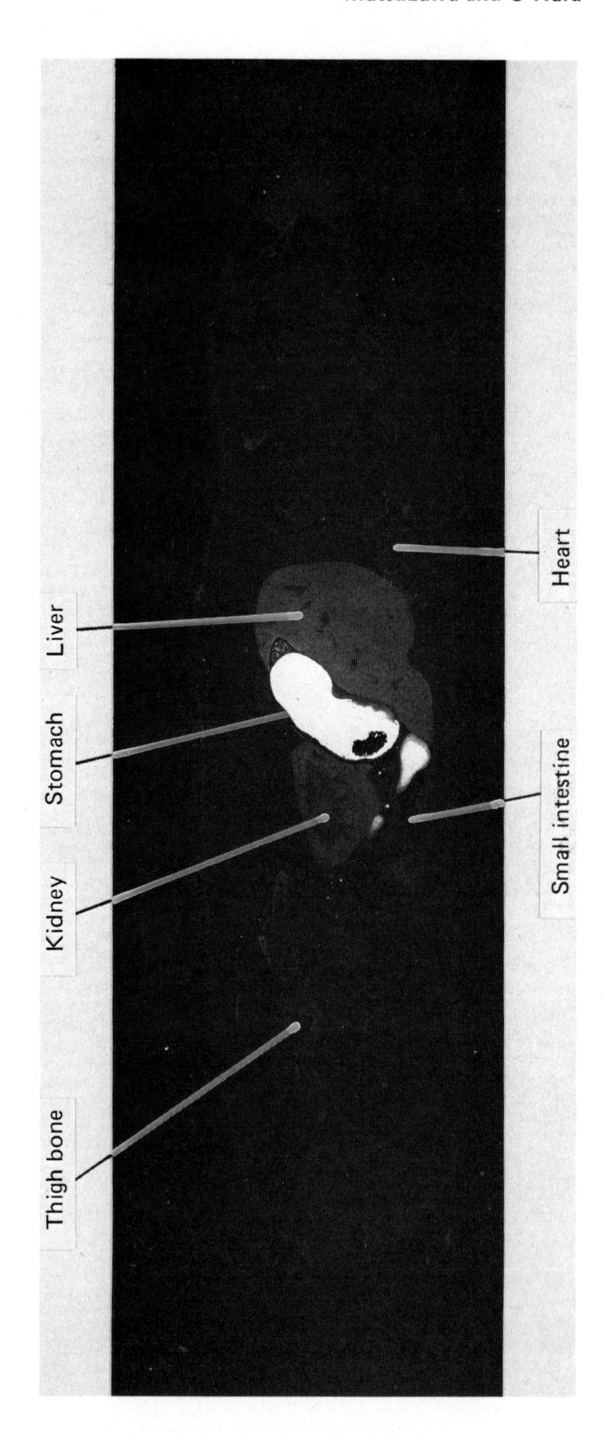
Heart
Liver
Stomach
Small intestine
Kidney
Thigh bone

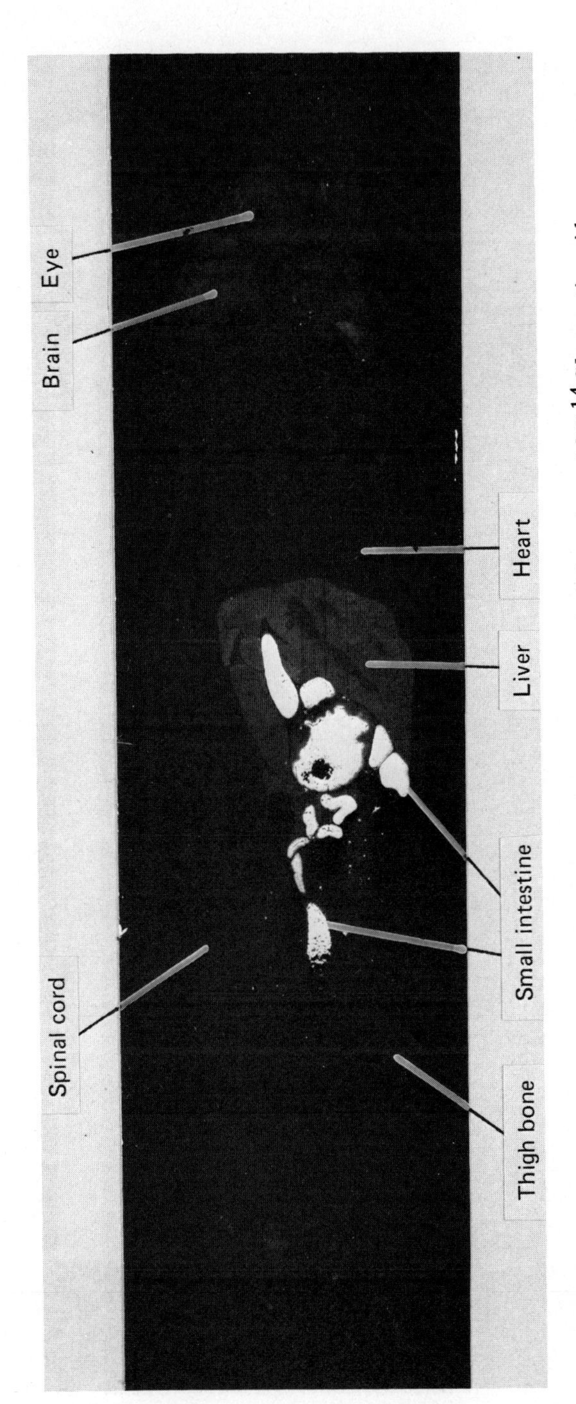

Figure 12 Whole-body autoradiogram of a male rat 30 min after oral administration of [U-^{14}C] aspartic acid.

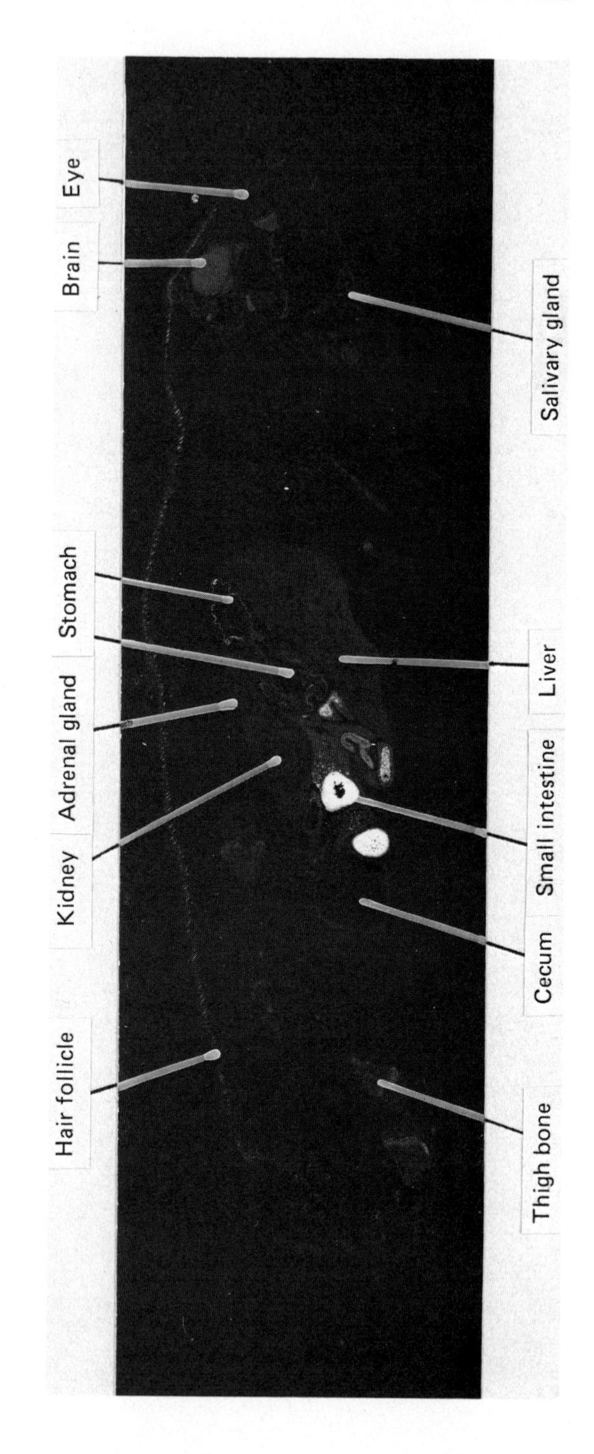
Eye
Brain
Salivary gland
Stomach
Liver
Adrenal gland
Small intestine
Kidney
Cecum
Hair follicle
Thigh bone

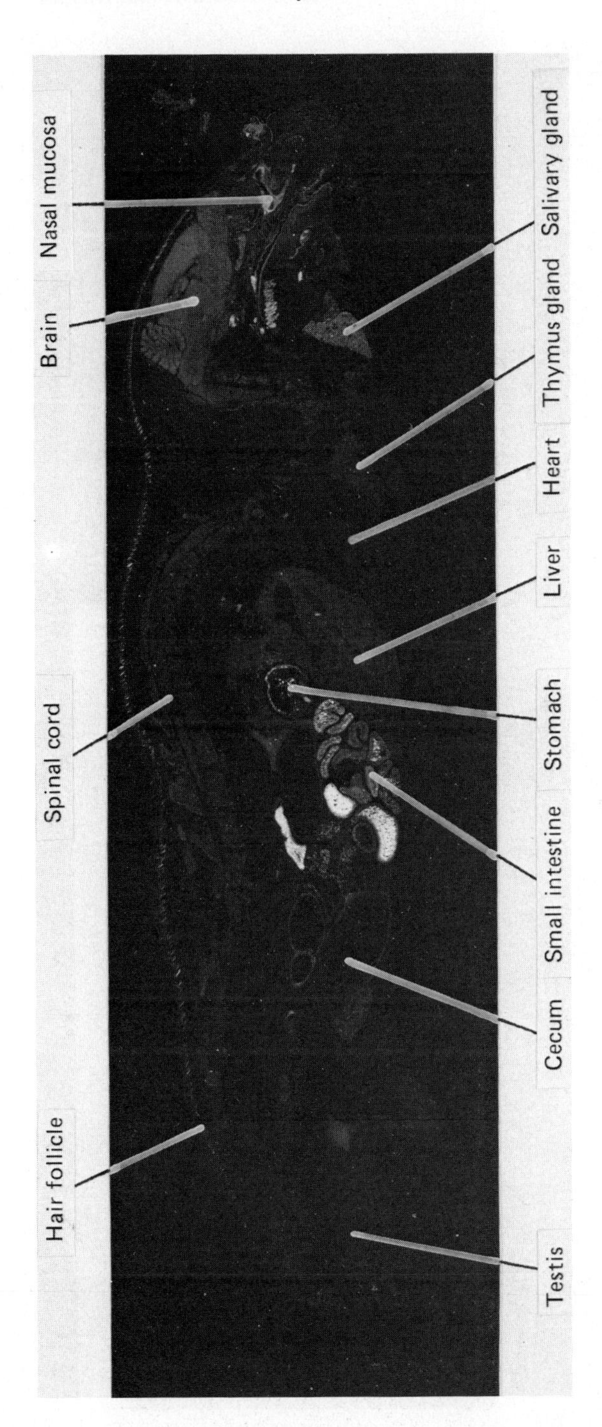

Figure 13 Whole-body autoradiogram of a male rat 2 hr after oral administration of [U-^{14}C-Asp] aspartame.

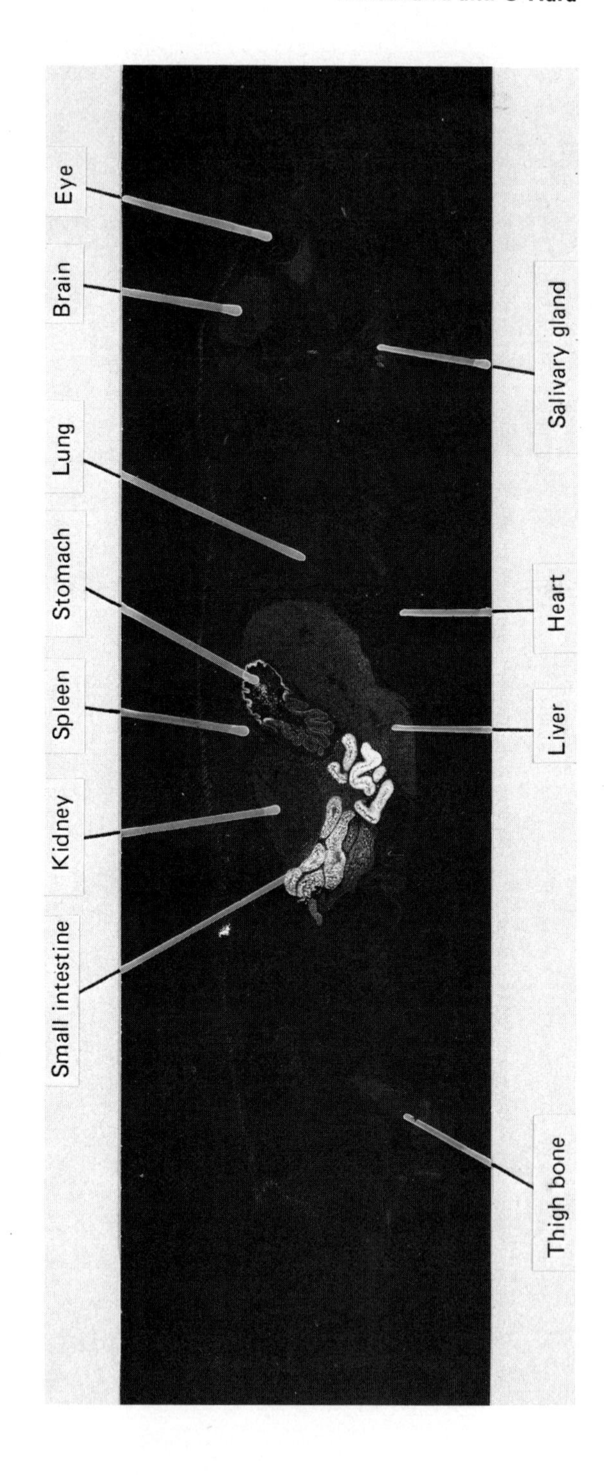
Eye
Brain
Lung
Stomach
Spleen
Kidney
Small intestine
Salivary gland
Heart
Liver
Thigh bone

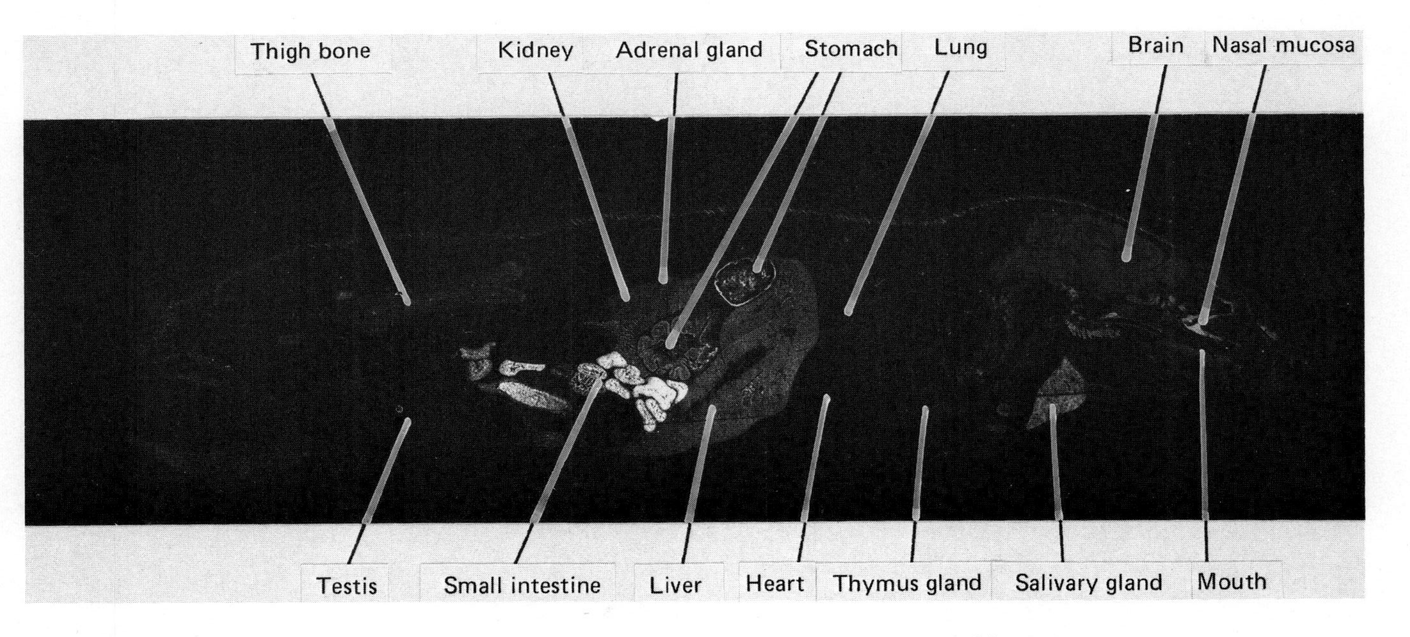

Figure 14 · Whole-body autoradiogram of a male rat 2 hr after oral administration of [U-^{14}C] aspartic acid.

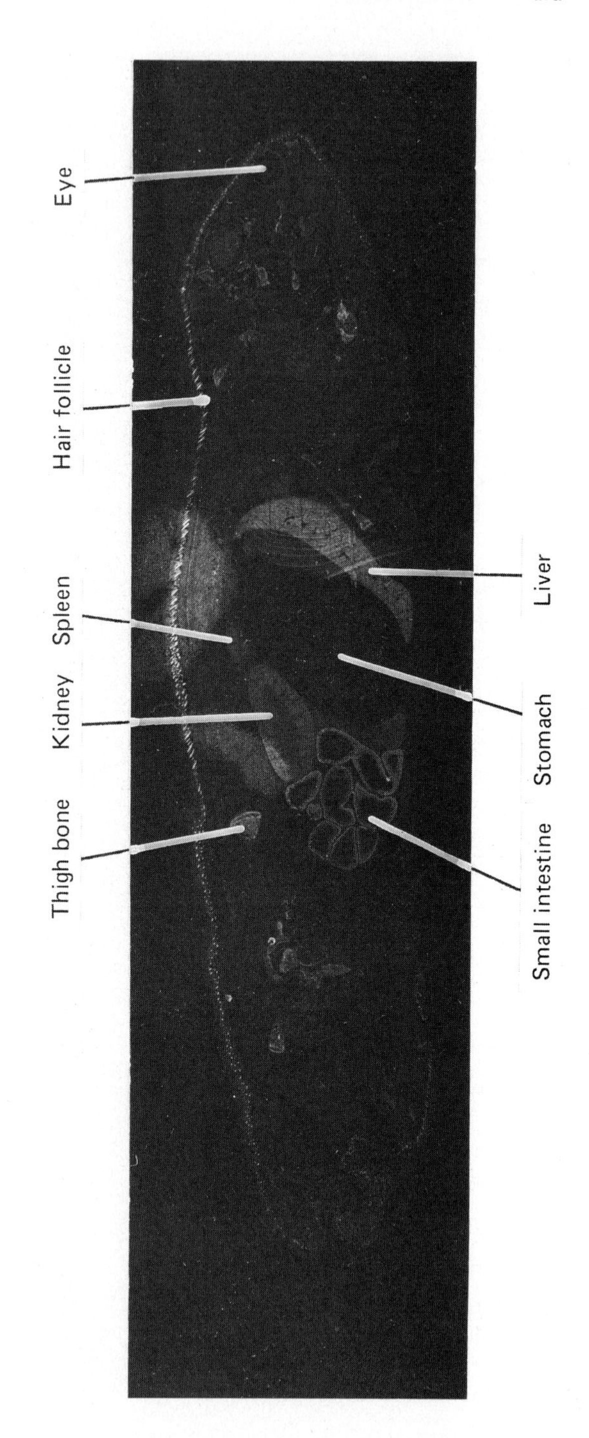
Eye
Hair follicle
Spleen
Kidney
Thigh bone
Liver
Stomach
Small intestine

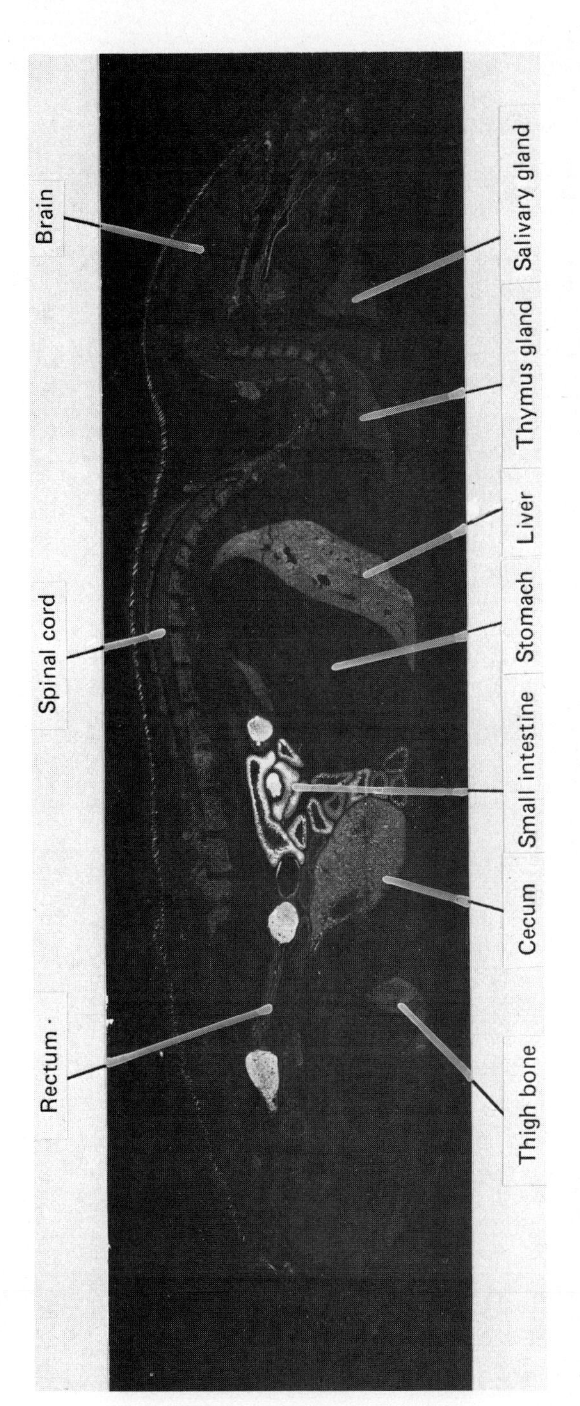

Figure 15 Whole-body autoradiogram of a male rat 6 hr after oral administration of [U-^{14}C-Asp] aspartame.

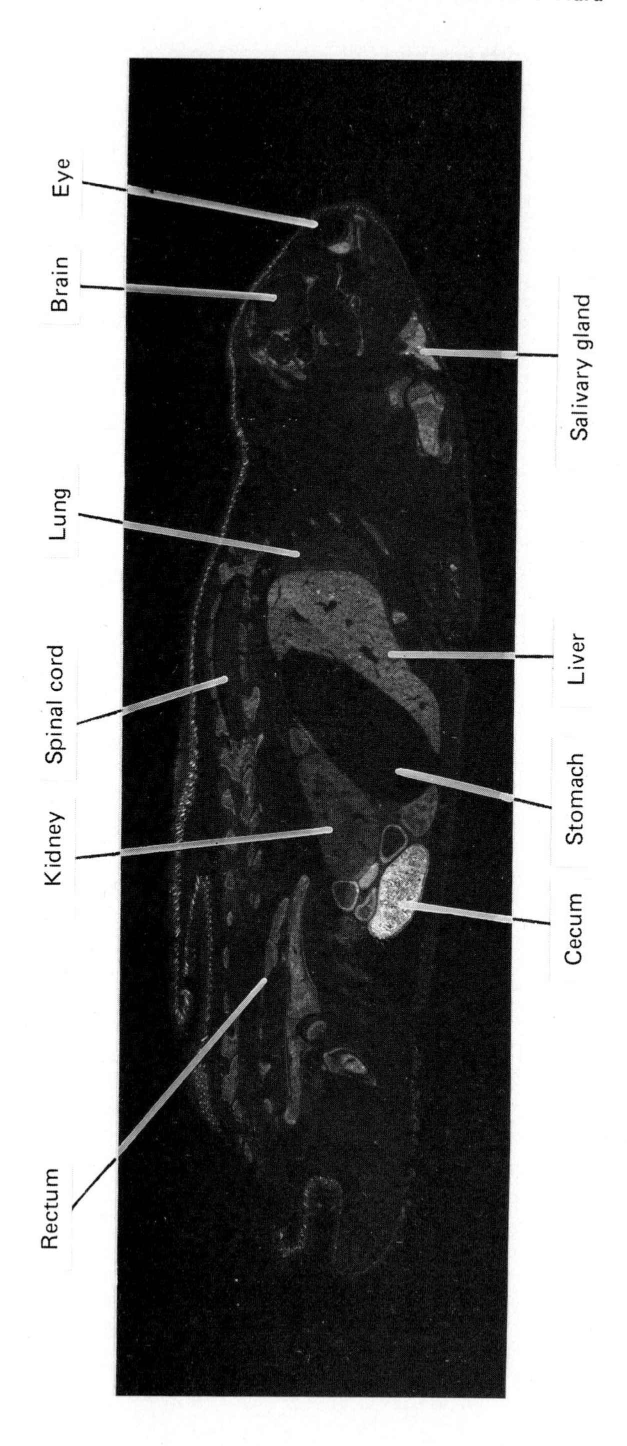
Eye
Brain
Salivary gland
Lung
Liver
Spinal cord
Stomach
Kidney
Cecum
Rectum

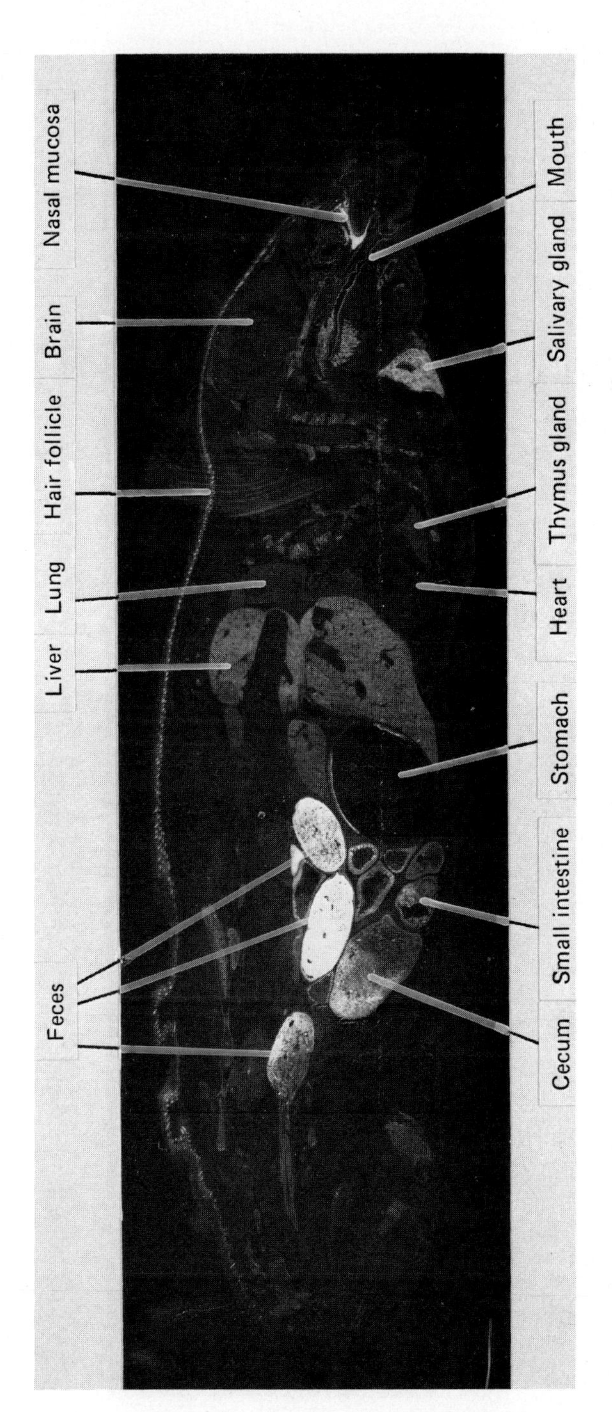

Figure 16 Whole-body autoradiogram of a male rat 6 hr after oral administration of [U-^{14}C] aspartic acid.

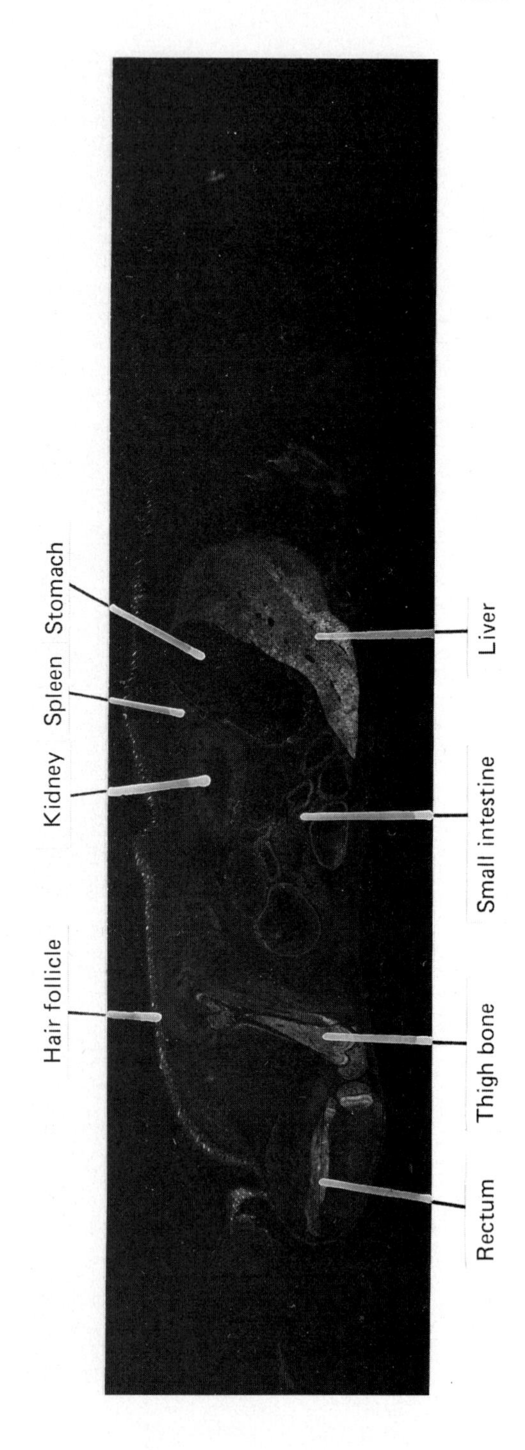
Stomach
Spleen
Kidney
Hair follicle
Liver
Small intestine
Thigh bone
Rectum

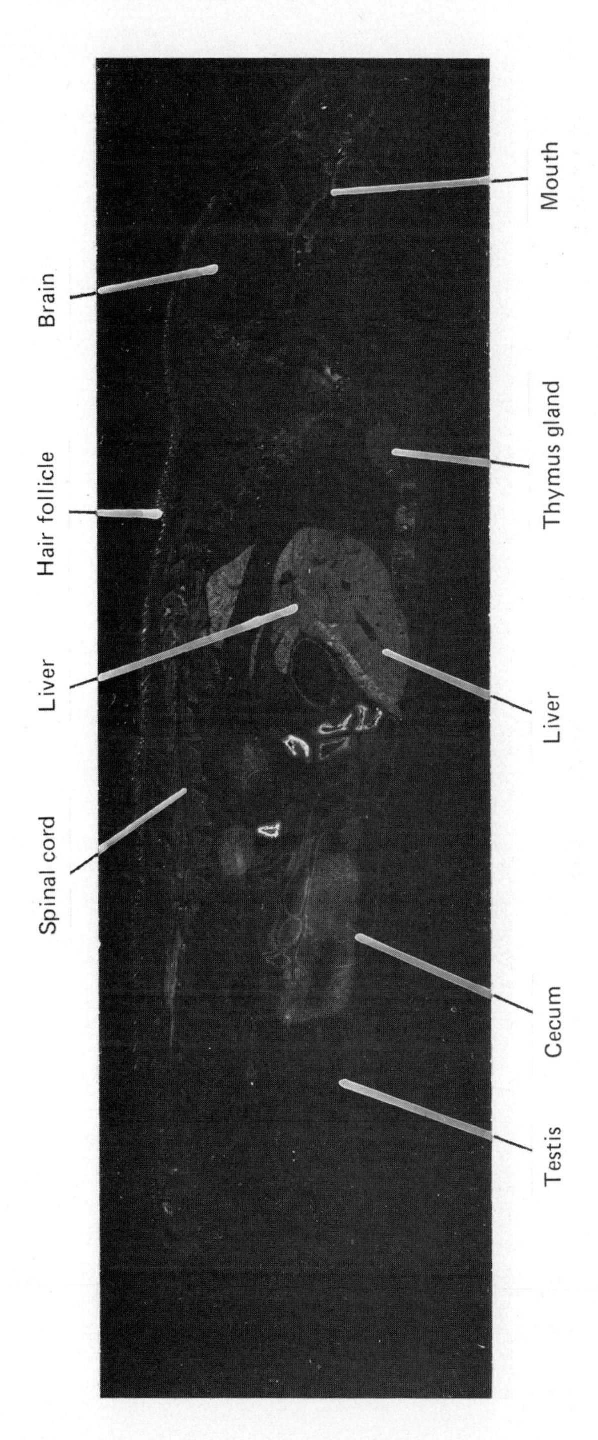

Figure 17 Whole-body autoradiogram of a male rat 24 hr after oral administration of [U-^{14}C-Asp] aspartame.

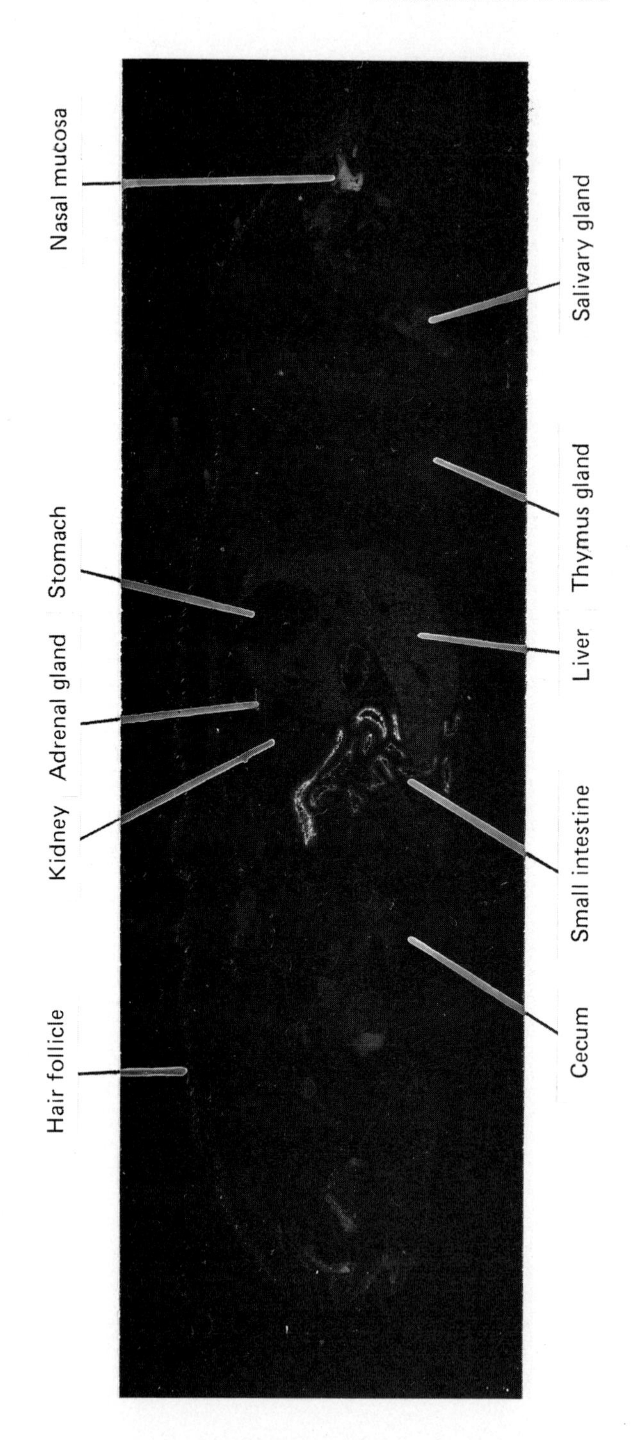
Nasal mucosa
Stomach
Adrenal gland
Kidney
Hair follicle
Salivary gland
Thymus gland
Liver
Small intestine
Cecum

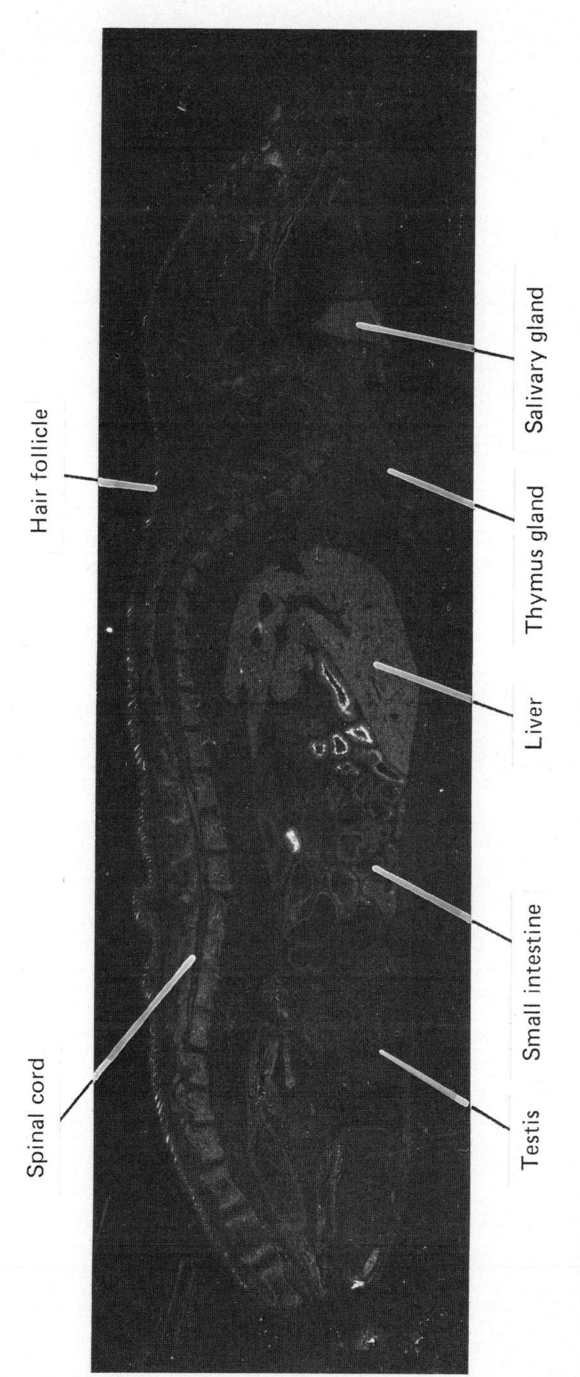

Figure 18 Whole-body autoradiogram of a male rat 24 hr after oral administration of [U-^{14}C] aspartic acid.

SENSORY AND DIETARY ASPECTS

9

Projected Aspartame Intake: Daily Ingestion of Aspartic Acid, Phenylalanine, and Methanol

Roberta Roak-Foltz and **Gilbert A. Leveille**
General Foods Corporation, White Plains, New York

The safety assessment of any food additive requires a knowledge of the pharmacology and toxicology of the additive and information regarding exposure. Population exposure is generally difficult to determine for a new compound and cannot be accurately established before its introduction. For this reason it is important to ensure that estimates of exposure be conservative. Usually this means consciously overestimating rather than underestimating intake exposure.

Elsewhere in this volume there is extensive discussion of the metabolism and toxicology of aspartame and its degradation products phenylalanine, aspartic acid, methanol, and diketopiperazine. These extensive studies demonstrate that high doses of aspartame are well tolerated. However, it is important to estimate the probable range of aspartame intake that might be anticipated.

We have used two approaches to estimate exposure to aspartame or its metabolites. The simplest involved the assumption that aspartame would replace the apparent per capita sugar intake. The per capita caloric sweetener intake was calculated, on the basis of disappearance, to be 156 g/day (1). Using a sweetener ratio of 180:1, this yields a daily estimated aspartame intake of 867 mg/day. Actual intake would be somewhat lower, since it is recognized that disappearance data overestimate consumption and not all of the sweetener applications can be replaced by aspartame.

The second approach used to project aspartame intake involved developing a menu containing generous amounts of added sugars and assuming the substitution of aspartame for the added sweeteners. This menu is shown in Table 1. In Table 2

Table 1 Daily Menu Used to Estimate Potential Aspartame Intake

Meals	Snacks
Breakfast	
6 fl oz breakfast beverage[a]	
1 oz sugar-sweetened cereal[a]	
½ banana	
½ cup milk	
1 slice toast	
1 teaspoon margarine	
1 cup coffee	
3 packets sugar[a]	1 fresh apple
Lunch	
½ cup pea soup	
1 sandwich	
3 slices bologna	
1 oz cheddar cheese	
2 slices bread	
1 teaspoon mustard	
2 lettuce leaves	
1½ oz potato chips	
8 fl oz soft drink[a]	2 sticks chewing gum[a]
½ cup vanilla pudding[a]	8 fl oz soft drink[a]
Dinner	
1 fried chicken leg	
½ cup peas and carrots	
½ cup mashed potatoes	
1 slice bread	
1 teaspoon margarine	
1 cup milk	
½ cup gelatin dessert[a]	
1 peach half	
2 tablespoons whipped topping[a]	
1 cup tea	
2 packets sugar[a]	8 fl oz soft drink[a]

[a]Foods and beverages in which aspartame could substitute for sucrose or corn sweeteners.

are shown the calculated values for the menu containing sweeteners or aspartame. This menu provided 2800 kcal and 260g of total sugars. Of this amount, 190 g was added sugar which conceivably could be substituted by aspartame. Using the 180:1 ratio of aspartame sweetness to sucrose sweetness, total substitution would result in a daily intake 1056 mg of aspartame. However, the sweetness ratio varies

Table 2 Nutrients Provided by Menu Before and After Replacement of Added Sweeteners by Aspartame

	Menu with sucrose	Menu with aspartame	Percent difference
Energy	2800 kcal	2200 kcal	-21
Protein	86 g	88 g	+2
Carbohydrate	396 g	225 g	-43
Total sugars	261 g	71 g	-73
Phenylalanine	4.0 g	4.4 g	+10
Aspartic acid	7.3 g	7.6 g	+4
Methanol	–	75 mg	–

from product to product. This menu provides about 750 mg of aspartame when the level typical for each product application is used for the potential aspartame-containing foods.

Both of these approaches yield similar values for aspartame intake and can be used to estimate the phenylalanine, aspartic acid, and methanol exposures. The metabolism of aspartame yields, on a weight basis, approximately 50% phenylalanine, 40% aspartic acid, and 10% methanol. Using the menu approach and the typical aspartame level for each food, the estimated intake of aspartame would result in 10 and 4% increased intakes of phenylalanine and aspartic acid, respectively, and an added methanol exposure of 75 mg.

On the basis of disappearance data, the estimated potential aspartame intake of 867 mg would translate to an increased daily consumption of 433 mg of phenylalanine, 347 mg of aspartic acid, and 87 mg of methanol. For comparison, the phenylalanine and aspartic acid daily intakes were estimated from data collected as part of the 1977-1978 U.S. Department of Agriculture Nationwide Food Consumption Survey (2). Amino acid levels were calculated for the average amount consumed for each of the 44 food groups reported in the survey. When a group represented several foods with different amino acid levels, an average was used (e.g., corn, oats, rice, wheat for cereal grains) or one form was selected as representative (e.g., chicken broiler or fryer, flesh only, roasted for all chicken).

Household measure equivalents were determined for the foods from the 44 groups using weights and measures from the U.S. Department of Agriculture (3-10). For this purpose the 44 groups were collapsed into 17 categories (Table 3). This approach yielded estimates of 3.6 and 6.8 g for daily phenylalanine and aspartic acid intakes, respectively. Combining these data, replacement of all sweeteners with aspartame would increase phenylalanine intake by 12% and aspartic acid intake by 5% and would add 87 mg of methanol to the diet.

It is clear from these estimates that aspartame is not likely to alter amino acid intake appreciably. Similarly, the added methanol burden is insignificant. Methanol, which is formed by enzymatic splitting of pectic substances, is a component

Table 3 Average Intake per Individual in a Day

Amount	Food
6.7 oz	Meat, poultry, or fish
1½ cups	Milk
½ oz	Cheese
½	Egg
1 oz	Legumes, nuts, or seeds
Equivalent of 4 slices	Bread (includes other baked goods)
½ oz	Ready-to-eat cereal
½ cup	Pasta or other grain mixtures
½	Potato
1 cup	Vegetables
½ cup	Fruit or fruit juice
2 teaspoons	Table fat or salad dressing
1 cup	Soft drinks or fruit drinks
¼ cup	Beer or ale
Equivalent of 2 tablespoons	Sugar, candy, or other sweets
1½ cups	Coffee (6 fl oz cup)
2/3 cup	Tea (6 fl oz cup)

Source: Adapted from *The USDA Nationwide Food Consumption Survey 1977-78, Preliminary Report* No. 2, Food and nutrient intakes of individuals in 1 day in the United States, Spring 1977, Tables 1.1a, 1.2a, 1.3a, 1.4a, and 1.5a.

of many fruits, vegetables, and wines. The amount of methanol contributed by these foods in the course of a day would likely exceed any contribution from aspartame (11-17).

It should be emphasized that these estimates are by design high. Actual intakes of aspartame will certainly be less, probably closer to 50% of the values we have estimated.

REFERENCES

1. Sugar and sweetener outlook and situation (1982). *USDA Economic Research Service SSR V7N4*, p. 11. U.S. Department of Agriculture, Washington, D.C.
2. Food and nutrient intakes of individuals in 1 day in the United States, Spring 1977 (1980). *USDA Nationwide Food Consumption Survey 1977-78 Preliminary Report* No. 2, pp. 45-59. U.S. Department of Agriculture, Washington, D.C.
3. Posati, L. P., and Orr, M. L. (1976). Composition of foods: Dairy and egg products. *USDA Agriculture Handbook* No. 8-1, U.S. Department of Agriculture, Washington, D.C.
4. Reeves, J. B., III, and Weihrauch, J. L. (1979). Composition of foods: Fats and oils. *USDA Agriculture Handbook* No. 8-4, U.S. Department of Agriculture, Washington, D.C.

5. Posati, L. P. (1979). Composition of foods: Poultry products. *USDA Agriculture Handbook* No. 8-5, U.S. Department of Agriculture, Washington, D.C.
6. Richardson, M., Posati, L. P., and Anderson, B. A. (1980). Composition of foods: Sausages and luncheon meats. *USDA Agriculture Handbook* No. 8-7, U.S. Department of Agriculture, Washington, D.C.
7. Douglass, J. S., Matthews, R. H., and Hepburn, F. N. (1982). Composition of foods: Breakfast cereals. *USDA Agriculture Handbook* No. 8-8, U.S. Department of Agriculture, Washington, D.C.
8. Gebhardt, S. E., Cutrufelli, R., and Matthews, R. H. (1982). Composition of foods: Fruits and fruit juices. *USDA Agriculture Handbook* No. 8-9, U.S. Department of Agriculture, Washington, D.C.
9. Cutrufelli, R., and Matthews, R. H. (1981). *Provisional Table on the Nutrient Content of Beverages, USDA*, U.S. Department of Agriculture, Washington, D.C.
10. Adams, C. F. (1975). Nutritive value of American foods in common units. *USDA Agriculture Handbook* No. 456, U.S. Department of Agriculture, Washington, D.C.
11. Lund, E. D., Kirkland, C. L., and Shaw, P. E. (1981). Methanol, ethanol, and acetaldehyde contents of citrus products. *J. Agric. Food Chem.* 29, 361-366.
12. Kazeniac, S. J., and Hall, R. M. (1970). Flavor chemistry of tomato volatiles. *J. Food Sci.* 35, 519-530.
13. Dyer, R. H. (1971). Comparison of GLC and colorimetric methods for determination of methanol in alcoholic beverages. *J. Assoc. Offic. Anal. Chem.* 54, 785-786.
14. Venturella, V. S., Graves, D., and Lang, R. E. (1974). Automated proof determination of liquors by gas-solid chromatography. *J. Assoc. Offic. Anal. Chem.* 57, 118-123.
15. Lee, C. Y., Acree, T. E., and Butts, R. M. (1975). Determination of methyl alcohol in wine by gas chromatography. *Anal. Chem.* 47, 747-748.
16. Kirchner, J. G., and Miller, J. M. (1957). Volatile water-soluble and oil constituents of Valencia orange juice. *J. Agric. Food Chem.* 5, 283-291.
17. Heatherbell, D. A., Wrolstad, R. E., and Libbey, L. M. (1971). Carrot volatiles: Characterization and effects of canning and freeze drying. *J. Food Sci.* 36, 219-224.

10

Comparison of Taste Properties of Aspartame with Other Sweeteners

Susan S. Schiffman
Duke University, Durham, North Carolina

INTRODUCTION

The purpose of this chapter is to provide an overview of the taste properties of aspartame and to compare them with other sweeteners. The potency of aspartame relative to sucrose depends on the delivery system and concentrations used. In general, however, its relative potency at moderate sweetness intensity is 150-200 times that of sucrose (1,2). A series of experiments are described that compare not only the intensive but also the qualitative and receptor properties of aspartame with a group of other sweeteners given in Table 1. The chemical structures of these sweeteners are shown in Figure 1. The chapter is organized around a series of questions that will provide a context for the data presented.

It is not possible at present to predict the taste properties of aspartame or any other sweetener from chemical structure. This is due to the fact that the physicochemical properties of molecules that lead to the perception of the sweet taste are not well understood. The most plausible theory to date that explains sweet taste is that of Shallenberger and Acree (3), who postulated that a necessary condition for sweetness is a pair of hydrogen bonds 3 Å apart. The stimulus molecule, according to this approach, possesses an AH-B configuration, where A and B are electronegative atoms and H is a hydrogen atom that is part of a polarized system A-H. The AH-B group in the stimulus molecule is presumed to interact with a complementary AH-B site in the taste receptor membrane to form two simultaneous hydrogen bonds. It has been suggested that a third lipophilic site may be

Table 1

Compound	Classification	Experiment			
		1	2	3	4
Acetosulfam (Acesulfame K)	Oxathiazinone dioxide (methyl derivative); 3,4-dihydro-6-methyl-1,2,3-oxathiazin-4-one-2,2-dioxide potassium salt	X	X	X	X
Aspartame	Dipeptide: L-aspartyl-L-phenylalanine methyl ester	X	X	X	X
Calcium cyclamate	Calcium cyclohexylsulfamate	X	X	X	X
Fructose	Monosaccharide ketohexose	X		X	X
Glucose	Monosaccharide aldohexose	X	X		
Maltose	Disaccharide	X			
Monellin	Protein (mol. wt. 10,700)	X		X	
Neohesperidin Dihydrochalcone	Dihydrochalcone glycoside	X	X	X	X
Rebaudioside A	Diterpene glycoside	X		X	X
Saccharin (sodium salt)	O-Sulfobenzimide: 1,2-benzothiazol-3(2H)-one-1,1-dioxide, Na^+ salt	X	X	X	X
Sorbose	Monosaccharide ketohexose	X			
Sorbitol	Polyhydric alcohol	X			
Stevioside	Diterpene glycoside	X		X	X
Thaumatin	Several distinct proteins (mol. wt. 18,000-21,000)	X		X	X
D-Tryptophan	D-Amino acid	X	X	X	X
Xylitol	Polyhydric alcohol	X	X		
Xylose	Monosaccharide aldopentose	X			

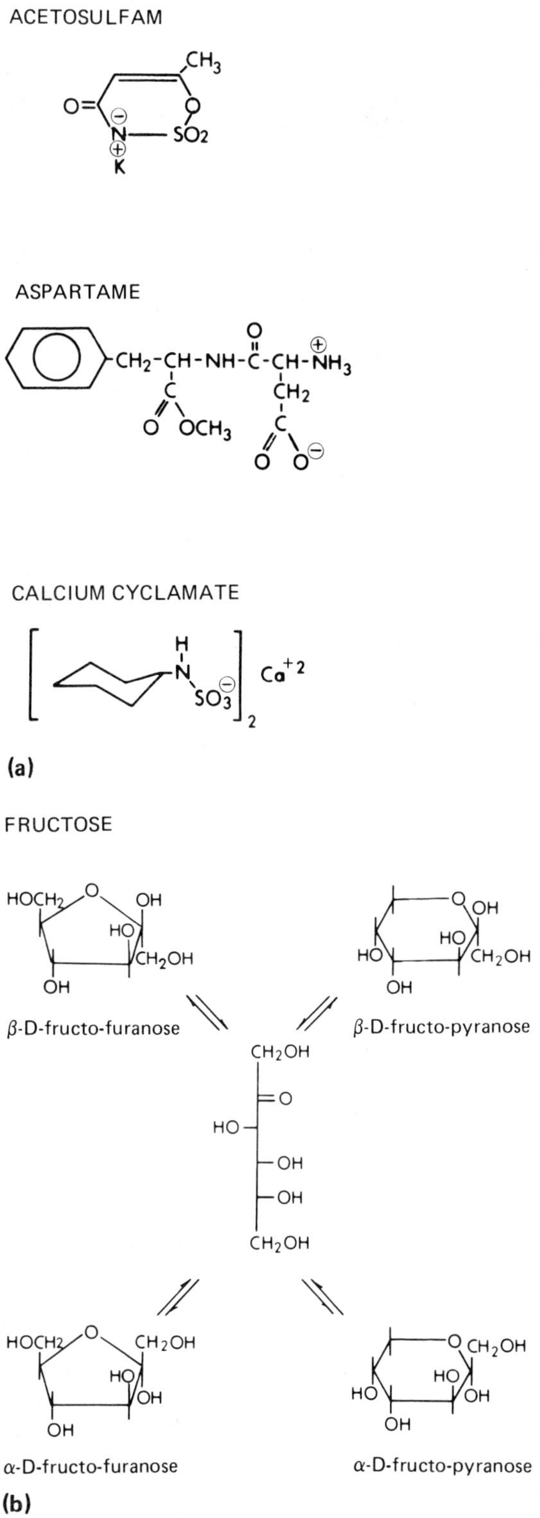

ACETOSULFAM

ASPARTAME

CALCIUM CYCLAMATE

(a)

FRUCTOSE

β-D-fructo-furanose

β-D-fructo-pyranose

α-D-fructo-furanose

α-D-fructo-pyranose

(b)

Figure 1 Chemical structures of the sweeteners employed. (From Ref. 7.)

GLUCOSE

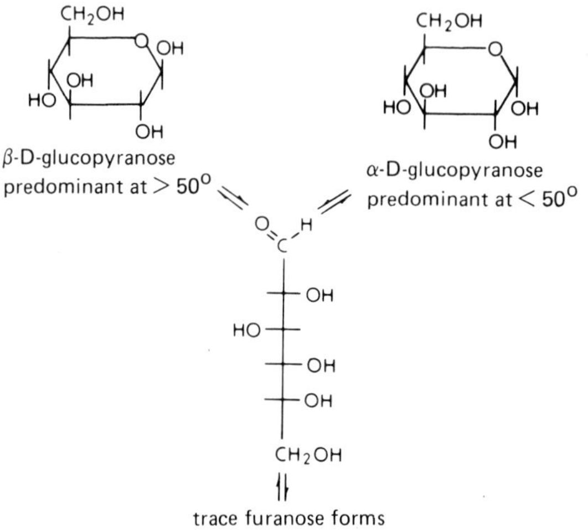

CH_2OH OH OH HO OH

β-D-glucopyranose
predominant at > 50°

CH_2OH OH HO OH OH

α-D-glucopyranose
predominant at < 50°

O H C OH HO OH OH CH_2OH

trace furanose forms

(c)

MALTOSE

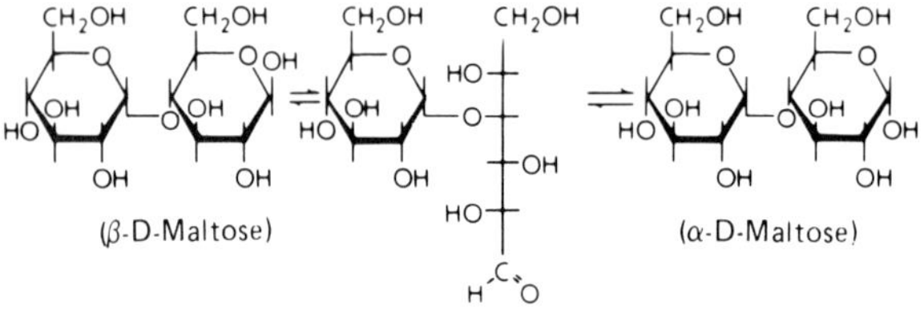

CH_2OH CH_2OH CH_2OH CH_2OH CH_2OH CH_2OH

(β-D-Maltose) (α-D-Maltose)

MONELLIN - PROTEIN

β- NEOHESPERIDIN DIHYDROCHALCONE

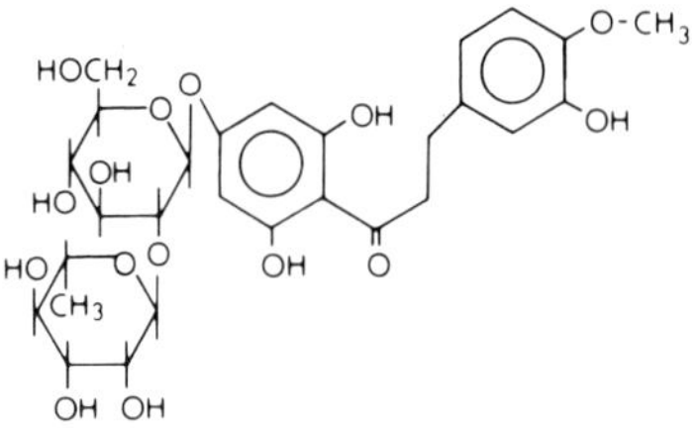

$HOCH_2$ O OH HO O HO CH_3 OH OH OH OH O O-CH_3 OH

(d)

Figure 1

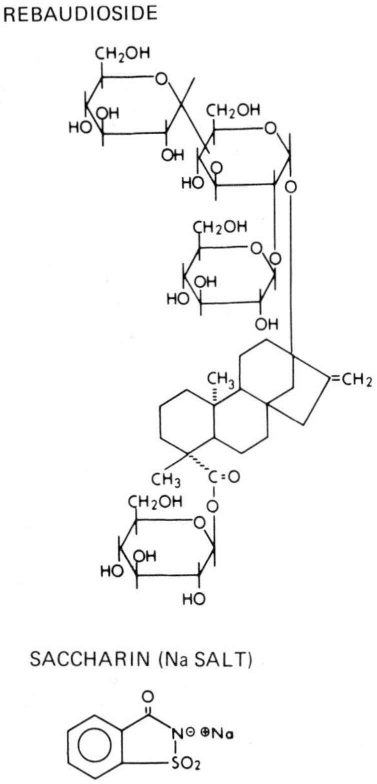

REBAUDIOSIDE

SORBITOL

SACCHARIN (Na SALT)

(e)

SORBOSE

cyclic hemiketal

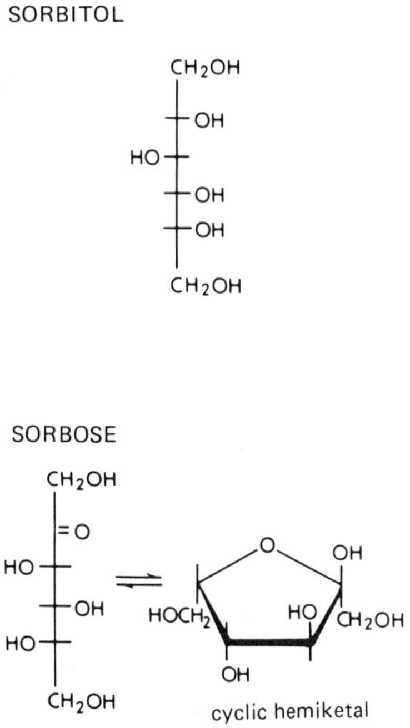

(f)

necessary for intense sweetness (4,5). Although this theory has been helpful in explaining sweetness after the fact, it has been distinctly unproductive in predicting a prioi these compounds that will taste sweet or how one sweetener differs from another.

HOW DOES ASPARTAME COMPARE IN QUALITY WITH OTHER SWEETENERS?

Traditional theories of taste have presumed that there are four so-called "primary" tastes: sweet, sour, salty, and bitter; all other taste sensations were held to be combinations of these four. Recent evidence, however, suggests that the qualitative range of taste is not only broader than originally assumed (6), but that sweetness itself may not be a simple unitary quality. In the experiment described here (from Ref. 7), aspartame is positioned in a spatial map along with 16 other sweeteners listed in experiment 1 in Table 1 by a mathematical technique called

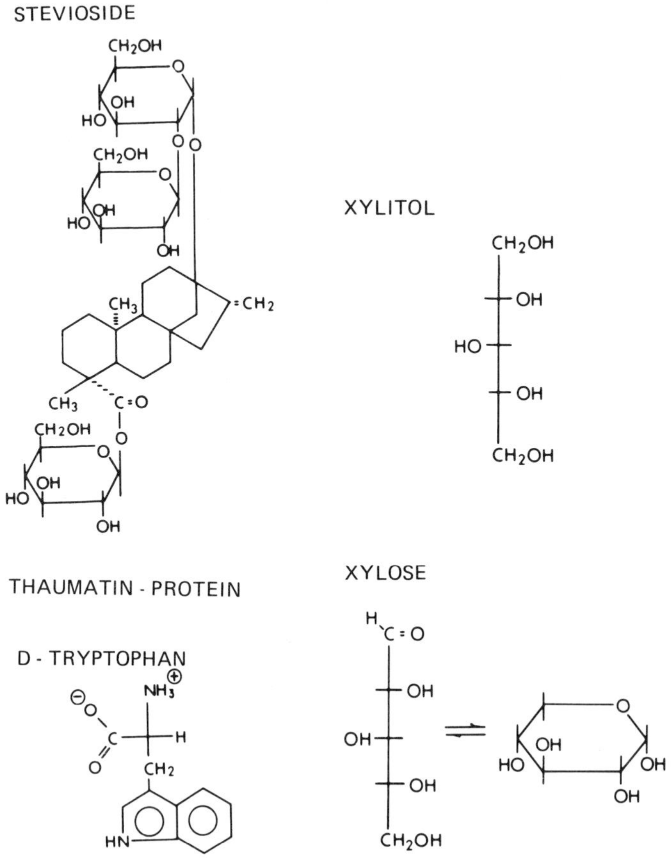

STEVIOSIDE

XYLITOL

THAUMATIN - PROTEIN

XYLOSE

D - TRYPTOPHAN

(g)

(h)

Figure 1 (continued)

multidimensional scaling (8). Sweeteners that were judged experimentally similar to one another are located near one another in the spatial arrangement. Sweeteners judged to be dissimilar are positioned distant from one another. The strength of the multidimensional scaling approach for studying sweeteners is that it permits an ordering of chemicals on the basis of their similarity in taste in spite of the fact that there are no well-defined verbal descriptors or well-understood physicochemical parameters by which to arrange them. All that is required for input to multidimensional scaling procedures is measures of similarity among all or most pairs of the stimuli tested.

The experiment was divided into three parts: (a) intensity matching, (b) similarity judgments, and (c) adjective ratings. In the intensity sessions subjects judged the overall intensity of a range of dilutions for each sweetener and recorded the total perceived intensity along a 5-in. line as shown here:

Weak————————————————————Strong

Intensity measures were transcribed such that 0 corresponded to the "weak" end of the scale and 100 corresponded to a rating of "strong." The concentrations of sweeteners for which the geometric mean intensity ratings over subjects fell between 55 and 65 were considered to be approximately equal in intensity.

Similarly judgments were obtained by taking 10 ml each of two sweeteners in succession into the mouth and rating the similarity of taste along another 5-in. line:

Same————————————————————Different

Similarity ratings were also coded from 0 to 100 in a manner analogous to the intensity judgments. Each stimulus was compared with every other, resulting in a total of 136 similarity judgments for each subject. In order to minimize adaptation effects, at least 5 min were allowed between presentations of pairs. Twenty 1-hr sessions were required to complete testing. After all the similarity judgments were obtained, the sweeteners were rated on a series of adjective scales previously used in our taste experiments (9-14). Four additional adjectives not previously employed were also included: "syrupy," "taste fades fast," "delayed sweetness," and "aftertaste." Like intensity and similarity judgments, the adjectives (such as "good" and "bad") defined the ends of a 5-in. line.

The similarity data were analyzed by an individual differences multidimensional scaling procedure called INDSCAL (15). This approach not only produces a multidimensional space that represents the perceived similarity of the sweeteners in a spatial configuration, but also provides weights for each subject on each of the dimensions of the multidimensional space, revealing individual differences in perception.

The three-dimensional solution common for all 12 subjects is shown in Figure 2. Three dimensions were considered adequate, because higher-dimensional solutions revealed no further new relationships among the stimuli. It can be seen that aspartame falls closer to the sugars than the other artificial sweeteners tested.

Cross sections through the space in Figure 2 are shown in Figure 3 in order to further illustrate relationships among the stimuli. Dimension I plotted versus dimension II in cross section represents the floor of the model in Figure 2. Fructose, sorbose, and glucose tend to group together along with xylose and xylitol as seen in both Figure 2 and the upper left-hand quadrant of Figure 3a. Aspartame, sorbitol, and maltose are located close by. Separate from the fructose area are monellin, thaumatin, and neohesperidin dihydrochalcone, which are found in the lower left quadrant of Figure 3a. Acetosulfam, rebaudioside, stevioside, and sodium saccharin are located near one another, with D-tryptophan a little further away. Calcium cyclamate is located between the sugars and the positions occupied by sodium saccharin, rebaudioside, and acetosulfam.

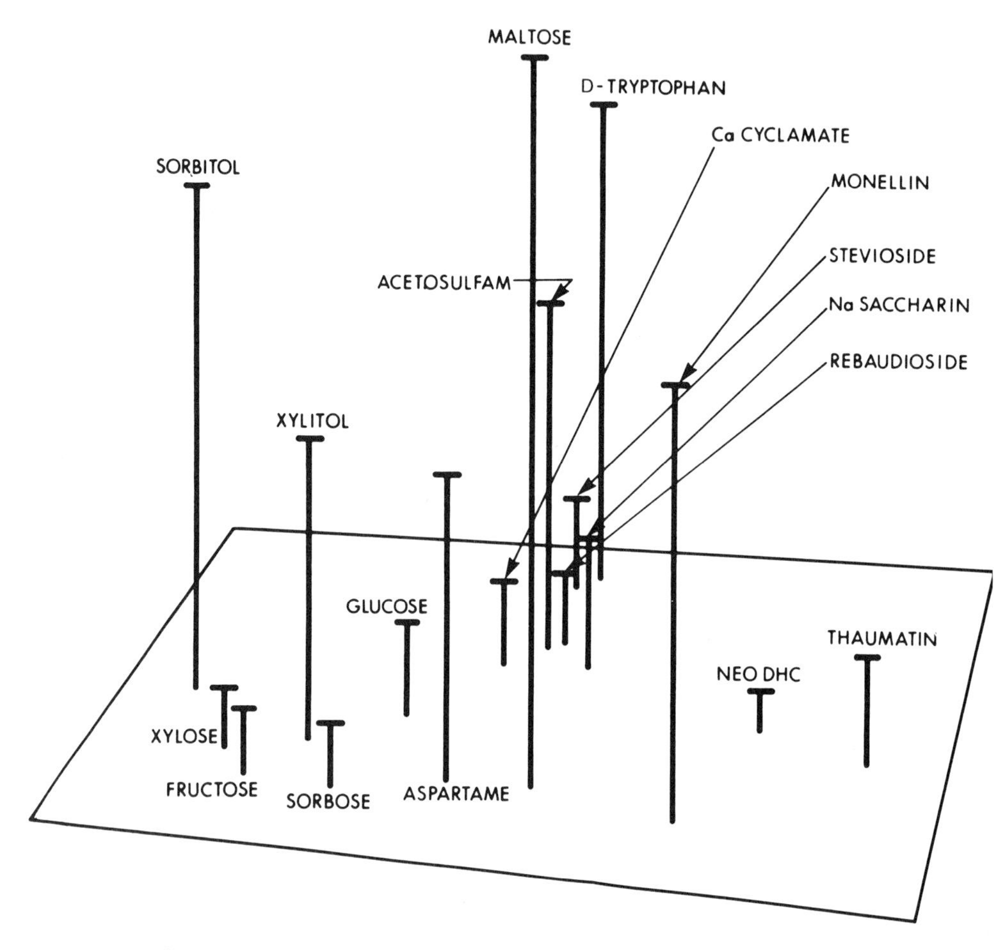

Figure 2 Three-dimensional arrangement of sweeteners achieved by INDSCAL. Stimuli that were judged similar to one another experimentally are arranged close to one another in the space. Stimuli judged dissimilar are located distant from one another. (From Ref. 7.)

Individual subject differences are revealed in the weight spaces shown in Figure 4. Weight spaces reflect idiosyncratic emphasis on the individual dimensions of a space and are interpreted in the following way. As an example, subject 3 weighted dimension I more than II, and thus for subject 3 the appropriate geometric representation of perceptual similarity among sweeteners is similar to that in Figure 2, but stretched out in an elliptical fashion along dimension I. Analogously, the appropriate multidimensional arrangement for subject 8 would be stretched out elliptically along dimension II. Dimension I will be shown below to be a "sweet-bitter" dimension and dimension II will be related to "aftertaste."

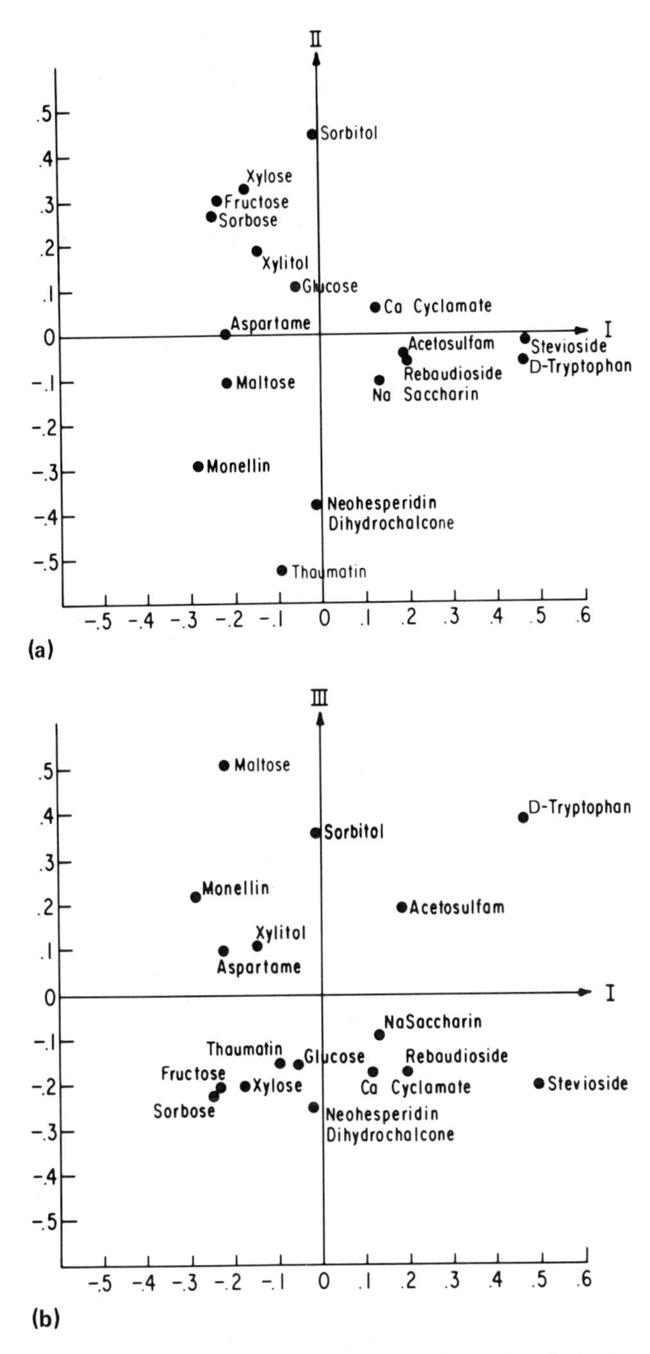

Figure 3 Cross sections of the three-dimensional arrangement in Figure 2. (a) dimension I versus dimension II (the floor of the model in Fig. 2) and (b) dimension I versus dimension III. (From Ref. 7.)

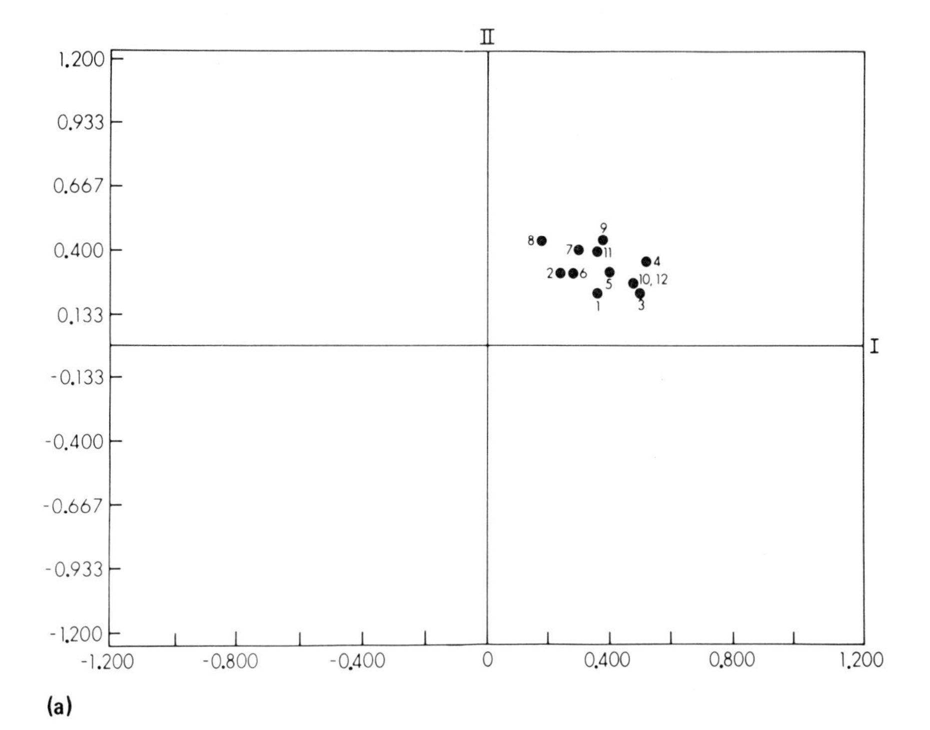
II
I
1.200
0.933
0.667
0.400
0.133
-0.133
-0.400
-0.667
-0.933
-1.200
-1.200
-0.800
-0.400
0
0.400
0.800
1.200
8
7
9
11
4
2
6
5
10, 12
1
3
(a)

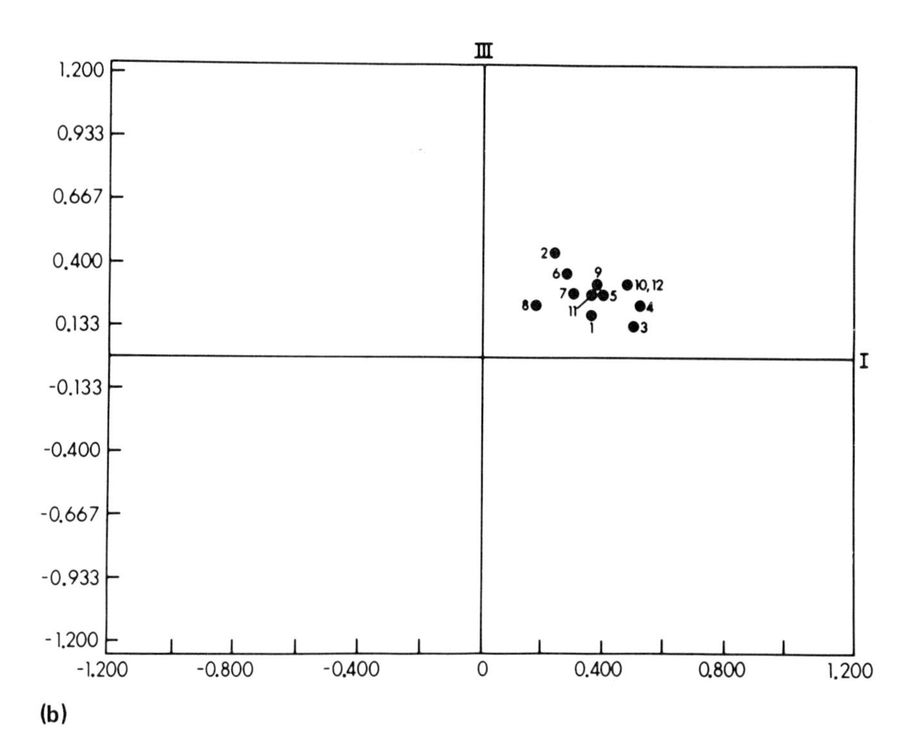
III
I
1.200
0.933
0.667
0.400
0.133
-0.133
-0.400
-0.667
-0.933
-1.200
-1.200
-0.800
-0.400
0
0.400
0.800
1.200
2
6
9
10, 12
7
5
8
4
11
1
3
(b)

The adjective ratings were examined both formally, by means of a procedure called PREFMAP (16), and informally to better understand the relations among the stimuli and the meaning of the dimensions in the spatial arrangement in Figure 2. Only 10 of the adjective scales were found to be relevant to the sweeteners tested. Histograms of the ratings for all 12 subjects on these 10 adjective scales are shown in Figure 5. It can be seen that there is considerable individual variation in these ratings. For example, one of the 12 subjects' rating for aspartame was transcribed from 1 to 10 on the "good-bad" scale. Four subjects' ratings were transcribed from 21 to 30. None were transcribed from 71-100. The rating "good" corresponds to the left side of the scale, and "bad" to the right side. Wider individual variation was found with language-based adjective ratings than with similarity judgments.

Although adjective ratings are not the best way to strictly quantify the relationships among individual sweeteners, they can be helpful in understanding the spatial arrangement in Figure 2 that is based on quantitative measures of similarity. The adjective ratings for the saccharides and alcohols that fell close together in the three-dimensional arrangement suggested that their tastes developed fairly fast with less aftertaste than the other stimuli. Glucose, fructose, and sorbitol were rated, in general, as good, natural, and somewhat syrupy sweet tastes. Aspartame and calcium cyclamate had relatively good sweetness, although calcium cyclamate was intermediate between sodium saccharin and the saccharides in pleasantness. Maltose was considered quite syrupy. Both xylose and xylitol had slight unpleasant components that were difficult to characterize verbally, although some subjects rated them as sour. While the tastes of thaumatin and neohesperidin dihydrochalcone developed slower than other stimuli, the taste of monellin developed fast. The aftertastes of monellin, thaumatin, and neohesperidin dihydrochalcone tended to linger, and many subjects found the aftertaste of the latter two unpleasant. Sodium saccharin, rebaudioside, stevioside, and D-tryptophan had the highest bitter and metallic ratings.

A quantitative approach using the technique called PREFMAP (relating the adjective ratings to the multidimensional sweetener space) confirmed the informal analysis. Vectors regressed through the spatial arrangement revealed that the first dimension is predominantly a sweet-bitter dimension and dimension II is related to aftertaste.

The finding that aspartame compares favorably with natural sugars has also been confirmed in studies that incorporate sweeteners into food products. Larson-Powers and Pangborn (17) evaluated the sensory properties of beverages and gelatins containing sucrose or one of three synthetic sweeteners, that is, aspartame

---

Figure 4 Weight spaces derived by INDSCAL that indicate the idiosyncratic stretching of the dimensions of the space in Figure 2 by individual subjects (see text for explanation): (a) dimension I versus dimension II and (b) dimension I versus dimension III. (From Ref. 7.)

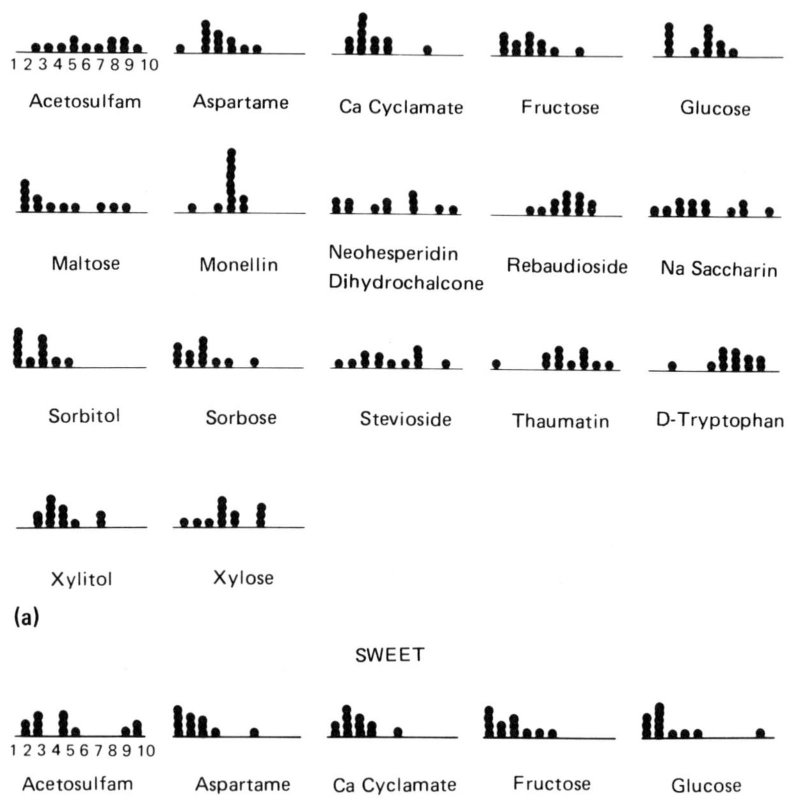
GOOD
1 2 3 4 5 6 7 8 9 10
Acetosulfam
Aspartame
Ca Cyclamate
Fructose
Glucose
Maltose
Monellin
Neohesperidin Dihydrochalcone
Rebaudioside
Na Saccharin
Sorbitol
Sorbose
Stevioside
Thaumatin
D-Tryptophan
Xylitol
Xylose

(a)

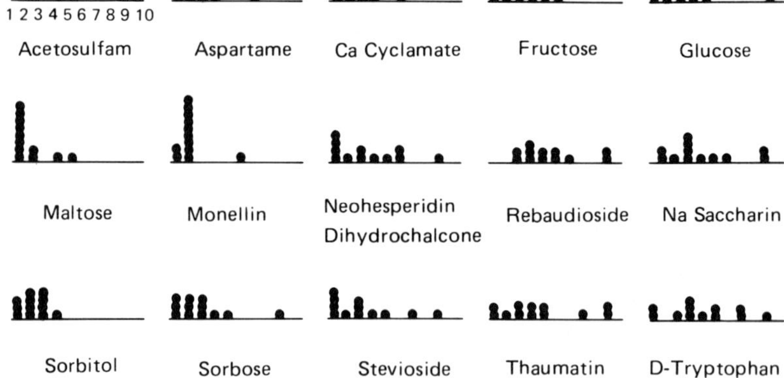
SWEET
1 2 3 4 5 6 7 8 9 10
Acetosulfam
Aspartame
Ca Cyclamate
Fructose
Glucose
Maltose
Monellin
Neohesperidin Dihydrochalcone
Rebaudioside
Na Saccharin
Sorbitol
Sorbose
Stevioside
Thaumatin
D-Tryptophan
Xylitol
Xylose

(b) **Figure 5**

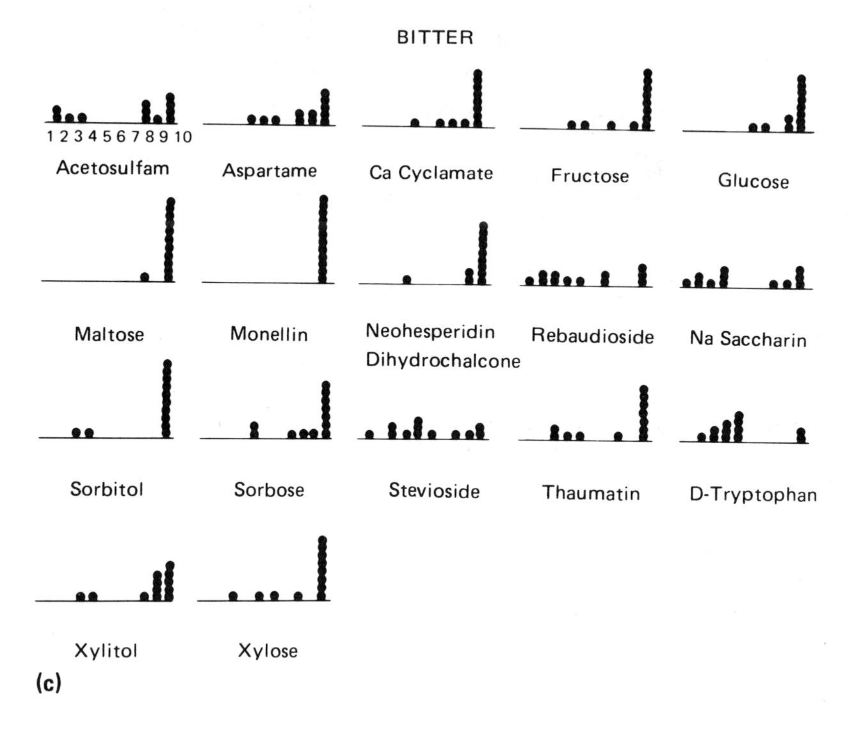
BITTER
1 2 3 4 5 6 7 8 9 10
Acetosulfam
Aspartame
Ca Cyclamate
Fructose
Glucose
Maltose
Monellin
Neohesperidin Dihydrochalcone
Rebaudioside
Na Saccharin
Sorbitol
Sorbose
Stevioside
Thaumatin
D-Tryptophan
Xylitol
Xylose

(c)

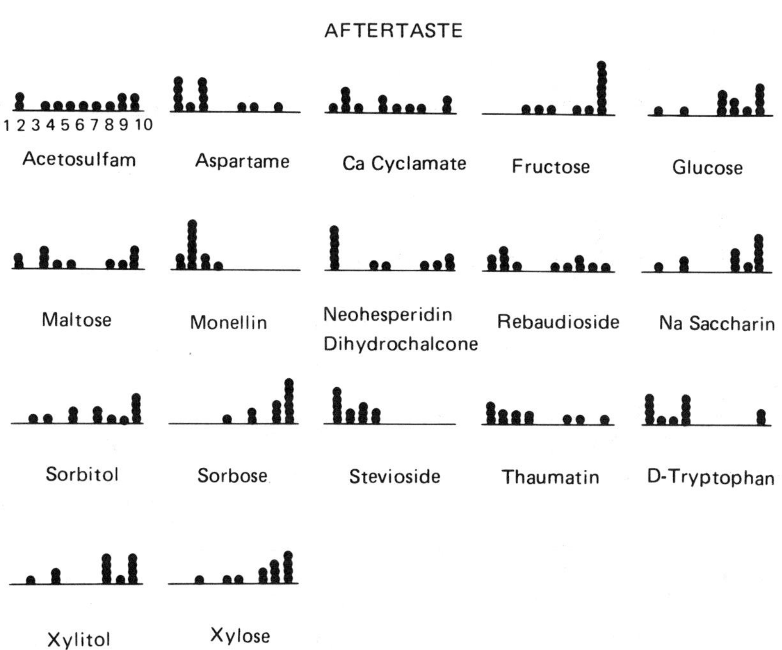
AFTERTASTE
1 2 3 4 5 6 7 8 9 10
Acetosulfam
Aspartame
Ca Cyclamate
Fructose
Glucose
Maltose
Monellin
Neohesperidin Dihydrochalcone
Rebaudioside
Na Saccharin
Sorbitol
Sorbose
Stevioside
Thaumatin
D-Tryptophan
Xylitol
Xylose

(d)

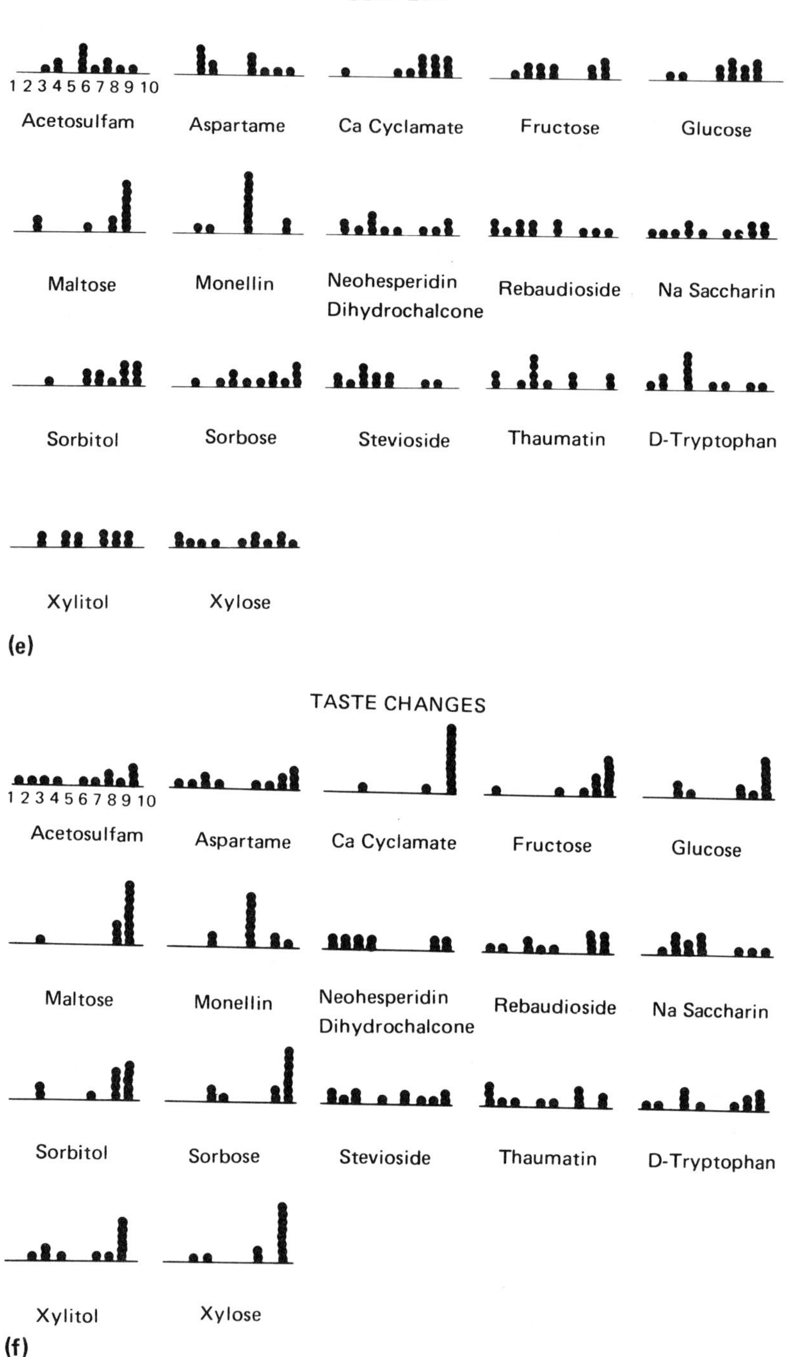

Figure 5 (continued)

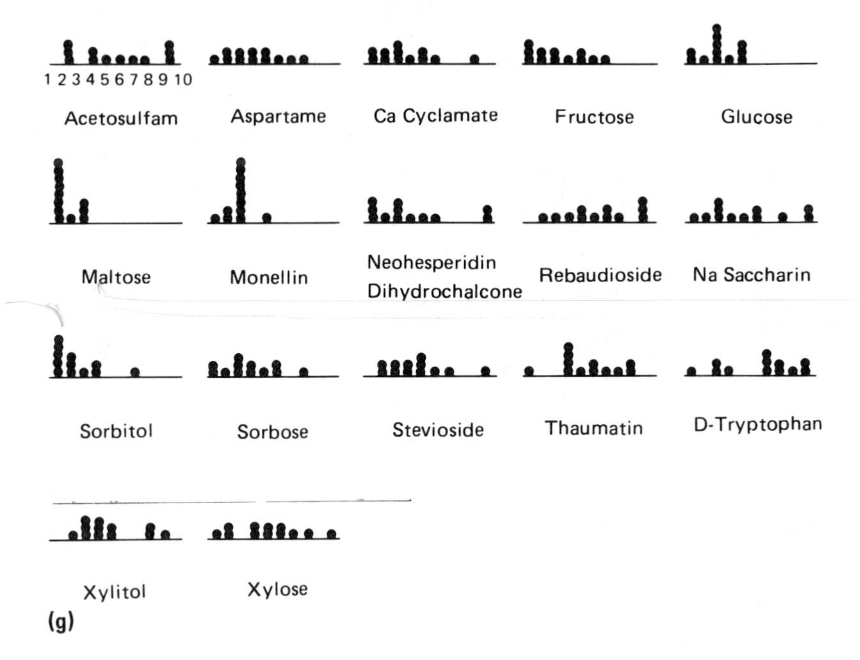
SMOOTH
1 2 3 4 5 6 7 8 9 10
Acetosulfam
Aspartame
Ca Cyclamate
Fructose
Glucose
Maltose
Monellin
Neohesperidin Dihydrochalcone
Rebaudioside
Na Saccharin
Sorbitol
Sorbose
Stevioside
Thaumatin
D-Tryptophan
Xylitol
Xylose
(g)

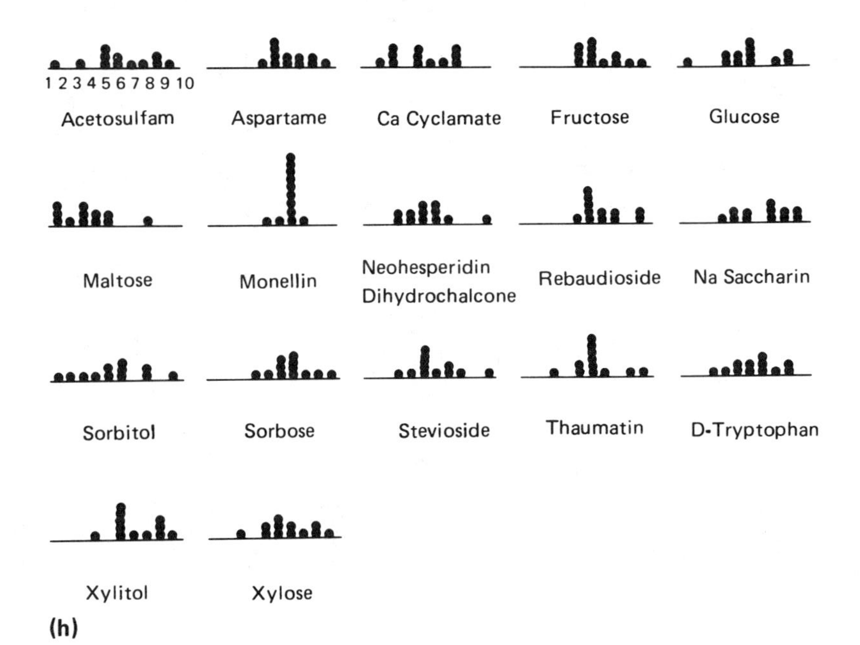
WARMING
1 2 3 4 5 6 7 8 9 10
Acetosulfam
Aspartame
Ca Cyclamate
Fructose
Glucose
Maltose
Monellin
Neohesperidin Dihydrochalcone
Rebaudioside
Na Saccharin
Sorbitol
Sorbose
Stevioside
Thaumatin
D-Tryptophan
Xylitol
Xylose
(h)

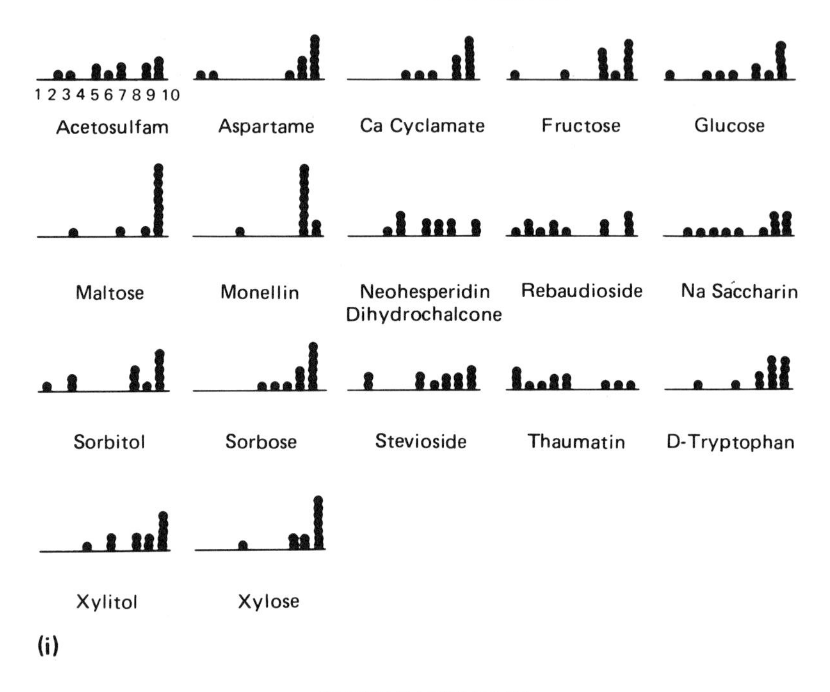

(i)

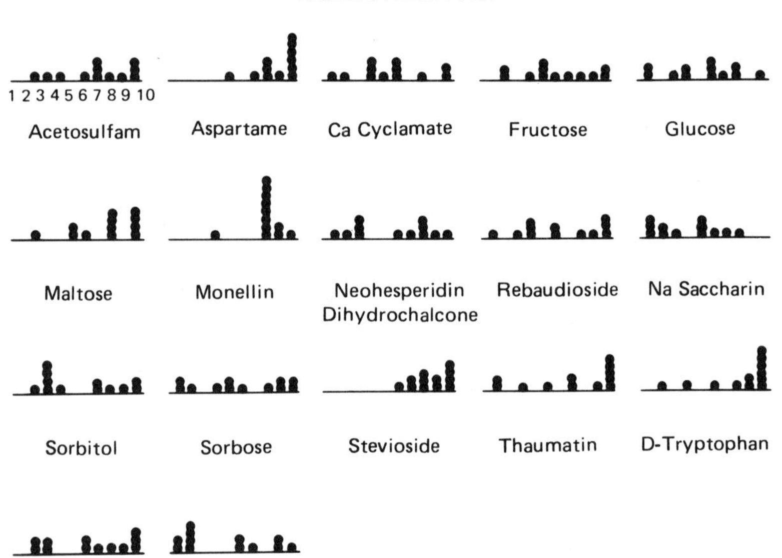

(j)

Figure 5 (continued)

calcium cyclamate, and sodium saccharin. Products sweetened with aspartame deviated least from those containing sucrose. Samples containing sodium saccharin differed the most, while those with calcium cyclamate were intermediate. Drinks containing sucrose or aspartame were generally regarded as having a sweet-clean taste, while those sweetened with calcium cyclamate or sodium saccharin had bitter or chemical components that were shown to persist markedly in time-intensity studies (18).

Conclusion

Aspartame is the artificial sweetener found to taste most similar to sugars. Sweeteners with long aftertastes such as monellin, thaumatin, and neohesperidin dihydrochalcone were perceived to deviate the most. Surprisingly, sweeteners with the highest metallic and bitter ratings, including acetosulfam, sodium saccharin, rebaudioside, and stevioside, were considered more similar to sugars than sweeteners with long aftertastes.

IS THE SWEETNESS OF ASPARTAME MEDIATED BY THE SAME RECEPTOR SITE(S) AS OTHER SWEETENERS?

The evidence to date suggests that sweetness is mediated by more than one receptor mechanism. Nonhomogeneous variability in human sensitivity found in the intensity-matching experiments (7), as well as threshold measurements for a range of compounds varying in chemical structure (19), suggests that there are several sweet receptor types. It has also been noted that although pronase E (20), semialkaline phosphatase (20), and gymnemic acid (21) all appear to selectively inhibit the sweet taste, their effects are not uniform across sweeteners. Alloxan has been found to selectively depress the response to sugars, but has no effect on artificial sweeteners such as sodium saccharin (22).

Evidence is presented in the following cross-adaptation experiment taken from Schiffman et al. (23) that further supports the view that sweetness is mediated by more than one receptor site. The experiment is based on the supposition that when adaptation to one sweetener results in a decreased response to another sweetener, the two stimuli may bind to a common receptor type. However, if adaptation does not decrease the sensation of the other sweetener, a possible implication is that different receptors code these stimuli.

Figure 5 Histograms for 10 adjective scales that represent the range of ratings for the 12 subjects. For example, one subject's rating for acetosulfam on the "good-bad" scale was transcribed as a rating between 11 and 20; one subject's rating was transcribed between 21 and 30. The left side of the scale corresponds with the rating "good"; the right side corresponds with the rating "bad." (From Ref. 7.)

All combinations of pairs of the eight stimuli listed under experiment 2 of Table 1 were presented to the subjects as triadic combinations in an ABA design. First, the subjects took 15 ml of a solution providing stimulus A at room temperature (72°F) into their mouths and immediately rated its preadaptation sweetness on a 5½-in. line labeled "very sweet" at one extreme and "not sweet" at the other. After thoroughly rinsing their mouths, subjects waited 4 min before holding 15 ml of the adapting solution B in their mouths for 60 sec. Solution B was then emptied from the mouth, and solution A, presented in a third cup, was promptly tasted without an intervening water rinse. Stimulus A was then rated again for postadaptation sweetness on a 5½-in. line. In addition, the impact of adaptation to a sweetener on water alone was measured. This involved adapting the tongue to each sweetener for 60 sec and rating the taste of 15 ml of deionized water alone for sweetness. An example of an experimental stimulus presentation is shown in Figure 6.

The reason for determining the impact of adaptation to sweeteners on the taste of water is that adaptation to certain compounds is known to induce particular taste qualities in water. For example, studies have revealed that water, after adaptation to a bitter stimulus, can acquire a sweet taste (24,25). This is an important point, because the water diluent may contribute a sweet taste to a test stimulus after adaptation to compounds with a bitter component.

The pre- and postadaptation ratings along the 5½-in. line were transcribed from 0 to 100, with "not sweet" corresponding to a transcription of 0 and "very sweet" respresented by 100. The difference between the pre- and postadaptation intensities over subjects for each combination of eight stimuli is given in Table 2. A positive magnitude of change indicates that the test solution's sweetness was reduced in intensity following adaptation; a negative value corresponds to an enhancement of the sweetness of the test solution. In order to determine whether the difference in the means was statistically significant, t tests were applied.

If there were only a single receptor site type for sweetness, the results in any one column in Table 2 might be expected to show the same result after adaptation

XYLITOL 3	ASPARTAME 6	DEIONIZED WATER 8	DEIONIZED WATER 10
GLUCOSE 2	SACCHARIN 5	GLUCOSE 7	SACCHARIN 9
XYLITOL 1	ASPARTAME 4		

Figure 6 An example of a stimulus arrangement for presentation to subjects. (From. Ref. 23.)

Table 2 Difference in Perceived Sweetness after Adaptation[a]

Test solution	Adapting solution							
	Acetosulfam	Aspartame	Calcium cyclamate	Glucose	Neohesperidin dihydrochalcone	Sodium saccharin	D-Tryptophan	Xylitol
		10.0	8.0	5.4	−18.6	43.8	30.6	25.2
Acetosulfam		—	—	—	—	0.001	0.01	0.02
	4.0		−7.4	13.8	−15.4	−13.0	40.0	23.8
Aspartame	—		0.05	0.05	0.05	0.05	0.001	0.01
	−11.0	0.7		20.6	14.6	4.2	−7.8	6.2
Calcium cyclamate	0.05	—		0.05	0.01	—	—	—
	−12.0	−9.3	9.1		−6.6	−17.4	6.0	15.2
Glucose	0.20	—	0.20		—	0.001	—	0.10
Neohesperidin	−11.5	6.6	−5.7	0.8		−21.6	15.8	1.7
dihydrochalcone	0.20	—	—	—		0.001	0.10	—
	29.7	2.2	15.4	16.4	−14.3		10.2	2.0
Sodium saccharin	0.001	—	0.20	0.10	—		0.20	—
	−13.6	36.0	−14.0	27.8	−5.8	−5.6		22.0
D-Tryptophan	0.10	0.001	0.10	0.05	—	—		0.05
	−9.2	−1.4	9.2	10.5	−19.4	−3.7	9.8	
Xylitol	0.10	—	0.20	0.20	0.001	—	—	

[a] t Tests were applied. The level of statistical significance is given below the difference of the means.

Table 3 Sweet Taste Induced in a Water Stimulus by Adaptation to Sweeteners

Adapting stimulus	Number of subjects	Arithmetic mean water taste
Acetosulfam	20	7.4
Aspartame	8	7.0
Calcium cyclamate	18	12.2
Glucose	27	3.0
Neohesperidin dihydrochalcone	20	9.0
Sodium saccharin	10	11.8
D-Tryptophan	9	2.6
Xylitol	35	9.6

to one specific stimulus; that is, the results in any one column should all be adaptations, enhancements, or all no changes. In addition, the magnitudes of change should ideally also be identical in any one column. However, this is not the case. There are fairly large, highly statistically significant inconsistencies in a given column that are not compatible with the existence of only one receptor type for sweetness. For example, adaptation to sodium saccharin produced a statistically significant sweetness reduction in acetosulfam of 43.8 units on a 100-unit rating scale. However, adaptation to sodium saccharin also enhanced the sweetness of neohesperidin dihydrochalcone by 21.6 units, also a statistically significant change. It is highly unlikely that such large and significant discrepancies within a column are compatible with a single receptor type for sweetness.

In order to further assess the possibility of multiple receptor sites for sweetness, an evaluation of the effect of the water taste is helpful. Table 3 gives the intensity of sweetness perceived for a water stimulus after adaptation to each of the eight stimuli employed in the study. Let us assume that, following adaptation to one stimulus, the taste of a second stimulus is the sum of the taste produced by the second compound and the water taste induced by adaptation to the first stimulus, minus any cross-adaptation between the two stimuli. By this reasoning, it can be seen that the values for the sweet water taste induced by adaptation to the eight stimuli are insufficient in numerous instances to account for the enhancement effects shown in Table 2. For example, the sweetness of xylitol was significantly enhanced by 19.4 units after adaptation to neohesperidin dihydrochalcone. The water taste, however, contributed only 9 units and is thus insufficiently large to account for the observed enhancement effect.

Table 4 adjusts the adaptation results in Table 2 to account for any sweetness contributed by the water diluent. It can be seen that this correction made the magnitudes of the cross-adaptations larger and the magnitudes of the enhance-

Table 4 Modification of Table 2 After Correcting for Water Taste

Test solution	Adapting solution							
	Acetosulfam	Aspartame	Calcium cyclamate	Glucose	Neohesperidin dihydrochalcone	Sodium saccharin	D-Tryptophan	Xylitol
Acetosulfam		17.0	20.2	8.4	-9.6	55.6	33.2	34.8
Aspartame	11.4		4.8	16.8	-6.4	-1.2	42.6	33.4
Calcium cyclamate	-3.6	7.7		23.6	23.6	16.0	-5.6	15.8
Glucose	-4.6	-2.3	21.3		2.4	-5.6	8.6	24.8
Neohesperidin dihydrochalcone	-4.1	13.6	6.5	3.8		-9.8	18.4	11.3
Sodium saccharin	37.1	9.2	-3.2	19.4	-5.3		12.8	11.6
D-Tryptophan	-6.2	43.0	-1.8	30.8	3.2	6.2		31.6
Xylitol	-1.8	5.6	21.4	13.5	-10.4	8.1	12.4	

ments smaller. In some cases the water taste correction changes an enhancement in Table 2 to a cross-adaptation in Table 4. In spite of the correction for the sweetness contribution from the water diluent after adaptation, some combinations of stimuli, especially those where acetosulfam, sodium saccharin, and neohesperidin dihydrochalcone are adapting stimuli, still show enhancement, further suggesting that more than one type of receptor codes sweetness.

Furthermore, a careful examination of Table 2 provides evidence for at least two different sweet receptor mechanisms. Aspartame and D-tryptophan appear to operate through a similar receptor site that is different from the receptor site shared by sodium saccharin and acetosulfam. Cross-adaptation between a pair of stimuli is not evidence for a common receptor type unless the cross-adaptation is reciprocal, that is, adaptation results when either stimulus of the pair is employed as the adapting solution. Adaptation to D-tryptophan significantly reduced the sweetness of aspartame by 40 units, while adaptation to aspartame significantly reduced the sweetness of D-tryptophan by 36 units. Also, adaptation to sodium saccharin significantly reduced the sweetness of acetosulfam by 43.8 units, while there was a significant degree of cross-adaptation with a magnitude of 29.7 units when acetosulfam served as the adapting stimulus. These results suggest that aspartame and D-tryptophan share similar receptor sites, while sodium saccharin and acetosulfam employ common sites. However, the receptors responsible for the sweetness of aspartame and D-tryptophan appear to be different from those of sodium saccharin and acetosulfam, since interpair stimulus combinations do not reveal consistent cross-adaptation.

The results of cross-adaptation are also shown pictorially in Figure 7. The cross-adaptation data, corrected for the water taste, were analyzed by the KYST multidimensional scaling procedure (26). The two-dimensional metric solution reveals, as expected, that aspartame and D-tryptophan are located close to one another, suggesting that they may share common sites. Acetosulfam and sodium saccharin fall near one another as well.

Hydrogen bonding has been suggested as the mechanism by which sweeteners bind to receptor sites, and for this reason the results of the cross-adaptation experiments were examined from this point of view. The possible AH-B systems that may be involved for each sweetener are given in Table 5. It is striking to note in Table 2 that adaptation to glucose and xylitol produced a reduction in sweetness intensity for all test solutions. Thus at least some of the AH-B receptor sites utilized by OH,OH stimulus systems may be shared by the AH-B systems of other stimuli. Sodium saccharin and acetosulfam, which mutually cross-adapt, have identical possible AH-B systems. Aspartame and D-tryptophan share one type in common. Cross-adaptation among calcium cyclamate (the only artificial sweetener with a single possible AH-B system), xylitol, and glucose is reciprocal, suggesting that at least some of the AH-B receptor sites used by these sweeteners are the same. An analogous argument holds for the mutually adaptive triad of D-tryptophan, xylitol, and glucose.

Table 5 Possible AH-B Systems

Stimulus	Number	Type
Acetosulfam[a,b]	2	NH,O (S→O) and/or NH,O (C=O)
Aspartame	2	NH–O (C=O) and/or NH_3^+,COO^-
Calcium cyclamate	1	NH,O (S→O)
Glucose	1	OH,OH
Neohesperidin dihydrochalcone	3	OH,OH and/or OH,⟨benzene ring⟩ and/or OH,OCH_3 (sugar units)
Saccharin[b] (sodium salt)	2	NH,O (S→O) and/or NH,O (C=O)
D-Tryptophan	2	NH_3^+,COO^- and/or NH,⟨ring⟩
Xylitol	1	OH,OH

[a]S→O refers to the fact that the S is an electron-donating atom. It does not form a true covalent bond with the O, but donates electrons, making the O electronegative and indicating that it has an unshared pair of electrons ($\ddot{O}$).
[b]Strictly speaking, the N in acetosulfam and sodium saccharin should be N^- because it is the salt form (Na^+ for saccharin, K^+ for acetosulfam) that is normally tasted. However, the "nonsalt" form (NH) is also sweet.
Source: Ref. 23.

Conclusion

Cross adaptation experiments suggest that more than one receptor type mediates sweetness. The receptor site for aspartame appears to be different from that of sodium saccharin.

HOW DOES THE THRESHOLD FOR ASPARTAME COMPARE WITH THRESHOLDS FOR OTHER SWEETENERS? IS THERE A CHANGE WITH AGE?

Thresholds for aspartame and 10 other sweeteners listed under experiment 3 of Table 1 were determined for 12 young (19-24 years) and 12 elderly (75-81 years) subjects by a forced choice technique described as follows. Subjects were presented with 13 trials of three unmarked cups, one of which contained 13 ml of a sweetener dissolved in deionized water. The other two cups contained 13 ml each of deionized water. Trials progressed from weaker to stronger concentrations,

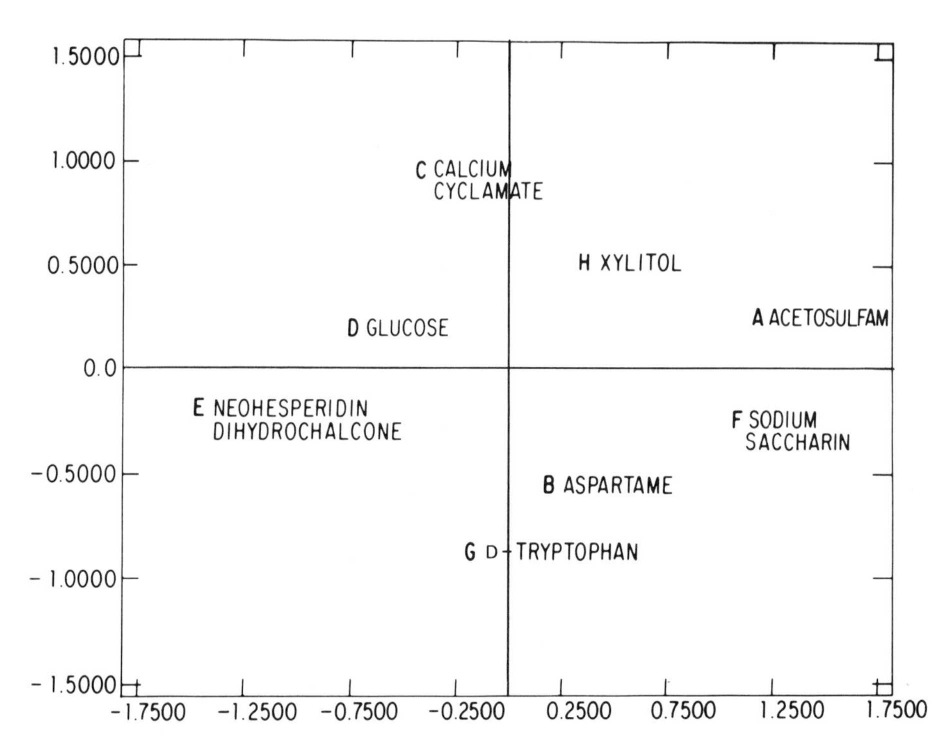

Figure 7 Two-dimensional arrangement derived by KYST based on cross-adaptation results in Table 4. Stimuli found proximate to one another in the space are assumed to use similar receptor mechanisms. (From Ref. 23.)

with the cup position for the sweetener relative to the water controls randomized over trials and subjects. The concentrations of sweeteners included 13 serial dilutions that differed from one another by a factor of 2. The concentration ranges are given in Table 6 in both molarity and weight percentage.

Subjects tasted the liquid in each of the three unmarked cups of a triad and indicated which cup they believed to contain the sweetener. If they were uncertain, they were instructed to make a guess. A taste detection threshold was considered to be established when a subject correctly discriminated the sweetener from the two water controls at three consecutive increasing concentrations. The most dilute of the three correct identifications was designated as the detection threshold for an individual subject. Recognition thresholds for sweetness in addition to detection-thresholds were determined. Subjects were requested to label the taste of the sample that was different from water blanks on each of the consecutive trials. The sweetness recognition threshold was designated as the most dilute concentration of three correct consecutive "sweet" identifications.

Table 6 Concentration Ranges Used for Each Sweetener for the Threshold Experiment

	Minimum concentration in molarity and weight percentage	Maximum concentration in molarity and weight percentage
Acetosulfam	5.81×10^{-7} M 0.0000117%	2.38×10^{-3} M 0.0480%
Aspartame	2.97×10^{-6} M 0.0000876%	1.22×10^{-2} M 0.359%
Calcium cyclamate	1.15×10^{-5} M 0.000232%	4.73×10^{-2} M 0.951%
Fructose	1.98×10^{-4} M 0.00393%	8.12×10^{-1} M 16.107%
Monellin	2.28×10^{-9} M 0.00000244%	9.35×10^{-6} M 0.0100%
Neohesperidin dihydrochalcone	1.91×10^{-7} M 0.0000117%	7.84×10^{-4} M 0.0480%
Rebaudioside	1.51×10^{-7} M 0.0000146%	6.19×10^{-4} M 0.0600%
Sodium saccharin	4.76×10^{-7} M 0.00000976%	1.95×10^{-3} M 0.0400%
Stevioside	1.81×10^{-7} M 0.0000146%	7.45×10^{-4} M 0.0600%
Thaumatin	1.36×10^{-9} M 0.00000244%	5.55×10^{-6} M 0.0100%
D-Tryptophan	3.57×10^{-6} M 0.0000730%	1.46×10^{-2} M 0.299%

Source: Ref. 19.

The detection thresholds are given in Table 7. Geometric means are reported here because individual thresholds were not found to be normally distributed. The geometric mean threshold for young and elderly subjects respectively are given in molarity in columns A and B and in weight percentage in columns C and D. The ratio of columns D and C [detection threshold (elderly)/ detection threshold (young)] is given in column E. Columns F and G give the ranges of the detection thresholds for sweeteners in terms of weight percentage for young and elderly, respectively. Column H indicates whether the young and elderly groups differed significantly as determined by Mann-Whitney U tests. N.S. indicates that the difference in detection thresholds for the two groups was not statistically

Table 7 Detection Thresholds for Young and Elderly[a]

	A	B	C	D	E	F	G	H
Acetosulfam	4.44×10^{-5} M	7.47×10^{-5} M	0.000891	0.00150	1.68	0.0000937-0.003	0.0000468-0.048	N.S.[b]
Aspartame	2.24×10^{-5} M	9.13×10^{-5} M	0.000661	0.00269	4.07	0.000350-0.00140	0.000991-0.00561	<0.05
Calcium cyclamate	2.66×10^{-4} M	4.12×10^{-4} M	0.00537	0.00830	1.55	0.000464-0.01485	0.000464-0.0297	N.S.
Fructose	4.39×10^{-3} M	10.1×10^{-3} M	0.0871	0.200	2.30	0.00786-0.503	0.0157 4.03	N.S.
Monellin	1.95×10^{-8} M	9.13×10^{-8} M	0.0000209	0.0000977	4.67	0.00000488-0.000156	0.00000976-0.000625	<0.05
Neohesperidin dihydrochalcone	2.20×10^{-6} M	4.60×10^{-6} M	0.000135	0.000282	2.90	0.0000234-0.0001875	0.0000468-0.0030	N.S.
Rebaudioside	4.61×10^{-6} M	13.0×10^{-6} M	0.000447	0.00126	2.82	0.0000293-0.000937	0.000468-0.00375	<0.05
Sodium saccharin	1.47×10^{-5} M	4.24×10^{-5} M	0.000302	0.000871	2.88	0.0000390-0.00125	0.000312-0.00250	<0.05
Stevioside	5.31×10^{-6} M	16.0×10^{-6} M	0.000427	0.00129	3.02	0.0000586-0.00375	0.000468-0.00375	<0.05
Thaumatin	7.16×10^{-8} M	13.3×10^{-8} M	0.000129	0.000240	1.86	0.00000488-0.000625	0.0000195-0.0100	N.S.
D-Tryptophan	1.09×10^{-4} M	3.22×10^{-4} M	0.00224	0.00661	2.95	0.000292-0.00467	0.00117-0.299	N.S.

[a]Column A, geometric mean detection thresholds in molarity for young subjects; column B, geometric mean detection thresholds in molarity for elderly subjects; column C, geometric mean detection thresholds in weight percentage for young subjects; column D, geometric mean detection thresholds in weight percentage for elderly subjects; column E, ratio of columns D and C, i.e., detection thresholds (elderly)/detection thresholds (young); column F, range of the detection thresholds for young subjects in weight percentage; column G, range of the detection thresholds for elderly subjects in weight percentage; column H, level of significance of difference for detection thresholds for young and elderly subjects as determined by Mann-Whitney U tests.

[b]N.S. indicates "not significant."

Table 8 Comparison of Detection and Sweetness Recognition Thresholds for Young Subjects[a]

	A	B	C	D	E
Acetosulfam	4.44×10^{-5} M	16.1×10^{-5} M	0.000891	0.00324	3.64
Aspartame	2.24×10^{-5} M	4.49×10^{-5} M	0.000661	0.00132	2.00
Calcium cyclamate	2.66×10^{-4} M	13.3×10^{-4} M	0.00537	0.0269	5.01
Fructose	4.39×10^{-3} M	16.6×10^{-3} M	0.0871	0.331	3.80
Monellin	1.95×10^{-8} M	6.76×10^{-8} M	0.0000209	0.0000724	3.46
Neohesperidin dihydrochalcone	2.20×10^{-6} M	5.28×10^{-6} M	0.000135	0.000324	2.40
Rebaudioside	4.61×10^{-6} M	13.6×10^{-6} M	0.000447	0.00132	2.95
Sodium saccharin	1.47×10^{-5} M	4.97×10^{-5} M	0.000302	0.00102	3.38
Stevioside	5.31×10^{-6} M	23.7×10^{-6} M	0.000427	0.00191	4.47
Thaumatin	7.16×10^{-8} M	20.1×10^{-8} M	0.000129	0.000363	2.81
D-Tryptophan	1.09×10^{-4} M	5.46×10^{-4} M	0.00224	0.0112	5.00

[a]Column A, geometric mean detection thresholds in molarity for young subjects; column B, geometric mean recognition thresholds in molarity for young subjects; column C, geometric mean detection thresholds in weight percentage for young subjects; column D, geometric mean recognition thresholds in weight percentage for young subjects; column E, ratio of columns D and C, recognition threshold (young)/detection threshold (young).

Table 9 Relationship of Rank Order of Detection Thresholds for Young Subjects and Possible Type and Number of AH–B Systems

Sweetener	Threshold in molarity	Number	Possible AH–B systems
Monellin	1.95×10^{-8} M	Many possible	1. NH,O (C=O) (peptide linkages). 2. NH_3^+,COO^- (N terminal and C terminal residues, the normal system of most amino acids). 3. NH_2,O (C=O) (e.g., asparagine and glutamine. 4. OH (e.g., hydroxyl groups of serine, threonine, and tyrosine) could be AH, and the OH oxygen atom possible B. 5. ϵ-Amino, guanido, and imidazole groups of lysine and arginine, respectively, could be AH. 6. Carboxyl oxygen atoms of aspartic and glutamic acid could be B. 7. Center of unsaturation (e.g., phenylalanine, tryptophan, and tyrosine) could be B. 8. Also, there is a shape factor in that there is a loss of sweetness if heated. Sweetness returns on cooling when the preferred conformation is reestablished.
Thaumatin	7.16×10^{-8} M	Many possible	As monellin
Neohesperidin dihydrochalcone	2.20×10^{-6} M	3	1. OH,OH (sugar units) 2. OH,(benzene ring) 3. OH,OCH_3
Rebaudioside	4.61×10^{-6} M	3	1. OH,OH (sugar units) 2. OH,O (C=O) (sugar hydroxyl group and carbonyl oxygen)

Table 9 (continued)

Sweetener	Threshold in molarity	Number	Possible AH–B systems
			3. OH,= CH_2 (sugar hydroxyl group and center of unsaturation)
Stevioside	5.31×10^{-6} M	3	As rebaudioside
Saccharin (sodium salt)	1.47×10^{-5} M	2	1. NH,O (S→O)[a,b] 2. NH,O (C=O)
Aspartame	2.24×10^{-5} M	2	1. NH,O (C=O) 2. NH_3^+,COO^-
Acetosulfam	4.44×10^{-5} M	2	1. NH,O (S→O)[b] 2. NH,O (C=O)
D-Tryptophan	1.09×10^{-4} M	2	1. NH_3^+,COO^- 2. NH,(aromatic ring)
Calcium cyclamate	2.66×10^{-4} M	1	1. NH,O (S→O)
Fructose	4.39×10^{-3} M	1	1. OH,OH

[a]S→O refers to the fact that the S is an electron-donating atom. It does not form a true covalent bond with the O, but donates electrons.
[b]Strictly speaking, the N in acetosulfam and sodium saccharin should be N^- because it is the salt form (Na^+ for saccharin, K^+ for acetosulfam) that is normally tasted. However, the acid form (NH) is also sweet.

significant. All mean detection thresholds were higher for elderly subjects than for young ones, with the average ratio in column E of Table 7 being 2.72. The detection threshold for aspartame in elderly subjects was 4.07 times higher than for young subjects.

The geometric mean detection and sweetness recognition thresholds for young subjects are compared in Table 8. Columns A and B give detection and recognition thresholds, respectively, in molarity, and columns C and D in weight percentage. The ratio of recognition to detection threshold for the sweeteners is found in column E. The ratio for aspartame is 2; that is, the recognition threshold for aspartame is twice as high as the detection threshold. The mean ratio for column E is 3.54. Recognition thresholds were not consistently determined for elderly subjects because the older persons were frequently unable to assign labels to tastes in spite of continued requests to do so.

Table 9 lists the detection thresholds for sweeteners for young subjects in rank order from lowest to highest along with the number and type of possible AH–B systems. The rank order for elderly subjects is almost identical to that for the young. It is striking that the lowest thresholds have the greatest possible number of AH–B types, while the highest thresholds are for sweeteners with only one possible type. It is possible that the decline in detection thresholds with age is related to a decrement in the number of papillae (and ultimately the number of complementary AH–B systems).

Threshold data, like cross-adaptation data, can thus be related to intermolecular hydrogen bonding. A compound interacting with one of the multiple AH–B units in the receptor membrane may be sweet; with two concertedly, quite sweet; with three simultaneously, intensely sweet; and with four or more, overwhelmingly sweet. It should be noted, however, that although hydrogen bonding is a highly probable mechanism in the perception of sweetness, an optimum balance between the strength of bonding and the number of AH–B types as well as lipophilicity may ultimately be found to be important as well.

Conclusion

The detection threshold in young subjects for aspartame is 195 times lower than that of fructose and 1148 times higher than monellin. The recognition threshold for aspartame is twice that of the detection threshold. The detection threshold for aspartame in elderly subjects is 4.07 times that for young subjects.

HOW DOES THE THRESHOLD OF THE DIPEPTIDE ASPARTAME COMPARE WITH THE THRESHOLDS OF AMINO ACIDS?

The detection thresholds for both D- and L-amino acids, along with their taste qualities, are given in Table 10 (see Ref. 27). It can be seen that the threshold for aspartame, 2.24×10^{-5} M, is lower than any of the D- or L-amino acids, including aspartame's constituent amino acids, L-aspartic acid and L-phenylalanine, with the exception of L-cysteine HCl. The sweet L forms tend to have the highest thresholds, while thresholds for sweet D forms tend to span the range for D-amino acids from high (D-serine, 6.48×10^{-2} M) to relatively low (D-tryptophan, 0.048×10^{-2} M).

Conclusion

The detection threshold for aspartame, 2.24×10^{-5} M, is lower than for all the amino acids except L-cysteine HC1. Detection thresholds for the constituent amino acids of aspartame, L-aspartic acid and L-phenylalanine, are both higher than aspartame itself.

Table 10 Comparison of Detection Thresholds and Taste Qualities for D- and L-Amino Acids Listed in Rank Order of Thresholds

Amino acids	D Forms taste quality	Threshold	Amino acids	L Forms taste quality	Threshold
D-Serine	Sweet, smooth, fresh, dilute, possibly tingling	6.48×10^{-2} M	Glycine	Sweet, pleasant smooth, refreshing	3.09×10^{-2} M
D-Proline	Sharp, unpleasant, bitter, minerally, metallic	6.04×10^{-2} M	L-Threonine	Flat to sweet; possibly bitter, sour, or fatty	2.57×10^{-2} M
D-Threonine	Somewhat tasteless; simple, weak, possibly sweet	3.37×10^{-2} M	L-Serine	Flat to sweet; possibly sour, complex	2.09×10^{-2} M
Glycine	Sweet, pleasant, smooth refreshing	3.09×10^{-2} M	L-Alanine	Sweet; possibly complex with bitter aftertaste	1.62×10^{-2} M
D-Isoleucine	Flat, bitter, minerally, metallic; possibly salty and smoky	1.25×10^{-2} M	L-Proline	Sweet, possibly complex with salty or sour components	1.51×10^{-2} M
D-Alanine	Sweet; simple with no bitter components	1.12×10^{-2} M	L-Glutamine	Flat, sweet, meaty, somewhat unpleasant	0.977×10^{-2} M
D-Asparagine	Sweet, simple, refreshing	0.977×10^{-2} M	L-Isoleucine	Flat to bitter	0.741×10^{-2} M
D-Leucine	Smooth, soft, moderately sweet	0.501×10^{-2} M	L-Phenylalanine	Bitter; possibly complex and strangling	0.661×10^{-2} M
D-Methionine	Alkaline, stale, bitter, minerally, meaty, sour, sweet components	0.501×10^{-2} M	L-Leucine	Flat to bitter (virtually indistinguishable from L-isoleucine)	0.645×10^{-2} M
D-Glutamine	Good, sweet, flavorous, smooth	0.347×10^{-2} M	L-Valine	Flat to bitter; slightly sweet	0.416×10^{-2} M
D-Valine	Somewhat tasteless; weak, alkaline, minerally; possibly sweet	0.295×10^{-2} M	L-Methionine	Flat to bitter; possibly sulfurous, meaty or sweet	0.372×10^{-2} M
D-Histidine	Sweet, flavorous, refreshing, fruity	0.186×10^{-2} M	L-Tryptophan	Flat to bitter	0.229×10^{-2} M

Table 10 (continued)

D-Arginine HCl	Bitter, alkaline, complex with salty and sour elements, minerally	0.162×10^{-2} M	L-Asparagine	Flat to bitter	0.162×10^{-2} M
D-Phenylalanine	Indistinct; possibly sweet, minerally; possibly bitter and metallic	0.155×10^{-2} M	L-Histidine	Flat to bitter, minerally	0.123×10^{-2} M
D-Lysine HCl	Bitter, minerally, poisonous, alkaline; metallic, soapy components	0.133×10^{-2} M	L-Arginine HCl	–	0.123×10^{-2} M
D-Aspartic acid	Strong, sour, salty, slightly bitter yet flavorous	0.0741×10^{-2} M	L-Arginine	Flat to bitter; alkaline complex	0.120×10^{-2} M
D-Tryptophan	Smooth, sweet, possibly bitter, minerally	0.048×10^{-2} M	L-Lysine	–	0.0708×10^{-2} M
D-Histidine HCl	Complex; sweet, sour, possibly salty; pungent	0.025×10^{-2} M	L-Lysine HCl	Bitter, complex, salty, sweet	0.0447×10^{-2} M
D-Cysteine	Obnoxious, repulsive, slightly bitter with persistent aftertaste	0.0085×10^{-2} M	L-Aspartic acid	Flat, sour, slightly bitter	0.0182×10^{-2} M
D-Glutamic acid	Sour, constant, pungent, possibly salty	0.0076×10^{-2} M	L-Histidine HCl	–	0.00794×10^{-2} M
			L-Glutamic acid	Unique, possibly meaty, salty, bitter, sour, complex	0.0063×10^{-2} M
			L-Cysteine	Sulfurous, obnoxious	0.0063×10^{-2} M
			L-Cysteine HCl	Sulfurous, obnoxious, concentrated, complex, poisonous	0.0016×10^{-2} M

HOW DOES THE GROWTH IN PERCEIVED INTENSITY FOR ASPARTAME COMPARE WITH OTHER SWEETENERS? IS THERE ANY CHANGE WITH AGE?

The method of magnitude estimation was used to determine the growth in perceived intensity with concentration for the 10 sweeteners listed under experiment 4 of Table 1 for 12 subjects ranging in age from 18 to 26 years and 12 elderly subjects aged 74-82 years. The concentration ranges and numbers of serial dilutions for the sweeteners are given in Table 11. The lower end of the range was determined by the average taste threshold for elderly subjects. The serial dilutions differed from one another by a factor of 2. Subjects assigned numbers to each dilution such that the ratios of the numbers reflected the ratios of the perceived taste intensities.

When the logs of the concentrations (C) of the sweeteners were plotted against the logs of the magnitude estimates of the sensations (S), a regression line could be fit to the points, indicating that a simple power function $S = kC^n$ (or $\log S = \log k + n \log C$) could describe the data. The slopes n for the regression lines for both young and elderly as well as quantification of the decrement with age are given in Table 12. In every case the slope for the young was greater than that for the elderly with a mean ratio of 2.060.

It can be seen that the mean slope for aspartame for young subjects is 0.454. This slope produces a much flatter line than the slope for D-tryptophan, which is 0.905. This is represented pictorially in Figure 8. The greatest age-related

Table 11 Concentration Ranges and Number of Dilutions for the Magnitude Estimation Experiment

	Range in weight percentage		
	Minimum	Maximum	Number of dilutions
Acetosulfam	0.00150	1.5360	11
Aspartame	0.00269	1.3773	10
Calcium cyclamate	0.00830	2.1248	9
Fructose	0.200	51.200	9
Neohesperidin dihydrochalcone	0.000282	0.072192	9
Rebaudioside	0.00126	0.32256	9
Sodium saccharin	0.00871	0.22298	9
Stevioside	0.00129	0.33024	9
Thaumatin	0.000240	0.03072	8
D-Tryptophan	0.00661	0.84608	8

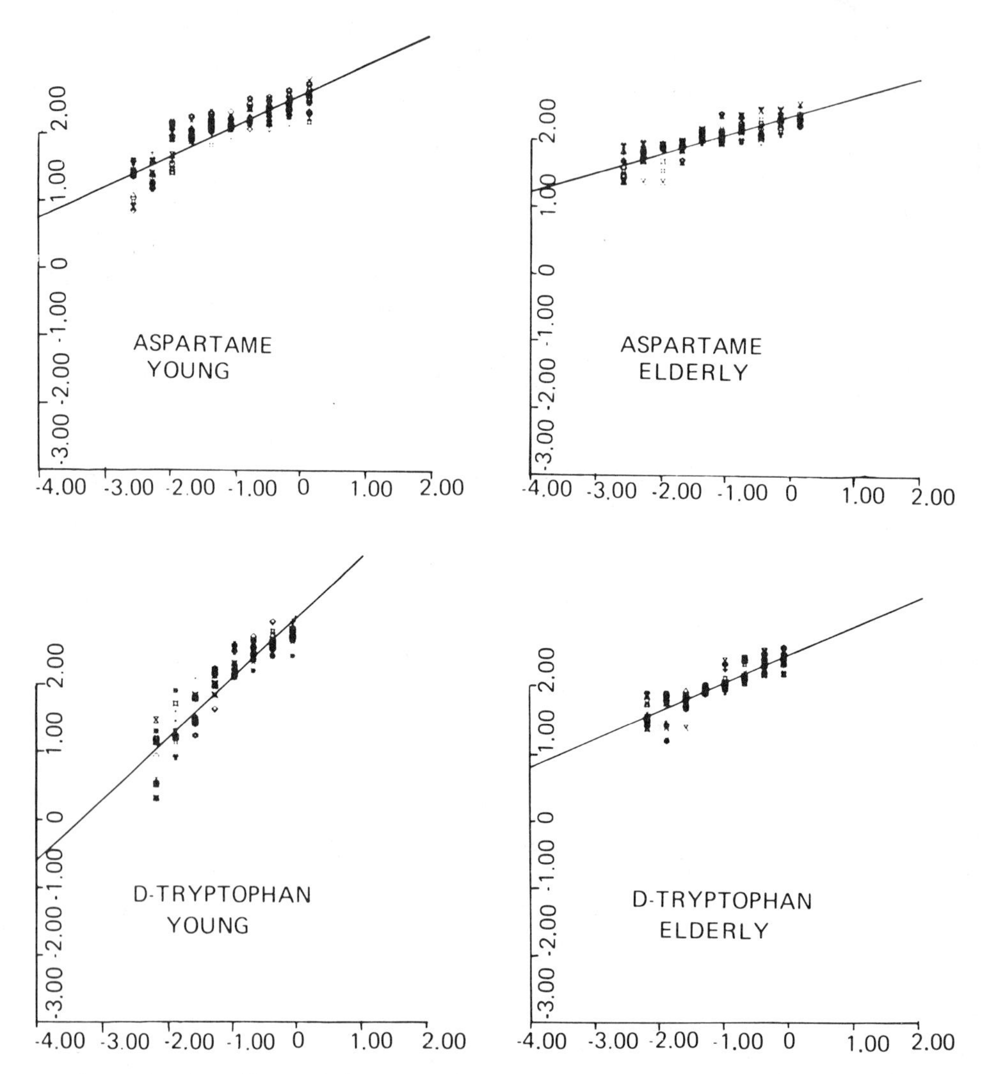

Figure 8 Logarithms of the concentrations of the sweeteners are plotted on the abscissa. Logarithms of the normalized magnitude estimates for young and elderly subjects are plotted on the ordinate. The slopes relating concentration and perceived intensity are flatter in all cases for elderly subjects. The locations of the functions on the ordinate are irrelevant because the magnitude estimates were normalized. (From Ref. 23.)

Table 12 Slopes for Young and Elderly Subjects as Well as Their Ratios[a]

	A	B	C	D	E
Acetosulfam	0.410	0.284	1.444	0.126	30.7%
Aspartame	0.454	0.284	1.599	0.170	37.4%
Calcium cyclamate	0.548	0.366	1.497	0.182	33.2%
Fructose	0.677	0.334	2.027	0.343	50.7%
Neohesperidin dihydrochalcone	0.528	0.231	2.286	0.297	56.3%
Rebaudioside A	0.704	0.300	2.347	0.404	57.4%
Sodium saccharin	0.552	0.282	1.957	0.270	48.9%
Stevioside	0.806	0.402	2.005	0.404	50.1%
Thaumatin	0.422	0.133	3.173	0.289	68.5%
D-Tryptophan	0.905	0.417	2.170	0.488	54.0%

[a]Column A, slope (young); column B, slope (elderly); column C, slope (young)/slope elderly; column D, difference of slope (young) − slope (elderly); column E, difference as a percentage of the slope (young).

depressions in the slopes were for thaumatin, rebaudioside, and neohesperidin dihydrochalcone, suggesting that the possibilities for concerted intermolecular hydrogen bonding may decline in the elderly.

Conclusion

The growth in the perceived intensity for aspartame is less steep than for the other sweeteners tested, with the exception of acetosulfam and thaumatin. Although the reasons for slope differences are not known, it is possible that receptors for sweeteners with steep slopes do not saturate as quickly as receptors for sweeteners with flatter slopes.

HOW DOES THE GROWTH IN PERCEIVED INTENSITY WITH CONCENTRATION FOR THE DIPEPTIDE ASPARTAME COMPARE WITH INDIVIDUAL AMINO ACIDS?

The slopes of the psychophysical functions relating concentrations and perceived intensity for both L- and D-amino acids have been determined by Schiffman and Clark (28) and Schiffman et al. (29) using the method of magnitude estimation. The values for the slopes for both L- and D-enantiomers, as well as their ratios are given in Table 13. The slopes for L-aspartic acid (0.498) and L-phenylalanine (0.796) are both greater than that for aspartame (0.454).

Conclusion

L-Threonine is the only sweet-tasting amino acid with a slope less than that of aspartame.

Table 13

Chemical grouping	Amino acid	Slopes[a] L	D	Direction	D/L
With neutral side chains (aliphatic)	Glycine	0.561			
	Alanine	0.603	0.713	D > L	1.182
	Valine	0.726	0.625	L > D	0.861
	Leucine	0.475	0.734	D ≫ L	1.546
	Isoleucine	0.675	0.538	L > D	0.797
(aromatic)	Phenylalanine	0.796	1.02	D > L	1.282
	Tryptophan	0.667	0.759	D > L	1.138
(amides)	Asparagine	0.323	0.761	D ≫ L	2.358
	Glutamine	0.495	0.553	D > L	1.117
With side chains containing basic (amine) groups	Arginine HCl	0.645	0.623	L ≈ D	0.966
	Lysine HCl	0.476	0.475	L ≈ D	0.998
	Histidine	0.415	0.758	D ≫ L	1.828
	Histidine HCl	0.574	0.563	L ≈ D	0.981
With side chains containing hydroxylic groups	Serine	0.788	0.671	L > D	0.852
	Threonine	0.396	0.448	D > L	1.131
With side chains containing acidic groups	Aspartic acid	0.498	0.652	D > L	1.309
	Glutamic acid	0.305	0.600	D ≫ L	1.969
With side chains containing sulfur atoms	Cysteine	0.277	0.373	D > L	1.346
	Cysteine HCl	0.513	0.594	D > L	1.158
	Methionine	0.414	0.679	D ≫ L	1.639

[a]The slopes for D-amino acids determined in this study are compared with those for L forms. "Direction" indicates whether the value found for one enantiomer is greater than (>), approximately equal to (≈), or less than (<) its mirror image. The actual ratios for the values for the slopes (D/L) are given as well. The symbol (≫) for "direction" signifies "very much greater than" and is used to indicate ratios greater than 1.5.

DOES THE TASTE OF CONSTITUENT AMINO ACIDS PREDICT THE TASTE OF DIPEPTIDES?

The taste of the constituent amino acids of a dipeptide is not predictive of the taste quality of the dipeptide itself (12). In a study of 46 dipeptides, Schiffman and Engelhard (12) noted that most were either weak or bitter in taste. Weak-tasting dipeptides tended to have constituent amino acids that are pleasant tasting, contain hydroxyl groups, or have aliphatic side chains. The dipeptides tested in this study that exhibited a sweet component had sweet-tasting amino acids at their NH_2 terminal position, with the exception of L-leucyl-L-tryptophan. This was true, however, for many bitter-tasting amino acids as well. In general, with one exception, dipeptides with a sour component contained amino acids with acidic groups. No clear trends were found for dipeptides with bitter or salty components.

The sequence of amino acids in a dipeptide clearly is important in determining taste quality. Of the seven pairs of dipeptides containing the same residues, but in reversed sequence, none had identical tastes. In addition, dipeptides with two identical constituent amino acids do not have the same taste as the amino acid alone (see Table 14).

Table 14 Brief Description of the Predominant Tastes of Dipeptides Consisting of Two Identical Constituent Amino Acids

Dipeptide	Taste of amino acid	Predominant taste of dipeptide
L-Alanyl-L-alanine	Alanine (sweet; possibly complex with bitter aftertaste	Weak
Glycylglycine	Glycine (sweet)	Bitter
L-Leucyl-L-leucine	Leucine (flat to bitter)	Sour, fairly salty and bitter
L-Phenylalanyl-L-phenylalanine	Phenylalanine (bitter; possibly complex and strangling)	Bitter-sour
L-Valyl-L-valine	Valine (flat to bitter; slightly sweet)	Weak

Conclusion

The taste of a dipeptide cannot be predicted at present from the taste of its constituent amino acids. L-Aspartic acid is flat, bitter, and sour and L-phenylalanine is bitter, while aspartame, L-aspartyl-L-phenylalanine methyl ester, is sweet.

WHAT IS THE ROLE OF SODIUM CHANNELS IN THE PERCEPTION OF SWEETENERS?

The diuretic amiloride, a potent inhibitor of sodium transport in a wide variety of epithelial systems (30), has been found to reduce the perceived intensity of a range of sweeteners, including aspartame, on the human tongue (31). These results suggest that either sodium transport is directly involved in the perception of sweet taste or the AH–B group in the amiloride molecule interferes with the binding of sweeteners.

In this study, one-half of the tongue was adapted to 5×10^{-4} M amiloride dissolved in deionized water and the other half to a deionized water control. Amiloride impaired the perception of all sweet tastes, including compounds ranging widely in chemical structure, such as sweet-tasting saccharides, glycosides, dipeptides, proteins, and amino acids.

Conclusion

Sodium transport may be involved in the perception of sweet tastes, including that of aspartame.

FINAL COMMENT

Aspartame closely approximates the taste of natural sugars and does not have the prominent metallic, bitter taste of many other artificial sweeteners such as sodium saccharin. The differences in taste quality found among artificial sweeteners are related to the fact that there appear to be multiple receptor sites for sweet-tasting compounds. Assessment of the type and number of AH–B systems in a sweetener molecule have some bearing on its taste properties, including thresholds. The detection threshold for aspartame in young subjects is 2.24×10^{-5} M; there is an increase in threshold by a factor of 4.07 in the elderly. The detection threshold for aspartame is slightly higher than that of sodium saccharin, but is lower than that of fructose by a factor of 195. Sodium transport may play a role in the transduction process for aspartame and other sweeteners.

REFERENCES

1. Mazur, R. H. (1977). Aspartame—A sweet surprise. *J. Toxicol. Environ. Health* 2, 243-249.

2. Mazur, R. H. (1979). Peptide-based sweeteners. In *Developments in Sweetness–1* (Hough, C. A. M., Parker, K. J., and Vlitos, A. J., eds.), Applied Science, London, pp. 125-134.
3. Shallenberger, R. S., and Acree, T. E. (1971). Chemical structure of compounds and their sweet and bitter taste. In *Handbook of Sensory Physiology* (Beidler, L. M., ed.), Springer-Verlag, Berlin, pp. 221-227.
4. Deutsch, E. W., and Hansch, C. (1966). Dependence of relative sweetness on hydrophobic bonding. *Nature* 221, 75.
5. Kier, L. B. (1972). A molecular theory of taste. *J. Pharm. Sci.* 61, 1394-1397.
6. Schiffman, S. S., and Erickson, R. P. (1980). The issue of primary tastes versus a taste continuum. *Neurosci. Biobehav. Rev.* 4, 109-117.
7. Schiffman, S. S., Reilly, D. A., and Clark, T. B. (1979). Qualitative differences among sweeteners. *Physiol. Behav.* 23, 1-9.
8. Schiffman, S. S., Reynolds, M. L., and Young, F. W. (1981). *Introduction to Multidimensional Scaling : Theory, Methods, and Applications.* Academic Press, New York.
9. Schiffman, S. S., and Erickson, R. P. (1971). A psychophysical model for gustatory quality. *Physiol. Behav.* 7, 617-633.
10. Schiffman, S. S., and Dackis, C. (1975). Taste of nutrients: Amino acids, vitamins, and fatty acids. *Percept. Psychophys.* 17, 140-146.
11. Schiffman, S. S., Moroch, K., and Dunbar, J. (1975). Taste of acetylated amino acids. *Chem. Senses*, 1, 387-401.
12. Schiffman, S. S., and Engelhard, H. H. (1976). Taste of dipeptides. *Physiol. Behav.* 17, 523-535.
13. Schiffman, S. S. (1977). Food recognition by the elderly. *J. Gerontol.* 32, 586-592.
14. Schiffman, S. S., Musante, G., and Conger, J. (1978). Application of multidimensional scaling to ratings of foods for obese and normal weight individuals. *Physiol. Behav.* 21, 417-422.
15. Carroll, J. D., and Chang, J. J. (1970). Analysis of individual differences in multidimensional scaling via an n-way generalization of "Eckhart-Young" decomposition. *Psychometrika* 35, 283-319.
16. Carroll, J. D. (1972). Individual differences and multidimensional scaling. In *Multidimensional Scaling: Theory and Applications in the Behavioral Sciences* (Shepard, R. N., Romney, A. K., and Nerlove, S., eds.), Academic, New York, pp. 105-155.
17. Larson-Powers, N., and Pangborn, R. M. (1978). Descriptive analysis of the sensory properties of beverages and gelatins containing sucrose or synthetic sweeteners. *J. Food Sci.* 43, 47-51.
18. Larson-Powers, M., and Pangborn, R. M. (1978). Paired comparison and time-intensity measurements of the sensory properties of beverages and gelatins containing sucrose or synthetic sweeteners. *J. Food Sci.* 43, 41-46.
19. Schiffman, S. S., Lindley, M. G., Clark, T. B., and Makino, C. (1981). Molecular mechanism of sweet taste: Relationship of hydrogen bonding to taste sensitivity for both young and elderly. *Neurobiol. Aging* 2, 173-185.

20. Hiji, Y. (1975). Selective elimination of taste responses to sugars by proteolytic enzymes. *Nature* 256, 427-429.
21. Bartoshuk, L. M., Dateo, G. P., Vandenbelt, D. J., Buttrick, R. L., and Long, L. (1969). Effects of *Gymnema sylvestre* and *Synsepalum dulcificum* on taste in man. In *Olfaction and Taste* (Pfaffmann, C., ed.), Rockefeller University Press, New York, pp. 436-444.
22. Zawalich, W. S. (1972). Comparison of taste and pancreatic beta cell receptor systems. In *Olfaction and Taste* IV (Schneider, D., ed.), Wissenschaftliche Verlagsgesellschaft MBH, pp. 280-286.
23. Schiffman, S. S., Cahn, H., and Lindley, M. G. (1981). Multiple receptor sites mediate sweetness: Evidence from cross adaptation. *Pharmacol. Biochem. Behav.* 15, 377-388.
24. McBurney, D. H., and Shick, T. R. (1971). Taste and water taste of twenty-six compounds for man. *Percept. Psychophys.* 10, 249-252.
25. Bartoshuk, L. M. (1968). Water taste in man. *Percept. Psychophys.* 3, 69-72.
26. Kruskal, J. B., Young, F. W., and Seery, J. B. (1978). *How to Use Kyst-2A, a Very Flexible Program to Do Multidimensional Scaling and Unfolding*, Bell Laboratories, Murray Hill, N.J.
27. Schiffman, S. S., Sennewald, K., and Gagnon, J. (1981). Comparison of taste qualities and thresholds of D- and L-amino acids. *Physiol. Behav.* 27, 51-59.
28. Schiffman, S. S., and Clark, T. B. (1980). Magnitude estimates of amino acids for young and elderly subjects. *Neurobiol. Aging* 1, 81-91.
29. Schiffman, S. S., Clark, T. B., and Gagnon, J. (1982). Influence of chirality of amino acids on the growth of perceived taste intensity with concentration. *Physiol. Behav.* 28, 457-465.
30. Benos, D. J. (1982). Amiloride: A molecular probe of sodium transport in tissues and cells. *Am. J. Physiol.* 242, C131-C145.
31. Schiffman, S. S., Lockhead, E., and Maes, F. W. (1983). Amiloride reduces the taste intensity of Na^+ and Li^+ salts and sweeteners. *Proc. Natl. Acad. Sci. (USA)* 80, 6136-6140.

11

Aspartame: Implications for the Food Scientist

Barry E. Homler
NutraSweet Group, G. D. Searle & Co., Skokie, Illinois

INTRODUCTION

Sweetness has always been one of man's most desired pleasures. The sweet taste is by far the most pleasing of the four basic "tastes" (sour, salt, bitter, and sweet) that we experience daily. However, when sweetness is ingested as carbohydrate, such pleasure carries with it an energy content of 4 kcal/g. Thus the pleasurable sensation normally obtained from sugar ingestion must be balanced against the amount of energy ingested if adiposity is to be avoided.

During this decade, the light and low-calorie food and beverage segments of the market have been major growth areas for the food industry. With the recent availability of aspartame,* these segments of the market have the potential for further growth. That growth will be limited only by the creativity and ingenuity of food manufacturers and food technologists.

The availability of aspartame extends the choice of sweetening agents available to the manufacturer. The choice of a sweetener (both caloric and noncaloric) for a food or beverage product depends on a variety of factors, including regulatory limitations, the quality of the product formulated, and the need for bulking agents. Aspartame's sweet, clean taste makes it a very desirable sweetener (2-8); however, aspartame's technical properties require that the formulation of new

Aspartame has been approved for use as a food additive in the United States, Canada, and several other countries. A full review of the food additive petition is contained in the *Federal Register* (1).

products involve more than a simple substitution of one sweetener for another. Reformulation and objective evaluation will often be required before new products can be marketed.

Before attempting to formulate new products, the food technologist should be aware of some of aspartame's characteristics that will affect the development of aspartame-sweetened food products. This chapter discusses the use of aspartame in food systems and provides information for typical applications.

CHEMISTRY

The structure of aspartame (N-L-aspartyl-L-phenylalanine 1-methyl ester is shown in Figure 1. Its physical properties are listed in Table 1. Aspartame is an odorless, white crystalline powder having a clean, sweet taste. It is slightly soluble in water (about 1.0% at 25°C) and is sparingly soluble in alcohol. Aspartame is not soluble in fats and oils. Being a peptide, the compound is amphoteric. The negative logs of its dissociation constants are 3.1 and 7.9 at 25°C, and its isoelectric point is 5.2.

Stability

Aspartame is an O-methyl ester. Under certain moisture, temperature, and pH conditions the ester bond is hydrolyzed, forming the dipeptide aspartylphenylalanine and methanol. Alternatively, methanol may be eliminated by the cyclization of aspartame to form its diketopiperazine. The diketopiperazine in turn can be hydrolyzed to form aspartylphenylalanine. Ultimately aspartylphenylalanine can be hydrolyzed to its individual amino acids, aspartate and phenylalanine. Aspartylphenylalanine, aspartate, phenylalanine, and the diketopiperazine are not sweet. When these compounds are formed in food products, a loss of sweetness is perceived; however, there is no offtaste or color. The primary reactions involved are shown in Figure 2.

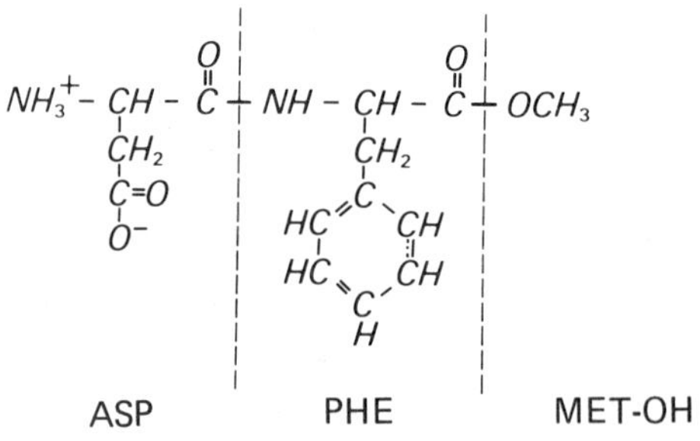

NH_3^+ – CH – C(=O) – NH – CH – C(=O) – OCH_3
CH₂ (ASP side chain): CH_2 – C=O – O^-
(PHE side chain): CH_2 – C_6H_5 (HC, CH, HC, CH, CH ring)

ASP PHE MET-OH

Figure 1 Chemical structure of aspartame.

Table 1 Properties of Aspartame

Color	White
Odor	None
Form	Powder
Taste	Clean, sweet
Solubility	Slightly soluble in water, sparingly soluble in alcohol
Diketopiperazine	NMT[a] 1.5%
Heavy metals	NMT 10 ppm
Loss on drying	NMT 4.5%

[a]NMT, not more than.

Dry Stability

The stability of aspartame in dry products is good. When used in dry applications, aspartame's stability is similar to that of the pure compound. Figure 3 shows aspartame stability under conditions far more severe than normal handling. At 105°C there is less than 5% conversion to the diketopiperazine.

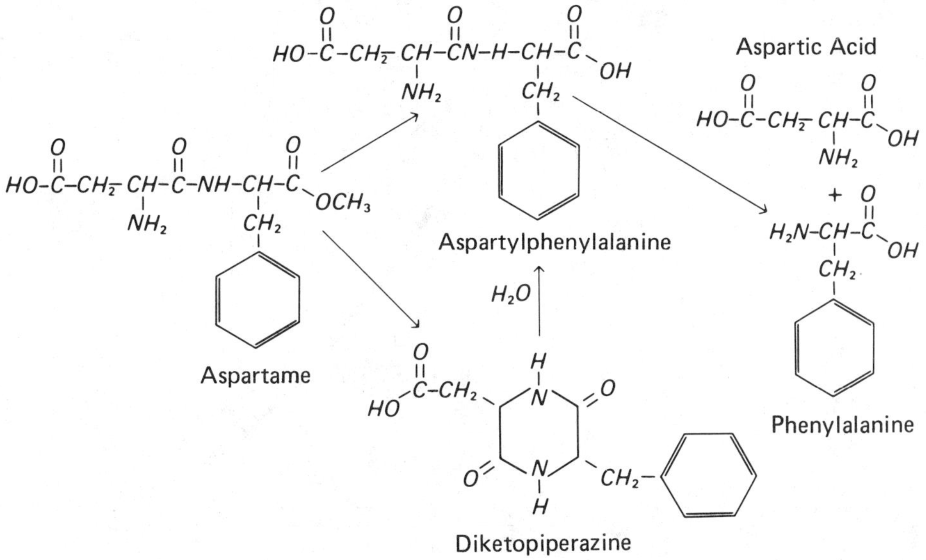

Figure 2 Typical chemical reactions by which aspartame is converted to non-sweet compounds.

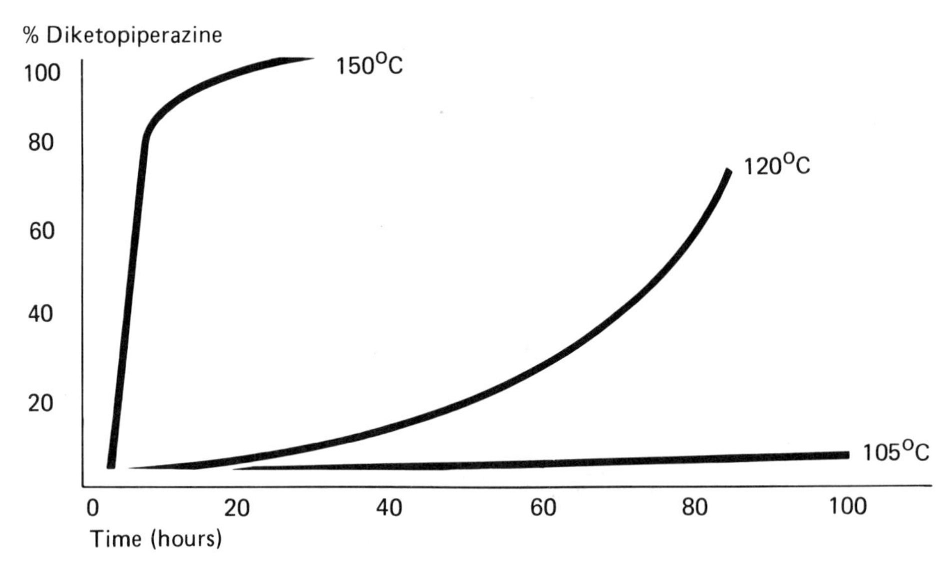

Figure 3 Rates of aspartame decomposition to form diketopiperazine at 105, 120, and 150°C.

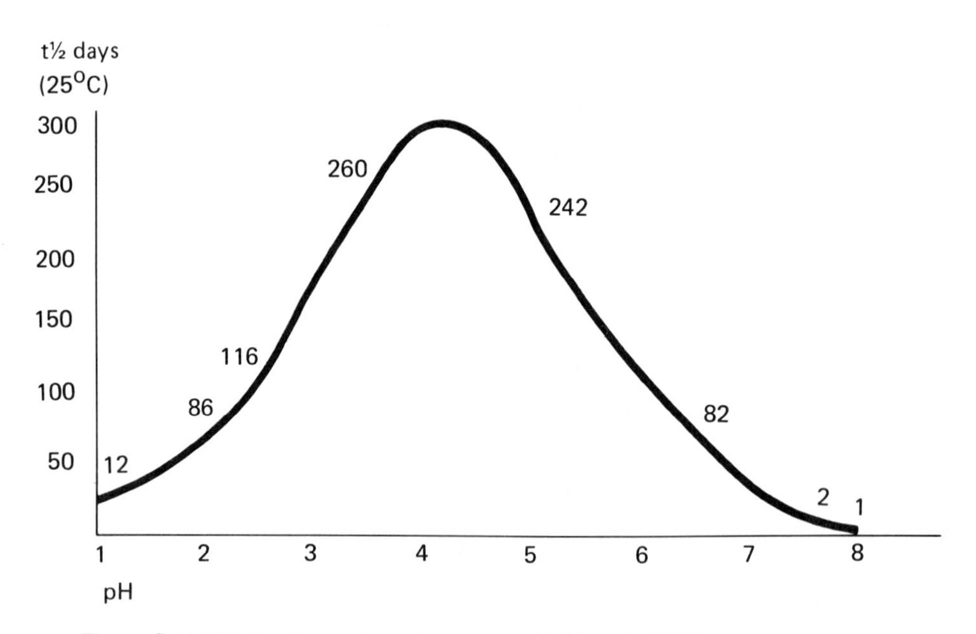

Figure 4 Stability of aspartame in aqueous buffers at 25°C.

Stability in Solution

The stability of aspartame in solution is a function of time, temperature, pH, and available moisture. Decomposition appears to follow simple first-order kinetics. At 25°C the maximum stability occurs at approximately pH 4.3 (Fig. 4). Aspartame is most stable between pH values of 3 and 5.

Figure 5 shows the stability of aspartame at 40°C in aqueous buffers ranging in pH from 2 to 8. A horizontal line represents almost complete stability. The more vertical the line, the less stability. Figures 6 and 7 show similar data obtained at 55 and 80°C. respectively. These figures show reasonable stability of aspartame in the weakly acid pH range (pH 2-5) present in many food systems, even with increasing temperature. The data obtained at 80°C are of particular interest, since they approximate conditions of high-temperature, short-time pasteurization. Use of aspartame in these applications is feasible, and no significant loss of the sweetener is noted. Aspartame will also withstand the processing used for aseptics.

Initial evaluation of these stability data might suggest that products with pH values outside the acidic range are poor prospects for aspartame sweetening. However, successful formulation of a wide range of products can be achieved by controlling the length of time a product is exposed to high temperatures during processing and then holding the product at low temperatures for distribution and sale. For example, ice cream has a pH of 6.5 to over 7. If the mix is pasteur-

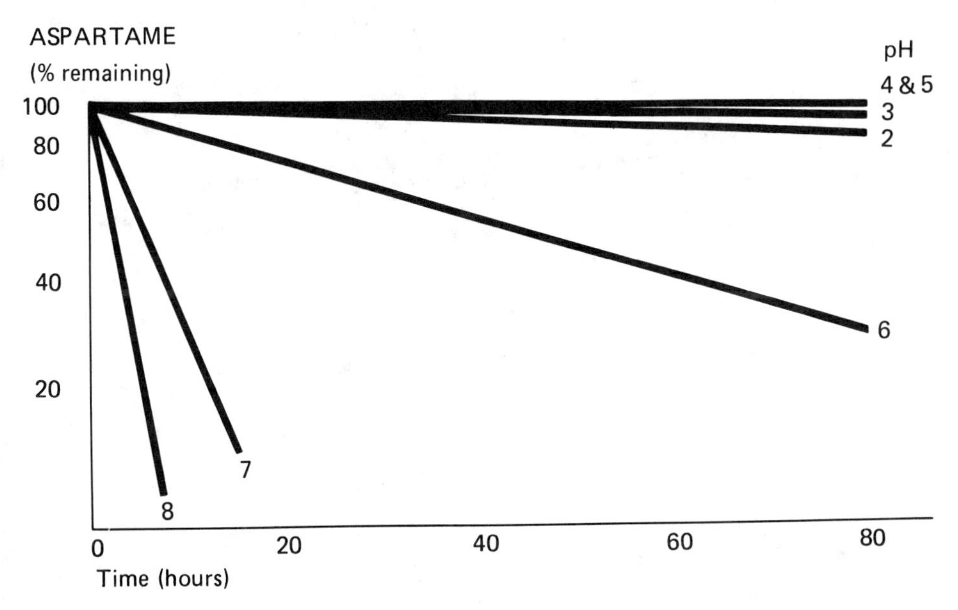

Figure 5 Stability of aspartame in aqueous buffers at 40°C.

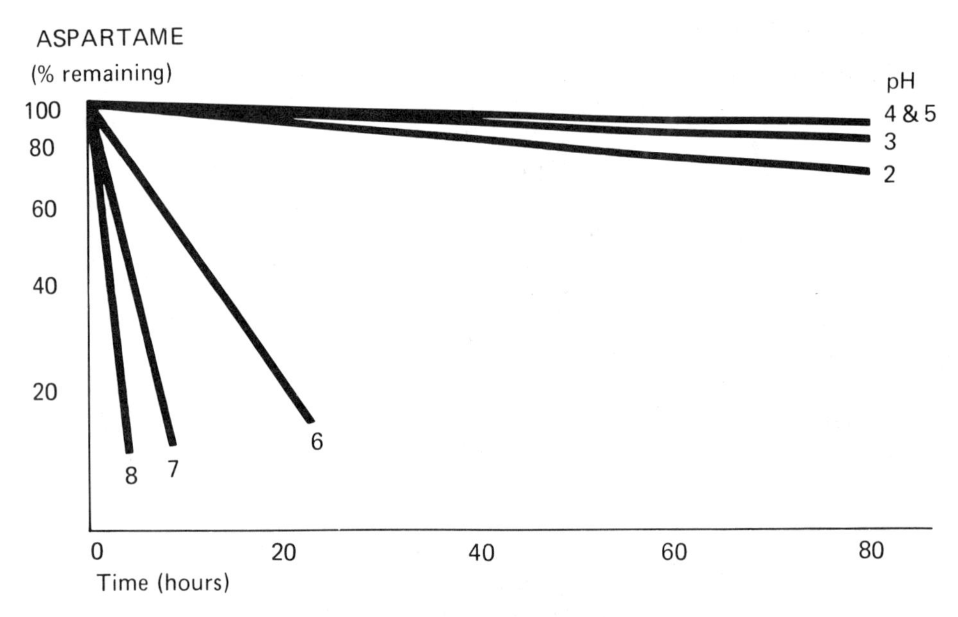

Figure 6 Stability of aspartame in aqueous buffers at 55°C.

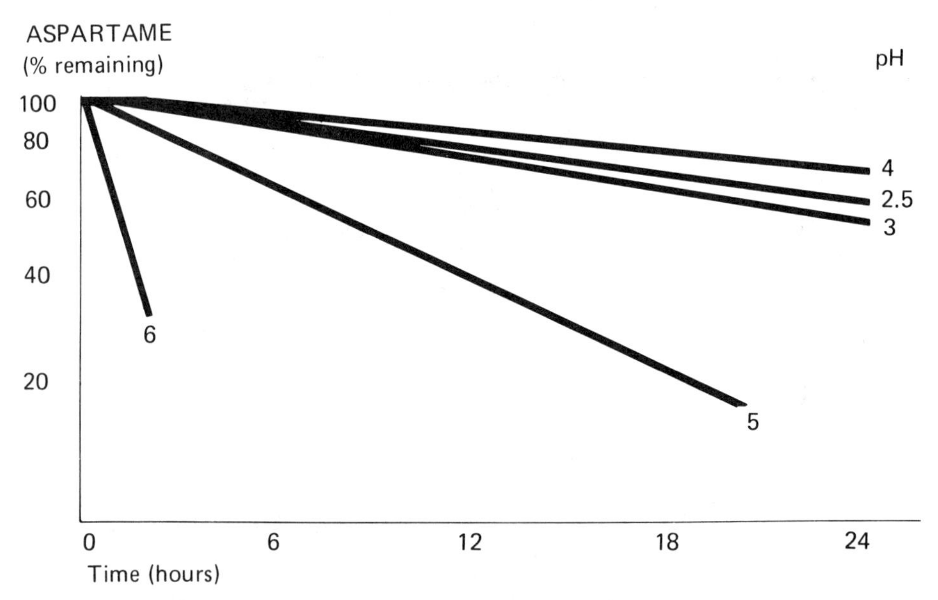

Figure 7 Stability of aspartame in aqueous buffers at 80°C.

ized by high-temperature, short-time conditions (180-190°F for under 1 min), virtually no loss of aspartame will be detected. Because the product is distributed and sold in the frozen state, the rate of aspartame loss is dramatically reduced (first-order reaction), probably owing to the reduction in free moisture. In this application, the stability of the product can exceed the normal desired shelf life of 6-8 months.

APPLICATIONS

General Comments

Although aspartame can provide sugarlike sweetness, both as an ingredient (called NutraSweet) and as a tabletop sweetener (Equal), its intensity of sweetness and light weight do not make it suitable as a sugar substitute for all food-processing applications. Other functional differences between the properties of sucrose and aspartame prevent its use as a "blanket" sugar substitute.

Unlike nutritive carbohydrate sweeteners, aspartame provides only sweetness. It cannot impart other physical properties to foods which are often associated with, and/or expected of, carbohydrate sweeteners. Some examples of the properties supplied by carbohydrate sweeteners include (9) texture, bulk, noncrystallization in some foods, solubility, humectant properties, preservation action, and caramelization.

The effect of removing sugar and replacing its sweetness with a high-potency sweetener in one typical food product, cake, produces a product that is no longer a cake, since the physical properties of the sugar in terms of bulk (volume), structure, and tenderness have been lost. To make a high-potency sweetener acceptable for this kind of use, a bulking agent to replace the sugar is needed. Preferably, the bulking agent should provide the benefits of sugar without the calories.

Aspartame Use Properties

Caloric Reduction

Aspartame provides a sweet, clean, sugarlike taste to food and beverages. Owing to its highly intense sweetness, food products using aspartame may have a much lower energy content. An example of this is shown in Table 2.

Aspartame is approximately 160-200 times sweeter than sugar. Its exact potency relative to sucrose is dependent on the food system in which it is used (Table 3). In water, a potency of 400 is observed for aspartame when it is tested at the threshold level for detection of sucrose sweetness; that is, aspartame is 400 times as sweet as sucrose at this level. At a sweetness level equal to that of a 10% sucrose-water solution, the potency declines to 133. By contrast, when the comparison is made in a lemonade mix, the potency rises to 180 at a 10% sucrose level.

Table 2 Comparison of the Energy Content of Aspartame- and Sugar-Sweetened Foods[a]

Food	Calories per serving	
	With sugar	With Aspartame
Lemonade made from mix (8 oz)	86	5
Gelatin dessert made from mix (½ cup)	81	10
Chocolate pudding made from mix (½ cup) with skim milk	150	75
Hot chocolate made from mix with water (8 oz)	116	63
Whipped topping made from mix (1 tablespoon)	14	5
Instant milk shake mix with skimmed milk (1 cup)	189	70

[a]These are examples of potential reductions in energy. Final formulations may not exhibit the same level of energy.
Source: Ref. 10.

In general, use levels for aspartame will vary from about 0.01% for a flavor modification role to 0.6% for sweetening in a finished product. Exact levels will depend on the product formulation, pH, temperature, and the specific flavor characteristics desired.

Enhancement and Extension of Flavor

In addition to its use as a sweetening agent, aspartame also enhances and extends some food and beverage flavors, particularly acidic fruit flavors. This property should be considered in product formulation. Any lingering sweetness from aspartame can be modified by use with other sweeteners or certain salts (11). As with all sweeteners, the flavor perceived may also be altered by temperature, other flavorings, pH, viscosity, and total solids, as well as by other factors.

Increasing Nutrient Density

In certain products, aspartame can be used to reduce the bulk provided by sugar. This makes it possible to increase the amount of other, more nutritious ingredients. For example, if some or all of the sugar in a presweetened cereal is replaced with aspartame sweetness, the amount of grain in the box can be increased by 10-40%.

The same is true of a peanut butter which has a standard of identity requiring a 90% minimum peanut content. When 2% stabilizer and 1% salt are added to peanut butter, the amount of sweetener is limited to 7%. Adding 0.12% aspartame as the sweetener would make the product twice as sweet, leaving room for a 97% total peanut product.

Table 3 Reported Potency Values for Aspartame in Various Systems

	Potency[a]	Sucrose in the control medium (%)	Tasting temperature °C	°F
Water	400	0.34	22	72
	215	4.3	22	72
	133	10	22	72
	100	15	22	72
	182	2	8	46
	160	4	8	46
	100	12	8	46
	43	30	8	46
	61	10	3	37
	54	10	22	72
Buffer-citrate	160	4	8	46
Phosphate at pH 3.2	100	12	8	46
Powdered soft drink mixes	133	10	8	46
	143	10	3	37
	180	11	8	46
Chewing gum	100-199	69	22	72
Gelatin	160-172	18	3	37
Pudding mix	200	17	8	46
Presweetened cereal	180	37	–	–
Soft drinks	180	10	8	46
Coffee/tea	180	4-6	70	
Flavored yogurt	175-220	17 (13 added)	8	46
Canned fruit	175	20	Chilled	

[a]Potency is equal to the concentration of sucrose divided by the concentration of aspartame at equal sweetness.

Bulk Reduction

A powdered beverage mix used to make 2 qt of final product weighs 200 g when made with sucrose and 14 g when made with aspartame; this represents a 93% reduction in net weight, and a significant reduction in the cost of packaging and shipping.

Synergy

By using aspartame in combination with other carbohydrate and high-potency sweeteners (sucrose, dextrose, fructose, saccharin), a variety of reduced- or low-calorie products can be formulated. Aspartame also exhibits synergy with many sweeteners, thereby allowing lower use levels of total sweetener. Synergy, simply put, is where 1 + 1 = 3.

Product Categories

Tabletop Sweeteners

High-potency sweeteners must be combined with diluents before they are suitable for home use in tablets, granular form, or in sachet packets. Diluents such as maltodextrin, dextrose, and lactose, along with agents to keep the material free flowing, are the most commonly used materials. In general, each tablet provides equivalent sweetness to one teaspoon of sucrose, while the packets provide the sweetness of two teaspoons of sucrose.

Dry Mix Products

This class of foods includes cold beverage mixes, milk shake mixes, hot chocolate mixes, milk amplifiers, instant presweetened tea, gelatin dessert mixes, pudding and pie filling mixes, breakfast drinks, dairy analog whipped topping, yogurt flavorings, and salad dressing mixes.

These products are generally easy to formulate, requiring only the dry blending of individual ingredients such as flavoring agents, colors, acidulants, gums, and other carriers. The rate of dissolution can be enhanced by preblending aspartame with a carrier such as maltodextrin and then blending this mix with the remaining ingredient. A ratio of 3-4 parts maltodextrin to 1 part of aspartame provides sufficient bulk to make handling easier and also makes dispersion and ultimate dissolution of aspartame in cold liquids more efficient. Packaging considerations should include the use of materials that serve as a moisture barrier to prevent caking of the dry mix while in distribution.

Excellent beverage mixes have been prepared, and a formulation for a typical product is shown in Table 4.

Table 4 Dry Mix Formulation: Lemonade Beverage[a]

Ingredient	Percent
Citric acid	55.00
Maltodextrin	19.93
Sodium citrate	8.45
Aspartame	5.74
Tricalcium phosphate	4.22
Ascorbic acid (vitamin C)	3.06
Thickener	2.41
Clouding agent	1.91
Flavoring	To suit
Coloring	To suit

[a]To serve, 16 g of the dry mix are dissolved in 2 qt of cold water. The final product provides approximately 4 kcal per 8-oz serving. In contrast, sugar-sweetened products provide 90-100 kcal.

A typical gelatin formulation is presented in Table 5. If more body is desired, maltodextrin, Polydextrose from Pfizer Chemicals or crystalline fructose may be used with the aspartame.

Puddings are usually set with starches or alginate systems. A formulation for a cooked pudding is shown in Table 6. A similar formulation for an instant pudding is shown in Table 7.

Carbonated Beverages

Products in this category represent a large potential market for high-potency sweeteners. Aspartame can be adapted to use in carbonated beverages by controlling pH, time (age), distribution time, and aspartame concentration. Products manufactured with aspartame have shown acceptance over the normal shelf life

Table 5 Dry Mix Formulation: Gelatin Dessert[a]

Ingredient	Percent
Gelatin (225 Bloom)	67.90
Adipic acid	17.00
Fumaric acid	1.40
Sodium citrate	6.40
Aspartame	3.80
Flavor	To suit
Color	To suit

[a]Product preparation: (a) In a mixing bowl, dissolve 12 g of mix in one cup of hot water; (b) mix in one cup cold water, and (c) pour into dessert cups and chill until set. The energy content is 12-13% of the comparable sugar-sweetened product.

Table 6 Dry Mix Formulation: Chocolate Pudding, Cooked[a]

Ingredients	Percent
Cornstarch	54.31
Modified cornstarch	23.05
Aspartame	15.00
Salt	2.00
Carob powder	1.35
Color	0.60
Flavor	To suit

[a]Product preparation: (a) stir 28 g of mix into two cups of milk (or skim milk) in a saucepan; (b) bring to a boil while stirring rapidly; (c) remove from heat; and (d) pour into dessert cups and chill till set. The energy content is approximately 50% of the sugar-sweetened product.

Table 7 Dry Mix Formulation: Chocolate Pudding, Instant[a]

Ingredient	Percent
Modified starch	70.00
Aspartame	2.60
Trisodium phosphate	5.50
Sodium dihydrophosphate	2.35
Calcium acetate	1.90
Salt	0.60
Carrageenan	4.35
Flavor	To suit
Caramel powder	2.20
Cocoa powder (10-12% fat)	8.75

[a]Product preparation: Add 25 g of the above instant pudding mix to two cups of milk, mix well, and chill. The energy content is approximately 50% of the sucrose-sweetened product.

of the product, and tests conducted on various flavors of these beverages show that loss of 40% of the aspartame is required before sweetness perception is no longer acceptable.

Typical carbonated soft drinks sweetened with carbohydrate sweeteners have sugar concentrations ranging from 8 to 15%. When substituted for sugar, the amount of aspartame that will give acceptable sweetness varies over a broad range attributable to the pH range found in the products, the storage life desired for the product, and flavor and flavor potentiation factors observed.

Table 8 shows typical levels for aspartame use in soft drinks. The process required to manufacture carbonated drinks with aspartame is essentially the same as that used in preparing sugar-sweetened drinks, except that the aspartame should be predissolved in a portion of the mix water in a separate step.

Cereals

Aspartame can be used to replace all or part of the sugar in cereals. As mentioned earlier, this can increase the amount of grain in a box of cereal, providing more

Table 8 Carbonated Beverages: Use Levels

Flavor	pH range	Percent aspartame used
Cola	2.4-3.1	0.055-0.068
Lemon-lime	3.0-3.1	0.030-0.060
Orange	3.1-3.4	0.055-0.090
Root beer	4.0-4.4	0.055-0.070

nutrients. Energy content is not reduced, since the added grain replaces the energy content of the removed sugar. Aspartame may be dusted onto the cereal prior to drying or incorporated into some sugar and sprayed on the cereal surface.

Chewing Gum

Because of the flavor enhancing properties of aspartame, its use in chewing gum is a promising application. Flavor and sweetness can be extended to close to 30 min versus the 5-6 min normally noted with a sugar-sweetened gum. Sugar is replaced by sorbitol and the bulk of the gum is retained.

Dairy Products

Flavored milks where sugar is replaced by aspartame are also potential applications. Since these products are distributed refrigerated with a "best when used by" date of about 2-3 weeks, products with aspartame retain full acceptable sweetness even though the product pH may be 7. Use levels of 0.04-0.06% aspartame will provide good sweetness.

Frozen desserts are another application for aspartame use. These include nonstandardized ice cream and ice milks, sherbets, puddings, and novelty products. Because these products are stored and distributed in the frozen condition, the products retain sweetness. Sugar in these products can either be reduced or replaced with sorbitol or Polydextrose to maintain volume, freezing point depression, and texture. Use levels of 0.05-0.08% aspartame provide acceptable sweetness.

Use of aspartame to sweeten yogurt is an additional category within the dairy product grouping. Aspartame can be used with flavor mixes and added to plain yogurt without affecting the caloric value. If some of the added sugar in yogurt is replaced, this will increase the product's protein content. A typical yogurt formulation is shown in Table 9.

Table 9 Aspartame-Based Yogurt, Swiss Style

Base for yogurt	
Skim milk[a]	100 gal
Nonfat milk solids	44 lb
Stabilizer[b]	5 lb
Yogurt culture[c]	Amount needed for 100 gal
Finished product	
Yogurt base	99.88%
Aspartame	0.07%
Flavor	To suit
Color	To suit

[a]8.5-9.0% solids.
[b]225 Bloom gelatin.
[c]*Lactobacillus bulgaricus* and *Streptococcus thermophilus* in a 1:1 ratio.

Refrigerated/Frozen Drinks

These beverage products are sold refrigerated and contain 0-50% juice. When refrigerated, these products are sold with an 8-week shelf life. This is attainable with aspartame owing to the refrigerated temperature and the acidic pH of these products (between 3.0 and 4.0). Aspartame sweetness likewise fits in well with frozen drinks, where shelf lives of over 1 year are attainable.

Confections

Candies and other confections are potential areas for aspartame use. However, confections contain a large proportion of sugar, and total replacement without a suitable bulking agent is unlikely. Use of a high-potency sweetener in such products will be limited until the technological problems of replacing sugar function are solved.

Other Applications

Product categories with potential for aspartame use include pickles (along with a crisping agent to replace that provided by sugar), creme fillings, glazes, jams, jellies, preserves, and fruit fillings.

While some technological problems remain, many can be overcome by balancing the various processing conditions to achieve the required stability, or by maintaining the pH and moisture content in a region giving greatest aspartame stability. Additional references of potential interest to the reader are listed in the References (12-24).

SUMMARY

Aspartame can help food producers exploit one of the major trends in the food market today: the trend toward "light foods." Although not a blanket substitute for sugar and other carbohydrate sweeteners, aspartame does provide many opportunities for formulating new products. These opportunities reflect aspartame's ability to be blended with other sweeteners, its unique flavor-enhancing properties, and its clean sugarlike taste. When new compatible bulking agents become available, even more opportunities will arise.

The formulation of products with aspartame is not a simple substitution of one sweetener for another, but requires reformulation and objective evaluation. However, technological ingenuity in reformulation coupled with creative new marketing strategies can result in many new and acceptable food and beverage products.

REFERENCES

1. Aspartame (1981). Commissioner's final decision. *Fed. Reg.* 46, 38284-38308.
2. Baldwin, R. E., and Korschgen, B. M. (1979). Intensification of fruit-flavors by aspartame. *J. Food Sci.* 44, 938-939.

3. Cloninger, M. R., and Baldwin, R. E. (1970). Aspartylphenylalanine methyl ester: A low-calorie sweetener. *Science* 170, 81-82.
4. Cloninger, M. R., and Baldwin, R. E. (1974). L-Aspartyl-L-phenylalanine methyl ester (aspartame) as sweetener. *J. Food Sci.* 39, 347-349.
5. Larson-Powers, N., and Pangborn, R. M. (1978). Paired comparisons and time-intensity measurements of the sensory properties of beverages and gelatins containing sucrose or synthetic sweeteners. *J. Food Sci.* 43, 41-46.
6. Larson-Powers, N., and Pangborn, R. M. (1978). Descriptive analysis of the sensory properties of beverages and gelatins containing sucrose or synthetic sweeteners. *J. Food Sci.* 43, 47-51.
7. McPherson, B. A., McGill, L. A., and Bodyfelt, F. W. (1978). Effect of stabilizing agents and aspartame on the sensory properties of orange sherbert. *J. Food Sci.* 43, 935-939.
8. Schiffman, S. S. (1984). Comparison of the taste properties of aspartame with other sweeteners. In *Aspartame: Physiology and Biochemistry* (Stegink, L. D., and Filer, L. J., Jr., eds.), Marcel Dekker, New York, Chap. 10.
9. Beck, C. I. (1978). Application potential for aspartame in low calorie and dietetic food. In *Low Calorie and Special Dietary Foods* (Dwivedi, B. K., ed.) CRC Press, West Palm Beach, Fla., pp. 59-114.
10. Kraus, B. (1971). *Calories and Carbohydrates*, Grosset & Dunlap. New York.
11. Schade, H. R. (1976). Taste modifier for artificial sweeteners, U.S. Patent No. 3,934,047, General Foods Corp., White Plains, New York, January 20.
12. Bahosy, B. J., Klose, R. E., and Nordstrom, H. A., (1976). Chewing gums of longer lasting sweetness and flavor, U.S. Patent No. 3,943,258, General Foods Corp., White Plains, New York, March 9.
13. Beck, C. I. (1974). Sweetness, character, and applications of aspartic acid-based sweeteners. In *ACS Sweetener Symposium* (Inglett, G. E., ed.), AVI Publishing, Westport, Ct., pp. 164-181.
14. Beck, C. I. (1978). Application potential for aspartame in low calorie and dietetic food. In *Low Calorie and Special Dietary Foods* (Dwivedi, B. K., ed.), CRC Press, West Palm Beach, Fla., pp. 59-114.
15. Food Chemical Codex, 3rd ed. (1981). National Academy Press, Washington, D.C., pp. 28-29.
16. Kobayashi, N., Kamimura, M., and Yamane, T. (1976). On aspartame, a new sweetener. *Seito Gijutsu Kenkyukaishi* 26, 7-17.
17. Larson, N. L. (1975). Sensory properties of flavored beverages and gelatins containing sucrose or synthetic sweeteners, M.S. thesis, University of California, Davis, California.
18. Mazur, R. H., Schlatter, J. M., and Goldkamp, A. H. (1969). Structure-taste relationships of some dipeptides. *J. Am. Chem. Soc.* 91, 2684-2691.
19. Mazur, R. H., and Ripper, A. (1979). Peptide-based sweeteners. In *Development in Sweeteners* (Hough, C. A. M., Parker, K. J., and Vlitos, A. J., eds.) Applied Science Publishers, London, pp. 125-134.
20. McPherson, B. A., McGill, L. A., and Bodyfelt, F. W. (1978). Effect of stabilizing agents and aspartame on the sensory properties of orange sherbert. *J. Food Sci.* 43, 935-939.

21. Moskowitz, H. R., and Bubose, C. (1977). Taste intensity, pleasantness and quality of aspartame, sugars, and their mixtures. *Can. Inst. Food Sci. Technol. J.* 10, 126-131.
22. Porikos, K. R., Booth, G., and Van Itallie, T. B. (1977). Effect of covert nutritive dilution on the spontaneous food intake of obese individuals: A pilot study. *Am. J. Clin. Nutr.* 30, 1638-1644.
23. Schiffman, S. S., Reilly, D. A., and Clark, T. B. III. (1979). Qualitative differences among sweeteners. *Physiol. Behav.* 23, 1-9.
24. Schiffman, S. S., Cahn, H., and Lindley, M. G. (1981). Multiple receptor sites mediate sweetness: Evidence from cross adaptation. *Pharmacol. Biochem. Behav.* 15, 377-388.

12

Role of Sugar and Other Sweeteners in Dental Caries

William H. Bowen
University of Rochester School of Medicine and Dentistry,
Rochester, New York

INTRODUCTION

An abundance of evidence has accumulated over several decades that leads any reasonable person to conclude that the frequent ingestion of sugars by humans is associated with the development of dental caries. The evidence has been gathered from epidemiological studies, clinical trials, and animal studies.

EPIDEMIOLOGICAL STUDIES

Although epidemiological studies at best can establish associations, after examining all the available results, one is inevitably led to the conclusion that there is a very strong relationship between the ingestion of sugars and the prevalence of dental caries.

A study conducted in Hopewood House (Australia) showed clearly that children who resided in the institution, and whose diet excluded such items as cookies, candies, and highly sugared foods, had substantially fewer carious lesions than did their cohorts whose diet was unrestricted (1). It is of interest to note that as children from the institution were allowed more frequent access to the nearby city, the prevalence of caries increased.

Weiss and Trithart (2) have observed that children who snack frequently were more likely to develop rampant caries than those who snack less frequently. In addition, they observed that a large proportion of children who on the average snacked fewer than three times daily were caries free.

There are a number of studies that report a strong association between rampant dental caries and protracted consumption of sweetened beverages by nursing infants (3,4). Children who have nursing bottle syndrome usually display severe caries along the palatal surfaces of the upper molars and upper incisors. This pattern of caries results from close contact between the nipple delivering the beverage and the teeth.

A strong association has been reported by Palmer (5) between ingestion of snacks and prevalence of dental caries. Children partaking of bedtime snacks developed significantly more caries than did children who did not so indulge.

It has been shown that the children of dentists have a much lower prevalence of caries than other children (6). This difference has been attributed to good dietary practices among the families of dentists.

The strict rationing of foods which occurred during and immediately following World War II had a profound effect on the incidence of dental caries in Europe and Japan (7). For example, studies conducted on Norwegian children aged 7-14 showed a reduction in prevalence of caries of 50-75% between 1940 and 1949 (8). Comparable reductions were observed in Great Britain (9); however, when candy became readily available by 1953, there was an immediate and dramatic increase in the prevalence of caries.

Research conducted over many years on the inhabitants of the island of Tristan da Cunha (10) revealed that when sugar and other refined carbohydrates were introduced to the island, there was a large increase in the prevalence of dental caries. Following an earthquake on the island, a large number of inhabitants emigrated to Great Britain, where their dental health was monitored. A dramatic increase in the prevalence of caries was observed over a period of years (11).

In contrast to these studies, Walker et al. (12) were unable to show a relationship between *total* sugar intake and the prevalence of caries in black children; similar observations were reported by Zita et al. (13), except they did find a strong positive correlation with between-meal sugar and prevalence of caries.

Studies conducted on patients who suffer from hereditary fructose intolerance and therefore must avoid sucrose have revealed that these persons are virtually caries free (14,15). It should be noted that such persons usually avoid not only sucrose, but all sweet-tasting foods.

CLINICAL STUDIES

For ethical and practical reasons very few clinical trials have been conducted in humans to determine the cariogenicity of sucrose or other sweetening agents. However, the few studies that have been carried out reveal that the onset of caries is associated with the ingestion of sugar.

In an elegant clinical investigation, Gustaffson et al. (16) demonstrated that the incidence of caries is related to the frequency of intake of carbohydrates and not to the total amount consumed. For example, patients who in 1 year consumed 94 kg of sugar included with their meals had fewer new carious lesions than patients who consumed 85 kg, 15 kg of which was taken between meals.

In a heroic study, Scheinin et al. (17) investigated the effects of substituting xylitol for sucrose in the diet of several hundred subjects. Control subjects consumed their regular diet. Subjects who ate the sucrose-free diet developed few carious lesions over a 2-year period when compared to controls.

Investigations conducted in humans to determine the effects of additives and sugar substitutes in chewing gum have revealed that sugar-containing gum is highly cariogenic (18,19).

ANIMAL STUDIES

Results of studies conducted in animals show that caries does not develop in the absence of sugars. Kite et al. (20) observed that caries-susceptible rats receiving their entire diet by gastric intubation remained caries free. This observation has been confirmed and extended by Bowen et al. (21), who also showed that sucrose is considerably more cariogenic than starch.

The overwhelming majority of diets used in animal experiments to study dental caries contain considerable levels of sucrose (22). When fed ad libitum to susceptible animals, all lead to rapid development of caries in a few weeks.

FACTORS AFFECTING CARIOGENICITY OF SUGAR

Although sucrose is highly cariogenic, there are several ways whereby its cariogenicity can be modified. For example, sugar taken in liquid form is substantially less cariogenic than powdered sugar (20). Particle size too can influence the cariogenicity of sugar (23). Large particles, because they do not become impacted in tooth fissures, are less cariogenic than sugar in powdered form. Sugar given in an adhesive form is usually highly cariogenic, and, in general, the longer sugar remains in the mouth, the greater the likelihood that caries will develop.

OTHER SUGARS

There is no evidence that suggests sugars such as glucose, fructose, or corn syrups are less cariogenic in humans than sucrose. Indeed, sucrose is probably believed to be the most cariogenic sugar in humans because it is undoubtedly the most frequently used sugar (24).

The results from studies conducted in rodents suggest that glucose and fructose may be slightly less cariogenic than sucrose (25,26). However, the results are by no means unequivocal and the differences observed are usually confined to smooth surfaces. The vast majority of carious lesions occur on occlusal surfaces and the incidence is relatively unaffected by the sugar fed.

The use of corn syrups (high fructose and high glucose) is increasing rapidly. Evidence gleaned from studies using corn syrup solids in rats (27) and a mixture of glucose and fructose (28) in monkeys suggests that little dental benefit would be derived by substituting these sweeteners for sucrose.

CARIOGENICITY OF SUGARS

Dental caries results from the action of specific bacteria which colonize the tooth surface and metabolize particular components of the diet, forming acid. Sugar, however, apparently has a broad range of effects which may result in enhancing the susceptibility of teeth to caries attack. In addition, frequent ingestion of sugar creates a cariogenic environment which tends to enhance the effect of each ingestion of sugar.

Newly erupted rat teeth are usually hypomineralized. Mineralization progressively increases posteruptively and is enhanced in the presence of fluoride (29). Available evidence indicates that sucrose, even in the absence of a direct cariogenic challenge, inhibits posteruptive mineralization (30), thereby presumably rendering teeth more susceptible to carious attack.

The presence of sucrose apparently promotes the colonization of tooth surfaces by *Streptococcus mutans*. Although *S. mutans* can become established in humans and in animals in the absence of sucrose, the minimal infective dose is greatly reduced in animals fed sucrose (31). In addition, the evidence suggests that ingestion of sucrose results in "irreversible" binding of *S. mutans* to tooth surfaces. Some strains of *S. mutans* are more dependent than others on the presence of sucrose to facilitate binding (32).

All serotypes of *S. mutans* synthesize glucosyl transferase, and some have the capacity to make fructosyl transferase (33). Using sucrose as substrate, these enzymes synthesize glucan and fructan. Essentially two types of glucan are formed. One is water soluble and is comprised mostly of α1-6 linked glucose. Most of the glucan formed is α1-3 linked glucose and is water insoluble (34). Glucan enhances the ability of microorganisms to adhere to tooth surfaces and in addition entraps organisms that might not otherwise colonize the tooth surface. Glucan synthesized in vivo apparently carries a charge (possibly bound lipoteichoic acid) which effectively prevents the diffusion of ions into or from dental plaque (35). Substances such as bicarbonate are prevented from diffusing into plaque, and acid produced within plaque cannot diffuse out. Uncharged materials such as sugars can readily diffuse into plaque.

Relatively high concentrations of fructan can be detected in plaque following ingestion of sucrose. Levels, however, rapidly decline, because a large number of microorganisms in dental plaque have the capacity to metabolize fructan (36).

Many microorganisms in plaque have the capacity to synthesize intracellular polysaccharide from a range of carbohydrates (37). This polysaccharide stains readily with iodine and is apparently of the amylopectin type. It can be catabolized by microorganisms in plaque during periods when extraneous sources of carbohydrate are lacking. This catabolism is probably responsible for the comparatively low pH values observed in plaque around carious lesions, even in patients who have been fasting (38). Available evidence suggests that the number of intracellular polysaccharide-forming organisms is positively correlated with the number of carious lesions (38). Both the synthesis and catabolism of intracellular polysaccharide are sensitive to the presence of fluoride (39).

Although all of the phenomena described thus far result in the creation of a cariogenic environment, dissolution of the tooth results from production of acids from sugar by bacteria in dental plaque. Each ingestion of sugar results in a rapid fall in pH values of dental plaque. Values as low as 4.0 are not uncommon. It is generally believed that enamel does not begin to dissolve rapidly until a pH of 5.5 is reached. The pH gradually returns to normal in the absence of additional sugar (40).

A large variety of acids can be detected in dental plaque from subjects who are highly caries active. These include acetic, lactic, propionic, and butyric acids (41). A constant relationship between the concentration of lactic acid in plaque and pH decrease is lacking (42). In some instances it appears that the concentration of lactic, propionic, and acetic acids accounted for less than 50% of titratable acidity. It has been observed that "fasting" plaque, that is, plaque that has not been exposed to carbohydrate for several hours, may contain 3×10^{-5} mmol of acid per milligram wet weight. Five minutes following exposure to sugar the acid concentration had increased to 5×10^{-5} mmol. A fivefold increase in D(−)-lactate and an eightfold increase in D(+)-lactate (43) are the major change observed.

SUGAR ALCOHOLS

The effects of polyols such as sorbitol, mannitol, and xylitol on plaque formation and caries development have been determined in animals (44-46) and to a lesser extent in humans (47,48). Ingestion of sorbitol in monkeys was followed initially by the formation of plaque with a syrupy consistency (44). The population of *S. mutans* declined sharply when sorbitol was substituted for sucrose, even though *S. mutans* ferments sorbitol. Prolonged ingestion of sorbitol by monkeys or humans did not lead to the development of a flora with enhanced ability to ferment sorbitol (44,47). All the evidence indicates that sorbitol and mannitol have negligible caries activity in humans and animals.

The effects of xylitol have been studied extensively in man (17) and in animals (46,49). Results of a clinical trial conducted over 2 years showed that xylitol is noncariogenic in humans (17). Studies conducted in humans and animals indicate that xylitol may possess cariostatic properties (49). Protracted use of xylitol in the absence of other sugars leads to remineralization of early carious lesions. Ingestion of xylitol leads to enhanced levels of lactoperoxidase and amylase in saliva. In addition, xylitol blocks the uptake of glucose by *S. mutans* (50,51).

Widespread substitution of sucrose by sugar alcohols is unlikely to be acceptable, because in many humans even moderate doses of sugar alcohols may have a cathartic effect. However, xylitol, which is about as sweet as sucrose, is less likely to induce catharsis than mannitol or sorbitol.

OTHER SWEETENERS

Although a large number of other sweeteners such as monellin and dihydrochalcones have been developed, their effect on dental caries has not been determined. In general, however, if they are not carbohydrates or can be used in low concentration, they are unlikely to be cariogenic. The effects of several sweeteners on plaque metabolism and caries development have been determined. These may be summarized as follows:

Saccharin

Growth of *S. mutans* is inhibited by saccharin (52), and, in addition, cell-free enzyme glucoysl-transferase is inhibited. The addition of saccharin to a cariogenic diet reduces the incidence of caries in rats. Although saccharin is apparently a weak carcinogen, its use in low doses in combination with cariogenic snack foods could make a contribution to dental health (53).

Aspartame

This dipeptide sweetener has been shown to be noncariogenic in rats; in addition, it appears to have a cariostatic effect on the free smooth surfaces (54). However, this latter observation has not been confirmed by Tanzer (personal communication).

It has also been observed that the inclusion of aspartame in saliva-glucose mixtures prevents a fall in pH, in contrast to those pH changes observed in control mixtures (55). The mechanism of action is unclear; however, several dipeptides and tripeptides that naturally occur in saliva are associated with a rise in pH in plaque following ingestion of sugar. It is believed that such peptides are broken down by plaque bacteria with the liberation of ammonia (56).

Lylose (Iso-Maltulose, Palatinose)

This substance is formed enzymatically from sucrose and is about 50% as sweet as sucrose. Recent evidence shows that it is not readily metabolized by plaque bacteria, and in addition it inhibits glucosyl transferase. Lylose is noncariogenic, and it appears that it may also be cariostatic (57,58).

CONCLUSION

It appears that no single sweetener is an ideal substitute for sucrose. Those substitutions which are several hundred times sweeter than sucrose lack the necessary bulk to be effective substitutes for sucrose. Other agents have undesirable side effects or aftertaste. It appears that a combination of sweeteners, some possessing cariostatic effects, could provide an ideal alternative for sucrose.

REFERENCES

1. Harris, R. M. (1963). Biology of the children of Hopewood House. *J. Dent. Res.* 42, 1387-1391.
2. Weiss, R. L., and Trithart, A. H. (1960). Between meal eating habits and dental caries experience in preschool children. *Am. J. Public Health* 50, 1097-1104.
3. Syrrist, A., and Selander, P. (1953). Some aspects of comforters and dental caries. *Odont. Revy.* 61, 237-240.
4. Goose, D. H. (1967). Infant feeding and caries of the incisors: An epidemiological approach. *Caries Res.* 1, 167-170.
5. Palmer, J. D. (1971). Dietary habits at bedtime in relation to dental caries in children. *Br. Dent. J.* 130, 288-293.
6. Ludwig, T. G., Denby, G. C., and Struthers, W. H. (1900). Dental health 1. Caries prevalence amongst dentists' children. *N. Z. Dent. J.* 50, 174-181.
7. Mandel, I. D. (1970). Effects of dietary modifications on dental caries in humans. *J. Dent. Res. Suppl.* 49, 1201-1211.
8. Sognnaes, R. F. (1948). An analysis of a war-time reduction of dental caries in European children. *Am. J. Dis. Child.* 75, 792-796.
9. James, P. M. C. (1965). The problem of dental caries. *Br. Dent. J.* 119, 295-300.
10. Sognnaes, R. F. (1965). In *Caries-Resistant Teeth* (Wolstenholm, G. E. W., and O'Connor, M., eds.), General discussion, p. 114, Churchill, London.
11. Holloway, P. J., James, P. M., and Slack, G. H. (1963). Dental disease in Tristan da Cunha. *Br. Dent. J.* 115, 19-25.
12. Walker, A. R. P., Dison, E., Duvenhage, A., Walker, B. F., Friedlander, I., and Aucamp, V. (1981). Dental caries in South African black and white high school pupils in relation to sugar intake and snack habits. *Community Dent. Oral Epidemiol.* 9, 37-43.
13. Zita, A. C., McDonald, R. E., and Andrews, A. L. (1959). Dietary habits and the dental caries experience in 200 children. *J. Dent. Res.* 38, 860-865.
14. Marthaler, T. M., and Froesch, E. R. (1967). Hereditary fructose intolerance. *Br. Dent. J.* 129, 597-599.
15. Hoover, C. I., Newbrun, E., Mettraux, G., and Graf, H. (1980). Microflora and chemical composition of dental plaque from subjects with hereditary fructose intolerance. *Infect. Immun.* 28, 853-859.

16. Gustafsson, B. E., Quensel, C. E., Lanke, L., Lundquist, C., Grannen, H., Bonbow, B. E., and Krasse, B. (1954). The Vipeholm dental caries study. The effect of different levels of carbohydrate intake on caries activity in 436 individuals observed for 5 years. *Acta. Odontol Scand.* 11, 232-388.
17. Scheinin, A., Makinen, K., and Ylitalo, K. (1975). Turku sugar studies V. Final report on the effect of sucrose fructose and xylitol on the caries incidence in man. *Acta Odont. Scand. Suppl.* 70, 67-104.
18. Makinen, K. K., and Scheinin, A. (1975). Turku sugar studies. XIX. Salivary peroxidase and invertase-like activity in relation to 1 year use of sucrose and xylitol chewing gum. *Acta Odont. Scand. Suppl.* 70, 60-67.
19. Glass, R. L. (1981). Effects on dental caries incidence of frequent ingestion of small amounts of sugars and stannous EDTA in chewing gum. *Caries Res.* 15, 256-262.
20. Kite, O. W., Shaw, J. H., and Sognnaes, R. F. (1950). The prevention of experimental tooth decay by tube-feeding. *J. Nutr.* 42, 89-96.
21. Bowen, W. H., Amsbaugh, S. M., Monell-Torrens, S., Brunelle, J., Kuzmiak-Jones, H., and Cole, M. F. (1980). *J. Am. Dent. Assoc.* 100, 677-681.
22. Navia, J. M. (1970). Evaluation of nutritional and dietary factors that modify animal caries. *J. Dent. Res. Suppl.* 49, 1213-1227.
23. Konig, K. G. (1970). Feeding regimes and caries. *J. Dent. Res. Suppl.* 49, 1327-1332.
24. Bowen, W. H. (1978). Role of carbohydrates in dental caries. In *Sweeteners and Dental Caries* (Shaw, J. H., and Roussos, G. G., eds.), Special Supplement, Feeding, Weight and Obesity Abstracts, pp. 147-156.
25. Green, R. M., and Hartles, R. L. (1969). The effect of diets containing different mono- and disaccharides on the incidence of dental caries in the albino rat. *Arch. Oral Biol.* 14, 235-245.
26. Shaw, J. H., Krumins, I., and Gibbons, R. J. (1967). Comparison of sucrose, lactose, maltose and glucose in the causation of experimental oral diseases. *Arch. Oral Biol.* 12, 755-768.
27. Bowen, W. H. (1983). The effect of substituting sucrose by corn syrup solids on caries in rats (unpublished data).
28. Colman, G., Bowen, W. H., and Cole, M. F. (1977). The effects of sucrose, fructose and a mixture of glucose and fructose on the incidence of dental caries in monkeys. *Br. Dent. J.* 142, 217-221.
29. Briner, W. H. (1981). Rodent model systems in dental caries research: Rats, mice and gerbils. In *Animal Models in Cariology* (Tanzer, J. M., ed.), Special Supplement, Microbiology Abstracts, pp. 111-119.
30. Konig, K. G. (1967). Caries induced in laboratory rats: Post-eruptive effect of sucrose and of bread of different degrees of refinement. *Br. Dent. J.* 132, 585-589.
31. van Houte, J. (1981). Experimental odontopathic infections effect of inoculation methods, dietary carbohydrate and host age. In *Animal Models in Cariology* (Tanzer, J. M., ed.), Special Supplement, Microbiology Abstracts, pp. 231-238.
32. Thomson, L. A., Bowen, W. H., Little, W. A., Kuzmiak-Jones, H., and Gomez, I. M. (1979). Simultaneous implantation of five serotypes of *Streptococcus mutans* in gnotobiotic rats. *Caries Res.* 13, 9-17.

33. Ciardi, J., Hageage, G., and Wittenberger, C. (1976). Multicomponent nature of the glycosyl transferase system of *Streptococcus mutans*. *J. Dent. Res.* 55, 87-96.
34. Trautner, K., Birkhead, D., and Svensson, S. (1982). Structure of extracellular glucans synthesized by *Streptococcus mutans* of serotypes a-e in vitro. *Caries Res.* 16, 81-89.
35. Rolla, G., Opperman, R. V., Bowen, W. H., Ciardi, J. E., and Knox, K. W. (1980). High amounts of lipoteichoic acid in sucrose-induced plaque in vivo. *Caries Res.* 14, 235-238.
36. Da Costa, T., and Gibbons, R. J. (1968). Hydrolysis of levan by human plaque streptococci. *Arch. Oral Biol.* 13, 609-615.
37. Houte, van J., Winkler, K. C., and Jansen, H. M. (1969). Iodophilic polysaccharide synthesis acid production and growth in oral streptococci. *Arch. Oral Biol.* 14, 45-50.
38. Stephan, R. M. (1940). Changes in hydrogen ion concentration on tooth surfaces and in carious lesions. *J. Am. Dent. Assoc.* 27, 718-723.
39. Hamilton, I. R. (1976). Intracellular polysaccharide synthesis by cariogenic microorganisms. In *Microbial Aspects of Dental Caries* (Stiles, H. M., Loesche, W., and O'Brien, T., eds.), Special Supplement, Microbiology Abstracts, pp. 683-701.
40. Graf, H., and Muhlemann, H. R. (1966). Telemetry of plaque pH from interdental area. *Helv. Odont. Acta* 10, 94-101.
41. Cole, M. F., Bowden, G., Korts, D., and Bowen, W. H. (1978). The effect of pyridoxine, phytate and invert sugar on production of plaque acids *in situ* in the monkey (*M. fascicularis*). *Caries Res.* 12, 190-210.
42. Gilmour, M. N., Green, G. C., Zah, L. M., Sparmann, C. D., and Pearlman, J. (1976). The C_1-C_4 monocarboxylic and lactic acids in dental plaques before and after exposure to sucrose in vivo. In *Microbial Aspects of Dental Caries* (Stiles, H. M., Loesche, W., and O'Brien, T., eds.), Special Supplement, Microbiology, pp. 539-556.
43. Geddes, D. A. (1972). The production of L(+) and D(−) lactic acid and volatile acids by human dental plaque and effect of plaque buffering and acidic strength on pH. *Arch. Oral Biol.* 17, 537-545.
44. Cornick, D. E. R., and Bowen, W. H. (1972). The effect of sorbitol on the microbiology of the dental plaque in monkeys (*Macaca irus*). *Arch. Oral Biol.* 17, 1637-1648.
45. Shaw, J. H. (1976). Inability of low levels of sorbitol and mannitol to support caries activity in rats. *J. Dent. Res.* 55, 376-382.
46. Leach, S. A., and Green, R. M. (1980). Effect of xylitol-supplemented diets on the progression and regression of fissure caries in the albino rat. *Caries Res.* 14, 16-23.
47. Moller, I. J., and Poulsen, S. (1973). The effect of sorbitol-containing chewing gum on the incidence of dental caries, plaque and gingivitis in Danish school children. *Community Dent. Oral Epidemiol.* 1, 58-67.
48. Makinen, K. K. (1978). The use of xylitol in nutritional and medical research with special reference to dental caries. In *Sweeteners and Dental Caries* (Shaw, J. H., and Roussos, G. G. eds.), Special Supplement, Feeding, Weight and Obesity Abstracts, pp. 193-224.

49. Leach, S. A., and Green, R. M. (1981). Reversal of fissure caries in the albino rat by sweetening agents. *Caries Res.* 15, 508-511.
50. Bird, J. L., Baum, B. J., Makinen, K. K., Bowen, W. H., and Longton, R. W. (1977). Xylitol associated changes in amylase and protein content of monkey parotid saliva. *J. Nutr.* 107, 1763-1767.
51. Makinen, K. K., Bowen, W. H., Dalgard, D., and Fitzgerald, G. (1978). Effect of peroral administration of xylitol on exocrine secretions of monkeys. *J. Nutr.* 108, 779-789.
52. Linke, H. A. (1977). Growth inhibition of glucose-grown cariogenic and other streptococci by saccharin in vitro. *Z. Naturforsch.* 32, 839-843.
53. Ciardi, J., Bowen, W. H., and Amsbaugh, S. M. (1983). Unpublished results.
54. Bowen, W. H. (1975). Unpublished results.
55. Mishiro, Y., and Kaneho, H. (1977). Effect of a dipeptide aspartame on lactic acid productions in human whole saliva. *J. Dent. Res.* 56, 1427.
56. Kleinberg, I., Kanapka, J. A., Chatterjee, R., Craw, D., A'Angelo, N., and Sandham, H. J. (1979). Metabolism of nitrogen by oral mixed bacteria. In *Saliva and Dental Caries* (Kleinberg, I., Ellison, S. A., and Mandel, I. D., eds.), Special Supplement, Microbiology Abstracts, pp. 357-377.
57. Ooshima, T., Izumitani, A., Sobue, S., Okahashi, N., and Hamadu, S. (1983). Non-cariogenicity of the disaccharide palatinose in experimental caries of rats. *Infect. Immun.* 39, 43-49.
58. Ciardi, J., and Bowen, W. H., Rolly, G., and Nagronski, K. (1983). Effect of sugar substitutes on bacterial growth, acid production and glucose synthesis. Abstract No. 110, Am. Assoc. Dent. Res. annual session.

13

Efficacy of Low-Calorie Sweeteners in Reducing Food Intake: Studies with Aspartame

Katherine P. Porikos* and Theodore B. Van Itallie
Columbia University College of Physicians and Surgeons, St. Luke's-Roosevelt Hospital Center, New York, New York

INTRODUCTION

The advent of synthetic sweeteners such as saccharin, cyclamate, and more recently aspartame has made it possible to offer people sweet taste in their diet without the calories that a diet high in sucrose often implies. However, stringent government regulations about food additives shown to be carcinogenic in animals have led to the withdrawal of Food and Drug Administration approval for food use of cyclamate and a potential ban on the use of saccharin. For any food additive there needs to be an assessment of benefits weighed against the potential risks from the use of such compounds. One group who would stand to benefit by the availability of low-calorie food analogs are obese people. However, there are very few data on the efficacy of low-calorie food analogs in reducing energy intake and promoting weight loss. Those which do exist are based on indirect measures, that is, self-reports of caloric intake and sweetener usage.

McCann et al. (1) questioned 247 obese people about their use of low-calorie sweeteners and found no relationship between the consumption of artificial sweeteners and whether or not the subjects lost weight. In that study acceptability of noncaloric sweeteners was apparently a problem in that only 26% of users reported liking artificially sweetened products. Parham and Parham (2) surveyed college students and found that the use of saccharin was correlated with

---

**Present affiliation*: the University of Calgary and Foothills Hospital, Calgary, Alberta, Canada

lower energy intake. However, the saccharin users in that study may have been restraining overall food intake because the degree to which their intake was suppressed was greater than could be explained by the substitution of saccharin for sugar. Another possibility is that their use of a noncaloric sweetener facilitated adherence to a diet by satisfying some need for sweets. However, when Farkas and Forbes (3) looked more systematically at whether noncaloric sweeteners facilitated dietary compliance in 100 diabetics, they found that adherence to a carbohydrate-restricted diet was independent of the use of low-calorie sweeteners. In a review of these studies Rosenman (4) concluded that "there is no evidence to show that artificial sweeteners are useful in weight reduction."

STUDIES OF FOOD INTAKE FOLLOWING CALORIC DILUTION WITH ASPARTAME

In a controlled laboratory setting we have examined energy intake of volunteers eating foods sweetened with sucrose or the low-calorie sweetener aspartame. We wanted to determine whether persons would defend their habitual level of energy intake in response to covert caloric dilution of their diet. In these studies we used aspartame as the caloric diluent, because human taste tests showed it tasted more like sucrose than many other sugar substitutes and had very little aftertaste (5,6). The use of a high-quality low-calorie sweetener was important because we did not want our subjects to detect the replacement of sucrose-sweetened products with low-calorie analogs. As long as the substitution was covert, any observed spontaneous adjustment of food intake could be attributed to a response to physiological signals of energy depletion. Such studies are directly relevant to the question of whether an artificial sweetener in the diet is effective in reducing overall energy intake.

We have studied a total of 24 people, 13 obese and 11 nonobese, while housed on a metabolic ward over periods of 15-30 days. This approach allowed for precise measurement of food intake, determination of various laboratory measurements, and maintenance of relatively constant energy output.

Methodology

The response to covert caloric dilution with aspartame was measured in three separate studies (7). The technique of recruiting subjects and the platter service method were the same throughout all studies. The subjects were healthy adults, aged 21-63 years, recruited as paid volunteers from advertisements in newspapers in the New York City area. All were screened by a physician, a psychologist, and a nutritionist to determine their eligibility for the study. To qualify, subjects had to be in good health, be willing to live as inpatients for the duration of the study, and have a history of using sucrose and sucrose-sweetened products in their customary diets. To prevent subjects from deliberately trying to adjust their energy intake during the course of the study, the subjects were not told the true purpose of the experiment until a debriefing session at the end. Instead, subjects were given an elaborate rationale for the experiment, which included the necessary information about aspartame to satisfy informed consent requirements but implied that we were studying the metabolic breakdown of aspartame. Subjects received a multivitamin tablet every morning with breakfast, which they believed to contain the aspartame. Blood samples were drawn regularly to further reinforce the alleged purpose of the study.

Subjects were offered ad libitum a varied diet of conventional foods by means of a platter service developed in our laboratory and described in detail in previous papers (8,9). Briefly, subjects could select from multiple servings of all menu items at meals and snacks. The latter were available 24 hr/day in a bedside refrigerator. All menu items were weighed covertly before serving, and the remaining portions reweighed after each meal. Snacks were weighed once a day. Grams consumed were converted by computer analysis to total daily intake of energy and grams of protein, fat, total carbohydrate, and sucrose.

Energy Intake

Study 1

The subjects in this 15-day study (8) were six obese women and two men, 24-134% overweight. A conventional sucrose-containing diet was fed on days 1-3 and 10-15. The energy-diluted diet containing aspartame was fed on days 4-9. During periods when the diet was sweetened with sucrose, subjects selected approximately 25% of their total energy intake as sucrose. When aspartame replaced sucrose, energy intake decreased significantly (see Fig. 1), and this decrease could be accounted for almost totally by the missing sucrose calories. For the first 3 days on the aspartame diet, energy intake averaged 77% of the baseline and there was no change from baseline in the quantity (in grams) of sweetened products consumed. Over the next 3 days energy intake increased by 310 kcal/day* to 86% of baseline. This increase, which represented a replacement of 41%† of the missing sucrose calories, was not statistically significant. Although these data show that the use of low-calorie food analogs decreased total energy intake by 14%, the fact that the subjects increased their energy intake during the second half of the aspartame substitution underscored the need for a longer-term study.

*When values for groups are given, they represent group means.

†The data for all the studies discussed in this paper were analyzed by either one- or two-way analyses of variance (ANOVA), with repeated measures on one factor. The two-way ANOVA was used in those studies which included both a normal-weight and an obese group. Wherever the effect for experimental periods was significant, specific comparisons of individual cell means were made either by F tests or by the Newman-Keuls procedure (10). Differences were considered significant if they reached a probability level of less than 0.05.

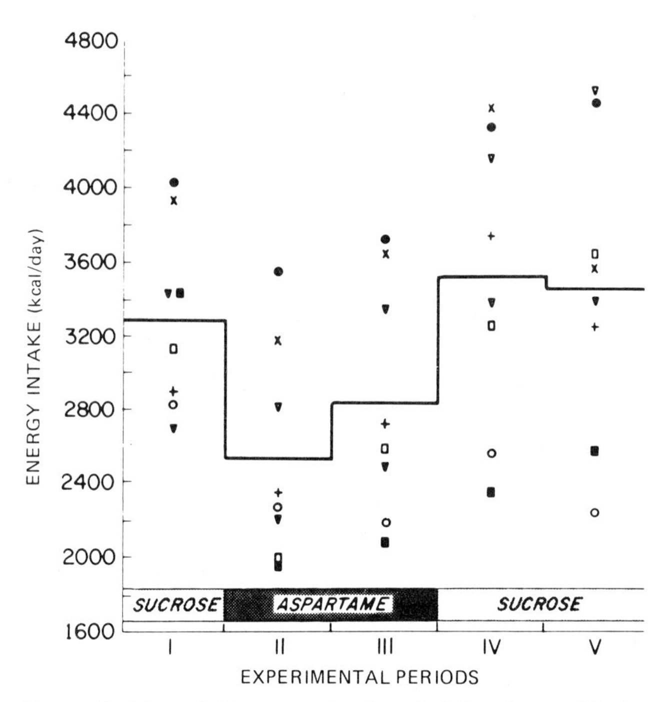

Figure 1 Mean daily energy intake of eight obese subjects as a function of experimental periods. Subjects consumed the sucrose-sweetened diet during periods I, IV, and V and the aspartame-sweetened diet during periods II and III. The solid line represents the average daily intake of the group for each consecutive 3-day period. The symbols represent the same data for the subjects individually. (From Ref. 8.)

Study 2

In this 24-day study the subjects were six men of normal body weight with no history of obesity or attempt at dieting (11). They were within 10% of desirable body weight. The study design was the same ABA format as in study 1, but the initial base-line period and the aspartame periods were doubled in length, making the design 6 days of base-line, 12 days of aspartame substitution, and a final 6 days of base-line. Once again the substitution of aspartame-sweetened analogs produced a significant decrease in energy intake (see Fig. 2). Once again intake was lowest in the first 3-day period after the substitution of aspartame for sucrose, 76% of base-line level. By days 4-6 intake of the diluted diet intake had increased by 354 kcal/day. However, there were no further adjustments in energy intake, and subjects stabilized their intake at 85% of base line. The adjusted energy intake replaced 40% of the diluted sucrose calories over the last 9 days on the

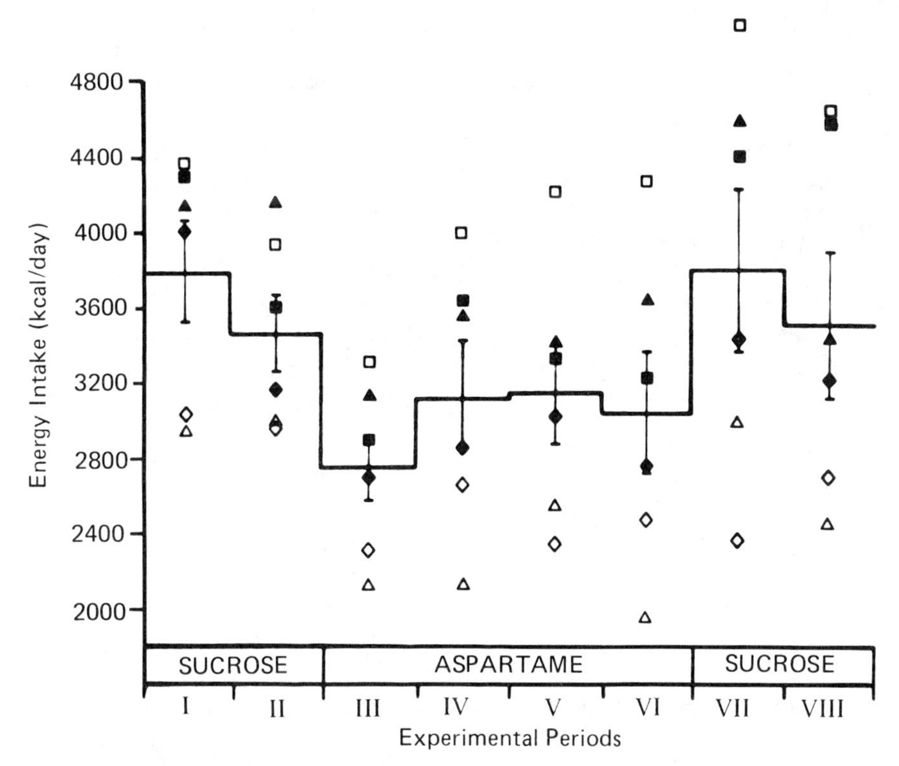

Figure 2 Mean daily energy intake (±SEM) of six normal-weight subjects as a function of experimental periods. Subjects consumed the sucrose-sweetened diet during periods I, II, VII, and VIII and the aspartame-sweetened diet during periods III-VI. The solid line represents the group mean and the symbols represent the individuals' data for each consecutive 3-day period as in Fig. 1. (From Ref. 11.)

aspartame diet. These data are important in two regards. First, they show that during aspartame substitution, energy intake does not return to base-line levels even when subjects are given almost 2 weeks to make the adjustment. In fact, whatever adjustment is going to take place seems to occur during the first 4-6 days on the altered diet. Second, the data suggest that normal-weight people do not regulate energy intake precisely and that low-calorie food analogs could be helpful to people interested in weight maintenance, as well as those people interested in weight loss. A third study was undertaken to look at the response of obese people over a 12-day dilution period and to compare it to the response of normal-weight people following the same protocol.

Study 3

In this study the subjects were eight normal-weight and five obese men.* Once again the control subjects were within 10% of desirable body weight.† The obese men were 18-45% overweight. The design was the same as in study 2, with an additional 6 days of base-line diet at the end of the study, for a total duration of 30 days. Both obese and normal-weight subjects showed a significant decrease in energy intake on the aspartame-sweetened diet. The normal-weight subjects consumed an average of 3418 kcal/day on the base-line diet, and 2880 kcal/day on the diluted diet. The obese consumed an average of 20% more kilocalories per day compared to the nonobese subjects throughout the study, but showed the same pattern of adaptation. They ingested 4131 kcal/day during the base-line period, and 3464 kcal/day while consuming the aspartame-sweetened diet. Substitution of aspartame for sucrose produced a sustained 15% decrease in total energy intake for both groups.

Summary of Human Intake Data

Our data indicate that, even under conditions where energy output remains relatively constant, people do not regulate their energy intake precisely. When the caloric density of a palatable and conventional diet was covertly altered by the substitution of an artificial sweetener for sucrose, people stabilized their intake at 85% of baseline.§ At least in the short term, there was no physiological drive to replace completely all calories diluted from the diet. These data suggest that low-calorie food analogs can offer an effective new approach to dieting. They allow for a reduction in energy intake without alteration in taste and only minor changes in volume of diet. A dietary regimen which includes low-calorie versions of people's favorite foods, particularly sweets, should encourage compliance.

CLINICAL LABORATORY FINDINGS

Another measure of the effectiveness of low-calorie food analogs in promoting health is their impact on medically relevant blood chemistries. In our studies we were interested in whether the substitution of aspartame for sucrose could have any effect on serum triglycerides, cholesterol, or glucose. While no significant changes in any of these parameters were noted in study 1 (with its short 3- to 6-day dietary periods), significant changes were found in serum triglycerides in studies 2 and 3. Some unexpected changes in routine measurements of serum transaminase levels led us to look at clinical laboratory tests of liver function. We

*Three of the eight nonobese subjects in study 3 had participated in study 2 for the standard 24 days followed immediately by an additional 6 base-line days. Their total 30-day data are included here together with the data of five new normal-weight subjects.

†Based on the midpoint of the range of desireable weights for medium frame published by the Metropolitan Life Insurance Company (12).

§Across all three studies (n = 24), intake of aspartame averaged 1.6 g/day (range = 1.1-2.5) and 20 mg/kg (range = 8-36).

studies 2 and 3. Some unexpected changes in routine measurements of serum transaminase levels led us to look at clinical laboratory tests of liver function. We found significant elevations of serum glutamic pyruvic transaminase (SGPT) and glutamic oxaloacetic transaminase (SGOT) in studies 2 and 3, but not in study 1. These data have been reported in detail elsewhere (13) and are summarized briefly for study 3 below.

The eight nonobese and five obese subjects who completed study 3 showed a mean serum triglyceride level at the upper limit of normal, 130 mg/dl, on the base-line diet. By the end of 12 days on the aspartame-sweetened diet, triglyceride levels had fallen significantly to 87 mg/dl. To investigate the cause of the changes in triglyceride levels, we correlated changes in triglyceride values at the end of each dietary manipulation with intake of total energy, sucrose, carbohydrate, fat, and protein and with changes in body weight during the same period. The changes in serum triglyceride levels did not vary systematically with any of these parameters, suggesting that the decrease in serum triglyceride concentration during the aspartame substitution was multiply determined.

There was a significant change in the composition of the diet when aspartame was substituted for sucrose. Whereas the self-selected base-line diet contained 56% carbohydrate, 32% fat, and 12% protein, the substitution of aspartame for sucrose led to a voluntarily chosen diet which contained 39% carbohydrate, 43% fat, and 18% protein. Serum cholesterol levels were unchanged by the alteration in diet composition and the removal of sucrose. The same was true for blood glucose levels in these nondiabetic subjects. There were no differences in the responses of obese and nonobese subjects.

During the base-line diet SGPT and SGOT levels rose significantly above admission levels with return to these initial levels during ingestion of aspartame. This phenomenon, more marked in the obese, was statistically significant in both groups of subjects. For the obese, SGPT values reached a high of 85 units/liter on day 6 compared to 49 units/liter at the start of the study (normal range = 45 units/liter or less). Comparable values for the nonobese were 40 and 23 units/liter, respectively. The SGOT changes were less marked, but no less statistically significant. Admission values were 21 units/liter for the obese and 19 units/liter for the nonobese. Those values rose to highs of 39 units/liter for the obese and 27 units/liter for the nonobese on day 6 (normal range = 40 units/liter or less). Correlations of changes in enzyme levels with total energy intake, major macronutrients, and changes in body weight revealed significant correlations between SGPT and SGOT changes and intake of total energy and sucrose.

STUDIES ON CALORIC DILUTION WITH ASPARTAME IN RATS

There has been virtually no work in animals that has looked at the regulation of energy intake by specifically altering energy supplied by dietary sucrose. Typically, energy dilutions have been made across the board, that is, water has been used to dilute a sweet-tasting liquid formula, or cellulose or kaolin has been added to powdered diet (14-16). Therefore, with the collaboration of two of our students, K. Ayoob and C. Semrad, we carried out a study on energy dilution with water, can, in some as yet unexplained fashion, interfere with caloric adaptation.

Methodology

A total of 15 naive female Charles River Sprague-Dawley rats were randomly assigned to one of two groups, that is, the dietary obese or control. The eight animals in the dietary obese group were fed a high-fat Purina chow diet for 2 weeks and then switched to a "supermarket" diet when their rate of weight gain was judged to be too slow. For 67 days they received the "supermarket" diet consisting of sweetened cereals, peanut butter, salami, bologna, cheese, and various crackers. Purina pellets were continuously available to these animals. The seven animals in the control group received only Purina chow. One control animal died during this period, and another ate virtually nothing later during one of the dilution periods. As a result, data for only five control animals are presented. At the start of the dilution experiment the dietary obese rats were 37% overweight relative to controls. They weighed 457 g as compared to the controls, which weighed only 334 g.

To facilitate substitution of aspartame for sucrose during the dilution manipulation, all animals were switched to a sweet-tasting liquid formula diet which contained 16% of energy as protein, 58% as carbohydrate, and 26% as fat, with all carbohydrate in the form of sucrose (the base-line diet). During the dilution periods enough sucrose was removed so as to produce a 20 or 40% dilution of total energy.* Aspartame replaced sucrose in quantities sufficient to produce diets equivalent in putative taste to the base-line diet. Preference testing in a pilot experiment had indicated what the appropriate levels of aspartame should be.

Animals were adapted to the base-line liquid diet for 30 days before the actual dilution experiment began. The dilution experiment was divided into five 10-day experimental periods. All animals received the base-line diet during periods I, III, and V. The animals received either the 20% or the 40% diluted diets during periods II and IV in a counterbalanced design. The diets were freshly prepared and changed once a day in the middle of the day. Energy intakes represent 24-hr values.

Energy Intake

The results of this study are shown in Figure 3. There were no significant differences in energy intake during the first 5 days as compared to the last 5 days

---

*The base-line diet was composed of 65 g of CHO-Free liquid formula, 4.3 g of sodium caseinate, 24 g of sucrose, and 6.7 g of water per 100 grams. The diluted diets contained the same amounts of CHO-Free and sodium caseinate as the base-line diet. In addition, the 20% diluted diet contained 3 g of aspartame table sweetener, 12 g of sucrose, 15.7 g of water, and 0.04 g of carageenan per 100 grams. The 40% diluted water contained 6 g of aspartame table sweetener, 2 g of sucrose, 22.7 g of water, and 0.075 g of carageenan per 100 grams.

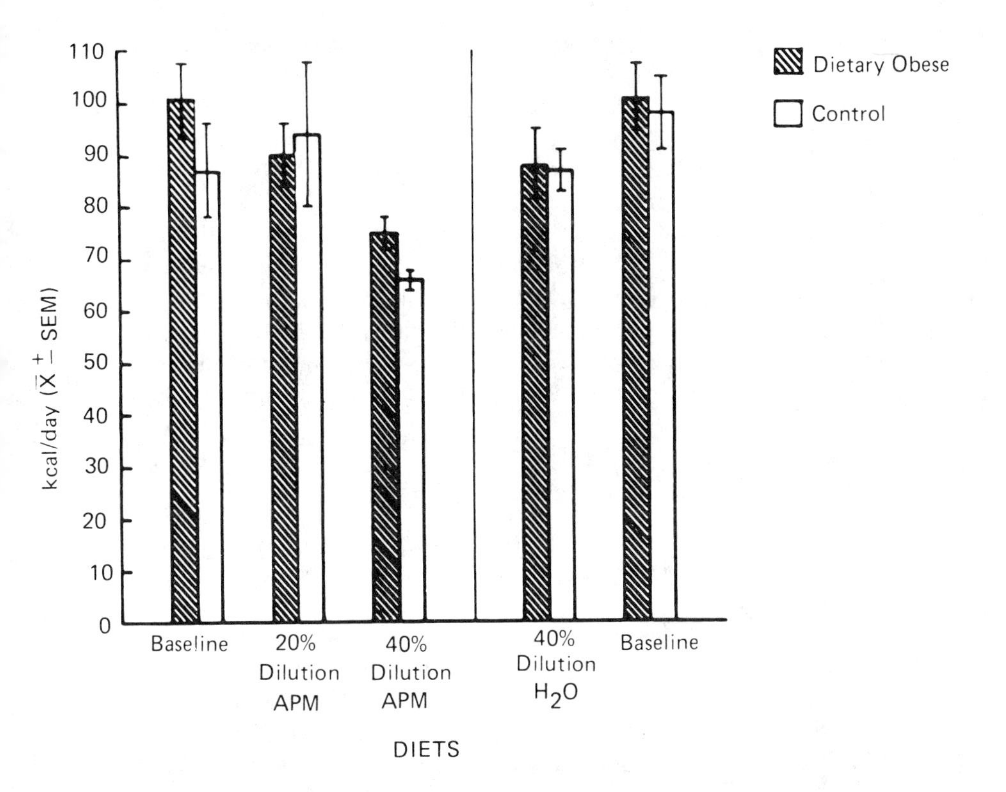

Figure 3 Mean daily energy intake (±SEM) of eight dietary obese and five control rats as a function of diet condition. The bars represent average daily intake per 10-day diet period.

on any diet for either the dietary obese or normal-weight animals. Therefore the data have been graphed on the basis of 10-day periods. Since data for the three 10-day base-line periods did not differ from one another, they have been combined and are shown on the left side of Figure 3.

On the diet diluted by 20%, intake did not differ significantly from base-line levels for either the dietary obese or the control animals. On the diet diluted by 40% both groups showed a significant decrease in caloric intake below baseline and 20% dilution levels, with normal-weight controls consuming 76% and obese rats 74% of their base-line calories. There were no significant differences in intake between the obese and normal-weight animals.

These results show poorer maintenance of caloric intake in response to a 40% dilution of the diet than is typically reported in the literature. Whereas our rats stabilized their intake at 75% of base-line, other studies using dilutions of 30 to 50% have found that rats generally maintain their intake at 85 to 90% of baseline

(14-17). None of these other studies has employed this type of dilution, that is, using an artificial sweetener to maintain the sweetness of the diet and removing only carbohydrate energy. Therefore we decided to add yet another test period at the end of the study. This condition involved a 40% dilution with water, where all macronutrients—protein, fat, and carbohydrate—were diluted and the sweetness of the diet changed. Neither aspartame nor any other low-calorie sweetener was added to restore the level of sweet taste diminished by water dilution. This final dilution manipulation period was followed by another 10-day base-line period. These results are shown in the two sets of bars on the right of Figure 3.

Under these conditions rats consumed 88% of base-line calories. This level of intake, consonant with the earlier literature, is statistically equivalent to base-line levels and significantly higher than intake on the diet diluted 40% with aspartame.

Summary of Animal Intake Data

To summarize, both obese and normal-weight control rats essentially maintained their habitual level of energy intake when 20% of base-line energy was removed by the substitution of aspartame for sucrose. However, when aspartame replaced sucrose in sufficient quantities to produce a 40% dilution of the diet, neither group increased its volume of intake sufficiently to maintain energy intake at base-line levels. The animals' failure to adjust volume of intake adequately is not due to a ceiling effect, that is, an inability to increase volume of intake any further. These animals managed to maintain their energy intakes at close to base-line levels in response to a 40% dilution of the diet when water alone was the diluent.

One could argue that some kind of "toxicity" at higher levels of aspartame intake held down the animals' intake. However, in a subsequent pilot study we demonstrated that the addition of the component parts of aspartame—aspartic acid, phenylalanine, and methanol—to the base-line diet did not produce any decrement in intake. These results decrease the likelihood that aversive effects of aspartame could explain the results in the 40% dilution study using aspartame. We also have preliminary data indicating that a 40% dilution with a saccharin-based sweetener produces results similar to those obtained with aspartame.

One intriguing interpretation of these data suggests that there may be a ceiling to the amount of sweet solution an organism will ingest in order to meet its energy requirements. If so, low-calorie sweet-tasting foods may have utility as a barrier to overeating at the end of a meal.

BLOOD CHEMISTRY DATA IN RATS

Serum triglyceride, cholesterol, glucose, SGPT, and SGOT levels were also measured in the rat study, since these data had proven of interest in our human studies. Three values from each animal were used in the statistical analyses: a base-line value after an average of 7 weeks on the supermarket diet, a second value

Table 1 Serum lipids, glucose and liver transaminases in rats as a function of diet condition[a]

	Dietary obese			Control			
	Supermarket diet, 7 weeks	Sucrose diet, 2 weeks	Sucrose diet, 15 weeks	Supermarket diet, 7 weeks	Sucrose diet, 2 weeks	Sucrose diet, 15 weeks	p (diet condition)
Triglycerides (62-276 mg/dl)[b-d]	240 ± 42	230 ± 31	634 ± 206	171 ± 29	209 ± 20	332 ± 31	$p = 0.025$
Cholesterol (40-124 mg/dl)	95 ± 8	115 ± 8	144 ± 16	109 ± 22	114 ± 8	136 ± 26	$p = 0.001$
Glucose, (150-421 mg/dl)	196 ± 13	186 ± 7	160 ± 9	158 ± 7	180 ± 10	170 ± 16	N.S.
SGPT (24-245 IU/liter)	62 ± 4	54 ± 7	67 ± 7	76 ± 10	57 ± 1	69 ± 12	$p < 0.05$
SGOT (97-281 IU/liter)	102 ± 7	116 ± 8	100 ± 7	111 ± 13	122 ± 8	96 ± 5	$p < 0.05$

[a]Values listed as the mean ± SEM.
[b]These clinical analyses were performed by Vetpath Laboratories, Teterboro, New Jersey.
[c]These values are for nonfasted animals feeding ad libitum.
[d]Reference ranges for rats as supplied by Vetpath Laboratories.

after an average of 2 weeks on the high sucrose diet*, and a final value after 15 weeks on the high sucrose diet.† Table 1 shows the average values (±SEM) for the dietary obese and control animals. There were no statistically significant differences between the dietary obese and control groups on any of the five serum factors.

Diet condition had the greatest effect on serum triglycerides. Dietary obese animals had average serum triglyceride levels of 634 mg/dl after 15 weeks on the high-sucrose diet, as compared to a value of 230 mg/dl after only 2 weeks on the sucrose diet or a value of 240 mg/dl while on the supermarket diet. Two dietary obese rats actually had serum triglycerides as high as 1938 and 1032 mg/dl at the end of the 15 weeks on the high-sucrose diet. The triglyceride response of the control rats to the different diet conditions, although less dramatic, was still statistically significant.

Serum cholesterol levels rose significantly at the end of the high-sucrose diet, but serum glucose levels did not change. The SGPT and SGOT activity levels did show statistically significant changes in response to diet condition, but these changes were not of a magnitude to be clinically significant. At no point did SGPT or SGOT activity levels approach the upper limit of the normal range in rats.

CONCLUSIONS

The studies in this paper are the first attempt of which we are aware to measure directly the effects of low-calorie sweeteners on spontaneous energy intake. The results provide support for the efficacy of low-calorie analogs in lowering overall caloric intake. The subjects in our studies adjusted their volume of intake to replace some but not all of the missing sucrose energy. Since the substitution of aspartame for sugar was covert, the adaptation that occurred was necessarily a response to physiological signals and was incomplete.

However, two additional questions must be answered before the efficacy of low-calorie sweeteners in the management of human obesity can be better established. First, studies in which people know that they are using low-calorie products need to be carried out. In a free-living situation people would be aware that they were using artificial sweeteners or other low-calorie food analogs. Several studies have shown the importance of cognitive factors as determinants of human food intake (18,19). It is possible that people would use the known energy savings from low-calorie products as an excuse to increase intake of other foods. This

---

*Blood samples were collected after 5 and 9 weeks on the supermarket diet and after 1 and 3 weeks on the high sucrose diet. The two sets of values for each diet condition were then averaged to produce the data for columns 1, 2, 4 and 5 in Table 1.

†The blood samples after 15 weeks on the high sucrose diet followed the end of the final baseline period. Thirty days out of this 15-week period were days on diluted diets.

situation is frequently depicted in cartoons showing someone ordering a high-calorie dessert with a calorie-free soft drink. Our studies have shown that there is no physiological drive to replace all energy removed from the diet when sugar sweetness is replaced with a low-calorie sweetener. However, there could be a psychological or cognitive drive to do so. This hypothesis requires testing.

The other issue needing direct study is that of actual weight loss with the use of low-calorie products. A carefully controlled long-term study is needed to look at the effects of using low-calorie food analogs on body weight. If, as our studies suggest, a sustained reduction in caloric intake of about 15% can be obtained by replacing sucrose with a noncaloric sweetener, an individual normally ingesting 2500 kcal/day could theoretically lose 25 lb in a year. However, such a degree of weight loss cannot be assumed in view of the many homeostatic mechanisms that might be involved. If sufficient weight were lost, the body might undergo metabolic adaptations so as to conserve energy.

Additional data on the efficacy of artificial sweeteners would be very valuable in allowing for rational regulatory policy decisions. Given the sizable role that low-calorie foods have in the American diet, it is surprising how little information exists concerning their effectiveness in promoting weight loss. The reason is no doubt the difficulty in collecting such data. By default, policy decisions on cyclamate and saccharin have been made on the basis of evaluations of risk only. Reliable data on the efficacy of low-calorie food analogs in promoting weight loss would facilitate better decision-making in these situations, with benefits weighed against risks.

ACKNOWLEDGMENTS

The aspartame-sweetened products and their sucrose counterparts were supplied by the General Foods Corporation, Tarrytown, New York. This work was supported in part by grants from the National Institutes of Health (Obesity Center AM-17624) and the General Foods Corporation, Tarrytown, New York.

REFERENCES

1. McCann, M. B., Trulson, M. F., and Stulb, S. C. (1956). Non-caloric sweeteners and weight reduction. *J. Am. Diet. Assoc.* 32, 327-330.
2. Parham, E. S., and Parham, Jr., A. R. (1980). Saccharin use and sugar intake by college students. *J. Am. Diet. Assoc.* 76, 560-563.
3. Farkas, C. S., and Forbes, C. E. (1965). Do non-caloric sweeteners aid patients with diabetes to adhere to their diets? *J. Am. Diet. Assoc.* 46, 482-484.
4. Rosenman, K. (1978). Benefits of saccharin: A review. *Environ. Res.* 15, 70-81.
5. Schiffman, S., Reilly, D. A., and Clark, III, T. B. (1979). Qualitative differences among sweeteners. *Physiol. Behav.* 23, 1-9.

6. Schiffman, S. S. (1984). Comparison of the taste properties of aspartame with other sweeteners. In *Aspartame: Physiology and Biochemistry* (Stegink, L. D., and Filer, L. J., Jr., eds.), Marcel Dekker, New York, Chap. 10.
7. Porikos, K. P. (1981). Control of food intake in man: Caloric dilution of a conventional and palatable diet. In *The Body Weight Regulatory System: Normal and Disturbed Mechanisms* (Cioffi, L. A., James, W. P. T., and Van Itallie, T. B., eds.), Raven Press, New York, pp. 83-87.
8. Porikos, K. P., Booth, G., and Van Itallie, T. B. (1977). Effect of covert nutritive dilution on the spontaneous food intake of obese individuals: A pilot study. *Am. J. Clin. Nutr.* 30, 1638-1644.
9. Porikos, K. P., Sullivan, A. C., McGhee, B., and Van Itallie, T. B. (1980). An experimental model for assessing effects of anorectics on spontaneous food intake of obese subjects. *Clin. Pharmacol. Ther.* 27, 815-822.
10. Winer, B. J. (1962). *Statistical Principles in Experimental Design*, McGraw-Hill, New York.
11. Porikos, K. P., Hesser, M. F., and Van Itallie, T. B. (1982). Caloric regulation in normal-weight men maintained on a palatable diet of conventional foods. *Physiol. Behav.* 29, 293-300.
12. Metropolitan Life Insurance Company. (1959). New weight standards for men and women. *Stat. Bull. Metropol. Life Insur. Co.* 40, 1-4.
13. Porikos, K. P., and Van Itallie, T. B. (1983). Diet-induced changes in serum transaminase and triglyceride levels in healthy adult men: role of sucrose and excess calories. *Am. J. Med.* 75, 624-630.
14. Williams D. R., and Teitelbaum, P. (1959). Some observations on the starvation resulting from lateral hypothalamic lesions. *J. Comp. Physiol. Psychol.* 52, 458-465.
15. Kennedy, G. C. (1950). The hypothalamic control of food intake in rats. *Proc. R. Soc.* 137, 535-549.
16. Janowitz, H. D., and Grossman, G. I. (1949). Effect of variations in nutritive density on intake of food of dogs and rats. *Am. J. Physiol.* 158, 184-193.
17. Friedman, M. I. (1978). Hyperphagia in rats with experimental diabetes mellitus: a response to a decreased supply of utilizable fuels. *J. Comp. Physiol. Psychol.* 92, 109-117.
18. Wooley, S. C. (1972). Physiologic versus cognitive factors in short term food regulation in the obese and nonobese. *Psvchosem. Med.* 34, 62-68.
19. Nisbett, R. E., and Storms, M. D. (1974). Cognitive and social determinants of food intake. In *Thought and Feeling: Cognitive Alteration of Feeling States* (London, H., and Nisbett, R. E., eds.). Aldine Press, Chicago, 190-208.

PRECLINICAL STUDIES

14

Preclinical Studies of Aspartame in Nonprimate Animals

Samuel V. Molinary
PepsiCo, Inc., Valhalla, New York

INTRODUCTION

Aspartame, [3-amino-N-(2-carboxyphenethyl)succinamic acid methyl ester], the methyl ester of L-aspartyl-L-phenylalanine, and its principal decomposition product, 3-carboxymethyl-6-benzyl-2,5-diketopiperazine (diketopiperazine, or DKP) have been extensively studied for their toxicological potential for the past 15 years. Over 112 studies concerning the metabolism, pharmacology, and toxicology of these two compounds were filed by G. D. Searle & Company with the Food and Drug Administration (FDA) as proof of the safety of aspartame (APM) as a food additive.* In approving APM as a sweetener in 1981, Commissioner Hayes said (1), "The safety evaluation of aspartame has been a long and arduous process, spanning the tenure of several FDA commissioners. Few compounds have withstood such detailed testing and repeated, close scrutiny, and the process through which aspartame has gone should provide the public with additional confidence of its safety." While APM may not be the most studied food additive, it probably has had the largest number of studies carried out *prior to* FDA approval.

Of these 112 safety studies submitted to the FDA, 93 are or pertain to reports of studies of aspartame and its DKP in animals. Many of these studies

---

These studies are contained in the so-called "E-File" at the Food and Drug Administration, Hearing Clerk File, Administrative Record, Aspartame 75F-0355, Rockville, Maryland.

have been reviewed and evaluated by the Joint Expert Committee on Food Additives (JEC/FA) of the Food and Agricultural Organization of the United Nations and the World Health Organization (FAO/WHO) in 1975, 1976, 1977, and 1980. A monograph on the toxicological evaluation was written following the 1980 JEC/FA meeting in Rome (2). This review will be confined to those animal studies carried out early in the development of aspartame (1968-1974). Some of these studies, especially the carcinogenicity studies in the rat, became the focal point for questions that were considered by the Public Board of Inquiry in 1980; the interpretation of these is discussed in depth elsewhere in this book (3-5).

In proposing a novel compound as a food additive, especially a sweetener which will be consumed by all segments of the population, it is important that the sponsor prove that this compound is safe to a reasonable certainty. The development of a data base for product safety assessment typically starts in three scientific areas: metabolism (including absorption, tissue distribution, pharmacokinetics, and elimination), pharmacology, and toxicology. Metabolism studies are carried out in a number of different species, ultimately including man, and the data obtained are used to design and interpret studies in the other two areas. In this specific case pharmacology studies are performed to discover if the compound has any biological activity other than that of creating the perception of a sweet taste. Pharmacological activity is not a desirable property for a food additive. Toxicology studies are carried out in two parts, acute and chronic exposure, and are undertaken to determine what harm may accrue as a result of ingesting the food additive, especially under conditions of high intake. Each of these areas will be considered as they relate to the studies of animals administered aspartame or its DKP.

The metabolism of aspartame and its diketopiperazine have been extensively studied in mice, rats, rabbits, dogs, monkeys, and man (2,6-19). The major point to emerge from the metabolism studies is that in all species tested, aspartame is broken down in the gastrointestinal tract to its constitutents L-phenylalanine, L-aspartic acid, and methanol. These findings suggest that outside the gastrointestinal tract, any toxicological activity of aspartame would result from systemic imbalances caused by increases in the plasma concentration of these two amino acids and methanol. This finding also suggests that if the dipeptide (aspartyl-phenylalanine) per se had any pharmacological activity, it would be in the gastrointestinal tract itself. This was considered possible since the dipeptide sequence aspartyl-phenylalanine is found at the carboxy terminal end of the gastrointestinal hormones gastrin and cholecystokinin (20).

Studies were undertaken by Bianchi et al. (21) to delineate any pharmacological effects that aspartame might have on the gastrointestinal system. Rats were given 200 mg/kg body weight of aspartame intragastrically and observed for effects on appetite suppression, inhibition or stimulation of gastric secretion, acid secretion, and proteolytic activity. No compound-related effects were observed.

In addition to tests on the gastrointestinal system, other pharmacological tests were carried out in a variety of in vivo and in vitro model systems (22). As pointed out by Potts et al. (23), "some of these studies were done to test specific hypotheses. . .while others. . .were done simply in the interest of safety." An evaluation of central nervous system effects (22,23) in rats fed a diet providing aspartame at 9% (11 g/kg body weight per day) sought to compare high-dose effects of phenylalanine with high doses of aspartame on learning in the rat. These authors concluded that on a mole-for-mole basis, there was no difference between aspartame and L-phenylalanine on the learning behavior of rats. No attempt was made in this study to measure the serum phenylalanine or tyrosine levels.

Possible effects of aspartame ingestion on the cardiovascular system were also studied (22); no pharmacological activity was observed. Aspartame was also tested in mice, rats, and rabbits for endocrine-like activity (24,25): None was found. In all, Searle studied the pharmacological activity of aspartame in 32 different animal studies, all with essentially negative results (22,24,26).

Toxicology studies were conducted on both aspartame and its DKP in a variety of species for various durations of exposure. These tests are typically done in a sequential fashion, with the data from each study being used to design the next experiment. The usual study sequence is

1. Acute toxicity (LD_{50})
2. Chronic toxicity, short-term (90-day study)
3. Chronic toxicity, long-term (52-week study)
4. Carcinogenicity (104 weeks study)

In addition, a series of teratology and reproduction studies are carried out concurrently with the chronic toxicity studies, as are mutagenicity and other special studies. Since the toxicology-teratology studies carried out on aspartame and its DKP are essentially negative and are too numerous to review in detail, only a few selected studies will be discussed.

Acute toxicity studies were carried out on aspartame and its DKP to establish the concentration that is lethal to 50% of the test animals (LD_{50}). Three species were studied: mice, rats, and rabbits (27,28). The compounds were administered by several routes, with the oral route being most pertinent to food additive safety evaluation. When administered orally at doses up to 5 g/kg body weight, no deaths occurred in any animal, nor was any abnormal behavior noted with either aspartame or its DKP. Therefore the minimum lethal dose for these three species is greater than 5000 mg/kg body weight.

Chronic short-term toxicity studies, 39-90 days in duration, were carried out in mice and rats with aspartame (26,29-31) and with aspartame's DKP (32-34) and in dogs with aspartame (35). No treatment-related or consistent dose-related effects were observed in any of these studies, other than a decrease in body weight at the highest dose, 10 g/kg body weight per day (30). Hematology, urinanalysis, organ weights, and gross and microscopic pathology were reported to be unremarkable in all studies. The principal purpose of these studies was to establish the dose ranges to be used in the chronic long-term and carcinogenicity studies.

CHRONIC TOXICITY STUDIES OF ASPARTAME

Chronic long-term toxicity studies of aspartame, 52-110 weeks in duration, were carried out in mice (36), rats (37-40), hamsters (41-43), and dogs (44). The study design and findings of these studies are detailed below. In addition, Ishii et al. have studied aspartame and aspartame-DKP mixtures fed to Wistar rats for 104 weeks (3,45). Studies E-33/34 (37,38) and E-70 (40) are also discussed elsewhere in this volume (3-5) with respect to the brain tumor issue.

Mouse

Study E-75 (36) was one of 12 Searle safety studies validated by Universities Associated for Research and Education in Pathology, Inc. prior to the Public Board of Inquiry on Aspartame* (46). This was a 110-week study. Groups of 36 male and 36 female mice (72 mice per dose group) were fed aspartame at 1, 2, and 4 g/kg body weight per day as a diet admix for 110 weeks. A group of 72 males and 72 females served as controls. Mean body weights of treated animals were comparable to those of the control group, even though food consumption was observed to decrease with increased aspartame dosage. During the 110 weeks on test, there were no observed effects on the survival, appearance, or behavior of the test animals. Ophthalmological findings were not remarkable. Although there were sporadic incidences of statistically significant differences in some of the measured hematological and clinical chemistry variables, there were no overall compound- or dose-related effects attributable to aspartame administration. Gross and microscopic pathology studies following necropsy gave no indication of tumorigenic or nontumorigenic changes with respect to aspartame administration.

Rat

In study E-33/34 (37,38) (see also Refs. 39 and 46) groups of 40 male and 40 female rats (80 rats per dose group) were given aspartame at 1, 2, 4, or 8 g/kg body weight per day as a diet admix for 104 weeks. A group of 60 males and 60 females served as controls. For the 104-week duration of the study, there was a slight reduction in mean body weight and in mean food consumption in the

*A detailed history of aspartame's course through the regulatory process in the U.S. is detailed in Ref. 1.

4 g/kg body weight per day dose group; these variables were markedly decreased in animals fed aspartame at 8 g/kg body weight per day. No effects on growth or food consumption were reported at the lower two doses. The 2-year survival for *all* groups was poor, especially the male control group. The investigators attribute the decreased survival to spontaneous disease. Females in the 4 and 8 g/kg body weight per day groups had a survival of only 54% of females in the control group. With the exception of persistent and increased numbers of red and white blood cells in the urine of animals in the 8 g/kg body weight per day group, no treatment-related effects were reported on hematological and clinical chemistry parameters. The authors report that no gross or microscopic pathology findings were treatment related. Brain tumor findings in these animals are discussed elsewhere (4,5).

In addition, Searle conducted a two-generation long-term toxicity study of aspartame (39,40). For this study, groups of 40 male and 40 female rats received aspartame at 2 and 4 g/kg body weight per day as a diet admix for 104 weeks. A group of 60 male and 60 female rats served as controls. All test animals were randomly retested from the F_1 litters of a multigeneration study in which the P_1 animals had been exposed to the corresponding dietary aspartame levels for 60 days prior to mating. The P_1 females were kept on diet during gestation and lactation. For the duration of the study, no treatment-related effects were reported with respect to the survival, behavior, or appearance of the animals. Decreased weight gain and food consumption were reported for the animals fed aspartame at 4 g/kg body weight per day. There was no clear dose- or treatment-related differences, that is, heart-to-body weight ratios were significantly decreased in treated males and liver weight was increased in all treated females. A detailed histopathology review of the brains, livers, and pituitary glands of control and treated rats was done. The types and numbers of neoplasms were comparable between the control and treated groups (see Refs. 4 and 5). An increased incidence of hyperplastic nodules was noted in the livers of treated females; however, it was concluded that these hyperplastic nodules were nonneoplastic and not treatment related.

Hamster

A long-term chronic toxicity study of aspartame in hamsters was started (41, see also Refs. 42 and 43). In this study, groups of five male and five female hamsters were fed aspartame in the diet at 1, 2, 4, and 12 g/kg body weight per day. Groups of 10 male and 10 female hamsters served as controls. All groups were replicated seven times. Erratic decreases in body weights and food consumption were noted early in the study. This was followed by the presence of an unidentified infection (later thought to be "wet tail") in the entire colony. This led to a greater than 50% mortality and subsequent termination of this study at 46 weeks.

Dog

Groups of five male and five female beagle dogs were fed 0, 1, 2, and 4 g/kg body weight per day of aspartame in the diet for 106 weeks (44). Depressed weight gain was reported at all levels of aspartame ingestion. A statistically significant lowering of the hemoglobin, hematocrit, and total red blood cell count was reported in the high-dose male animals. No other consistent effects were noted in the hematology studies. Bromsulfophthalein (BSP) clearance time was increased in the males fed aspartame at 2 and 4 g/kg body weight per day at 78 and 116 weeks. Other liver function tests were not remarkable. Gross and histopathological examination did not reveal any treatment-related effects.

Monkey

Infant monkeys fed large doses of aspartame or phenylalanine for the first 9 months of life were also studied. These data are reviewed by Reynolds et al. in Chap. 20 (19). Neuropathology studies in infant primates administered aspartame (2 g/kg body weight) or aspartame (2 g/kg body weight) plus monosodium L-glutamate (1 g/kg body weight) were negative (17,18).

CHRONIC TOXICITY STUDIES OF DKP

The principal decomposition product of aspartame, its DKP, was studied independently in chronic, long-term toxicity studies in the mouse (47) and rat (48,49).

Mouse

In study E-76 (47), groups of 36 male and 36 female mice were given 0.25, 0.5, and 1 g/kg body weight per day of DKP as a diet admix for 110 weeks. A group of 72 male and 72 female mice served as controls. No treatment-related effects on weight gain, food consumption, appearance, behavior, or survival were reported. No adverse effects were reported for the hematology and clinical chemistry measurements. The only suggestion of a treatment effect in the pathology report was a significant increase in thyroid weight and in the ratio of thyroid to body weight. There were no other tumorigenic or nontumorigenic changes reported that were attributable to administration of aspartame's DKP.

Rat

In this study (48,49), groups of six male and six female rats were fed 0.75, 1.5, and 3 g/kg body weight per day of DKP in the diet for 115 weeks. A group of 12 males and 12 females served as the control. Each group was replicated six times (36 rats per gender per group). No effects were reported with respect to survival, behavior, or appearance; however, animals of both sexes in the 1.5 and

3 g/kg body weight per day groups were shown to have a dose-related, statistically significant increase in food consumption in the high-dose group males. Hematology and clinical chemistry measurements reportedly show no difference from controls. Gross and microscopic pathology studies did not indicate the presence of tumorigenic or nontumorigenic changes related to DKP ingestion.

In addition to these routine chronic toxicity studies, both aspartame and its DKP were studied for specific tumorigenic potential in the urinary bladders of mice (50,51). The same study design was used for each compound. Groups of 200 albino mice (60-90 days of age) had pellets (80% cholesterol and 20% aspartame or 80% cholesterol and 20% DKP) surgically implanted in their urinary bladders. The negative control group animals were implanted with pellets composed of cholesterol and the 8-methyl ether of xanthurenic acid. Each study was 56 weeks in duration, during which time the following parameters were measured or noted: morbidity, mortality, motor and behavioral activity, growth, general external features, and digital palpation of protruding tissue masses. At the termination of the study, all animals were necropsied and examined histopathologically for bladder tumors. Any animals dying during the study were examined in the same way. In neither study was there an increased incidence of bladder tumors over that of the negative controls. These data are discussed in detail by Bryan in Chap. 16 (52).

In addition to studies of toxicity and carcinogenicity, aspartame and its DKP were tested for any effects they might exert with respect to reproduction, reproductive performance, and teratology. For purposes of safety evaluation, these studies are typically done in segments. These segments and their purposes were as follows:

Segment I. To evaluate effects of the test substance on fertility, conception, and implantation. These studies also provide an overview of fetal development and growth.

Segment II. To evaluate the test compound for any embryotoxic and teratogenic effects occurring during the first half of pregnancy.

Segment III. To evaluate the test compound for effects on late pregnancy, parturition, and lactation.

Multigeneration. To evaluate the test compound with respect to the total reproductive performance of two consecutive generations. This includes observations on the offspring during neonatal growth.

Aspartame and its DKP were examined for reproduction and teratological effects in 24 such studies. One study was done in a chick embryo system (53); all others were done using rats or rabbits. In the chick embryo study (53), aspartame doses of 0.25 mg and 0.5 mg per egg, along with appropriate controls, were studied. No morphological abnormalities were reported in the embryos or hatched chicks of the aspartame treatment groups.

A segment I study of aspartame was performed using Charles River CD rats (54). Two dose groups composed of 14 male and 30 female rats each were given 2 and 4 g of aspartame per kilogram body weight per day, respectively, in the diet. The control group consisted of 14 male and 48 female rats fed a basal chow diet. All treated animals received aspartame in the diet during the premating and mating period; the females continued to receive aspartame during gestation and lactation. Some animals were sacrificed at the end of the mating period. Fifty percent of the dams were sacrificed at day 13 of gestation; the ovaries, uterus, and uterine contents were examined. The remaining dams were allowed to go through gestation, delivery, and lactation. The pups were examined for evidence of malformation at birth and allowed to go to weaning (21 days), at which time they were sacrificed. Aspartame was reported to have no effects on paternal survival rate, mating performance, fertility, or body weight gain, even though paternal food consumption was depressed. With the exception of a few random fluctuations, maternal survival, body weight gain, and food consumption were unremarkable. No evidence of an aspartame effect was reported after examination of the data obtained from the hysterotomy, litter examination, or neonatal examination.

A multigeneration reproduction study was also performed in Charles River CD rats (55). As in the above experiment (54), aspartame was given in the diet at 0, 2, and 4 g/kg body weight per day throughout the study to groups of 12 male and 24 female rats. Aspartame administration began 9 weeks prior to mating the P_1 animals. All F_1 litters were reduced to 10 pups within 24 hr of delivery; these remaining animals were used as the P_2 group, 20 males and 60 females per group. The P_2 groups were mated at 9 weeks of age. Of the F_2 litters, five litters per group were killed at birth and used in preparing a separate clinical pathology report (56). The remaining 15 F_2 litters per group were followed to weaning (21 days) and sacrificed. Mean food consumption, body weight, survival, physical appearance, and behavior were recorded. Evaluation of the data indicated no difference between treated and control groups, with the exception of reduced mean body weight of weanling rats in both the F_1 and F_2 high-dose litters. Evaluation of the data relating to fertility, gestation, live births, and litter size at weaning indicated comparable findings between treated and control groups.

A segment II teratology study was done in Charles River CD rats (57). Aspartame was given as a diet admix to 24 mated females in each of three dosage groups (0, 2, and 4 g/kg body weight per day) during gestation days 6-15. On gestation day 20, all dams were sacrificed, the ovaries and uterus were removed, and the uterine contents examined. Fetuses were examined externally and preserved intact for subsequent soft tissue and skeletal examination. Forty-seven litters comprised of 589 fetuses were examined; no evidence of soft tissue or skeletal anomalies was observed. In addition, no differences were reported between control and treated groups with respect to the mean number of resorption sites, fetuses per pregnant female, viable fetuses, fetal sex distribution, fetal body weight and length, and crown-rump distance. There were no effects reported on maternal survival or body weight. It was concluded that oral administration of up to 4 g/kg body weight per day aspartame during the 6th to 15th gestational day was neither embryotoxic nor teratogenic.

Four segment III teratology studies were done using Charles River CD rats (58-61). These studies were of the same design and yielded comparable results; therefore they will be reviewed as a unit. Three groups of 24 pregnant female rats each were given 0, 2.5, and 4.4 g/kg body weight per day aspartame during gestation days 14-21 (parturition) and 0, 3.6, and 6.8 g/kg body weight per day aspartame on postpartum days 1-21 (weaning). There were 21 litters (246 pups) from the control females, 20 litters (236 pups) from the low-dose females, and 22 litters (289 pups) from the high-dose females. All pups were examined at birth. No differences were reported between control and treatment groups with respect to maternal food consumption, behavior, mobidity, and mortality. Mean maternal body weights were depressed for the high-dose groups during the lactation period. Live birth data, litter size, and data from the physical examination of the pups were not remarkable. However, weanling pup survival was reported to be depressed for those pups in the high-dose group. In addition, 3% of the high-dose pups were reported to have incompletely opened eyelids, and two pups from the same high-dose litter had "grossly observable lens opacities" (58).

One other segment II study was carried out in rats fed a 3:1 mixture of aspartame and its DKP (62). Four groups each of 30 mated females were given the aspartame:DKP mixture in the diet at levels of 0, 1, 2, and 3 g/kg body weight per day from gestational days 6 to 14. On gestation day 19, all dams were sacrificed and their uterine contents examined as described above in study E-5 (57). Eighty-two litters (1026 term fetuses) from treated females were examined for soft tissue and skeletal anomalies. None were observed. No differences were noted between control and treated groups with respect to maternal survival, body weight, food consumption, mean number of implantation sites, resorption sites, number of fetuses per pregnant female, live fetuses, dead fetuses, fetal sex distribution, fetal body weight, or crown-rump distance.

Aspartame was studied in a total of six segment II teratology studies using New Zealand white rabbits. In five of these studies aspartame alone was fed (63-67), while in the sixth a mixture of aspartame and its DKP was fed in a 3:1 ratio (68). The design of these studies was identical; therefore they will be reviewed as a unit. In all studies, the female rabbits were artificially inseminated using pooled sperm specimens. Aspartame was administered by gavage at 2 g/kg body weight per day during gestation days 6-18 (63,64). The aspartame:DKP mixture was given by gavage at 3 g/kg body weight per day during gestation days 6-18 (68). In studies E-54, E-55, and E-79 (65-67), aspartame was administered as a pelleted diet at mean dosages of 1.1 and 1.9 g/kg body weight per day. All animals were sacrificed at term and fetuses were examined externally. Approximately one-half of the fetuses were processed for soft tissue examination and one-half processed for skeletal examination. In addition, observations were made on maternal survival rates, conception rates, body weights, hysterotomy findings, litter size and viability, fetal size, and sex distribution. These reports concluded that there were no differences between the control and treated groups and that there were no compound-related embryotoxic or teratogenic effects.

Aspartame's diketopiperazine derivative was independently studied in a series of reproduction and teratology studies. Three studies (segments I-III) were carried out in Charles River CD rats (58,69), and one segment II study was carried out in New Zealand white rabbits (70). In the segment I rat study (58), aspartame's DKP was administered in the diet (14 males and 28 females in each dosage group) at 0.45, 0.9, and 1.8 g/kg body weight per day. A group of 14 male and 60 females served as controls. The diet was given throughout premating, mating, gestation, and lactation. The DKP was reported to have no effect on parental survival, food consumption, mating performance, fertility, or parental body weight gain. At the high-dose level, dam body weights were depressed during the midgestation period. It was concluded that DKP fed continuously had no adverse effects on the sires, dams, or pups.

File study E-37 (58) also contains the report of a segment III study of DKP in Charles River CD rats. In this study, four groups of 20 pregnant rats were fed DKP in a diet admix at 0, 0.7, 1.3, and 2.5 g/kg body weight per day from gestation day 14 to postpartum day 21 (weaning). The report concluded that there were no adverse effects attributable to DKP administration in this study.

The segment II DKP study (69) in rats consisted of four groups of 24 mated females fed aspartame's DKP at 0, 1, 2, or 4 g/kg body weight per day, respectively, from the 6th through the 15th day of gestation. In addition to evaluation for possible soft tissue and skeletal anomalies, observations were made on maternal survival, conception rate, body weight, food consumption, number of resorption sites, mean number of fetuses per pregnant female, fetal sex distribution, fetal body weight, and crown-rump distance. A significant decrease in resorption sites was reported for the high-dose group. No treatment-related anatomical alterations were observed.

A segment II study of DKP was performed using New Zealand white rabbits (70). Four groups of 21 artificially inseminated rabbits received 0, 0.5, 1, or 2 g/kg body weight per day of DKP by gavage from gestation days 6 to 18. While no embryotoxic or teratogenic effects were discerned, there were insufficient data from the high-dose group to permit a complete evaluation.

Aspartame and its DKP were subjected to a series of mutagenicity tests, ranging from in vitro to in vivo assays. Both aspartame and its DKP were tested twice each in the Ames *Salmonella*/microsome assay, with no mutagenic activity being observed in any assay (71-74). Both compounds were studied separately in the dominant lethal assay using Charles River CD rats (75,76), in the host mediated assay using Purina Caesarean rats (77,78), and in in vivo cytogenetic assays using Holtzman and Purina Caesarean rats (79-81). In none of these assays was there evidence of genetic activity by either aspartame or its DKP.

CONCLUSIONS

The preclinical studies enumerated above, as well as clinical studies discussed elsewhere in this volume, formed the basis for deliberation by the Bureau of Foods of the U.S. Food and Drug Administration (FDA) and, subsequently, by the Commissioner of the FDA who made the final decision to allow aspartame as a food additive (1). This data base was also evaluated by the Joint FAO/WHO Expert Committee on Food Additives (JEC/FA) and was the basis on which that body established an allowable daily intake (ADI) of 0-40 mg/kg body weight per day for aspartame and an ADI of 0-7.5 mg/kg body weight per day for aspartame's DKP (2). The acceptable daily intake established the the JEC/FA is based on no-observed-effect levels in the preclinical studies enumerated above of 4 g/kg body weight per day for aspartame and of 750 mg/kg body weight per day for aspartame's DKP. A safety factor of 100 was applied to these to arrive at the respective ADIs (2). Commissioner Hayes of the Food and Drug Administration, based on clinical data and projected consumption data provided by the G. D. Searle Co., took the 99th percentile of human consumption to be 34 mg/kg body weight per day (1). This figure, arrived at by a line of reasoning different from that of the JEC/FA, is remarkably close to the ADI for aspartame established by the JEC/FA.

The data from preclinical studies, such as those enumerated above, cannot by themselves assure that a food additive is completely safe for all segments of the consuming population. When integrated with the clinical data discussed elsewhere in this volume and with projected consumption data, there is reasonable certainty that aspartame will cause no harm to the public who consume it. Only use, concommitant with regulatory surveillance, can lead to a final determination of whether aspartame is safe for use by all segments of the population.

REFERENCES*

1. Aspartame (1981). Commissioner's final decision. *Fed. Reg.* 46, 38285-38308.
2. Aspartame (1980). In *Toxicological Evaluation of Certain Food Additives,* Twenty-fourth report of the Joint FAO/WHO expert Committee on Food Additives, WHO Technical Report Series No. 653, Food and Agriculture Organization of the United Nations and the World Health Organization, Rome, pp. 18-86.

---

*In the "E-file" reports by G. D. Searle to the Food and Drug Administration, SC-18862 refers to aspartame (Searle Compound 18862) and SC-19192 refers to aspartame's diketopiperazine (DKP) derivative (Searle Compound SC-19192).

3. Ishii, H. (1984). Chronic feeding studies with aspartame and its diketopiperazine. In *Aspartame: Physiology and Biochemistry* (Stegink, L. D. and Filer, L. J., Jr., eds.), Marcel Dekker, New York. Chap. 15.
4. Koestner, A. (1984). Aspartame and brain tumors. 1. Pathology issues. In *Aspartame: Physiology and Biochemistry* (Stegink, L. D. and Filer, L. J., Jr., eds.), Marcel Dekker, New York, Chap. 22.
5. Cornell, R. G., Wolfe, R. A., and Saunders, P. G. (1984). Aspartame and brain tumors: Statistical Issues. In *Aspartame: Physiology and Biochemistry* (Stegink, L. D., and Filer, L. J., Jr., eds.), Marcel Dekker, New York, Chap. 23.
6. File E-15 (1972). *SC-18862: The Metabolism of Aspartame*, Vol. 1, Parts I-XIV, submitted by G. D. Searle & Co. to the Food and Drug Administration, Hearing Clerk File, Administrative Record, Aspartame 75F-0355, Food and Drug Administration, Rockville, Md.
7. File E-17 (1972). *SC-18862: The Metabolism of Aspartame*, Vol. 2, Parts XV-XIX, submitted by G. D. Searle & Co. to the Food and Drug Administration, Hearing Clerk File, Administrative Record, Aspartame 75F-0355, Food and Drug Administration, Rockville, Md.
8. File E-18 (1972). *SC-18862: The Metabolism of Aspartame*, Vol. 3, Parts XX-XXIII, submitted by G. D. Searle & Co. to the Food and Drug Administration, Hearing Clerk File, Administrative Record, Aspartame 75F-0355, Food and Drug Administration, Rockville, Md.
9. File E-80 (1974). *SC-18862: The Metabolism of Aspartame*, Vol. 4, Parts XXIV-XXXI, submitted by G. D. Searle & Co. to the Food and Drug Administration, Hearing Clerk File, Administrative Record, Aspartame 75F-0355, Food and Drug Administration, Rockville, Md.
10. File E-104 (1980). *Developmental Assessment of Infant Macaques Receiving Dietary Aspartame or Phenylalanine*, submitted by G. D. Searle & Co. to the Food and Drug Administration, Hearing Clerk File, Administrative Record, Aspartame 75F-0355, Food and Drug Administration, Rockville, Md.
11. File-105 (1980). *Aspartame Administration to the Infant Monkey: Hypothalamic Morphology and Blood Amino Acid Levels*, submitted by G. D. Searle & Co. to the Food and Drug Administration, Hearing Clerk File, Administrative Record, Aspartame 75F-0355, Food and Drug Administration, Rockville, Md.
12. Oppermann, J. A. (1984). Aspartame metabolism in animals. In *Aspartame: Physiology and Biochemistry* (Stegink, L. D., and Filer, L. J., Jr., eds.), Marcel Dekker, New York, Chap. 7.
13. Oppermann, J. A., Muldoon, E., and Ranney, R. E. (1973). Effect of aspartame on phenylalanine metabolism in the monkey. *J. Nutr.* 103, 1460-1466.
14. Ranney, R. E., Opperman, J. A., Muldoon, E., and McMahon, F. G. (1976). Comparative metabolism of aspartame in experimental animals and humans. *J. Toxicol. Environ. Health* 2, 441-451.
15. Ranney, R. E., and Oppermann, J. A. (1979). A review of the metabolism of

the aspartyl moiety of aspartame in experimental animals and man. *J. Toxicol. Environ. Health* 2, 979-985.
16. Oppermann, J. A., and Ranney, R. E. (1979). The metabolism of aspartame in infant and adult mice. *J. Environ. Pathol. Toxicol.* 2, 987-998.
17. Reynolds, W. A., Parsons, L., and Stegink, L. D. (1984). Neuropathology studies following aspartame ingestion by infant nonhuman primates. In *Aspartame: Physiology and Biochemistry* (Stegink, L. D., and Filer, L. J., Jr., eds.), Marcel Dekker, New York. Chap. 18.
18. Reynolds, W. A., Stegink, L. D., Filer, L. J., Jr., and Renn, E. (1980). Aspartame administration to the infant monkey: hypothalamic morphology and plasma amino acid levels. *Anat. Rec.* 198, 73-85.
19. Reynolds, W. A., Baumann, A. F., Stegink, L. D., Renn, E., Filer, L. J., Jr., and Naidu, S. (1982). Developmental Assessment of Infant macaques receiving dietary aspartame or phenylalanine. In *Aspartame: Physiology and Biochemistry* (Stegink, L. D., and Filer, L. J., Jr., eds.), Marcel Dekker, New York, Chap. 20.
20. White, A., Handler, P., and Smith, E. (1973). *Principles of Biochemistry,* 5th ed., McGraw-Hill, New York, pp. 914-918.
21. Bianchi, R. G., Muir, E. T., Cook, D. L., and Nutting, E. F. (1980). The biological properties of aspartame. II. Actions involving the gastrointestinal system. *J. Environ. Pathol. Toxicol.* 3, 355-362.
22. File E-1 (1972). *A Sweetening Agent Pharmacological Studies,* submitted by G. D. Searle & Co. to the Food and Drug Administration, Hearing Clerk File, Administrative Record, Aspartame 75F-0355, Food and Drug Administration, Rockville, Md.
23. Potts, W. S., Blass, J. L., and Nutting, E. F. (1980). Biological properties of aspartame. I. Evaluation of central nervous system effects. *J. Environ. Pathol. Toxicol.* 3, 341-353.
24. File E-19 (1972). *SC-18862: A Sweetening Agent: Endocrine Studies,* submitted by G. D. Searle & Co. to the Food and Drug Administration, Hearing Clerk File, Administrative Record, Aspartame 75F-0355, Food and Drug Administration, Rockville, Md.
25. Saunders, F. J., Pautsch, W. F., and Nutting, E. F. (1980). The biological properties of asparatame. III. Examination for endocrine-like activities. *J. Environ. Pathol. Toxicol.* 3, 363-373.
26. File E-4 (1972). *SC-18862: Nine Week Oral Toxicity Study in the Rat,* submitted by G. D. Searle & Co. to the Food and Drug Administration, Hearing Clerk File, Administrative Record, Aspartame 75F-0355, Food and Drug Administration, Rockville, Md.
27. File E-45 (1973). *SC-19192: Acute Toxicity Studies in the Rat, Mouse and Rabbit,* submitted by G. D. Searle & Co. to the Food and Drug Administration, Hearing Clerk File, Administrative Record, Aspartame 75F-0355, Food and Drug Administration, Rockville, Md.
28. File E-46 (1973). *SC-18862: Acute Toxicity Studies in the Rat, Mouse and Rabbit,* submitted by G. D. Searle & Co. to the Food and Drug Administration, Hearing Clerk File, Administrative Record, Aspartame 75F-0355, Food and Drug Administration, Rockville, Md.

29. File E-2 (1972). *SC-18862: Four Week Oral Tolerance Study in the Mouse*, PT 815S69, submitted by G. D. Searle & Co. to the Food and Drug Administration, Hearing Clerk File, Administrative Record, Aspartame 75F-0355, Food and Drug Administration, Rockville, Md.
30. File E-3 (1972). *SC-18862: Four Week Oral Tolerance Study in the Rat*, PT 814S69, submitted by G. D. Searle & Co. to the Food and Drug Administration, Hearing Clerk File, Administrative Record, Aspartame 75F-0355, Food and Drug Administration, Rockville, Md.
31. File E-20 (1972). *SC-18862: Two Month Oral Administration–Rats*, PT 719H68, Final Report, submitted by G. D. Searle & Co. to the Food and Drug Administration, Hearing Clerk File, Administrative Record, Aspartame 75F-0355, Food and Drug Administration, Rockville, Md.
32. File E-6 (1972). *SC-19192: Two Week Oral Toxicity Study in the Mouse*, PT 855S70, submitted by G. D. Searle & Co. to the Food and Drug Administration, Hearing Clerk File, Administrative Record, Aspartame 75F-0355, Food and Drug Administration, Rockville, Md.
33. File E-7 (1972). *SC-19192: Two Week Oral Toxicity Study in the Rat*, PT 884S70, submitted by G. D. Searle & Co. to the Food and Drug Administration Hearing Clerk File, Administrative Record, Aspartame 75F-0355, Food and Drug Administration, Rockville, Md.
34. File E-8 (1972). *SC-19192: Five Week Oral Toxicity Study in the Rat*, PT 972S71, submitted by G. D. Searle & Co. to the Food and Drug Administration, Hearing Clerk File, Administrative Record, Aspartame 75F-0355, Food and Drug Administration, Rockville, Md.
35. File E-21 (1972). *SC-18862: Two Month Oral Toxicity–Dogs*, PT 720H68, submitted by G. D. Searle & Co. to the Food and Drug Administration, Hearing Clerk File, Administrative Record, Aspartame 75F-0355, Food and Drug Administration, Rockville, Md.
36. File E-75 (1974). *SC-18862: 104-Week Toxicity Study in the Mouse*, PT 984H73, submitted by G. D. Searle & Co. to the Food and Drug Administration, Hearing Clerk File, Administrative Record, Aspartame 75F-0355, Food and Drug Administration, Rockville, Md.
37. File E-33 (1973). *SC-18862: Appendix: Two Year Toxicity Study in the Rat*, PT 838H71, submitted by G. D. Searle & Co. to the Food and Drug Administration, Hearing Clerk File, Administrative Record, Aspartame 75F-0355, Food and Drug Administration, Rockville, Md.
38. File E-34 (1973). *SC-18862: Two Year Toxicity Study in the Rat*, PT 838H71, submitted by G. D. Searle & Co. to the Food and Drug Administration, Hearing Clerk File, Administrative Record, Aspartame 75F-0355, Food and Drug Administration, Rockville, Md.
39. File E-87 (1975). *SC-18862: A Supplemental Evaluation of Rat Brains from Two Tumorigenicity Studies*, PT 1227, submitted by G. D. Searle & Co. to the Food and Drug Administration, Hearing Clerk File, Administrative Record, Aspartame 75F-0355, Food and Drug Administration, Rockville, Md.
40. File E-70 (1974). *SC-18862: Lifetime Toxicity Study in the Rat*, PT 892H72, Final Report, submitted by G. D. Searle & Co. to the Food and Drug

Administration, Hearing Clerk File, Administrative Record, Aspartame 75F-0355, Food and Drug Administration, Rockville, Md.
41. File E-27 (1973). *SC-18862: 46-Week Oral Toxicity–Hamster*, PT 852S72, submitted by G. D. Searle & Co. to the Food and Drug Administration, Hearing Clerk File, Administrative Record, Aspartame 75F-0355, Food and Drug Administration, Rockville, Md.
42. File E-35 (1973). *SC-18862: 46-Week Oral Toxicity Study in the Hamster, Supplement*, No. 1, Part I, PT 852S72, submitted by G. D. Searle & Co. to the Food and Drug Administration, Hearing Clerk File, Administrative Record, Aspartame 75F-0355, Food and Drug Administration, Rockville, Md.
43. File E-36 (1973). *SC-18862: 46-Week Oral Toxicity Study in the Hamster, Supplement*, No. 1, Part II, PT 852S72, submitted by G. D. Searle & Co. to the Food and Drug Administration, Hearing Clerk File, Administrative Record, Aspartame 75F-0355, Food and Drug Administration, Rockville, Md.
44. File E-28 (1973). *SC-18862: 106-Week Oral Toxicity Study in the Dog,* submitted by G. D. Searle & Co. to the Food and Drug Administration, Hearing Clerk File, Administrative Record, Aspartame 75F-0355, Food and Drug Administration, Rockville, Md.
45. Ishii, H. Koshimizu, T., Usami, S., and Fujimoto, T. (1981). Toxicity of aspartame and its diketopiperazine by Wistar rats by dietary administration for 104 weeks. *Toxicology* 21, 91-94.
46. File E-102 (1978). *Authentication Review of Selected Materials submitted to the Food and Drug Administration Relative to Application of Searle Laboratories to Market Aspartame*, Vols. 1-3, report prepared by Universities Associated for Research and Education in Pathology, report submitted to the Food and Drug Administration, Hearing Clerk File, Administrative Record, Aspartame 75F-0355, Food and Drug Administration, Rockville, Md.
47. File E-76 (1974). *SC-19192: 110-Week Toxicity Study in the Mouse*, PT 985H73, Final report, submitted by G. D. Searle & Co. to the Food and Drug Administration, Hearing Clerk File, Administrative Record, Aspartame 75F-0355, Food and Drug Administration, Rockville, Md.
48. File E-77 (1974). *SC-19192: 115-Week Oral Tumorigenicity Study in the Rat*, Vol. 1, PT 988S73, submitted by G. D. Searle & Co. to the Food and Drug Administration, Hearing Clerk File, Administrative Record, Aspartame 75F-0355, Food and Drug Administration, Rockville, Md.
49. File E-78 (1974). *SC-19192: 115-Week Oral Tumorigenicity Study in the Rat*, Vol. 2, Postmortem evaluation, PT 988S73, submitted by G. D. Searle & Co. to the Food and Drug Administration, Hearing Clerk File, Administrative Record, Aspartame 75F-0355, Food and Drug Administration, Rockville, Md.
50. File E-58 (1973). *SC-18862: A 26 Week Urinary Bladder Tumorigenicity Study in the Mouse by the Intravesical Pellet Implant Technique*, PT 1031ot72, submitted by G. D. Searle & Co. to the Food and Drug Administration, Hearing Clerk File, Administrative Record, Aspartame 75F-0355, Food and Drug Administration, Rockville, Md.
51. File E-59 (1973). *SC-19192: A 26 Week Urinary Bladder Tumorigenicity*

study in the Mouse by the Intravesical Pellet. Implant Technique Final Report, PT 1032ot72, submitted by G. D. Searle & Co. to the Food and Drug Administration, Hearing Clerk File, Administrative Record, Aspartame 75F-0355, Food and Drug Administration, Rockville, Md.

52. Bryan, G. T. (1984). Artificial sweeteners and bladder cancer: Assessment of potential urinary bladder carcinogenicity of aspartame and its diketopiperazine derivative in mice. In *Aspartame: Physiology and Biochemistry* (Stegink, L. D., and Filer, L. J., Jr., eds.), Marcel Dekker, New York, Chap. 16.
53. File E-22 (1972). *SC-18862: Chicken Embryo Study of SC-18862, Calcium Cyclamate and Sucrose*, PT 870H70, submitted by G. D. Searle & Co. to the Food and Drug Administration, Hearing Clerk File, Administrative Record, Aspartame 75F-0355, Food and Drug Administration, Rockville, Md.
54. File E-10 (1972). *Toxicological Evaluation of SC-18862: Evaluation of Reproductive Performance*, PT 857S70, submitted by G. D. Searle & Co. to the Food and Drug Administration, Hearing Clerk File, Administrative Record, Aspartame 75F-0355, Food and Drug Administration, Rockville, Md.
55. File E-11 (1972). *Two Generation Reproduction Study in Rats*, PT 867H71, submitted by G. D. Searle & Co. to the Food and Drug Administration, Hearing Clerk File, Administrative Record, Aspartame 75F-0355, Food and Drug Administration, Rockville, Md.
56. File E-9 (1972). *Toxicological Evaluation in the Neonatal Rat*, PT 893H71, submitted by G. D. Searle & Co. to the Food and Drug Administration, Hearing Clerk File, Administrative Record, Aspartame 75F-0355, Food and Drug Administration, Rockville, Md.
57. File E-5 (1972). *Evaluation of Embryotoxic and Teratogenic Potential in the Rat*, PT 851S70, submitted by G. D. Searle & Co. to the Food and Drug Administration, Hearing Clerk File, Administrative Record, Aspartame 75F-0355, Food and Drug Administration, Rockville, Md.
58. File E-37 (1973). *SC-19192: Evaluation of Reproductive Performance in the Rat: Segment I of the Teratology Reproduction Profile*, PT 996S72, submitted by G. D. Searle & Co. to the Food and Drug Administration, Hearing Clerk File, Administrative Record, Aspartame 75F-0355, Food and Drug Administration, Rockville, Md.
59. File E-47 (1973). *SC-18862: A Study of the Pregnant and Lactating Rat and of Her Offspring*, PT 858S70, submitted by G. D. Searle & Co. to the Food and Drug Administration, Hearing Clerk File, Administrative Record, Aspartame 75F-0355, Food and Drug Administration, Rockville, Md.
60. File E-48 (1973). *SC-18862: A Study of the Pregnant and Lactating Rat and Her Offspring, Segment III*, PT 896S70, submitted by G. D. Searle & Co. to the Food and Drug Administration, Hearing Clerk File, Administrative Record, Aspartame 75F-0355, Food and Drug Administration, Rockville, Md.
61. File E-49 (1973). *SC-18862: A Study of the Pregnant and Lactating Rat and of Her Offspring, Segment III. Comparison by Feeding of Equimolar Quantities of L-Phenylalanine and/or L-Aspartic Acid*, PT 898S70, submitted by G. D. Searle & Co. to the Food and Drug Administration, Hearing Clerk File, Administrative Record, Aspartame 75F-0355, Food and Drug Administration, Rockville, Md.

62. File E-56 (1973). *SC-18862 and SC-19192: 3:1 Ratio Segment II Teratology Study in the Rat*, PT 1001H72, submitted by G. D. Searle & Co. to the Food and Drug Administration, Hearing Clerk File, Administrative Record, Aspartame 75F-0355, Food and Drug Administration, Rockville, Md.
63. File E-52 (1973). *SC-18862: Segment II–An Evaluation of the Teratogenic Potential in the Rabbit*, PT 1045H72, submitted by G. D. Searle & Co. to the Food and Drug Administration, Hearing Clerk File, Administrative Record, Aspartame 75F-0355, Food and Drug Administration, Rockville, Md.
64. File E-53 (1973). *SC-18862: An Evaluation of the Embryotoxic and Teratogenic Potential in the Rabbit. Segment II Study*, PT 968S71, submited by G. D. Searle & Co. to the Food and Drug Administration, Hearing Clerk File, Administrative Record, Aspartame 75F-0355, Food and Drug Administration, Rockville, Md.
65. File E-54 (1973). *SC-18862: An Evaluation of the Embryotoxic and Teratogenic Potential in the Rabbit. Segment II Study*, PT 859S70, submitted by G. D. Searle & Co. to the Food and Drug Administration, Hearing Clerk File, Administrative Record, Aspartame 75F-0355, Food and Drug Administration, Rockville, Md.
66. File E-55 (1973). *SC-18862: Segment II Teratology Study in the Rabbit,* PT 941H71, submitted by G. D. Searle & Co. to the Food and Drug Administration, Hearing Clerk File, Administative Record, Aspartame 75F-0355, Food and Drug Administration, Rockville, Md.
67. File E-79 (1974). *SC-18862: Segment II Evaluation of the Teratogenic Potential in the Rabbit*, PT 1062H73, submitted by G. D. Searle & Co. to the Food and Drug Administration, Hearing Clerk File, Administrative Record, Aspartame 75F-0355, Food and Drug Administration, Rockville, Md.
68. File E-29 (1973). *SC-18862 and SC-19192: 3:1 Ratio Segment II Teratology Study. Rabbit*, PT 1002H72, submitted by G. D. Searle & Co. to the Food and Drug Administration, Hearing Clerk File, Administrative Record, Aspartame 75F-0355, Food and Drug Administration, Rockville, Md.
69. File E-38 (1973). *SC-19192: An Evaluation of the Embryotoxic and Teratogenic Potential in the Rat: Segment II of the Teratology Reproduction Profile*, PT 997S72, submitted by G. D. Searle & Co. to the Food and Drug Administration, Hearing Clerk File, Administrative Record, Aspartame 75F-0355, Food and Drug Administration, Rockville, Md.
70. File E-57 (1973). *SC-19192: Segment II Teratology Study in the Rabbit*, PT 1003H72, submitted by G. D. Searle & Co. to the Food and Drug Administration, Hearing Clerk File, Administrative Record, Aspartame 75F-0355, Food and Drug Administration, Rockville, Md.
71. File E-97 (1978). *SC-18862: An Evaluation of Mutagenic Potential Employing the Ames* Salmonella*/Microsome Assay*, S.A. 1377, submitted by G. D. Searle & Co. to the Food and Drug Administration, Hearing Clerk File, Administrative Record, Aspartame 75F-0355, Food and Drug Administration, Rockville, Md.
72. File E-101 (1978). *An Evaluation of the Mutagenic Potential of SC-18862 Employing the Ames* Salmonella*/Microsome Assay*, S.A. 1385, submitted by G. D. Searle & Co. to the Food and Drug Administration, Hearing Clerk

File, Administrative Record, Aspartame 75F-0355, Food and Drug Administration, Rockville, Md.
73. File E-98 (1978). *SC-19192: An Evaluation of Mutagenic Potential Employing the Ames* Salmonella/*Microsome Assay*, S.A. 1378, submitted by G. D. Searle & Co. to the Food and Drug Administration, Hearing Clerk File, Administrative Record, Aspartame 75F-0355, Food and Drug Administration, Rockville, Md.
74. File E-106 (1980). *An Evaluation of the Mutagenic Potential of SC-19192 Employing the Ames* Salmonella/*Microsome Assay*, S.A. 1384, submitted by G. D. Searle & Co. to the Food and Drug Administration, Hearing Clerk File, Administrative Record, Aspartame 75F-0355, Food and Drug Administration, Rockville, Md.
75. File E-40 (1973). *SC-18862: An Evaluation of the Mutagenic Potential in the Rat Employing the Dominant Lethal Assay*, PT 868S70, submitted by G. D. Searle & Co. to the Food and Drug Administration, Hearing Clerk File, Administrative Record, Aspartame 75F-0355, Food and Drug Administration, Rockville, Md.
76. File E-41 (1973). *SC-18862: An Evaluation of the Mutagenic Potential in the Rat Employing the Dominant Lethal Assay*, PT 1007S72, submitted by G. D. Searle & Co. to the Food and Drug Administration, Hearing Clerk File, Administrative Record, Aspartame 75F-0355, Food and Drug Administration, Rockville, Md.
77. File E-31 (1973). *SC-19192: Evaluation of Mutagenic Potential Employing the Host-Mediated Assay–Rat*, PT 1029H72, submitted by G. D. Searle & Co. to the Food and Drug Administration, Hearing Clerk File, Administrative Record, Aspartame 75F-0355, Food and Drug Administration, Rockville, Md.
78. File E-44 (1973). *SC-18862: Evaluation of Mutagenic Potential Employing the Host Mediated Assay in the Rat*, PT 1028H72, submitted by G. D. Searle & Co. to the Food and Drug Administration, Hearing Clerk File, Administrative Record, Aspartame 75F-0355, Food and Drug Administration, Rockville, Md.
79. File E-12 (1972). *SC-18862: Mutagenic Study in Rats*, PT 869H70, submitted by G. D. Searle & Co. to the Food and Drug Administration, Hearing Clerk File, Administrative Record, Aspartame 75F-0355, Food and Drug Administration, Rockville, Md.
80. File E-43 (1973). *SC-18862: An Evaluation of Mutagenic Potential Employing the In Vivo Cytogenetics Method in the Rat*, PT 1026H72, submitted by G. D. Searle & Co. to the Food and Drug Administration, Hearing Clerk File, Administrative Record, Aspartame 75F-0355, Food and Drug Administration, Rockville, Md.
81. File E-30 (1973). *SC-19192: Evaluation of Mutagenic Potential Employing the In Vivo Cytogenics Method in the Rat*, PT 1027H72, submitted by G. D. Searle & Co. to the Food and Drug Administration, Hearing Clerk File, Administration Record, Aspartame 75F-0355, Food and Drug Administration, Rockville, Md.

15

Chronic Feeding Studies with Aspartame and Its Diketopiperazine

Hiroyuka Ishii
Life Science Laboratory, Central Research Laboratories,
Ajinomoto Company, Inc., Yokohama, Japan

INTRODUCTION

Since aspartame (APM) forms a diketopiperazine (DKP) under certain conditions of storage or food processing (1), toxicological studies were conducted on aspartame's diketopiperazine. The results of chronic feeding studies in rats and dogs fed aspartame or its diketopiperazine are reviewed in this chapter.

STUDIES IN THE RAT

Chronic toxicity studies of APM and its DKP in the rat have been reported by Ishii et al. (2). A total of 860 SLC Wistar rats were divided into five groups comprising 86 male and 86 female animals per group. Control animals were fed a basal diet whose composition is given in Table 1. Aspartame or an APM-DKP (3:1) mixture were added to the basal diet to provide approximately 1, 2, and 4 g/kg body weight per day of APM or 4 g/kg body weight per day of the 3:1 APM-DKP mixture. The experimental diet was fed from 6 to 110 weeks of life. The quantity of test materials added to the basal diet was adjusted every week according to changes in mean body weight and food consumption. Each group was subdivided into a secondary group of 60 male and 60 female animals, with smaller satellite groups for interim clinical and postmortem studies (i.e., 10 male and 10 female animals studied at 26 weeks of feeding, and 16 male and 16 female animals studied after 52 weeks of feeding).

Table 1 Composition of the Basal Diet

Component	Amount (%)
Water	7.8
Crude protein	23.6
Crude fat	4.4
Crude fiber	4.9
Crude ash	6.6
Nitrogen-free extract	52.7

Animals were evaluated with respect to physical appearance, behavior, body weight, food and water consumption, survival, clinical laboratory data, and at postmortem for organ weight as well as for gross and microscopic pathology.

Mean consumption of APM or the 3:1 APM-DKP mixture was within 5% of projected intake for all groups. There were no dose-related changes in physical appearance, behavior, or mortality (Figs. 1 and 2). A dose-dependent suppression of body weight gain was observed in male animals receiving APM at 2 and 4 g/kg body weight and those receiving the APM-DKP mixture at 4 g/kg body weight. Slightly lowered mean body weights were noted among female animals of all groups from approximately 20 weeks of feeding (Figs. 3 and 4). Mean food consumption and the ratio of food consumption per unit body weight were lower in animals fed APM or the APM-DKP mixture than in controls throughout the study. Reduced body weight gain during the 78 weeks of feeding was correlated

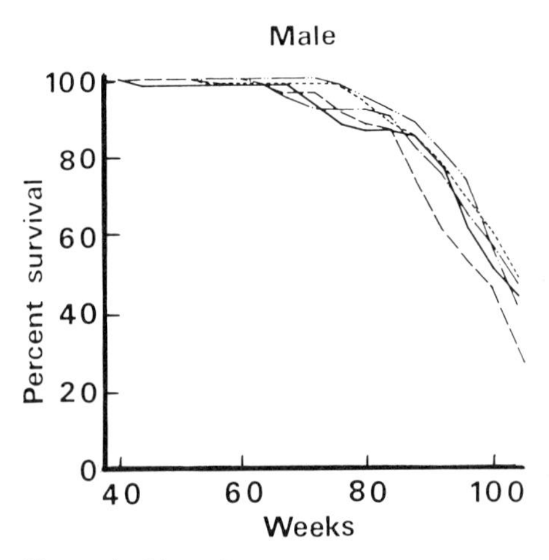

Figure 1 Mortality in male rats fed APM for 104 weeks: control (——), 1 g/kg body weight APM (— —), 2 g/kg body weight APM (— · —), 4 g/kg body weight APM (— · · —), and 4 g/kg body weight APM + DKP (3:1) (- - -).

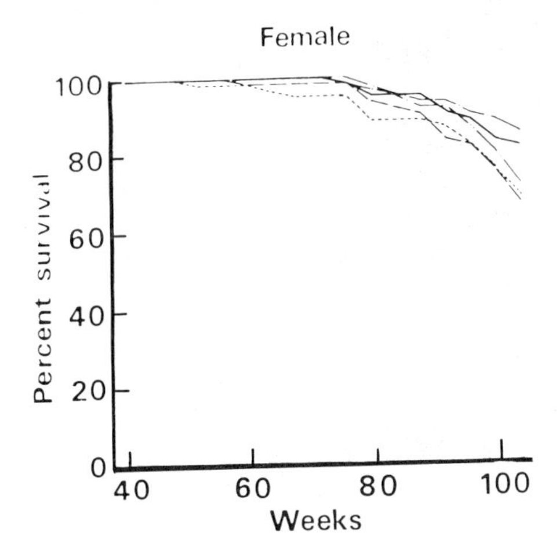

Figure 2 Mortality in female rats fed APM for 104 weeks: control (——), 1 g/kg body weight APM (— —), 2 g/kg body weight APM (— · —), 4 g/kg body weight APM (— · · —), and 4 g/kg body weight APM + DKP (3:1) (- - -).

with the decrease in food consumption. The decreased food consumption probably resulted from an amino acid imbalance created by APM ingestion at these levels. The addition of excessive amounts of individual amino acids such as glycine, phenylalanine, and tryptophan has been shown to decrease food consumption and depress body weight gain (3-6).

Both male and female animals fed the APM-DKP mixture showed a statistically significant increase in urine specific gravity and a decrease in urinary pH at 26 and 52 weeks (Table 2). The decrease in urinary pH was less pronounced in these animals at 104 weeks of study. The urinary pH of female animals fed 4 g/kg body weight APM was slightly decreased at 26 and 52 weeks of study. Similar changes were noted in female animals fed 2 g/kg body weight APM at 52 weeks of study. The changes in urinary specific gravity and pH were within normal limits. A dose-related increase in excretion of urinary calcium was observed in animals of both genders fed APM at 2 and 4 g/kg body weight and those fed the APM-DKP mixture (4 g/kg body weight). This effect is not unique for APM, since other investigators have reported that rats fed high-protein diets increase both calcium absorption and urinary excretion (7,8).

There were no hematological changes attributable to APM or its DKP. A decrease in total serum cholesterol concentration was recorded at 52 and 104 weeks in animals of both sexes fed the APM-DKP mixture and at 104 weeks in animals of both genders fed APM at 4 g/kg body weight. This was not considered to be an adverse effect.

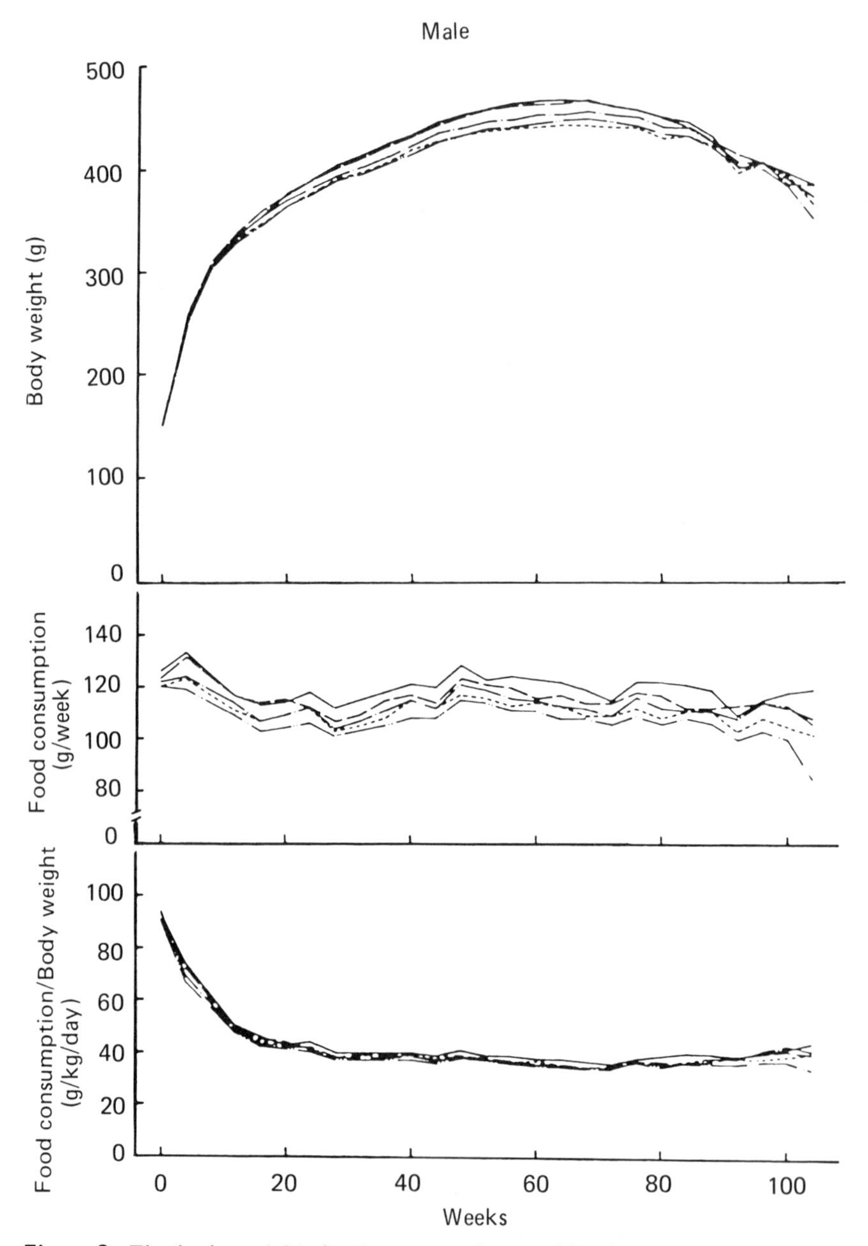

Figure 3 The body weight, food consumption, and food consumption per body weight ratio in female rats fed APM for 104 weeks: control (——), 1 g/kg body weight APM (— —), 2 g/kg body weight APM (— · —), 4 g/kg body weight APM (— · · —), and 4 g/kg body weight APM + DKP (3:1) (- - -).

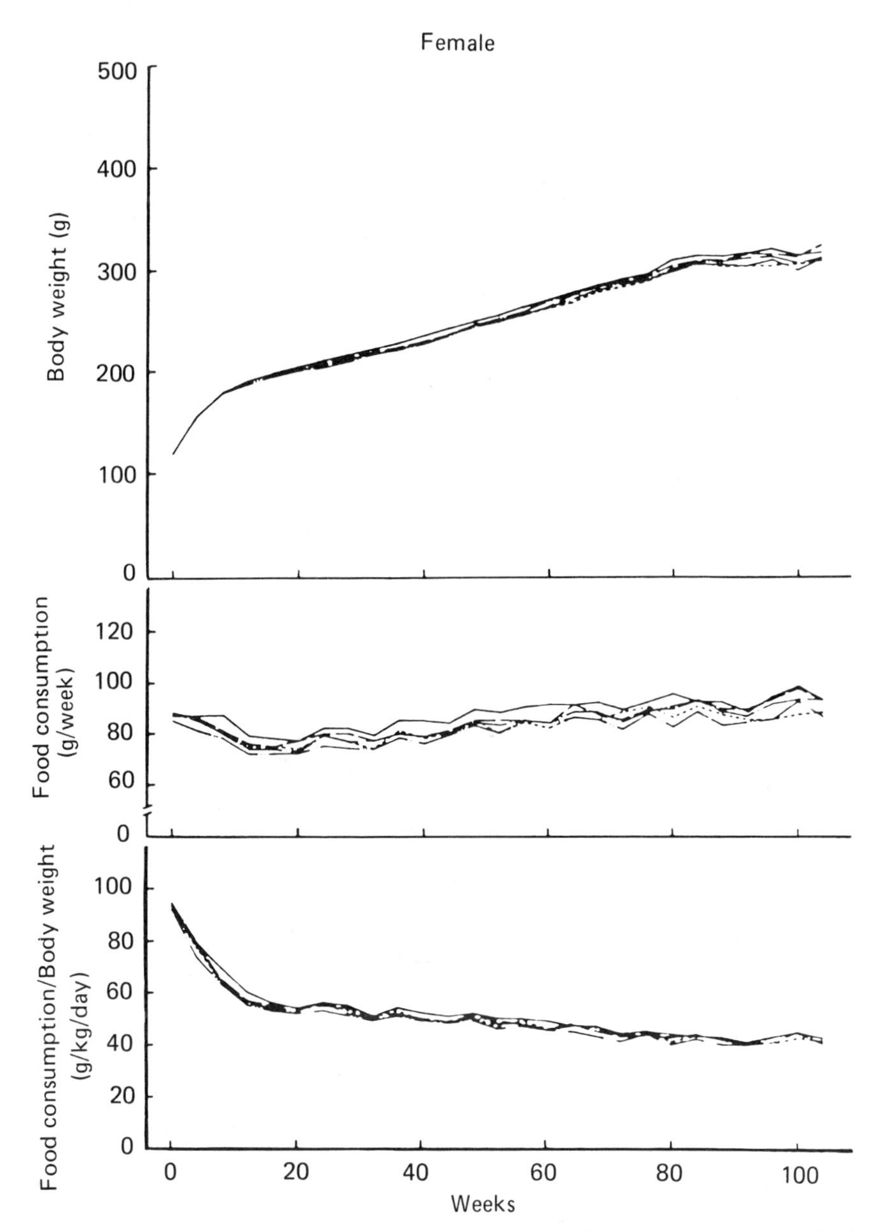

Figure 4 The body weight, food consumption, and food consumption per body weight ratio in female rats fed APM for 104 weeks: control (——), 1 g/kg body weight APM (— —), 2 g/kg body weight APM (— · —), 4 g/kg body weight APM (— · · —), and 4 g/kg body weight APM + DKP (3:1) (- - -).

Table 2 Urine Specific Gravity and pH in Rats Fed APM and DKP (Mean ± SD)

Gender	Treatment	Specific gravity			pH		
		26 weeks	52 weeks	104 weeks	26 weeks	52 weeks	104 weeks
Male							
	Control						
	APM, 1 g/kg body weight	1.040 ± 0.008	1.034 ± 0.007	1.034 ± 0.006	6.64 ± 0.14	6.54 ± 0.38	6.14 ± 0.30
	APM, 2 g/kg body weight	1.041 ± 0.012	1.027 ± 0.005[a]	1.032 ± 0.005	6.68 ± 0.26	6.50 ± 0.28	6.04 ± 0.22
	APM, 4 g/kg body weight	1.042 ± 0.006	1.032 ± 0.007	1.030 ± 0.004[a]	6.58 ± 0.13	6.35 ± 0.11	6.13 ± 0.43
	APM+DKP	1.040 ± 0.005	1.033 ± 0.006	1.032 ± 0.007	6.58 ± 0.11	6.53 ± 0.26	6.23 ± 0.41
	(3:1), 4 g/kg body weight	1.048 ± 0.011[a]	1.042 ± 0.012[a]	1.034 ± 0.008	6.28 ± 0.18[b]	6.30 ± 0.14[a]	5.97 ± 0.23
Female							
	Control						
	APM, 1 g/kg body weight	1.026 ± 0.009	1.030 ± 0.007	1.032 ± 0.010	7.02 ± 0.68	6.74 ± 0.40	6.34 ± 0.36
	APM, 2 g/kg body weight	1.032 ± 0.007	1.028 ± 0.010	1.032 ± 0.009	6.71 ± 0.55	6.75 ± 0.52	6.46 ± 0.42
	APM, 4 g/kg body weight	1.028 ± 0.004	1.031 ± 0.005	1.031 ± 0.008	6.80 ± 0.30	6.44 ± 0.26[a]	6.34 ± 0.36
	APM+DKP	1.033 ± 0.006	1.027 ± 0.004	1.029 ± 0.007	6.51 ± 0.19[a]	6.48 ± 0.31[a]	6.44 ± 0.35
	(3:1), 4 g/kg body weight	1.034 ± 0.005[a]	1.035 ± 0.005[a]	1.034 ± 0.008	6.47 ± 0.34[a]	6.34 ± 0.22[a]	6.22 ± 0.24

[a]Significant difference from control at $p < 0.05$.
[b]Significant difference from control at $p < 0.01$.

A slight increase in relative spleen weight was observed at 26 and 52 weeks of study in male animals fed either APM at 4 g/kg body weight or the APM-DKP mixture (Table 3). Similar changes in spleen weight were noted at 26 weeks of study in female animals fed APM at 4 g/kg body weight, at 52 weeks of study in male animals fed APM at 2 g/kg body weight, and at 52 weeks of study in female animals fed the APM-DKP mixture (4 g/kg body weight). A similar trend in relative kidney weight was observed at 26 and 52 weeks in male and female groups fed the APM-DKP mixture (4 g/kg body weight).

Histopathological examination of the kidneys of rats killed at 104 weeks showed a dose-related increase in the number of animals with focal mineralization. The foci were found as minute mineral deposits in a few epithelial cells, unrelated to epithelial hyperplasia or inflammatory response. These foci may reflect the increase in urinary calcium excretion by these animals.

There were no gross histopathological findings attributable to APM or its DKP in any other organ, including the spleen. Since the issue of brain tumorigenicity received considerable discussion at the aspartame Public Board of Inquiry, the brains of animals fed APM or APM + DKP were carefully examined (9). There was no significant difference in the incidence of brain tumors between control and treatment groups. On the basis of this study, it can be concluded that a daily oral intake of 4 g/kg body weight of APM or 4 g/kg body weight of the 3:1 APM-DKP mixture for 104 weeks is not associated with toxicological changes in the rat.

Searle has reported to the Food and Drug Administration the results of a 2-year toxicity study in rats carried out by Hazleton Laboratories (10). Aspartame was administered in the diet to groups of 40 male and 40 female Charles River albino rats at levels of 1, 2, 4, and 6-8 g/kg body weight per day for 104 weeks. The intake of the very high-dose group, started at a dose of 6 g/kg body weight per day, was increased from 6 to 7 g/kg body weight per day at week 16 and then increased from 7 to 8 g/kg body weight per day at week 44. A fifth group of 60 male and 60 female rats served as controls and received only the basal laboratory diet (Purina Laboratory Chow). An ex vitro degradation product of APM (1), DKP, was present in various amounts throughout the study. It is estimated that the very high-dose group ingested about 80 mg/kg body weight per day of DKP throughout the last 60 weeks of study.

Criteria used to evaluate safety included physical appearance, behavior, growth, food consumption, survival, clinical laboratory data, ophthalmologic examinations, organ weights, tumor incidence, and gross and microscopic pathology. There was no evidence that the administration of APM at any level produced any effect upon the physical appearance and behavior of the animals. Growth rates were significantly reduced from that of control animals in rats fed APM at an intake of 6-8 g/kg body weight per day. Analyses of food consumption data for the first 52 weeks of study showed significantly lower intake

Table 3 Relative Organ Weights in Rats Fed APM and DKP (Mean ± SD)

Gender	Treatment (g/kg body weight)	Organ weight/body weight X 10^4								
		Spleen			Left kidney			Right kidney		
		26 weeks	52 weeks	104 weeks	26 weeks	52 weeks	104 weeks	26 weeks	52 weeks	104 weeks
Male	Control	17.1 ± 1.1	16.8 ± 1.1	62.8 ± 47.7	28.2 ± 1.2	29.7 ± 1.2	44.5 ± 8.2	27.7 ± 1.0	29.8 ± 1.5	45.5 ± 8.1
	APM, 1 g/kg	17.1 ± 0.5	17.9 ± 2.3	68.4 ± 50.9	27.2 ± 1.0	29.7 ± 1.2	45.1 ± 8.4	27.1 ± 0.8	29.3 ± 1.5	46.0 ± 9.0
	APM, 2 g/kg	18.1 ± 1.2	18.0 ± 1.7[a]	64.9 ± 45.9	28.3 ± 0.8	30.2 ± 0.9	50.4 ± 10.6[a]	28.1 ± 1.0	29.8 ± 1.1	51.0 ± 11.1[a]
	APM, 4 g/kg	18.3 ± 0.7[a]	18.8 ± 1.3[b]	54.8 ± 28.7	28.1 ± 0.9	30.4 ± 1.0	48.0 ± 11.6	27.2 ± 1.0	29.6 ± 0.9	49.3 ± 12.4
	APM+DKP (3:1) 4 g/kg	18.8 ± 1.2[b]	19.1 ± 2.5[b]	66.9 ± 42.4	29.5 ± 1.4[a]	31.8 ± 1.3[b]	47.0 ± 8.8	29.6 ± 1.2[b]	31.7 ± 1.4[b]	47.7 ± 8.7
Female	Control	20.3 ± 1.2	19.4 ± 2.8	28.8 ± 29.5	31.2 ± 0.9	32.6 ± 2.6	35.7 ± 4.3	30.7 ± 1.9	32.5 ± 2.2	35.7 ± 4.5
	APM, 1 g/kg	21.1 ± 1.3	18.5 ± 1.4	25.5 ± 24.3	31.6 ± 1.9	32.2 ± 0.8	34.1 ± 4.0	31.5 ± 1.2	32.1 ± 1.5	34.1 ± 4.0
	APM, 2 g/kg	21.3 ± 1.3	19.2 ± 1.6	26.4 ± 17.9	31.4 ±0.8	32.5 ± 1.8	35.6 ± 5.6	30.9 ± 1.3	32.4 ± 2.2	35.4 ± 6.4
	APM, 4 g/kg	22.2 ± 1.4[b]	21.0 ± 1.6	35.9 ± 38.7	31.7 ± 1.3	32.9 ± 1.9	34.8 ± 5.8	32.6 ± 1.8[a]	32.2 ± 1.8	33.9 ± 5.7
	APM+DKP (3:1), 4 g/kg	21.2 ± 0.8	21.3 ± 1.7[a]	27.6 ± 14.7	32.2 ± 1.5	34.1 ± 1.8	34.6 ± 4.7	32.3 ± 1.4	34.0 ± 2.1	34.4 ± 5.1

[a]Significant difference from control at $p < 0.05$.
[b]Significant difference from control at $p < 0.01$.

for males fed APM at 6-8 g/kg body weight per day and for females fed APM at 4 and 6-8 g/kg body weight per day.

Survival rates after 104 weeks among rats fed APM were comparable to those of control animals, except for female rats fed the highest level of APM.

The results of clinical laboratory tests obtained periodically throughout the study from all treated groups revealed no consistent statistically significant changes from the normal values reported for this species and strain of rat.

Periodic ophthalmoscopic examination of animals at all treatment levels showed no indication of compound-related effects. Similarly, organ-to-body weight ratios showed no changes attributable to treatment.

Tumor data were statistically analyzed by gender employing an actuarial (life-table) technique. Tumor formation in either gender at any of the treatment levels for any tumor types studied did not differ from that of controls. Except for renal changes, gross and microscopic pathologic findings at postmortem were unremarkable.

Deposits of hemosiderin in renal tubular or pelvic epithelial cells of male survivors and nonsurvivors were increased at the higher treatment levels. Females at all treatment levels and males administered lower doses were not affected. The incidence of focal hyperplasia within the renal pelvic epithelium was increased in male animals fed APM at 6-8 g/kg body weight. Male and female animals at other treatment levels were sporadically affected. Tubular degeneration was increased in male animals (both survivors and nonsurvivors) fed 6-8 g/kg body weight APM. Female animals were not affected. Intratubular proteinaceous material was observed with greater frequency among male survivors fed APM. The dose relationship, however, was an inverse one. Male nonsurvivors and female survivors and nonsurvivors fed APM were not affected. Neither renal function nor longevity were affected, suggesting that the above effects were of limited biological significance.

Administration of APM to male and female rats for a period of 2 years did not increase the incidence of neoplasms, even when given at levels of 6-8 g/kg body weight. Aspartame intakes of 4 g/kg body weight per day did not affect survival rates of male or female rats, with the exception of the renal changes observed in male rats fed very high levels of APM for 2 years. There was little evidence of treatment-related nonneoplastic changes in other organs or tumors.

Lifetime chronic feeding studies on the rats have also been conducted for Searle by Hazleton Laboratories (11). Adult female Charles River albino rats were fed APM at levels of 2 and 4 g/kg body weight per day for 60 days prior to mating. They continued to receive APM at this level of intake throughout mating, gestation, and nursing. Weanling rats (40 male and 40 females for each dose level) continued to receive APM at 2 and 4 g/kg body weight per day for 104 weeks postweaning. An ex vitro degradation product of APM (1), DKP, was present in various amounts throughout the study; DKP intake was approximately

20 and 40 mg/kg body weight per day for the low- and high-dose groups, respectively. A third group of 60 male and 60 female weanling rats fed the basal laboratory diet (Purina Laboratory Chow) served as controls.

Criteria used for safety evaluation were physical appearance, behavior, growth, food consumption, survival, clinical laboratory data, ophthalmologic examination, organ weights, tumor incidence, and gross and microscopic pathology. There was no evidence that the administration of APM at any level produced any effect upon the physical appearance and behavior of the animals. The growth rate of male rats fed the high level intake of APM was significantly lower than that of control animals. Analyses of food consumption data for the first 52 weeks of study revealed significantly lower intakes by these animals.

Statistical analysis of survival calculated by the life-table technique indicated that survival after 104 weeks was comparable in both treatment groups to that of control animals irrespective of gender.

While random fluctuations occurred within the clinical laboratory data and occasionally reached statistical significance, these changes were sporadic, and the values generally remained within normal limits.

Periodic ophthalmoscopic examination of all animals showed no indication of a compound-related effect. Likewise, gross observations at necropsy failed to reveal compound-related changes in any organ or tissue. Specific organ weignts and the ratio of organ weight to body weight showed no statistically significant differences by treatment group.

Liver phenylalanine hydroxylase activity was determined on five rats of each gender per group upon termination of the study (Table 4). Statistical analysis showed a significant compound-related increase in enzyme activity for male rats fed low levels of APM and for both male and female animals fed a high-level intake of APM.

Tumor incidence data, by gender, were analyzed statistically using an acturial (life-table) technique. There was no increase in frequency of occurrence of any tumor type by gender or treatment group over tnat tound in control animals. Histopathological examination of multiple organs was unremarkable.

Table 4 Liver Phenylalanine Hydroxylase Activity

Treatment (g/kg body weight)	Liver phenylalanine hydroxylase activity, μmol tyrosine/g/30 min: Male	Female
Control	1.36 ± 0.39	1.24 ± 0.54
APM, 2 g/kg	3.21 ± 0.42[a]	1.21 ± 0.46
APM, 4 g/kg	3.13 ± 1.35[a]	2.52 ± 0.92[a]

[a]Significant difference from control at $p < 0.05$.

On the basis of these results, it can be concluded that rats exposed to 2 or 4 g/kg body weight per day of APM and/or its metabolic products throughout the full prenatal period by placental transfer, throughout the nursing period via maternal milk and/or dry diet, and throughout the 104-week postweaning period by direct dietary administration exhibit no adverse effects as measured by survival rate, incidence of neoplasms, or treatment-related nonneoplastic changes in any organ or tissue from animals of either gender.

STUDIES IN THE DOG

Rao and co-workers reported chronic toxicity studies in the dog (12). This was a 106-week chronic toxicity study of APM conducted in 5-month-old purebred beagle dogs. Treated dogs received daily doses of APM admixed with a powdered diet at levels to provide APM at 1,2, and 4 g/kg body weight per day. These levels of intake are 0, 33, 67, and 133 times greater than the calculated maximum daily intake by man. Control animals received the APM-free basal diet (Powdered KASCO dog basal diet, raw meal form). Five animals of each gender were assigned to each feeding group. The various lots of commercial grade APM employed in this study contained up to 1% of the DKP of aspartame. Thus dogs in the high-dose APM group ingested approximately 40 mg/kg body weight per day of APM's DKP.

Physical examinations were performed at 4-week intervals, and ophthalmoscopic examinations at 24, 52, 78, and 104 weeks of treatment. Hematological examinations, clinical biochemical analyses, and urinalysis were performed periodically on all dogs. All animals were promptly necropsied, and representative tissues from control and all treated animals processed for microscopic examination.

Survival rate was 100% in all groups. No compound-related variations in food consumption or body weight gain were observed. There were no changes in motor or behavioral activity and in general physical appearance. Ophthalmoscopic examination of the dogs beginning at 24 weeks of treatment did not reveal any consistent treatment-related changes. However, when the initial ophthalmoscopic examination of all dogs was performed at 24 weeks of treatment, two dogs, one each from the medium-dose group and one from the high-dose group exhibited bilateral anterior subcapsular and nuclear cataracts. Retrospective examination of the breeding records revealed that these two dogs were littermates. The parental sire and dam were remated, and unilateral cataract involving the temporal aspect of the lens cortex was found in one dog from two full sibs at approximately 6 months of age. It was concluded that the cataracts observed in the two treated dogs entered into the study were congenital in origin and unrelated to treatment.

Hematological and biochemical studies revealed no meaningful treatment-related changes, although transient but significant differences in several parameters were observed. Most values however, remained within normal limits and no consistent dose-related pattern was evident.

Some dogs given 4 g/kg body weight per day of APM showed a significant increase in excretion of urinary phenylketones at weeks 2, 4, and 26 of treatment. However, these dogs were negative with respect to phenylketonuria at all other intervals (8, 14, 52, 78, and 106 weeks). Little biological significance is attached to this transient and/or sporadic increase in urinary phenylketones. Fasting serum phenylalanine levels remained normal throughout the 106 weeks of treatment.

At postmortem, gross anatomical and microscopic findings were not remarkable. Relative organ weights were not affected.

It was concluded that continuous dietary administration of APM at daily intake levels of 1, 2, and 4 g/kg body weight to 5-month-old male or female beagle dogs for 106 consecutive weeks caused no biologically meaningful alterations in body weight, food consumption, physical findings, clinical laboratory results, or postmortem examination, including gross anatomical and microscopic studies.

SUMMARY

To evaluate possible adverse effects of orally administered aspartame (APM) and its diketopiperazine (DKP), two toxicity studies of 104 weeks duration and one lifetime toxicity study were performed in rats and a single toxicity study of 106 weeks duration was performed in dogs.

The results of these studies indicate that the oral intake of APM at levels up to 4 g/kg body weight or of the 3:1 APM-DKP mixture at a level of 4 g/kg body weight did not cause adverse effects in rats, and that dietary administration of APM at daily intake levels up to 4 g/kg body weight did not cause any biologically meaningful alterations in dogs.

REFERENCES*

1. Homler, B. (1984). Aspartame: Implications for the food scientist. In *Aspartame: Physiology and Biochemistry*. (Stegink, L. D., and Filer, L. J., Jr., eds.), New York, Marcel Dekker, Chap. 11.
2. Ishii, H., Koshimizu, T., Usami, S., and Fugimoto, T. (1981). Toxicity of aspartame and its diketopiperazine for Wistar rats by dietary administration for 104 weeks. *Toxicology* 21, 91-94.
3. Harper, A. E. (1956). Amino acid imbalances, toxicities and antagonisms. *Nutr. Rev.* 14, 225-227.

*In the "E-file" reports by G. D. Searle to the Food and Drug Administration, SC-18862 refers to aspartame (Searle Compound 18862).

4. Sauberlich, H. E. (1961). Studies on the toxicity and antagonism of amino acids for weanling rats. *J. Nutr.* 75, 61-72.
5. Muramatsu, K., Odagiri, H., Morishita, S., and Takeuchi, H. (1971). Effect of excess levels of individual amino acids on growth of rats fed casein diets. *J. Nutr.* 101, 1117-1125.
6. Rogers, Q. R., and Leung, P. M. B. (1973). The influence of amino acids on the neuroregulation of food intake. *Fed. Proc.* 32, 1709-1719.
7. Bell, R. R., Engelmann, D. T., Sie, T. L., and Draper, H. H. (1975). Effect of a high protein intake on calcium metabolism in the rat. *J. Nutr.* 105, 475-483.
8. McCance, R. A., Widdowson, E. M., and Lehmann, H. (1942). The effect of protein intake on the absorption of calcium and magnesium. *Biochem. J.* 36, 686-691.
9. Ishii, H. (1981). Incidence of brain tumors in rats fed aspartame. *Toxicol. Lett.* 7, 433-437.
10. Study E-34 (1973). *SC-18862: Two-Year Toxicity Study in the Rat*, PT-838H71, report filed by G. D. Searle & Co. regarding food additive petition for aspartame, Food Additive Master File 134, U.S. Department of Health and Human Services, Food and Drug Administration, Hearing Clerk File, Administrative Record, Aspartame 75F-0355, Food and Drug Administration, Rockville, Md.
11. Study E-70 (1974). *SC-18862: Lifetime Toxicity Study in the Rat*, PT-892H72, report filed by G. D. Searle & Co. regarding food additive petition for aspartame, Food Additive Master File 134, U.S. Department of Health and Human Services, Food and Drug Administration, Hearing Clerk File, Administrative Record, Aspartame 75F-0355, Food and Drug Administration, Rockville, Md.
12. Rao, K. S., Mauro, J., and McConnell, R. G. (1972). *E-28, SC-18862: 106 Week Oral Toxicity Study in the Dog*, report filed by G. D. Searle & Co. regarding food additive petition for aspartame, Food Additive Master File 134, U.S. Department of Health and Human Services, Food and Drug Administration, Hearing Clerk File, Administrative Record, Aspartame 75F-0355, Food and Drug Administration, Rockville, Md.

16

Artificial Sweeteners and Bladder Cancer: Assessment of Potential Urinary Bladder Carcinogenicity of Aspartame and Its Diketopiperazine Derivative in Mice

George T. Bryan
Wisconsin Clinical Cancer Center, University of Wisconsin Center for Health Sciences, Madison, Wisconsin

INTRODUCTION

Toxicologic studies of nonnutritive or low caloric sweetening agents have on several occasions demonstrated tumorigenic effects for the mammalian lower urinary tract, that is, renal pelvis, ureter, urinary bladder, and urethra (1-6). For example, dulcin [(4-ethoxyphenyl)urea], a nonnutritive sweetener first prepared in 1883, was reported (3) to produce papillomas of the renal pelvis and urinary bladder in rats. Although the significance of these tumors is not yet known, dulcin has not been approved for use by the United States Food and Drug Administration (1). Urinary bladder tumorigenic effects have been reported in experimental animals treated with saccharin and cyclamates in a variety of ways (2). Because of concerns for human safety, cyclamates were banned in the United States, and saccharin use was permitted only by special legislation.

The urinary bladder as a target tissue for carcinogenic effects of cyclamates (4) and saccharin (5,6) was first demonstrated through use of the mouse intravesical pellet implantation technique. This technique has been advocated as a valid method of assessing bladder carcinogenicity when properly employed (7-9). The objectives of the studies described here were to test the possible bladder tumorigenic effects of aspartame (N-L-α-aspartyl-L-phenylalanine methyl ester) and its major decomposition product, 3-phenylmethyl-2,5-diketopiperazine-6-acetic acid (DKP), by the mouse intravesical pellet implantation technique.

MATERIALS AND METHODS

Chemicals

Aspartame (Lot No. 5606 7B), [ring-U-^{14}C]aspartame ([^{14}C]aspartame) (specific activity 0.45 mCi/mmol), DKP (Lot No. IR A6 906), and [ring-U-^{14}C]DKP ([^{14}C]DKP) (specific activity 0.46 mCi/mmol) were gifts of G. D. Searle Co., Chicago, Illinois. L-Alanine-1-^{14}C ([^{14}C]alanine) (specific activity 11.4 mCi/mmol; Lot No. SG 5842) and L-phenylalanine-1-^{14}C ([^{14}C]phenylalanine) (specific activity 3.7 mCi/mmol; Lot No. G 5830-2) were purchased from Volk Radiochemical Co., Burbank, California. Xanthurenic acid 8-methyl ether (XAE) was prepared as described by Price and Dodge (10).

Animals and Animal Care

Female Swiss albino mice (29-35 g), 60-90 days old (Rolfsmeyer Co., Madison, Wisconsin) were utilized for all experiments. For studies using ^{14}C-labeled chemicals, mice were individually housed in raised stainless steel, screen-bottomed cages or in all-glass metabolism cages which permitted the separate collection of urine, feces, and CO_2 (ethanolamine:methanol, 20:80 vol/vol) as described (11-14). For 26- and 56-week bladder tumorigenicity studies, mice were housed five mice or less per raised stainless steel, screen-bottomed cage. Animals were maintained in rooms with controlled light, temperature, and humidity. Mice received pelleted diet (Wayne Lab-Blox, Allied Mills, Inc., Chicago, Illinois) and tap water ad libitum.

Assessment of Stability and Disappearance Rate of Aspartame or DKP Administered Intravesically

Pellets of 20- to 22-mg mass and 0.4-cm diameter, composed of one part of powdered [^{14}C]aspartame or [^{14}C]DKP mixed with four parts of recrystallized (3 times) powdered cholesterol (Sigma Chemical Co., St. Louis, Missouri) were fashioned as described previously (15) in a Colton pellet press. Lots composed of 80 pellets were prepared to ensure uniformity and reproducibility of chemical composition. All pellets were weighed following preparation, and those exceeding tolerance limits were discarded. Pellets retained were placed in individually labeled small glass vials for storage at 22°C prior to animal administration. Storage in this manner was for no more than 7 days prior to animal placement.

Mice were individually anesthesized with pentobarbital (Nembutal sodium, Abbott Laboratories, North Chicago, Illinois) and diethyl ether. Each study mouse had a pellet surgically placed into the urinary bladder lumen by the technique of Jull (16) as modified by Allen et al. (5). Our use of these techniques in other studies has been reported in detail (4, 6, 9, 11-13, 15).

The ^{14}C content of pellets was assayed by dissolving the pellets in 10 ml of chloroform followed by quantitative dilution to a final volume of 100 ml with absolute ethanol. Following mixing, aliquots were quantitatively transferred to vials, scintillation fluid (14) was added, and the samples were counted (68-74% efficiency) in a Nuclear Chicago Mark I scintillation counter. Efficiency corrections were made using automatic external standardization with the channel ratio method.

Aliquots of the chloroform-ethanol solutions containing either [^{14}C]aspartame or [^{14}C]DKP were concentrated under a stream of air and chromatographed on 0.25-mm silica gel thin-layer plates with fluorescent indicator (Sil G-25 UV_{254}, Brinkman Inst., Inc., Des Plaines, Illinois) employing a saturated atmosphere solvent system consisting of (a) n-butanol:glacial acetic acid:water (8:2:2 vol/vol) (R_f aspartame = 0.77, R_f DKP = 0.89), or (b) ethanol:water (7:3 vol/vol) (R_f aspartame = 0.43, R_f DKP = 0.55). Following development radioactivity was monitored with a Varian Series 6000 radiochromatogram scanner.

The in vivo elution of [^{14}C]aspartame or [^{14}C]DKP was measured using procedures described previously (4, 6, 9, 11, 13, 15). Briefly, pellets, randomly selected from those prepared, were weighed and the content of [^{14}C]aspartame or [^{14}C]DKP calculated. Following surgical introduction into the mouse bladder lumen, mice were housed individually in steel cages. At intervals of 1, 2, 4, 6, or 8 hr following surgery, two to four mice were killed and the pellets recovered. The quantity of [^{14}C]aspartame or [^{14}C]DKP remaining in each pellet was determined, and, from this and the calculated content of the test chemical present prior to insertion of the pellet into the mouse bladder, an estimate of the amount of compound that had disappeared from the pellet during the time it was in the bladder in vivo was made. The percentage of [^{14}C]aspartame or [^{14}C]DKP remaining in a pellet after x hours intravesically was converted to its common logarithm. When the analytic results, representing several pellets remaining in the mouse bladders for various periods of time, were plotted graphically versus the number of hours each pellet had been exposed to urine, a linear relationship was apparent. The slope b of the regression equation best representing this linear relationship was computed by the method of least squares, tested for statistical significance by the F test, and a correlation coefficient was calculated (17). A specific elution rate constant K was calculated for [^{14}C]-aspartame or [^{14}C]DKP by multiplying the slope b by (–) 2.303. The time required for 50% of each chemical to disappear from a pellet, $t_{1/2}$, was computed from K and was used as an index of the first-order elution time. Coordinates (x, y) lying on the regression line were computed from the regression equation, and a graphic representation of this line was drawn as described previously (4, 6, 15, 18).

Absorption by and Interaction of Chemicals with the Mouse Urinary Bladder

Pellets of 20- to 22-mg mass and 0.4-cm diameter composed of [^{14}C] aspartame or [^{14}C] DKP were constructed, weighed, and surgically placed into the urinary bladder lumina of female Swiss mice (4, 6, 9, 11-13, 15).

The method used for the introduction of aqueous solutions into the mouse bladder has been published (11-13). Briefly, after anesthetization (see above), a low abdominal midline incision was made, the bladder and urethra were exposed, and the urethra was carefully dissected away from the dorsal-lying vagina. Two small, round, wooden sticks (3 mm in diameter) were inserted between the urethra and vagina to facilitate the entry of the needle (syringe, 0.05 ml, with needle, no. 705, Hamilton Co., Inc., Whittier, California) into the urethra about 0.2-0.3 mm distal to the bladder. Two ligatures were placed around the urethra, but they were not tightened until the needle had been passed into the urethra and inserted into the bladder lumen. The ureters were identified next and ligated bilaterally about midway between the bladder and kidneys. After injection of 0.05 ml of the 0.9 N sterile saline solution containing the test chemical, the needle was carefully withdrawn past each ligature. The ligatures were successively tightened to prevent leakage of the solution as the needle was removed. The abdomen was then sutured.

Following direct pellet placement, by per urethram instillation in the urinary bladder lumen, or after intraperitoneal administration of test chemicals, mice were placed individually into an all-glass metabolism chamber. They were not fed, but were permitted distilled water ad libitum. Respiratory CO_2, urine, and feces were collected until the mice were killed (11-13).

Chemical dose and specific activity were as follows: for intravesical pellet implantation, [^{14}C] aspartame, 4.0- 4.4 mg per mouse, 0.93 ± 0.18 μCi per mouse; [^{14}C] DKP, 4.0- 4.4 mg per mouse, 0.94 ± 0.08 μCi per mouse; for intraurethral administration, [^{14}C] aspartame, 0.66 mg per mouse, 1.0 μCi per mouse; [^{14}C] DKP, 0.57 mg per mouse, 1.0 μCi per mouse; for intraurethral and intraperitoneal administration, [^{14}C] alanine, 4.9 μg per mouse, 0.63 μCi per mouse; [^{14}C] phenylalanine, 27.9 μg per mouse, 0.63 μCi per mouse.

Mice bearing intravesical pellets were killed by cervical fracture at specified times, and pellets were removed and analyzed as described above. The radioactivity of aliquots of the ethanolamine carbonate solution used to trap carbon dioxide, urine, or homogenized feces was determined using liquid scintillation techniques (14, 19). Liver, lung, kidney, and urinary bladder were dissected, weighed, and individually digested in 2 N KOH:ethanol:toluene (10:5:1 vol/vol) for 24 hr and then brought to volume with ethanol. The bladder was washed and processed separately from the bladder contents. Stomach and intestinal contents were pooled with feces prior to homogenization. The carcass was digested, as were the other tissues, and then filtered to remove bone. Duplicate 0.5-ml samples of the different tissues, urine, feces, or CO_2 were counted in vials containing 20 ml of scintillation fluid for up to 10 min each. Background counts for each tissue were determined by testing two to four mice of the same age processed individually through the above procedures using unlabeled administered chemical.

The mean ± SD percentage of recovered ^{14}C was calculated using four to seven mice for each time period studied. The mean ± SD tissue specific activity, expressed as disintegrations per minute (dpm) per milligram wet weight, was computed. Statistical procedures used were described previously (11-13).

A 26-Week Urinary Bladder Tumorigenicity Study in the Mouse of Aspartame and DKP Administered by the Intravesical Pellet Implantation Technique

A total of 400 female 60- to 90-day-old Swiss albino mice were randomly allocated to one of four treatment groups: group 1 (negative control), exposed only to intraluminal pellets of 20- to 22-mg mass composed of purified cholesterol; group 2, exposed to cholesterol pellets containing 4.0- 4.4 mg of aspartame; group 3, exposed to cholesterol pellets containing 4.0- 4.4 mg of DKP; and group 4 (positive control), exposed to cholesterol pellets containing 4.0- 4.4 mg of XAE. Pellets were fashioned as described previously (15) in lots of 130-140 for each group to encompass the needs of the study and to ensure uniformity and reproducibility of the chemical composition of the pellets. All pellets were weighed, those exceeding tolerance limits were discarded, and those retained were placed in individually labeled small glass vials for storage at 22°C for no more than 7 days prior to surgical intraluminal administration as described (4, 6, 9, 11-13, 15, 16).

Mice were inspected twice daily for morbidity, mortality, and motor and behavioral activity. Individual body weights were recorded weekly up to 4 weeks after surgery, biweekly for the next 8 weeks, once every 4 weeks thereafter, and at death. Pertinent observations, including general external features and digital palpation of protruding tissue masses, were recorded.

Mice found dead or killed by ether anesthesia at the end of the study were weighed and necropsies were performed under the supervision of a pathologist. All tissues in the thoracic, abdominal, and pelvic cavities were examined, as well as the skin. Representative tissues and all gross abnormalities were sampled for histological inspection. All preserved tissues, except the urinary bladders, were fixed in 10% neutral buffered formalin. All urinary bladders were distended with Bouin's fixative inserted intraurethrally (4, 6, 20). Fixed bladders were bisected in the midsagittal line and inspected grossly. The bisected halves of each bladder were imbedded and six intermittent sagittal sections of each half were prepared at 5-μm thickness and stained with hematoxylin and eosin.

Statistical evaluation of tumors was restricted to a comparison of the incidence of bladder carcinomas observed in animals surviving more than 175 days. The comparison was made between the incidence of carcinomas related to the introduction of pellets of cholesterol containing a test chemical (groups 2-4) with the cholesterol alone (negative control) group (group 1). Probabilities of statistical significance were computed by the exact method for 2 X 2 tables (21).

A 56-Week Urinary Bladder Tumorigenicity Study in the Mouse of Aspartame and DKP Administered by the Intravesical Implantation Technique

Details of the pellet preparation; operative technique; selection, care, and observations of the mice; preparation and staining of bladder and other tissue specimens for histological examination; histological criteria and neoplasia staging; and statistical procedures were described above and published in detail (4-6, 15, 20, 22-24). In this study, 800 female 60- to 90-day-old Swiss mice were randomly allocated to duplicate (A and B) treatment groups 1-4 composed each of 100 mice. Treatment groups were group 1 (negative control), exposed only to intraluminal pellets of 20- to 22-mg mass composed of purified cholesterol; group 2, exposed to cholesterol pellets containing 4.0-4.4 mg of aspartame; group 3, exposed to cholesterol pellets containing 4.0-4.4 mg of DKP; and group 4 (positive control), exposed to cholesterol pellets containing 4.0-4.4 mg of XAE. Each mouse was provided with a unique code number so that none of the investigators were aware of its treatment allocation until experimental data had been analyzed. The in vivo study duration was prospectively established as 56 weeks. All mice surviving to this time were killed by ether anesthesia and evaluated as described above.

RESULTS

Assessment of Stability and Disappearance Rate of Aspartame or DKP Administered Intravesically

Analyses of five pellets not inserted into mouse bladders demonstrated a mean (±SD) radioactivity content of [^{14}C]aspartame of 0.93 ± 0.18 μCi and a mean (±SD) radioactivity content of [^{14}C]DKP of 0.94 ± 0.08 μCi. Thin-layer radiochromatography of these solubilized pellets demonstrated only one spot corresponding to the chemical known to be present in the pellets.

The loci of the experimentally determined coordinates; the linear elution curves; the 50% elution times, $t_{1/2}$; and the elution rate constants K for aspartame (Fig. 1) and DKP (Fig. 2) are presented. The $t_{1/2}$ and K for both were 4.8 hr and 0.13 hr^{-1}, respectively. The correlation coefficient for aspartame was (−) 0.94, and that for DKP was (−) 0.81. The slopes of the computed regression lines for aspartame and DKP were highly significant ($p < 0.01$). Thin-layer radiochromatograms of the solubilized pellets that had been in vivo for variable periods revealed only one spot corresponding to the chemical known to be present in the pellets. These data were used to select a chemical, XAE, with similar $t_{1/2}$ and K

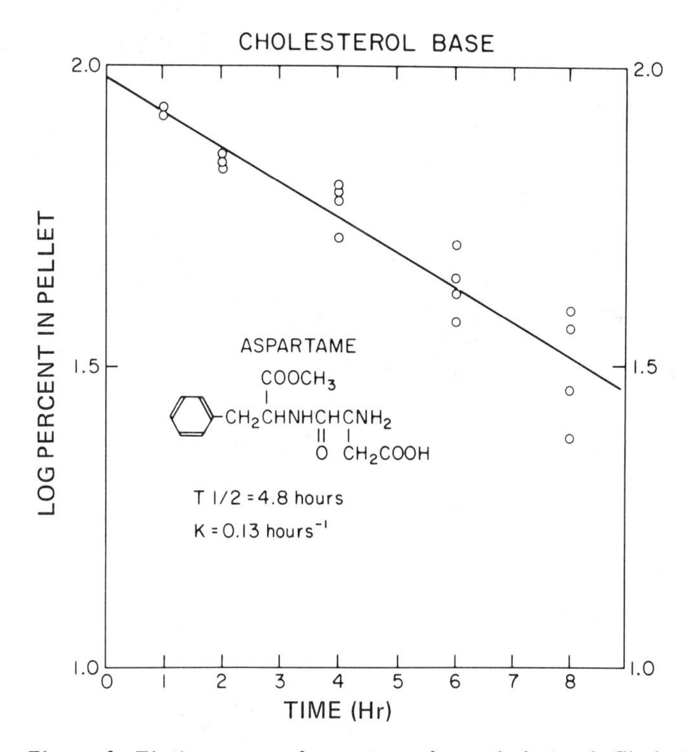

Figure 1 Elution curve of aspartame from cholesterol. Cholesterol pellets (20-22 mg) containing 20% [^{14}C]aspartame were surgically placed into urinary bladder lumina of female Swiss mice. At 1, 2, 4, 6, and 8 hr following surgery, mice were killed, and removed pellets were analyzed for radioactivity content as described in Materials and Methods. Each open circle indicates one mouse at the designated time. The regression line, $t_{1/2}$, and K were computed as described.

characteristics (15,18) as a positive control (20) for the 26- and 56-week bladder tumorigenicity studies of aspartame and DKP.

Absorption by and Interaction of Chemicals with the Mouse Urinary Bladder

The percentage of total radioactivity recovered and the tissue specific activity measured at 4.8 ($t_{1/2}$), 9.6 ($2t_{1/2}$), and 14.3 ($3t_{1/2}$) hr following implantation of cholesterol pellets containing [^{14}C]aspartame or [^{14}C]DKP into the vesical lumen of mice are presented in Table 1. Radioactivity for both compounds disappeared from pellets and was detected in all tissues and biological samples measured. Tissue incorporation of radioactivity 4.8 hr ($t_{1/2}$) after intraluminal

Table 1 Percentage of Total Radioactivity (dpm) Recovered and Tissue Specific Activity After Implantation of Cholesterol Pellets Containing [^{14}C] Aspartame or DKP into the Urinary Bladders of Mice

		Animal hours of survival					
		4.8 ($t_{1/2}$)		9.6 ($2t_{1/2}$)		14.4 ($3t_{1/2}$)	
Compound	Sample	Percent ± SD[a]	dpm/mg ± SD[b]	Percent ± SD[a]	dpm/mg ± SD[b]	Percent ± SD[a]	dpm/mg ± SD[b]
Aspartame	$^{14}CO_2$	3.3 ± 2.3	–	12.5 ± 3.1	–	7.4 ± 2.3	–
	Urine	2.2 ± 1.5	–	9.1 ± 2.0	–	13.9 ± 12.3	–
	Feces[c]	0.2 ± 0.2	–	4.5 ± 1.1	–	3.2 ± 4.3	–
	Bladder	0.2 ± 0.01	136 ± 64	0.8 ± 0.4	372 ± 194	0.7 ± 0.1	277 ± 65
	Carcass	12.6 ± 3.1	12.8 ± 6.0	17.3 ± 7.0	7.6 ± 2.4	7.0 ± 1.8	4.7 ± 1.8
	Liver	2.7 ± 0.2	45.5 ± 15.2	8.4 ± 1.9	63.7 ± 15.4	4.2 ± 1.2	52.4 ± 25.7
	Lung	0.3 ± 0.07	17.0 ± 4.1	0.8 ± 0.2	28.3 ± 22.7	0.5 ± 0.08	27.3 ± 12.2
	Kidney	0.6 ± 0.06	31.7 ± 8.3	1.6 ± 0.5	46.3 ± 14.2	1.0 ± 0.2	40.2 ± 16.7
	Pellet	77.8 ± 4.1	–	45.1 ± 14.0	–	62.2 ± 14.9	–
DKP	$^{14}CO_2$	0.2 ± 0.07	–	0.6 ± 0.4	–	0.9 ± 0.9	–
	Urine	10.8 ± 8.4	–	19.9 ± 4.8	–	29.0 ± 7.3	–
	Feces[c]	1.4 ± 1.6	–	16.9 ± 9.6	–	13.3 ± 8.3	–
	Bladder	0.4 ± 0.06	124 ± 28	0.5 ± 0.2	335 ± 153	1.8 ± 0.5	635 ± 152
	Carcass	19.7 ± 7.7	8.5 ± 3.1	10.4 ± 1.6	5.5 ± 0.8	6.9 ± 6.7	3.5 ± 25
	Liver	0.2 ± 0.06	1.6 ± 0.4	0.3 ± 0.07	2.8 ± 1.3	0.2 ± 0.2	2.7 ± 1.1
	Lung	0.04 ± 0.01	1.3 ± 0.4	0.08 ± 0.03	3.5 ± 2.0	0.04 ± 0.03	1.1 ± 0.7
	Kidney	0.3 ± 0.09	7.6 ± 3.3	0.9 ± 0.4	27.5 ± 14.5	0.1 ± 0.1	3.2 ± 2.2
	Pellet	67.8 ± 3.8	–	50.5 ± 8.8	–	49.3 ± 11.1	–

[a]Percentage ± SD of four mice at each time period.
[b]Tissue specific activity (disintegrations per milligram wet weight) ± SD.
[c]Including intestinal contents.

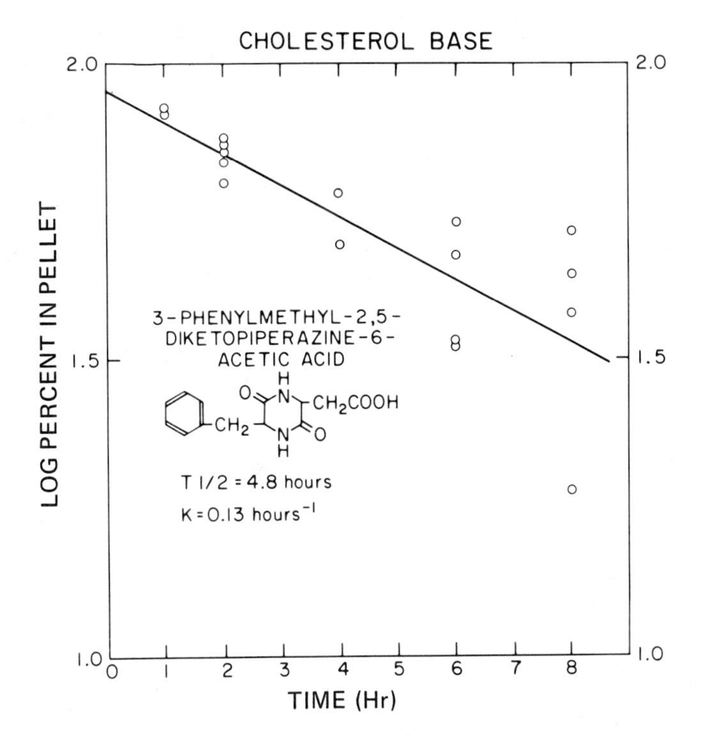

Figure 2 Elution curve of DKP from cholesterol. Cholesterol pellets (20-22 mg) containing 20% [^{14}C] DKP were surgically placed into urinary bladder lumina of female Swiss mice. At 1, 2, 4, 6, and 8 hr following surgery, mice were killed, and removed pellets were analyzed for radioactivity content as described in Materials and Methods. Each open circle indicates one mouse at the designated time. The regression line, $t_{1/2}$, and K were computed as described.

exposure to either [^{14}C] aspartame or [^{14}C] DKP is presented in Figure 3. Bladder incorporation of radioactivity was significantly greater ($p < 0.05$) than for other tissues studied at this time. Distribution of radioactivity into bladder was also greater than other tissues at 9.6 ($2t_{1/2}$) and 14.4 hr ($3t_{1/2}$) (Table 1). With [^{14}C] aspartame pellets, metabolism of aspartame occurred following bladder absorption as demonstrated by the significant amount of $^{14}CO_2$ exhaled (25).

The percentage of total radioactivity recovered and the tissue specific activity measured at 4.8 ($t_{1/2}$) hr following intraurethral installation of aqueous solutions containing [^{14}C] aspartame or [^{14}C] DKP into the isolated urinary bladder lumen of mice are displayed in Table 2. Within the brief observation period of 4.8 hr, about 30 and 25%, respectively, of ^{14}C label from aspartame and DKP had passed into or through the bladder and was distributed in other tissues or expired as CO_2.

Table 2 Percentage of Total Radioactivity (dpm) Recovered and Tissue Specific Activity 4.8 hr ($t_{1/2}$) After Intraurethral Instillation of Aqueous Solution Containing [^{14}C]Aspartame or DKP into the Urinary Bladder Lumen of Mice

	Aspartame 4.8 hr ($t_{1/2}$) after instillation		DKP 4.8 hr ($t_{1/2}$) after instillation	
Sample	Percent ± SD[a]	dpm/mg ± SD[b]	Percent ± SD[a]	dpm/mg ± SD[b]
$^{14}CO_2$	2.9 ± 1.5	–	0.03 ± 0.01	–
Bladder contents	70.6 ± 14.6	–	75.9 ± 11.9	–
Intestinal contents and feces	0.01 ± 0.01	–	0.07 ± 0.11	–
Bladder	6.3 ± 1.2	1085 ± 175	0.3 ± 0.08	56.6 ± 32.6
Carcass	15.9 ± 11.0	5.5 ± 3.8	13.3 ± 6.6	14.2 ± 5.1
Liver	3.1 ± 1.7	19.9 ± 9.6	2.7 ± 1.9	24.4 ± 14.7
Lung	0.3 ± 0.2	10.2 ± 3.9	0.6 ± 0.4	29.3 ± 13.8
Kidney	1.0 ± 0.5	15.9 ± 6.3	7.2 ± 3.5	151.2 ± 64.4

[a] Percentage ± SD of four mice.
[b] Tissue specific activity (disintegrations per milligram wet weight) ± SD.

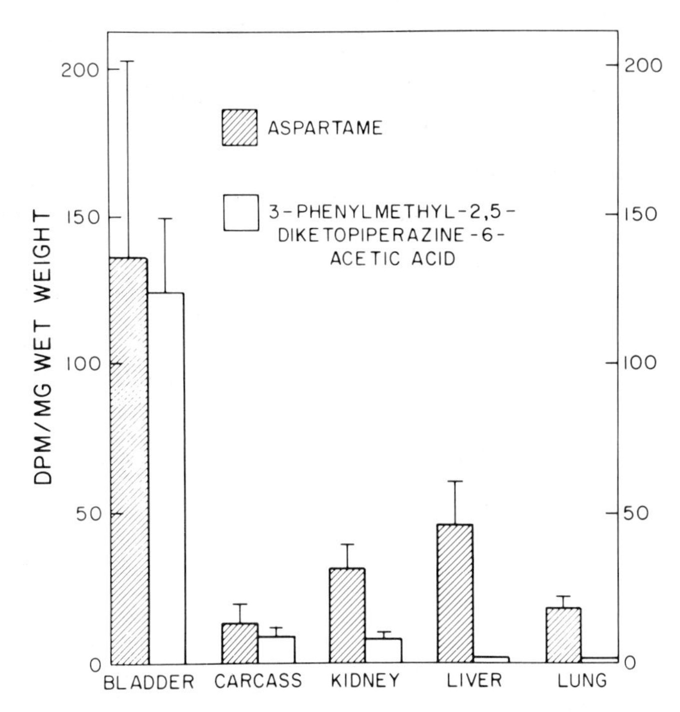

Figure 3 Comparison of the mean radioactivity (dpm/mg ± SD) of washed tissues 4.8 hr ($t_{1/2}$) following surgical introduction of cholesterol pellets containing either [^{14}C]aspartame or [^{14}C]DKP into the urinary bladder lumina of female Swiss mice. At 4.8 hr, four mice exposed to each chemical were killed by cervical fracture, and tissues removed and analyzed for radioactivity as described in Materials and Methods.

The intraurethral absorption of aspartame was compared with the intraurethral and intraperitoneal absorptions of [^{14}C]phenylalanine, a metabolic product of aspartame (25), and [^{14}C]alanine. Data from these experiments are presented in Figures 4-7. $^{14}CO_2$ was detected in the breath of mice 1 hr after the intraurethral instillation of [^{14}C]phenylalanine (Fig. 4) or [^{14}C]alanine (Fig. 6). At the study termination after 24 hr, most of the ^{14}C had disappeared from the bladder contents and was present in $^{14}CO_2$ or animal carcass and feces. Carbon-14 distributions following intraperitoneal administration of [^{14}C]phenylalanine (Fig. 5) or [^{14}C]alanine (Fig. 7) were similar to those following intraurethral administration.

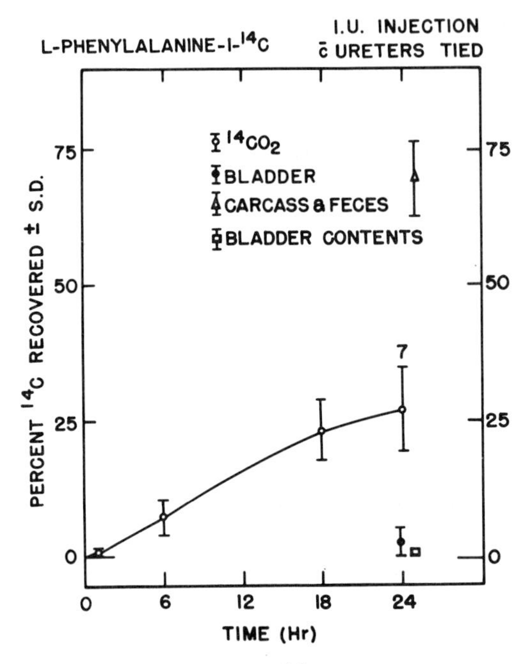

Figure 4 Distribution of ^{14}C in seven female Swiss mice injected intraurethrally with an aqueous solution of [^{14}C] phenylalanine. Cumulative $^{14}CO_2$ was measured at 1, 6, 18, and 24 hr following intraurethral injection. At 24 hr, mice were killed by cervical fracture, and tissues removed and analyzed for radioactivity as described in Materials and Methods. Data shown as mean ± SD.

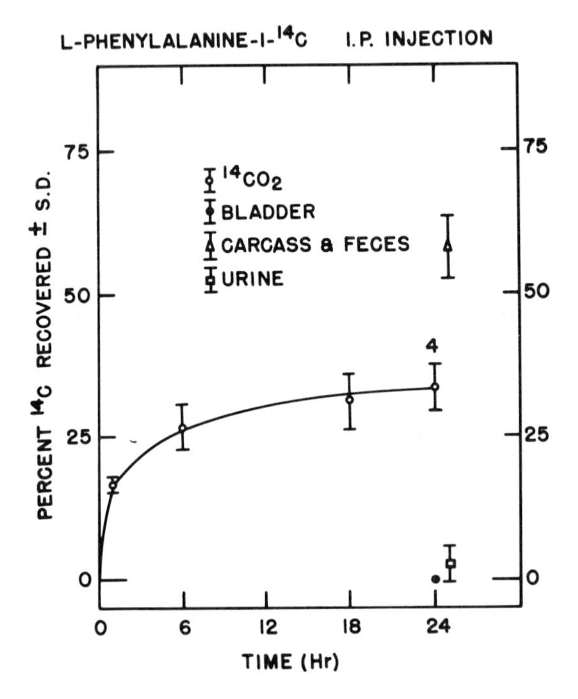

Figure 5

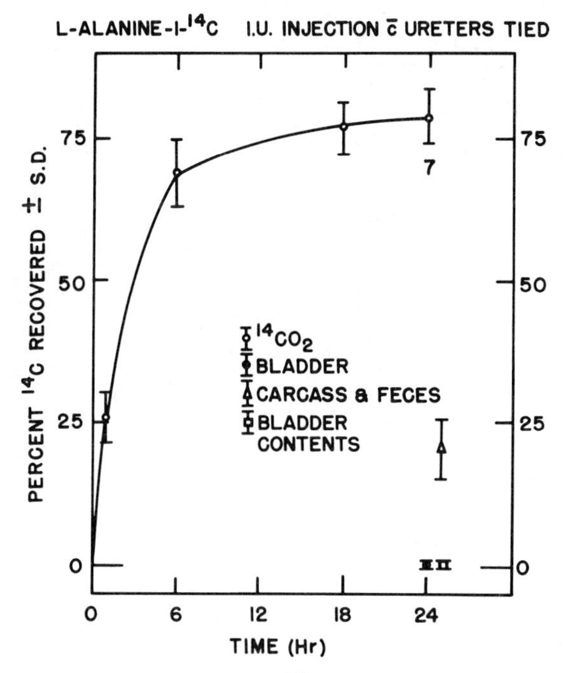

Figure 6 Distribution of ^{14}C in seven female Swiss mice injected intraurethrally with an aqueous solution of [^{14}C]alanine. Cumulative $^{14}CO_2$ was measured at 1, 6, 18, and 24 hr following intraurethral injection. At 24 hr, mice were killed by cervical fracture, and tissues removed and analyzed for radioactivity as described in Materials and Methods. Data shown as mean ± SD.

A 26-Week Urinary Bladder Tumorigenicity Study in the Mouse of Aspartame and DKP Administered by the Intravesical Pellet Implantation Technique

The number and percentage of surviving mice, and their mean weights, following surgical placement of pellets into urinary bladder lumina are presented in Table 3 for groups 1-4. Survival was comparable for mice of groups 1-3, but mice of group 4 exposed to XAE-containing pellets exhibited a consistently significant mortality compared to mice of group 1 ($p < 0.01$) at all time periods after the fourth week. Weights of mice in groups 1-4 were comparable at each time period.

Interval survival and incidences of bladder metaplastic and neoplastic lesions present in mice surviving more than 176 days postsurgery are shown in Table 4.

Figure 5 Distribution of ^{14}C in four female Swiss mice injected intraperitoneally with an aqueous solution of [^{14}C]phenylalanine. Cumulative $^{14}CO_2$ was measured at 1, 6, 18, and 24 hr following intraperitoneal injection. At 24 hr, mice were killed by cervical fracture, and tissues removed and analyzed for radioactivity as described in Materials and Methods. Data shown as mean ± SD.

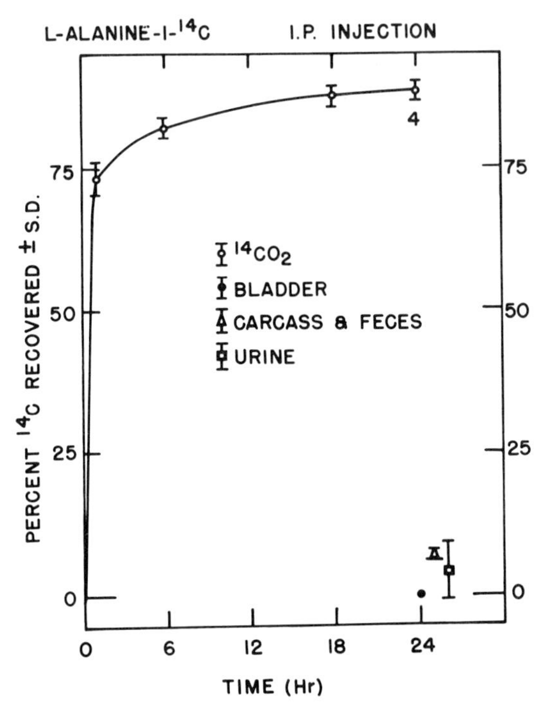

Figure 7 Distribution of ^{14}C in four female Swiss mice injected intraperitoneally with an aqueous solution of [^{14}C]alanine. Cumulative $^{14}CO_2$ was measured at 1, 6, 18, and 24 hr following intraperitoneal injection. At 24 hr, mice were killed by cervical fracture, and tissues removed and analyzed for radioactivity as described in Materials and Methods. Data shown as mean ± SD.

Bladder lesions possessing cellular characteristics compatible with epithelially derived neoplasia and with extension into the bladder submucosa were classified as stage I; those that additionally were seen to invade muscle as stage II; and those that additionally presented evidence of serosal spread, local pelvic metastases, regional, nodal, or distant metastases as stage III. No urinary bladder neoplasms were found in mice subject to study by the intravesical pellet implantation technique dying or killed prior to 175 days following surgery in these or previous studies (4,6,15,20,22-24). Thus only those mice surviving a minimum of 176 days were included for incidence tabulations and statistical analyses (Table 4). The incidences of bladder neoplasia varied from 7.2 (group 1) to 17.2% (group 3), but those of groups 2-4 were not significantly different from that of group 1.

A summary of tumors observed in other organs is presented in Table 5. No statistically significant augmentation of incidence of a single tumor type or of

Table 3 Summary of Mean Weights of Mice Subjected to a 26-Week Urinary Bladder Tumorigenicity Study of Aspartame or DKP by the Intravesical Pellet Implantation Technique

	Weeks					
Group	1	4	8	12	20	26
1. Cholesterol (negative control)						
Number examined	97	85	80	78	74	71
Percent surviving	97	85	80	78	74	71
Mean weights (g)	30.8	37.4	38.9	42.2	43.1	44.1
2. Aspartame						
Number examined	94	75	72	71	68	68
Percent surviving	94	75	72	71	68	68
Mean weights (g)	32.2	37.4	40.2	42.1	43.1	44.2
3. DKP						
Number examined	97	75	72	70	68	66
Percent surviving	92.4	71.4[a]	68.6	66.7	64.8	62.9
Mean weights (g)	32.2	38.6	41.6	42.9	44.6	45.3
4. XAE (positive control)						
Number examined	87	68	59	55	50	46
Percent surviving	87	68[a]	59[a]	55[a]	50[a]	46[a]
Mean weights (g)	34.5	40.2	42.0	44.9	45.2	46.1

[a] $p < 0.01$, calculated by the chi-squared test of the treated group versus the negative control group for the same week.

the aggregated tumor incidence was detected for any chemical treatment group (groups 2-4) compared with the negative control (group 1).

A 56-Week Urinary Bladder Tumorigenicity Study in the Mouse of Aspartame and DKP Administered by the Intravesical Pellet Implantation Technique

The number of surviving mice and their mean ± SEM weights following surgical placement of pellets into urinary bladder lumina are presented in Table 6 for groups 1-4 and their paired duplicate subgroups A and B. Survival of mice at each time interval was comparable except for group 2B (aspartame) beyond the 32nd week, group 3A (DKP) at the 44th week, and group 4B at the 44th and 56th weeks. Weights of mice in groups 1-4A and 4B were comparable at each time period.

Table 4 Summary of Survival and Incidence of Bladder Neoplasia Observed in a 26-Week Urinary Bladder Tumorigenicity Study of Aspartame or DKP in the Mouse by the Intravesical Pellet Implantation Technique

	Survival (days)						Incidence of bladder changes					
								Neoplasia stage (all 176-200 days)				
Group	0-50	51-100	101-150	151-175	176-200	Unknown	Metaplasia	I	II	III	Total	p value[a]
1. Cholesterol (negative control)	6	3	4	1	69		1	3	2	0	5/69	–
NA	3	0	1	0	2	11						
2. Aspartame	14	1	3	0	66		4	2	3	0	5/66	>0.9
NA	5	0	0	0	2	9						
3. DKP	19	2	2	2	64		2	2	9	0	11/64	0.07
NA	1	0	0	0	2	13						
4. XAE (positive control)	21	2	2	0	44		1	2	5	0	7/44	0.12
NA	8	1	1	1	2	18						

[a]p value, calculated by the exact method for the 2 X 2 table, compares the total bladder neoplasia incidence of groups 2, 3, or 4 versus group 1.
[b]NA, animal cannibalized, too autolyzed, or lost, or bladder tissue not available for microscopic inspection.

Table 5 Summary of Tumors Observed in Organs Other Than Urinary Bladder in a 26-Week Urinary Bladder Tumorigenicity Study of Aspartame or DKP in the Mouse by the Intravesical Pellet Implantation Technique

Group	Number of mice	Lesions in evaluated mice surviving over 176 days			
		Leukemia	Mammary adenofibroma	Pulmonary adenoma	Stomach papilloma
1. Cholesterol (negative control)	69	4	2	1	
2. Aspartame	66	1		1	
3. DKP	64	3	1		1
4. XAE (positive control)	44	2		2	

Table 6A Summary of Survival and Mean Weights of Mice Subjected to a 56-Week Urinary Bladder Tumorigenicity Study of Aspartame or DKP by the Intravesical Pellet Implantation Technique

	Cholesterol (negative control)				Aspartame			
	Group A		Group B		Group A		Group B	
Time (weeks)	Number examined	Mean weight ± SEM (g)	Number examined	Mean weight ± SEM (g)	Number examined	Mean weight ± SEM (g)	Number examined	Mean weight ± SEM (g)
0	100	31.9 ± 3.1	100	30.9 ± 3.4	100	33.2 ± 3.4	100	32.4 ± 3.3
4	90	36.4 ± 3.6	91	36.1 ± 3.7	88	36.2 ± 4.1	87	36.5 ± 4.3
10	88	39.3 ± 3.9	89	38.9 ± 4.1	84	39.9 ± 4.5	82	40.1 ± 4.6
20	82	42.5 ± 4.4	82	42.3 ± 4.5	78	43.1 ± 4.2	76	43.8 ± 4.5
32	73	45.2 ± 5.6	73	45.1 ± 5.7	61	45.7 ± 5.8	56[a]	46.1 ± 6.3
44	65	47.1 ± 6.3	65	46.9 ± 6.2	51	46.8 ± 6.7[b]	42[a]	47.1 ± 7.3
56	52	47.6 ± 7.6	48	47.7 ± 7.8	39	49.5 ± 7.6	33[a]	49.9 ± 7.9

[a] $p < 0.01$, calculated by the chi-squared test of the treated group versus the appropriate negative control group for the same week.
[b] $0.05 > p > 0.01$, calculated by the chi-squared test of the treated group versus the appropriate negative control group for the same week.

Table 6B Summary of Survival and Mean Weights of Mice Subjected to a 56-Week Urinary Bladder Tumorigenicity Study of Aspartame or DKP by the Intravesical Pellet Implantation Technique

	DKP				XAE (Positive Control)			
	Group A		Group B		Group A		Group B	
Time (weeks)	Number examined	Mean weight ± SEM (g)	Number examined	Mean weight ± SEM (g)	Number examined	Mean weight ± SEM (g)	Number examined	Mean weight ± SEM (g)
0	100	33.5 ± 3.5	100	33.1 ± 3.2	100	32.6 ± 3.7	100	32.9 ± 3.5
4	80	37.5 ± 3.7	80	37.8 ± 3.8	84	37.9 ± 4.1	83	38.2 ± 4.3
10	73	42.3 ± 3.9	79	42.1 ± 4.1	78	41.9 ± 4.3	80	42.3 ± 4.4
20	67	43.9 ± 5.1	76	44.2 ± 5.3	72	44.9 ± 4.8	70	45.2 ± 4.8
32	61	45.4 ± 5.3	68	45.7 ± 5.2	61	46.4 ± 5.4	60	46.9 ± 5.5
44	51[b]	46.7 ± 5.9	63	46.9 ± 5.6	53	46.9 ± 5.9	50[b]	47.2 ± 6.0
56	39	49.6 ± 7.3	47	49.9 ± 7.5	36	49.5 ± 7.1[b]	27[a]	49.9 ± 7.8

[a] $p < 0.01$, calculated by the chi-squared test of the treated group versus the appropriate negative control group for the same week.
[b] $0.05 > p > 0.01$, calculated by the chi-squared test of the treated group versus the appropriate negative control group for the same week.

Table 7 Summary of Survival and Comparison of Incidence of Neoplasia Observed in a 56-Week Urinary Bladder Tumorigenicity Study of Aspartame or DKP in the Mouse by the Intravesical Pellet Implantation Technique

		Survival (days)			Incidence of bladder changes							
						Neoplasia stage (all 176 to end days)						
Treatment	Group	0-175	176 to termination	Mean[a]	Metaplasia	I	II	III	Total	Percent	Chi squared	p value[b]
1. Cholesterol	A											
(negative control)	Assessed	10	77	368	4	3	4	0	7/77	9.1	–	–
	NA[c]	9	4									
	B											
	Assessed	10	78	362	4	8	2	0	10/78	12.8	0.556	0.5
	NA	9	3									
	Total											
	Assessed	20	155	365	8	11	6	0	17/155	10.6	–	–
	NA	18	7									
2. Aspartame	A											
	Assessed	19	65	358	3	3	5	0	8/65	12.3	–	–
	NA	10	6									
	B											
	Assessed	25	58	348	3	3	2	0	5/58	8.5	0.439	0.5
	NA	9	8									
	Total											
	Assessed	44	123	353	6	6	7	0	13/123	10.6	–	–
	NA	19	14									

3. DKP	A											
	Assessed	17	57	372	3	6	3	0	9/57	15.8	–	–
	NA	17	9									
	B											
	Assessed	7	68	376	5	4	4	0	8/68	11.8	0.429	0.5
	NA	21	4									
	Total											
	Assessed	24	125	374	8	10	7	0	17/125	13.6	–	–
	NA	38	13									
4. XAE	A											
(positive control)	Assessed	18	59	368	17	12	11	0	23/59	39.0	–	–
	NA	14	9									
	B											
	Assessed	26	52	366	15	6	10	1	17/52	32.7	0.471	0.5
	NA	10	12									
	Total											
	Assessed	44	111	367	32	18	21	1	40/111	36.0	–	–
	NA	24	21									

[a]Mean days of survival of assessed animals surviving over 176 days.
[b]p value, calculated by the chi-squared test, compares the total bladder neoplasia incidence of group A with that of group B.
[c]NA, animal cannibalized, too autolyzed, or lost, or bladder tissue not available for microscopic inspection.

Table 8 Summary of Survival and Incidence of Neoplasia Observed in a 56-Week Urinary Bladder Tumorigenicity Study of Aspartame or DKP in the Mouse by the Intravesical Pellet Implantation Technique

		Survival (days)			Incidence of bladder changes							
						Neoplasia stage (all 176 to end days)						
Group	Treatment	0-175	176 To termination	Mean	Metaplasia	I	II	III	Total	Percent	Chi squared	p Value[a]
A	1. Cholesterol (negative control)	10	77	368	4	3	4	0	7/77	9.1	–	–
	NA[b]	9	4									
	2. Aspartame	19	65	358	3	3	5	0	8/65	12.3	0.388	0.6
	NA	10	6									
	3. DKP	17	57	372	3	6	3	0	9/57	15.8	1.4	0.2
	NA	17	9									
	4. XAE (positive control)	18	59	368	17	12	11	0	23/59	39.0	17.8	<0.001
	NA	14	9									
B	1. Cholesterol (negative control)	10	78	362	4	8	2	0	10/78	12.8	–	–
	NA	9	3									
	2. Aspartame	25	58	348	3	3	2	0	5/58	8.5	0.598	0.5
	NA	9	8									
	3. DKP	7	68	376	5	4	4	0	8/68	11.8	0.0348	0.9
	NA	21	4									
	4. XAE (positive control)	26	52	366	15	6	10	1	17/52	32.7	7.45	$0.01 > p > 0.001$
	NA	10	12									

[a]p value, calculated by the exact method for the 2 X 2 table, compares the total bladder neoplasia incidence of treatments 2, 3, or 4 versus treatment 1 within each group.

Table 9 Summary of Survival and Incidence of Neoplasia Observed in a 56-Week Urinary Bladder Tumorigenicity Study of Aspartame or DKP in the Mouse by the Intravesical Pellet Implantation Technique (Pooled Groups A and B)

	Survival (days)			Incidence of bladder changes							
					Neoplasia stage (all 176 to end days)						
Treatment	0-175	176 to termination	Mean	Metaplasia	I	II	III	Total	Percent	Chi squared	p value[a]
1. Cholesterol (negative control)	20	155	365	8	11	6	0	17/155	10.6	–	–
NA[b]	18	7									
2. Aspartame	44	123	353	6	6	7	0	13/123	10.6	0.0113	0.9
NA	19	14									
3. DKP	24	125	374	8	10	7	0	17/125	13.6	0.453	0.5
NA	38	13									
4. XAE (positive control)	44	111	367	32	18	21	1	40/111	36.0	24.8	<0.001
NA	24	21									

[a]p value, calculated by the exact method for the 2 X 2 table, compares the total bladder neoplasia incidence of treatments 2, 3, or 4 with that of treatment 1.

[b]NA, animal cannibalized, too autolyzed, or lost, or bladder tissue not available for microscopic inspection.

Table 10 Summary of Tumors Observed in Organs Other Than Urinary Bladder in a 56-Week Urinary Bladder Tumorigenicity Study of Aspartame or DKP in the Mouse by the Intravesical Pellet Implantation Technique

			Lesions in evaluated mice surviving over 176 days							
Treatment	Groups	Number of Mice	Mammary fibroadenoma	Leukemia	Pulmonary adenoma	Stomach papilloma	Intestinal fibroma	Ovarian cyst	Liver cholangioma	Other
1. Cholesterol	A	77	4	2	4	1				
(negative control)	B	78	2	2	5		1			
2. Aspartame	A	65	4	2	1					Uterus fibroma Colon adenoma
	B	58	1	1	1				1	Submaxillary gland adenoma
3. DKP	A	57	2	2	1			2		
	B	68	2		3			1	3	Submaxillary gland adenoma (2)
4. XAE	A	59	10[a]	3	4					
(positive control)	B	52	8[b]	2	3		2			

[a] $0.05 > p > 0.01$, calculated by the chi-squared test of the treated group versus the appropriate negative control group.
[b] $p < 0.01$, calculated by the chi-squared test of the treated group versus the appropriate negative control group.

Survival of mice from 0 to 175 days after surgery and for 176 days to study termination; mean survival; and incidences of bladder metaplastic and neoplastic lesions in mice surviving more than 176 days are shown for treatment groups 1-4, for duplicated subgroups A and B of each treatment group, and for the composite total in Table 7. The scoring criteria used for bladder neoplastic lesions were described above. The incidences of bladder neoplasia for subgroups A and B of treatment groups 1-4 were comparable and not statistically different. These data are again presented in Table 8, where analytical emphasis has been placed on the incidences of bladder neoplastic lesions present in duplicated subgroups A and B for treatment groups 1-4. Statistical comparisons of bladder tumor incidences within subgroups A and B showed that mice exposed to aspartame (group 2) or DKP (group 3) did not exhibit significantly higher incidences of bladder neoplasia than did the negative control (group 1) mice exposed only to cholesterol pellets. Conversely, the incidences of vesical tumors associated with treatment by the positive control, XAE (group 4), were highly statistically significantly greater than those for the negative control (group 1) mice. Summaries for the composite total of survival, incidences of bladder lesions in combined subgroups A and B for treatment groups 1-4, and statistical analyses of comparative bladder neoplastic lesions are presented in Table 9. In these analyses, mice exposed to aspartame (group 2) or DKP (group 3) did not reveal statistically significantly higher bladder neoplasia incidences compared to group 1, while mice exposed to XAE (group 4) did demonstrate a highly statistically significant incidence of bladder tumors.

A summary of tumors detected in other organs of mice in treatment groups 1-4 and for subgroups A and B is presented in Table 10. The incidences of tumors of other organs in treatment groups 1-3 ranged from 8.6 (group 2B, aspartame) to 16.2% (group 3B, DKP) and were not statistically significantly different from those of negative control group 1 ($p < 0.5$). Mice of treatment group 4 (XAE, positive control) in subgroups A and B exhibited a 28.8% incidence of tumors at other sites that was significantly greater ($p < 0.05$) than that of negative control (group 1) mice. This difference was entirely related to a statistically higher incidence of mammary fibroadenoma present in both subgroups A and B of treatment group 4.

DISCUSSION

Aspartame and DKP did not demonstrate urinary bladder carcinogenic effects when tested by the mouse bladder pellet implantation technique in a 26-week study (Table 4) or in a duplicated 56-week study (Tables 7-9). It was shown that aspartame and DKP disappeared from cholesterol pellets implanted intravesically with a $t_{1/2}$ of 4.8 hr (Figs. 1 and 2) and that they were incorporated significantly into urinary bladder and other murine tissues (Table 1, Fig. 3).

Aqueous solutions of [^{14}C]aspartame and [^{14}C]DKP placed intraurethrally into the bladder lumina of mice readily passed into and through the bladder and were detected in other tissues and excreta (Table 2). Thus both aspartame and DKP did disappear from the cholesterol pellets in vivo, both reached the bladder epithelium, and both had the opportunity to be metabolically converted into putative active forms by urothelium, satisfying prerequisite conditions for a valid negative test of bladder carcinogenicity by this bioassay (9). In this test, cholesterol pellets were always found intact in the bladder lumina at necropsy. As these pellets themselves provide concurrent urothelial stimulatory activity (9) and additionally evoke bladder hypertrophy (13), this bioassay is capable of detecting weak bladder carcinogenic activity of a test compound (9).

In contrast, XAE, previously demonstrated as a mouse bladder carcinogen by this method (20), was selected as a positive control chemical for this study because of its relatively rapid disappearance in vivo from cholesterol pellets placed intravesically (15,18). The compound was shown (13) to disappear from cholesterol pellets in vivo, to reach bladder epithelium, and to pass through the bladder under conditions similar to those used in the present study. In this investigation, XAE did not significantly increase the incidence of bladder neoplasia in the 26-week urinary bladder tumorigenicity study (Table 4), but it did significantly augment bladder tumor incidences in duplicate groups of mice in the 56-week bioassay (Tables 7-9). In addition, XAE-treated mice had significantly greater incidences of mammary fibroadenoma (Table 10). In a prior study (26), XAE administered to female Swiss mice by repeated subcutaneous injections significantly increased the incidence of malignant lymphoreticular tumors compared to that of unmedicated control mice. The mechanism(s) by which XAE participates in the carcinogenic process is not known.

ACKNOWLEDGMENTS

I thank Doctors G. M. Lower, Jr., Y. Kotake, and W. A. Croft for assistance with aspects of this study; Doctors A. M. Pamukcu and E. Ertürk for a review of many of the histopathological specimens and providing valuable advice; R. Clingan, D. Scholl, A. Healy, D. B. Headley, E. Grunden, J. Ross, and D. Moseley for providing assistance with mouse surgery, animal care, and histotechnology; and S. Pertzborn for editorial assistance with manuscript preparation. This work was supported in part by Grants CA 10017, CA 14520, and CA 14524 from the National Cancer Institute, USPHS, the latter grant through The National Bladder Cancer Project, and by a grant from G. D. Searle Co., Chicago, Illinois. A portion of this work was conducted as a Career Development Awardee of the National Cancer Institute, USPHS (1-K4-CA-08245).

REFERENCES

1. International Agency for Research on Cancer (1976). Some carbamates, thiocarbamates and carbazides. In *IARC Monographs on the Evaluation of Carcinogenic Risk of Chemicals to Man*, Vol. 12, International Agency for Research on Cancer, Lyon, pp. 97-105.
2. International Agency for Research on Cancer (1980). Some non-nutritive sweetening agents. In *IARC Monographs on the Evaluation of the Carcinogenic Risk of Chemicals to Humans*, Vol. 22, International Agency for Research on Cancer, Lyon, pp. 1-185.
3. Griepentrog, F. (1959). Tumoren der Harnwege und Harnsteine in chronischen Versuchen mit dem Süssstoff p-Phenetylcarbamid. *Arzneim. Forsch.* 9, 123-125.
4. Bryan, G. T., and Ertürk, E. (1970). Production of mouse urinary bladder carcinomas by sodium cyclamate. *Science* 167, 996-998.
5. Allen, M. J., Boyland, E., Dukes, C. E., Horning, E. S., and Watson, J. G. (1957). Cancer of the urinary bladder induced in mice with metabolites of aromatic amines and tryptophan. *Br. J. Cancer* 11, 212-228.
6. Bryan, G. T., Ertürk, E., and Yoshida, O. (1970). Production of urinary bladder carcinomas in mice by sodium saccharin. *Science* 168, 1238-1240.
7. Boyland, E., Busby, E. R., Dukes, C. E., Grover, P. L., and Manson, D. (1964). Further experiments on implantation of materials into the urinary bladder of mice. *Br. J. Cancer* 18, 575-581.
8. Clayson, D. B., Pringle, J. A. S., Bonser, G. M., and Wood, M. (1968). The technique of bladder implantation: Further results and an assessment. *Br. J. Cancer* 22, 825-832.
9. Bryan, G. T., and Yoshida, O. (1971). Artificial sweeteners as urinary bladder carcinogens. *Arch. Environ. Health* 23, 6-12.
10. Price, J. M., and Dodge, L. W. (1956). Occurrence of the 8-methyl ether of xanthurenic acid in normal human urine. *J. Biol. Chem.* 223, 699-704.
11. Bryan, G. T., Morris, C. R., and Brown, R. R. (1965). Absorption of ^{14}C-labeled 3-hydroxy-L-kynurenine and 3-hydroxyanthranilic acid from the mouse urinary bladder under carcinogenic conditions. *Cancer Res.* 25, 1432-1437.
12. Morris, C. R., and Bryan, G. T. (1966). Absorption of C^{14}-labeled tryptophan, its metabolites, glycine, and glucose by the urinary bladder of the mouse. *Invest. Urol.* 3, 577-585.
13. Bryan, G. T., and Morris, C. R. (1968). Distribution of ^{14}C-labeled 8-methyl ether of xanthurenic acid in the mouse following bladder luminal, subcutaneous, and intraperitoneal administration under carcinogenic conditions. *Cancer Res.* 28, 186-191.
14. Lower, G. M., Jr., and Bryan, G. T. (1969). The metabolism of the 8-methyl ether of xanthurenic acid in the mouse. *Cancer Res.* 29, 1013-1018.
15. Bryan, G. T., Brown, R. R., Morris, C. R., and Price, J. M. (1964). In vivo elution of tryptophan metabolites and other aromatic nitrogen compounds

from cholesterol pellets implanted into mouse bladders. *Cancer Res.* 24, 586-595.

16. Jull, J. W. (1951). The induction of tumours of the bladder epithelium in mice by the direct application of a carcinogen. *Br. J. Cancer* 5, 328-330.
17. Steel, R. G. D., and Torrie, J. H. (1960). *Principles and Procedures of Statistics, with Special Reference to the Biological Sciences,* McGraw-Hill, New York, pp. 161-182.
18. Bryan, G. T. (1969). Pellet implantation studies of carcinogenic compounds. *J. Nat. Cancer Inst.* 43, 255-261.
19. Cohen, S. M., Alter, A., and Bryan, G. T. (1973). Distribution of radioactivity and metabolism of formic acid 2-[4-(5-nitro-2-furyl)-2-^{14}C-2-thiazolyl] hydrazide following oral administration to rats and mice. *Cancer Res.* 33, 2802-2809.
20. Bryan, G. T., Brown, R. R., and Price, J. M. (1964). Mouse bladder carcinogenicity of certain tryptophan metabolites and other aromatic nitrogen compounds suspended in cholesterol. *Cancer Res.* 24, 596-602.
21. Fisher, R. A. (1958). *Statistical Methods for Research Workers*, 13th ed., Hafner, New York, pp. 96-97.
22. Bryan, G. T., and Springberg, P. D. (1966). Role of the vehicle in the genesis of bladder carcinomas in mice by the pellet implantation technic. *Cancer Res.* 26, 105-109.
23. Bryan, G. T. (1969). Role of tryptophan metabolites in urinary bladder cancer. *Am. Ind. Hygiene Assoc. J.* 30, 27-34.
24. Bryan, G. T. (1971). The role of urinary tryptophan metabolites in the etiology of bladder cancer. *Am. J. Clin. Nutr.* 24, 841-847.
25. Oppermann, J. A., Muldoon, E., and Ranney, R. E. (1973). Metabolism of aspartame in monkeys. *J. Nutr.* 103, 1454-1459.
26. Bryan, G. T. (1968). Neoplastic response of various tissues to the systemic administration of the 8-methyl ether of xanthurenic acid. *Cancer Res.* 28, 183-185.

17

Aspartate-Induced Neurotoxicity in Infant Mice

Arnold E. Applebaum, Tahia T. Daabees*, and Lewis D. Stegink
University of Iowa College of Medicine, Iowa City, Iowa

Michael W. Finkelstein
University of Iowa College of Dentistry, Iowa City, Iowa

INTRODUCTION

There is no doubt that administration of large amounts of the dicarboxylic amino acids glutamate and aspartate to neonatal rodents produces hypothalamic neuronal necrosis (1-9). Although the glutamate-induced lesion has been studied extensively, aspartate is less well studied.

Glutamate is not considered neurotoxic to the infant nonhuman primate; however, this subject is somewhat controversial. Olney and colleagues report neuronal necrosis in infant nonhuman primates given large doses of glutamate (10,11). Other research groups, however, are unable to produce a lesion in the infant nonhuman primate (12-22), even when plasma glutamate and aspartate concentrations are grossly elevated (17,20). These other groups generally had no difficulty in producing the glutamate-induced lesion in infant rodents. Studies on aspartate neurotoxicity in the nonhuman primate have not been reported.

The controversy over nonhuman primate data relative to the use of glutamate as a food additive led to studies to determine the threshold plasma concentration of glutamate and aspartate associated with neuronal necrosis in the infant mouse. These data have been valuable in evaluating glutamate safety in man. Although the neonatal mouse is highly sensitive to glutamate-induced neuronal necrosis, it tolerates large doses of glutamate and marked elevation of plasma glutamate and aspartate concentration without developing neuronal necrosis. Our data (23) and those of Takasaki and colleagues (24,25) indicate that the plasma glutamate plus

**Present affiliation*: University of Alexandria, Alexandria, Egypt.

aspartate concentration must reach 60-100 μmol/dl (5-10 times normal) in infant mice before neuronal necrosis is noted. When glutamate dissolved in water is given orally, the lowest dose producing neuronal necrosis varies from 500 mg/kg body weight (6,19,20) to 700 mg/kg body weight (24,25). In contrast, infant nonhuman primates do not develop neuronal necrosis even when plasma glutamate and aspartate concentrations are grossly elevated (17,21). Thus plasma glutamate and aspartate concentrations of the magnitude produced by nontoxic doses of glutamate in the sensitive infant mouse are probably not toxic to the infant nonhuman primate.

Studies to determine the threshold plasma concentration of aspartate and glutamate associated with aspartate-induced neuronal necrosis have only recently been carried out (26). These studies have become more important with the introduction of aspartame into the food supply. Aspartame (L-aspartyl-L-phenylalanine methyl ester) contains 40% aspartate by weight. It is hydrolyzed in the intestinal mucosa to its component amino acids, which in turn are handled in a manner similar to those arising from dietary protein. Early questions about aspartame safety arose in part because of concern about potential toxic effects of its aspartate content. Olney (27-29) and Reif-Lehrer (30) suggested that aspartame should not be added to the food supply because of interactions between the aspartate content of aspartame and the monosodium L-glutamate (MSG) that may be added to certain foods.

Although Olney and Ho (6) reported that large doses of aspartate were neurotoxic to infant mice, few attempts were made to determine the threshold dose of aspartate required to produce neurotoxicity. Similarly, few neuropathological studies have been carried out in animals administered both glutamate and aspartate. Olney and Ho (6) reported that the neurotoxic effects of glutamate and aspartate appeared to be additive in the infant mouse, but did not report plasma amino acid concentrations. Thus our objectives were to determine the lowest dose of aspartate that was neurotoxic for the neonatal mouse and to determine the plasma concentration of aspartate and glutamate associated with that dose.

One puzzling finding about glutamate-induced neurotoxicity in the infant mouse is glutamate's relative lack of neurotoxicity when administered with food as part of the diet (31,32). This protective effect probably reflects the carbohydrate content of the diet. In normal human subjects and young pigs the administration of MSG with carbohydrate produces much lower plasma glutamate levels than when an equivalent dose is ingested with water (33-36). Takasaki (37) demonstrated that mono- and disaccharides administered with glutamate decreased the extent of neuronal necrosis in infant mice when compared to the extent of neuronal necrosis observed in animals administered MSG alone. Takasaki and Yugari (38) demonstrated a protective effect of pharmacological doses of insulin when administered prior to MSG dosing. However, neither study reported plasma glutamate and aspartate concentrations. Thus our studies also tested the effects of carbohydrate and insulin on aspartate-induced neuronal necrosis and plasma glutamate and aspartate concentrations.

METHODS

Eight-day-old Swiss-Webster mice, maintained on a 12-hr light-dark cycle, remained with their dam until the morning of an experiment. One hour prior to oral dosing the pups were removed from their mother and weighed. Body temperature of the animals was maintained with a thermostatically controlled heating pad and lamp. Test and control solutions were administered by gavage using a flexible polyethylene tube attached to a microliter syringe.

Amino acids were dissolved in distilled water prior to administration. Aspartic acid, neutralized with solid sodium bicarbonate, was administered as a 10% (wt/vol) solution at a dose of 1 g/kg body weight. As a control, isotonic saline was given in a volume equivalent to that of a 10% aspartic acid solution. In some experiments both glutamate and aspartate were administered as a 10% solution with each amino acid present at 5%. Polycose (Ross Laboratories, Columbus, Ohio), a partially hydrolyzed starch preparation, was used to study the effect of orally administered carbohydrate on aspartate-induced neuronal necrosis. In these experiments both Polycose and aspartate were administered in a solution providing each substance at 1 g/kg body weight. The effect of prior administration of insulin was evaluated by giving 0.02 units of regular insulin subcutaneously 4 hr prior to oral dosing with sodium aspartate.

Neuropathological studies were performed on animals allowed to survive 5 hr post oral dosing. During the survival period animals remained separated from their dam but were kept warm. Animals were killed by intracardiac perfusion of a mixed aldehyde fixative. The brains were removed and blocked for embedding in JB-4 (Polysciences) plastic. Serial sections (2 μm) were made throughout the extent of the hypothalamic arcuate nucleus and stained for Nissl substance using either cresyl violet or toluidine blue. All sections were examined with the light microscope. When present, the extent of the aspartate-induced lesion was noted and the number of necrotic neurons per 2-μm section counted. The maximum number of necrotic neurons per section was used as the count for each animal. In some animals the arcuate nuclear region was embedded in Epon to allow both light- and electron-microscopic examination. A more detailed description of these methods has been published (26,39).

Separate groups of animals, treated in the same manner as those used in the neuropathological studies, were used for the determination of plasma amino acid levels. Blood was obtained by decapitation and processed for amino acid analysis following the method previously described by Finkelstein et al. (39).

RESULTS

Figure 1 shows sections through the arcuate nucleus of neonatal mice receiving either saline (Fig. 1a) or aspartate (Fig. 1b) at a dose of 1000 mg/kg body weight. The arcuate nucleus of the treated animal (Fig. 1b) has the characteristic "Swiss cheese" appearance associated with glutamate-induced damage. In addition, many neurons with pyknotic nuclei are seen throughout the field. At higher magnifica-

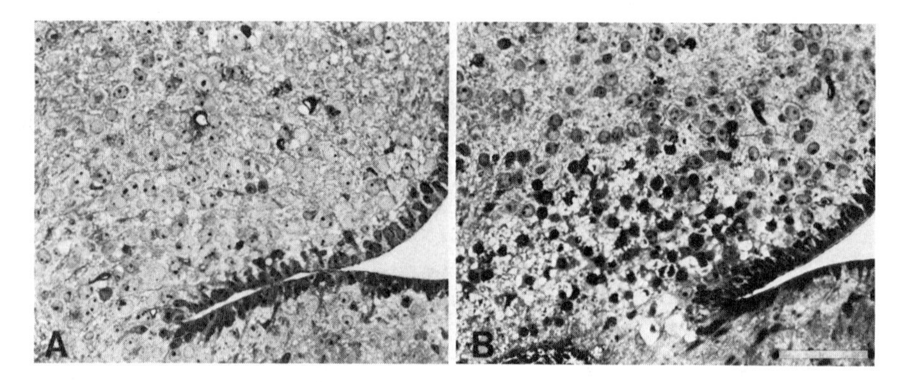

Figure 1 Arcuate nucleus of the hypothalamus 5 hr following oral administration of (A) saline or (B) 1000 mg/kg body weight aspartic acid. In (B) numerous neurons with pyknotic nuclei are easily identified (calibration bar = 40 μm).

tion, swelling of neuronal cytoplasm and organelles is evident. These changes are more obvious at the electron-microscopic level (Fig. 2).

A summary of the results of our neuropathological studies of neonatal mice receiving aspartate alone is shown in Table 1. No animal receiving saline or aspartate at dose levels of 250 or 500 mg/kg body weight developed neuronal necrosis in the arcuate nucleus. However, 3 of 10 mice given aspartate at a level of 650 mg/kg body weight developed neuronal necrosis. All mice given aspartate at 750 or 1000 mg/kg body weight sustained extensive damage to the arcuate nuclear region of the hypothalamus. As can be seen from Table 1, the mean (±SD) peak plasma aspartate plus glutamate level associated with the onset of neuronal necrosis is 227 ± 55.3 μmol/dl. In contrast, a mean (±SD) peak plasma aspartate plus glutamate level of 127 ± 76 μmol/dl does not result in neuronal necrosis. While we have not tested aspartate dosages between 500 and 650 mg/kg, the relatively small number of necrotic neurons encountered at the latter level suggest that the threshold is near 650 mg/kg. This assumption is strengthened by the fact that, once the neurotoxic threshold is reached, the extent of neuronal necrosis is proportional to the dose of aspartate administered.

Other investigators have reported that the neurotoxic effects of glutamate and aspartate are additive (6). To test this hypothesis, we compared the results from studies employing aspartate alone to results obtained when mice were given both aspartate and glutamate. The data from these experiments are summarized in Table 2. At a total dose of 1000 mg/kg body weight glutamate plus aspartate, infant mice had a mean (±SD) of 38.3 ± 30.4 necrotic neurons per section and a mean (±SD) peak plasma aspartate plus glutamate level of 377 ± 32.3 μmol/dl. In contrast, administration of 1000 mg/kg body weight of aspartate alone led to a mean (±SD) of 80.8 ± 17.3 necrotic neurons per section and a mean (±SD) peak plasma level of 430 ± 161 μmol/dl. Interestingly, the administration of aspartate

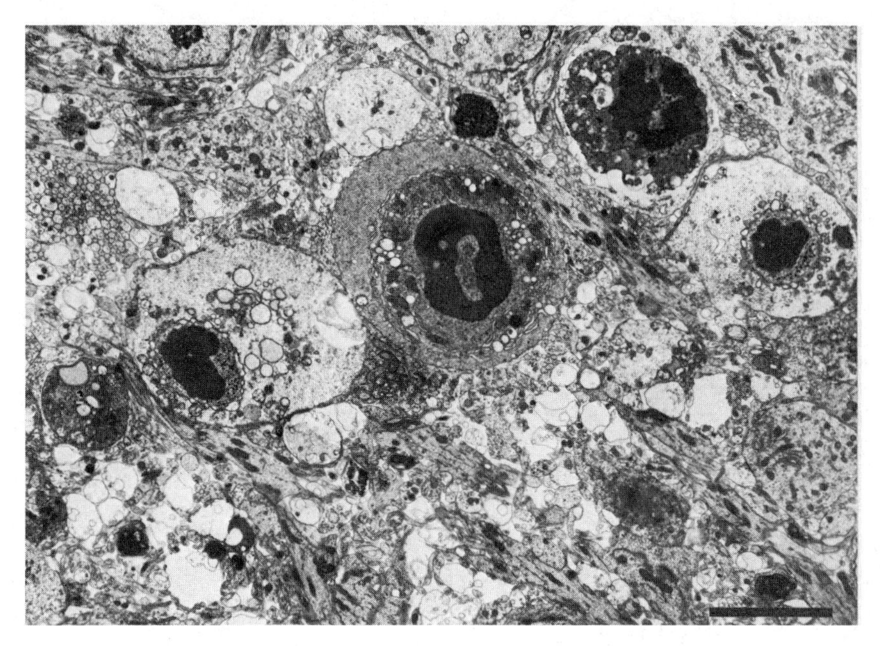

Figure 2 Electron micrograph of necrotic neurons within the arcuate nucleus of a neonatal mouse 5 hr following oral administration of 1000 mg/kg body weight aspartic acid (calibration bar = 5.0 μm).

Table 1 Plasma Glutamate Plus Aspartate Concentrations and the Degree of Neuronal Necrosis in Infant Mice Given Graded Doses of Aspartate

Aspartate dose (mg/kg body weight)	Peak plasma aspartate + glutamate level (μmol/dl)[a]	Affected animals[b]	Necrotic neurons per section[c]
0	14 ± 1.6	0/8	0
250	49 ± 31	0/12	0
500	127 ± 76	0/19	0
650	227 ± 55	3/10	7.3 ± 1.52
750	329 ± 119	12/12	45.9 ± 7.20
1000	430 ± 161	18/18	80.8 ± 17. 3

[a]Data expressed as the mean ± SD.
[b]Values reported as the number of affected animals per total number studied.
[c]Mean ± SD for affected animals only.

Table 2 Plasma Glutamate Plus Aspartate Concentrations and the Degree of Neuronal Necrosis in Infant Mice Given Aspartate Plus Glutamate

Aspartate dose (mg/kg body weight)	Glutamate dose (mg/kg body weight)	Peak plasma aspartate plus glutamate concentrations[a] (μmol/dl)	Affected animals[b]	Necrotic neurons per section[c]
500	0	127 ± 76	0/19	0
250	250	169 ± 23	2/12	12.0 ± 1.0
1000	0	430 ± 161	18/18	80.8 ± 17.3
500	500	377 ± 32	25/32	38.3 ± 30.4

[a]Data expressed as mean ± SD.
[b]Values reported as the number of affected animals per total number studied.
[c]Mean ± SD for affected animals only.

plus glutamate at a total dose of 500 mg/kg body weight resulted in neuronal necrosis in 2 of the 12 animals tested. No lesions were observed in animals given 500 mg/kg body weight aspartate alone (see Table 1).

Both insulin and carbohydrate have a mitigating effect upon the neurotoxic effects of glutamate (37,38). When partially hydrolyzed starch (Polycose) was administered with aspartate, the maximum number of necrotic neurons encountered within the arcuate nucleus was reduced (Table 3). A dose of 1000 mg/kg body weight aspartate plus 1 g/kg body weight partially hydrolyzed starch produced a mean (±SD) peak plasma aspartate plus glutamate concentration of 279 ± 76.6 μmol/dl. Despite the high plasma aspartate plus glutamate level, simul-

Table 3 Plasma Glutamate Plus Aspartate Concentrations and Degree of Neuronal Necrosis in Infant Mice Given 1000 mg/kg Aspartate ± Insulin or Polycose

Treatment	Peak plasma aspartate plus glutamate levels[a] (μmol/dl)	Affected animals[b]	Maximum number of necrotic neurons per section[c]
Aspartate, 1000 mg/kg	430 ± 161	18/18	80.8 ± 17.3
Aspartate, 1000 mg/kg, plus prior insulin injection	355 ± 76.6	13/14	28.4 ± 12.6
Aspartate, 1000 mg/kg, plus 1 g/kg Polycose	279 ± 120	9/10	30.1 ± 14.2

[a]Data expressed as mean ± SD.
[b]Values reported as the number of affected animals per total number studied.
[c]Mean ± SD for affected animals only.

taneous administration of Polycose with aspartate reduced the maximum number of necrotic neurons encountered in arcuate nucleus to a mean (±SD) of 30.1 ± 14.2. This value is significantly lower than the value observed (80.8 ± 17.3 necrotic neurons per section) following administration of aspartate alone. Prior injection of insulin also reduced the observed number of necrotic neurons within the arcuate nucleus from that observed following a 1000 mg/kg body weight dose of aspartate (Table 3). Although carbohydrate and insulin decreased the extent of neuronal necrosis and the mean peak plasma level of aspartate plus glutamate, the decrease in plasma glutamate plus aspartate concentration was not as great as had been expected on the basis of human studies of glutamate loading with and without carbohydrate (35).

DISCUSSION

The oral administration of aspartate in doses of 750 or 1000 mg/kg body weight to 8-day mice produced extensive lesions of the arcuate nucleus in all of the animals tested. Aspartate dosing at 650 mg/kg body weight produced lesions in 3 of the 10 animals studied. None of the animals administered 250 or 500 mg/kg body weight aspartate demonstrated hypothalamic neuronal necrosis. Thus the threshold dose of aspartate associated with arcuate damage in this study was between 500 and 650 mg/kg body weight. These results are similar to those observed following the oral administration of glutamate at 700 and 1000 mg/kg body weight (1,2,6). Studies using glutamate doses of 500 mg/kg body weight have shown variable results. For example, Olney and Ho (6) found lesions in 52% of mice administered 500 mg/kg body weight glutamate, while Reynolds et al. (19,20) found arcuate lesions in 22% of mice at this dose. In contrast, Takasaki et al. (24) and O'hara and Takasaki (25) reported that infant mice administered glutamate orally at 500 mg/kg body weight did not develop lesions. However, these authors reported that a 700-mg/kg body weight glutamate dose produced lesions in 25% of the infant mice. The variation in sensitivity may reflect the different strains of mice used in these experiments. Lemkey-Johnston and Reynolds (2) reported strain-related variation in the sensitivity of infant mice to glutamate-induced neuronal necrosis.

In the present study administration of 500 mg/kg body weight aspartate resulted in a mean (±SD) peak plasma glutamate plus aspartate concentration of 127 ± 76 μmol/dl but did not cause arcuate lesions. A dose of 650 mg/kg body weight aspartate produced a mean (±SD) peak plasma glutamate plus aspartate concentration of 227 ± 55 μmol/dl and resulted in arcuate lesions in some animals. Thus the threshold peak plasma glutamate plus aspartate concentration associated with neuronal necrosis must lie between 127 and 227 μmol/dl. However, if we consider data obtained from the loading studies with glutamate and aspartate (Table 2) and assume that the effects of glutamate and aspartate are additive as suggested by Olney and Ho (6), then the threshold plasma aspartate plus glutamate level associated with toxicity must lie between 127 and 169 μmol/dl. Both values are higher than the threshold plasma aspartate plus glutamate concentration estimated in earlier experiments using either a protein hydrolysate or glutamate dosing. Stegink et al. (23) measured plasma amino acid levels in infant mice injected with a protein hydrolysate solution containing glutamate and aspartate, correlating their results with neuropathological data reported by Olney et al. (9). Their data suggested that plasma glutamate plus aspartate concentrations greater than 60 μmol/dl were associated with neurotoxicity. However, the protein hydrolysate injected contained many other amino acids and produced a severe generalized hyperaminoacidemia (23). The resulting hyperosmolarity may make the infant animal more sensitive to dicarboxylic amino acid-induced neuronal necrosis. The latter possibility is also suggested by the data of McCall and colleagues (40). O'hara and Takasaki (25) studied the threshold plasma glutamate concentration associated with neuronal necrosis in infant mice administered MSG. Extreme changes is osmolarity were not present in these animals. In these experiments the threshold plasma glutamate value was estimated to be 100 μmol/dl, a value approximately twice that estimated from studies with protein hydrolysates. O'hara and Takasaki (25) did not report plasma aspartate values.

The threshold aspartate dose required to produce aspartate-induced neurotoxicity in the infant mouse is somewhat higher than the threshold dose of aspartate associated with neuronal necrosis in studies reported by Okaniwa et al. (7) in young rats. These investigators administered potassium aspartate orally to 7-day old rats at doses of 0, 220, 440, 910, and 1900 mg/kg body weight. Rats administered aspartate at 0 or 220 mg/kg did not develop neuronal necrosis. Two of 19 rats administered aspartate at 440 mg/kg body weight, 9 of 12 animals administered aspartate at 910 mg/kg body weight, and all animals administered aspartate at 1900 mg/kg body weight developed neuronal necrosis. The extent of neuronal necrosis was not quantitated, and plasma amino acid levels were not reported. These data suggest that an oral aspartate dose of approximately 440 mg/kg body weight is required for lesions to develop in the rat. Our data in the mouse suggest a threshold aspartate dose between 500 and 650 mg/kg body weight.

Itoh et al. (41) reported plasma glutamate and aspartate concentrations in 7-day-old rats administered potassium aspartate orally at either 220 or 910 mg/kg body weight. The authors reported a mean peak plasma glutamate plus aspartate concentration of 41 μmol/dl at an aspartate dose of 220 mg/kg body weight. This value is similar to the mean peak plasma glutamate plus aspartate level observed in infant mice administered 250 mg aspartate/kg body weight (49 ± 31 μmol/dl). However, the mean peak plasma glutamate plus aspartate concentration reported by Itoh et al. (41) at an aspartate dose of 910 mg/kg is very small (66 μmol/dl) when compared to values they observed at the lower doses of aspartate studied (220 mg/kg body weight). By contrast, we observed a mean (±SD) peak plasma glutamate plus aspartate concentration of 329 ± 131 μmol/dl in infant mice administered aspartate at 750 mg/kg and a mean (±SD) value of 430 ± 189 μmol/dl in animals administered aspartate at 1000 mg/kg body weight. The difference between findings is difficult to understand, but may be related to the fact that Itoh et al. (41) allowed their animals to suckle after dosing. In humans and young pigs the simultaneous ingestion of food with administered glutamate decreases plasma glutamate levels when compared to plasma values noted after ingestion of an equivalent glutamate dose in water (33-36). A similar finding may occur with aspartate.

Carbohydrate (Polycose) produced a significant protective effect against neuronal necrosis when administered simultaneously with 1000 mg/kg body weight aspartate. These findings are in agreement with earlier studies. Takasaki (31) and Anantharaman (32) reported that mice can ingest very large quantities of glutamate as part of their diet without developing neuronal necrosis. These data have led to considerable speculation about the nature of the "factors" involved in the protective effect associated with simultaneous ingestion of food. Stegink et al. (34) reported that plasma glutamate plus aspartate concentrations were depressed in adult human subjects ingesting large glutamate doses with Sustagen (Mead Johnson, Evansville, IN) when compared to plasma values noted after ingestion of the equivalent quantity of glutamate in water. These data suggest that the simultaneous ingestion of food decreases the extent of neuronal necrosis by decreasing peak plasma glutamate plus aspartate concentration. Subsequent studies in humans and young pigs implicated carbohydrate as the meal component affecting plasma glutamate levels (35,36). These experiments demonstrate that simultaneous ingestion of metabolizable carbohydrate with glutamate significantly decreases plasma glutamate plus aspartate concentrations when compared to plasma values noted when glutamate was ingested without added carbohydrate.

Takasaki (37) reported that simultaneous ingestion of mono- and disaccharides with MSG significantly reduced the severity of arcuate lesions in mice administered 2 g/kg body weight MSG. However, carbohydrate did not completely prevent glutamate-induced neurotoxicity, and no plasma amino acid levels were reported.

Polycose administration had a much smaller effect on the mean peak plasma glutamate plus aspartate concentrations in aspartate-loaded mice than expected from studies in man using glutamate. Stegink et al. (35) gave normal adult subjects 150 mg/kg body weight loads of glutamate with and without added Polycose (1 g/kg body weight). When the glutamate load was administered in water, the mean peak plasma glutamate concentration was 59.4 ± 46.5 μmol/dl. When glutamate was administered with Polycose, the mean peak plasma glutamate concentration was only 7.18 ± 3.48 μmol/dl. Thus Polycose's smaller effect on plasma glutamate and aspartate levels in mice was surprising. It is not clear whether this difference represents a species effect (mouse versus man) or whether this results from a difference between the metabolism of glutamate and the metabolism of aspartate.

Takasaki and Yugari (38) also reported that injection of neonatal mice with 0.02 units of insulin prior to oral administration of 2000 mg/kg body weight glutamate markedly reduced the extent of hypothalamic neuronal necrosis as compared to animals receiving glutamate without preinjection of insulin. They suggested that insulin's protective effect is due to facilitation of tissue uptake of dicarboxylic amino acids, thus lowering plasma amino acid levels. In our study using aspartate, injection of the mice with insulin prior to administration of 1000 mg/kg body weight aspartate resulted in decreased arcuate damage. However, prior injection of insulin had a relatively small effect upon the mean peak plasma glutamate plus aspartate concentration and values were still well above the toxic threshold. Therefore our data suggest that insulin pretreatment may have an effect on aspartate-induced neuronal necrosis beyond that associated with a reduction in peak plasma glutamate and aspartate levels.

We were surprised to note that administration of 500 mg/kg body weight glutamate plus 500 mg/kg body weight aspartate resulted in significantly less arcuate damage, in terms of both percent of animals with lesions and mean maximum number of necrotic neurons, than administration of 1000 mg/kg body weight aspartate. Olney and Ho (6) emphasized the similarities between these two amino acids and reported that glutamate and aspartate were additive in their neurotoxic effect, since animals treated orally with 500 mg/kg body weight glutamate plus 500 mg/kg body weight aspartate developed a degree of hypothalamic damage essentially similar to that seen in animals treated with either agent at 1000 mg/kg body weight. The neuropathological data in our study can be explained by the plasma glutamate plus aspartate concentration. Peak plasma glutamate plus aspartate concentration correlated well with observed neuronal necrosis. The peak plasma glutamate plus aspartate concentration and number of necrotic neurons per section observed in animals administered 500 mg/kg body weight glutamate plus 500 mg/kg body weight aspartate were similar to values observed in animals administered 750 mg/kg body weight aspartate. The values for these two groups of animals were less than those observed in the group given 1000 mg/kg body weight aspartate.

It is probable that both glutamate and aspartate are neurotoxic, each to different groups of neurons. Current data suggest that glutamate and aspartate act as neurotransmitters at different receptor sites (42,43). Thus high plasma glutamate concentration might result in neuronal necrosis in one population of neurons, while elevated plasma aspartate concentration might result in neuronal necrosis in a different neuronal population.

The neuropathological data reported in this study are consistent with available data on aspartame toxicity in infant mice. Neurotoxic effects reported in mice administered large doses of aspartame are thought to result from the sweetener's aspartate content. Reynolds et al. (19) reported that 40% of animals administered aspartame at 2 g/kg body weight developed neuronal necrosis, while 13%

of the animals administered aspartame at 1 g/kg body weight exhibited lesions. No animals administered aspartame at 0.5 g/kg body weight developed neuronal necrosis. Olney (44) reported lesions in infant mice administered 2-2.5 g/kg body weight aspartame, but did not report a "no-effect" dose. It is assumed that aspartame-induced neurotoxicity results from the release of aspartate to the blood during aspartame metabolism. Thus the effect dose for aspartame and aspartate should be similar if equivalent plasma concentrations of dicarboxylic amino acid result. Neither Olney (44) nor Reynolds et al. (19) reported plasma amino acid concentrations after aspartame dosing.

The estimated neurotoxic doses of aspartate and the plasma amino acid concentrations associated with those doses can be compared with the expected intake of aspartame by humans and with plasma dicarboxylic amino acid concentrations at those intake levels. The 99th percentile of projected daily intake for aspartame is 34 mg/kg body weight, or about 14 mg/kg body weight aspartate (45). In infant mice aspartate doses as high as 500 mg/kg body weight did not lead to neuronal necrosis despite a mean peak plasma glutamate plus aspartate concentration of 127 ± 76 μmol/dl. Normal humans ingesting abuse doses of aspartame (200 mg/kg body weight, 80 mg/kg body weight aspartate content) had a mean peak plasma glutamate plus aspartate concentration of 6.66 μmol/dl (46), a value far below the toxic threshold value noted in the mouse. Furthermore, Reynolds et al. (19,21) failed to find neuronal necrosis in infant nonhuman primates administered aspartame (2 g/kg body weight) plus glutamate (1 g/kg body weight) despite mean peak plasma glutamate plus aspartate concentrations of 217 μmol/dl. Thus aspartame ingested at projected levels of intake for the sweetener appears to present little danger to man from aspartate-induced neurotoxicity.

REFERENCES

1. Olney, J. W. (1969). Brain lesions, obesity, and other disturbances in mice treated with monosodium glutamate. *Science* 164, 719-721.
2. Lemkey-Johnston, N., and Reynolds, W. A. (1974). Nature and extent of brain lesions in mice related to ingestion of monosodium glutamate. A light and electron microscope study. *J. Neuropathol. Exp. Neurol.* 33, 74-97.
3. Olney, J. W., Rhee, V., and de Gubareff, T. (1973). Neurotoxic effects of glutamate. *N. Engl. J. Med.* 289, 1374-1375.
4. Tafelski, T. J., and Lamperti, A. A. (1977). Effects of a single injection of monosodium glutamate on the reproductive neuroendocrine axis of the female hamster. *Biol. Reprod.* 17, 404-411.
5. Lamperti, A., and Blaha, G. (1976). The effects of neonatally-administered monosodium glutamate on the reproductive system of adult hamsters. *Biol. Reprod.* 14, 362-369.
6. Olney, J. W., and Ho, O.-L. (1970). Brain damage in infant mice following oral intake of glutamate, aspartate or cysteine. *Nature* 227, 609-611.

7. Okaniwa, A., Hori, M., Masuda, M., Takeshita, M., Hayashi, N., Wada, I., Doi, K., and Ohara, Y. (1969). Histopathological study on effects of potassium aspartate on the hypothalamus of rats. *J. Toxicol. Sci.* 4, 31-46.
8. Olney, J. W., Labruyere, J., and de Gubareff, T. (1980). Brain damage in mice from voluntary ingestion of glutamate and aspartate. *Neurobehav. Toxicol.* 2, 125-129.
9. Olney, J. W., Ho, O.-L., and Rhee, V. (1973). Brain damaging potential of protein hydrolysates. *N. Engl. J. Med.* 289, 391-395.
10. Olney, J. W., and Sharpe, L. G. (1969). Brain lesions in an infant rhesus monkey treated with monosodium glutamate. *Science* 166, 386-388.
11. Olney, J. W., Sharpe, L. G., and Feigin, R. D. (1972). Glutamate-induced brain damage in infant primates. *J. Neuropathol. Exp. Neurol.* 31, 464-488.
12. Reynolds, W. A., Lemkey-Johnston, N., Filer, L. J., Jr., and Pitkin, R. M. (1971). Monosodium glutamate: Absence of hypothalamic lesions after ingestion by newborn primates. *Science* 172, 1342-1344.
13. Abraham, R., Dougherty, W., Golberg, L., and Coulston, F. (1971). The response of the hypothalamus to high doses of monosodium glutamate in mice and monkeys. *Exp. Mol. Pathol.* 15, 43-60.
14. Newman, A. J., Heywood, R., Palmer, A. K., Barry, D. H., Edwards, F. P., and Worden, A. N. (1973). The administration of monosodium L-glutamate to neonatal and pregnant rhesus monkeys. *Toxicology* 1, 197-204.
15. Wen, C., Hayes, K. C., and Gershoff, S. N. (1973). Effects of dietary supplementation of monosodium glutamate on infant monkeys, weanling rats and suckling mice. *Am. J. Clin. Nutr.* 26, 803-813.
16. Abraham, R., Swart, J., Golberg, L., and Coulston, F. (1975). Electron microscopic observations of hypothalami in neonatal rhesus monkeys (*Macaca mulatta*) after administration of monosodium L-glutamate. *Exp. Mol. Pathol.* 23, 203-213.
17. Stegink, L. D., Reynolds, W. A., Filer, L. J., Jr., Pitkin, R. M., Boaz, D. P., and Brummel, M. C. (1975). Monosodium glutamate metabolism in the neonatal monkey. *Am. J. Physiol.* 229, 246-250.
18. Heywood, R., and James, R. W. (1979). An attempt to induce neurotoxicity in an infant rhesus monkey with monosodium glutamate. *Toxicol. Lett.* 4, 285-286.
19. Reynolds, W. A., Butler, V., and Lemkey-Johnston, N. (1976). Hypothalamic morphology following ingestion of aspartame or MSG in the neonatal rodent and primate: A preliminary report. *J. Toxicol. Environ. Health* 2, 471-480.
20. Reynolds, W. A., Lemkey-Johnston, N., and Stegink, L. D. (1979). Morphology of the fetal monkey hypothalamus after in utero exposure to monosodium glutamate. In *Glutamic Acid: Advances in Biochemistry and Physiology* (Filer, L. J., Jr., Garattini, S., Kare, M. R., Reynolds, W. A., and Wurtman, R. J., eds.), Raven Press, New York, pp. 217-229.
21. Reynolds, W. A., Stegink, L. D., Filer, L. J., Jr., and Renn, E. (1980). Aspartame administration to the infant monkey: Hypothalamic morphology and plasma amino acid levels. *Anat. Rec.* 198, 73-85.
22. Heywood, R., and Worden, A. N. (1979). Glutamate toxicity in laboratory animals. In *Glutamic Acid: Advances in Biochemistry and Physiology* (Filer,

L. J., Jr., Garattini, S., Kare, M. R., Reynolds, W. A., and Wurtman, R. J., eds.), Raven Press, New York, pp. 203-215.
23. Stegink, L. D., Shepherd, J. A., Brummel, M. C., and Murray, L. M. (1974). Toxicity of protein hydrolysate solutions: Correlation of glutamate dose and neuronal necrosis to plasma amino acid levels in young mice. *Toxicology* 2, 285-299.
24. Takasaki, T., Matsuzawa, Y., Iwata, S., O'hara, Y., Yonetani, S., and Ichimura, M. (1979). Toxicological studies of monosodium-L-glutamate in rodents: Relationship between routes of administration and neurotoxicity. In *Glutamic Acid: Advances in Biochemistry and Physiology* (Filer, L. J., Jr., Garattini, S., Kare, M. R., Reynolds, W. A., and Wurtman, R. J., eds.), Raven Press, New York, pp. 255-275.
25. O'hara, Y., and Takasaki, Y. (1979). Relationship between plasma glutamate levels and hypothalamic lesions in rodents. *Toxicol. Lett.* 4, 499-505.
26. Finkelstein. M. W. (1982). Aspartate-induced neuronal necrosis in infant mice. Effect of dose, insulin and carbohvdrate. M. S. Thesis in Oral Pathology, The University of Iowa, Iowa City, Iowa.
27. Olney, J. W. (1975). L-Glutamic and L-aspartic acids–A question of hazard? *Food Cosmet. Toxicol.* 13, 595-596.
28. Olney, J. W. (1975). Another view of aspartame. In *Sweeteners, Issues and Uncertainties*, Academy Forum, National Academy of Sciences, Washington, D.C., pp. 189-195.
29. Olney, J. W. (1981). Excitatory neurotoxins as food additives: An evaluation of risk. *Neurotoxicology* 2, 163-192.
30. Reif-Lehrer, L. (1976). Possible significance of adverse reactions to glutamate in humans. *Fed. Proc.* 35, 2205-2211.
31. Takasaki, Y. (1978). Studies on brain lesions after administration of monosodium L-glutamate to mice. II. Absence of brain damage following administration of monosodium L-glutamate in the diet. *Toxicology* 9, 307-318.
32. Anantharaman, K. (1979). In utero and dietary administration of monosodium L-glutamate to mice: Reproductive performance and development in a multigeneration study. In *Glutamic Acid: Advances in Biochemistry and Physiology* (Filer, L. J., Jr., Garattini, S., Kare, M. R., Reynolds, W. A., and Wurtman, R. J., eds.), Raven Press, New York, pp. 213-253.
33. Stegink, L. D., Filer, L. J., Jr., and Baker, G. L. (1982). Plasma and erythrocyte amino acid levels in normal adult subjects fed a high protein meal with and without added monosodium glutamate. *J. Nutr.* 112, 1953-1962.
34. Stegink, L. D., Baker, G. L., and Filer, L. J., Jr. (1983). Modulating effect of Sustagen on plasma glutamate concentration in humans ingesting monosodium L-glutamate. *Am. J. Clin. Nutr.* 37, 194-200.
35. Stegink, L. D., Filer, L. J., Jr., and Baker, G. L. (1983). Effect of carbohydrate on plasma and erythrocyte glutamate levels in humans ingesting large doses of monosodium L-glutamate in water. *Am. J. Clin. Nutr.* 37, 961-968.
36. Daabees, T. T., Andersen, D. W., Zike, W. L., Filer, L. J., Jr., and Stegink, L. D. (1984). Effect of meal components on peripheral and portal plasma glutamate levels in young pigs administered large doses of monosodium L-glutamate. *Metabolism* 33, 58-67.

37. Takasaki, Y. (1979). Protective effect of mono- and disaccharides on glutamate-induced brain damage in mice. *Toxicol. Lett.* 4, 205-210.
38. Takasaki, Y., and Yugari, Y. (1980). Protective effect of arginine, leucine, and preinjection of insulin on glutamate neurotoxicity in mice. *Toxicol. Lett.* 5, 39-44.
39. Finkelstein, M. W., Daabees, T. T., Stegink, L. D., and Applebaum, A. E. (1983). Correlation of aspartate dose, plasma dicarboxylic amino acid con-concentration, and neuronal necrosis in infant mice. *Toxicology* 29, 109-119.
40. McCall, A., Glaeser, B. S., Millington, W., and Wurtman, R. J. (1979). Monosodium glutamate neurotoxicity, hyperosmolarity, and blood brain barrier dysfunction. *Neurobehav. Toxicol.* 1, 279-283.
41. Itoh, H., Kishi, T., Iwasawa, Y., Kawashima, K., and Chibata, I. (1979). Plasma aspartate levels in rats following administration of monopotassium aspartate via three routes. *J. Toxicol. Sci.* 4, 337-388.
42. McLennan, H. (1981). On the nature of the receptors for various excitatory amino acids in the mammalian central nervous system. In *Glutamate as a Neurotransmitter* (Di Chiara, G., and Gessa, G. L., eds.), Raven Press, New York, pp. 253-262.
43. Davies, J., and Watkins, J. C. (1981). Pharmacology of glutamate and aspartate antagonists on cat spinal neurones. In *Glutamate as a Neurotransmitter* (Di Chiara, G., and Gessa, G. L., eds.), Raven Press, New York, pp. 275-284.
44. Olney, J. W. (1976). Brain damage and oral intake of certain amino acids. In *Transport Phenomena in the Nervous System* (Levi, G., Battistin, L., and Lajtha, A., eds.), Plenum Press, New York, pp. 497-506.
45. Stegink, L. D., Filer, L. J., Jr., and Baker, G. L. (1977). Effect of aspartame and aspartate loading upon plasma and erythrocyte free amino acid levels in normal adult volunteers. *J. Nutr.* 107, 1837-1845.
46. Stegink, L. D., Filer, L. J., Jr., and Baker, G. L. (1981). Plasma and erythrocyte concentrations of free amino acids in adult humans administered abuse doses of aspartame. *J. Toxicol. Environ. Health* 7, 291-305.

18

Neuropathology Studies Following Aspartame Ingestion by Infant Nonhuman Primates

W. Ann Reynolds
California State University, Long Beach, California

Linda Parsons
University of Illinois College of Medicine, Chicago, Illinois

Lewis D. Stegink
University of Iowa College of Medicine, Iowa City, Iowa

INTRODUCTION

The administration of large doses of the dicarboxylic amino acids glutamate and aspartate produces hypothalamic neuronal necrosis in the infant rodent. Large amounts of monosodium L-glutamate (MSG) administered to neonatal mice result in retinal (1) and brain lesions (2-4). It was initially believed that such brain lesions were restricted to the hypothalamic areas; however, we have demonstrated that at high glutamate doses multiple brain areas are affected, including the tectum, habenular nuclei, subfornical organ, dorsolateral surface of the thalamus, dentate-hippocampal gyri, and cerebral cortex. The lesion is characterized by neuronal death, resulting in pyknotic nuclei, dilated dendrites, and, ultimately, phagocytic events involving glial cells (4,5). Neonatal mice are sensitive to doses of MSG as low as 500 mg/kg body weight (6,7).

The administration of aspartate to infant mice also results in neuronal necrosis (6,8-10). The minimun dose of aspartate producing a lesion is in the range of 500-650 mg/kg body weight. Mice given an oral dose of MSG (500 mg/kg body weight) plus aspartate (500 mg/kg body weight) show the hypothalamic damage characteristic of animals given either substance at 1 g/kg body weight (6), indicating an additive neurotoxic effect of the two substances.

After considerable controversy, it appears that glutamate and aspartate do not induce damage to the fetal or neonatal primate brain, even when administered in large doses. Although Olney and colleagues (11,12) reported hypothalamic damage in neonatal nonhuman primates following glutamate administration, four other research groups were unable to find hypothalamic damage following MSG administration in spite of extensive and painstaking efforts (3,13-19). In general, these groups had no difficulty in demonstrating glutamate-induced neurotoxicity in infant rodents. To our knowledge, no systematic studies of aspartate neurotoxicity have been carried in nonhuman primates.

Olney has expressed concern that aspartame, because of its aspartate content, might pose a risk to the developing hypothalamus (20,21). He was also concerned about the potential additive effect of aspartame with the MSG present in some foods (20,21). Reynolds et al. have previously assessed aspartame neurotoxicity in infant mice administered aspartame at dosages ranging from 0.25 to 4 g/kg body weight (7). Hypothalamic lesions were encountered at aspartame dose levels equal to or exceeding 1 g/kg body weight. At comparable dosages per unit body weight, aspartame toxicity was less than that noted in mice treated with MSG. For example, at an aspartame dose of 1 g/kg body weight, neuronal damage was encountered in only a small proportion of the arcuate neuronal population in aspartame-treated animals, and brain damage was seldom encountered in regions other than the ventral hypothalamus. Olney (22) also reported aspartame-induced neuronal necrosis in infant mice.

The brain of the infant nonhuman primate more closely resembles that of the human infant than that of the infant mouse. The present study was designed to evaluate the effects of massive aspartame loads, with and without added MSG, upon the hypothalamus of the infant nonhuman primate. The details of this study have been published previously (23).

MATERIALS AND METHODS

Three species of macaques were used in these experiments: *Macaca mulatta, Macaca fascicularis,* and *Macaca arctoides*. All animals were delivered in a primate facility from timed pregnancies. Eleven infant monkeys served as controls. They received either no treatment, or water by stomach tube (Table 1). Eight infant monkeys received aspartame at a level of 2 g/kg body weight (Table 2), and six infant monkeys were given aspartame at a dose of 2 g/kg body weight plus 1 g/kg body weight of MSG (Table 3).

After dosing, each infant was observed constantly for 4 hr, so that vomiting, respiratory difficulties, cyanosis, or other problems could be observed. If a monkey regurgitated, it was observed until vomiting ceased and then redosed with the estimated volume of solution regurgitated.

Serial blood samples (1 ml) were obtained at 0, 20, 40, 60, 90, 120, 180, and 240 min in four animals administered aspartame and in six animals administered aspartame plus MSG for plasma amino acid analyses.

After dosing, each animal was placed on a warming pad to maintain body temperature at 37°C for the duration of the experiment. The treatment period

Table 1 Control Monkeys Receiving Nil per Os (NPO) or Water by Gastric Tube

Age (days)	Species	Sex	Body weight (g)	Dose
6	*M. fascicularis*	F	335	NPO
8	*M. mulatta*	M	450	NPO
9	*M. fascicularis*	M	310	H_2O
15	*M. fascicularis*	M	310	NPO
16	*M. mulatta*	F	590	H_2O
18	*M. fascicularis*	M	350	H_2O
19	*M. fascicularis*	F	370	H_2O
21	*M. fascicularis*	M	360	NPO
30	*M. arctoides*	M	520	NPO
51	*M. fascicularis*	M	410	NPO
120	*M. arctoides*	F	820	H_2O

Table 2 Monkeys Receiving 2 g/kg Body Weight Aspartame by Gastric Tube

Age (days)	Species	Sex	Body weight (g)
1	*M. arctoides*	F	430
1.5	*M. mulatta*	F	370
2	*M. arctoides*	M	430
3	*M. mulatta*	F	410
3	*M. arctoides*	F	470
5	*M. arctoides*	F	500
8	*M. fascicularis*	F	280
22	*M. mulatta*	M	530

Table 3 Monkeys Receiving 2 g/kg Body Weight Aspartame Plus 1 g/kg Body Weight Monosodium L-Glutamate by Gastric Tube

Age (days)	Species	Sex	Body weight (g)
1	*M. mulatta*	F	422
1	*M. arctoides*	M	460
1	*M. mulatta*	M	480
2	*M. mulatta*	F	370
2	*M. arctoides*	F	380
3	*M. mulatta*	M	380

selected was between 4 and 5 hr, a length of time sufficient for the development of any potential lesion. The infant mouse exhibits swollen dendrites with cellular damage 15 min after oral administration of MSG (4). Olney et al. (12) reported observing necrotic neurons with beginning phagocytosis in infant nonhuman primate brains perfused 30 min after dosing with MSG.

At the termination of each experiment, the animal was anesthetized and the brain perfused as described in previous publications (7,23). Following perfusion, the head was removed, the calvarium opened, and the head with exposed brain stored overnight in additional volumes of the perfusate. Subsequently, two or three samples (1 mm in thickness), including the arcuate-median eminence area, were cut. The slices were postfixed in osmium tetraoxide, dehydrated, and embedded in Durcapan ACM. Thin sections, cut with a diamond knife at 700 Å, were stained with uranyl acetate and lead citrate and examined with an electron microscope. In order to examine the entire hypothalamic region, 1-μm plastic sections were cut serially, stained with methylene blue and azure II, and studied by light microscopy.

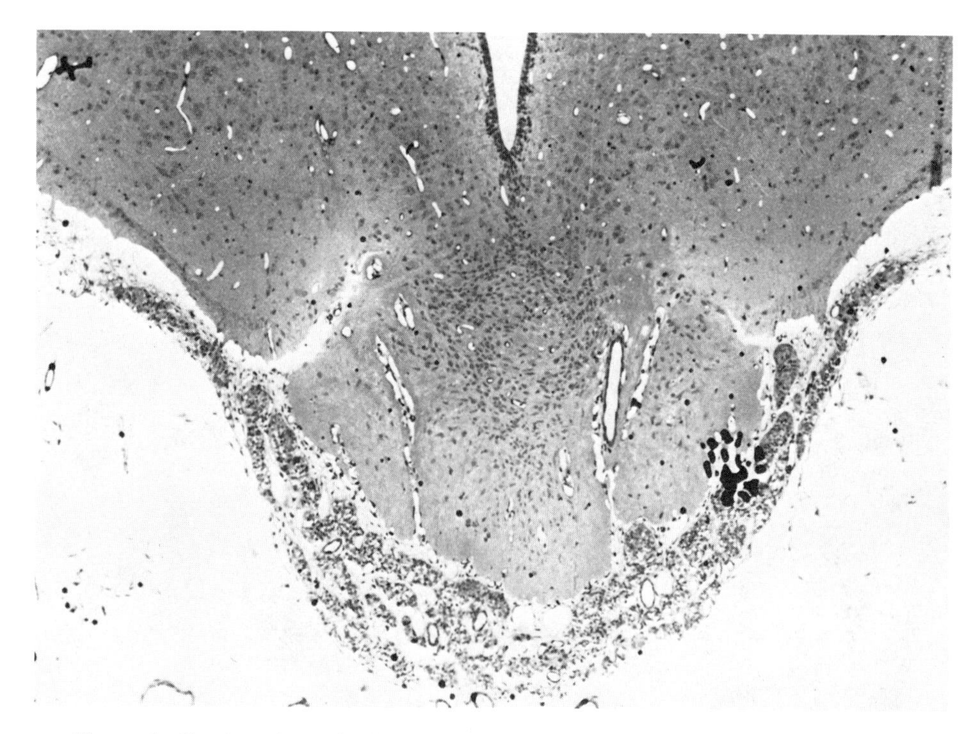

Figure 1 Section through the rostral portion of an infant monkey hypothalamus. This infant (#2747) received 2 g/kg body weight of aspartame by a gastric tube on the first day of life. The neuropil is normal in appearance (X58).

RESULTS

Histopathology

Because of localized regions of vasoconstriction, it is virtually impossible to perfuse a brain as large as that of the infant monkey with glutaraldehyde so as to achieve high-quality fixation throughout. Thus accurate detection of neuropathological events requires the study of a representative series of control brains so that perfusion artifacts do not become misleading. For this reason, the massive task of preparing 1-μm serial sections of the entire ventral hypothalamic region was undertaken in all of the animals listed in Tables 1-3.

The appearance of the neuropil, neurons, neuronal processes, and glia did not vary between normal control animals and those treated with either aspartame (2 g/kg body weight) or aspartame (2 g/kg body weight) plus MSG (1 g/kg body weight). Representative sections from the hypothalamic region of these animals (Figs.1-4) did not contain evidence of abnormal cell death or neuronal degeneration.

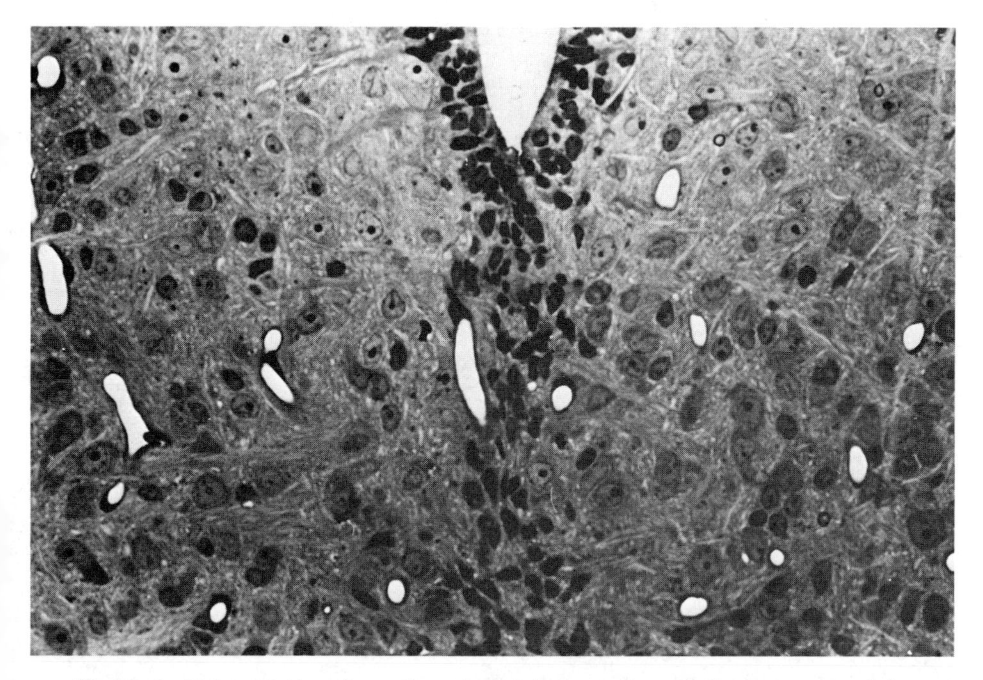

Figure 2 Enlarged view from the subventricular region of the same animal depicted in Figure 1. Neurons appear normal. Dilated capillaries have resulted from the perfusion process (X575).

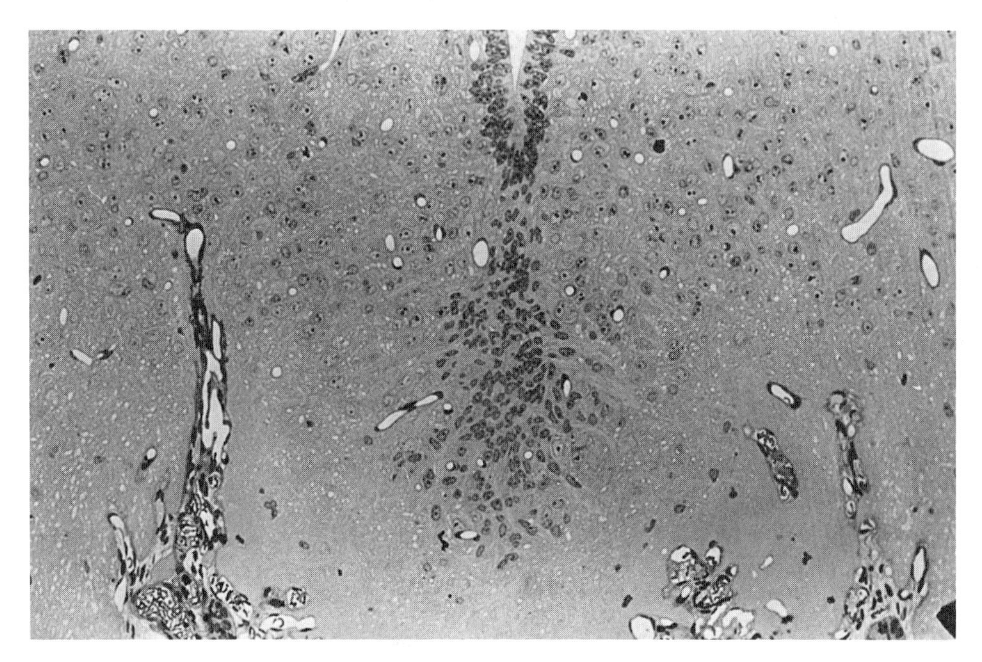

Figure 3 Section through the infundibular portion of a newborn monkey (#2823) receiving 2 g/kg body weight of aspartame plus 1 g/kg body weight of monosodium L-glutamate via a gastric tube at day 1 of age. Neurons and neuropil are normal in appearance (X144).

At the electron-microscopic level, there were no differences between experimental and control brain sections. The imperfection of even the best perfusion process became apparent in the 700 Å sections. Irrespective of whether material was obtained from control or treated animals, localized areas of poor perfusion exhibiting abnormal morphology were noted. Swollen processes and other evidence of poor fixation were often found in close juxtaposition to areas of otherwise well-fixed neuropil in the brains of both control and experimental animals (Fig. 4).

Detailed examination of the hypothalamic region of this large group of newborn monkeys consistently revealed supraependymal cellular blebs. The blebs, usually attached to the sidewalls of the third ventricle, were noticed caudal to the optic chiasm and rostral to the caudal end of the infundibular recess (Figs. 5 and 6). They ranged in size from 5 to 175 cells and covered from 6 to 105 μm when measured in the frontal plane. There were no differences in the appearance, size, or incidence of blebs in control or treated monkeys. Most of the blebs were cleanly distinct from the ependyma (Fig. 6), but occasionally, bleblike cells

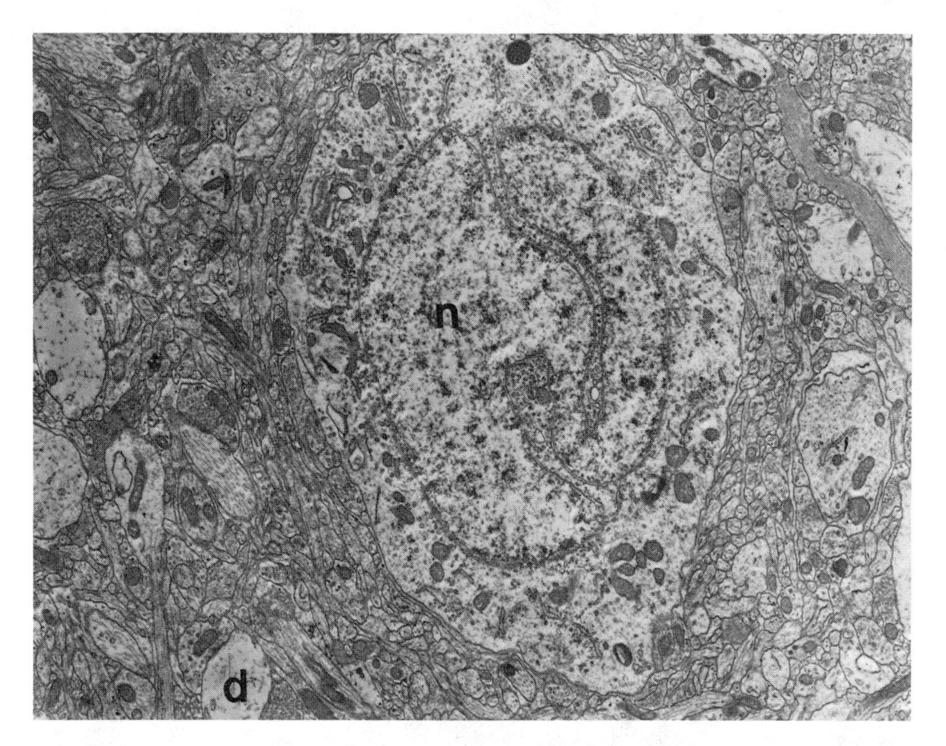

Figure 4 Electron-microscopic view of the subventricular portion of an infant monkey (#2858) receiving 2 g/kg body weight of aspartame on day 3 by a stomach tube. The cytoplasm and folded nucleus (n) are normal in appearance. The swollen dendrite (d) amid otherwise well-fixed neuropil emphasizes the challenge of achieving good perfusion throughout a primate brain (X8084).

appeared to have invaded the subependymal area (Fig. 5). Single supraependymal cells and small cell aggregates have been reported previously in the monkey third ventricle (24) and in the mouse and rabbit (25,26).

Biochemical Observations

Plasma aspartate, glutamate, phenylalanine, and tyrosine concentrations in animals administered aspartame with and without added MSG are shown in Figures 7-10. Considerable individual variation was noted in the shape of the plasma concentration-time curves for phenylalanine, tyrosine, glutamate, and aspartate after administration of either aspartame or aspartame plus MSG. This undoubtedly reflects differences in gastric emptying and the fact that aspartame was administered as a slurry to achieve a dose level of 2 g/kg body weight. Plasma

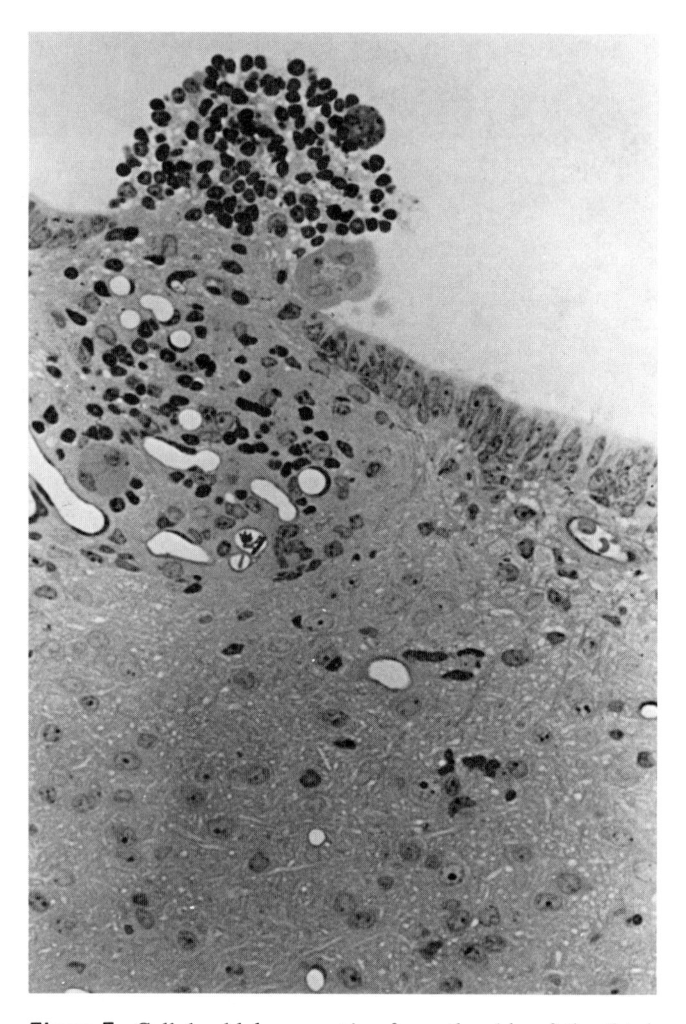

Figure 5 Cellular bleb emanating from the side of the third ventricle of a 1-day-old stumptail monkey (#2796) receiving 2 g/kg body weight aspartame plus 1 g/kg body weight monosodium L-glutamate. Note the disruption of ependymal cells and the presence of cells with dark nuclei (characteristic of cells populating blebs) beneath the ependymal surface. Numerous capillaries invade the subependymal bleb mass (X575).

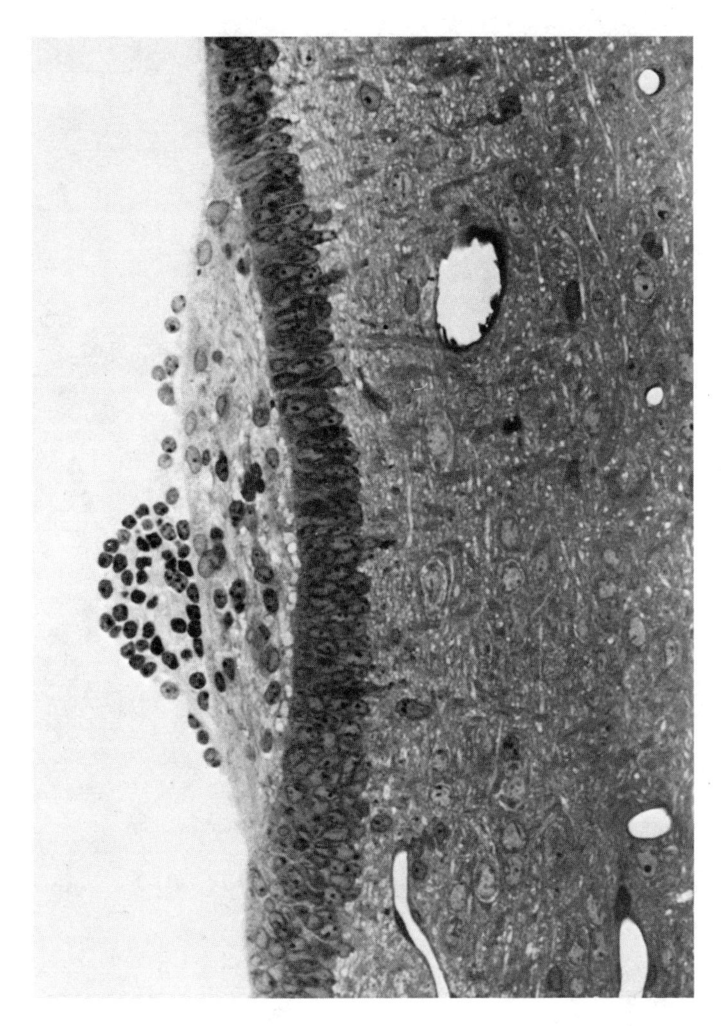

Figure 6 An uninterrupted layer of ependymal cells underlies this supraependymal bleb found in the third ventricle of a 3-day-old stumptail monkey (#2746). An aggregation of cells with small, dark nuclei protrudes above the basilar fibrous-appearing layer of cells (X575).

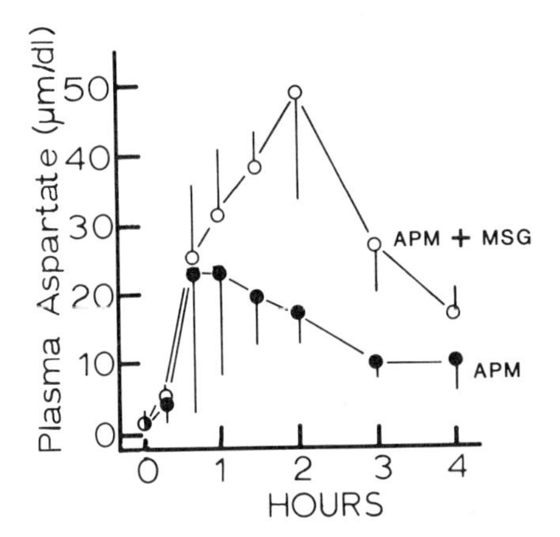

Figure 7 Mean (±SEM) aspartate concentrations (μmol/dl) in infant monkeys administered aspartame (APM) at 2 g/kg body weight (●——●) or aspartame (2 g/kg body weight) plus monosodium L-glutamate (MSG) at 1 g/kg body weight (○——○).

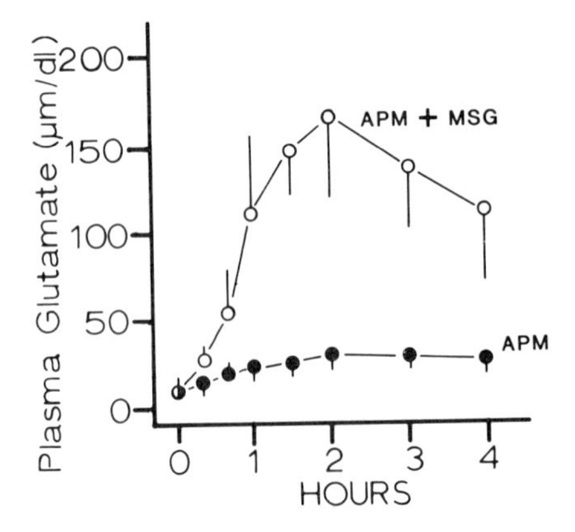

Figure 8 Mean (±SEM) plasma glutamate concentrations (μmol/dl) in infant monkeys administered aspartame (APM) at 2 g/kg body weight (●——●) or aspartame (2 g/kg body weight) plus monosodium L-glutamate (MSG) at 1 g/kg body weight (○——○).

aspartate and phenylalanine concentration-time curves in human subjects show similar variability after slurry administration of aspartame (27).

Administration of aspartame alone significantly ($p = 0.001$) increased plasma aspartate concentrations above mean (±SEM) base-line values (0.69 ± 0.22 μmol/dl), with mean (±SEM) peak levels of 23 ± 16 μmol/dl noted at 60 min (Fig. 7). Individual peak values for aspartate at any given time ranged from 21 to 84

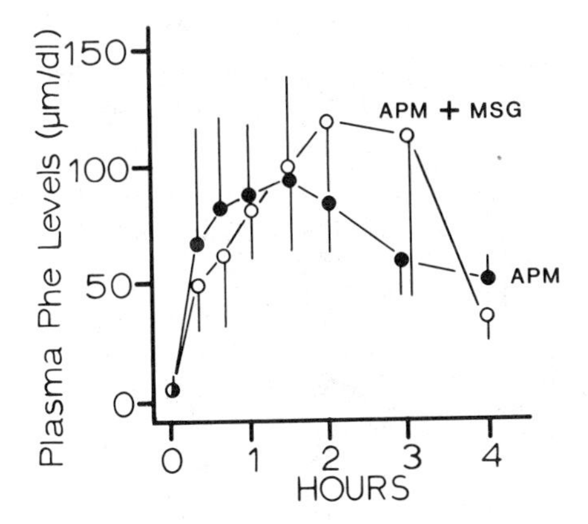

Figure 9 Mean (±SEM) plasma phenylalanine (Phe) concentrations (μmol/dl) in infant monkeys administered aspartame (APM) at 2 g/kg body weight (●—●) or aspartame (2 g/kg body weight) and monosodium L-glutamate (MSG) at 1 g/kg body weight (○—○).

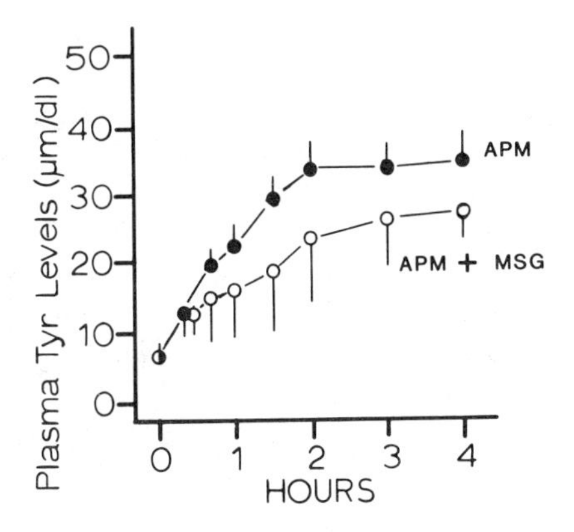

Figure 10 Mean (±SEM) plasma tyrosine (Tyr) concentrations (μmol/dl) in infant monkeys administered aspartame (APM) at 2 g/kg body weight (●—●) or aspartame (2 g/kg body weight) and monosodium L-glutamate (MSG) at 1 g/kg body weight (○—○).

μmol/dl. Plasma aspartate concentrations of animals administered aspartame plus MSG were significantly higher than animals administered aspartame alone. Plasma aspartate concentrations increased significantly over base line (1.36 ± 0.5 μmol/dl), reaching mean (±SEM) peak values of 49 ± 15 μmol/dl 120 min after dosing. Individual peak values ranged from 20 to 92 μmol/dl.

Aspartame loading also increased plasma glutamate concentration (Fig. 8). Plasma glutamate concentrations increased from a base-line level of 9.38 ± 0.86

μmol/dl to a mean peak value of 31.5 ± 6.8 μmol/dl 120 min postdosing. Individual peak values ranged from 14.4 to 43.5 μmol/dl. Plasma glutamate concentrations were significantly higher when MSG was administered with aspartame. Plasma glutamate concentrations increased significantly over base-line levels (8.0 ± 0.7 μmol/dl), reaching a mean (±SEM) peak level of 168 ± 44 μmol/dl 120 min after administration of the test load. Individual peak plasma glutamate levels ranged from 121 to 280 μmol/dl in these animals.

Mean plasma glutamate concentrations observed in infant monkeys administered glutamate plus aspartame were similar to those noted in infant monkeys administered an equivalent dose of MSG (1 g/kg body weight) without added aspartame in earlier studies (17).

Plasma phenylalanine concentrations (Fig. 9) in animals administered aspartame alone increased significantly from a mean (±SEM) base-line value of 5.93 ± 1.40 μmol/dl to a mean (±SEM) peak value of 95 ± 29 μmol/dl 90 min after aspartame loading. Individual peak values ranged from 37 to 215 μmol/dl. Plasma tyrosine concentrations (Fig. 10) increased significantly after aspartame dosing. Values increased from a mean (±SEM) base-line value of 6.7 ± 1.3 μmol/dl to a mean (±SEM) peak value of 35.5 ± 3.4 μmol/dl 4 hr after aspartame ingestion. The increase in plasma tyrosine concentration undoubtedly reflects conversion of phenylalanine to tyrosine.

The observed changes in plasma phenylalanine and tyrosine concentrations in animals administered aspartame alone are similar to those noted in animals administered aspartame plus MSG. In the latter group, plasma phenylalanine concentrations (Fig. 9) increased from a mean (±SEM) base-line value of 6.8 ± 1.1 μmol/dl to mean peak values of 120 ± 63 μmol/dl 120 min after aspartame loading. Individual peak values ranged from 45 to 380 μmol/dl. Plasma tyrosine concentrations (Fig. 10) increased over base-line values (6.6 ± 0.6 μmol/dl), reaching a mean (±SEM) peak level of 27.9 ± 4.88 μmol/dl 4 hr after administration.

DISCUSSION

The test compounds were clearly absorbed following dosing as indicated by the increased plasma aspartate, glutamate, and phenylalanine concentrations. Thus the absence of neuronal necrosis in animals administered aspartame or aspartame plus MSG was not due to a failure to elevate plasma glutamate and aspartate levels. Olney et al. (12) suggested this rationale for the failure of other research groups (3,13-19) to reproduce their report of glutamate-induced hypothalamic neuronal necrosis in neonatal nonhuman primates. The plasma glutamate and aspartate levels in neonatal monkeys administered 2 g/kg body weight aspartame plus 1 g/kg body weight MSG are similar to those noted in neonatal monkeys given MSG alone at a dose of 1 g/kg body weight (17).

The individual data from various animals showed a marked variability in the concentration-time curves for plasma aspartate, phenylalanine, and glutamate.

This variation presumably reflects administration of the compound in slurry form. Human subjects administered aspartame in slurry form (27) showed variations in plasma phenylalanine and aspartate absorption-metabolism curves similar to those noted in these monkeys. This variation did not occur when humans ingested the identical aspartame dose in solution (27).

We have previously reported that mice and monkeys metabolize glutamate loads more rapidly than humans (28). The present study suggests that monkeys also metabolize phenylalanine more rapidly than humans. In humans the administration of increasing doses of aspartame produces an increase in peak plasma phenylalanine levels proportional to dose (Fig. 11). Extrapolation from these human data (Fig. 11) to an aspartame dose of 2 g/kg body weight predicts much higher plasma phenylalanine levels than the values observed in monkeys at this dose. This higher rate of amino acid catabolism is consistent with the higher protein requirement (g protein/kg body weight) of the monkey compared to the human (33,34).

The administration of aspartame or aspartame plus MSG in this study was designed to resemble the situation of inadvertent ingestion of large amounts of aspartame by a young child. For example, a 1-year-old child (10 kg) who accidentally ingested the entire contents of aspartame coffee sweetener (100 tablets of 20 mg each) would ingest 200 mg/kg body weight aspartame. The low solubility of aspartame actually would result in ingestion as a slurry. Thus large doses of the test compounds were given as a slurry in water by stomach tube to fasted infant nonhuman primates. Even so, this massive load of aspartame or aspartame plus MSG did not produce histopathological changes in the hypothalamus.

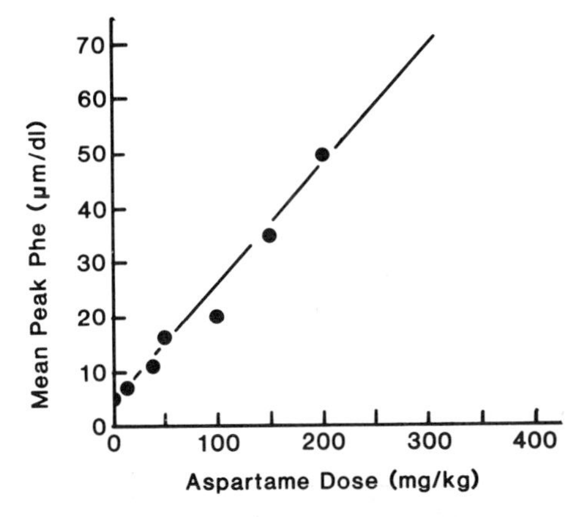

Figure 11 Mean peak plasma phenylalanine (Phe) concentrations (μmol/dl) in human subjects administered aspartame at 10, 34, 50, 100, 150, and 200 mg/kg body weight. (From Refs. 29-31.)

The absence of hypothalamic lesions in these 6 infant monkeys administered aspartame with MSG is consistent with previous studies of 47 infant monkeys administered acute oral loads of glutamate (19). These results differ from studies reported by Olney and colleagues (11,12). The inability of other laboratories (3, 13-19) to replicate the findings reported by Olney and colleagues in MSG-treated monkeys (11,12) implies that the latter investigators may have regarded perfusion artifacts as hypothalamic damage. The central nervous system damage originally reported by Olney et al. (12) following oral ingestion of MSG numbered two to three neurons per section. Damaged cells totaled 50-90 in number and were limited to the rostral subventricular portion of the infundibular nucleus. This contrasts with findings in the rodent, where lateral portions of the hypothalamus sustain the most obvious damage, involving large numbers of neurons per section (4,5). Developmental biologists have long recognized that cell death is an integral part of differentiation in the developing nervous system. During fetal brain maturation, the ependymal cells of the ventricles divide rapidly and produce the dense cellular population of the neuropil. Only a small portion of these cells survive into adulthood as mature neurons. It is believed that neuronal survival is based on successful connection either to peripheral neurons or to adjacent cells. This protective proliferative process of the central nervous system, whereby a plethora of cells is produced, suggests that early reports of MSG-induced neuronal death were observations of neurons passing through predictable cell death, perhaps enhanced by the dehydrated state of some of the infant monkeys studied.

From evolutionary, physiological, and neuroanatomical points of view, the Old World nonhuman primate is the available animal model most closely akin to the human infant. Exhaustive study of serial sections of brains from infant macaques demonstrates that aspartame ingestion at high dosage levels does not cause hypothalamic damage. Furthermore, the ingestion of MSG plus aspartame does not elicit abnormal hypothalamic morphology.

REFERENCES

1. Lucas, D. R., and J. P. Newhouse (1957). The toxic effect of sodium L-glutamate on the inner layers of the retina. *Arch. Ophthalmol.* 58, 193-201.
2. Olney, J. W. (1971). Glutamate-induced neuronal necrosis in the infant mouse hypothalamus. *J. Neuropathol. Exp. Neurol.* 30, 75-90.
3. Abraham, R., Dougherty, W., Golberg, L., and Coulston, F. (1971). The response of the hypothalamus to high doses of monosodium glutamate in mice and monkeys. *Exp. Mol. Pathol.* 15, 43-60.
4. Lemkey-Johnston, N., and Reynolds, W. A. (1974). Nature and extent of brain lesions in mice related to ingestion of monosodium glutamate. *J. Neuropathol. Exp. Neurol.* 33, 74-97.
5. Lemkey-Johnston, N., and Reynolds, W. A. (1972). Incidence and extent of brain lesions in mice following ingestion of monosodium glutamate (MSG). *Anat. Rec.* 172, 354.

6. Olney, J. W., and Ho, O. L. (1970). Brain damage in infant mice following oral intake of glutamate, aspartate or cysteine. *Nature* 227, 609-611.
7. Reynolds, W. A., Butler, V., and Lemkey-Johnston, N. (1976). Hypothalamic morphology following ingestion of aspartame or MSG in the neonatal rodent and primate: A preliminary report. *J. Toxicol. Environ. Health* 2, 471-480.
8. Inouye, M., and Murakami, U. (1973). Brain lesions in mouse infants and fetuses induced by monosodium L-aspartate. *Congenital Anomalies* 13, 235-244.
9. Finkelstein, M. W., Daabees, T. T., Stegink, L. D., and Applebaum, A. E. (1983). Correlation of aspartate dose, plasma dicarboxylic amino acid concentration and neuronal necrosis in infant mice. *Toxicology* 29, 109-119.
10. Applebaum, A. E., Finkelstein, M. W., Daabees, T. T., and Stegink, L. D. (1984). Aspartate-induced neurotoxicity in infant mice. In *Aspartame: Physiology and Biochemistry* (Stegink, L. D., and Filer, L. J., Jr., eds.), Marcel Dekker, New York, Chap. 17.
11. Olney, J. W., and Sharpe, L. G. (1969). Brain lesions in an infant rhesus monkey treated with monosodium glutamate. *Science* 166, 386-388.
12. Olney, J. W., Sharpe, L. G., and Feigin, R. D. (1972). Glutamate-induced brain damage in infant primates. *J. Neuropathol. Exp. Neurol.* 31, 464-488.
13. Reynolds, W. A., Lemkey-Johnston, N., Filer, L. J., Jr., and Pitkin, R. M. (1971). Monosodium glutamate: Absence of hypothalamic lesions after ingestion by newborn primates. *Science* 172, 1342-1344.
14. Newman, A. J., Heywood, R., Palmer, A. K., Barry, D. H., Edwards, F. P., and Worden, A. N. (1973). The administration of monosodium L-glutamate to neonatal and pregnant rhesus monkeys. *Toxicology* 1, 197-204.
15. Wen, C., Hayes, K. C., and Gershoff, S. N. (1973). Effects of dietary supplementation of monosodium glutamate on infant monkeys, weanling rats and suckling mice. *Am. J. Clin. Nutr.* 26, 803-813.
16. Abraham, R., Swart, J., Golberg, L., and Coulston, F. (1975). Electron microscopic observations of hypothalami in neonatal rhesus monkeys (*Macaca mulatta*) after administration of monosodium L-glutamate. *Exp. Mol. Pathol.* 23, 203-213.
17. Stegink, L. D., Reynolds, W. A., Filer, L. J., Jr., Pitkin, R. M., Boaz, D. P., and Brummel, M. C. (1975). Monosodium glutamate metabolism in the neonatal monkey. *Am. J. Physiol.* 229, 246-250.
18. Heywood, R., and James, R. W. (1979). An attempt to induce neurotoxicity in an infant rhesus monkey with monosodium glutamate. *Toxicol. Lett.* 4, 285-286.
19. Reynolds, W. A., Lemkey-Johnston, N., and Stegink, L. D. (1979). Morphology of the fetal monkey hypothalamus after in utero exposure to monosodium glutamate. In *Glutamic Acid: Advances in Biochemistry and Physiology* (Filer, L. J. Jr.. Garattini, S., Kare, M. R., Reynolds, W. A., and Wurtman, R. J., eds.), Raven Press, New York, pp. 217-229.
20. Olney, J. W. (1975). Another view of aspartame. In *Sweeteners, Issues and Uncertainties*, Academy Forum, National Academy of Sciences, Washington D.C., pp. 189-195.

21. Olney, J. W. (1975). L-Glutamic and L-aspartic acids—A question of hazard? *Food Cosmet. Toxicol.* 13, 595-600.
22. Olney, J. W. (1976). Brain damage and oral intake of certain amino acids. In *Transport Phenomena in the Nervous System* (Levi, G., Battistin, L., and Lajtha, A., eds.), Plenum, New York, pp. 497-506.
23. Reynolds, W. A., Stegink, L. D., Filer, L. J. Jr., and Renn, E. (1980). Aspartame administration to the infant monkey: Hypothalamic morphology and plasma amino acid levels. *Anat. Rec.* 198, 73-85.
24. Coates, P. W. (1973). Supraependymal cells: Light and transmission electron microscopy extends scanning electron microscopic demonstration. *Brain Res.* 57, 502-507.
25. Bleier, R. (1971). The relations of ependyma to neurons and capillaries in the hypothalamus: A Golgi-Cox study. *J. Comp. Neurol.* 142, 439-464.
26. Leonhardt, H. (1968). Bukettförmige Strukturen in Ependym der Regio hypothalamica des III. Ventrikels beim Kaninchen. *Z. Zellforsch.* 88, 297-317.
27. Stegink, L. D., Filer, L. J., Jr., Baker, G. L., and Brummel, M. C. (1979). Plasma and erythrocyte amino acid levels of adult humans given 100 mg/kg body weight aspartame. *Toxicology* 14, 131-140.
28. Stegink, L. D., Reynolds, W. A., Filer, L. J., Jr., Baker, G. L., Daabees, T. T., and Pitkin, R. M. (1979). Comparative metabolism of glutamate in the mouse, monkey and man. In *Glutamic Acid: Advances in Biochemistry and Physiology* (Filer, L. J., Jr., Garattini, S., Kare, M. R., Reynolds, W. A., and Wurtman, R. J., eds.), Raven Press, New York, pp. 85-102.
29. Stegink, L. D., Filer, L. J., Jr., and Baker, G. L. (1981). Plasma and erythrocyte concentrations of free amino acids in adult humans administered abuse doses of aspartame. *J. Toxicol. Environ. Health* 7, 291-305.
30. Stegink, L. D., Filer, L. J., Jr., and Baker, G. L. (1979). Plasma, erythrocyte and human milk levels of free amino acids in lactating women administered aspartame or lactose. *J. Nutr.* 109, 2173-2181.
31. Stegink, L. D., Baker, G. L., and Filer, L. J., Jr. (1983). Effect of repeated ingestion of aspartame-sweetened beverages upon plasma aminograms in normal adults. *Am. J. Clin. Nutr.* 36, 704.
32. Stegink, L. D., Filer, L. J., Jr., and Baker, G. L. (1977). Effect of aspartame and aspartate loading upon plasma and erythrocyte free amino acid levels in normal adult volunteers. *J. Nutr.* 107, 1837-1845.
33. National Academy of Sciences, National Research Council (1974). *Recommended Dietary Allowances*, 8th rev. ed., Washington, D.C.
34. National Academy of Sciences, National Research Council (1978). *Nutrient Requirements of Domestic Animals, Nutrient Requirements of Nonhuman Primates* No. 14, Washington D.C.

19

Behavioral Testing in Rodents Given Food Additives

Richard E. Butcher
Western Behavioral Sciences Institute, La Jolla, California

Charles V. Vorhees
Psychoteratology Laboratory, Institute for Developmental Research, Children's Hospital Research Foundation, University of Cincinnati College of Medicine, Cincinnati, Ohio

INTRODUCTION

This chapter is devoted to an examination of the experimental procedures by which food additives have been tested for adverse behavioral effects in immature animals (developmental psychotoxicology) and a review of the results produced by these methods. The inclusion of such a review in this volume is a reflection of recent heightened interest in the effects of environmental agents on behavior, particularly the behavior of children.

Research in this specific area is part of a larger movement to consider functional departures from normality as indices of a toxicological effect and to include developing as well as adult animals in toxicology research programs. To better understand this context and for purposes of a discussion of the developmental psychotoxicology of food additives, two assumptions which underlie the recent enthusiasm for examining toxicity in immature subjects with functional end points should be made more explicit. First, it is generally held that the developing organism is uniquely vulnerable to toxic effects. The immature physiology of the test animal is different in many respects from that of the adult and hence may respond in adverse ways to influences which the mature subject can withstand without permanent injury. The second working hypothesis is the notion that functional examinations are more "sensitive" than are physical examinations, and perhaps more "meaningful" as well. Irrespective of the appearance of a tissue, its ability to perform its adaptive function in the intact organism is the true determinant of its integrity.

The interest in developmental psychotoxicology is a logical outgrowth of these ideas. The behavior of test animals is preeminently a functional measure and is demonstrably influenced by a large number of physical, physiological, and environmental variables. In addition to the currency of these ideas, the recent growth of interest in developmental psychotoxicology has been further encouraged by the reports of Feingold (1), which suggested a link between ingestion of food additives and hyperactivity on children and the series of studies by Fernstrom (2) on the influence of dietary constituents on neurotransmitter levels. Recently, mental retardation has been identified as a sign of the fetal alcohol syndrome (for a review see Refs. 3-5), and mental deficits have been associated with prenatal exposure to anticonvulsants (6).

BACKGROUND: TEST METHODS

In the early 1970s the U.S. Food and Drug Administration (FDA) began a systematic review of food additives on the so-called GRAS (Generally Recognized as Safe) list, using the improved methods of toxicological assessment developed in recent years. As part of the FDA's overall toxicological review of these compounds, and in recognition of the emerging importance of psychotoxicological phenomena, the Bureau of Foods of the FDA sponsored a research program with two objectives. The first was to develop a psychotoxicological test battery suitable for use in the evaluation of food additives in developing laboratory rodents. The second objective was to use the new battery of tests to screen a number of GRAS list food additives and related compounds of interest to the FDA and the toxicological community for their potential for producing developmental psychotoxicity.

The FDA sponsorship took the form of a contract awarded to the Cincinnati Children's Hospital Research Foundation (7). A principal mission of the investigators was the need to design a testing procedure that was comprehensive, sensitive, and usable. There was an acknowledged need to survey the largest number of functions possible and examine aspects of physical, sensory, locomotor, motivational, and learning capacities in the test subjects. Because almost all the additives to be tested were "regarded as safe," all aspects of the investigations had to be measured against the expectation that "positive" results from test agent exposure would be unlikely–this expectation led to the inclusion of positive control groups and the use of large numbers of animals in the studies. It can be seen that a consequence of the need for comprehensiveness and sensitivity was a strong practical demand for a test system that, insofar as possible, would provide clear end points and which could be administered to a large number of animals.

Historically, the origins of the test battery developed with the FDA may be traced to a number of early workers in developmental psychobiology. The general strategy may be traced to Bolles and Woods (8), who, in 1964, published the most comprehensive effort up to that time to map the normal ontogeny of the laboratory rat. In the same year Fox (9) published data on the reflex ontogeny of laboratory mice. By 1968, Irwin (10) had published a wide-range test system to evaluate toxicological effects in mice by examining the behavioral, neurological, and general status of laboratory mice, and in the early 1970s behavioral test batteries of several types were in use. One of the best known of these, offered by Altman and his associates, mapped normal ontogeny (11), as well as the ontogeny of animals exposed to low-protein diets or X-irradiation during early postnatal life (12,13). An equally well-known test battery appeared in the classic work by Smart and Dobbing (14,15) on abnormal reflex and behavioral development in rats subjected to early undernutrition. Neither of these test batteries, however, were used or offered as standardized toxicological screening instruments but, rather, were used as descriptive research devices to document impaired behavioral development in rats subjected to treatments already known to impair brain and behavioral development.

In addition to those researchers engaged in developing wide-range assessment devices, others were involved in the development of specific tasks that might be of special value in developmental psychotoxicology. In this area the most influential early efforts came from Butcher and his associates (16-20) on the use of the Biel water maze (21) as a measure of problem solving in developmentally brain-damaged rats; Vorhees (22) on the use of the Y maze as an improved measure of avoidance and discrimination learning in rats exposed to a prenatal insult; and Shapiro et al. (23), who proposed a new method to evaluate abnormal swimming ontogeny in rats. This list is representative, rather than comprehensive, but illustrates that some limited methodological research was going on. Most of the early research in developmental psychotoxicology, however, was not focused upon the development of new tasks to detect this form of behavioral toxicity but, rather, relied upon the use of already published behavioral methods drawn from the psychopharmacology, physiological psychology, and animal learning fields.

GOVERNMENTAL AND REGULATORY CONTEXT

The single most important influence that affected the planners of the test battery developed under FDA sponsorship was the work of the Committee on Postnatal Evaluation of Animals Subjected to Insult During Development. Although the final report of this committee was not issued until 1977 (24), its work was widely known to those actively engaged in developmental psychotoxicology at the time of its deliberations (1974-1977). The first version of the test battery that we developed for use with food additives was greatly influenced by this committee's findings and by those consulted by the committee for advice. Those consulted from outside the committee were drawn from a wide range of areas in both psychology and teratology and formed the basis of a consensus viewpoint.

The committee concluded its work by proposing a test battery which consisted of the following: (a) assessment of physical landmarks of development, specifically pinna detachment, incisor eruption, eye opening, testis descent, and vaginal patency; (b) assessment of reflex development, specifically righting reflex (surface and air righting), auditory startle, stages of locomotor ontogeny, and visual placing; and (c) assessment of adult behavioral competence, specifically rotorod performance, appetitive position discrimination, discriminative operant performance, avoidance learning, spontaneous locomotor activity (running wheel and open-field types), and aggressive behavior. The match between this proposal and the initial version of the test battery we used was quite close. The early physical landmarks were included, but vaginal patency was not added until later, and testis descent was judged to be too variable to be useful. Among the reflex tests, surface righting was included, but air righting was not included based on a presumed overlap of functions assessed by these two tasks; auditory startle, stages of locomotor development, and visual placing were all included as proposed by the committee. Other early behavioral measures that were added to these to widen the scope of the battery still further were tests of cliff avoidance and pivoting, after the work of Altman and Sudarshan (11), and swimming ontogeny, after the work of Schapiro et al. (23).

For the tests of adult performance, our test battery matched the committee's recommendations closely by the inclusion of rotorod, running wheel, open-field, appetitive position discrimination, and both active and passive avoidance testing. There were some differences, however, in the specific tasks chosen, even though the same types of tasks were included. For example, in our test battery the open-field test of activity and the passive avoidance test of memory/reactivity were conducted as separate tests, using well-established procedures taken from the literature, rather than the committee's unusual suggestion of combining these two tasks into a single test paradigm. For active avoidance the present battery used a wheel-turn variable-interval task rather than the Y maze proposed by the committee. In retrospect, the Y maze may have proven more satisfactory, but at the time the present battery was assembled, the data demonstrating the unique merits of the Y maze were just appearing in print and it was not immediately apparent that this task would translate well into a toxicological assessment setting. Indeed the first experimental evidence using the Y maze as a toxicological assessment technique did not appear until 1974 (22), too late to modify the battery we were assembling. For appetitive position discrimination the committee recommended an appetitive version of the Y maze avoidance test, but since there were no published data on this test as a measure of toxicity, our battery used a double-ended T maze based on the published data of Howard et al. (25). The committee also recommended an S+/S− operant discrimination test and the Lindsey test of aggressive behavior. Neither of these tests were included in the initial test battery we assembled, the first because it was felt to be impractical for between-group study designs, as well as the time required to obtain a reasonable assessment of operant discrimination behavior, and the second because its value in behavioral toxicology was unknown.

These were not the only behaviors the battery did not cover. In terms of measuring complex problem solving behavior, the Biel water maze would have been an obvious choice, but practical constraints kept it out of the battery until later revisions eliminated less useful tests, thereby making an opening in the battery for this test. In addition, the first version did not include assessments of negative geotaxis, olfactory discrimination, social behavior, or schedule-maintained operant behavior not of the S+/S− type. Negative geotaxis and olfactory tests were later added, but social behavior and non-S+/S− operant performance remain absent, primarily owing to the lack of methods in these areas that meet crucial criteria for use in a standardized test battery.

A central issue in the development of a behavioral test battery for use in a regulatory context has been the status of tests administered early in the life of the subject. Such early testing, if found to have validity, has many assets in screening for neurobehavioral impairment in safety evaluation studies. The individual tests are usually rapidly administered, permitting the examination of a fairly large number of animals, and are fairly wide in scope, permitting the testing of several functions. In addition, tests of this sort are, by definition, administered shortly after the birth of the test animals. If these tests are sufficient by themselves, the costs involved in housing subjects for long periods of time would be reduced. It is reasonable to suspect, on both theoretical and empirical grounds, however, that early test results often indicate deficits that are transitory. If early testing has this characteristic, it cannot by itself serve as a safety assessment procedure.

In the mid-1970s, the Japanese, British, and French governments revised their teratology and reproductive premarket safety assessment guidelines and included requirements for the evaluation of behavioral teratogenicity. (Equivalent terms used in the literature for this field are psychoteratogenicity, developmental psychotoxicity, and developmental neurobehavioral toxicity.) These regulatory changes were first applied to pharmaceuticals, but their impact on all of behavioral teratology, particularly in industry, has been substantial. These regulatory actions have, more than any other single event, launched behavioral teratology from the realm of basic research into the arena of large-scale day-to-day use (26). The United State's FDA has yet to issue similar requirements, but the implementation of behavioral guidelines in other countries has unquestionably increased the likelihood that similar regulations will eventually be enacted here. Thus, from the time we began working on the test battery, larger events were occurring which caused the outcome of this endeavor to take on greater significance. For example, in 1980 the Environmental Protection Agency (EPA) discussed the possible use of this test battery as part of the toxicity data they wanted industry to provide on chlorinated benzenes. In addition, several major international pharmaceutical companies have set up developmental psychotoxicity testing programs for their use modeled after this test battery. Thus it is becoming increasingly clear that momentum toward the routine inclusion of developmental psychotoxicity and adult behavioral toxicity assessments of food additives as well as of pharmaceuticals and industrial products is gaining acceptance. As it does, the face of toxicology will be changed in a substantial way.

CINCINNATI TEST BATTERY

The test battery we developed, ultimately named the Cincinnati Psychoteratogenicity Screening Test Battery (27), was never static. It has evolved through nine versions. The key idea, however, is that it did evolve, and as in most evolutionary processes, the changes occurred gradually, in a series of small steps. Moreover, the revisions were always made with several crucial considerations in mind. First, no test was discarded unless it was proven to be insensitive to a known neurotoxic/psychoteratogenic treatment, or unless it produced data that was too variable to hold out much promise that it could be improved to a point where it might ever be useful. Second, tests were not revised unless either a clear improvement in objectivity could be realized or the existing procedures had problems when used in large-scale testing that were not anticipated by their use in the traditional settings from which they were drawn. Third, new tests were not introduced into the battery unless they offered considerable promise of improving the overall sensitivity of the battery.

Several design features were common to all the food additive experiments using the Cincinnati Psychoteratogenicity Screening Test Battery, except for the caffeine experiment, which will be described separately. All of the food additives were administered to Sprague-Dawley rats (Laboratory Supply Co., Indianapolis, Indiana) mixed in Purina rat chow meal. The doses are described as the percentage of the diet supplemented with the test food additive for each treatment group. Dosages, calculated on a gram per kilogram body weight per day basis as determined from food consumption measurements, were obtained on animals in each phase of each experiment. In all of these experiments the additives were administered to adult male and female rats for 2 weeks prior to pairing for breeding. The dietary treatments were then continued throughout the mating period. After conception, males were eliminated and the dietary treatment continued to the dams during gestation and lactation. Dams were eliminated from the experiment at weaning (postpartum day 21). The diets continued directly to the offspring from weaning through to the end of the experiment at approximately 110 days of age. The end of each experiment varied slightly because animals in avoidance testing required variable lengths of time to master each phase of this test.

Uniform procedures in breeding and handling were also followed. Pregnancy in the females was determined by daily vaginal lavages during breeding, each specimen being examined microscopically for the presence of sperm. The day a female was discovered as being sperm positive was counted as day 0 of gestation. Birth was counted as day 0 of postnatal age of the offspring, and examinations and adjustments of the litter were conducted 24 hr after birth.

Litters were examined on postnatal day 1 and litter size was adjusted to 8-12, balancing for equal numbers of males and females in each litter as much as possible. All offspring were examined for external malformations, sexed, and weighed. Dead offspring were also counted. Two males and two females were then randomly selected and marked with indelible ink for preweaning neurobehavioral testing. Weaning occurred at 21 days of postnatal age. At 24 days of age littermates were separated from one another and rehoused in individual cages. The four animals used in preweaning testing and four additional animals not previously tested were kept for postweaning testing. Any additional offspring were then eliminated from the study. Of the eight offspring kept from each litter on day 24, four (two having had preweaning testing and two not having had preweaning testing) were used for postweaning testing. The remaining animals were held in reserve and used only if the designated test animal died or was mistested. These reserve animals were also the ones used for day-90 brain, eye, and body weights. Litters with fewer than eight live offspring on day 1 were eliminated from the study.

The experiments were designed so that two litters per treatment group were enrolled in the study each week. The food additives were tested in sets of two and shared control groups. In all experiments two kinds of control groups were prepared: (a) a negative control group that received Purina rat chow meal unsupplemented by any food additive and (b) a positive control group that received no dietary supplement, but which was injected either prenatally or postnatally with a known neuroembryotoxic drug. This latter group was intended to provide a reference treatment that could be used as a concurrent measure of the sensitivity of the test battery. The value of such an approach has been discussed elsewhere (27). Three experimental groups given different doses of the test food additive were prepared for each study.

Weekly or biweekly body weight measurements were recorded during each phase of the experiment: prebreeding, gestation, lactation (maternal), preweaning (offspring), juvenile, and adult (day 90). Food consumption was also measured on a subset of the animals during these same phases. Regional brain weights and eye weights were taken from selected animals at 90 days of age, and in some experiments neurohistological examinations were conducted on selected offspring at 21 and/or 90 days of age.

Few food additives have received any psychotoxicity testing. Those that we have examined using the Cincinnati Battery and for which adequate data exist to render a reasonable review include aspartame, monosodium glutamate (MSG), calcium carrageenan, butylated hydroxytoluene (BHT), butylated hydroxyanisole (BHA), FD & C Red Dyes Nos. 3 and 40, brominated vegetable oil, sodium nitrite, and caffeine. Among these, only BHT, BHA, FD & C Red Dye No. 3, and caffeine have been investigated by others for their possible psychotoxic effects. Aside from these, the only other food additive evaluated for its psychotoxic effects appears to be FD & C Yellow Dye No. 5 (28).

Details of the offspring evaluations have been published elsewhere. Because the reader may refer to the specific manuscripts for greater detail, this manuscript will discuss only the overall outcome of each experiment, beginning with aspartame. The studies reported below were done on 10 food additives, each with at least three dose levels and a control group, and each treatment group had approximately 15 dams, each producing litters with 8-12 offspring. In turn, the offspring were tested in about 15 tests, most with multiple dependent measures of behavior. Thus the amount of data generated ranged in the hundreds of thousands of measurements, and it is worth remembering that because of this, some type I statistical errors (false positives) are bound to be present among the findings. Given this, what may be most surprising about the data that follow is how few compounds actually produced interpretable patterns of behavioral toxicity.

BEHAVIORAL EFFECTS OF ASPARTAME

One of the initial compounds nominated for study was the artificial sweetener aspartame (APM). The investigation of this compound was unique in two ways: First, APM was the only additive in the study not then approved for use and available through commercial sources. Our supplies of APM were therefore obtained directly from the research departments of G. D. Searle & Co. The second unique feature of the study of APM was the availability of a considerable amount of background literature—not on the compound itself, but on one of its two principal catabolites, phenylalanine (29).

In an attempt to mimic the psychological impairments of phenylketonuric children, Waisman and Harlow (30) and Polidora et al. (31,32) had administered large amounts of phenylalanine to monkeys and rats. Although the resemblance of the physiological and behavioral symptoms of these animals to the conditions present in the human phenylketonuric was incisively questioned in Karrer and Cahilly's (33) review, these studies did permit several conclusions about phenylalanine's effect in test animals. For example, it was clear from these studies that animals will not accept large amounts of phenylalanine in their diets and respond by reducing their food intake. More importantly for the study of APM, the literature indicates that substantial amounts of phenylalanine do impair behavioral test performance.

The specifications for the APM study called only for the administration of the additive at 0, 2, 4, and 6% of the diet. In view of the effects of phenylalanine described in the studies above, however, it seemed necessary to include, in addition to a positive control group, an additional diet group fed 3% phenylalanine to distinguish the unique effects of APM from the previously identified consequences of phenylalanine administration.

In order to provide an idea of the actual dosage received by the different groups, we measured food consumption and body weights throughout the experiment and calculated mean (±SEM) intake of aspartame. The 6% aspartame group dams consumed 5.0 ± 0.2 g of aspartame per kilogram body weight per day during pregnancy and 9.5 ± 0.6 g/kg per day during lactation. Their male offspring consumed 8.5 ± 0.4 g/kg per day, and female offspring 9.2 ± 0.4 g/kg per day during a sampling period in which food consumption was measured between postnatal days 24-42. The 4% aspartame dams consumed 3.2 ± 0.1 g/kg per day during pregnancy and 6.8 ± 0.5 g/kg per day of aspartame during lactation. Their offspring consumed 5.7 ± 0.1 (males) and 5.7 ± 0.2 (females) g/kg per day of aspartame during the postweaning sampling period. Finally, the 2% aspartame dams consumed 1.6 ± 0.1 g/kg per day during pregnancy and 3.8 ± 0.1 g/kg per day of aspartame during lactation. Their offspring consumed 3.2 ± 0.2 (males) and 2.9 ± 0.1 (females) g/kg per day of aspartame during the postweaning sampling period.

The details of this study are reported elsewhere (29). The general procedures were as described above (29) and the postnatal tests are shown in Table 1. Both the 6% APM and 3% phenylalanine animals showed the deleterious effects of these diets in almost every one of the preweaning observations, including measures of growth, physical development, and early behavioral development. There were substantially fewer indications of a toxic effect in the postweaning tests, a somewhat curious result in view of the consistency of the early deficits. This apparent "recovery of function" in older animals will receive further discussion below.

Table 1 Cincinnati Psychoteratogenicity Screening Test Battery used to Evaluate Aspartame in Rats

Preweaning tests[a]	Postweaning tests[a]
Pinna detachment (PN1-crit.)	Running wheel (PN30-51)
Incisor eruption (PN8-crit.)	Open field 36 (PN40-42)
Eye opening (PN10-crit.)	Rotorod (PN60-61)
Surface righting (PN3-crit.)	Wheel-turn active avoidance (PN70-crit.)
Cliff avoidance (PN3-crit.)	Double T-maze appetitive position discrimination (PN65-crit.)
Forward locomotion (PN6-crit.)	Passive avoidance (PN110-112)
Pivoting locomotion (PN7)	
Auditory startle (PN10-crit.)	
Swimming ontogeny (PN6-20 even days)	
Open field 18 (PN15-17)	

[a]Postnatal ages are designated as PN followed by the offspring's age in days; crit. = the criterion behavior being tested for and used as the basis to discontinue further observations for the behavior being studied.

The essential result of the initial APM investigation was a determination that chronic exposure to 6% APM in the diet has a substantial psychotoxicological effect, but that this effect is not distinguishable from treatment with an equivalent amount of phenylalanine (3%). More specifically, the 6% APM and 3% phenylalanine groups both produced a large increase in offspring mortality and reduced maternal and offspring body weight. These groups also exhibited delayed physical and reflex development, including impaired eye opening, surface righting, and swimming development. Both groups were also significantly hypoactive using an open-field test of preweaning spontaneous locomotor activity. The 6% APM groups also showed significantly delayed auditory startle and forward quadripedal locomotor development. The 3% phenylalanine group showed similar trends on these two tests, but they were not as large and not statistically significant. Few effects were seen after weaning. In part, this is certainly due to the high mortality that occurred in these two groups. By 30 days of age 58% of the 6% APM group and 83% of the 3% phenylalanine group had died. This severely limited the number of animals available for postweaning testing in these groups. While one might have expected graded effects at the lower doses of APM, this generally did not occur. While significant delays in swimming development occurred in the 4 and 2% APM groups, this was the only test on which reliable effects occurred in these groups. The other behavioral tests, as well as measures of body weight gain and mortality rates, were all within normal limits at the lower doses of APM. We also performed a smaller study using only the 6% APM group, but employing a cross-fostering design to separate the influence of prenatal from postnatal factors. We found that all of the adverse effects we had found previously in the original 6% APM and 3% phenylalanine groups were present in the group that received APM only after birth; those receiving it only prenatally exhibited no adverse signs.

Based on these observations, the following conclusions seem warranted: (a) At high doses APM is capable of producing adverse effects on growing rats which are indistinguishable from the effects of a diet containing equivalent amounts of L-phenylalanine; (b) The effect of APM on developing rats shows a very steep dose-response curve. This effect is consistent with the phenylketonuria animal model literature, in which it has been demonstrated that symptoms only occur in immature rodents if the dose of phenylalanine is large enough to surpass their liver's phenylalanine-hydroxylating capacity (34). Since the phenylalanine hydroxylase enzyme is still developing postnatally in rats, it is possible to feed young rats more phenylalanine that they can metabolize. This overload then causes serum phenylalanine levels to rise to toxic levels, thought to be analogous to humans born with phenylketonuria, a disease characterized by an inborn deficiency of phenylalanine hydroxylase (34). At slightly lower doses, however, below the threshold of phenylalanine hydroxylase's metabolizing capacity in young rats, no effects are observed. Finally, (c) the cross-fostering experiment demonstrated that no adverse effects of APM, even in high doses, occurred from prenatal exposure. Again this probably relates to the phenylalanine-hydroxylating capacity, since the dam's capacity is substantial. It has been repeatedly demonstrated that feeding phenylalanine alone, even at what appear to be extremely high doses, is insufficient to overload the phenylalanine-hydroxylating capacity of an adult rat to the point of adversely affecting her fetuses (16). This can only be accomplished by simultaneously inhibiting the rats' phenylalanine-metabolizing capacity using a phenylalanine hydroxylase inhibitor (16,34). This explains why no adverse outcome was seen in the APM group treated only prenatally. In conclusion, our data support the concept that phenylalanine overload may be deleterious to developing organisms, regardless of the source of phenylalanine, be it in natural foods or in the artificial sweetener aspartame. But unless doses are very high, the only population that would be at significant risk from an added source of dietary phenylalanine would be individuals with the inherited disease of phenylketonuria.

In 1980 Potts et al. (35) reported the results of another study on the behavioral effects of exposure to aspartame. Potts et al. (35) performed two types of experiments, acute studies using large doses of aspartame (up to 1000 mg/kg) administered by gavage and chronic feeding experiments not unlike those in our experiments reported above. The acute studies were done primarily in mice, and included tests of rotorod coordination, hexabarbital sleeping time, hot plate and tail clip tests of analgesia, electroconvulsive and pentylenetetrazol tests for modification of seizure effects, and an eyelid ptosis test of antidepressant activity using Ro4-1284. One acute test was done in rats (F344 strain), and it involved shuttle-box shock avoidance learning. Potts et al. (35) found no consistent effects of acute APM administration on any of these endpoints.

The chronic dietary studies by Potts et al. (35) used rats, again of the F344 strain, and doses of 4.5 or 9.0% APM. The tests were of photocell locomotor activity, shuttle-box shock avoidance and unsignaled, or so-called Sidman shock avoidance learning. As in our experiment, Potts et al. (35) included groups given phenylalanine in concentrations equivalent to that contained in the phenylalanine moiety of APM. Aspartame feeding in the experiments of Potts et al. (35), however, did not begin until the animals were weanlings. As a result, Potts et al. saw none of the weight gain impairments or increases in mortality we observed (29), even though their high-dose group received more APM than our high-dose group (9 versus 6%, respectively). This is not surprising in view of the previous discussion on the maturation of the hepatic phenylalanine hydroxylase system. Behaviorally, Potts et al. (35) found a few non-dose-dependent increases in locomotor activity in the low-dose phenylalanine and high-dose APM groups, but these are of doubtful significance because they were not consistent. They also found a significant reduction in male shuttle-box avoidance acquisition in the high-dose APM and high-dose phenylalanine groups. This effect appears to be accounted for by the effect of the phenylalanine in the diet, since the effect was no larger in the APM group than in the phenylalanine group. No effect was observed on this task among females, due in part at least, to their greater variability in acquiring this response. Finally no overall effect was seen in either sex in any group in the Sidman avoidance test. The high-dose male phenylalanine group showed a significant increase in shocks received at the 2-hr test point out of 6 hr of testing, but this effect was not confirmed in either of the APM groups.

Overall, our data (29) and those of Potts et al. (35) converge in finding little psychotoxicity in rodents from aspartame exposure, except perhaps at very high doses. Moreover, the psychotoxicity that is seen at high doses may be fully accounted for by the known psychotoxicity of equivalent doses of L-phenylalanine.

BEHAVIORAL EFFECTS OF OTHER FOOD ADDITIVES

We now turn to a discussion, in more abbreviated form, of studies on the developmental psychotoxic effects of other food additives in rats. We will begin with a discussion of the food additives monosodium glutamate (MSG) and calcium carrageenan (CC) (26).

The MSG groups received diets containing 1.7, 3.4, or 5.1% MSG, and the CC group received diets containing 0.45, 0.9, or 1.8% CC. The positive control group (C+) was injected on day 12 of gestation with 550 mg/kg hydroxyurea. Prior to breeding, the 5.1% MSG group had an average MSG intake of 5.15 g/kg per day, the 3.4% MSG group averaged 3.45 g/kg per day, and the 1.7% MSG group averaged 1.85 g/kg per day for both males and females. During the same period, the 1.8% CC group averaged 1.9 g/kg per day, the 0.9% CC group 0.9 g/kg per day, and the 0.45% CC group 0.45 g/kg per day.

Neither MSG nor CC produced any significant effects on paternal or maternal weight, food consumption, or reproductive performance. A small but significant increase in offspring mortality was seen in both the MSG and CC mid-dose groups. Neither food additive significantly reduced offspring weight gain or food consumption. Monosodium glutamate at all doses and the 0.9% CC group showed a significant delay in incisor eruption, but no effects were seen on pinna detachment or eye-opening development. Monosodium glutamate had no effect on righting, cliff avoidance, auditory startle, visual placing or locomotor development. Calcium carrageenan had no effect on most of these tests, but the 1.8% CC group showed a significant delay in the appearance of pivoting locomotion, and the 0.9% CC group a significant acceleration in the appearance of pivoting. On swimming development, several significant effects were observed in both MSG and CC groups, but they were not dose dependent and were split between being delays and accelerations of swimming competence, suggesting that these differences were not biologically significant. Calcium carrageenan had no significant effects on any of the postweaning behavioral tests. The only MSG group that showed postweaning behavioral effects that were noteworthy occurred in the 5.1% MSG group, which showed decreased open-field rearing, slower active avoidance acquisition, and longer 48-hr passive avoidance response latencies than negative controls. Neither MSG nor CC reduced regional brain weights, eye weights, H and E neuronal cell counts, or Golgi-Cox dendritic spine counts. The positive control group showed several clear effects, including delayed startle and swimming development and reduced open-field activity.

Based on the outcome of this experiment it was possible to conclude the CC produced no evidence of developmental psychotoxicity. The evidence concerning MSG was less clear. The high-dose MSG group showed several significant effects, but they were generally small. The mid- and low-dose MSG groups showed no consistent pattern of effects. We concluded, therefore, that there was no evidence of developmental psychotoxicity from dietary MSG at doses of 3.4% of the diet or below. This conclusion was disputed by Olney (36). As we noted, both in this article (26) and later (37), our data were consistent with the evidence of Takasaki (38,39), that the route of administration produces a major shift in the toxicity of MSG. Bolus intraperitoneal, subcutaneous, or intravenous injections of MSG can be quite neurotoxic in developing rodents, particularly in mice (40), but is many times less toxic when administered continuously in the diet (see discussion in Ref. 26).

The next food additive we evaluated was the antitoxidant butylated hydroxytoluene (BHT) (41). This compound was administered in the diet in concentrations of 0.125, 0.25, and 0.5%. The positive control procedure was again hydroxyurea injected on day 12 of gestation in a dose of 550 mg/kg. Butylated hydroxytoluene proved to be quite toxic to developing rats at these doses. The high-dose BHT group exhibited significantly reduced maternal and offspring body weights, this effect being most pronounced during early postnatal development. Offspring mortality was 39% up to day 30 of age. This dose also delayed eye opening, surface righting, swimming development, and reduced preweaning open-field activity. However, this dose produced no significant effects on postweaning physical or behavioral development. The mid-dose group exhibited significant mortality up to day 30 of 23%, while the low-dose group exhibited no significant increase in mortality. Neither the mid- nor low-dose of BHT had any significant deleterious effects on pre- or postweaning behavioral development. The positive control group again exhibited considerable evidence of toxicity. This group showed delayed eye opening, locomotor development, swimming development, and reduced growth and day-90 brain weights.

Based on these data we were able to conclude that BHT exhibited significant developmental toxicity in rats at doses of 0.25-0.5% of the diet, with marginal effects appearing at 0.125%. The behavioral data expanded the picture of BHT's toxicity by demonstrating the functional consequences of its effects, and in that way added significant information to the standard developmental toxicological profile. Moreover, the data indicated that the central nervous system was not the primary target organ of BHT's effects (41).

As a result of the findings with BHT, it seemed warranted to also evaluate the closely related antioxidant food additive butylated hydroxyanisole (BHA) (42).

The same doses were used in this experiment with BHA as were used in the BHT experiment, that is, doses of 0.125, 0.25, and 0.5% of the diet. Hydroxyurea was again used as the positive control treatment, but this time it was administered directly to the offspring by daily subcutaneous injections on days 2-10 of life at a dose of 50 mg/kg per day.

The salient finding from this experiment was that BHA was significantly less toxic to developing rats than BHT. Specifically, the high-dose BHA group showed significantly increased mortality; 13.5% of the offspring in this group died between day 1 and 30 of life. However, this increase in mortality was substantially less than the 39% mortality for the same period observed with BHT administration at the same dose. A marginal increase ($p<0.06$) in offspring mortality occurred in the mid-dose BHA group (8.3%), but no noteworthy increase occurred in the low-dose BHA group. Behaviorally, the high- and mid-dose BHA groups exhibited significantly delayed auditory startle development, and nonsignificant but suggestive trend toward increased diurnal running wheel activity.

We concluded from this experiment that BHA is less developmentally toxic than BHT, and that while its toxic effects clearly produce functionally significant changes in the offspring, BHA is not a potent central nervous system toxin (42).

We have recently reported the results of three experiments on food dyes (43, 44). Two of the experiments were on FD & C Red Dye No. 3 (Red 3), and the other was on FD & C Red Dye No. 40 (Red 40). In both experiments using Red 3 the dye was administered at doses of 0.25, 0.5, and 1.0% of the diet. In the Red 40 experiment the doses were 2.5, 5.0, and 10.0% of the diet. For the first Red 3 experiment and the Red 40 experiment the positive control groups were treated with 50 mg/kg per day of hydroxyurea administered on days 2-10 of postnatal life as in the BHA experiment. No positive control group was included in the second Red 3 experiment, which was done with an aim toward resolving some non-dose-dependent effects observed in the first experiment with Red 3.

Red 3 (43) produced no reductions in parental or offspring body weight or food consumption. However, the high- and mid-dose groups in the first Red 3 experiment showed increased preweaning mortality. This effect was not observed in the second Red 3 experiment. Behaviorally, Red 3 significantly increased open-field activity (first experiment only) and had mixed effects on swimming development (both experiments). We concluded from the two Red 3 experiments that this food coloring produced no consistent evidence of developmental toxicity or psychotoxicity.

Red 40 (44) produced a clearer picture of developmental toxicity. Red 40 impaired reproductive success rates among breeding pairs, reduced parent and offspring body weight, reduced offspring brain weight, increased mortality (primarily in the high-dose group), and delayed the onset of vaginal patency in the female offspring. Behaviorally, Red 40 produced significantly decreased running wheel activity and increased open-field rearing behavior. The evidence with Red 40 indicated that this additive was both a general developmental toxin and a behavioral toxin. It is important to bear in mind, however, that these clear toxic effects from Red 40 exposure occurred at doses that were 10 times those used in the two Red 3 experiments; therefore it may not be too surprising that overt toxicity was observed in the case of Red 40.

The next food additive investigated was one of a class of additives collectively referred to as the brominated vegetable oils (BVOs) (45,46). These compounds are thought to be relatively similar; therefore only one was evaluated, with the intent that it would be representative of the group. For this purpose brominated soybean oil was chosen. The doses of BVO used were 0, 0.25, 0.5, 1.0, and 2.0% of the diet. The C+ treatment was 50 mg/kg per day of hydroxyurea administered to the offspring by subcutaneous injection on days 2-10 of life. The 2.0% BVO dose completely blocked reproduction, making this group unavailable for postnatal evaluation. At 1.0% of the diet, BVO reduced reproductive success (percentage of bred females littering) by more than 50%. The two lower doses also reduced littering rates significantly, but not as severely, both of these doses having reduced littering rates by 25%. The 1.0% BVO group also increased the number of small litters delivered and severely impaired offspring viability. Only 26% of the offspring in this group lived beyond 24 days of postnatal life. The 0.5% BVO group also increased offspring mortality rates, but the effect was much less severe than in the 1.0% BVO group, amounting to 13.3% of the offspring having died by postnatal day 24. At 0.25% of the diet, BVO had no significant effect on offspring viability; BVO significantly impaired offspring body growth as reflected by weight, and this effect was dose dependent. Behaviorally, BVO at 1.0% of the diet impaired the development of all of the early behaviors tested, save only the righting reflex. At 0.5% BVO significantly delayed all of the early tests except righting and pivoting; BVO at 0.25% of the diet significantly delayed the development of pivoting, negative geotaxis, normal open-field ambulation, and swimming. There were not enough surviving offspring in the 1.0% BVO group to test after weaning. Among the remaining groups, both showed significantly reduced running wheel activity and delayed onset of vaginal patency. On many of these tests BVO showed evidence of greater toxicity than was seen in the C+ group.

We concluded that BVO is a very toxic food additive, and we suggested that safe levels of this additive should undoubtedly be set very low. We also concluded that the inclusion of measures of behavioral capacities altered assumptions about the dose-response curve of this food additive. We found evidence of neurobehavioral toxicity at doses heretofore thought to be at or below the no-effect level. This conclusion was valid even though the central nervous system appears not to be the primary target organ for this additive's toxic effects, indicating that functional measures may also be informative indicators of non-central nervous system perturbations. The data suggested that germ cells and rapidly dividing cells are the most susceptible to the toxic effects of BVO, and that behavioral measures appear to be a sensitive method for detecting some of these effects (45).

The final food additive we have reported thus far using this test battery is sodium nitrite ($NaNO_2$) (47). This additive was administered in the diet in concentrations of 0, 0.0125, 0.025, and 0.05% of the diet. The C+ group was treated with 4 mg/kg of 5-azacytidine on day-16 of gestation. Sodium nitrite produced no significant reductions in parental weight or food consumption, nor did it adversely affect reproductive performance. Offspring mortality and weight gain were significantly reduced in the two highest dose groups during the preweaning period, but not thereafter. Sodium nitrite at the two highest doses also significantly delayed swimming development. The mid-dose sodium nitrite group also reduced open-field activity and 90-day eye weights. The C+ group exhibited marked evidence of developmental toxicity, including reduced litter size, increased offspring mortality and pivoting, and delayed startle and swimming development. Postnatally, the C+ group exhibited altered open-field performance, delayed vaginal patency development, and reduced active avoidance acquisition.

From the data we concluded that sodium nitrite showed signs of moderate developmental toxicity which included manifestations involving altered behavioral performance. Nevertheless, the developing central nervous system did not appear to be the primary target organ of sodium nitrite's toxicity (47).

Using essentially the same test battery, but a route of additive administration through the drinking water rather than the diet, we also investigated the developmental toxicity and psychotoxicity of caffeine (48). In this experiment caffeine was administered in one of two forms, at one of two dose levels. Rats received water with either 0.056 or 0.014% caffeine dissolved in it, or brewed coffee at a strength containing essentially the same amounts of caffeine. A C+ group was administered 40,000 IU/kg per day of vitamin A by gavage on days 7-20 of gestation. The caffeine and coffee treatments began to the parent animals 60 days prior to breeding and continued through gestation and lactation to the females. Treatment was discontinued at weaning (postnatal day 21).

All caffeine and coffee groups consumed significantly more fluid during pregnancy than controls, but the two high-dose (the high-dose caffeine and the high-dose coffee) groups weighed significantly less during pregnancy than did the controls as did their offspring. No reproductive abnormalities or malformations were found. Several tests revealed evidence of delayed development. These included delayed incisor eruption and swimming development, and increased open-field activity at the high doses. After weaning the high-dose groups showed effects on only one test, the open field, showing decreased activity and increased defecation frequency. The two low-dose groups did not show much in the way of behavioral or physical developmental effects. The C+ group showed delayed incisor eruption, increased preweaning and postweaning open-field activity, and decreased running wheel circadian activity rhythms.

We concluded, based on our data and those of others (49), that most of the neonatal effects we observed from caffeine exposure were consistent with the expected effects of concurrent administration of this agent, and that we found no persuasive evidence of long-term developmental toxicity or psychotoxicity (48).

In 1977 Sobotka et al. (28) published data on the developmental toxic and psychotoxic potential of dietarily administered FD & C Yellow Dye No. 5 (tartrazine). Pregnant rats were fed diets containing 0, 1, and 2% tartrazine throughout gestation and lactation, and then the dye was continued in the diets of the offspring for 90 days after weaning. Prior to weaning the offspring were evaluated for righting, clinging, startle, placing, eye opening, and locomotor activity. After weaning animals were evaluated behaviorally on tests of activity and acquisition of an avoidance response. In addition, some animals after weaning were used to assay brain protein, DNA content, and cholesterol. Tartrazine at these dietary levels did not affect viability, physical milestones, or behavioral indices of development, save a small increase in the clinging in females in the tartrazine groups. No postweaning effects were noted. Brain weights and protein, DNA, and cholesterol measurements, as well as other organ weights, revealed no effects from tartrazine exposure. Tartrazine, did however, significantly depress body weights, thymus weights, and slightly increased red blood cell counts and hemoglobin. Overall, tartrazine did not appear to be a developmental toxin, and certainly did not appear to be a psychotoxin.

SUMMARY OF RESULTS

The results of all of our experiments on the behavioral effects of food additives reviewed above are summarized in Table 2. An overview of the functional testing of these compounds indicates an absence of strong developmental psychotoxic effects, which is consistent with the status of these additives as GRAS list substances. The noteworthy exception may be BVO. Another consistent element in the repeated use of the Cincinnati Test Battery with food additives is the occurrence of abnormalities in preweaning testing which are not evident in tests administered after weaning. This phenomenon will be discussed later.

CONCLUSIONS AND IMPLICATIONS

The findings decribed in the previous section resulted from the continuous exposure of rats to moderate to high levels of additives mixed into the diet. Given the nature of the materials tested and the expected pattern of human use, this exposure pattern seems reasonable. In the context of developmental psychotoxicity, however, this technique places some limitations on the interpretation of findings, and each of the elements in this exposure procedure warrants a discussion of its possible relationship to the results obtained.

First, the inclusion of additives in the diet obviously assumes the appropriateness of the oral route of administration. While this assumption with food additives seems warranted, it is noteworthy that at least one of these additives, monosodium glutamate, apparently has different toxicological consequences when administered without food (38,39).

Table 2 Summary of Behavioral Effects of Food Additives

Measure	Aspartame	MSG	Carrageenan	BHT
Dose (% of diet)	2,4,6	1.7,3.4, 5.1	0.45,0.9, 1.8	0.125,0.25, 0.5
Preweaning tests				
Physical milestones	↓eye at 6[d]	↓incisors all doses	↓incisors all doses	↓eye at 0.5
Surface righting	↓at 6			↓at 0.5
Cliff avoidance				
Forward locomotion	↓at 6			
Negative geotaxis				
Pivoting			↓at 1.8 and 0.9	
Auditory startle	↓at 6			
Swimming	↓at 6		↑↓[c]	↓at 0.5
Open field 18	↓at 6			↓at 0.5
Postweaning tests				
Running wheel				
Open field 36		↓at 5.1		
Rotorod				
Active avoidance		↓at 5.1		
Double T maze				
Passive avoidance latency		↑at 5.1		

[a]Blocked reproduction; therefore no litters were obtained at this dose.
[b]Owing to a testing error, no data were available at the 1% BVO dose level on this one test.
[c]Statistically significant, but not consistent and therefore not considered reliable.
[d]Change noted at the listed % of compound present in the diet.

The adulteration of a test animal's normal food supply with a surplus dietary constituent, no matter how agreeable the substance might seem, often produces a significant reduction in the gross food intake of the subject (diet rejection). This phenomenon is a well-known consequence of feeding animals diets that are toxic, and appears with regularity in feeding phenylalanine to animals in an attempt to produce an animal model of phenylketonuria (34). The usual control manipulation for this effect is that of "pair feeding" a group of control animals with quantities of normal diet matched in quantity to that consumed each day by the experimental diet group.

In our use of food additives, this phenomenon has been observed several times, and, inasmuch as the protocol used did not include provision for a pair-fed control, the degree to which results are attributable to the animal's self-imposed partial starvation is uncertain. In most cases it is known that the tests are not

BHA	Red 3	Red 40	BVO	Sodium nitrite	Caffeine
0.125,0.25, 0.5	0.25,0.5, 1.0	2.5,5,10	0.25,0.5 1.0,2.0[a]	0.0125,0.025, 0.05	0.014,0.056
			↓patency all doses		↓incisors at 0.056
			↓at 0.5 and 1.0		
			↓all doses		
			↓all doses		
			↓at 0.5[b]		
↓at 0.25 and 0.5		↓at 2.5	↓all doses	↓at 0.025 and 0.05	↓at 0.056
	↑↓[c]				
			↓all doses	↓at 0.025	↑at 0.056
		↓all doses	↓all doses		
		↑at 5 and 10			↓at 0.056
	↑_[c]				
		↓at 2.5			
	↑_[c]				

influenced much by physical size alone. The tests were chosen, in fact, with this attribute in mind.

It may be that the most significant element of the dietary route technique is the continuous nature of the exposure. In these experiments the exposure period was continued throughout all of the tests and examinations. Where results indicating an impairment are found, therefore, no sure determination can be made of whether the deficit is the result of a permanent defect induced by the test agent or a temporary reduction in the subject's ability to perform which is induced by the immediate effects of the experimental diet.

The general finding in the studies reviewed above, early developmental deficits which are not followed by indications of impairment in adult testing may be related to the length of treatment with the test article. This pattern of behavioral results is consistent with a "recovery of function" in these animals in which there is gradual functional compensation for a handicap incurred in early life. This apparent "recovery," observed rather frequently in our data, stimulated a comparison of the statistical sensitivity of our early- and later-age tests (46). In view of the importance of this phenomenon in the interpretation of investigations directed at safety evaluation (see above), a wider examination of the developmental psychotoxicity literature relevant to this topic seemed justified. Therefore we attempted a complete survey of the developmental psychotoxicology literature to determine the extent to which the results of juvenile testing are reliable predictors of lifelong impairment. In this view, 30 studies were identified in which comparisons between pre- and postweaning results were possible (22, 29,50-77). The review of reports in which testing of both juvenile and adult animals was administered indicated that only a very few of these reports used parallel or identical test procedures at the several ages tested. Any reported indication of developmental delay in the early period (<30 days) was therefore compared and contrasted to results from later testing, irrespective of the modality or function being examined. Thus the question was asked, "Did the observation of an early deficit predict *any* subsequent change in behavioral test results?" Surprisingly, in view of our results, only 9 of the 30 reports surveyed reported developmental defects which were unconfirmed in later testing. The majority of the studies report evidence of behavioral abnormality which is consistent in early and adult testing.

This literature can be interpreted with reference to the experimental literature on recovery of function. This research area offers several principal elements to be considered in studying recovery: (a) the age (in terms of neurophysiological and neuroanatomical development) at which the insult is experienced, (b) the magnitude of the lesion and the region of the central nervous system that is damaged, (c) the time allowed for recovery (the time between insult and observation), (d) the function(s) being tested, and (e) the availability of specific tests for latent deficits (for a review see Ref. 77). These considerations are strikingly similar to those applied by the developmental psychotoxicologist. Each of the variables has its analog in developmental toxicology experimentation, and indeed, such investigations could easily be redefined as recovery of function research with the lesion applied very early in brain development.

A final element to be added to the list of variables stated above is the rate at which the brain injury occurs—which has been described by Finger (77) as lesion "staging" (78). Although there are inconsistencies in the literature, research in the staging of brain lesioning suggests (all other things being equal, including the ultimate size of the lesion) that recovery from serial lesions is more complete than that from a single, traumatic event, and, furthermore, that the longer the interval between the lesion stages, the slower the momentum and the more complete the recovery.

If it may be assumed that a prolonged period of toxicant administration is like slow lesion momentum; the consideration of the staging factor permits an intriguing reinterpretation of the literature surveyed. It is noteworthy that all but one of the reports in which early deficits were not confirmed in adult testing utilized extended exposure periods, the majority using continuous exposure and in the others the shortest exposure period used being 13 days. It is tempting to speculate that continuous exposure has an effect akin to a many-staged lesion, and, accordingly, a more complete restoration of normal function is possible.

In a regulatory context this interpretation, if found to have general application, implies that the exposure period for a given test compound should not be prolonged beyond that which is consistent with the need to reflect human exposure patterns and other regulatory considerations.

ACKNOWLEDGMENTS

The work done here was supported in part by a research contract from the Bureau of Foods of the FDA (223-75-2030), a contract from the Bureau of Drugs of the FDA (223-76-3026), and for the study on caffeine by a contract from the National Coffee Association. These sources of support and support received from the Children's Hospital Research Foundation of Cincinnati and the Western Behavioral Sciences Institute are gratefully acknowledged.

REFERENCES

1. Feingold, B. F. (1975). Behavioral disturbances, learning disabilities, and food additives. *Chemtech*, 7, 264-267.
2. Fernstrom, J. D. (1976). The effect of nutritional factors on brain amino acid levels and monoamine synthesis. *Fed. Proc.* 35, 1151-1156.
3. Abel, E. L. (1980). Fetal alcohol syndrome: Behavioral teratology. *Psychol. Bull.* 87, 29-50.
4. Abel, E. L. (1981). Behavioral teratology of alcohol. *Psychol. Bull.* 90, 564-581.
5. Streissguth, A. P., Landsman-Dwyer, S., Martin, J. C., and Smith, D. W. (1980). Teratogenic effects of alcohol in humans and laboratory animals. *Science* 209, 353-361.
6. Smith, D. W. (1977). Teratogenicity of anticonvulsive medications. *Am. J. Dis. Child.* 131, 1337-1339.
7. Butcher, R. E., Vorhees, C. V., Wootten, V., Brunner, R. L., and Sobotka, T. J. (1980). A survey of early tests for the developmental psychotoxicity of food additives and related compounds. In *Effects of Food and Drugs on the Development and Function of the Nervous System: Methods for Predicting Toxicity* (Gryder, R. M., and Frankos, V. H., eds.), Office of Health Affairs, Food and Drug Administration, Department of Health and Human Services, Washington, D.C., pp. 62-69.
8. Bolles, R. C., and Woods, P. J. (1964). The ontogeny of behavior in the albino rat. *Anim. Behav.* 12, 427-441.
9. Fox, W. M. (1964). Reflex-ontogeny and behavioral development of the mouse. *Anim. Behav.* 13, 234-241.

10. Irwin, S. (1968). Comprehensive observational assessment: Ia. A systematic, quantitative procedure for assessing the behavioral and physiologic state of the mouse. *Psychopharmacologia* 13, 222-257.
11. Altman, J., and Sudarshan, K. (1975). Postnatal development of locomotion in the laboratory rat. *Anim. Behav.* 23, 896-920.
12. Altman, J., Sudarshan, K., Das, G. D., McCormich, N., and Barnes, D. (1971). The influence of nutrition on neural and behavioral development: III. Development of some motor, particularly locomotor patterns during infancy. *Dev. Psychobiol.* 4, 97-114.
13. Altman, J., Brunner, R. L., Belut, F. G., and Sudarshan, K. (1974). The development of behavior in normal and brain-damaged infant rats, studied with homing (nest-seeking) as motivation. In *Drugs and the Developing Brain* (Vernadakis, A., and Weiner, N., eds.), Plenum, New York, pp. 321-248.
14. Smart, J. L. (1971). Long-lasting effects of early nutritional deprivation on the behavior of rodents. *Psychiatr. Neurol. Neurochem.* 74, 443-452.
15. Smart, J. L., and Dobbing, J. (1971). Vulnerability of developing brain. II. Effects of early nutritional deprivation on reflex ontogeny and development of behavior in the rat. *Brain Res.* 28, 85-95.
16. Butcher, R. E. (1970). Learning impairment associated with maternal phenylketonuria in rats. *Nature* 226, 555-556.
17. Butcher, R. E., Brunner, R. L., Roth, T., and Kimmel, C. A. (1972). A learning impairment associated with maternal hypervitaminosis-A in rats. *Life Sci.* 11, 141-145.
18. Butcher, R. E. (1976). Behavioral testing as a method for assessing risk. *Environ. Health Perspect.* 18, 75-78.
19. Butcher, R. E., Vorhees, C. V., and Kimmel, C. A. (1972). Learning impairment from maternal salicylate treatment in rats. *Nature New Biol.* 236, 211-212.
20. Vorhees, C. V., Brunner, R. L., McDaniel, C. R., and Butcher, R. E. (1978). The relationship of gestation age to vitamin A induced postnatal dysfunction. *Teratology* 17, 271-276.
21. Biel, W. C. (1940). Early age differences in the maze performance of the albino rat. *J. Genet. Psychol.* 56, 439-453.
22. Vorhees, C. V. (1974). Some behavioral effects of maternal hypervitaminosis A in rats. *Teratology* 10, 269-274.
23. Schapiro, S., Salas, M., and Vukovich, K. (1970). Hormonal effects on ontogeny of swimming ability in the rat: Assessment of central nervous system development. *Science* 168, 147-151.
24. Kimmel, C. A., Brunner, R. L., Butcher, R. E., Scott, W. J., Wilson, J. G., Rodier, P. M., Grant, L. D., Routh, D. K., Damstra, T., Falk, H. L., and Staples, R. E. (1977). Final report of the Committee on Postnatal Evaluation of Animals Subjected to Insult During Development. Report to the National Institute of Environmental Health Sciences, Research Triangle Park, North Carolina.
25. Howard, J., Grant, L. D., and Breese, G. R. (1974). Effects of intracisternal 6-hydroxydopamine treatment on acquisition and performance of rats in a double T-maze. *J. Comp. Physiol. Psychol.* 86, 995-1007.

26. Vorhees, C. V., and Butcher, R. E. (1982). Behavioral teratogenicity. In *Developmental Toxicology* (Snell, K., ed.), Praeger, New York, pp. 249-297.
27. Vorhees, C. V., Butcher, R. E., Brunner, R. L., and Sobotka, T. J. (1979). A developmental test battery for neurobehavioral toxicity in rats: A preliminary analysis using monosodium glutamate, calcium carrageenan and hydroxyurea. *Toxicol. Appl. Pharmacol.* 50, 267-282.
28. Sobotka, T. J., Brodie, R. E., and Spaid, S. L. (1977). Tartrazine and the developing nervous system of rats. *J. Toxicol. Environ. Health* 2, 1211-1220.
29. Brunner, R. L., Vorhees, C. V., Kinney, L., and Butcher, R. E. (1979). Aspartame: Assessment of developmental psychotoxicity in a new artificial sweetener. *Neurobehav. Toxicol.* 1, 79-86.
30. Waisman, H. A., and Harlow, H. F. (1965). Experimental phenylketonuria in infant monkeys. *Science* 147, 685-695.
31. Polidora, V. J., Cummingham, R. F., and Waisman, H. A. (1966). Dosage parameters of a behavioral deficit associated with phenylketonuria in rats. *J. Comp. Physiol. Psychol.* 61, 436-441.
32. Polidora, V. J., Cunningham, R. F., and Waisman, H. A. (1966). Phenylketonuria in rats: Reversibility of behavioral deficit. *Science* 151, 219-221.
33. Karrer, R., and Cahilly, G. (1965). Experimental attempts to produce phenylketonuria in animals: A critical review. *Psychol. Bull.* 64, 52-64.
34. Vorhees, C. V., Butcher, R. E., and Berry, H. K. (1981). Progress in experimental phenylketonuria: A critical review. *Neurosci. Biobehav. Rev.* 5, 177-190.
35. Potts, W. J., Bloss, J. L., and Nutting, E. F. (1980). Biological properties of aspartame. I. Evaluation of central nervous system effects. *J. Environ. Pathol. Toxicol.* 3, 341-353.
36. Olney, J. W. (1980). Letter to the editor on Vorhees et al. study: A developmental test battery for neurobehavioral toxicity in rats: A preliminary analysis using monosodium glutamate, calcium carrageenan, and hydroxyurea. *Toxicol. Appl. Pharmacol.* 53, 177.
37. Vorhees, C. V., Butcher, R. E., Brunner, R. L., and Sobotka, T. J. (1980). Letter to the editor: A reply to J. W. Olney's letter on Vorhees et al. study: A developmental test battery for neurobehavioral toxicity in rats: A preliminary analysis using monosodium glutamate, calcium carrageenan, and hydroxyurea. *Toxicol. Appl. Pharmacol.* 53, 177-178.
38. Takasaki, Y. (1978). Studies on brain lesion by administration of monosodium L-glutamate to mice. I. Brain lesions in infant mice caused by administration of monosodium glutamate. *Toxicology* 9, 293-305.
39. Takasaki, Y. (1978). Studies on brain lesion by administration of monosodium L-glutamate to mice. II. Absence of brain damage following administration of monosodium L-glutamate in the diet. *Toxicology* 9, 307-318.
40. Pizzi, W. J., and Barnhart, J. E. (1976). Effects of monosodium glutamate on somatic development, obesity and activity in the mouse. *Pharmacol. Biochem. Behav.* 5, 551-557.
41. Vorhees, C. V., Butcher, R. E., Brunner, R. L., and Sobotka, T. J. (1981). Developmental neurobehavioral toxicity of butylated hydroxytoluene in rats. *Food Cosmet. Toxicol.* 19, 153-162.

42. Vorhees, C. V., Butcher, R. E., Brunner, R. L., Wootten, V., and Sobotka, T. J. (1981). Developmental neurobehavioral toxicity of butylated hydroxyanisole (BHA) in rats. *Neurobehav. Toxicol. Teratol.* 3, 321-329.
43. Vorhees, C. V., Butcher, R. E., Brunner, R. L., Wootten, V., and Sobotka, T. J. (1983). A developmental toxicity and psychotoxicity evaluation of FD & C red dye No. 3 (erythrosine) in rats. *Arch. Toxicol.* 53, 253-264.
44. Vorhees, C. V., Butcher, R. E., Brunner, R. L., Wootten, V., and Sobotka, T. J. (1983). Developmental toxicity and psychotoxicity of FD & C red dye No. 40 (allura red AC) in rats. *Toxicology* 28, 207-217.
45. Vorhees, C. V., Butcher, R. E., Wootten, C., and Brunner, R. T. (1983). Behavioral and reproductive effects of chronic developmental exposure to brominated vegetable oil in rats. *Teratology* 28, 309-318.
46. Butcher, R. E., Wootten, V., and Vorhees, C. V. (1980). Standards in behavioral teratology testing: Test variability and sensitivity. *Teratogen. Carcinogen. Mutagen.* 1, 49-61.
47. Vorhees, C. V., Butcher, R. E., Brunner, R. L., and Wootten, C. (1984). Developmental toxicity and psychotoxicity of sodium nitrite in rats. *Fd. Chem. Toxicol.* 22, 1-6.
48. Butcher, R. E., Vorhees, C. V., and Wootten, V. (1984). Behavioral and physical development of rats chronically exposed to caffeinated fluids. *Fund. Appl. Toxicol.* 4, 1-13.
49. Sobotka, T. J., Spaid, S. L., and Brodie, R. E. (1979). Neurobehavioral teratology of caffeine exposure in rats. *Neurotoxicology* 1, 403-416.
50. Koeter, H. W. B. M., and Rodier, P. M. (1981). Functional development after pre- and postnatal exposure to inhalant anesthetics. *Teratology* 24, 56A.
51. Ordy, J. M., Samorajski, T., Collins, R. L., and Rolsten, C. (1966). Prenatal chlorpromazine effects on liver, survival and behavior of mice offspring. *J. Pharmacol. Exp. Ther.* 151, 110-125.
52. Bornschein, R. L., Hastings, L., and Manson, J. M. (1980). Behavioral toxicity in the offspring of rats following maternal exposure to dichloromethane. *Toxicol. Appl. Pharmacol.* 52, 29-37.
53. Spyker, J. M., and Avery, D. L. (1977). Neurobehavioral effects of prenatal exposure to the organophosphate Diazinon in mice. *J. Toxicol. Environ. Health* 3, 989-1002.
54. Coyle, I. R. (1975). Changes in developing behavior following prenatal administration of imipramine. *Pharmacol. Biochem. Behav.* 3, 799-807.
55. Olson, K., and Boush, G. M. (1975). Decreased learning capacity in rats exposed prenatally and postnatally to low doses of mercury. *Bull. Environ. Contam. Toxicol.* 13, 73-79.
56. Werboff, J., Havlena, J., and Sikov, M. R. (1962). Effects of prenatal X-irradiation on activity, emotionality, and maze-learning ability in the rat. *Radiat. Res.* 16, 441-462.
57. Lee, M. H., Rabe, A., Haddad, R., and Dumas, R. (1980). Developmental delays in the progeny of rats consuming ethanol during pregnancy. *Teratology* 21, 53A.
58. Shaywitz, B. A., Griffieth, G. G., and Warshaw, J. B. (1979). Hyperactivity and cognitive deficits in developing rat pups born to alcoholic mothers: An

experimental model of the expanded fetal alcohol syndrome (EFAS). *Neurobehav. Toxicol.* 1, 113-122.
59. Reiter, L., Anderson, G. E., Ash, M. E., and Gray, L. E. (1977). Locomotor activity measurements in behavioral toxicology: Effects of lead administration on residential maze behavior. In *Behavioral Toxicology: An Emerging Discipline* (Zenick, H., and Reiter, R. W., eds.), U.S. Environmental Protection Agency, Research Triangle Park, N.C., pp. 6-1-6-18.
60. Rodier, P. M. (1977). Correlations between prenatally-induced alterations in CNS cell populations and postnatal function. *Teratology* 16, 253-346.
61. Sorbrian, S. K. (1977). Prenatal morphine administration alters behavioral development in the rat. *Pharmacol. Biochem. Behav.* 7, 285-288.
62. Zagon, I. S., McLaughlin, P. J., and Thompson, C. I. (1979). Development of motor activity in young rats following perinatal methadone exposure. *Pharmacol. Biochem. Behav.* 10, 743-749.
63. Hutchings, D. E., Feraru, E., Gorinson, H. S., and Golden, R. R. (1979). Effects of prenatal methadone on the rest-activity cycle of the pre-weanling rat. *Neurobehav. Toxicol.* 1, 33-40.
64. Hutchings, D. E., Towey, J. P., Gorinson, H. S., and Hunt, H. F. (1979). Methadone during pregnancy: Assessment of behavioral effects in the rat offspring. *J. Pharmacol. Exp. Ther.* 208, 106-112.
65. Jensh, R. P., Ludlow, J., Weinberg, I., Vogel, W. H., Rudder, T., and Brent, R. L. (1978). Studies concerning the postnatal effects of protracted low dose prenatal 915 MHz microwave radiation. *Teratology* 17, 21A.
66. Goodwin, P. J., Perez, V. J., Eatwell, J. C., Palet, J. L., and Jaworski, M. J. (1980). Phencyclidine: Effects of chronic administration in the female mouse on gestation, maternal behavior, and the neonates. *Psychopharmacology* 69, 63-67.
67. Borgen, L. A., Davis, W. M., and Pace, H. B. (1973). Effects of prenatal Δ^9-tetrahydrocannabinol on the development of rat offspring. *Pharmacol. Biochem. Behav.* 1, 203-206.
68. Riley, E. P., Lochry, E. A., Shapiro, N. R., and Baldwin, J. (1979). Response perseveration in rats exposed to alcohol prenatally. *Pharmacol. Biochem. Behav.* 10, 255-259.
69. Ata, M. M., and Sullivan, F. M. (1977). Effect of prenatal phenytoin treatment on postnatal development. *Br. J. Pharmacol.* 59, 494P.
70. Anandam, N., Felegi, W., and Stern, J. M. (1980). *In utero* alcohol heightens juvenile reactivity. *Pharmacol. Biochem. Behav.* 13, 531-535.
71. Grant, L. D., Kimmel, C. A., West, G. L., Martinez-Vargas, C. M., and Howard, J. L. (1980). Chronic low level lead toxicity in the rat. *Toxicol. Appl. Pharmacol.* 56, 42-58.
72. Robertson, R. T., Majka, J. A., Peter, C. P., and Bokelman, D. L. (1980). Effects of prenatal exposure to chlorpromazine on postnatal development and behavior of rats. *Toxicol. Appl. Pharmacol.* 53, 541-549.
73. Olson, C. L., Boush, G. M., and Matsumura, F. (1980). Pre- and postnatal exposure to dieldrin: Persistent stimulatory and behavioral effects. *Pestic. Biochem. Physiol.* 13, 20-33.
74. Ostertag, B. (1969). Behavioral, motility and anatomical findings in mice

surviving prenatal irradiation. In *Proceedings of the Ninth Handford Biology Symposium*, Richland, Wash., U.S. Printing Office.

75. Eccles, C. V., and Annau, Z. (1978). Long-term change in behavior after prenatal exposure to methyl mercury in rats. *Pharmacologist* 20, 222.
76. Reiter, L. W., Anderson, G. E., Laskey, J. W., and Cahill, D. F. (1975). Developmental and behavioral changes in the rat during chronic exposure to lead. *Environ. Health Perspect.* 12, 119-123.
77. Finger, S. (1978). Lesion momentum and behavior. In *Recovery from Brain Damage: Research and Theory* (Finger, S., ed.), Plenum, New York, pp. 135-164.
78. Johnson, D., and Almli, C. R. (1978). Age, brain damage, and performance. In *Recovery from Brain Damage: Research and Theory* (Finger, S., ed.), Plenum, New York, pp. 115-134.

20

Developmental Assessment of Infant Macaques Receiving Dietary Aspartame or Phenylalanine

W. Ann Reynolds
California State University, Long Beach, California

Anne F. Bauman
University of Illinois at Chicago, Chicago, Illinois

Lewis D. Stegink and L. J. Filer, Jr.
University of Iowa College of Medicine, Iowa City, Iowa

Sakkubai Naidu
Loyola University Medical Center, Maywood, Illinois

Questions about aspartame safety have arisen because of concern about the potential toxic effects of its constituent amino acids, aspartate and phenylalanine. Like many chemical substances, these amino acids may exert toxic effects when administered at very high dose levels, although species and age susceptibility vary greatly.

When administered in high doses, aspartic acid and aspartame have been shown to be neurotoxic to newborn mice (1,2). However, the administration of equally high doses of aspartame (2 g/kg body weight) to infant monkeys did not produce hypothalamic damage (3).

The potential toxicity of the phenylalanine moiety of aspartame is also of interest because of the metabolic disorder phenylketonuria. In this genetic disease, the ability of the patient to convert phenylalanine to tyrosine is defective, and phenylalanine accumulates in blood and tissues. These high phenylalanine concentrations are associated with abnormal electroencephalograms (EEGs) and profound mental retardation (4,5). To study the mechanism of phenylalanine-induced damage, animal models have been developed. In most of these models, animals are fed large doses of phenylalanine with or without compounds that are inhibitors of the phenylalanine hydroxylase system. The elevated blood levels of phenylalanine, produced by feeding large amounts of dietary phenylalanine, are associated with a variety of central nervous system and behavioral abnormalities in the weanling rat and infant monkey (5-7).

Waisman and Harlow (7) reported that infant monkeys fed a diet providing phenylalanine at 3 g/kg body weight per day exhibited grand mal seizures and persistent intellectual deficits. The effects of chronic ingestion of aspartame in high doses by nonhuman primates had not been studied in detail prior to our studies. In 1971 Waisman et al. initiated a pilot study to evaluate the effects of chronic aspartame ingestion upon growth and development in the infant macaque. Unfortunately, only incomplete data are available from this study, which was terminated prematurely by Dr. Waisman's untimely death (8). This study was further complicated by a lack of adequate numbers of monkeys, the inclusion of one monkey with severe birth defects, and persistent infection in two of the treatment groups with *Shigella sp*. throughout the duration of the feeding trial.

Since aspartame's use as a food additive makes it available in the diets of toddlers and children, we felt that a test of aspartame's safety in developing nonhuman primates was advisable. The Old World macaque provides an excellent biomedical model for studies involving human growth and development, metabolic parameters, brain maturation, endocrine regulation, and other indices. Dietary requirements in the macaque have been well defined (9).

The present study provided for the protracted intake of aspartame and phenylalanine by infant monkeys. A special feeding protocol was designed to avoid a problem inherent in the earlier primate feeding studies of Waisman and colleagues (7,8), that is, that of administering special diets to infant monkeys. The infant monkeys studied by Waisman and colleagues were hand held and fed from a nursing bottle on four occasions during a typical 8-hr technician working day. Thus the desired dose of aspartame for the entire day was in fact administered over an 8-hr period. As a result, the infant monkeys were exposed to an extraordinarily high aspartame intake during that 8-hr period, and then not exposed to any aspartame during the remaining 16 hrs of each day. The present study spread the aspartame dose over a greater time period of each day.

MATERIALS AND METHODS

Infant *Macaca arctoides* were selected for the study. All infants were born in the closed *M. arctoides* colony in the Primate Facility of the Biologic Resources Laboratory, University of Illinois at the Medical Center. *Macaca arctoides* is closely related to *Macaca mulatta*, the rhesus monkey, and interbreeds with this species in the wild.

The doses of aspartame chosen were 1.0, 2.0, and 3.0 g/kg body weight per day. Phenylalanine was given at a level of 1.65 g/kg body weight per day, approximately equimolar to the phenylalanine intake from a 3 g/kg body weight aspartame load. Four infants were assigned to each of these four treatment groups, and four infants were assigned to a control group (Table 1). The sexes of the animals in each group varied randomly, as animals were assigned in sequence to a group as they were born. Weight gains per group did not vary during the period of study in spite of different sex ratios in the treatment groups.

Following birth, each infant was allowed to remain with its mother for a minimum of 1 week to ensure stability and good sucking ability. Infants delivered by cesarean section were nursery reared and hand fed from birth. Infants ranged in age from 17 to 42 days at the beginning the feeding program and were able to self-feed. Each infant was then weaned by being placed in an incubator and taught to nurse itself from a plastic bottle with a rubber nipple (Fig. 1). At the beginning of the study period, each infant was assigned randomly to one of five diets. Similac with Iron (Ross Laboratories, Columbus, Ohio) was fed to the control group, while others received this formula with added aspartame or phenylalanine. These experimental diets were fed for the next 9 months (270 days). Additional water was available at all times. When the monkeys reached the point that the incubator no longer provided comfortable accommodations, the animals were moved into larger individual activity cages. In these cages, water and formula were initially available from plastic bottles with rubber nipples. When milk intake became greater than the capacity of the plastic bottles, or when the infants began to eat the rubber nipples, formula and water was supplied from large glass bottles suspended on the outside cage walls with a metal drinking spigot projecting into the cage. Aspartame and pure phenylalanine are not readily soluble in milk or water; thus it was technically difficult to achieve solution of the 3 g/kg body weight dose of aspartame and the 1.65 g/kg body weight dose of phenylalanine. Every attempt was made to keep the aspartame and phenylalanine in solution or in slurry form such as by hand shaking the bottles at intervals and using mechanical shaking devices on suspended bottles. Nipple holes were enlarged so that the slurry of aspartame plus formula could pass through.

Each week the prior week's intake of formula was assessed individually in order to estimate the next week's daily intake. Aspartame or phenylalanine was

Table 1 Study Groups

Group	Treatment	Dose (g/kg body weight per day)	Number of infants
1	Control		3 male, 1 female
2	Phenylalanine	1.65[a]	3 male, 1 female
3	Aspartame	1.0	3 female, 1 male
4	Aspartame	2.0	3 male, 1 female
5	Aspartame	3.0	4 male, 0 female

[a]Approximately equimolar to the phenylalanine in a 3 g/kg body weight dose of aspartame.

Figure 1 Infant *M. arctoides* in an incubator. Water and formula or formula plus aspartame or phenylalanine were available on a 24-hr basis.

added to the formula in sufficient quantities each day so as to achieve the desired dose. Each morning, after cage cleaning, fresh formula and water were substituted for the previous day's allotment. Daily records were kept of milk and water intake. This procedure resulted in a more normal feeding situation over each 24-hr period than did the protocol used by Waisman and colleagues (7,8). Because the liquid diet was to be fed for 270 days, the usual daily addition of one piece of fruit (apple, orange, or banana) and a weekly vitamin sandwich was delayed until the animals attained 3 or more months of age. After the cessation of the liquid diets, all monkeys were fed a diet of monkey chow (Ralston Purina, St. Louis, Missouri).

Blood Analyses

Blood samples were obtained after a 4-hr morning fast from the saphenous vein of all monkeys at 2, 4, 6, 8, and 9 months of treatment. On days that blood samples were to be drawn, the infants were not allowed access to formula from 0600 until 1000. Since infant monkeys will suckle or eat almost continually while awake, these samples would essentially represent a fasted state. Samples were analyzed for serum electrolytes and osmolality, blood glucose, a complete blood count, and plasma concentrations of free amino acids.

Urine

Urine was checked at 2, 4, 6, 8, and 9 months of age by Labstix for pH, occult blood, protein, glucose, ketones, and bilirubin, and by Phenistix for phenylketones.

Growth and Behavior

Each infant was weighed on a weekly basis. Once a month for the first 5 months, and then at 2-month intervals, each infant was measured with respect to crown-rump length,crown-heel length and head circumference and examined for various developmental milestones, which included extent of teething, ability to vocalize, alertness, tractability, and general behavior. The infant monkeys socialized with the two research assistants involved in this study.

Electroencephalograms

Electroencephalograms (EEGs) were obtained on all monkeys prior to or shortly after going on the diet. Electroencephalograms were also obtained at months 4 and 9 while the animals were fed the test diets. After the monkeys had completed the 9-month period of experimental feeding, eight animals were continued on the control diet (infant formula without additives) for 1 month with repeat EEGs. Repeat EEGs were also obtained on randomly selected animals at 4-month intervals following termination of experimental feeding. All EEGs were obtained using techniques similar to those used for newborn human infants by the same technician in the Department of Neurology where they were analyzed without knowledge of group assignment. Each animal was restrained with towels before being lightly sedated with 10% chloral hydrate for the EEG studies.

Postfeeding Behavioral Studies

Human infants with phenylketonuria fed normal diets during early development are mentally retarded. Such mental retardation is not reversed if these children are later put on a low-phenylalanine diet. Waisman and Harlow (7) reported similar findings in infant monkeys fed high levels of phenylalanine (3 g/kg body weight) during development. Because of these observations, all animals in this study were sent to the Primate Behavorial Laboratory at the University of Wisconsin for learning test performance studies after completion of the studies reported here. The results of these learning test performance studies on these animals are summarized by Sumoi in the following chapter (10).

RESULTS

Aspartame and Phenylalanine Intake

The average daily intake of aspartame or phenylalanine closely approached the planned dose level (Fig. 2). Animals receiving aspartame had differing levels of intake appropriate to the quantity fed, with values for each group being uniform throughout the study. The standard error was less than 1% of the mean intake of all monthly values. However, animals in group 5, scheduled to consume 3 g/kg body weight per day of aspartame typically consumed only 2.5-2.7 g/kg body weight per day.

Formula and Water Intakes

Daily formula (Table 2) and water intake (Table 3) increased with age. Analysis of variance showed significant differences with respect to formula intake during months 4-7 of the study between animals fed aspartame at 3 g/kg body weight and animals fed phenylalanine. With exception of the first 2 months of study (periods 1 and 2), monkeys fed 1.65 g/kg body weight of phenylalanine per day consumed significantly less formula than others (Table 2). The monkeys fed 3 g/kg body weight of aspartame per day consumed significantly more water than monkeys fed aspartame at lower doses during months 3, 5, and 9 (Table 3).

The total daily intake of formula plus water by group is shown in Figure 3. No significant differences were noted between groups when volumes of formula and water were combined. When a ratio of mean formula intake to mean water intake was calculated (Fig. 4), some interesting relationships emerged. Monkeys

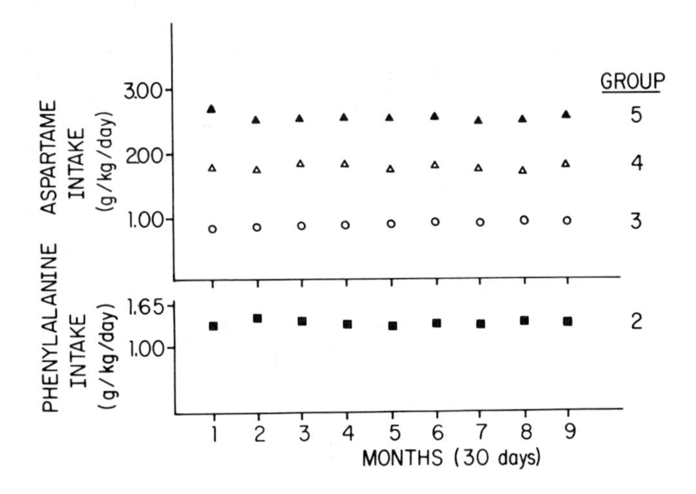

Figure 2 Aspartame or phenylalanine intake in test animals. The upper panel shows the average daily intake of aspartame by month in group 3 animals (●), group 4 animals (Δ), and group 5 animals (▲). The lower panel shows the average daily phenylalanine intake by month in group 2 animals (■). Control animals (group 1, not shown) received formula alone.

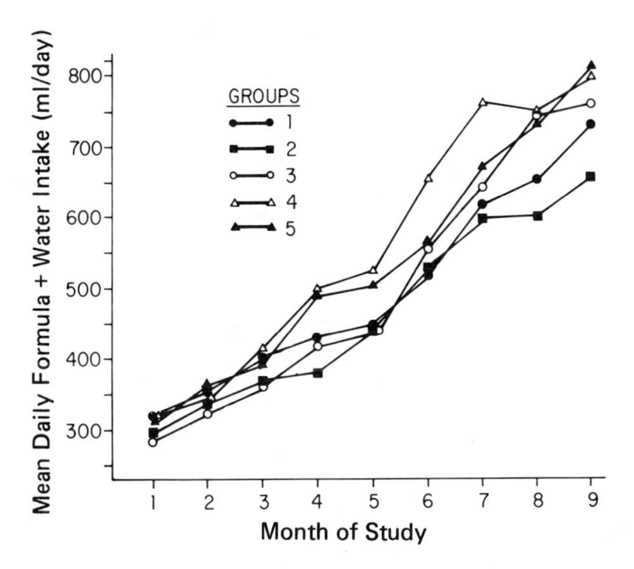

Figure 3 Mean total daily liquid intake (formula plus water) for each group over time. No significant differences between groups were observed. Group 1 animals (●) are controls. Group 2 animals (■) were fed phenylalanine at 1.65 g/kg body weight. Animals in groups 3 (○), 4 (Δ), and 5 (▲) were fed aspartame at 1, 2, and 3 g/kg body weight, respectively.

Table 2 Daily Formula Intake[a]

Group	Month of study								
	1	2	3	4	5	6	7	8	9
Control	277 ± 10	308 ± 8	354 ± 8	378 ± 10	395 ± 11	431 ± 16	507 ± 20	546 ± 16	581 ± 16
Phenylalanine, 1.65 g/kg body weight	237 ± 11	266 ± 11	306 ± 12	316 ± 13[b]	341 ± 16[b]	413 ± 18[b]	455 ± 22[b]	463 ± 20	493 ± 21
Aspartame, 1 g/kg body weight	265 ± 10	284 ± 8	327 ± 10	369 ± 13	394 ± 13	484 ± 12	527 ± 13	592 ± 12	604 ± 14
Aspartame, 2 g/kg body weight	257 ± 8	295 ± 8	372 ± 10	449 ± 9[c]	469 ± 16[c]	561 ± 15[c]	625 ± 16[c]	603 ± 17	613 ± 19
Aspartame, 3 g/kg body weight	220 ± 8	282 ± 9	314 ± 8	409 ± 37	409 ± 12	453 ± 15	498 ± 16	513 ± 16	539 ± 19

[a]Values shown as the mean ± SEM, in milliliters per day.
[b,c]Values with these superscripts in each vertical column differ significantly at $p < 0.05$.

Table 3 Daily Water Intake[a]

Group	Month of study								
	1	2	3	4	5	6	7	8	9
Control	53 ± 6	47 ± 5	49 ± 5	49 ± 5	60 ± 7	97 ± 10	124 ± 11	124 ± 11[b]	152 ± 14[b]
Phenylalanine, 1.65 g/kg body weight	62 ± 5	76 ± 7	65 ± 6	70 ± 7	98 ± 10[c]	126 ± 9	138 ± 12	138 ± 11	161 ± 16[b]
Aspartame, 1 g/kg body weight	58 ± 5	45 ± 6	41 ± 5[b]	46 ± 7	40 ± 6[b]	72 ± 9	114 ± 10	149 ± 13	157 ± 20[b]
Aspartame, 2 g/kg body weight	67 ± 5	46 ± 6	42 ± 6[b]	53 ± 6	69 ± 7	102 ± 10	147 ± 11	148 ± 10	181 ± 18
Aspartame, 3 g/kg body weight	91 ± 6	84 ± 7	74 ± 8[c]	81 ± 6	102 ± 8[c]	126 ± 10	178 ± 15	212 ± 16[c]	270 ± 24[c]

[a]Values shown as the mean ± SEM in milliliters per day.
[b,c]Values with these superscripts in each vertical column differ significantly at $p < 0.05$.

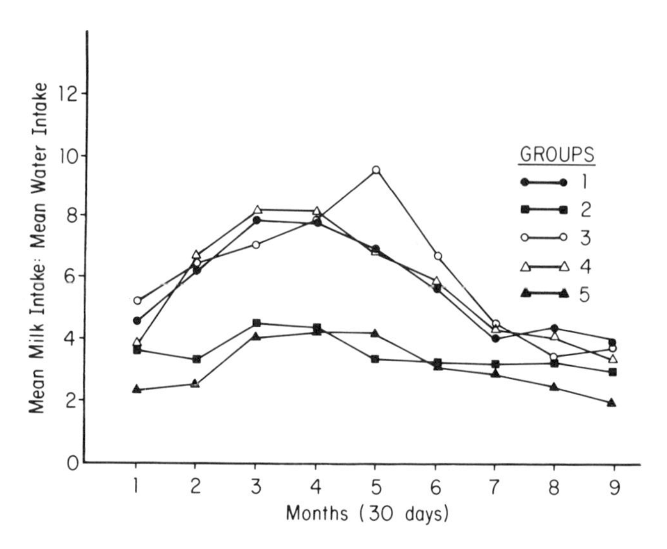

Figure 4 Ratio of the mean daily formula intake to water intake as a function of the duration of study. Group 1 animals (●) are controls. Group 2 animals (■) were fed phenylalanine at 1.65 g/kg body weight. Animals in groups 3 (○), 4 (△), and 5 (▲) were fed aspartame at 1, 2, and 3 g/kg body weight, respectively. Monkeys in group 3, given 1 g of aspartame per kilogram body weight per day, were consistently the best formula drinkers after 3 months. Monkeys in group 5 given 3 g of aspartame per kilogram body weight per day consistently consumed the most water in relationship to formula.

fed 1 g/kg body weight of aspartame per day had the largest formula consumption in relation to water consumed after 4 months of study. This difference was statistically significant at 5 months of study from that of the other groups ($p < 0.05$). The two groups exhibiting the largest water intake in relationship to formula intake were those animals receiving the most aspartame or phenylalanine added to the diet, 3 g/kg body weight of aspartame or 1.65 g/kg body weight of phenylalanine.

Growth Rates

Whether analyzed on an individual basis, or as groups (Fig. 5), incremental weight gain was remarkably uniform. Coefficients of linear regression expressing growth with time were similar. Animals receiving the diet providing 1.65 g/kg body weight of phenylalanine showed a somewhat reduced growth rate from that of the other four groups, whose rates were essentially identical. Crown-heel growth rates were also uniform for all five groups, with no differences noted between the control animals and any of the four experimental groups.

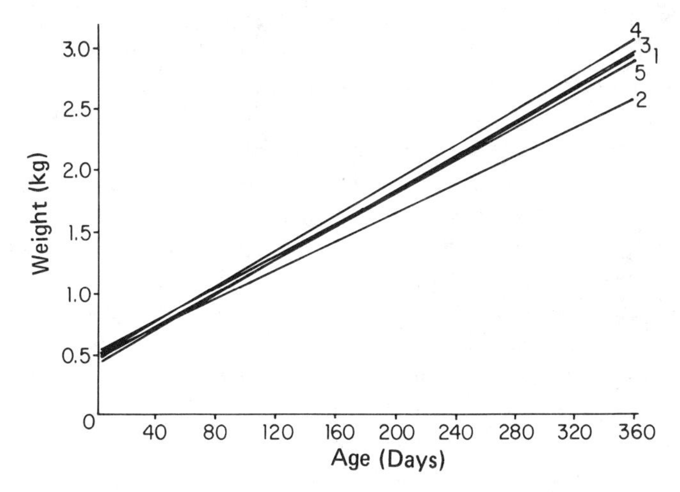

Figure 5 Weight gain for all animals studied. The increase in weight over the study period was essentially identical for all five groups. Group 1 animals are controls. Group 2 animals were fed phenylalanine at 1.65 g/kg body weight. Animals in groups 3, 4, and 5 were fed aspartame at 1, 2, and 3 g/kg body weight, respectively. The b values for groups 1, 2, 3, 4, and 5 are 0.00639, 0.00571, 0.00697, 0.00738, and 0.00701, respectively.

Blood Indices

Hematocrit, hemoglobin, and white and red cell counts were normal for all groups throughout the study. Blood glucose levels ranged from 33 to 100 mg/dl, well within the normal range observed in the infant macaque. Serum sodium, potassium, and chloride concentrations and serum osmolality were all within normal range.

Urine Analyses

Urine specimens collected at intervals from each animal throughout the study had normal pH values without significant levels of protein, glucose, or ketones.

EEG Studies

Knox (11) reported that EEGs were severely abnormal in the majority of untreated human infants with phenylketonuria. Since phenylketonuric patients have been treated with low-phenylalanine diets, their clinical course and EEG patterns have improved dramatically, and EEGs traditionally have been used to monitor the effect of therapy and neurological development in children treated for phenylketonuria (11). Electroencephalogram abnormalities are present early in the phenylketonuric infant. Blaskovics et al. (12) reported abnormal EEGs in

19 of 161 infants with a confirmed diagnosis of phenylketonuria within days of the diagnosis. With dietary treatment, the initially abnormal EEGs normalized by 1 year of age in all but two of these patients (who had minimal and mild abnormalities, respectively).

Abnormal EEGs and seizures have been reported in infant monkeys administered high doses of phenylalanine (13). Seizure patterns were never observed in any of the EEGs obtained on the 20 infant macaques in this study. Fortuitously, an infant rhesus monkey residing within the primate colony at the University of Illinois, but not involved in this study, had grand mal seizures during the time of this study. An extensive series of electroencephalograms obtained on this infant monkey exhibited the typical spike and wave discharge that characterizes human subjects with phenylketonuria or idiopathic epilepsy. Thus this monkey provided us with a positive control.

Plasma Amino Acid Analyses

Overall mean fasting plasma phenylalanine, tyrosine, glutamate, and aspartate levels for the five groups of animals are given in Table 4. No significant differences were found in fasting plasma aspartate and glutamate levels among groups. These data indicate rapid metabolism of the aspartate portion of aspartame when administered with diet.

Fasting plasma phenylalanine levels were increased in all groups ingesting either the diet with added phenylalanine or the diets with added aspartame. Fasting plasma phenylalanine levels in animals ingesting aspartame at 1 g/kg body weight were only slightly higher (8.88 ± 5.15 μmol/dl) than values noted in control animals receiving no aspartame or phenylalanine (5.49 ± 1.49 μmol/dl). Similarly, fasting plasma tyrosine levels were only slightly higher (9.38 ± 5.38 μmol/dl) than values in control animals (7.61 ± 2.49 μmol/dl). This undoubtedly reflects the rapid metabolism of aspartame at this level of intake when ingested over the course of an entire day, and its rapid metabolism and clearance during a 4-hr period of fasting prior to blood sampling.

Fasting plasma phenylalanine levels in animals ingesting aspartame at 2 g/kg body weight were higher (28.7 ± 48.8 μmol/dl) than values in control animals. The elevation in the mean fasting level of plasma phenylalanine suggests that plasma phenylalanine levels were extremely high during the postprandial period. A plasma phenylalanine level measured 2 hr postprandially in one animal fed aspartame at 2 g/kg body weight was 227 μmol/dl, indicating that postprandial plasma phenylalanine levels were considerably higher than fasting levels. Fasting plasma tyrosine levels in animals fed aspartame at 2 g/kg body weight were also elevated (13.1 ± 8.58 μmol/dl) above levels noted in controls (7.61 ± 2.49 μmol/dl). These increased tyrosine levels reflect the conversion of phenylalanine to tyrosine.

Fasting plasma phenylalanine levels in animals ingesting aspartame at 3 g/kg body weight were significantly higher (66.2 ± 83.3 μmol/dl) than levels in con-

Table 4 Mean (±SD) Fasting Plasma Phenylalanine, Tyrosine, Aspartate, and Glutamate Levels (μmol/dl) in Infant Monkeys Fed Diets Supplemented with Aspartame or Phenylalanine

	Plasma levels while on diet			
Animal group	Phenylalanine	Tyrosine	Aspartate	Glutamate
Control	5.49 ± 1.94	7.61 ± 2.49	1.08 ± 0.66	6.06 ± 2.31
Aspartame (1 g/kg body weight)	8.88 ± 5.15	9.38 ± 5.33	1.23 ± 0.90	4.48 ± 1.31
Aspartame (2 g/kg body weight)	26.7 ± 48.8	13.1 ± 8.58	1.61 ± 1.21	5.33 ± 1.59
Aspartame (3 g/kg body weight)	66.2 ± 83.3	23.2 ± 11.4	1.78 ± 1.43	5.90 ± 2.41
Phenylalanine (1.65 g/kg body weight)	54.4 ± 64.9	19.3 ± 6.37	1.08 ± 0.54	4.96 ± 1.35

trol animals. Considerable variability was noted in the plasma phenylalanine levels of these animals, reflecting the differences in length of time since the monkeys had last eaten. Fasting plasma tyrosine levels were also elevated (23.3 ± 11.4 μmol/dl), indicative of considerable conversion of phenylalanine to tyrosine.

Relative differences in fasting plasma phenylalanine levels at aspartame intakes of 1, 2, and 3 g/kg body weight indicate that animals fed formula calculated to provide the 3 g/kg body weight load ingested doses close to this level in spite of aspartame's lack of solubility at this dose level.

The diet with phenylalanine added at 1.65 g/kg body weight provides a quantity of phenylalanine approximately equivalent to that found in the diet providing 3 g/kg body weight aspartame. As expected, fasting plasma phenylalanine and tyrosine levels were similar in animals fed these two diets. Mean fasting plasma phenylalanine levels in animals fed phenylalanine at 1.65 g/kg body weight were 54.4 ± 65 μmol/dl, and mean plasma tyrosine levels were 19.3 ± 6.37 μmol/dl.

General Observations

At 2-3 months of age, all of the infants were troubled with chronic mild diarrhea. The first five animals enrolled in the study became infected with *Salmonella* sp. that responded quickly to Keflex (Eli Lilly & Co., Indianapolis, Indiana). Subsequently, repeated stool samples were cultured and no common enteric bacteria that result in diarrhea could be identified. After the study was nearly complete, it was learned that Old World monkeys tend to become lactose intolerant at about 3 months of age (14). Thus the animals were experiencing chronic mild diarrhea secondary to a decrease in intestinal lactase activity. The good weight gains achieved in the study, however, indicate that this lactose intolerance did not interfere with nutrient absorption or utilization.

Time of tooth eruption and changes in pelage were recorded, and comprehensive anecdotal behavioral records were maintained. No significant differences were noted between groups.

DISCUSSION

The growth and maturational data were essentially identical for all 20 infants, regardless of aspartame or phenylalanine intake. No adverse effects associated with aspartame or phenylalanine ingestion at these levels of intake were noted. Although all of the monkeys thrived on their respective diets, some differences in intake of formula and water emerged between groups.

The negative effects of aspartame and/or phenylalanine feeding in the present study contrast with those reported by Waisman and colleagues (7,8). Those investigators reported a variety of adverse effects in infant monkeys fed phenylalanine at 3 g/kg body weight. The differences between our study and those of Waisman and colleagues may reflect the lower level of phenylalanine fed in our study (1.65 g/kg body weight). In their preliminary and unfinished study of aspartame feeding, Rao et al. had planned to study an aspartame dose of 4-6 g/kg body weight (8). However, the monkeys would not consume this load, presumably owing to its intense sweetness. For this reason the actual dose administered in his preliminary study approximated 3.6 g/kg body weight.

Interpretation of both Waisman studies (7,8) is further complicated by an uncertainty of whether animals in those studies had ad libitum access to water. If not, serious questions can be raised with respect to adverse effects associated with increased blood osmolality and the general metabolic state of the animals. Convulsions, seizures, and alterations in brain electrical activity are common sequelae of high solute loads (15,16). The infant monkeys in the Waisman studies were observed to convulse primarily when being handled. This would correspond to the time of feeding. Thus the question arises whether the adverse effects noted in the infant monkeys studied by Waisman reflected hyperosmolality, or generalized hyperaminoacidemia resulting from dehydration, rather than elevated phenylalanine levels. In the present study close attention was paid to providing free access to water as well as monitoring water intake in relation to formula intake. The ratio of milk to water intake was decreased in the animals receiving 3 g/kg body weight of aspartame (Fig. 4). Thus simple access to water may be the key element accounting for the contradictory findings between the present study and the prior work of Waisman and colleagues. Ausman (17) has observed a high mortality rate in infant *Cebus* and squirrel monkeys raised on infant formula without free access to water. His observations again suggest the adverse effects of hyperosmolality.

Fasting plasma phenylalanine and tyrosine values indicate rapid metabolism of aspartame and phenylalanine by the infant nonhuman primate. An intake of aspartame of 1 g/kg body weight resulted in only a minimal elevation in fasting plasma phenylalanine level despite the large quantity of aspartame fed. Fasting plasma phenylalanine levels increased in response to an increase in aspartame or phenylalanine load. Ingestion of phenylalanine in dipeptide form (as aspartame) rather than as the free amino acid did not seem to affect fasting plasma phenylalanine values. Fasting plasma phenylalanine levels were similar in animals given 1.65 g/kg body weight phenylalanine to values observed in animals given an approximately equivalent amount of phenylalanine as aspartame (3 g/kg body weight).

The variation noted in fasting plasma phenylalanine levels of animals ingesting diets with added aspartame and phenylalanine probably reflects variable periods of fasting. Food was removed from the cages at 0600, and blood samples were obtained after a minimum 4-hr fast, usually at 1000 to 1030, sometimes slightly later. However, no data were available on when the animals had last ingested food. In some cases, no food remained at 0600, while in other cases food was still available to the animals at 0600. Thus the length of fasting prior to blood sampling was at least 4-hr, but undoubtedly varied from animal to animal. Animals with a longer fasting period would be expected to show lower plasma phenylalanine levels. Postprandial plasma amino acid concentrations were not measured in these animals. It is clear, however, that postprandial values must be considerably higher than the levels noted in fasted animals.

The control of food intake and the correlated regulation of body energy balance relative to plasma amino acid levels has been studied for many years. Recently, considerable interest has developed around the relationship of diet to plasma and brain amino acid levels and brain neurotransmitter levels (18-22). The importance of amino acids in neurotransmitter synthesis in the central nervous system has been known for some time. However, only recently has it been shown that physiological changes in plasma amino acids can alter neurotransmitter synthesis in the brain, indicating that peripheral metabolism may have a major influence on central metabolism.

For example, the concentration of tyrosine, and hence catecholamines, in the brain has been shown to depend upon the relative plasma concentrations of tyrosine and the other large neutral amino acids (20).

Anderson and Ashley (18,19,21,22) have postulated that the plasma ratio of tyrosine to phenylalanine may be a major factor in regulating energy intake in the rat. In their experiments, weanling rats were allowed to self-select from two diets that differed only in protein content. After food selection patterns were established, the rats were allowed to eat overnight, and blood was collected between 0900 and 1100. Changes in plasma amino acid patterns were correlated with protein and energy intake by the various groups. Anderson and Ashley (21,22) reported that the plasma tyrosine-to-phenylalanine ratio, but not the tyrosine-to-large neutral amino acid ratio, was correlated consistently with energy intake. Anderson (18) has suggested "that changes in plasma tyrosine/phenylalanine ratio reflect, or stimulate, at least in part a mechanism operating via the CNS to control energy intake."

The dietary treatment in the present study caused a significant change in the plasma tyrosine-to-phenylalanine ratio. Such changes allow us the opportunity to evaluate the tyrosine-to-phenylalanine ratio to energy intake hypothesis in the nonhuman primate. This is particularly important, since most feeding studies have been carried out in the rat, and it is not certain that the nonhuman primate will have a similar response.

The data in Table 5 correlate the mean fasting plasma tyrosine-to-phenylalanine ratio in these monkeys with average daily milk intake for the 270-day study. Tyrosine-to-phenylalanine ratios in fasting plasma varied from 1.38 in control animals to 0.35 in animals ingesting either 3 g/kg aspartame or 1.65 g/kg phenylalanine as part of the diet. Milk or energy intake of the monkeys with tyrosine-to-phenylalanine ratios of 0.35 was reduced, significantly so for the animals fed phenylalanine at 1.65 g/kg body weight. However, no changes were noted with less extreme changes in the tyrosine-to-phenylalanine ratio.

It must be pointed out that the imbalance in the plasma tyrosine-to-phenylalanine ratio as measured in the fasting sample is likely to be less than values

Table 5 Comparison of Fasting Plasma Tyrosine to Phenylalanine Ratio with Milk Intake of Infant Monkeys Fed Diets Supplemented with Aspartame or Phenylalanine

Animal group	Tyrosine/Phenylalanine	Average daily milk intake (ml)
Control	1.38	422
Aspartame (1 g/kg body weight)	1.05	428
Aspartame (2 g/kg body weight)	0.49	472
Aspartame (3 g/kg body weight)	0.35	389
Phenylalanine (1.65 g/kg body weight)	0.35	366

noted in postprandial plasma samples. Preliminary data indicate much higher postprandial plasma phenylalanine concentrations in another group of infant monkeys fed the aspartame or phenylalanine diets. Thus, the postprandial tyrosine-to-phenylalanine ratio in animals fed aspartame or phenylalanine must be even lower than that indicated by the fasting plasma levels.

SUMMARY

Newborn monkeys *(M. arctoides)* were reared for 9 months on infant formula diets containing added aspartame to provide intakes of 1, 2, or 3 g/kg body weight per day. Control monkeys received only infant formula, and one group received formula plus phenylalanine at a level of 1.65 g/kg body weight per day, approximately equivalent to the phenylalanine content of the 3 g/kg body weight intake of aspartame. The protocol allowed self-selection of formula intake plus free access to water at all times.

The infant monkeys were monitored for formula and water intake, weight, and crown-heel length increases. Blood samples were taken at intervals and analyzed for electrolytes, osmolality, differential cell count, hematocrit, hemoglobin, and glucose and amino acid levels. Urine samples were checked for protein, glucose, and ketones. Each animal was handled often and evaluated carefully for feeding habits and behavioral milestones. Electroencephalograms were obtained at regular intervals.

The growth rates for all five groups of animals were indistinguishable, both with respect to weight gain and increments in crown-heel length. Blood chemistries and urine analyses were normal in all five groups throughout the dietary period and did not differ between groups. Water intake in relation to formula consumed was elevated for the groups receiving 3.0 g/kg body weight aspartame or 1.65 g/kg body weight phenylalanine. Animals fed formula providing 3.0 g/kg body weight aspartame consumed the most water. Animals fed 1.65

g/kg body weight phenylalanine became the most avid fruit eaters. The bitter taste imparted by phenylalanine and the excessive sweetness of the high levels of intake of aspartame in addition to the increase in intake of amino acids by these animals may have been responsible for these observations.

The electroencephalographic findings did not vary from group to group and resembled those characteristic of newborn human infants. No infant monkey in any group exhibited convulsions, seizures, shudders, or any sort of abnormal neurological behavior during administration of the experimental diet or, in the year following, while fed a more traditional monkey diet.

Fasting plasma phenylalanine levels were dose related to the daily intake of aspartame and/or phenylalanine. Plasma phenylalanine levels were similar in monkeys ingesting aspartame at 3 g/kg body weight to values noted in animals ingesting an equivalent dose of phenylalanine (1.65 g/kg body weight). The variation in fasting plasma values probably reflects the duration of the fast prior to obtaining the blood sample.

It was not possible to confirm earlier work suggesting that neurological deficits can be induced in infant monkeys by adding large amounts of phenylalanine to the diet. Free access to water and self-selection of formula characterized the present study. Hyperosmolality (and/or a generalized hyperaminoacidemia related to hyperosomality) induced by forced feeding may have been responsible for the nonspecific neurological abnormalities reported previously. However, earlier studies also administered phenylalanine at higher dose levels. Large intakes of aspartame as part of the diet appear to have no effect upon developmental parameters of the infant macaque.

REFERENCES*

1. Olney, J. W., Ho, O. -L., and Rhee, V. (1971). Cytotoxic effects of acidic and sulphur containing amino acids on the infant mouse central nervous system. *Exp. Brain Res.* 14, 61-76.
2. Reynolds, W. A., Butler, V. and Lemkey-Johnston, N. (1971). Hypothalamic morphology following ingestion of aspartame or MSG in the neonatal rodent and primate: A preliminary report. *J. Toxicol. Environ. Health* 2, 471-480.
3. Reynolds, W. A., Stegink, L. D., Filer, L. J., Jr., and Renn, E. (1980). Aspartame administration to the infant monkey: Hypothalamic morphology and plasma amino acid levels. *Anat. Rec.* 198, 73-85.
4. Tourian, A., and Sidbury, J. B. (1983). Phenylketonuria and hyperphenylalaninemia. In *The Metabolic Basis of Inherited Disease*, 5th ed. (Stanbury, J. B., Wyngaarden, J. B., Fredrickson, D. S., Goldstein, J. L., and Brown, M. S., eds.), McGraw-Hill, New York, pp. 270-286.
5. Vorhees, C. V., Butcher, R. E., and Berry, H. K. (1981). Progress in experi-

*In the "E-file" reports by G. D. Searle to the Food and Drug Administration, SC-18862 refers to aspartame (Searle Compound 18862).

mental phenylketonuria: A critical review. *Neurosci. Biobehav. Rev.* 5, 177-190.
6. Kerr, G. R., and Waisman, H. A. (1967). Dietary induction of hyperphenylalaninemia in the rat. *J. Nutr.* 92, 10-18.
7. Waisman, H. A., and Harlow, H. P. (1965). Experimental phenylketonuria in infant monkeys. *Science* 147, 685-695.
8. Rao, K. S., McConnell, R. G., and Waisman, H. A. (1972). *File E-32. SC-18862: 52 Week Oral Toxicity Study in the Infant Monkey*, a report by the Department of Biological Research, G. D. Searle & Co., submitted to the Food and Drug Administration, Hearing Clerk File, Administration Record, Aspartame 75F-0355, Food and Drug Administration, Rockville, Md.
9. *Nutrient Requirements of Nonhuman Primates* (1978). Nutrient Requirements of Domestic Animals, No. 14 The National Research Council, National Academy of Sciences, Washington, D.C.
10. Suomi, S. J. (1984). Effects of aspartame on the learning test performance of young stumptail macaques. In *Aspartame: Physiology and Biochemistry* (Stegink, L. D., and Filer, L. J., Jr., eds.), Marcel Dekker, New York, Chap. 21.
11. Knox, W. E. (1970). Retrospective study of phenylketonuria. *Phenylketonuria Newslett. N. Engl. Deaconess Hosp.* March, pp. 1-3.
12. Blaskovics, M., Engel, R., Podosin, R. L., Azen, C. G., and Friedman, E. G. (1981). EEG pattern in phenylketonuria under early initiated dietary treatment. *Am. J. Dis. Child.* 135, 802-808.
13. Cadell, T. E., Harlow, H. F., and Waisman, H. A. (1962). EEG changes in experimental phenylketonuria. *Electroencephalogl. Clin. Neurophysiol.* 14, 540-543.
14. Hart, N. A., Reeves, J., Dalgard, D. W., and Adamson, R. H. (1980). Lactose malabsorption in two sibling rhesus monkeys (*Macaca mulatta*). *J. Med. Primatol.* 9, 309-313.
15. Gauthier, B., Freeman, R., and Beveridge, J. (1969). Accidental salt poisoning in a hospital nursery. *Aust. Paediatr. J.* 5, 101-105.
16. Simmons, M. A., Adcock, E. W., Bard, H., and Battaglia, F. C. (1974). Hypernatremia and intracranial hemorrhage in neonatates. *N. Engl. J. Med.* 291, 6-10.
17. Ausman, L. M. (1979). Personal communication.
18. Anderson, G. H. (1977). Regulation of protein intake by plasma amino acids. *Adv. Nutr. Res.* 1, 145-166.
19. Anderson, G. H. (1979). Control of protein and energy intake: Role of plasma amino acids and brain neurotransmitters. *Can. J. Physiol. Pharmacol.* (in press).
20. Fernstrom, J. D. (1976). The effect of nutritional factors on brain amino acid levels and monamine synthesis. *Fed. Proc.* 35, 1151.
21. Anderson, G. H., and Ashley, D. V. M. (1976). Correlation of the plasma tryptophan and tyrosine to neutral amino acid ratios with protein and energy intakes in the self-selecting weanling rat. *Proc. Can. Fed. Biol. Soc.* 19, 24.
22. Anderson, G. H., and Ashley, D. V. M. (1977). Correlation of the plasma tyrosine to phenylalanine ratio with energy intake in self-selecting weanling rats. *Life Sci.* 21, 1227-1234.

21

Effects of Aspartame on the Learning Test Performance of Young Stumptail Macaques

Stephen J. Suomi*
University of Wisconsin, Madison, Wisconsin

INTRODUCTION

Consideration of the learning capabilities of animals has attracted the attention of scientists, philosophers, theologians, naturalists, and interested layman for many centuries. Particular emphasis has centered on questions regarding the degree to which the cognitive capabilities of various animal species resemble those of human beings. Here opinions have been as diverse as the background of the investigators choosing to pursue these questions. For example, the seventeenth-century philospher Descartes boldly asserted that all animals, lacking the rationality of the human mind, behaved like machines driven only by simple instincts. At the other extreme, the nineteenth-century German educator von Osten believed that his horse, Clever Hans, possessed mathematical abilities comparable to those of most educated people of his time. We now know that Descartes and von Osten were both quite wrong. Pfungst (1) convincingly demonstrated that Clever Hans was clever not in mathematical ability but, rather, in his ability to recognize cues (probably unintentional) from his proud human owner signifying when to stop "counting" (2). On the other hand, Descartes' limited view of animal intelligence has been convincingly proven false for most animals by almost a century of systematic scientific research documenting the effects of experience on behavior (3-10).

*Present affiliation: National Institute of Child Health and Human Development, National Institutes of Health, Bethesda, Maryland

The scientific study of animal learning has become increasingly diverse and sophisticated over the past 50 years. One major research emphasis in this area has centered on the discovery and subsequent elucidation of basic principles or "laws" regarding learning processes that generalize across a broad range of different species and situations. Two exceedingly widespread and robust forms of learning have been identified by such research efforts: "classical" (Pavlovian) conditioning and "instrumental" (operant) conditioning (5,7,11,12). Both classical, and instrumental conditioning can be demonstrated in virtually all organisms that possess even the simplest of nervous systems, and in these and all species with more advanced nervous systems, including humans, the empirical rules governing the acquisition and retention of conditioned behavior are remarkably similar.

Some investigators have focused their research efforts on documenting and understanding *differences* between species in their respective learning capacities, and in these endeavors they have gone far beyond the two basic conditioning paradigms by developing sensitive comparative tests of cognitive capabilities. These research efforts have met with considerable success over the past 25 years. It is now well established that major differences exist in the learning capacities of animals of different taxa, and there appear to be systematic relationships between various aspects of brain structure in different species and the cognitive capabilities of representative members of those species (13,14). Of particular interest is the fact that, by and large, the greater the anatomical similarity of the characteristic brain structure of a given species to that of the human brain, the more closely that species will resemble human beings with respect to its characteristic learning capacity. Old World monkeys, including the genus *Macaca*, and (especially) the apes (e.g., chimpanzees and gorillas) have brains whose anatomical structure more closely resembles that of the human brain than any other living organisms (15). Thus, while there is little argument that *Homo sapiens* possess unique cognitive abilities, especially with respect to language, it is also true that they share more capabilities with advanced nonhuman primates than with any other taxonomic group (16).

Compelling evidence that advanced nonhuman primates have patterns of cognitive functioning that are characteristic of normal humans but not of most, if not all, nonprimate animal species comes from two basic lines of research. The first involves the comparative study of performance on learning tests that require complex cognitive representations and/or strategies for successful problem solution. Typically, these tests cannot be reliably solved via simple conditioning practices, more complicated response chains, or rote memorization; instead, some degree of concept formation, abstract representation, and/or development of flexible response strategies is needed to achieve optimal performance on these tasks. On tests of this sort normal adult humans and advanced nonhuman primates will readily obtain solutions with little practice, whereas nonprimate subjects will require extensive practice to reach equivalent levels of performance or not be able to solve the problem at all (8,14,17).

The second body of evidence for parallel cognitive functioning in human and advanced nonhuman primates comes from comparative studies of ontogenic development of intelligence in these species. One of the best-documented sets of findings in the entire field of human developmental research is that human infants do not possess the full intellectual capabilities of normal adults, but, instead, develop such capabilities in an orderly and highly predictable fashion over the first 12-16 years of life. Such developmentally based increments provide the primary rationale for construction of most current standardized intelligence tests for human children (e.g., the Weschler Scales for Intelligence and the Stanford-Binet), and they form the empirical basis for the widely accepted theoretical views of human cognitive development associated with Piaget (18,20). Recent findings from studies of cognitive development in advanced nonhuman primates have demonstrated that not only do macaques, orangutans, and chimpanzees (and probably all other advanced nonhuman primate species) display dramatic and systematic increases in their learning abilities from birth to adolescence, but also that they go through the very same developmental stages, in the same sequence, and in roughly the same relative time span, as do human children during at least their first 2-3 years of life (21-23). Thus there appear to be strong parallels, if not outright homologies, between cognitive development in young human children and in young members of advanced nonhuman primate species (24).

The presence of such strong parallels has made it possible to develop compelling *animal models* of cognitive functioning in human infants and young children. Using batteries of standarized tests that are age sensitive for advanced nonhuman primate subjects, investigators have been able to carry out systematic, carefully controlled studies that simulate phenomena which are of considerable practical interest to investigators, parents, and policy makers, but which are difficult to study experimentally in human children for ethical and practical reasons. Some prominent examples of such animal models that involve young nonhuman primate subjects include studies of the effects of various types of brain lesions (25,26), diets deficient in specific essential nutrients (27), exposure to various environmental toxins, including PCBs, PBBs, and dioxin (28), and rearing in socially deprived environments (29) on the development of learning in these subjects.

The present research project was conducted for the purpose of evaluating the learning performance of juvenile-age stumptail macaques (*Macaca arctoides*) who had been exposed during infancy to a diet containing one of three levels of aspartame, a diet containing phenylalanine, or a control diet containing no additives. Learning performance was assessed by a battery of discrimination learning tests that have been standardized for macaques over the past four decades.

The battery included tests of two-choice object discrimination, two-dimensional pattern discrimination, six trial learning set problems, and six trial oddity problems. Various tests in the battery previously had been shown to differentiate macaque subjects on the basis of age, social rearing history, test environment, relative size and location of various cortical lesions, and type of rearing diet. In addition, each subject was administered a test of its hearing capabilities.

METHODS

Subjects

The subjects for the study were 20 stumptailed macaques (*Macaca arctoides*), each approximately 1½ years of age at the beginning of the testing described below. All subjects had been born and raised at the University of Illinois Medical School Animal Resources Laboratory. During infancy, each subject was maintained on one of five different diets for 270 days, with equal number of subjects (n = 4) assigned to each dietary condition. Three of the diets (A1, A2, A3) provided aspartame at 1, 2, and 3 g/kg body weight, respectively, a fourth diet provided phenylalanine at 1.65 g/kg body weight (P), while the fifth diet contained no additives and served as the controlled condition (C). Details concerning the actual levels of additives, the preparation procedures, and the schedule of feeding for each diet group are presented in the preceding chapter by Reynolds et al. (30).

The 20 infant monkeys were placed randomly in a dietary regimen as born (4 animals per dietary group). All 20 animals were housed individually during the 270 day period while fed the test diets. At the end of the 270 day test feeding period, animals were fed standard monkey chow while awaiting transfer to the University of Wisconsin. The first 10 animals to finish the feeding study were housed in a social group for approximately 3 months before being transferred to quarantine facilities at the University of Wisconsin (see below). The second group of 10 animals coming off the feeding studies were housed in individual cages for several months before transport to Wisconsin.

Transport and Quarantine Procedure

When the first group of 10 subjects were approximately 1¼ years old, they were placed in individual compartments of a shipping cage and driven by van from the University of Illinois Medical School Animal Resources Laboratory (Chicago) to the University of Wisconsin Health Sciences Center (Madison), where they were individually quarantined for a 90-day period in an 0.8 X 0.6 X 0.8 m wire-mesh cage where the animal could see and hear, but not contact, the other monkeys in the group. After quarantine all animals were moved to the University of Wisconsin Department of Psychology Primate Laboratory, where they were housed individually in cages identical in size and design as before. They remained housed in this manner throughout the entire period of adaptation and testing; during this time each subject was maintained on the standard Primate Laboratory macaque diet, consisting mainly of Purina monkey chow (Ralston Purina, St. Louis, Missouri), fruit and vegetables, and vitamin additives. After completion of the final test, subjects were placed in shipping cages and returned by van to the University of Illinois.

Members of the second group of 10 monkeys went through an identical transport and quarantine procedure, except that this procedure began (and ended) approximately 1 calendar year later. While in the Wisconsin Primate Laboratory, the animals were maintained under precisely the same living and test conditions as were members of the first group of subjects, with one exception: The second group of monkeys spent the last 60 days of the 90-day quarantine period physically housed in the Primate Laboratory rather than spending the entire quarantine period at the Health Sciences Center, as did members of the first group. Inasmuch as they were housed in the same-sized cages in approximately the same-sized housing room and maintained on exactly the same diet in both facilities, it seems likely that this difference was inconsequential with respect to the results of all the learning and hearing tests.

Test Apparatus and Adaptation Procedure

All tests of learning performance and of hearing were carried out in a Wisconsin General Test Apparatus (WGTA), a device that has been used for tests of primate learning for the past 40 years. The WGTA was first described by Harlow and Bromer (31); it is pictured in Figure 1, and details of its construction and operation can be found in Ref. 32. All tests were administered by four different investigators throughout the course of the study; all had extensive previous experience in the testing of macaques in the WGTA. All investigators were blind to the dietary history of the subjects.

All subjects in the present study were first adapted to the WGTA according to the standard three-step procedure described by Harlow et al. (26) prior to the start of the formal learning test battery. Adaptation began by training each subject to obtain food from the centered well of a one-hole stimulus tray. Grapes, small marshmallows, Fruit Loops, and pieces of apple were used for food, according to each subject's particular preference. Each subject was given 25 trials per day until it would respond within 5 sec of presentation on each trial. At this point an unpainted wood cube was placed behind the baited well and, on succeeding trials, moved forward until it covered the well (and the food reward within the well). This procedure was continued until the subjects responded by displacing the block consistently within 5 sec on all 25 trials of 2 consecutive days.

The second stage of adaptation began when criterion had been reached on stage 1. Here the subject was presented with a two-hole stimulus board, with the unpainted wood cube covering one of the two food wells; the covered food well was always baited, while the uncovered food well was never baited. The subject

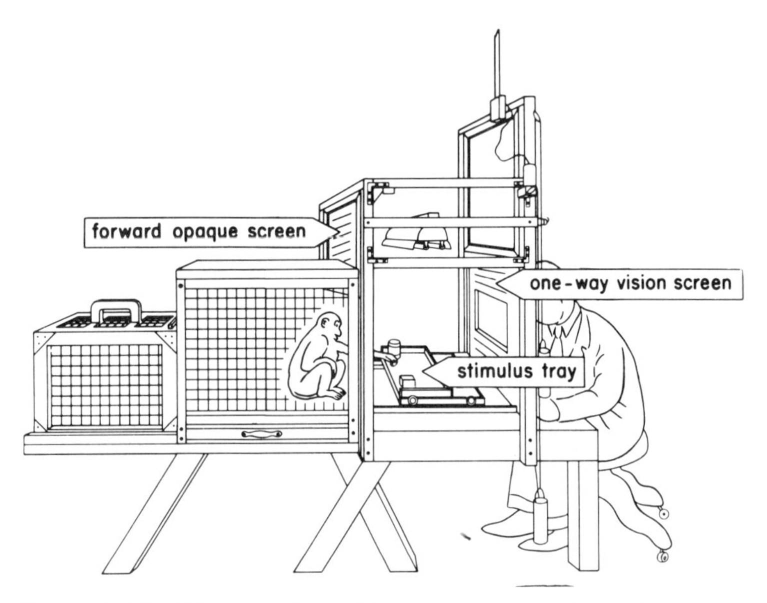

Figure 1 The Wisconsin General Testing Apparatus (WGTA).

was given 25 trials per day for a total of 5 days; the length of time needed to complete the 25 trials was recorded each day.

The final stage of adaptation also utilized the two-hole stimulus board, with a single stimulus object covering the baited food well. However, instead of the previously used unpainted wood cube, a different three-dimensional stimulus object was used on each day for 25 trials. Ten days of this stage of adaptation were presented to each subject; as before, the length of time required to complete each day's 25 trials was recorded.

Learning Test Procedures

After each subject had successfully completed the adaptation procedure, it was run through the formal battery of learning tests, which consisted of 10 40-trial object discrimination problems, 10 40-trial pattern discrimination problems, 600 6-trial object discrimination learning set problems, and 256 6-trial oddity problems. The tests were run consecutively for each subject, that is, all the problems for one test were run before any problems for the next test were presented. The entire test battery was completed on the first group of 10 subjects (see above) prior to the beginning of the adaptation procedure for the second group of 10 monkeys. A detailed description of the task, procedure, and schedule for each of the four tests comprising the overall battery is presented below.

Object Discrimination

In this test each subject was presented with two three-dimensional objects covering the two food wells on each trial. One object ("correct") always covered a baited food well, while the other object ("incorrect") always covered an empty food well. The rewarded food well position was shifted from left to right in random order, balancing the two positions over each day of testing (40 trials), with the additional condition that the same position ("left" or "right") would not be correct for more than three trials in succession. A noncorrection procedure was used for this and all subsequent tests.

Each subject was presented with a single two-choice problem each day for 40 trials. A different 40-trial problem, involving a new pair of object stimuli, was presented on each day, for a total of 10 days. Stimuli were always three-dimensional, and they differed at least in color and shape. Previous research has shown that 1½-year-old macaques rapidly learn to solve object discrimination problems and typically achieve an asymptotic performance of 80-95% correct on later problems (21,29).

Pattern Discrimination

Following the test of object discrimination, each subject was given 10 problems of two-dimensional pattern discrimination. These problems were very similar to those of object discrimination, except that the two stimuli (one always "correct," the other always "incorrect") were two-dimensional patterns, differing in shape and color, each pasted on a 2 × 2 in. white card mounted on a rectilinear block of wood.

Each problem was presented for 40 trials; only a single problem was presented on each of the 10 days of pattern discrimination testing. The same procedure (i.e., randomized left-right positions of correct patterns, noncorrection procedure) was followed for pattern discrimination problems. Previous research has demonstrated that normal 1½-year-old macaque subjects can solve these problems, but that their asymptotic performance is typically lower than their performance on three-dimensional object discrimination problems (29). In contrast, monkeys with lesions in the temporal cortex typically perform at chance levels on pattern discrimination problems (25).

Object Discrimination Learning Set

Tests of learning set involve presenting subjects with a long series of six-trial discrimination problems (8). In the present test battery, each subject was given a total of 600 six-trial problems; each problem presented a new pair of three-dimensional objects that differed in color and shape. Eight six-trial problems were presented each day for a total of 75 test days. The same procedure that had been used in the previous two-choice discrimination problems was employed.

Considerable previous research has demonstrated that macaque subjects who learn to solve learning set problems typically show progressive improvement from problem to problem. More importantly, over successive blocks of problems the form of intraproblem trial-to-trial performance changes, such that at asymptotic levels trial 1 performance is nearly at chance levels, trial 2 performance is nearly perfect, and performance on trials 3-6 is similar to that of trial 2. Monkeys "acquiring a learning set" show that they have learned a strategy that enables them to solve the problem once they obtain feedback regarding the outcome of their trial 1 choice; i.e., they learn an abstract concept, or, in Harlow's words, they "learn to learn" (8). Previous research has shown that many nonprimate species (14) and infant macaques under 1 year of age (21) do not acquire learning sets, but normal adolescent subjects perform at nearly adult levels, with trial 2 performance typically reaching asymptote at 85-95% correct (29,33). Other work has shown the test to be sensitive to cortical lesions (34,35) and to long-term diets initiated at birth that contain excessive levels of phenylalanine (36) or low levels of protein (27). Thus the test is sensitive to phylogenetic, developmental, and dietary factors.

Oddity Learning Set

The final test in the battery was that of oddity. In this test, subjects were presented problems that contained three stimuli, two of which were identical and one that was different. The subject's task was to select the "odd" stimulus on each trial.

Because oddity problems involve the simultaneous presentation of three, rather than two, stimuli on each trial, it was necessary to adapt each subject to a three-hole stimulus board prior to presentation of the test. Therefore, after each subject had completed its final object discrimination learning set problem, it began a 5-day adaptation procedure similar to the previously described two-hole adaptation procedure. On each of the 5 days, a different three-dimensional object covered a baited food well for 25 trials each day. The position of the object was shifted from trial to trial; the remaining food wells were left uncovered (and unbaited) on each trial. For each day of adaptation the experimenter recorded the length of time the subject took to complete the 25 trials.

Formal oddity testing began on the sixth day. Eight problems were presented each succeeding day until 256 problems (32 days) had been completed. Each problem consisted of six trials with the same stimuli, two of which were identical and the third "odd." The odd stimulus differed in color and shape from the other two stimuli.

The oddity test requires that a subject form "same" and "different" con ceptual catagories; in addition, the present format involves a learning set mode of presentation (many problems, each with a small number of trials), so that both interproblem and intraproblem learning can be measured. Note that a sophisticated subject should be able to achieve near-perfect performance on all trials of a given problem. Nevertheless, oddity learning set problems tax the cognitive capacity of most primate subjects, while nonprimate subjects cannot solve oddity problems reliably (14,17). Among the macaque subjects, the task discriminates subjects by maturational status [subjects under 1 year cannot solve oddity problems, and asymptotic performance is not achieved before late adolescence-early adulthood (21)], presence of cortical lesions (26), early diets (27), and early rearing history [socially deprived subjects display inferior performance (29)]. Thus the oddity test represents the most difficult and discriminative test in the present overall battery.

Hearing Test Procedure

After completion of the formal battery of learning tests, each subject was run through an additional test of its hearing capability. For this test a pair of speakers was mounted behind the subject's holding cage in the WGTA apparatus, so that sound emanating from either speaker during WGTA testing would be perceived as coming from the right or left rear of the subject as it was oriented toward the WGTA stimulus tray and the tester. The speakers were connected to an audio tape recorder that could be operated by the tester as he or she was running the subject through WGTA trials.

A tape recording had been made of 1000-Hz pure tones of 1-sec duration each, played at six different intensities (68, 72, 77, 79, 82 and 88 dB), with an interstimulus interval of 16-43 sec (X = 31 sec). The order of different intensities was randomized on the tape; two different orders were followed, for a total of 12 stimulus presentations.

Subjects were each brought into the WGTA test room and run through trials of simple two-choice object discrimination problems. During testing the level of ambient white noise in the test room was set at 63-66 dB, so that the various sound stimuli differed from the ambient noise level by 2-25 dB, depending on the particular stimulus tone. The recording of the 1000-Hz sound stimuli was played over the speakers beginning 3 min after the subject had started the WGTA tests. The observer noted whether or not the subject displayed an orienting response to each presentation of the pure tone; such an orientation involved turning away from the WGTA stimulus tray and the tester, and as such could easily be detected and recorded by the tester (who could see when the tones were presented via movement of the sound level meters on the tape recorder, independent of his or her ability to hear the tones when they were presented.

Data Analysis

Data from both the adaptation sessions and each of the four tests in the formal learning test battery were analyzed statistically by means of analyses of variance (ANOVA). For the initial stage of adaptation, the number of days required to reach criterion for each subject was entered into a 5 (diet group) × 2 (first versus second test group) two-way ANOVA. For the other two stages of adaptation and for the oddity adaptation, the total length of time required for completion of the stage under analysis for each subject was likewise entered into a 5 X 2 two-way ANOVA. For those ANOVA tests that yielded significant main effects and/or interactions ($p < 0.05$), Duncan tests were applied to the relevant cell means.

Three-way repeated measures of ANOVA were used on the data from each of the four formal learning tests. For both the object discrimination and the pattern discrimination tests, each subject's total number of correct responses was summed over 2-day periods (total possible correct = 50), and these sums were entered into three-way repeated measures of ANOVA, with the diet group (1-5) and subject group (first group of 10 versus the second group of 10) as independent variables and with the five 2-day sums as the repeated measure. Duncan tests were subsequently performed whenever significant main effects and/or interactions were detected.

For the object discrimination learning set test, two different ANOVA tests were performed. The first involved summing each subject's scores for trials 2-6 over blocks of 100 problems (6 blocks in all), and then subjecting these sums to a three-way repeated-measures ANOVA, with the diet group and subject test group as independent variables and with the six 100-problem blocks as a repeated measure. In the second ANOVA, only trial 2 scores were used; these were likewise summed over blocks of 100 trials and then cast into a comparable three-way repeated measures ANOVA. As before, Duncan tests were subsequently performed when significant main effects and/or interactions were disclosed.

Scores from the oddity tests were subjected to two separate repeated ANOVA tests. The first involved summing subject scores on trials 1-6 over blocks of eight problems (four blocks in all); these sums were entered into a three-way repeated-measures ANOVA, with the diet group and subject test group as independent variables and with problem blocks as a repeated measure. A second, comparable ANOVA was performed on the trial 1 data only. Again, Duncan tests were subsequently supplied where appropriate.

Finally, scores from the hearing test were inspected. Because all subjects detected each level of the sound stimuli at least once (see Results below), no formal statistical analyses were performed on these data.

RESULTS

The results of the present experiment revealed that (a) all 20 subjects were successfully adapted to the WGTA, all subjects displayed significant learning on all of the tests in the formal battery, and no subject displayed any deficits on the hearing test; (b) no significant differences were found between any of the five diet groups with respect to adaptation to the WGTA, performance on any of the formal learning tests, or performance on the hearing test; and (c) significant differences between the first and second groups of 10 subjects each were found on the adaptation measures and on the trials 1-6 measure of performance on the oddity learning set test. Details of these findings are presented below.

Adaptation Measures

As mentioned above, all of the subjects successfully completed all three stages of the adaptation procedure, but the second test group of 10 subjects took longer to adapt than the first test group of 10 subjects. There were no significant diet group differences or associated interactions disclosed for any of the adaptation measures.

The first stage of adaptation to the WGTA (for both groups of 10 subjects) was completed in an average of 7.25, 8.0, 7.25, 9.25, and 7.0 days for diet groups A1, A2, A3, P, and C, respectively; no differences between any of these groups achieved statistical significance. However, the first test group of 10 subjects completed this stage of adaptation in significantly fewer days than did the second test group of 10 subjects (5.6 versus 10.0 days; $F = 20.61; df = 1,10; p < 0.0005$). The interaction between diet groups and subject test groups was not significant.

There were no significant differences between diet groups in the amount of time required to complete the two-hole board adaptation state ($\overline{X}$'s = 224, 217, 200, 222, and 220 sec per adaptation day for diet groups A1, A2, A3, P, and C, respectively; $F < 1$; $df = 4,10$). There was, however, a significant difference between the first and second test groups of subjects ($\overline{X}$'s = 176 and 257 sec per adaptation day for test group 1 and test group 2, respectively; $F = 25.54$; $df = 1,10$; $p < 0.0005$). The diet group X test group interaction was not significant ($F = 1.27; df = 4,10$).

For the final stage of adaptation, the ANOVA revealed no significant effect of diet group ($\overline{X}$'s = 198, 203, 179, 182, and 209 sec per adaptation day for diet groups A1, A2, A3, P, and C, respectively; $F = 1.86$; $df = 4,10$). However, a significant effect of test group was obtained ($\overline{X}$'s = 149 and 239 sec per adaptation day for the test of group 1 and the test of group 2, respectively; $F = 113.58$; $df = 1,10; p < 0.0005$). The diet group X test group interaction was not significant.

Learning Test Battery

All 20 subjects displayed significant improvement on each of the four learning tests over repeated trials, displaying levels of performance that were consistent with those displayed by normal rhesus and stumptail macaques of comparable age (21, 33).

Object Discrimination

Levels of performance attained by each diet group are presented in Figure 2. The ANOVA for object discrimination test scores revealed a significant effect of the repeated measure ($F = 5.28$; $df = 4,40$; $p < 0.0005$), reflecting a gradual

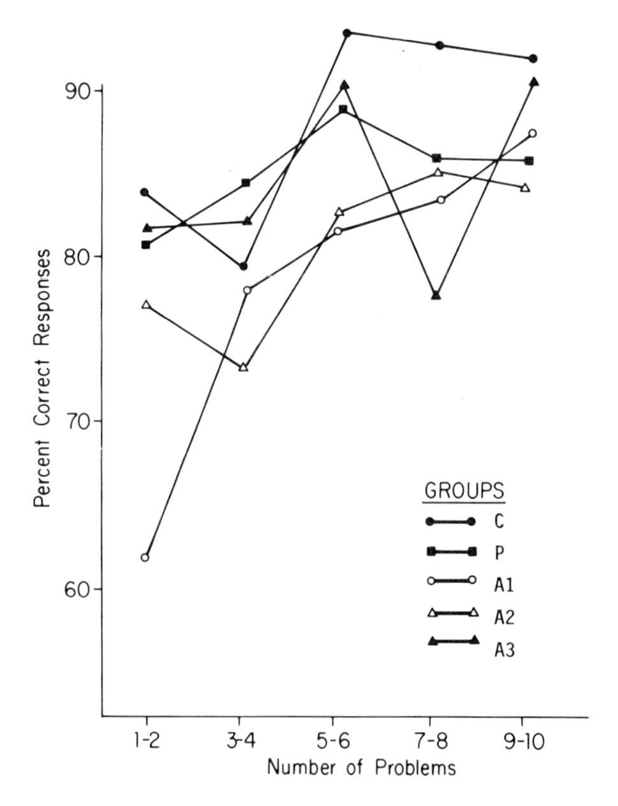

Figure 2 Object discrimination test performance. Group C animals (● — ●) were controls fed formula without any additions. Group P animals (■ — ■) received formula providing phenylalanine at 1.65 g/kg body weight per day. Animals in groups A1 (○ — ○), A2 (△ — △), and A3 (▲ — ▲) received formula providing aspartame at 1, 2, and 3 g/kg body weight per day, respectively.

improvement in performance over the 10 days of testing. No other significant main effects or interactions were disclosed by the ANOVA.

Pattern Discrimination

Levels of performance of test group 1 and test group 2 for each diet group are presented in Figure 3. All subjects showed improvement over the 10-day test period, as indicated by the highly significant effect of the repeated measure in the ANOVA ($F = 10.46; df = 4{,}40; p < 0.0005$). The ANOVA yielded no other significant main effects or interactions, although the test group X days interaction approached statistical significance ($F = 1.89; df = 4{,}10; p < 0.10$). In general, the results of the pattern discrimination test paralleled those of the object discrimination test, although scores for the former were slightly lower

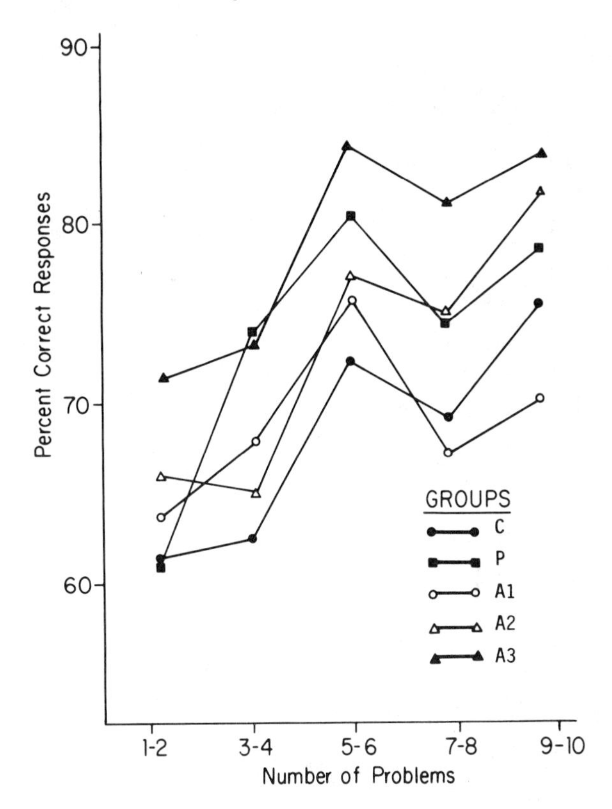

Figure 3 Pattern discrimination test performance. Group C animals (●——●) were controls fed formula without any additions. Group P animals (■——■) received formula providing phenylalanine at 1.65 g/kg body weight per day. Animals in Groups A1 (○——○), A2 (△——△) and A3 (▲——▲) received formula providing aspartame at 1, 2, and 3 g/kg body weight per day, respectively.

than those for the latter, consistent with previous findings for normal 1- to 2-year-old macaques (29).

Object Discrimination Learning Set

Cell means for trials 2-6 of the object discrimination learning set are presented in Figure 4, while cell means for trial 2 only are presented in Figure 5. As indicated in the figures, all subgroups acquired a learning set over repeated blocks of problems; indeed, trial 2 performance during the final block of 100 problems ranged from 84 to 95% correct across all subgroups, again consistent with levels reported for normal macaque subjects of comparable age (29,33). Trials 2-6 and trial 2 ANOVA tests all yielded highly significant effects of repeated problem blocks (F's = 57.86 and 44.52; df's = 5,50; $p < 0.0005$ for trials 2-6 and trial 2, respectively), but no other significant main effects or interactions.

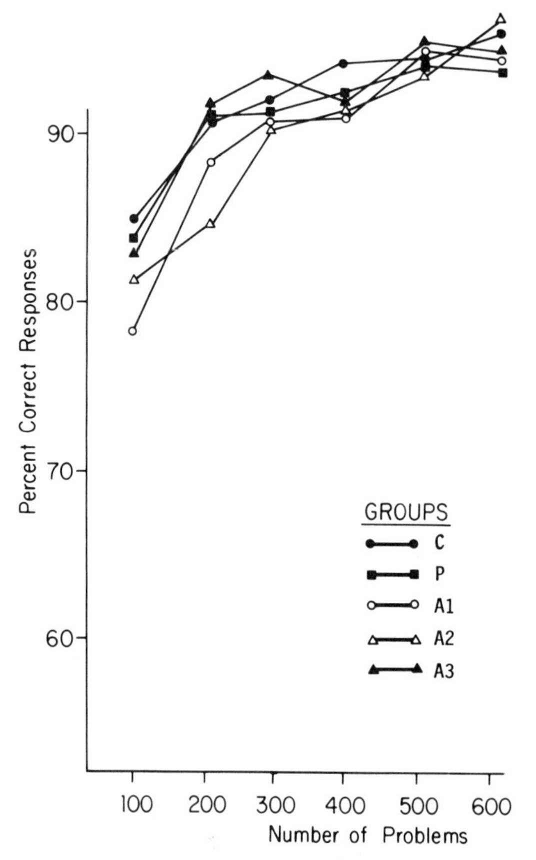

Figure 4 Object discrimination learning set: performance on trials 2-6. Group C animals (●—●) were controls fed formula without any additions. Group P animals (■—■) received formula providing phenylalanine at 1.65 g/kg body weight per day. Animals in groups A1 (○—○), A2 (△—△), and A3 (▲—▲) received formula providing aspartame at 1, 2, and 3 g/kg body weight per day, respectively.

Oddity Learning Set

As mentioned in the Methods section, all subjects were adapted to three-hole stimulus boards for 5 days prior to the start of formal oddity testing. The ANOVA of the total time required by each subject to complete each day's 25 adaptation trials revealed that the second test group of 10 subjects took significantly longer than did the first test group of 10 ($\bar{X}$'s = 205 versus 260 sec/day; $F = 20.57$; $df = 1,10$; $p < 0.005$. No other significant main effects or interactions were revealed.

Results of the actual oddity tests are presented in Figure 6, which illustrates performance levels on trials 1-6. The levels of performance shown in Figure 6

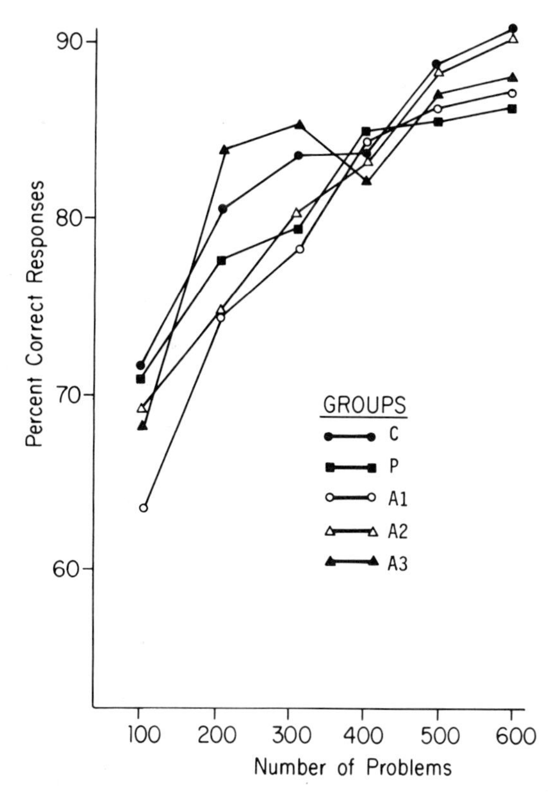

Figure 5 Object discrimination learning set: performance on trial 2 only. Group C animals (●——●) were controls fed formula without any additions. Group P animals (■——■) received formula providing phenylalanine at 1.65 g/kg body weight per day. Animals in groups A1 (○——○), A2 (△——△), and A3 (▲——▲) received formula providing aspartame at 1, 2, and 3 g/kg body weight per day, respectively.

indicate that all subgroups of subjects learned the oddity task and formed adequate learning sets; asymptomatic performance levels were consistent with those reported in previous studies of physically normal young macaque subjects (21,29).

The ANOVA for trials 1-6 data revealed a highly significant effect of the repeated measure ($F = 48.18$; $df = 3{,}30$; $p < 0.0005$), which reflected the improvement in performance displayed by subjects over problem blocks. The ANOVA also yielded a significant main effect of subject test group ($F = 5.76$; $df = 1{,}10$; $p < 0.05$), with the first test group of subjects displaying superior performance relative to that of the second test group of 10 subjects. No other significant main effects or interactions were disclosed by the trials 1-6 ANOVA. The ANOVA for data from trial 1 alone yielded only a significant main effect of the repeated

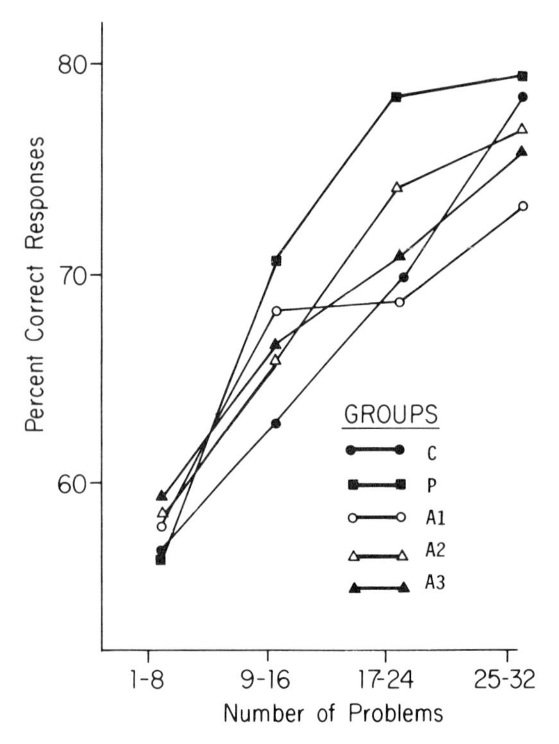

Figure 6 Oddity learning set: performance on trials 1-6. Group C animals (●——●) were controls fed formula without any additions. Group P animals (■——■) received formula providing phenylalanine at 1.65 g/kg body weight per day. Animals in groups A1 (○——○), A2 (△——△), and A3 (▲——▲) received formula providing aspartame at 1, 2, and 3 g/kg body weight per day, respectively.

measure ($F = 24.84$; $df = 3{,}30$; $p < 0.0005$), reflecting improved performance over blocks of problems.

Hearing Tests

The results of the hearing test provided no evidence of major hearing deficits in any of the 20 stumptail macaque subjects. All subjects oriented toward each of the six levels of auditory stimulation at least one of the two times the stimulus was presented. In fact, all but three subjects oriented toward the speakers on all trials. Those three subjects (#3346, #3593, and #3585) failed to orient on the second presentations of the 70-, 79-, and 88-dB stimuli, respectively. The facts that these were not the weakest auditory stimuli presented and that the failures to orient occurred late in the presentation schedule suggest that habituation, rather than an auditory deficit, accounted for the failure to orient on those trials.

DISCUSSION

The overall results of the present experimentation did not provide any evidence of deficits in either the learning performance or the hearing capabilities of any of the 20 juvenile stumptail macaques tested. The performance of these monkeys on every learning test was quite consistent with performance levels for physically normal macaques of comparable age previously reported for several different primate laboratories. The hearing test, while clearly not an exhaustive examination of each subject's auditory capabilities, failed to reveal any major deficits in the range of auditory stimulation to which macaques are maximally sensitive (37). In addition, informal reports of the testers indicated that all of the present subjects were responsive to various laboratory noises (e.g., animal calls, cage-washing machines, conversation between laboratory personnel) occurring outside of the test room during WGTA sessions, despite an ambient white noise level of 63-66 db inside of the test room. Thus the occurrence of gross hearing deficits, at least within the "normal" range of auditory stimulation, seems highly unlikely for any of these stumptail macaques.

The failure to find any statistically significant diet group differences (or associated interactions) on any of the learning test measures was somewhat unexpected, given that one of the diet conditions involved dietary addition of phenylalanine and that previous studies from Wisconsin (36) had disclosed some learning deficits associated with diets high in phenylalanine content. However, a rereading of the original studies performed at Wisconsin in the 1960s revealed that (a) learning deficits had been disclosed only on the most taxing learning tests rather than throughout the whole battery; (b) learning deficits were found only in some of the subjects exposed to high-phenylalanine diets and those subjects showing the deficits were not always those who had the highest blood levels of phenylalanine; thus it was *not* the case that a direct relationship between phenylalanine levels in blood and degree of learning deficit existed; and (c) in those cases where learning deficits had been found, the subjects had been chronically maintained on the phenylalanine diet not only during infancy but throughout the entire learning test period as well. Moreover, the Waisman-Harlow (36) subjects also displayed clear-cut physical abnormalities. In contrast, the present juvenile stumptail macaque subjects were maintained on normal diets during the WGTA testing periods, and they displayed no gross physical abnormalities or high susceptibility to contagious diseases during their stay at Wisconsin.

At any rate, the fact remains that no significant diet group differences or associated interactions were found for any of the present measures. One can only conclude that the dietary conditions to which these subjects had been exposed early in life did not yield long-term effects on their performance on a WGTA test battery that previously had been shown to be sensitive to age difference, size and location of cortical lesions, form of early rearing history, some chronic dietary conditions, and exposure to various environmental toxins.

On the other hand, significant differences between the first and second test groups of subjects were found for all stages of adaptation and for the trial 1-6 measure on the oddity learning set test. While the first and second test groups of subjects were tested at different times and were exposed to slightly different quarantine procedures, it seems unlikely that these alone could have produced the above group differences, inasmuch as the test procedures were identical and the quarantine differences were limited to the external facility in which the animals were housed (the actual cages were the same for both groups, and the physical characteristics of their respective housing rooms were highly similar). Instead, the group differences most likely resulted from differences in the subjects' early social histories, with the second test group reared under more socially isolated (or at least restricted) conditions than the first test group. While such a conclusion is speculative to some degree, it is consistent with previous findings regarding socially reared versus isolation-reared macaques (e.g., see Refs. 29 and 38). These findings revealed that isolates generally are slower to adapt to the WGTA than are socially reared monkeys on all tasks *except oddity*, for which they show a slight but statistically significant deficit. The fact that the testers also reported the second test group of subjects to be generally less tractable, more aggressive (toward the testers), more difficult to handle during transport to and from the WGTA, and more easily disturbed than the first test group is also consistent with the fact that the two groups of subjects had different social rearing histories.

The results obtained in monkeys are consistent with data obtained in rats by Potts et al. (39). They carried out behavioral tests (activity studies, conditioned avoidance response, Sidman avoidance response) in weanling rats fed diets providing no additions, 2.5 and 5% phenylalanine, and 4.5 and 9% aspartame for 90 days, with behavioral tests carried out while the animals were fed the diets. No significant effects were noted between control animals and those fed diets providing 2.5% phenylalanine or 4.5% aspartame (equivalent to an aspartame dose of 5.5 g/kg body weight per day). However, small effects were noted in animals fed diets providing 5% phenylalanine or 9% aspartame (equivalent to an aspartame dose of 11 g/kg body weight per day). These data suggest that phenylalanine-induced behavioral effects require administration of very high doses of phenylalanine and/or aspartame.

In summary, the present study of learning test performance by juvenile-age stumptail macaque monkeys, while revealing some effects apparently due to differential early social rearing histories, nevertheless failed to disclose any obvious adaptation, hearing, or learning test performance deficits that could be attributed to early exposure to various levels of aspartame in their diet. Instead, all subjects displayed age-normative levels of performance on all portions of the standardized learning test battery. Of course, failure to reject a null hypothesis does not carry the same scientific weight as a successful direct test of that same hypothesis.

The original rationale for conducting the present study was to test an animal model for the effects of aspartame consumption early in life on cognitive functioning during childhood. It should be remembered that cross-species generalizations always involve certain risks, especially when the generalizations in question are simulative rather than substitutive in nature (24). While there is clearly nothing in the present findings to suggest in any way that aspartame consumption at reasonable levels by young human children might pose some risk for subsequent normal cognitive development, one would be wise to keep in mind that stumptail macaques are not furry little humans with stumpy tails but, instead, members of a different (albeit closely related) species. Nonetheless, they are an appropriate animal model to study behavioral effects of high doses of test compounds.

REFERENCES

1. Pfungst, O. (1904). Research cited in Ref. 2 (by Sparks), pp. 145-146.
2. Sparks, J. (1982). *The Discovery of Animal Behavior*, Collins, London.
3. Thorndike, E. L. (1898). Animal intelligence. *Psych. Rev.* 2, 1-109.
4. Watson, J. B. (1913). Psychology as the behaviorist views it. *Psych. Rev.* 20, 158-177.
5. Pavlov, I. P. (1927). *Conditioned Reflexes*, Oxford University Press, London.
6. Lorenz, K. (1935). Der Kumpan in der Umwelt des Vogels. *J. Ornithol.* 83, 137-213.
7. Skinner, B. F. (1938). *The Behavior of Organisms*, Appleton-Century-Crofts, New York.
8. Harlow, H. F. (1949). The formation of learning sets. *Psychol. Rev.* 56, 51-66.
9. Seligman, M. E. P. (1970). On the generality of the laws of learning. *Psychol. Rev.* 77, 406-418.
10. Kandel, E. R. (1978). *A Cell-Biological Approach to Behavior*, Grass Lecture Monograph 1, Society for Neuroscience, Bethesda, Md.
11. Hilgard, E. R., and Marquis, D. G. (1940). *Conditioning and Learning*, Appleton-Century-Crofts, New York.
12. Rescorla, R. A. (1967). Pavlovian conditioning and its proper control procedures. *Psychol. Rev.* 74, 71-80.
13. Bitterman, M. E. (1960). Toward a comparative psychology of learning. *Am. Psychol.* 15, 704-711.
14. Warren, J. M. (1965). Primate learning in comparative perspective. In *Behavior of Nonhuman Primates*, Vol. 1 (Schrier, A., Harlow, H., and Stollnitz, F., eds.), Academic, New York.
15. Fulton, J. F. (1955). *A Textbook of Physiology*, Saunders, Philadelphia.
16. Rumbaugh, D. M., and Pate, J. R. (1983). Toward a comparative framework for studies in intelligence. In *Developmental Behavior Genetics and Behavior Analysis* (Gray, D., Plomin, R., and Johnston, J., eds.), Lawrence Erlbaum, Hillsdale, N.J.

17. Rumbaugh, D. M., and Gill, T. U. (1973). The learning skills of great apes. *J. Hum. Evol.* 2, 171-179.
18. Piaget, J. (1960). *The Child's Conception of the World*, Littlefield-Adams, Patterson, N.J.
19. Flavel, J. H. (1977). *Cognitive Development*, Prentice-Hall, Englewood Cliffs, N.J.
20. Sameroff, A. J., and Cavanaugh, P. J. (1979). Learning in infancy: A developmental perspective. In *Handbook of Infant Development* (Osofsky, J., ed.), Wiley, New York.
21. Harlow, H. F. (1959). The development of learning in the rhesus monkey. *Am. Sci.* 47, 459-479.
22. Wise, K. L., Wise, L. A., and Zimmermann, R. R. (1974). Piagetian object permanence performance in the infant rhesus monkey. *Dev. Psychol.* 10, 429-437.
23. Chevalier-Skolnikoff, S. (1977). A Piagetian model for describing and comparing socialization in monkey, ape, and human infants. In *Primate Bio-Social Development* (Chevalier-Skolnikoff, S., and Poirier, F., eds.), Garland Press, New York.
24. Suomi, S. J., and Immelmann, K. (1983). On the process and product of cross-species generalization. In *Observing Man Observing Animals* (Rajecki, D. W., ed.), Lawrence Erlbaum, Hillsdale, N.J.
25. Mishkin, M., and Pribram, K. H. (1954). Visual discrimination performance following partial ablations of the temporal lobe. I. Ventral vs. lateral. *J. Com. Physiol. Psychol.* 47, 167-174.
26. Harlow, H. F., Blomquist, A. J., Thompson, C. I., Schiltz, K. A., and Harlow, M. K. (1968). Effects of induction age and size of frontal lobe lesions on learning in rhesus monkeys. In *The Neuropsychology of Development* (Isaacson, R. L., ed.), Wiley, New York, pp. 79-120.
27. Zimmermann, R. R., Strobel, D. A., Steere, P., and Geist, C. R. (1975). Behavior and malnutrition in the rhesus monkey. In *Primate Behavior*, Vol. 4 (Rosenblum, L. A., ed.), Academic, New York, pp. 241-302.
28. Bushnell, P. J., and Bowman, R. E. (1979). Effects of chronic lead ingestion on social development in infant rhesus monkeys. *Neurobehav. Toxicol.* 1, 207-219.
29. Harlow, H. F., Gluck, J. P., and Schiltz, K. A. (1973). Differential effect of early enrichment and deprivation on learning in the rhesus monkey (*Macaca mulatta*). *J. Comp. Physiol. Psychol.* 84, 598-604.
30. Reynolds, W. A., Bauman, A. F., Naidu, S., Stegink, L. D., and Filer, L. J., Jr. (1984). Developmental assessment of infant macaques receiving dietary aspartame or phenylalanine. In *Aspartame: Physiology and Biochemistry* (Stegink, L. D., and Filer, L. J., Jr., eds.), Marcel Dekker, New York, Chap. 20.
31. Harlow, H. F., and Bromer, J. (1938). A test apparatus for monkeys. *Psychol. Rec.* 2, 434-436.
32. Meyer, D. R., Treichler, F. R., and Meyer, P. M. (1965). Discrete-trial training techniques and stimulus variables. In *Behavior of Nonhuman Primates,*

Vol. 1 (Schrier, A. M., Harlow, H. F., and Stollnitz, F., eds.), Academic, New York, pp. 1-49.

33. Schrier, A. M. (1965). Learning-set formation by three species of macaque monkeys. *J. Comp. Physiol. Psychol.* 61, 490-495.
34. Riopelle, A. J., Alper, R. G., Strong, P. N., and Ades, H. W. (1953). Multiple discrimination and patterned string performance of normal and temporal lobectomized monkeys. *J. Comp. Physiol. Psychol.* 46, 145-149.
35. Raisler, R. L., and Harlow, H. F. (1965). Learned behavior following lesions of the posterior association cortex in infant, immature, and preadolescent monkeys. *J. Comp. Physiol. Psychol.* 60, 167-174.
36. Waisman, H. A., and Harlow, H. F. (1965). Experimental phenylketonuria in infant monkeys. *Science* 147, 685-695.
37. Stebbins, W. C. (1978). Hearing of the primates. In *Recent Advances in Primatology*, Vol. 1 (Chivers, D., and Herbert, J., eds.), Academic, New York, pp. 705-720.
38. Harlow, H. F., Harlow, M. K., Schiltz, K. A., and Mohr, D. J. (1971). The effect of early adverse and enriched environments on the learning ability of rhesus monkeys. In *Cognitive Processes of Nonhuman Primates* (Jarrard, L. E., ed.), Academic, New York, pp. 121-148.
39. Potts, W. J., Bloss, J. L., and Nutting, E. F. (1980). Biological properties of aspartame. I. Evaluation of central nervous system effects. *J. Environ. Pathol. Toxicol.* 3, 341-353.

22

Aspartame and Brain Tumors: Pathology Issues

Adalbert Koestner
Michigan State University, East Lansing, Michigan

INTRODUCTION

The purpose of this presentation is to assess the prospective neuro-oncogenic potential of aspartame in rats. The possibility that aspartame had neuro-oncogenic potential was one of the major issues at the Public Board of Inquiry evaluating the safety of aspartame (1,2). My assessment of this question will be based primarily upon two 2-year studies conducted by G. D. Searle & Co. and Hazelton Laboratories. These studies are designated as E-33/34 and E-70 (3). In this assessment the brain tumors encountered in these two studies will be compared with the natural incidence and spectrum of central nervous system (CNS) tumors in rats and with the incidence and characteristics of brain tumors induced by known neurocarcinogens.

An increased awareness of the neurocarcinogenic potential of a variety of chemicals has resulted in more careful histological examination of all toxicological studies, even where no clinical history of neurological signs were noted or where detectable gross CNS lesions were not observed at necropsy. This concern resulted in an evaluation of the data from rats fed aspartame in the two Searle studies designated E-33/34 and E-70. Surprisingly, a higher incidence of brain tumors than previously reported in nontreated animals was suggested by this review. Since the chance of equal distribution of brain tumors in a randomly selected and grouped population of rats (groups of 40-100) is low, large variations in brain tumor incidences among such small groups of animals should be expected.

The chances are always greater for one of several experimental groups to be randomly affected by a higher incidence of tumors than they are for the single control group. The results become alarming to the pathologist if the highest incidence of brain tumors arbitrarily coincides with the highest-dose group. Statistical analysis in such cases may at times be equivocal, and convincing solutions can only be achieved by a thorough biological analysis. We shall present criteria for such a biological analysis using aspartame as an example. A statistical evaluation of the data from the Searle studies is presented in the chapter by Cornell et al. (4).

ASPARTAME STUDIES E-33/34 AND E-70

The E-33/34 study employed four experimental groups and one control group of Charles River CD rats. The experimental rats received daily aspartame doses of 1000, 2000, 4000, and 6000-8000 mg/kg body weight orally in the diet for 104 days following weaning. The E-70 study had one control group and two experimental groups. In this study pregnant animals in the experimental groups received 2000 or 4000 mg/kg aspartame daily during pregnancy and lactation. This feeding was continued in the offspring thereafter. As will be discussed, this latter study is particularly important for an accurate assay of neurocarcinogenicity, because aspartame was administered during pregnancy, thus exposing the animals to the test substance and/or its metabolites during fetal life, as well as during adolescent and adult life. Both experiments were terminated when the rats were 2 years old, at which time all survivors were killed. All rats were necropsied after premature death or at the termination of the experiment. A total of eight brain sections from five coronal blocks were examined histologically. The brain tumor incidence in these two studies is listed in Tables 1 and 2.

As can be seen in Tables 1 and 2, the brain tumor incidence in study E-33/34 was low in the control animals and animals fed the high dose of aspartame, and highest in animals fed a medium dose of aspartame. In study E-70 the greatest incidence of brain tumors was observed in the control group. The incidence of

Table 1 E-33/34 Aspartame 2-Year Study (Postnatal Exposure)

	Control	1000 mg/kg	2000 mg/kg	4000 mg/kg	6000-8000 mg/kg	Total
Number of brains examined[a]	119	80	80	80	80	439
Number without tumors	118	77	79	75	78	427
Number with tumors[b]	1	3	1	5	2	12 (2.7%)

[a]Eight coronal sections per brain. A Searle audit of Hazelton's slides revealed 119 instead of 120 control preparations.
[b]From p. 838 of UAREP Authentication Report (HCFV-112) (Ref. 28).

Table 2 E-70 Aspartame 2-Year Study (Transplacental and Postnatal Exposure)

	Control	2000 mg/kg	4000 mg/kg	Total
Number of brains examined[a]	119	78	79	272
Number without tumors	111	75	78	264
Number with tumors[b]	4	3	1	8 (2.9%)

[a]Eight coronal sections per brain. Numbers from report E-70, p. 14, confirmed Searle audit of Hazelton's slides.
[b]From p. 838 of UAREP Authentication Report (HCFV-112) (Ref. 28).

brain tumors in all aspartame treated animals combined was 3.1%, and 2.1% in control rats. The overall brain tumor incidence in these experiments was 2.8%, well within the range of naturally occurring brain tumors in Charles River CD rats (Table 3) (5-7). A statistical analysis of these data is presented by Cornell et al. (4). The biological analysis presented here is based upon criteria which are recognized as characteristic features of neurocarcinogenic agents:

1. Capability to reliably and consistently increase the tumor incidence
2. Transpolation of the appearance of brain tumors to younger age
3. Demonstration of a dose-effect relationship
4. Affection of embryonal and fetal brain cells greater than that of adult neuro-ectodermal cells
5. Shift to more anaplastic tumor types

Each of these criteria will be examined in turn relative to the data observed in Searle studies E-33/34 and E-70.

Table 3 CNS Tumor Incidence in Charles River CD Rats

Reference	Histopathology	Strain	Tumors (%)
5	Restricted to gross tumors	Charles River CD	1.2
6	Several sections	Charles River CD	2.1
7	Serial sections	Charles River CD	3.2
3	Eight sections	Charles River CD	2.8

CAPABILITY TO RELIABLY AND CONSISTENTLY INCREASE BRAIN TUMOR INCIDENCE

To assess this criterion we need to briefly discuss the natural brain tumor incidence in rats. There is great variation in the incidence of tumors of the nervous system in different animal species. Primary tumors of the CNS in humans account for about 1.2% of all autopsied deaths and approximately 9% of all primary human tumors (8). The dog closely shares man's environment and, when permitted to live its whole life-span, demonstrates a brain tumor incidence similar to that of adult human beings (9). The incidence of CNS tumors is often considered to be low in most other animal species; however, statements of this sort need to be interpreted with caution, since they depend greatly upon (a) the availability of prospective brain tumor material to the pathologist; (b) specific interest in a particular anatomical area; and (c) thoroughness in the examination.

The recent development of the rat as a reliable brain tumor model (10-18) increased interest in the spontaneous incidence of CNS tumors in that species. Brain tumors in various rat strains have been recorded by commercial breeding companies, individual investigators, and governmental agencies. However, many surveys of naturally occurring brain tumors in rats were either carried out in animals that were too young, or were restricted to grossly visible tumors. In such surveys microtumors were not detected; thus the true tumor incidence is not reflected in the reports. This is also the reason why reported incidences of experimentally produced brain tumors differ between reports, in spite of identical experimental designs with the same strain of rats and identical doses of the same carcinogen.

Dagel et al. (5) reported a 1.2% brain tumor incidence in the Charles River CD strain of rats, 1.9% in the Osborn-Mendel strain, and 1.3% in the Wistar strain. In their survey histological examination of the brain was restricted to animals with neurological signs or those with gross brain lesions. As a result, microtumors were definitely missed. This tumor incidence in Charles River CD rats compares with a 2.1% brain tumor incidence reported from a National Cancer Institute survey (6) where several sections of every brain were histologically examined (Table 3). In comparison, Sumi et al. (16) reported a brain tumor incidence of 7.3% (including microtumors) in old Wistar rats (up to 798 days old) based upon histological examination of eight sections from eight coronal blocks obtained from every brain.

A considerable number of publications are available on naturally occurring brain tumors in rats; however, most are difficult to evaluate. Some surveys suffer from incomplete information with respect to the age of the animal at the time of necropsy used to calculate the statistics (17-21). In other surveys, only brains with macroscopic lesions were histologically examined (5,22), or only one to two sections were prepared. Since brain tumors are an age-related lesion in rats, and may only be present as microtumors at an age of 2 years (the time most studies are terminated), both age and thoroughness of histological examination significantly influence the incidence of brain tumors.

In summary, the reported incidence of naturally occurring brain tumors in rats varies considerably, depending upon the specific method of investigation. The two most important parameters determining detection of brain tumors are the age at which the assessment is made and the thoroughness with which the examination is performed. Brain tumors in rats are a late life event. If examination is done too early (between 1 and 2 years), tumors may be too small to be detected (microtumors) and may even be missed microscopically if too few brain segments are examined. Because of these factors, the reported incidence of brain tumors in rats ranges from less than 1% (if only grossly detectable tumors were counted) to over 7% when numerous brain sections of all aged rats are examined microscopically. Based on the most reliable reports, the rat exceeds all mammalian species, including man, in brain tumor incidence.

A comparison of brain tumors of aspartame-fed rats in the Searle studies (E-33/34 and E-70) with the reported natural brain tumor incidence in rats reveals that the brain tumor incidence in the Searle studies is well within the range of the natural incidence of brain tumors in CD rats. This conclusion is further supported by a chronic feeding study (E-77-78) with daily doses of 0, 750, 1500, and 300 mg/kg of diketopiperazine, a degradation product of aspartame (23).

TRANSPOLATION OF TUMOR APPEARANCE TO A YOUNGER AGE

This criterion basically suggests that exposure of an animal to an authentic neurocarcinogen should cause tumor appearance at a young age. This contrasts with the normal situation where tumors occur late in life. For example, methylnitrosourea is a potent neurocarcinogen. When administered repeatedly to young adult rats, 97% of the exposed animals developed tumors of the nervous system, with an average survival time of 339 days (12,24). The great majority of the resulting tumors killed the animals before the first year of life. The latency period was further reduced if embryos were exposed transplacentally to one single dose of 50 mg/kg ethylnitrosourea. In that case there was a 100% neurogenic tumor incidence. The average survival time was 211 days with the first tumor clinically recognized at the 141st day of life (12,14,25).

In contrast, all tumors observed in the Searle-Hazelton studies were mostly microtumors at the time of necropsy, and only incidentally detected by histological examination. The exceptions to this were five astrocytomas found in rats killed at 2 years of age, and a medulloblastoma which were recognized on gross examination. The medulloblastoma is also the only tumor that occurred early in life, namely, before the first year of age, a typical feature of this tumor in animals and man (26). Because of the inherent early occurrence of this type of tumor, and the fact that it occurred in study E-33/34, where administration of the test substance was not begun until after weaning (the tumor killed the rat at 16 weeks of age), no relationship between this tumor and the administration of the test compound could be established. Medulloblastomas are induced congenitally in most cases. If aspartame was capable of inducing such a tumor, it should have produced a significant number of medulloblastomas in E-70, where the compound was administered throughout gestation, thus exposing the fetal brain to the test compound or its metabolites.

All other tumors in the Searle-Hazelton study occurred after the first year of life and 9 of the 14 astrocytomas were still microtumors at termination of the 2-year experiments.

Thus aspartame does not meet the second criterion of a neurocarcinogen, that is, causing the transpolation of tumors to a younger age.

DOSE-EFFECT RELATIONSHIP

Authentic neurocarcinogens should increase tumor incidence as the dose administered increases. For example, a clear dose-effect relationship can be demon-

Table 4 Transplacental Tumor Induction with Ethylnitrosourea in SPF Sprague-Dawley Rats on the 20th Day of Gestation[a]

Single intravenous dose (mg/kg)	Number of dams	Number of offspring	Number of rats with tumors	Average survival time (days)[a]
1	4	46	13	487 ± 119
5	3	24	14 (58%)	368 ± 149
20	2	17	16 (94%)	269 ± 77
50	2	25	25 (100%)	211 ± 70

[a]Mean SD.
Source: From Ref. 12.

Table 5 Tumor Incidence in Rats Fed Aspartame (2 years)

Dose (mg/kg)	Number of rats (E-33/34)	Tumors	Dose (mg/kg)	Number of rats (E-70)	Tumors
Control	119	1	Control	115	4
1000	80	3	–	–	–
2000	80	1	2000	78	3
4000	80	5	4000	79	1
6000-8000	80	2	–	–	–
Total	439	12 (2.7%)	Total	272	8 (2.9%)

strated using ethylnitrosourea; as the dose is increased, the tumor incidence increases and survival time is shortened (Table 4).

This was not the case with aspartame. Tumors in the Searle study appeared randomly; in E-70 the highest incidence was in controls, and in E-33/34 the highest incidence was in an experimental group exposed to a middle dose of aspartame. The tumor incidence in both studies was low in groups fed the highest dose (Table 5).

Thus aspartame does not meet the third criterion for a carcinogenic agent, that is, one that produces a dose-effect relationship.

HIGHER SUSCEPTIBILITY OF FETAL BRAIN CELLS TO CARCINOGENS AS COMPARED TO ADULT BRAIN CELLS

In studies with ethylnitrosourea, fetuses were 50-100 times more susceptible to neurogenic tumor induction by N-nitroso compounds than adults. One single intravenous inoculation of 50 mg/kg ethylnitrosourea into pregnant rats resulted in 100% neurogenic tumors in the offspring. This is an important observation, since a high brain tumor incidence is noted in children (27). Because of this greater susceptibility, studies in which the test compound is administered in utero, while the developing brain is most vulnerable and susceptible, are a sensitive and accurate means to measure neurocarcinogenicity. Multiple inoculations with N-nitroso compounds are always required to produce neurogenic tumors in adult rats.

If aspartame were a neurocarcinogen, one would have expected an increase in brain tumor incidence in the E-70 study compared to the E-33/34 study, since in E-70 fetuses were exposed throughout gestation to aspartame and its metabolites. However, as may be calculated from the data in Table 5, a lower incidence of brain tumors was noted in aspartame-treated rats in study E-70 (2.5%) than in study E-33/34 (3.4%).

SHIFT TO MORE ANAPLASTIC TUMOR TYPES

If we examine the tumor types produced in rats exposed to nitrosourea compounds, an increased number of anaplastic tumors and a decreased incidence of differentiated astrocytomas can be noted (Table 6).

In contrast, the spectrum of tumors noted in the Searle-Hazelton study is comparable to surveys of spontaneous brain tumors in CD rats, which upon histological examinations consist primarily of mature and differentiated tumors of the glial cell population (Table 7). Thus aspartame does not meet this criterion either.

Table 6 Classification of Tumors Produced in Sprague-Dawley Rats with Methylnitrosourea (Incidence of Nervous System tumors 97%) and Ethylnitrosourea (Incidence 100% in Offspring)

Tumor type	Methylnitrosourea, 5 mg/kg i.v. weekly for 30 weeks	Ethylnitrosourea, 50 mg/kg i.v. to pregnant rats on 20th day of gestation
Astrocytoma	5	4
Oligodendroglioma	18	32
Mixed glioma	14	19
Anaplastic	14	5
Gliaependymoma	0	4
Ependymoma	0	10
Gliosarcoma	7	0
Meningioma	0	1
Total CNS	58	75
Neurinoma	7	3
Anaplastic neurinoma	6	24
Total peripheral nervous system	13	27
Extraneural	8	7
Total neoplasia	79	109

Source: From Ref. 12.

Table 7 Comparison of Tumor Types in the Searle-Hazelton Studies with Those Reported by Dagel et al.

Type	Searle-Hazelton	Dagel et al.[c]
Astrocytome	14	19
Oligodendroglioma	1	1
Glioma, not specified, gliomatosis	1	1
Meningioma	1	3
Ependymoma	3	4
Reticulosis	0	2
Pinealoma	0	2
Medulloblastoma	1	0
Total number of brain tumors	21 (2.9%)[a]	32 (1.4%)
Total number of rats	711	2242[b]

[a]Medulloblastoma included.
[b]Several rat strains in addition to CD rats.
[c]Data from Ref. 5.

CONCLUSION

In the two lifetime aspartame feeding studies conducted by Searle-Hazelton, the number of brain tumors was well within the range of reported brain tumor incidences in rats (2.8%), particularly since the incidence rate was established by microscopic examination of eight sections from every brain. There was no transpolation of tumors to an earlier age, no higher effect rate in fetal brain cells, and no dose-effect relationship between the amount of aspartame administration and the incidence of brain tumors. Furthermore, there was no shift in the type of tumor from that normally seen in rats.

Significantly, in the most rigorous test for brain tumorigenicity, when treatment was begun in utero, control animals showed more tumors than the treatment group. Since aspartame did not fulfill any of the criteria of a neurocarcinogenic agent and did not cause an increased incidence of brain tumors in rats, even at extremely high doses and administration over a long period of time, aspartame cannot be classified as a neurocarcinogenic agent in rats.

REFERENCES

1. Nauta, W. J. H., Lampert, P. W., and Young, V. R. (1980). Aspartame. Decision of the Public Board of Inquiry, Department of Health and Human Services, Food and Drug Administration, Docket No. 75F-0355, Washington, D.C. (September).
2. Aspartame. Commissioner's Final Decision (1981). Docket No. 75F-0355. *Fed. Reg.* 46, 38285-38308.
3. Studies E-33/34 and E-70. Report filed by G. D. Searle. Docket No. 75F-0355, Department of Health and Human Services, Food and Drug Administration, Washington, D.C.
4. Cornell, R. G., Wolfe, R. A., and Sanders, P. G. (1984). Aspartame and brain tumors: Statistical issues. In *Aspartame: Physiology and Biochemistry.* (Stegink, L. D., and Filer, L. J., Jr., eds.), Marcel Dekker, Inc., New York, Chapter 23.
5. Dagel, G. E., Zwicker, G. M., and Renne, R. A. (1979). Morphology of spontaneous tumors in the rat. *Vet. Pathol.* 16, 318-324.
6. Gart, J. J., Chu, K. C., and Tarone, R. E. (1979). Statistical issues in interpretation of chronic bioassay tests for carcinogenicity. *J. Natl. Can. Inst.* 62, 957-984.
7. Thompson, S. W., and Hunt, R. D. (1963). Spontaneous tumors in the Sprague-Dawley rat: Incidence rates of some types of neoplasms as determined by serial section versus single section technics. *Ann. N.Y. Acad. Sci.* 108, 832-845.
8. Rubinstein, L. J. (1972). Tumors of the central nervous system. AFIP Fascicle 6, American Registry of Pathology, Washington, D.C.
9. Luginbühl, H., Fankhauser, R., and McGrath, J. T. (1968). Spontaneous neoplasms of the nervous system in animals. *Neurol. Surg.* 2, 85-164.
10. Druckrey, H., Ivankovic, S., and Preussman, R. (1964). Selektive Erzeugung

von Hirntumoren bei Ratten durch Methylnitrosoharnstoff. *Naturwissenschaften* 51, 144.
11. Janisch, W., and Schreiber, D. (1977). In *Experimental Tumors of the Central Nervous System* (Bigner, D. D., and Swenberg, J. A., eds.), Upjohn Co., Kalamazoo, Mi.
12. Koestner, A. (1978). Animal model of human disease: N-nitrosourea-induced neurogenic tumors in the rat. *Comp. Pathol. Bull.* 10, 2-3.
13. Koestner, A., Swenberg, J. A., and Denlinger, R. H. (1979). Host factors affecting perinatal carcinogenesis by resorptive alkylnitrosoureas in rats. *Nat. Cancer Inst. Monogr.* 51, 211-217.
14. Swenberg, J. A., Wechsler, W., and Koestner, A. (1972). The sequential development of transplacentally induced neuroectodermal tumors. *J. Neuropathol. Exp. Neurol.* 31, 202-203.
15. Zimmerman, H. M., and Arnold, H. (1943). Experimental brain tumors. I. *Cancer Res.* 1, 919-938; II. *Am. J. Pathol.* 19, 939-955.
16. Sumi, N., Stavrou, D., Frohberg, H., and Jochmann, G. (1976). The incidence of spontaneous tumours of the central nervous system of Wistar rats. *Arch. Toxicol.* 35, 1-13.
17. Fitzgerald, J. E., Schardein, J. L., and Kurtz, S. M. (1974). Spontaneous tumors of the nervous system in albino rats. *J. Nat. Cancer Inst.* 52, 265-273.
18. MacKenzie, W. F., and Garner, F. M. (1973). Comparison of neoplasms in six sources of rats. *J. Nat. Cancer Inst.* 50, 1243-1257.
19. Mawdesley-Thomas, L. E., and Newman, A. J. (1974). Some observations on spontaneously occurring tumors of the central nervous system of Sprague-Dawley derived rats. *J. Pathol.* 112, 107-116.
20. Newman, A. J., and Mawdesley-Thomas, L. E. (1944). Spontaneous tumors of the central nervous system of laboratory rats. *J. Comp. Pathol.* 84, 39-50.
21. Schardein, J. L., Fitzgerald, J. E., and Kaump, D. H. (1968). Spontaneous tumors in Holtzman-source rats of various ages. *Pathol. Vet.* 5, 238-252.
22. Gilbert, C., and Gillman, J. (1958). Spontaneous neoplasms in the albino rat. *S. Afr. J. Med. Sci.* 23, 257-272.
23. Ishii, H. (1984). Effects of excessive aspartame on the learning test performance of young stumptail macaques. In *Aspartame: Physiology and Biochemistry* (Stegink, L. D., and Filer, L. J., Jr., eds.), Marcel Dekker, New York Chap. 15.
24. Swenberg, J. A., Koestner, A., and Wechsler, W. (1972). The induction of tumors of the nervous system with intravenous methylnitrosourea. *Lab. Invest.* 26, 74-85.
25. Koestner, A., Swenberg, J. A., and Wechsler, W. (1971). Transplacental production with ethylnitrosourea of neoplasms of the nervous system in Sprague-Dawley rats. *Am. J. Pathol.* 63, 37-50.
26. Zülch, K. J. (1965). *Brain Tumors: Their Biology and Pathology*, 2nd American ed., Springer, New York.
27. Wechsler, W., and Koestner, A. (1977). Developmental biology related to oncology. In *Principles of Surgical Oncology* (Raven, R., ed.), Plenum, pp. 93-112.

28. Authentication review of selected materials submitted to the Food and Drug Administration relative to application of Searle Laboratories to market aspartame (1978). Report prepared by Universities Associated for Research and Education in Pathology (UAREP) (3 vols.). Filed as Study E-102, Aspartame Petition Docket No. 75F-0355, Department of Health and Human Services, Food and Drug Administration, Washington, D.C. (November 18).

23

Aspartame and Brain Tumors: Statistical Issues

Richard G. Cornell and Robert A. Wolfe
University of Michigan, Ann Arbor, Michigan

Paul G. Sanders
G. D. Searle & Co., Skokie, Illinois

INTRODUCTION

Modern society demands that government provide protection against possible harm from the results of technological improvement. Very often the process of regulating the introduction of a new substance relies on research which requires careful statistical design and analysis. It is widely recognized that a careful blending of statistical and scientific reasoning is required to reach a proper evaluation and balance of the many conflicting value judgments involved in such regulation (1-3).

Most readers are familiar with the widespread public controversy that has attended regulation of food additives (cyclamate, saccharin, Red No. 2), insecticides (malathion, heptachlor), and naturally occurring substances (aflatoxin). Often the overriding concern in regulating test substances has been to limit human exposure to carcinogenic compounds. While great strides are being made in the understanding, diagnosis, and treatment of cancer, there is still no certain way to predict the carcinogenic potential of a compound.

This chapter describes the statistical analysis of the brain tumor data from long-term feeding studies on aspartame. Much of the statistical work associated with aspartame was motivated from a concern for its carcinogenic potential. However, for many of those who had been concerned with the safety of aspartame during its development, the Public Board of Inquiry's decision (4) concerning aspartame's carcinogenic potential came as a surprise. The ubiquity of the component amino acids aspartic acid and phenylalanine suggests that the likelihood of carcinogenicity is small. However, from among the many studies exploring carcinogenic potential by both in vitro and in vivo methods, one study designated E-33/34 suggested a possible association of aspartame with an increase in incidence of brain tumors in rats. By 1980 several independent groups of qualified experts had reviewed this particular aspect of the aspartame safety record and concluded that there were no compound-related effects.

The analysis of data relative to the ingestion of aspartame by rats and the subsequent development of brain tumors is centered on data on 711 rats summarized in 32 frequencies in Table 1 in the next section. This table was formed by cross-tabulating frequencies of rats with brain tumors by dosage level, including a zero control dose. This is typical of tabular data in problems assigned to students in basic statistics courses on tests of independence, in this case independence of aspartame exposure and tumor development. Such an assignment would involve a choice among common statistical techniques for comparing rates. However, despite the apparent simplicity, closer scrutiny has shown that several major issues are involved. The resolution of these issues was crucial to the interpretation of the data on tumorigenicity in rats for aspartame. More generally, these issues arise in the evaluation of the carcinogenicity of any food additive, drug, or substance in the environment.

The statistical issues are embodied in the following questions concerning the aspartame tumorigenicity studies:

1. What data should be utilized in the analysis?
2. Are the data sufficiently accurate and reliable to form the basis for scientific inference?
3. Should comparisons be made with data on controls from studies other than those involving aspartame?
4. Should the fact that a particular anatomical site is singled out for more intensive study (from the many sites initially studied for tumors) be taken into account in the analysis?
5. Should each dose level of aspartame be compared separately to control, and how should trends over dosages be examined?
6. How should information from different experiments on aspartame be incorporated into a single conclusion?
7. Should the power of the experiments to detect a dosage effect influence the conclusions drawn?
8. Should times from aspartame ingestion to the discovery of brain tumors, or to natural death or sacrifice without a brain tumor, be analyzed in addition to the rate of tumor development?

An affirmative response to the eighth question would lead to reconsideration of the first seven questions with a view to the analysis of tumor-free survival times. It would also lead to other questions, the most important of which are the following:

9. Should tumors discovered at sacrifice at the end of the experiment be analyzed differently from tumors discovered at the time of "early" death, that is, death of a rat before the end of an experiment?
10. Should brain tumors found at an "early" death be viewed as the cause of death, or only as incidental to death?

In the sections which follow, these questions will be addressed in the context of the aspartame data. A description of the studies involved is presented followed by the various statistical analyses and interpretations that were presented to the Public Board of Inquiry. The issues discussed do not cover nearly all of the difficult problems in the design and analysis of studies for the detection of potential carcinogens; however, they do cover a number of more common problems and they give strong insight into the necessary intertwining of statistical and biological inference in a complex regulatory matter. A fuller discussion of statistical issues is presented by Gart et al. (5), and an extensive guide to the literature on carcinogenic risk assessment is given by Krewski and Brown (6).

ANALYSIS OF TUMOR RATES

Description of the Experiments

Incidence Data

The data on brain tumor incidence relative to aspartame ingestion are presented in Table 1 separately for each experiment by gender and type of exposure. The ex-

Table 1 Number of Rats with Brain Tumors Detected by Innes and UAREP[a] and Total Number of Rats Examined by Exposure Group, Experiment Number, and Gender

Exposure group	Males			Females		
	Innes	UAREP	Total	Innes	UAREP	Total
E-33/34						
Control	0	1	59	0	0	60
Dose 1	2	1	40	2	2	40
Dose 2	1	1	40	0	0	40
Dose 4	4	4	40	1	1	40
Doses 6-8	0	0	40	2	2	40
E-70						
Control	3	3	58	1	1	57
Dose 2	2	2	39	1	1	39
Dose 4	1	1	39	1	0	40

[a]Universities Associated for Research and Education in Pathology.

periments are labeled E-33/34 and E-70. For each cross-classification, the number of rats examined and the number found to have brain tumors by two series of pathological examinations are given. The rats were necropsied at the time of early death or sacrifice at about 104 weeks, whichever came first. As part of the complete histopathological search for tumors routinely done in such studies, brains were sectioned and examined microscopically for tumors. The result of the examinations in the earlier study, E-33/34, led to expansion of the protocol to include additional sectioning to search for microscopic tumors in all animals in both studies. Brain sections were examined independently by Dr. Innes, a neuropathological consultant, and at a later time by the Universities Associated for Research and Education in Pathology (UAREP) group, with slightly different results. The UAREP review was carried out as part of the authentication of these and other studies. Both the Innes and UAREP findings are given in Table 1.

Note that the use of the number of rats with brain tumors in the analysis implies that the first question has been answered and that it has been decided to analyze the number of rats with site-specific tumors instead of alternative summary statistics such as the average number of tumors per rat, the average number of sites with tumors per rat, or the average number of site-specific tumors per rat. The rationale behind this choice is discussed in detail by Gart et al. (5), but, in brief, it is clear that the site of interest was identified in advance of our analysis, so site-specific data are needed. Also, the presence of even one tumor is by itself life threatening and is therefore a primary interest, as opposed to the number of tumors at a selected site. Thus only the presence or absence of a site-specific tumor per rat has been used in forming the frequencies in Table 1.

Time without detection of a site-specific tumor could form the basis for an alternative analysis, as discussed later in the answer to the eighth question. However, in the first analysis described in this section, survival times without detected brain tumors have not been utilized because the presence of many major competing causes of death, other than brain tumors, confounds the interpretation of the analysis of survival times.

Data Quality

The fact that two slightly different sets of frequencies are displayed in Table 1 would seem to lead naturally to the second question: Are the data sufficiently accurate and reliable to form the basis for scientific inference? The importance of data of high quality is shown by the fact that the Food and Drug Administration (FDA) requested that a second independent, scientific review be carried out to assess the accuracy and reliability of the data, which led to the UAREP study.

Bias, or lack of accuracy, could easily arise with data of this type. The likelihood of finding a tumor depends not only on the size and location of the tumor, but also on the number of sections of material examined for each site. If more sections were prepared from material from a given site for rats fed aspartame than for controls, for instance, the likelihood of finding a small tumor at that site among rats fed aspartame would be greater than for control rats, and a bias would have been introduced through the procedure of using different examination intensities. Unconscious bias on the part of a pathologist could also be introduced through a failure to blind the pathologist with respect to the diets (treatments) of the experimental animals.

Reliability refers to the extent to which the data are reproducible, regardless of their inherent bias. The use of different intensities of examination, if not related to treatment assignment, would lead to a lack of reliability without a lack of accuracy, that is, without bias. However, a lack of consistent procedures could lead to observations which are both inaccurate and unreliable.

The purpose here is not to fully describe the experimental procedures used to assess the accuracy and reliability of the resultant data; instead, the purpose is to underscore the need for procedures which yield accurate and reliable data. This is shown not only by the thoroughness with which the UAREP checked the quality of the Innes data and the high level of agreement between the two sets of readings, but also by the fact, as shall be seen, that a small discrepancy in results can lead to different decisions following tests of hypotheses.

Experimental Design

Before turning to data analysis, the experiments leading to the data in Table 1 will be described more fully. The exposure groups within each experiment included two control groups, one of each gender, which were fed chow with no aspartame supplement. Each of these control groups initially contained 60 rats, but 6 of them did not provide brains that could be examined. Careful review revealed no indication that the failure to obtain sections for microscopic examination for these rats was related to the presence or absence of a brain tumor.

In study E-33/34 four groups of 40 rats each were exposed to aspartame for each sex. The dosage levels for these groups are referred to as doses 1, 2, 4, and 6-8 in Table 1, where the dose number gives the approximate daily dosage of aspartame in multiples of 1000 mg/kg body weight (i.e., dose 2 is a dose of 2000 mg/kg body weight). Dietary administration of aspartame was initiated after weaning.

In study E-70 exposure to aspartame began with the onset of pregnancy. Feeding was continued through lactation and weaning until death. For each sex two dosage groups were used in addition to a control. As in the E-33/34 study, the groups that received aspartame are labeled dose 2 and dose 4 to indicate the approximate daily dosage of aspartame. In study E-70 data were obtained on 39 male rats at each of the two dosage levels. Two male rats, one from each dosage level, were necropsied at sacrifice, but their brain tissues were not sectioned. There was no reported gross evidence of a brain tumor in either of them. Data were obtained on 39 female rats for dose 2 and 40 female rats for dose 4.

Statistical Analysis

Use of External Controls

The analysis of the data in Table 1 will be discussed in considerable detail. Implicit in the consideration of these data are at least tentative answers to the third and fourth questions: Question 3 concerns the use of external controls. Since experimental animals of the same strain may be used in different experiments, perhaps in different laboratories at different times, it is natural to consider making comparisons with data on similar control animals from other experiments in the analysis of this experiment. However, the propensity for bias is great, since the detailed protocol for the pathological examination of brain tissue is unlikely to be the same in each experiment. The laboratory environment will also change. This is particularly true if historical controls are used, since methods for experimental observation, including the maintenance of a stable environment, are continually being refined. Thus in the analysis reported here statistical comparisons are made only with internal controls. In general, detailed comparisons with external controls are unlikely to be warranted, although the results of experiments should be interpreted in the light of experience on related studies.

Effect of Multiple Sites

The data in Table 1 are for the brain only; however, data were also gathered for many other sites. Should these data be taken into account (question 4)?

First, suppose that prior to the experiment the compound under investigation was thought to be as apt to produce tumors at one site as at another. Moreover, suppose there is no evidence to suggest a lack of independence in its action at different sites. Then, if the hypothesis of no differences between tumor rates between treatment and control groups were tested at 20 different sites with a significance level of 0.05 for each test, the expected number of sites at which a significant difference is found would be 1, even if the compound had *no* carcinogenic effect. Moreover, the probability of there being at least one significant result would be $1 - (1 - 0.05)^{20} = 0.64$. Thus, even if the significance for each site were only 0.05, the overall probability of rejection of a true null hypothesis of no carcinogenic treatment effect would be much higher. To avoid this problem, a lower significance level can be utilized for each site by using the Bonferroni inequality, which would lead to an acceptably low experiment-wide significance level for all sites combined (see Ref. 5).

However, such an adjustment in the significance level for each individual site is not necessarily appropriate. The problem is no doubt less severe than just indicated, because similar sites would likely produce similar responses to a compound, thus invalidating the assumption that the compound acts independently at the various sites. More importantly, likely target sites for most potential carcinogens can be determined in advance so that appropriate sites for careful scrutiny can often be identified in advance. In this case, a Bonferroni adjustment of significance levels should not be carried out for such a site, even though data are also analyzed for many other sites.

In the case of aspartame the brain has been singled out as a likely site for tumor activity before the hearing of the Public Board of Inquiry, but only after the data had been available for public scrutiny for several years. Nevertheless, because this site was of special concern at the time of the hearing apart from others initially investigated, a Bonferroni correction to take into account testing at multiple sites was not utilized.

Comparisons Over Dose Levels

It is common practice in the analysis of results, such as those from the aspartame experiments, to compare separately each dosage group receiving a compound being tested for carcinogenicity with the control group. This has the advantage of not requiring any assumptions about the functional form of a possible relationship of carcinogenic effect to dosage level. It also is a direct procedure for investigating if particular doses are "no-effect" doses. However, this approach has major drawbacks. It utilizes the same control group in each comparison, so separate comparisons of different dosage groups with the same control group do not yield independent information. Secondly, disparate results may be obtained for different experimental doses, which may only confuse the issue rather than lead to an overall conclusion. Thirdly, patterns of responses over the doses are not taken into account. Finally, by dividing the data into several separate analyses, the power of each test will be considerably less than that which could be achieved by simultaneous analysis of data for all levels of aspartame ingestion and controls. Thus it is preferable to investigate the possibility of trends over dosages and not confine the statistical analysis to separate comparisons of dosage groups with the control group. This answers the fifth question raised in the Introduction.

The procedure used in the analysis reported here was to first test for the possibility of a linear dose effect on tumor incidence rates among rats fed aspartame, not including controls. Logistic regression analysis was used for these calculations. This gave an iterative weighted least-squares analysis of the relationship between tumor incidence and the logarithm of the dose of aspartame for the treated rats. The incidence rates for groups of rats fed aspartame, which were calculated from the frequencies in Table 1 and then used in this analysis of trend, are given as percentaged for both Innes and UAREP by experiment number and sex in Table 2. The analysis was repeated for the Innes and UAREP data; males and females were considered separately and in combination, leading to 12 analyses in all. No significant trend in tumor incidence over positive dose levels was found in any of these analyses. In fact, the significance level (p) was greater than 0.20 for each analysis. The meaning of p will be brought up again later when the formulation of a statistical test of hypothesis is presented in more detail. It suffices now to

Table 2 Brain Tumor Incidence Rate in Percent Detected by Innes and UAREP for Rats Fed Aspartame by Exposure Group, Experiment Number, and Gender

Exposure group	Male		Female	
	Innes	UAREP	Innes	UAREP
E-33/34				
Dose 1	5.0	2.5	5.0	5.0
Dose 2	2.5	2.5	0	0
Dose 4	10.0	10.0	2.5	2.5
Doses 6-8	0	0	5.0	5.0
E-70				
Dose 2	5.1	5.1	2.6	2.6
Dose 4	2.6	2.6	2.5	0

recognize that p summarizes the extent to which the data are in accord with a null hypothesis. Since p is a probability, it cannot be negative or greater than 1. Values less than 0.05 or 0.01 are usually taken as evidence that the null hypothesis should be rejected.

Since no evidence of a linear trend was found in brain tumor incidence rates with increasing levels of aspartame exposure (Table 2), the hypothesis that brain tumor frequencies arose as independent random expressions of the same underlying rate of brain tumor development, regardless of dose, experiment or gender designation, was tested using the variance test based on a Poisson distribution. The resultant chi-square statistics were approximately equal to their expected value for both the Innes and UAREP data, so the hypothesis was accepted ($0.05 < p < 0.75$, Innes; $0.25 < p < 0.50$, UAREP). Thus not only was no trend observed in the incidence rates for rats fed different amounts of aspartame, but a complete lack of an effect of varying the level of aspartame ingestion was observed for rats fed aspartame. Moreover. no experiment or gender effect was exhibited for rats fed aspartame.

Because these analyses showed no differential effect of aspartame dosage on tumor incidence, the various groups corresponding to different doses of aspartame were pooled into one combined treatment group for further analysis. If a consistent trend has been observed, this would have been modeled and summarized by one or more statistics, such as a slope over increasing dosages for a linear trend, which would also have enabled all the data on animals fed aspartame, as well as the data on controls, to be incorporated into a coherent analysis over several doses. This would be more powerful than any single comparison of groups of treated and control animals.

The examination of trends in tumor incidence rates over increasing dosage levels was also central to the conclusions reached by the Public Board of Inquiry (4). However, their analysis did not utilize statistical methodology developed for this purpose, but was informal and heuristic, inadvertently showing the importance of a more thorough statistical approach. In their comments on the E-33/34 data, they stated (Ref. 4, p. 41) that "the two relatively low-dose (1000-2000 mg/kg/day) groups could be combined as well as the two high-dose groups. This would yield a figure of, respectively 5 brain tumors in 160 rats (3.1%), and 7 in 160 (4.3%), suggesting a possible dose-effect relationship." Later in their evaluation they concluded that "a further cause for concern in study E-33/34 is presented by the suggested dose-effect relationship: whereas the combined brain tumor incidence in the two lower-dose groups was 3.1%, it was 4.3% for the two higher-dose groups."

It is clear that the Public Board of Inquiry did not take into account testimony based on the analysis presented here that concluded that the observed frequencies of brain tumors among dosed rats were "independent random expressions of the same underlying rate of brain tumor development." Instead they inferred that there was a possible dose-effect relationship by noting that there were five tumors for the lower two dose groups and seven for the higher two groups among 160 rats for each pair of groups. The probability of frequencies that close together in the absence of a dose effect is very high.

The Public Board of Inquiry apparently reached their conclusion based on inspection of the data as opposed to carefully reasoned tests of hypotheses with probability calculations. However, even an informal analysis should have pointed to the conclusion reached here as opposed to theirs. They examined the Innes data for E-33/34 which are presented again in Table 3 for emphasis and ease of inspection.

The frequency of tumors varies from 1 through 5 per dosage group, and the highest frequencies are in the lowest and next to highest dose groups, but not in the highest dose group. The lack of trend is evident, but was masked by grouping dose levels before examination for a possible trend. It is appropriate to group

Table 3 Innes Data on the Frequency and Percentage of Brain Tumors and on the Number of Rats Examined by Dosage Groups for Experiment E-33/34

	Number of brain tumors	Number of rats examined	Percentage of rats with brain tumor
Dose 1	4	80	5.00
Dose 2	1	80	1.25
Dose 4	5	80	6.25
Doses 6-8	2	80	2.50

after observing the absence of a dose-response relationship, as was done in the analysis reported here, but not in order to show its presence. Arbitrary grouping may lead to biased or incorrect conclusions, as it did for the Public Board of Inquiry, especially if it is carried out after inspection of the data. Another illustration from these data would be to put groups 1 and 2 together to form a low-dose group but to leave the highest two dose groups separate. This could be supported on the basis of a possible confounding effect of dose saturation. This would lead to the highest percentage of brain tumors in the middle group (6.25 versus 3.13 and 2.50% for the low and high groups, respectively). Thus the data from these groupings give some evidence of a parabolic trend but any inference that a particular trend is indicated is inappropriate; it is caused by grouping data exhibiting random variation in an arbitrary manner.

In summary, for the data on the incidence of brain tumors among rats fed aspartame, there is no evidence of a dose-effect trend, even though there is natural variation from group to group which is not related to dose level in any apparent way.

Comparisons of Rates: Aspartame Versus Control

Next it is necessary to compare the data on rats fed aspartame, pooled together because of the lack of a dose-effect relationship, with the data on control rats which did not ingest aspartame. The frequency data are presented in Table 4; corresponding incidence rates as percentages are given in Table 5. As observed in the analysis of rates displayed in Table 2 for groups of animals fed aspartame, no experiment or gender effect exists for these rats. However, there is more variability in rates for control rats, so the data in Tables 4 and 5 are given separately for different experiment gender combinations.

The tumor incidence rates in the control and combined treatment groups were compared using the one-tailed Fisher's exact test for 2 X 2 contingency tables.

Table 4 Number of Rats with Brain Tumors Detected by Innes and UAREP and Total Number of Rats Examined for Control and Combined Aspartame Treatment Groups by Experiment Number and Gender

Exposure group	Male			Female		
	Innes	UAREP	Total	Innes	UAREP	Total
E-33/34						
Control	0	1	59	0	0	60
Treatment	7	6	160	5	5	160
E-70						
Control	3	3	58	1	1	57
Treatment	3	3	78	2	1	79

Table 5 Brain Tumor Incidence Rates in Percent as Detected by Innes and UAREP for Control and Combined Aspartame Treatment Groups by Experiment Number and Gender

Exposure group	Male		Female	
	Innes	UAREP	Innes	UAREP
E-33/34				
Control	0	1.7	0	0
Treatment	4.4	3.8	3.1	3.1
E-70				
Control	5.2	5.2	1.8	1.8
Treatment	3.8	3.8	2.5	1.3

For E-33/34 the rates for the treatment groups did not significantly exceed the rates for the corresponding control groups for either males ($p = 0.11$) or females ($p = 0.20$) using the Innes data. The same is true for the UAREP data ($p = 0.39$ and 0.20 for males and females, respectively). However, when the results for males and females are combined using one-sided Mantel-Haenszel statistics (see Ref. 7), the incidence rate for the treated rats is significantly higher than that for the control rats for the Innes data ($p < 0.029$), but not for the UAREP data ($p > 0.05$). Note that even though it was not deemed appropriate to pool the data by gender, the results have been incorporated into a single conclusion through use of the Mantel-Haenszel procedure. This same approach will later be used to form a single conclusion for all the data from both experiments and provide an answer for question 6.

Not a great deal of emphasis should be placed on the fact that the overall significance level for the comparison of treated and control rats for E-33/34 is less than the conventional critical probability level of 0.05 for the Innes data but not for the UAREP data. The difference in significance levels is due to minor differences in the evaluation of brain tissues from treatment rats and control rats. More specifically, among controls in E-33/34 Innes found no brain tumors, but UAREP found one. Since no dose-response relationship was found within the group of rats treated with aspartame and since other experimentation has shown that brain tumors sometimes occur in control animals, the Innes and UAREP results viewed together at most show a possible increase in tumor incidence rate among treated rats relative to controls. Further evidence than that given by study E-33/34 is needed to clarify the relationship between exposure to aspartame and the development of brain tumors.

Such additional evidence is given by study E-70. The data of E-70 give no indication of an effect of aspartame on tumor incidence with either the Innes or UAREP data for either males or females. The p values calculated with Fisher's exact test are 0.79 and 0.62 for the Innes data for males and females, respectively. The corresponding p values for the UAREP data are 0.79 and 0.66. None of these are even close to the critical value 0.05. For males the incidence rate for controls is higher than for experimental rats; the two incidence rates are essentially the same for females, with a reversal in order between the Innes and UAREP data brought about by another difference in the evaluation of brain tumor frequencies.

Statistical results for the E-33/34 and E-70 experiments have been combined over the four gender-experiment categories with the one-sided Mantel-Haenszel procedure. This method gives a single summary chi-squared statistic from both experiments for the comparison of the tumor incidence rates for control rats and rats treated with aspartame. The resultant statistic is not statistically significant for either the Innes ($p > 0.05$) or the UAREP ($p > 0.20$) data, so the null hypothesis of no treatment effect of aspartame on brain tumor incidence is sustained.

Approach to Scientific Inference

Before continuing with analyses of other data, it is helpful to clarify certain statistical issues in the approach to scientific inference embodied in the analysis just described. The statistical methodology has been tailored with one main goal in mind, namely, to achieve a high probability of detection of a carcinogenic effect of aspartame if such an effect exists. In statistical terms, this means that high power has been sought in the selection of a testing procedure. This has been facilitated by the integration of all the data into a single summary statement on the safety of aspartame. It has also been enhanced through the choice of one-sided tests of hypotheses. The hypothesis testing context of these terms is presented in this section and the importance of power in the interpretation of the results is explained in answer to question 7 raised in the Introduction.

The decision process in the application of statistical tests to scientific reasoning consists of the formation of a null hypothesis (H) of no difference between an experimental treatment and an appropriate control followed by the rejection, or the failure to reject H, on the basis of experimental data. This process is summarized in Table 6. The entries in the body of Table 6 denote the probabilities

Table 6 Probability of Decisions About a Null Hypothesis H Given the Status of H

Decision rule outcome	Status of hypothesis H: True	Status of hypothesis H: Not true
Reject H	α	$1 - \beta$ = power
Do not reject H	$1 - \alpha$	β
Total	1	1

determined by the decision rule used in the analysis of the data conditional upon the state of nature with respect to the truth of H (given as a column heading). The fact that the probabilities are calculated conditional upon a column heading is made clear by showing that the probabilities in each column sum to unity. A decision rule is traditionally chosen so that α is small; that is, the probability of making the error of rejecting H when it is true is small. This error is called a type 1 error. Then the observed data leads to the decision to reject H only if there is strong evidence that it does not hold. When H is rejected, it is often said that the difference between treatment and control procedures is statistically significant at the level of α chosen.

The choice of α, and the associated decision rule based on a statistic to be calculated from the experimental data, is ideally made before the experimental data are available. Alternatively, a p value is calculated after collection of the experimental data to summarize the extent to which the data are in an accord with the null hypothesis. The p value is the probability of a type 1 error when the boundary between rejection and acceptance is given by the value of the test statistic calculated using the experimental data. Thus p and α are on the same scale, and the decision rule can be stated in terms of a comparison of p and α. In particular, the null hypothesis would be rejected if $p < \alpha$, and retained otherwise.

When departures from H in only one direction lead to rejection of H, a test of hypothesis is said to be one-sided; that is, the motivational question framed by the null and alternative hypotheses is directional. In this case the motivational question could be phrased as follows: Is aspartame more carcinogenic than control? This is a one-sided question in contrast to the question Is aspartame different from control in its carcinogenic potential? Clearly the first question is the appropriate one here, since aspartame would not be banned because of carcinogenic potential if it were judged to be less carcinogenic than control. This is the reason one-sided hypothesis tests have been reported in this chapter.

When the data do not lead to a rejection of H, the situation is not as clear-cut as was indicated in the discussion of the decision of the Commissioner of the FDA (1) on cyclamate. Some confusion is evident in that discussion on the interpretation of $1-\alpha$, which, as can be seen from Table 6, is the probability of the data not leading to rejection of H conditional upon the truth of H. Thus $1-\alpha$ is the probability of a correct decision when H holds; it is calculated under the assumption that H holds and therefore contains no information on the extent to which H does not hold. Such information is given by probabilities calculated when H is not true, namely, β and $1-\beta$. Like α, β is the probability of an error in the decision process, the error of not rejecting H condition upon H not being true, commonly called a type II error. The probability $1-\beta$ is the probability of correctly rejecting H when it in fact does not hold and is called the power of the test.

One reason why the situation may not be clear-cut when H is not rejected is because the failure to reject H may be due to the fact that H is true, or because the power $1-\beta$ is small, that is, there is insufficient evidence to overturn H with the small value specified for α even though H is false. This indeterminacy can be avoided in the design of the experiment by choosing large enough sample sizes to assure that if H is not true, the experimental evidence will be likely, with probability $1-\beta$, to lead to rejection of H. When the power $1-\beta$ is close to 1, a failure to reject H can reasonably be attributed to the truth of H. However, no experimental design will assure that the power is close to 1 for all of the possible alternatives to the truth of H. Various alternatives to H correspond to different extents of departure from H, and the power $1-\beta$ will depend upon the alternative under which it is calculated. This problem is ideally handled by the specification of a threshold such that, for an extent of departure from H equal or greater than this threshold, it is essential that the decision rule based on experimental data have a high probability of rejecting H, that is, have high power. Once such a threshold is specified along with α and the decision rule to be applied to experimental data, then a sample size for a subsequent experiment can be chosen so that if H is not rejected based upon the experimental evidence, it can be accepted with high confidence. Alternatively, once the sample size for an experiment is chosen and even after an experiment is completed, the power of the experiment can be calculated for several threshold values to help ascertain if failure to reject H gives sufficient evidence to retain H.

Power of Tests for Experimental Data

Power probabilities for E-33/34 and E-70 have been calculated. They are reported in this section and lead to the conclusion that these experiments have sufficient power so that if H is not rejected, then the null hypothesis of no effect of adding aspartame to a rat's diet can be sustained with high confidence. In the previous section it was shown that the null hypothesis of no aspartame effect should not be rejected based on the data from these experiments. It was also shown that it is reasonable to group rats receiving different levels of aspartame into a single treatment group for comparison with control. This grouping is used in the power calculations presented in this section.

As already indicated, power generally increases as the sample sizes in experimental groups increase or as the size of the carcinogenic effect to be detected is taken to be larger. The approximate power of the design of E-33/34 and E-70 is recorded in Table 7 for several combinations of the tumor rate among control animals (π_1) and the increase in tumor rate due to aspartame dosage (δ). For example, if the *true* rate for control animals were 0.4% and the *true* rate for dosed animals were 3.4%, then the design of the experiments would lead to finding a significant ($p < 0.05$) difference in *observed* rates about 77% of the time. Since the true rates are not known, a range of possible values is considered in Table 7. The calculations reported in Table 7 are based on sample sizes for the control and dosed groups of 240 and 480, respectively. These sample sizes would result if the data from the two experiments were combined into just two groups,

Table 7 Power (%) to Detect the Difference Between Control (π_1) and Dosed ($\pi_2 = \pi_1 + \delta$) Tumor Rates[a]

	δ		
π_1 (%)	3%	4%	5%
0.4	77	92	98
1.0	66	85	95
1.4	61	81	92
2.0	54	75	88

[a]There were 240 rats in the control group and 480 rats in the dosed group. The power was generally less than 50% for δ less than 3% (one-sided test with $\alpha = 0.5$; E-33/34 and E-70 combined).

disregarding the gender, experiment number, and dose level of treated animals. The calculations are based on the fourth approximation recommended by Ury and Fleiss (8).

The values in Table 7 are not exact but represent close approximations to the combined power of the two experiments. Similar values for the design of experiment E-33/34 alone are reported in Table 8.

The conclusion that the null hypothesis of no aspartame effect should be retained is based upon powers greater than 50%, the power referred to by the Commissioner of the FDA in his discussion, for δ as small as 3% for a broad range of π_1 values for controls and, more specifically, on the considerably higher power of 77% for $\delta = 3\%$ for the small value of π_1 of 0.04%.

As an alternative to the power calculations just described after an experiment has been completed, confidence limits on a difference between treatment and control populations can be calculated. If a parameter which describes this differ-

Table 8 Power (%) to Detect the Difference Between Control (π_1) and Dosed ($\pi_2 = \pi_1 + \delta$) Tumor Rates for $\delta = 5\%$[a]

π_1 (%)	Power for $\delta = 5\%$ (%)
0.4	75
1.0	67
1.4	63
2.0	57

[a]There were 120 rats in the control group and 320 rats in the dosed group (one-sided test with $\alpha = 0.05$; E-33/34).

ence has been estimated closely as shown by a narrow interval, and if the interval includes a "no-difference" value of the parameter, then this is evidence that the null hypothesis should be sustained. For this calculation, experimental and control groups were compared on the basis of the logarithm of the odds ratio (7). This is a statistic which summarizes all of the information in a fourfold table given the frequencies of rats with and without brain tumors separately for the treatment and control groups.

First, a test of the null hypothesis was recalculated on the basis of 13 tumors found by UAREP and/or Innes among the E-33/34 rats and 8 tumors found by UAREP among the E-70 rats. The null hypothesis of no aspartame effect was not rejected. The observed evidence against the null hypothesis could be expected to occur by chance about 18% of the time if aspartame did not cause brain tumors, which is a higher probability than conventionally selected values of α. This calculation was done with a one-sided Mantel-Haenszel test combining evidence across both experiments and gender. Secondly, based on a 95% confidence interval for the logarithm of the odds ratio and data from both E-33/34 and E-70, the tumor rate for rats fed aspartame is no more than 3.4 times the control rate. Based on E-70 alone this factor is reduced to 2.4.

Both the power and confidence interval calculations show that in addition to not rejecting the hypothesis that aspartame is not carcinogenic based upon the E-33/34 and E-70 experiments, it is also reasonable to sustain the null hypothesis that in fact aspartame is not carcinogenic. This is in accord with but strengthens the conclusion previously reached. It also incorporates the approach to evaluating carcinogenesis on the basis of experimental data discussed by the Commissioner of the FDA (1) in his decision on cyclamate.

During the review of the brain tumor issue, concern was raised that Searle's studies, which had been designed more than 10 years previously, might not meet current regulatory standards for such studies. The analysis of statistical power proved useful in showing how Searle's experimental design compared to currently accepted rodent carcinogenesis protocols.

The Bureau of Foods at the FDA had indicated that current study design usually had four groups, each with 100 animals equally divided between males and females. One group is a control and the other three are carefully selected doses of the test compound. Searle studies differed from these in that the control group had 120 animals and each treated group had 80 animals. Table 9 shows how animals would be allocated in each type of study.

The Searle design reflected a common practice of assigning more animals to the control group because it is "used more" than the treated group. Under certain assumptions, the optimal number of animals in a control group exceeds the number in each treated group by a factor which is the square root of the number of treated groups. The increment in the Searle study was 50% (120 control and

Table 9 Allocation of Animals in the Searle Aspartame Study and in Currently Recommended Protocols

Study	Number of groups	Number of animals		
		Control	Treated	Total
Searle E-70	3	120	80	280
Current FDA recommendation	3	100	100	300
Searle E-33/34	5	120	80	440
Current FDA recommendation	5	100	100	500

80 per treatment), lying between the optimal values of 41% (E-70) and 100% (E-33/34).

In assessing the comparability of the two designs, we elected to ignore this multiple-treated-group factor and simply perform the comparisons for hypothetical studies with two groups. In one study, referred to as study A, there were 100 control and 100 treated; in the second study, study B, which is similar to Searle's, there were 120 controls and 80 treated rats. It was expected that the power of study B would be slightly less than the power of study A.

Using the method previously mentioned, due to Ury and Fleiss (8), power curves were calculated for the two studies as well as for a combined study (477 control and 234 per treatment), study C, that is almost identical to the study described in Table 7. A normal control incidence of 1% (π_1 in Tables 7 and 8) was assumed and a specially developed computer program was used. Figure 1 shows the comparisons of the power functions of the three studies. As expected, the power curves for study A and study B are close. Surprisingly, the curve for study B is higher than the curve for study A over a substantial range of detectable difference. This situation suggests the following conclusions:

1. The Searle study is at least as powerful as the current regulatory standard.
2. The Searle study achieves equivalent power with fewer total animals, constituting a more efficient design.

In addition to giving a conclusive answer to the question of comparability of Searle's and current standards, the power analysis has raised some interesting possibilities for the effective design of studies on very small incidence rates. Preliminary investigation of this result using an exact method, as opposed to the approximate methods used for generating Figure 1 and Tables 7 and 8, suggests that the conclusions indicated here are correct.

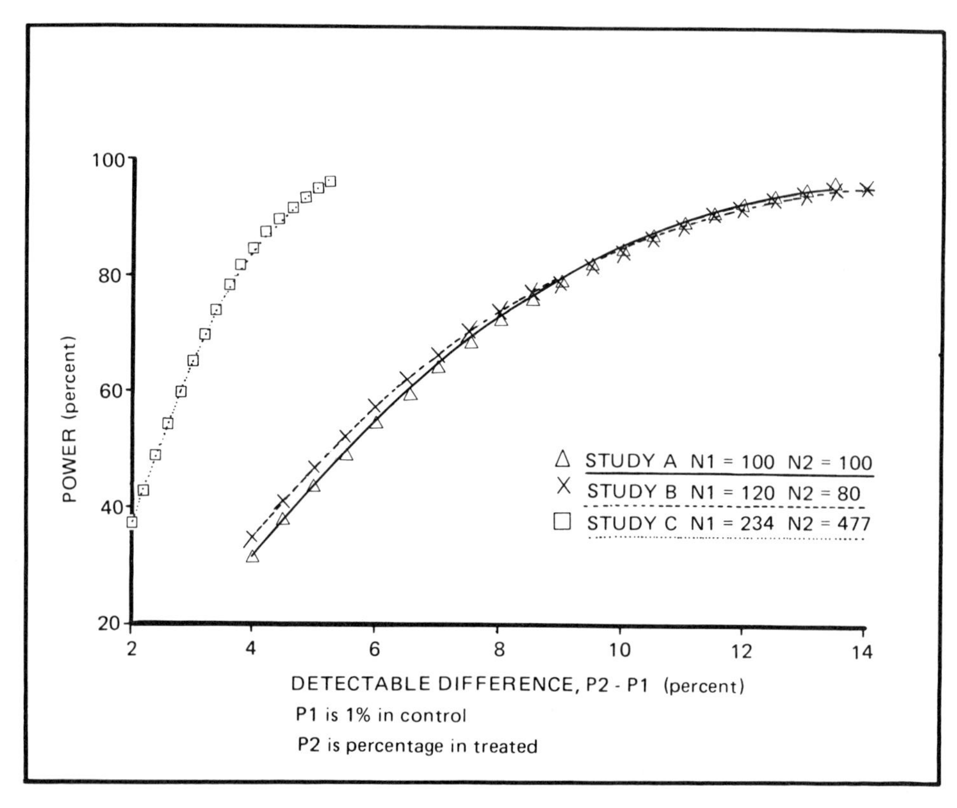

Figure 1 Comparisons of power functions of the three studies relevant to the brain tumor issue.

TIME TO EVENT CONSIDERATIONS

When comparing tumor incidence rates among various groups of animals, it is important to compare mortality patterns as well. This is because animals which are subject to higher mortality rates from other causes are likely to exhibit lower tumor incidence rates. The reason for this is that there is less time for the tumors to develop during the animals' lifetimes. Thus an affirmative answer should be given to question 8 raised in the Introduction, which leads to the discussion in this section of the issues raised by questions 9 and 10.

For the aspartame data the average survival times of the animals in the various treatment and control groups of E-33/34 and E-70 vary from each other only slightly, as can be seen in Table 10. Any adjustment of tumor incidence rates due to differing average survival times for the groups thus would be small. The potential effect of such an adjustment on the tumor incidence can be evaluated in a straightforward manner.

Table 10 Average Weeks of Survival by Exposure Group and Experiment Number

	Male	Female
E-33/34 (total)	89.1	91.0
Control	81.3	94.7
Treatment (combined)	92.1	89.6
Dose 1	89.6	94.4
Dose 2	90.7	91.2
Dose 4	95.2	87.3
Doses 6-8	92.8	85.6
E-70 (total)	92.5	93.5
Control	91.4	93.5
Treatment (combined)	93.3	93.4
Dose 2	93.5	92.1
Dose 4	93.1	94.6

The E-33/34 female control group average survival time was about 6% greater than the E-33/34 female treatment group average survival. In E-70 the average survival time of the female control rats was very close to the average survival time of the female treatment rats. Making the naive adjustment of increasing the tumor incidence in the E-33/34 female treatment group by 6% would lead to an increase in the number of tumors from 5 to 5.3 for that group. This would not greatly affect the comparison of treatment and control incidence rates. Even if six tumors had been reported in this group, the difference between control and treatment rates would not be significant ($p > 0.10$) and the overall Mantel-Haenszel statistic comparing control and treatment incidence rates for E-33/34 and E-70 females would not be significant ($p > 0.05$), one-sided).

The corresponding adjustment of tumor rates among the male rats would have the effect of lowering the treatment tumor incidence rate relative to the control rate. The single summary chi-square statistic reported earlier for males would also remain insignificant with such augmented data.

While the discussion so far has focused on the tumor rates observed during the course of an experiment, the issue of time to tumor is also relevant in examining the question of tumorigenicity. It is common for a tumorigenic substance to both increase the number of tumors and to decrease the time of occurrence of the tumors. Statistical methods called survival analysis have been developed which can be used to analyze times to tumor and time without tumor for data such as are available from the aspartame experiments (see Refs. 9-11).

Survival analysis methods allow the comparison of times to a specific event of interest, such as time to tumor, for animals in two or more groups. Survival analysis offers two major advantages over the simpler comparison of incidence rates which we have presented previously. First, survival analysis of times to tumor makes an appropriate adjustment for any differences which might be present in mortality patterns among the experimental groups. Thus, although such differences were not substantial for the aspartame experiments, survival analysis methods would have alleviated our concern for the potential effects of differences in mortality patterns among experimental groups. Second, survival analysis methods are sensitive to changes in times to tumor, as well as to numbers of tumors. Thus survival methods utilize more of the relevant data for evaluating tumorigenicity than would analysis of incidence rates alone. Survival analyses of time to tumor were carried out for the aspartame data and led to much the same conclusion as did the results reported for tumor incidence rates. While these methods are commonly used for analyzing data such as the aspartame data, we have not emphasized such methods because of questions regarding the appropriate interpretation of the time to tumor data.

The aspartame data give information about the time with no brain tumor and the time to detection of brain tumor for the animals without and with brain tumors, respectively. It is the interpretation of the times to tumor which raise substantial issues, because not all tumors were detected in the same fashion. Some tumors were found at the time of death and may well have been the leading cause of death. Other tumors were found at death, but may not have contributed to the death of the animal. These tumors may not have been detected until later (at sacrifice, say) had the animal not died from other causes. Still more tumors were found at sacrifice. These tumors would not have been discovered until later had the experiments been longer. These differences in the mode of detection of the tumors have implications for the appropriate interpretation and analysis of the recorded times to tumor. Peto et al. (3) have suggested that tumors be classified, through pathological examination, as to their likely relationship to the death of the animal. For example, tumors could be classified as the direct cause of death or as incidental to death. Statistical methodologies, such as those proposed by Peto et al. (3), which account for this classification of tumors, could then be used.

CONCLUSION

The analyses described here on comparisons of the incidence and timing of brain tumors in rats fed aspartame as opposed to a control diet have demonstrated that aspartame is not carcinogenic for this site, the only one for which concern has surfaced. In addition, issues which need to be addressed in the analysis and interpretation of experimental data on animal tumors for any chemical substances have been discussed. The conclusions reached have been hammered out in experiments in industrial laboratories, decisions of governmental agencies, deliberations of consumer groups, and reports from the scientific community.

They have led to the marketing of a new sweetener which will be beneficial to the health of many consumers. The procedures utilized form the basis for similar evaluations in the future.

REFERENCES

1. Commissioner of the Food and Drug Administration (1980). Cyclamate (cyclamic acid, calcium cyclamate, and sodium cyclamate); *Commissioner's Decision*. Department of Health and Human Services, Food and Drug Administration (Docket No. 76F-0392), *Fed. Reg.* 45, 61474 ff.
2. Interagency Regulatory Liaison Group (1979). Scientific bases for identification of potential carcinogens and estimation of risks. *J. Nat. Cancer Inst.* 63, 244-268
3. Peto, R., Pike, M. C., Day, N. E., Gray, R. G., Lee, P. N., Parish, S., Peto, J., Richards, S., and Wahrendorf, J. (1980). Guidelines for simple, sensitive significance tests for carcinogenic effects in long-term animal experiments. In *Long-Term and Short-Term Screening Assays for Carcinogens: A Critical Appraisal*, IARC Monographs on the Evaluation of the Carcinogenic Risk of Chemicals to Humans, Annex to Supplement 2. International Agency for Research on Cancer, Lyon, pp. 311-426.
4. Public Board of Inquiry (1980). Aspartame: Decision of the Public Board of Inquiry. Department of Health and Human Services, Food and Drug Administration (Docket No. 75F-0355), Washington, D.C.
5. Gart, J. J., Chu, K. C., and Tarone, R. E. (1979). Statistical issues in the interpretation of chronic bioassay tests for carcinogenicity. *J. Nat. Cancer Inst.* 62, 957-974.
6. Krewski, D., and Brown, C. (1981). Carcinogenic risk assessment: A guide to the literature. *Biometrics* 37, 353-366.
7. Fleiss, J. L. (1981). *Statistical Methods for Rates and Proportions*, 2nd ed., Wiley, New York.
8. Ury, H. K., and Fleiss, J. L. (1980). On approximate sample sizes for comparing two independent proportions with the use of Yates' correction. *Biometrics* 36, 347-352.
9. Kalbfleish, J. D., and Prentice, R. L. (1980). *The Statistical Analysis of Failure Time Data*, Wiley, New York.
10. Peto, R., Pike, M. C., Armitage, P., Breslow, N. E., Cox, D. R., Howard, S. V., Mantel, N., McPherson, K., Peto, J., and Smith, P. G. (1976). Design and analysis of randomized clinical trials requiring prolonged observation of each patient. Part I. *Br. J. Cancer* 34, 385-612.
11. Peto, R., Pike, M. C., Armitage, P., Breslow, N. E., Cox, D. R., Howard, S. V., Mantel, N., McPherson, K., Peto, J., and Smith, P. G. (1977). Design and analysis of randomized clinical trials requiring prolonged observation of each patient. Part II. *Br. J. Cancer* 35, 1-39.

24

Possible Neurohormonal Effects of Aspartame Ingestion

Frank M. Sturtevant
G. D. Searle & Co., Skokie, Illinois

INTRODUCTION

Prior to the approval of aspartame as a food additive by the Food and Drug Administration (1), the question had been raised whether the ingestion of aspartame, either alone or together with glutamate, posed a risk of contributing undesirable effects on neuroendocrine regulatory systems. It had been hypothesized that an infant or child gorging himself on aspartame might suffer a "silent lesion" in the hypothalamus that could give rise to neuroendocrine abnormalities in later life, but would sound no signal of distress at the time of acute exposure (2).

HORMONAL BLOOD LEVELS

One basis for the foregoing hypothesis was apparently set forth in a 1976 report by Olney et al. (3) in which a supposedly subneurotoxic, parenteral dose of 1000 mg/kg body weight of glutamate caused an acute disturbance in luteinizing hormone (LH) and testosterone serum levels in a single experiment on rats. The implication was made that the same consequence would follow oral ingestion of aspartame as a food additive by children.

A further extension of this theory was made by Olney (4) when he posited that "the routine intake of MSG [monosodium glutamate] and Aspartame several times per day by children throughout their formative years could possibly entail repetitive disturbances in several neuroendocrine axes (e.g., gonadotrophins, growth hormone and prolactin) and this would adversely influence somatosexual development."

Subsequent work by Olney and co-workers has shown LH spikes in rats immediately following the injection of other compounds known to cause neuronal necrosis, namely, kainic acid, N-methylaspartic acid, and homocysteic acid. The results of the work with these compounds are not relevant to aspartame, however. These foreign chemicals do not arise from the metabolism of aspartame, nor have they been shown to be handled by the body in the same manner as aspartate. If there is any relevance to aspartame in the LH experiments of Olney, one must turn to the one experiment involving glutamate.

This experiment (3) was a pharmacological study utilizing a subcutaneous dose of 1000 mg/kg body weight of glutamate. The working hypothesis was that a dose of glutamate, below that required to destroy arcuate neurons, "might nevertheless stimulate them to fire at increased rates and, thereby, disturb endocrine systems regulated by these neurons, one of which is thought to be the pituitary gonadal axis." Olney estimated that 1000 mg/kg body weight glutamate administered subcutaneously in 55-day-old rats would be such a dose, although this was not confirmed in this study by examination of the hypothalamus. Rats were killed 0.25, 0.5, 1. 2, 4, and 6 hr after glutamate injection and serum LH and testosterone determined.

Rats were always killed at the same time of day "to avoid possible circadian influences"; however, it is well known in chronobiology that such a procedure is ineffective for this purpose. Diametrically opposite results could have been obtained if the animals were killed at a different phase of the circadian system (see Ref. 5).

The LH results of Olney's experiment are not from serial determinations in individual rats, but are group averages. Since all of the groups were killed at the same time of day, this required that monosodium glutamate (MSG) injections be made at different times of the day, which assumes that the resulting plasma glutamate levels would not be affected by the circadian phase of the animals. There is good basis for doubting this (6); nevertheless, the assumption was not investigated by the authors. Interestingly, Olney's saline-injected control data, though not graphically presented, were said not to vary from the base line. In other words, the data presumably displayed no temporal rhythm during the 6-hr period of observation.

When the excursions in Olney's LH levels were compared with those reported by other investigators for noninjected control rats (5,7-11) it was obvious that significant, several-fold variability in LH levels with time of day occurred even in the absence of glutamate injection.

It is well recognized that LH secretion occurs in an episodic, ultradian fashion (12); however, this factor was not considered in Olney's work. In subsequent studies, Dyer et al. (13) found no effect of neonatal MSG treatment on pulsatile LH secretion or on sex steroid-induced LH surge in ovariectomized rats. On the other hand, MSG administration to neonatal rats has been reported to result in a disturbance of episodic growth hormone (14,15) and prolactin (14) secretion in the adult animal.

It is of interest to compare Olney's original LH results (3) to those of others who tried to duplicate his experiment. Yonetani and Matsuzawa (7) could not reproduce the finding of an LH elevation at 15 min postinjection, regardless of whether the rats were killed at the end of the light or dark phase of the circadian cycle. The authors commented (7) that the fluctuations observed by Olney following MSG administration "appear to be within the range of circadian variation obtained from the present study."

Nemeroff et al. (16) also attempted to reproduce Olney's findings because they had previously found (17) that the MSG-sensitive system does not directly participate in the acute regulation of LH release. In no experimental group did LH reach the detection level of 15-20 ng/ml, leading them to conclude that "the results do not support the suggestion that MSG when administered to adults stimulates arcuate neurons to fire resulting in activation of the pituitary axis" (16).

In addition to LH, Olney's 1976 paper (3) also reported on testosterone levels after the subcutaneous injection of 1000 mg/kg body weight of MSG. Keating and Tcholakian (18) showed that rats who have received no treatment displayed circadian variation of serum testosterone when measured by serial sampling. Olney's saline-injected rats apparently displayed no rhythm, but these data were not graphed in the publication.

Yonetani and Matsuzawa (7) failed to observe any acute elevation of serum testosterone following MSG administration, regardless of whether the rats were killed at the end of the light or dark phase of the circadian cycle.

Terry et al. (14) observed a suppression of rhythmic growth hormone secretion and a transient release of prolactin following 1000 mg/kg body weight of MSG (injected intraperitoneally) in adult rats. They concluded that these findings do not support the hypothesis of MSG excitation of tuberoinfundibular dopaminergic neurons.

With regard to aspartame itself, Mares and Berg (19) had investigated gonadotrophin levels in aspartame-treated rats. Serum gonadotrophin levels were determined in 30 male and 30 ovariectomized female rats 1-3 hr after the 10th daily dose of aspartame administered at 300 mg/kg body weight per day by gavage. Aspartame was administered between 0800 and 0900. Before-noon serum levels of LH, follicle-stimulating hormone (FSH), and prolactin, approximately 1-3 hr after dosage, were essentially unchanged on the 10th day of treatment (Table 1). Pituitary levels were also unaffected. Lennon et al. (20) found no effect on serum prolactin or gonadotrophins of 7000 mg/kg body weight per day of aspartame in the diet of lactating rats. Although such studies do not purport to duplicate Olney's experiment, they certainly do not provide any support for his hypothesis. In summary, the LH changes observed by Olney appear to fall within the range seen with normal circadian variation. His findings could not be reproduced by others. The most important criticism that should be made, however, is the following: The plasma levels of glutamate following 1000 mg/kg body weight injected subcutaneously cannot conceivably be related to plasma levels of aspartate

Table 1 Serum Gonadotrophin Levels (ng/ml) in Male and Ovariectomized Female Rats 1-3 hr After 10th Daily Dose of Aspartame[a] (19)

	LH	FSH	Prolactin
Females			
Control	379 ± 56	1409 ± 54	45 ± 9.7
Aspartame	315 ± 51	1443 ± 114	45 ± 14
Males			
Control	<50	302 ± 53	108 ± 24
Aspartame	<50	401 ± 38	120 ± 12

[a]300 mg/kg body weight per day, oral.

following routing use of aspartame as a food additive. In human beings, plasma levels are not elevated at many times the projected level of intake as a food additive, as has been convincingly shown by Stegink (21). Thus no mechanism exists for the stimulation of the hypothalamus in the first place.

REPRODUCTIVE PERFORMANCE IN RATS

There should be little question that a critical test of any adverse effect of chronic aspartame use on somatosexual development, as hypothesized by Olney, would be an assessment of reproductive performance.

Preliminary studies (20) had shown that intragastric administration of 300 mg/kg body weight for 5-7 days to hamsters and rats did not affect postcoital fertility. Furthermore, acute studies in various species had shown aspartame to be devoid of any sex hormone agonist or antagonist properties (22).

In one chronic study (23), Charles River CD rats divided into three groups were fed a diet containing aspartame at dose levels of 0, 1800, or 3700 mg/kg body weight per day for 9 weeks until they reached approximately 90-100 days of age. This generation can be denoted as the P_2 generation.

For the first breeding phase, one male and two females within each dosage group were bred to obtain a P_1 generation. This generation was observed at birth, during nursing, and at weaning. Animals of both sexes were then grouped and continued on the treatment diets to form a 2-year study (24), to be described later. Other animals from this generation were fed the test diets and bred as before to yield an F_1 generation. These F_1 animals were killed at age 21 days and reported separately (24).

There was no indication of a compound-related effect during either the first or second reproductive phase as assessed by indices of fertility, gestation, live birth, or nursing, or by the mean litter size born (Table 2). Thus subfertility or infertility was not observed in this two-generation aspartame feeding study.

Table 2 Reproductive Performance in Rats (23)

Indices	$P_2 \to P_1$		$P_1 \to F_1$	
	Control	High dose	Control	High dose
Fertility: percent pregnant postmating	87.5	95.8	78.9	95.0
Gestation: percent pregnancies → litters	100	100	100	100
Birth: percentage of live pups	99.6	99.6	98.9	99.6
Nursing: percentage of pups weaned	79.9	71.9	88.6	98.1
Mean litter size	12	10	11	12

VENTRAL BRAIN COMPRESSION

The foregoing discussion addressed Olney's hypothesis of possible effects on neuroendocrine regulatory systems that might result from subneurotoxic plasma levels of aspartate. Olney proposed possible endocrine dysfunctions upon other bases as well. One such basis is that of "ventral compression" of the brain. In his 1976 communication (4), Olney alleged that rodents treated with aspartate have ventral compression of the brain due to loss of neurons about the mediobasal hypothalamus. Furthermore, he suggested that females thus affected have, as a result, increased numbers of cystic or atretic ovarian follicles. This position is not supported by the facts.

The diagnosis of "ventral compression" of the brain refers to a *dorsad* compression of the ventral aspect of the brain in the region of the pituitary resulting from an expanded pituitary tumor (26). It is not related to a loss of hypothalamic neurons as stated by Olney.

The Searle chronic toxicity studies do not support Olney's thesis. First, cases of "ventral compression" were diagnosed microscopically, not grossly, as stated by Olney. In one 2-year rat study (27), there were 11 instances of "ventral compression" and 29 cases of cystic ovarian follicles. In only one case did the two conditions exist in the same rat. Therefore it is extremely difficult to perceive how Olney can allege that one caused the other.

In a second 2-year rat study (24), "ventral compression" was diagnosed more frequently in the controls than in the treated animals, which is opposite to what one would expect if this condition were a compound-induced effect. Cystic ovarian follicles were also diagnosed more frequently in control animals. Six of the 12 cases of "ventral compression" and 7 of the 16 cases of cystic follicles occurred in aspartame-treated rats. In only two rats did the two conditions occur in the same animal, and both of these were control animals.

One other 2-year study on rats was carried out by Ishii et al. (28). Dietary doses of aspartame up to 4000 mg/kg body weight per day did not yield any diagnosed occurrences of ventral brain compression or of cystic ovarian follicles in Wistar rats.

In brief, then, there can be no question that the issue of ventral compression raised by Olney is not valid.

MULTIPLE ENDOCRINE DISTURBANCES

Yet another basis for possible endocrine dysfunction was founded upon a 1969 report by Olney (29). Neonatal mice administered massive amounts of MSG parenterally developed a constellation of endocrine disturbances several months later that, while varying in their specifics according to different subsequent authors, generally involved the following: obesity with adiposity, skeletal stunting, reproductive disabilities, and depressed weights of the pituitary gland and gonads. Decreased pituitary content of prolactin and growth hormone was reported, as was increased serum prolactin, decreased serum growth hormone, and disturbances in pulsatile secretion. Somewhat paradoxically, there may be decreased or increased locomotor activity and an absence of hyperphagia. Sensitivity to this effect of MSG decreases with age. Aspartic acid at neurotoxic dose levels is reported to have similar properties.

The dose of MSG employed in published reports of these conditions has varied from a single injection of 2000 mg/kg body weight (30) to 10 daily injections in amounts increasing from 2200 to 4400 mg/kg body weight per day. When the brain was examined following such doses, typical acute, arcuate nuclear lesions, or subsequently decreased neuronal populations were observed. The severity of the acute lesion is correlated with the frequency of the subsequent obesity (31). Parenteral doses of MSG that failed to produce hypothalamic pyknotic nuclei, namely, 500 mg/kg body weight and less, did not cause endocrine disturbances even when the doses were administered repetitively on alternate days (30). Thus the somatosexual disturbances following parenteral administration of MSG to neonatal mice are apparently the consequence of hypothalamic damage induced by elevated plasma glutamate levels.

Similar observations have been made in newborn rats given parenteral doses of MSG (32-34), although some investigators have reported negative results in rats given equivalent doses (30,35-38). In any event, hypothalamic neuronal necrosis appears to be the necessary antecedent of somatosexual disturbance in the rat as well as the mouse.

When MSG was administered to rodents as a dietary component, somatosexual disturbances were not observed. It is well recognized that even massive doses of MSG so administered do not elevate plasma glutamate levels to the threshold required for hypothalamic damage. Therefore no subsequent somatosexual perturbations would be expected.

In spite of the near-universal acceptance of these facts, it is of interest to survey the results of Searle's animal experiments for evidence of signs relating to somatosexual disturbances.

STUDIES ON ASPARTAME

In one study (39), aspartame was administered in diet to groups of ICR Swiss mice at dose levels of 1000, 2000, and 4000 mg/kg body weight per day for 104 weeks, beginning at age 28 days. Each group consisted of 36 males and 36 females. A similar study was reported by Ito (40). This was a 95-week feeding study in weanling mice who received up to 5900-9800 mg/kg body weight per day of aspartame. Somatosexual disturbances were not observed in either study.

Groups of 40 male and 40 female Charles River CD weanling rats weighing 75-108 g were fed aspartame at dose levels of 1000, 2000, 4000, and 6000 mg/kg body weight per day in diet for 104 weeks (27). Aspartame intake of the animals fed the highest dosage was increased from 6000 to 7000 mg/kg body weight per day at week 16, and to 8000 mg/kg body weight per day at week 44.

Histological examination showed that the ovaries of treated rats had a greater incidence of cystic follicles than those of control animals (Table 3). Olney has alleged that this was due to "ventral compression" of the brain (4). However, as already indicated, these two conditions almost never coexisted in the same animal. Furthermore, since a dose-response relationship was not evident for cystic follicles, it may be concluded that the observed distribution merely reflects a subnormal frequency (5.6%) in control animals. Consistent with this conclusion is a 26% frequency of cystic follicles in control animals involved in a similar feeding study on the diketopiperazine of aspartame (41).

A second chronic study (24) was performed in a manner similar to the preceding one, except that exposure to aspartame began prior to conception, with extension through gestation until 2 years of age. After weaning, Charles River CD rats, 40 males and females per group, received aspartame in the diet at levels of 2000 and 4000 mg/kg body weight per day.

Histologically, ovarian cystic follicles were recorded in 18% of control animals and in 9-11% of the two aspartame groups (Table 4). Thus there is no confirmation of any increased incidence of cystic ovarian follicles in aspartame-treated rats. Furthermore, such follicles were not diagnosed at all in the 2-year Wistar rat study of Ishii et al. (28).

In studies on other species, hamsters received aspartame in the diet at doses up to 12,000 mg/kg body weight per day from age 20 days for 46 weeks, at which

Table 3 Histology of the Ovaries in Rats (27)

	Dose (mg/kg body weight per day)				
	0	1000	2000	4000	6000→8000
Cystic follicles					
Incidence	3/54	5/16	5/12	6/40	5/37
Percent	5.6	31.2	41.7	15.0	13.5

Table 4 Histology of the Ovaries in Rats (24)

	Dose (mg/kg body weight per day)		
	0	2000	4000
Cystic follicles			
Incidence	9/49	3/34	4/37
Percent	18.4	8.8	10.8

time the study was terminated because of an uncontrolled disease in the colony (42). Further, 5-month-old beagle dogs received 1000, 2000, or 4000 mg/kg body weight aspartame in the diet for 2 years (43). Finally, monkeys were employed in a controlled, chronic toxicity study of aspartame, described by Reynolds et al. (44). In none of these species were somatosexual disturbances observed.

In summary, Searle has carried out 2-year feeding studies with aspartame in mice, rats (two separate studies), and dogs. There has been a 10-month study in hamsters and a 9-month study in monkeys. Searle also carried out a two-generation reproduction study in rats fed aspartame. Japanese investigators carried out additional 2-year studies in mice and rats fed aspartame. The results were uniformly negative in regard to Olney's hypothesis.

CONCLUSIONS

The "silent lesion" hypothesis of Olney predicts that routine consumption of aspartame by children may cause repetitive neuroendocrine disturbances leading to adverse influences upon somatosexual development.

1. This "silent lesion" hypothesis is based, in part, upon a pharmacological experiment carried out in adult rats by Olney, involving a single 1000 mg/kg body weight parenteral injection of MSG, which he assumed to be subneurotoxic. Subsequent work with foreign chemicals such as N-methylaspartic and kainic acids is not relevant for aspartame. The size of the MSG dose employed has no relevance for the use of aspartame as a food additive.
2. Other, independent investigators could not reproduce Olney's experimental data on LH and testosterone in specific attempts to do so. Thus Olney's report fails the test of reproducibility, which is a scientific sine qua non.
3. Dietary aspartame does not elevate plasma aspartate, as shown by Stegink (21) elsewhere in this volume. Thus, even if one accedes to all of Olney's assumptions and extrapolations, no mechanism exists for dietary aspartame administration to exert an effect upon the hypothalamus, since aspartate levels are not elevated.

4. The characteristic signs of multiple endocrine or somatosexual disturbances were not observed in the many aspartame feeding studies performed by Searle and others. This is not unexpected, because neuronal necrosis of the arcuate nucleus appears to be an obligate antecedent for these disturbances, and the arcuate would remain unaffected in the absence of elevated plasma aspartate levels.
5. In view of all the foregoing, it is concluded that, contrary to the allegations of Olney, aspartame does not pose a risk for neuroendocrine regulatory systems.

REFERENCES

1. Hayes, A. H., Jr. (1981). Aspartame: Commissioner's final decision. *Fed. Reg.* 46, 38285-38308, 46394, 50947-50948.
2. Olney, J. W. (1974). Memorandum to the Commissioner, Food and Drug Administration, August 16.
3. Olney, J. W., Cicero, T. J., Meyer, E. R., and de Gubareff, T. (1976). Acute glutamate-induced elevations in serum testosterone and luteinizing hormone. *Brain Res.* 112, 420-424.
4. Olney, J. W. (1976). Letter to Richard Merrill, Chief Counsel, Food and Drug Administration, November 12.
5. Dunn, J. D., Hess, M., and Johnson, D. C. (1976). Effect of thyroidectomy on rhythmic gonadotropin release. *Proc. Soc. Exp. Biol. Med.* 151, 22-27.
6. Sturtevant, F. M. (1976). Chronopharmacokinetics of ethanol. I. Review of the literature and theoretical considerations. *Chronobiologia* 3, 237-262.
7. Yonetani, S., and Matsuzawa, Y. (1978). Effect of monosodium glutamate on serum luteinizing hormone and testosterone in adult male rats. *Toxicol. Lett.* 1, 207-211.
8. Kalra, P. S., and Kalra, S. P. (1977). Circadian periodicities of serum androgen, progesterone, gonadotropins and luteinizing hormone-releasing hormone in male rats: The effects of hypothalamic deafferentiation, castration and adrenalectomy. *Endocrinology* 101, 1821-1827.
9. Matsuzawa, Y., Yonetani, S., Takasaki, Y., Iwata, S., and Sekine, S. (1979). Studies on reproductive endocrine function in rats treated with monosodium L-glutamate early in life. *Toxicol. Lett.* 4, 359-371.
10. Dunn, J. D., Arimura, A., and Scheving, L. E. (1972). Effect of stress on circadian periodicity in serum LH and prolactin concentration. *Endocrinology* 90, 29-33.
11. Dunn, J. D. (1974). Circadian variation in adrenocortical and anterior pituitary hormones. In *Biological Rhythms in Neuroendocrine Activity* (Kawakami, M., ed.), Igaku Shoin, Tokyo, pp. 119-139.
12. Weitzman, E. D. (1976). Circadian rhythms and episodic hormone secretion in man. *Annu. Rev. Med.* 27, 225-243.
13. Dyer, R. G., Weick, R. F., Mansfield, S., and Corbet, H. (1981). Secretion of luteinizing hormone in ovariectomized adult rats treated neonatally with monosodium glutamate. *J. Endocrinol.* 91, 341-346.

14. Terry, L. C., Epelbaum, J., and Martin, J. B. (1981). Monosodium glutamate: Acute and chronic effects on rhythmic growth hormone and prolactin secretion, and somatostatin in the undisturbed male rat. *Brain Res.* 217, 129-142.
15. Millard, W. J., Martin, J. B., Jr., Audet, J., Sagar, S. M., and Martin, J. B. (1982). Evidence that reduced growth hormone secretion observed in monosodium glutamate-treated rats is the result of a deficiency of growth hormone-releasing factor. *Endocrinology* 110, 540-549.
16. Nemeroff, C. B., Bissette, G., Greeley, G. H., Mailman, R. B., Martin, J. B., Brazeau, P., and Kizer, J. S. (1978). Effects of acute administration of monosodium L-glutamate (MSG), atropine or haloperidol on anterior pituitary hormone secretion in the rat. *Brain Res.* 156, 198-201.
17. Greeley, G. H., Jr., Nicholson, G. F., Nemeroff, C. B., Youngblood, W. W., and Kizer, J. S. (1978). Direct evidence that the arcuate nucleus-median eminence tuberoinfundibular system is not of primary importance in the feedback regulation of luteinizing hormone and follicle-stimulating hormone secretion in the castrated rat. *Endocrinology* 103, 170-175.
18. Keating, R. J., and Tcholakian, R. K. (1979). *In vivo* patterns of circulating steroids in adult male rats III. Effect of total parenteral nutrition on the diurnal variation of testosterone. *Endocrinol. Res. Commun.* 6, 95-105.
19. Mares, S. E., and Berg, J. R. (1978). Effects of aspartame (SC-18862) on gonadotropin secretion in rats. Searle Biol. Rep. 78D1169, FDA Hearing Clerk Doc. 75F-0355, Appendix A to "Statement of Issues," G. D. Searle & Co., Skokie, Ill., 7/31/79, Aspartame Public Board of Inquiry.
20. Lennon, H. D., Metcalf, L. E., Mares, S. E., Smith, J. H., Nutting, E. F., and Saunders, F. J. (1980). The biological properties of aspartame. IV. Effects on reproduction and lactation. *J. Environ. Pathol. Toxicol.* 3, 375-386.
21. Stegink, L. D. (1984). Aspartame metabolism in human subjects: Acute dosing studies. In *Aspartame: Physiology and Biochemistry* (Stegink, L. D., and Filer, L. J., Jr., eds.), Marcel Dekker, New York, Chapter 26.
22. Saunders, F. J., Pautsch, W. F., and Nutting, E. F. (1980). The biological properties of aspartame. III. Examination for endocrine-like activities. *J. Environ. Pathol. Toxicol.* 3, 363-373.
23. Reno, F. E. (1971). Two-generation reproduction study—Rats. SC-18862, Final Report, P-T 867H71, September 8, Food Additive Petition 3A2885, Part E-11; Hearing Clerk's File Vol. 20, Docket No. 75F-0355, Hazleton Laboratories, Inc., Vienna, Va.
24. Trutter, J. A., and Reno, F. E. (1974). Lifetime toxicity study in the rat. SC-18862, Final Report, P-T No. 892H72, January 11, Food Additive Petition 3A2885, Part E-70; Hearing Clerk's File Vol. 80, Docket No. 75F-0355, Hazleton Laboratories, Inc., Vienna, Va.
25. Reno, F. E. (1972). Toxicological evaluation in the neonatal rat, SC-18862, Final Report, P-T 893H71, January 26, Food Additive Petition 3A2885, Part E-9; Hearing Clerk's File Vol. 18, Docket No. 75F-0355, Hazleton Laboratories, Inc., Vienna, Va.
26. Ferrell, J. F. (1978). Letter to Robert G. Bost, G. D. Searle & Co., November 29, Experimental Pathology Laboratories, Inc., Herndon, Va.

27. Reno, F. E. (1973). Two-year toxicity study in the rat. SC-18862, Final Report, P-T No. 838H71, January 12, Food Additive Petition 3A2885, Parts E-33 and E-34; Hearing Clerk's File Vol. 43, 44, Docket No. 75F-0355, Hazleton Laboratories, Inc., Vienna, Va.
28. Ishii, H., Koshimizu, T., Usami, S., and Fujimoto, T. (1981). Toxicity of aspartame and its diketopiperazine for Wistar rats by dietary administration for 104 weeks. *Toxicology* 21, 91-94.
29. Olney, J. W. (1969). Brain lesions, obesity, and other disturbances in mice treated with monosodium glutamate. *Science* 164, 719-721.
30. Matsuyama, S., Oki, Y., and Yokoki, Y. (1973). Obesity induced by monosodium glutamate in mice. *Nat. Inst. Anim. Health Q.* 13, 91-101.
31. Tanaka, K., Shimada, M., Nakao, K., and Kusunoki, T. (1978). Hypothalamic lesion induced by injection of monosodium glutamate in suckling period and subsequent development of obesity. *Exp. Neurol.* 62, 191-199.
32. Hansson, H. A. (1979). Scanning electron microscopic studies on the long term effects of sodium glutamate on the rat retina. *Virchows Arch. B.* 4, 357-367.
33. Redding, T. W., Schally, A. U., Arimura, A., and Wakabayaski, I. (1977). Effect of monosodium glutamate on some endocrine functions. *Neuroendocrinology* 8, 245-255.
34. Nikoletseas, M. M. (1977). Obesity in exercising, hypophagic rats treated with monosodium glutamate. *Physiol. Behav.* 19, 767-773.
35. Bunyan, J., Murrell, E. A., and Shah, P. P. (1976). The induction of obesity in rodents by means of monosodium glutamate. *Br. J. Nutr.* 35, 25-39.
36. Adamo, N. Y., and Ratner, A. (1970). Monosodium glutamate: Lack of effects on brain and reproductive function in rats. *Science* 169, 673-674.
37. Trentini, G. P., Botticelli, A., and Botticelli, C. S. (1974). Effect of monosodium glutamate on the endocrine glands and on the reproductive function of the rat. *Fertil. Steril.* 25, 478-483.
38. Lengvari, I. (1977). Effect of perinatal monosodium glutamate treatment on endocrine functions of rats in maturity. *Acta Biol. Acad. Sci. Hung.* 28, 133-141.
39. Reno, F. E. (1974). 104 week toxicity study in the mouse. SC-18862, Final Report, P-T No. 984H73, September 6, Food Additive Petition 3A2885, Part E-75; Hearing Clerk's File Vol. 87, Docket No. 75F-0355, Hazleton, Laboratories, Inc., Vienna, Va.
40. Ito, R. (1975). Toxicity of AP and DKP in dietary administration to mice for twenty-six, fifty-three and ninety-five weeks. January 25, Hearing Clerk's File Vol. 119, Docket No. 75F-0355.
41. Rao, K. S., Stejskal, R., and McConnell, R. G. (1974). 115 week oral tumorigenicity study in the rat. SC-19192, P-T No. 988S73, September, Food Additive Petition 3A2885, Part E-77; Hearing Clerk's File Vol. 89, Docket No. 75F-0355.
42. Rao, K. S., Mauro, J., and McConnell, R. G. (1972). 46 week oral toxicity study in the hamster. SC-18862, December 8, Food Additive Petition 3A2885, Part E-27; Hearing Clerk's File Vol. 37, Docket No. 75F-0355.

43. Rao, K. S., Mauro, J., and McConnell, R. G. (1972). 106 week oral toxicity study in the dog. SC-18862, August 1, Food Additive Petition 3A2885, Part E-28; Hearing Clerk's File Vol. 38, Docket No. 75F-0355.
44. Reynolds, W. A., Bauman, A. F., Stegink, L. D., Renn, E., and Filer, L. J., Jr. (1979). Developmental assessment of infant macaques receiving dietary aspartame or phenylalanine. FDA Hearing Clerk Doc. 75F-0355, Suppl. Info. Vol. III, Tab 27, submitted by G. D. Searle & Co., July 31.

STUDIES OF ASPARTAME METABOLISM IN HUMANS

25

Chronic Ingestion of Aspartame in Humans

Willard J. Visek
University of Illinois College of Medicine at Urbana-Champaign, Urbana, Illinois

INTRODUCTION

The Food Additives Amendment enacted in 1958 (Public Law 85-929) requires that manufacturers demonstrate the safety of a new food additive. The amendment includes the extremely important but controversial Delaney clause, which states that any food additive in any concentration found to cause cancer in animals or humans will be prohibited for food and beverage use. The intent of the Delaney clause is to protect the consuming public. However, the problems it raises in data interpretation, protocol evaluation, and extrapolation of the data to conditions of use in man have polarized scientists, legislative bodies, and the consuming public into proponents and opponents of the measure. These controversies were brought into particular focus in the case of artificial sweeteners when the cyclamates were banned (1) under the Delaney provision in 1969 and when the Food and Drug Administration (FDA) proposed to ban saccharin in 1977 (2). The proposed ban on the use of saccharin came at a time when it was the only general-use artificial sweetener approved in the United States. The majority of the 100,000 comments the FDA received concerning this issue were opposed to the ban. The public outcry was so intense that the U.S. Congress passed the Saccharin Study and Labeling Act (Public Law 95-203), which imposed an 18-month moratorium on the proposed ban to consider its impact. This latter law also required warning labels and notices for foods containing saccharin stating that saccharin has been shown to cause cancer in laboratory animals (3).

The intensity of the debate over the wisdom of the Delaney clause dramatized the need for an artificial sweetener for dieters and diabetics for whom obesity is a health problem. It was clear that an artificial sweetener without carcinogenic properties would receive wide public acceptance. The molecular structure of aspartame (L-aspartyl-L-phenylalanylmethyl ester) argues that it is rapidly converted in the body to metabolites which would not present a carcinogenic hazard, and extensive studies in animals support this postulate. However, concerns were expressed by a small number of individuals that the metabolic products might elicit other adverse responses under some circumstances. Attempts to address these concerns involved studies on over 290 human subjects from selected groups that included healthy children and adolescents, young persons during weight reduction, non-insulin-dependent diabetics, and phenylketonuric heterozygous adults (4-7).

STUDIES WITH APPARENTLY HEALTHY CHILDREN AND ADOLESCENTS

This double blind study involved 126 children and adolescents, chosen from an apparently healthy group of 255, ranging in age from 2 to 21 years (4). The subjects selected included an equal number of each sex. Selection for study began with a complete physical examination and medical history for each prospective subject. In subjects aged 2-12 years, specimens were obtained in a fasting state for the laboratory examinations of plasma phenylalanine concentrations, plasma tyrosine concentration, complete blood count, partial thromboplastin time, prothrombin time, creatinine concentration, direct, indirect and total bilirubin, serum glutamate oxalacetate transaminase (SGOT) activity, complete urinalysis, and urine tests for phenylpyruvic acid. Additional tests were done in the adolescent group (ages 13-21 years). These included plasma thyroxine, fasting blood glucose, uric acid, cholesterol (total and esters), serum triglycerides, and blood urea nitrogen. Individuals who were married, pregnant, or immunized within the 2 weeks preceding enrollment in the study were not accepted. Although not used as a basis of exclusion, acne vulgaris lesions about the forehead and facial area were also scored in the 13- to 21-year-old group. A total of 33 experimental subjects were selected at random for determinations of serum and urine methanol. Individuals whose prestudy methanol values fell outside of the normal laboratory range were eliminated. Ophthalmoscopic examinations were conducted on all subjects. In addition, the 7- to 21-year-old subjects underwent slit-lamp visual acuity and color vision evaluations. The 16- to 21-year-old subjects also had tonometry meaurements.

The age, weight, and sex distributions and other pertinent data are presented in Table 1. The experimental subjects received their aspartame or sucrose as supplements to their regular diet in the form of gelatin, milk, soft drink or pudding

Table 1 Age and Sex Distribution, Weight Range, and Aspartame Consumption of 126 Apparently Healthy Children and Adolescents Who Completed a 13-Week Study After They Had Been Assigned on a Double-Blind Basis to Aspartame or Sucrose Treatments

	Study group					Total
	A	B	C	D	E	
Age (years)	2-3	4-6	7-9	10-12	13-21	
Weight range (kg)	11.4-20.0	14.1-29.1	20.9-40.9	25.9-56.8	35-76.8	
Aspartame consumed						
Per 24 hr (g)	0.61	0.8	1.6	2.0	2.4	
Per kg/body weight/ 24 hr (mg)	53.7-30	56.8-27.5	76.5-39.1	77.2-35.2	68.6-31.2	
Sex distribution of 126 subjects completing study						
Males	7	10	10	12	23	62
Females	6	12	12	12	22	64
Number of subjects on aspartame or sucrose						
Aspartame	7	11	11	12	20	61
Sucrose	6	11	11	12	25	65

Source: Modified from Ref. 4.

mixtures, cream fillings, especially prepared cookies, packets for use in cereal, or in sucrose- or aspartame-containing capsules. Other details of the experimental protocol are published elsewhere (4).

Effects on Laboratory Values

Statistical analyses employed Student's t tests for comparing means for the data obtained at 13 weeks against base-line values. Statistically significant differences in the laboratory data are summarized in Table 2, but none were believed to be of clinical importance and were not seen in laboratory tests, which were all repeated 7 days after the end of the 13th week. The laboratory data obtained 7 days after the termination of the test did show statistically significant, but not clinically significant, differences ($p<0.05$) in blood hemoglobin and hematocrit between the aspartame and sucrose treatments in age group C (7- to 9-year-olds). Respective mean values were 14.2 and 13.4 mg/dl for hemoglobin and 41.5 and 39.5% for hematocrit. Additional laboratory determinations on the blood of adolescents (ages 13-21, Group E) for urea nitrogen, serum thyroxine, serum cholesterol, and serum triglycerides revealed no statistically significant differences.

Effects of Aspartame on Plasma Phenylalanine and Tyrosine Concentrations or Phenylalanine Metabolism

Plasma phenylalanine and tyrosine concentrations were determined on blood samples drawn on specific times and days of the week during week 0 (the week preceding the study) and weeks 1, 3, 5, 7, 9, 11, and 13. Both treatment groups in each age category showed slight decreases in phenylalanine over the course of the 13 weeks ($p<0.05$), but no significant differences between treatment groups were noted. No differences in plasma tyrosine concentrations were demonstrated

Table 2 Statistically Significant Differences Observed in Children and Adolescents Who Consumed Aspartame or Sucrose for 13 Weeks

		Mean Value		
Group	Test	Aspartame	Sucrose	Significance
A	No differences			
B	Hematocrit	36.8%	38.2%	$p<0.05$
C	Direct bilirubin	0.07 mg%	0.14 mg%	$p<0.05$
	Creatinine	0.37 mg%	0.51 mg%	$p<0.05$
	Prothrombin time	12.4 sec	11.7 sec	$p<0.02$
D	Indirect bilirubin	0.05 mg%	0.18 mg%	$p<0.01$
E	White blood cell count	6260 mm^3	7340 mm^3	$p<0.05$

Source: Ref. 4.

between or within treatment groups as a function of time or age category. Plasma phenylalanine-to-tyrosine ratios likewise showed no significant differences. Baseline ratios tended to be higher than those determined during the test period for all age groups.

All determinations for the presence of phenylpyruvic acid in the urine (positive values would be presumptive evidence of disturbances in phenylalanine metabolism) were negative before the study and at weeks 7 and 13.

Methanol Determinations, Physical Status, Eye Examinations, Acne Evaluations, and Subjective Complaints

Blood methanol determinations for 33 subjects of all age categories were negative. Pre- and posttest physical examinations, including blood pressure measurements, revealed no significant changes. As expected, all children tended to gain weight during the study, but the initial and final weights for each group showed no significant differences between experimental treatments. Ophthalmological examinations and acne evaluations likewise failed to show differences or abnormalities. The subjective complaints which were recorded in biweekly interviews were not clinically important.

EFFECTS OF ASPARTAME IN YOUNG PERSONS DURING WEIGHT REDUCTION

It is reasonable to expect that a potent sweetener with little caloric value would find particular use among weight-conscious individuals. Early in the evaluations of aspartame, investigations were undertaken to determine its possible toxicity for young persons actively engaged in a clinically supervised weight reduction program. During these studies particular attention was given to glucose metabolism and the behavior of the glucoregulatory hormones insulin and glucagon (5).

A total of 59 subjects were enrolled in these studies, and 55 completed them. Most were recruited from schools of nursing, and the remainder were referred by pediatricians for treatment of obesity; 51 were females. The ages ranged from 10 to 21, with a mean of 19.3 years. As a group, the subjects had a mean height of 166.1 cm and an average weight of 74.7 kg. Since the ideal average body weight for this group would be 56.4 kg (8), they averaged 33% above this standard. Their ideal average body weight ranged from 45.4 to 81.8 kg. Before assignment to the study, medical histories were obtained and physical examinations were carried out. Laboratory tests, ophthalmological examinations, and an estimation of facial acne were also a part of each subject's evaluation. None had diabetes or abnormalities of vision, blood clotting, or thyroid function.

Aspartame and the placebo were administered in 300-mg gelatin capsules. The substances were assigned and administered in a randomized double-blind fashion. Dieticians of the Thorndike Metabolic Ward, Boston City Hospital, Boston,

Massachusetts, obtained dietary histories on the initial visit and gave instructions for an individualized calorie-restricted diet which was ingested for the 13 weeks that aspartame or the placebo were consumed. The daily intake of aspartame was 2.79 g.

Random assignment to treatments resulted in tightly clustered groups that were virtually identical for average height, weight, and caloric intake. The aspartame subjects were 1.1 years older in average age, a difference which was highly statistically significant but of little practical importance. During the course of the 13 weeks, the 24 subjects taking aspartame lost approximately 3.1 kg, compared to 2 kg for the 33 subjects who consumed lactose (placebo), but this difference was not statistically significant. Blood pressure measurements in both groups showed a significant downward trend through week 7, but not through week 13. Assessment of acne and ophthalmological slit-lamp examinations showed no changes that could be related to either experimental regimen.

Hemoglobin concentration, hematocrit, white cell count, prothrombin time, and partial thromboplastin time showed no consistent differences. Average plasma glucose measured at weeks 7 and 13 in both groups was significantly below initial concentrations, as would be expected during weight loss, but there was no significant difference between the aspartame and placebo treatments. Measurements of blood urea nitrogen, creatinine, triglycerides, total cholesterol, cholesterol esters, direct, indirect and total bilirubin, total thyroxine, SGOT activity, and uric acid gave no basis for suspecting toxicity of aspartame. Conventional urinalyses at weeks 7 and 13 likewise gave no evidence of important differences between treatments. Urinary methanol determinations on random occasions were negative. Generally, plasma phenylalanine and tyrosine concentrations tended to be higher during aspartame administration in specimens obtained during weeks 0, 1, 3, 5, 7, 9, 11, and 13. However, only the difference for plasma phenylalanine concentration in week 11 proved to be statistically significant. Since blood was drawn 12 hr after the last dosage of the sweetener, the practical significance of this difference shown by statistical analyses is uncertain and may not represent prior intake of aspartame.

Body weight, plasma glucose, and immunoreactive insulin showed a downward trend during weight reduction through 7 weeks (Fig. 1). Even though weight reduction persisted, but at a slower rate, from week 7 to 13, plasma glucose and insulin tended to rise in the subjects consuming aspartame. Throughout the entire 13 weeks immunoreactive glucagon tended to rise in both groups. The investigators drew particular attention to the degree of weight loss being significant during weeks 0 and 7 and not thereafter. Plasma glucose and immunoreactive insulin were correspondingly lower at week 7, but no meaningful further change was seen at week 13. The authors concluded that weight reduction in these subjects caused a detectable metabolic shift in carbohydrate metabolism which was not affected by aspartame administration.

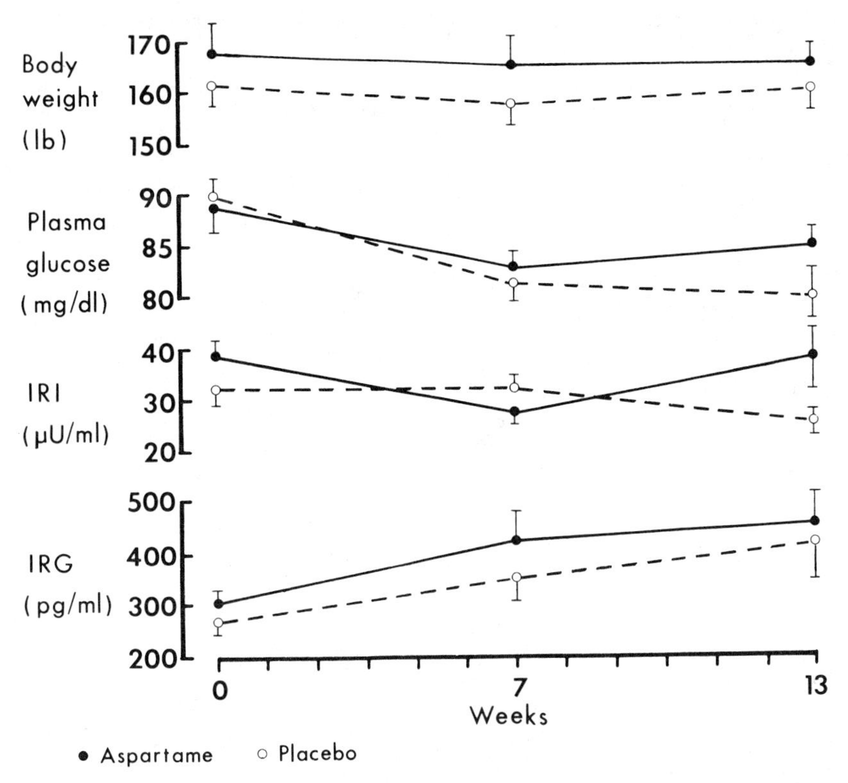

Figure 1 Effect of ingestion of aspartame (A) or placebo (P) on body weight, plasma glucose, immunoreactive insulin (IRI), and immunoreactive glucagon (IRG) during caloric restriction in young persons. At no time are A and P significantly different from each other. Within A and P groups, paired comparisons against the initial measurement (week 0) show significant differences at weeks 7 and 13. Significant reductions are seen in body weight at both weeks 7 and 13 ($p < 0.001$ for A and P in every instance). For glucose at week 7, significant reductions are seen in A ($p < 0.01$) and P ($p < 0.05$). At week 13 glucose remains significantly reduced only in P ($p < 0.001$). IRI is significantly reduced only at week 7: A ($p < 0.025$), P ($p < 0.01$). For IRG, significant elevations are seen at weeks 7 ($p < 0.001$ for both A and P) and 13: A ($p < 0.01$) and P ($p < 0.001$). (From Ref. 4.)

Complaints and Side Effects

An assessment of the complaints registered by the subjects of this study are instructive (Table 3). The most frequent complaints were bowel changes, febrile illnesses, and menstrual changes. This list emphasizes the importance of randomization and the need to include a placebo treatment whenever assessments are

Table 3 Complaints by Young Overweight Persons While Consuming Aspartame or Placebo for 13 Weeks

Side effect	Number of side effects	
	Aspartame (n = 22)	Placebo (n = 33)
Appetite change	5	8
Nausea	4	3
Bowel change	30	32
Fatigue	3	4
Headache	1	6
Dizziness	2	2
Abdominal cramps	1	6
Menstrual changes	16	12
Polyuria	5	1
Fainting	1	1
Nosebleed	1	2
Congestion	2	
Febrile illnesses	16	19
Pain in leg	1	
Cold hands	2	
Noticeable taste	2	
Prickly sensation	1	
Dry skin	1	3
Acne	5	8
Rash	3	1
Dry mouth	1	
Bumps in mouth		1
Crying spells		1
Anal fissure		1
Pills hard to swallow		2
Total	103	113

Source: Ref. 5.

made in which subjective responses may influence interpretation. Considering the number and nature of the complaints, one would be hard pressed to assess differences between placebo and experimental groups which rely upon such criteria.

The subjects of this study were primarily women in late adolescence. The aspartame dosage was 2.7 g/day, about four times the anticipated daily intake under conditions of expected ordinary use. The weight reductions, averaging 2-3 kg (4.5-6.5 lb), were comparable and were achieved mainly during the first 7 weeks, with little further progress by week 13. The authors reported that they had the subjective impression that weight loss corresponded to the degree of enthusiasm which the subjects showed for the program. This seemed high initially, and declined as the study progressed, although a number of determined individuals showed continued weight loss for the duration of the 13-week program. Other investigators have also observed a decline in enthusiasm by subjects during the progress of weight reduction studies (9). The changes in glucoregulatory hormones also corresponded to significant weight loss during weeks 0-7 when significant declines in blood glucose and immunoreactive insulin occurred accompanied by a rise in immunoreactive glucagon. By week 13 these values changed little or began to advance toward base line, probably reflecting increasing caloric intakes. The reciprocal changes in immunoreactive insulin and immunoreactive glucagon at week 7 reflect changes in gluconeogenesis, which is enhanced during decreased food intake to meet glucose needs of the brain and other tissues (10).

Similar shifts in glucose, insulin, and glucagon have been seen in other studies employing calorie restriction (11,12). However, aspartame produced no detectable changes in these indices, even though its metabolism yields aspartic acid and phenylalanine, both recognized as glucagon secretagogues (13). Thus it appears from these studies that aspartame has no synergistic action with caloric restriction to stimulate glucagon release. The same can be said about insulin, because aspartic acid and phenylalanine likewise are suggested to stimulate insulin secretion from the beta cells of the pancreas (13,14). It is noteworthy that the above observations are based upon blood drawn 12 hr after the last dose of aspartame. Although there were no consistent differences in plasma phenylalanine concentration, the possibility remains that stimulation of immunoreactive insulin and immunoreactive glucagon secretion may occur earlier following aspartame administration with or without a meal.

It is of interest that weight loss was achieved with detectable effects on carbohydrate metabolism, but with no lowering of plasma triglycerides or cholesterol. This may have occurred because the subjects were normolipidemic. The significant reduction in systolic blood pressure occurred only in week 7. Although several factors may have been responsible, including a simple diminution of stress, caloric restriction is known to cause urinary loss of sodium and water and their retention when carbohydrate and nitrogen intakes rise. Therefore attendant changes in body fluid volume may have been the basis for the changes in blood pressure seen at weeks 7 and 13.

EFFECTS OF ASPARTAME IN NON-INSULIN-DEPENDENT DIABETES

Aspartame is approximately 180 times as sweet as sucrose and supplies 0.5% of the caloric value for an equivalent amount of sweetening. These attributes were important in stimulating its evaluation in patients with diabetes mellitus. Treatment of this disorder requires control of caloric and sugar intake. Although this disease and obesity can be managed without artificial sweeteners, the availability of a suitable sweetening agent generally makes the patient's life more tolerable. Artificial sweeteners are particularly helpful to physicians in managing juvenile diabetics, who have difficulty restricting their intake of sweets. The potential benefits an artificial sweetener may have for the diabetic and its usefulness, real or psychological, for weight control are legitimate issues that deserve careful and sympathetic examination. Of even greater importance is that such agents are safe and do not interfere with other necessary therapy.

The experimental subjects of this study were 69 adults, between the ages of 21 and 70, with diabetes mellitus that was managed by diet and/or oral hypoglycemic agents (6). A complete medical history was obtained and a physical examination was performed at least 1 week before administration of aspartame or placebo. Only individuals showing normal results for the following tests were selected for the study: complete blood count, pregnancy test (women), partial thromboplastin time, blood urea nitrogen, creatinine, bilirubin, plasma phenylalanine, and plasma tyrosine. The following laboratory tests were also done at screening, but abnormal values were not used to preclude selection for participation: SGOT, serum glutamic pyruvate transaminase, lactic dehydrogenase, alkaline phosphatase, uric acid cholesterol (total and esters), triglycerides, fasting blood sugar, oral glucose tolerance (selected cases), and urinalysis.

Complete eye examinations including tonometry, fundoscopy, and slit-lamp examination were done before and at completion of the study.

The patients were instructed to follow their usual diet. They were randomly assigned in double-blind fashion to two groups receiving either two aspartame- or placebo-containing capsules three times daily. The daily consumption of aspartame was 1.8 g for 13 weeks. Laboratory tests were repeated at the end of weeks 1, 3, 5, 7, 9, 11, 13, and 14. Final clinical and ophthalmological examinations were also done during the 14th week, 1 week after the cessation of aspartame or placebo consumption.

A total of 43 of the diabetic patients who participated were under the care of the Department of Medicine, Tulane University School of Medicine, New Orleans, Louisiana, while 26 were patients of the Jewish Hospital and Medical Center, Brooklyn, New York. Of the 43 patients studied at Tulane University, 18 showed fasting blood sugar concentrations which were below 100 mg/dl for the majority of their analyses. Of these 18 subjects, 7 had consumed aspartame, and 11 placebo. There were 10 patients with a majority of their blood sugar values above 110 and below 130 mg/dl: 7 had consumed aspartame, and 3 placebo. The remaining 15 patients showed blood sugar concentrations above 130 mg/dl for a majority of their analyses: 9 had consumed aspartame, and 6 placebo. The remaining 26 patients, followed at the Jewish Hospital and Medical Center in Brooklyn, showed no statistically significant differences in blood sugar concentrations between aspartame and placebo groups at weeks 1, 5, 9, and 14 of the study. As with the 43 patients studied at Tulane, there was no evidence that aspartame influenced the control of diabetes.

There were no significant differences between treatment groups in mean body weight or blood pressure measurements. The laboratory values for individual patients fluctuated, but no abnormalities were observed that could be ascribed to aspartame or placebo. Plasma phenylalanine and tyrosine concentrations and the plasma phenylalanine-to-tyrosine ratios did not show appreciable changes, and significant changes were not observed between treatment groups. No ophthalmological or electrocardiographic effects were observed.

A patient dropped from the study during the 11th week underwent a partial gastrectomy for a reticulum cell sarcoma. No evidence was obtained to show that this lesion was linked to either experimental treatment. Other details of this study are available elsewhere (6).

LONG-TERM STUDIES OF ASPARTAME IN PHENYLKETONURIC ADULTS

Phenylketonuria (PKU) is a disease caused by an autosomal recessive gene carried by approximately 2% of the population. It is the most common of the aminoacidurias and occurs at an incidence of 1 in 10,000 births. Mental retardation associated with the disease is lessened by placing phenylketonuric infants on diets low in phenylalanine. Fortunately phenylalanine blood tests are now routinely conducted on newborn infants at most hospitals. In PKU, elevated serum phenylalanine concentrations greater than 20 mg/dl are associated with development of mental retardation. Moderate elevations of 5-20 mg/dl may be associated with perceptual difficulties and learning disorders.

An average U.S. adult consumes about 4.7 g of phenylalanine per day from all sources. It has been estimated that an average adult might consume from 0.34 to 3 g of phenylalanine daily in aspartame. Thus studies were conducted in parents of PKU patients, since by definition such parents must be either heterozygous (a carrier) or homozygous for the disease.

A total of 43 individuals (19 males and 25 males) with an average age of 35 years participated in a double-blind study. They were divided according to sex and randomly assigned to one of two treatment regimens. Each group consumed identical capsules containing either aspartame or placebo. Table 4 contains the schedule of capsule intake and the daily intake of phenylalanine in the group receiving the capsules containing aspartame. During the first 7 weeks (phase A) the intake of aspartame was increased each week from 3 to 37 capsules per day. For the ensuing 21 weeks (phase B) six capsules per day were ingested.

Each participant had a medical and ophthalmological examination to determine general health status at the beginning and at the completion of the study. During each week of the study period, blood pressures, body weights, intakes of

Table 4 Aspartame and Placebo Capsule Intake by Heterozygous Phenylketonuric Adults During a Course of 2 Weeks

	Phase A						Phase B:
	Week 1	Week 2	Week 3	Week 4	Week 5	Week 6	Weeks 7-27
Number of capsules taken with each meal (three times per day)	1	2	4	5	7	9	2
Amount of aspartame per capsule (g)	0.2	0.2	0.2	0.3	0.3	0.3	0.3
Total amount aspartame per day (g)	0.6	1.2	2.4	4.5	6.3	8.1	1.8
Total daily phenylalanine from aspartame (g)	0.34	0.67	1.35	2.53	3.54	4.55	1.01

Source: Ref. 7.

medication, and observations and complaints of the participants were recorded. Extensive laboratory studies included serum phenylalanine and tyrosine determinations, hematological and clotting studies, urinalysis, blood urea nitrogen, thyroxine, bilirubin direct and indirect, SGOT, alkaline phosphatase, uric acid, creatinine, cholesterol (total and esters), triglycerides, glucose tolerance, urine chromatography for phenylketones, and tests for pregnancy.

Fasting serum phenylalanine concentration was determined before the study and during weeks 4-6, 8, 11, 15, 19, 23, 27, and 28. Serum tyrosine concentration was determined weekly. The hematologic, urinalysis, and blood chemistry studies were done during screening and during weeks 7, 13, 19, 27, and 28. Glucose tolerance tests were done during weeks 3, 6, 7, 19, 23, 27, and 28.

No significant medical or biochemical changes were observed during this 28-week study. Aspartame was well tolerated, and serum phenylalanine and tyrosine concentrations remained within normal limits throughout the 28 weeks.

CONCLUDING REMARKS

The literature contains reports of studies with 290-295 human subjects, ranging in age from 2 to 70 years, who have consumed aspartame daily for 13-28 weeks. These included children and adolescents, young individuals during weight reduction, non-insulin-dependent adult diabetics, and phenylketonuric heterozygous adults. Daily intakes of aspartame in some of the studies have exceeded those that would be expected in daily practice. All of the studies can be criticized for their experimental design and because the results have not been subjected to sophisticated statistical analyses. Numerous questions remain unanswered, particularly with respect to the possible effects of aspartame on human beings who may consume the compound over years of daily use. Nonetheless, the observations made in susceptible individuals under frequent clinical and extensive laboratory surveillance are reassuring, because no adverse responses were reported and the investigators did not report even suggestive evidence of harmful effects.

REFERENCES

1. Anonymous (1969). Cyclamates barred as "food additives" under Delaney Clause. *Food Chem. News Spec. Suppl.*, October 20.
2. Anonymous (1978). Saccharin: Where do we go from here? *FDA Consumer* 12, 16-21.
3. Anonymous (1978). Saccharin warning notice required in stores. *FDA Consumer* 12, 4.
4. Frey, G. H. (1976). Use of aspartame by apparently healthy children and adolescents. *J. Toxicol. Environ. Health* 2, 401-415.
5. Knopp, R. H. (1976). Effects of aspartame in young persons during weight reduction. *J. Toxicol. Environ. Health* 2, 417-428.

6. Stern, S. B., Bleicher, S. J., Flores, A., Gombos, G., Recitas, D., and Shu, J. (1976). Administration of aspartame in non-insulin-dependent diabetes. *J. Toxicol. Environ. Health* 2, 429-439.
7. Koch, R., Shaw, K. N. F., Williamson, M., and Haber, M. (1976). Use of aspartame in phenylketonuric heterozygous adults. *J. Toxicol. Environ. Health* 2, 453-457.
8. Bierman, E. L., and Glomset, J. A. (1974). Disorders of lipid metabolism. In *Textbook of Endocrinology*, 5th ed. (Williams, R. H., ed.), Saunders, Philadelphia, pp. 890-937.
9. London, A. M., and Schreiber, E. D. (1966). A controlled study of the effects of group discussions and an anorexiant in outpatient treatment of obesity. *Ann. Int. Med.* 658, 80-92.
10. Cahill, G. F. (1971). Physiology of insulin in man. *Diabetes* 20, 758-799.
11. Marliss, E. B., Aoki, T. T., Unger, R. H., Soeldner, J. S., and Cahill, G. F. (1970). Glucagon levels and metabolic effects in fasting man. *J. Clin. Invest.* 49, 2256-2270.
12. Saudek, C. D., Boulter, P. R., and Arky, R. A. (1973). The naturiuretic effect of glucagon and its role in starvation. *J. Clin. Endocrinol. Metab.* 36, 761-765.
13. Rocha, D. M., Faloona, G. R., and Unger, R. H. (1972). Glucagon-stimulating activity of 20 amino acids in dogs. *J. Clin. Invest.* 51, 2346-2351.
14. Floyd, J. C., Fajans, S. S., Conn, J. W., Knopf, R. F., and Rull, J. (1966). Stimulation of insulin secretion by amino acids. *J. Clin. Invest.* 45, 1487-1502.

26

Aspartame Metabolism in Humans: Acute Dosing Studies

Lewis D. Stegink
University of Iowa College of Medicine, Iowa City, Iowa

Toxicology is based on the premise that all compounds are toxic at some dose. Salt, water, sugar, and even a mother's love produce deleterious effects when given in inappropriate amounts. Thus it is not surprising that very large doses of aspartame (Fig. 1) or aspartame's component parts (aspartate, phenylalanine, and methanol) produce deleterious effects in sensitive animal species. The critical question is whether the compound is potentially harmful at normal use and potential abuse levels.

Aspartame may be absorbed and metabolized in one of two ways (Fig. 2). It may be hydrolyzed in the intestinal lumen to aspartate, phenylalanine, and methanol by proteolytic and hydrolytic enzymes (1-5). These compounds are absorbed from the lumen and reach the blood in a manner similar to that of amino acids and methanol arising from dietary protein or polysaccharides. Alternatively, aspartame may be absorbed directly into mucosal cells by peptide transport mechanisms (4,5) with subsequent hydrolysis within the cell to aspartate, phenylalanine, and methanol. In either case, large doses of aspartame release aspartate, phenylalanine, and methanol to the portal blood, and these components must be metabolized and/or excreted.

Olney (6-9) and Reif-Lehrer (10) expressed concern about the safety of aspartame because of its aspartate content. Administration of high doses of aspartate to neonatal mice or rats results in elevated plasma aspartate concentrations (11-16) and hypothalamic neuronal necrosis (14-18). Aspartame administered in large amounts (1-2.5 g/kg body weight) to infant mice produces neuronal necrosis

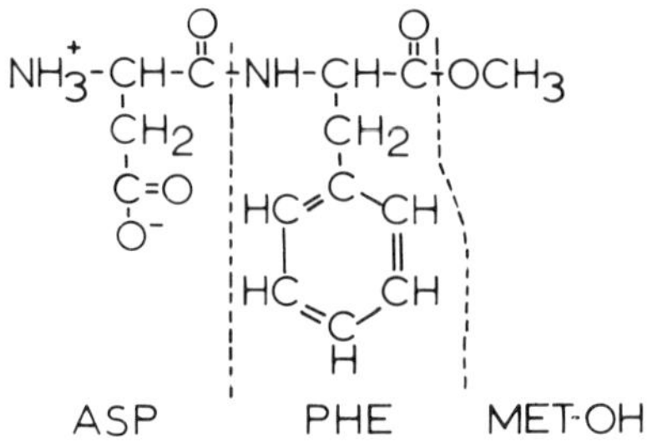

$NH_3^+-CH(CH_2-C(=O)-O^-)-C(=O)-NH-CH(CH_2-C_6H_5)-C(=O)-OCH_3$

ASP PHE MET-OH

Figure 1 Chemical structure of aspartame. The dotted lines divide the structure into its component parts: aspartate (ASP), phenylalanine (PHE), and methanol (MET-OH).

(19,20), presumably as a result of elevated blood aspartate concentrations. However, aspartame administration (2 g/kg body weight) to infant nonhuman primates, with or without added monosodium L-glutamate (1 g/kg body weight), did not produce neuronal necrosis, even though plasma aspartate and glutamate levels were elevated (20,21).

Turner objected to aspartame (22) because of its phenylalanine content. He speculated that aspartame ingestion would markedly elevate plasma phenylalanine concentration. Grossly elevated plasma phenylalanine concentrations, such as those found in children with classic phenylketonuria, are associated with mental retardation (23-26).

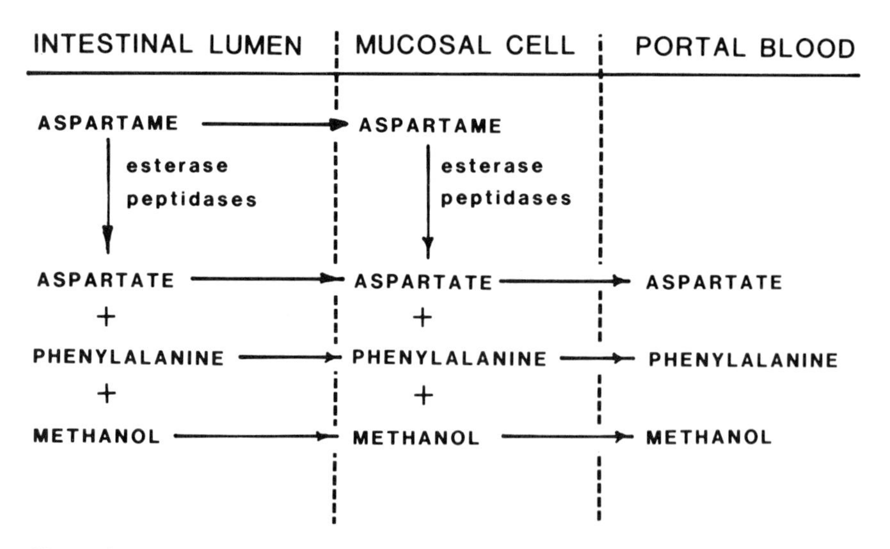

Figure 2 Aspartame hydrolysis occurs in both the intestinal lumen and in mucosal cells, releasing aspartate, phenylalanine, and methanol to the portal blood.

Since aspartame is a methyl ester, its metabolism releases methanol to the circulation. Ingestion of large doses of methanol is associated with adverse effects in sensitive species (27-30). Thus the potential toxicity of aspartame due to its methanol content must be considered.

In considering the toxicity of each of aspartame's three components, it is important to recognize that the blood concentration of each component (aspartate, phenylalanine, or methanol) must be markedly elevated to produce toxic effects. Certain facets of methanol toxicity, in fact are associated with formate rather than methanol accumulation (30,31).

Evaluation of potential toxicity of new food additives must deal with the question of sensitive species. Some species are more resistant than others to adverse effects resulting from the administration of a specific compound. Although rodents are commonly used in toxicology studies, nonhuman primates are often a better model of the human response. Investigators must deal with conflicting toxicity data obtained in different species and determine how these data relate to toxicity in humans. Tests evaluating the potential toxicity of aspartate, phenylalanine, and methanol in humans must be indirect; however, it is possible to examine the effect of large doses of aspartame on human blood amino acid and methanol concentrations to determine whether such doses produce the excessively high blood concentrations of these compounds associated with toxicity in animals.

This chapter reviews plasma and erythrocyte levels of amino acids as well as blood methanol and formate concentrations that result from the administration of large doses of aspartame to normal adults and individuals heterozygous for phenylketonuria.

The first section will discuss the effect of aspartame loading on plasma phenylalanine concentration, the second its effect on plasma aspartate concentration, and the third section its effect on blood methanol concentration.

PROJECTED ASPARTAME INTAKE

Projected levels of aspartame ingestion have been calculated by the Food and Drug Administration (FDA), the Market Research Corporation of America (MRCA), and our research group (32-35). Table 1 summarizes these data. If

Table 1 Summary of Projections for Aspartame Intake

Source	Aspartame totally replaces mean sucrose sweetness (mg/kg body weight)	Maximum (mg/kg body weight)
FDA (32,33)	Not calculated	22-28
MRCA (34)	3-11	25-34
Stegink et al. (35)	7-9	23-25

aspartame totally replaces estimated mean daily sucrose intake on a sweetness basis, aspartame intake will range between 3 and 11 mg/kg body weight. This amounts to 1.7-6.2 mg/kg body weight phenylalanine, 1.3-4.9 mg/kg aspartate, and 0.33-1.2 mg/kg body weight methanol. The highest daily aspartame ingestion, according to these calculations (32-35), would range from 22 to 34 mg/kg body weight. This is equivalent to ingesting 12-19 mg/kg body weight phenylalanine, 9.8-15.2 mg/kg body weight aspartate, and 2.4-3.7 mg/kg body weight methanol. According to the MRCA an aspartame intake of 34 mg/kg body weight represents the 99th percentile of projected daily ingestion.*

Next, let us compare the quantity of amino acids contributed by aspartame ingestion at these levels with normal intakes for aspartate and phenylalanine provided by dietary protein. Table 2 (taken from Ref. 33) lists the estimated dietary intake of phenylalanine and aspartate based upon average daily protein intake for various age groups. Values range from 52 to 229 mg/kg body weight for phenylalanine and from 80 to 395 mg/kg body weight for aspartate per day. Similar calculations for adults have been made by Roak-Foltz and Leveille (37). Intake per kilogram body weight is highest in children owing to the relative increase in protein intake needed for growth.

The data in Table 1 can be used to project the increase in phenylalanine and aspartate intake if aspartame replaces all sucrose in the diet. The normal protein intake of a 4-year-old child usually provides about 316 mg/kg body weight aspartate and 206 mg/kg body weight phenylalanine (Table 2). If aspartame was ingested at 34 mg/kg body weight, this level of aspartame ingestion would increase the mean total aspartate intake from 316 to 331 mg/kg body weight per day,

Table 2 Estimates of Age-Related Dietary Intake of Phenylalanine and Aspartate Based Upon Average Daily Protein Intake[a]

		Daily intake (mg/kg body weight)	
Age	Weight (kg)	Phenylalanine	Aspartate
8 months	8	229	395
4 years	16	206	316
9 years	28	144	219
15 years	50	100	170
34 years	60	52	80

[a]From Ref. 33. Estimates are based on a representative diet delivering the total average protein consumed by people in the age-weight categories shown in the left column. The source of the data is the *Food Intake and Nutritive Value of U.S. Diets* (36).

*Calculation carried out before approval of aspartame for use in carbonated beverages.

and the mean phenylalanine intake from 206 to 225 mg/kg body weight per day. Thus the projected intake levels for aspartame suggest that it will have a relatively small effect on aspartate or phenylalanine intake.

PHENYLALANINE

Plasma phenylalanine levels increase after ingestion of meals containing protein, with postprandial plasma phenylalanine levels reaching 12 μmol/dl, or about 2 mg % (38-40). Grossly elevated plasma phenylalanine levels are associated with mental retardation in persons with phenylketonuria (23-26). In this genetic disease, phenylalanine is metabolized poorly and accumulates in blood and tissues. In children with phenylketonuria (PKU) plasma phenylalanine levels range from 120 to 600 μmol/dl (20-100 mg %).

Such information can be used to evaluate aspartame safety by relating the plasma phenylalanine concentrations in these infants to those levels observed after administration of graded doses of aspartame. To address this point, high-use doses of aspartame were first administered to normal subjects; subsequently possible abuse doses were studied. Finally, use and abuse doses were evaluated in phenylketonuric heterozygotes. Our premise in all of these human studies was that if large quantities of aspartame are fed to human subjects without producing gross elevations of plasma phenylalanine levels, the risk of toxicity is minimal.

Normal Subjects Given Aspartame at 34 mg/kg Body Weight

Our first study investigated the effect of aspartame loading at 34 mg/kg body weight upon blood amino acid concentrations in normal humans (35). This dose represents the 99th percentile of projected ingestion of aspartame *for an entire day*.when aspartame replaces dietary sucrose on a sweetness basis.

Twelve normal adults were administered aspartame (34 mg/kg body weight) or an equimolar quantity of aspartate (13 mg/kg body weight) dissolved in 300 ml of orange juice in a randomized crossover design.

Figure 3 shows plasma phenylalanine and tyrosine levels in these subjects. Plasma phenylalanine levels decreased significantly ($p \leqslant 0.001$) from the fasting base-line level 90 min after the aspartate load and returned to baseline by 8 hr. After the aspartame load, plasma phenylalanine levels increased significantly ($p \leqslant 0.001$) from a normal base-line level of 5-6 μmol/dl to levels normally seen postprandially in orally fed infants and adults (38-40). A significant rise in plasma tyrosine levels, representing the conversion of phenylalanine to tyrosine, was also seen after aspartame loading.

It has been suggested that some amino acids are transported in the erythrocyte to a greater extent than in plasma under certain circumstances (41-45). Thus erythrocyte levels of free amino acids were also measured. Erythrocyte phenylalanine and tyrosine levels in these subjects are shown in Figure 4. The increase in erythrocyte phenylalanine follows the pattern noted in plasma (Fig. 5). Note

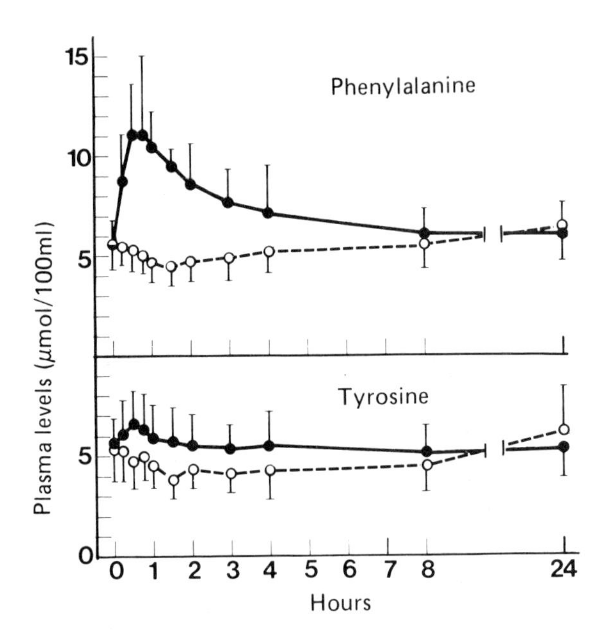

Figure 3 Mean (±SD) plasma phenylalanine and tyrosine concentrations (μmol/dl) in normal adults administered either 34 mg/kg body weight aspartame (●——●) or 13 mg/kg body weight of aspartate (○- - -○). (From Ref. 35.)

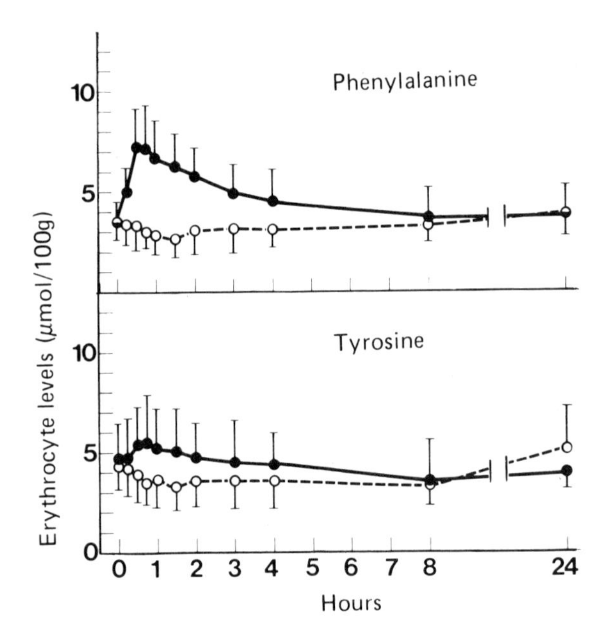

Figure 4 Mean (±SD) erythrocyte phenylalanine and tyrosine concentrations (μmol/100 g red cells) in normal adults administered either 34 mg/kg body weight aspartame (●——●) or 13 mg/kg body weight aspartate (○- - -○). (From Ref. 35.)

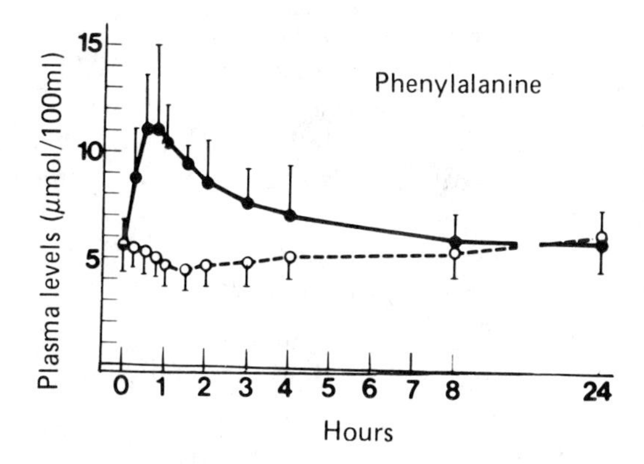

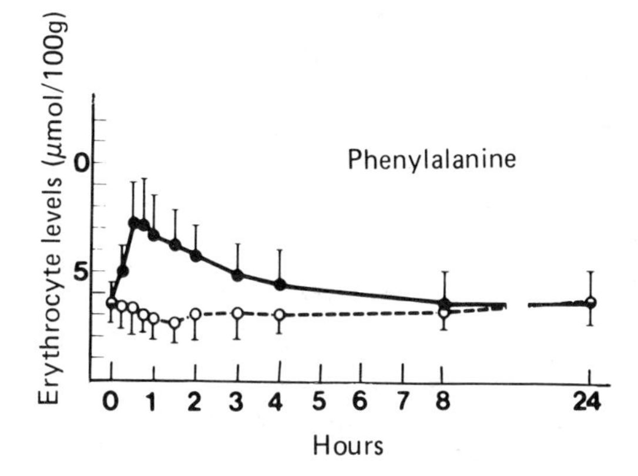

Figure 5 Comparison of plasma and erythrocyte phenylalanine levels in normal adults administered 34 mg/kg body weight of aspartame (●—●) or 13 mg/kg body weight aspartate (○- - -○) (data shown as the mean ±SD).

the rise in erythrocyte phenylalanine concentration as plasma level increases, with a corresponding decrease when plasma levels fall. Neither plasma nor erythrocyte phenylalanine levels exceed usual postprandial concentrations, despite ingestion in a single load of an aspartame dose calculated to be the 99th percentile projected for an entire day. These data indicate no risk to human subjects after administration of aspartame at this level.

Normal Subjects Given Aspartame at 50 mg/kg Body Weight

Six adult female subjects were given 50 mg/kg body weight of either aspartame or lactose in a randomized crossover design (46). A 50 mg/kg body weight aspartame load is approximately two times the 99th percentile of projected daily aspartame intake and can be considered a high-use dose. As expected, plasma phenylalanine levels (Fig. 6) increased significantly over baseline ($p \leqslant 0.001$), reaching a mean (±SD) peak value of 16.2 ± 4.9 μmol/dl after aspartame ingestion. Plasma phenylalanine levels were not affected by lactose loading. Plasma tyrosine levels increased slightly after aspartame loading to reach values in the normal postprandial range for this amino acid. Despite the large aspartame dose, plasma phenylalanine concentrations were only slightly higher than values found postprandially (12 ± 3 μmol/dl) in orally fed infants (38,39) or adults (40).

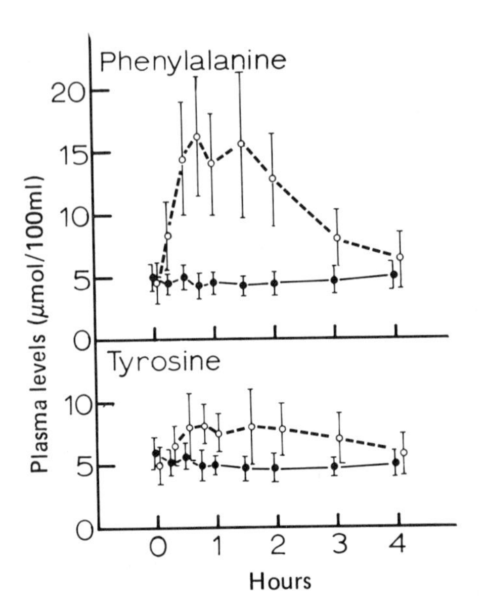

Figure 6 Mean (±SD) plasma phenylalanine and tyrosine concentrations (μmol/dl) in normal lactating females administered 50 mg/kg body weight of aspartame (○- - -○) or lactose (●——●). (From Ref. 46.)

Erythrocyte phenylalanine and tyrosine levels were also measured in these subjects. As shown in Figure 7, erythrocyte phenylalanine and tyrosine levels follow plasma levels after aspartame loading.

As shown in Figure 8, plasma phenylalanine levels increase proportionally to the aspartame dose. Peak levels were higher, and the plasma concentration-time curve broader in those subjects administered 50 mg/kg body weight aspartame than in subjects administered aspartame at 34 mg/kg body weight. The plasma phenylalanine levels observed in these subjects were far below peak levels routinely noted in subjects tested for heterozygosity for phenylketonuria by phenylalanine loading (47). In such subjects, peak values of 78 ± 45 μmol/dl were noted. Since 50 mg/kg body weight aspartame given as a single bolus only increased plasma phenylalanine levels into the high postprandial range, ingestion of 50 mg/kg body weight aspartame during the course of an entire day should represent no significant risk from its phenylalanine content.

Normal Subjects Given Abuse Doses of Aspartame

Following evaluation of blood amino acid data from the first two studies, the effects of abuse doses of aspartame were studied in normal adult volunteers (48).

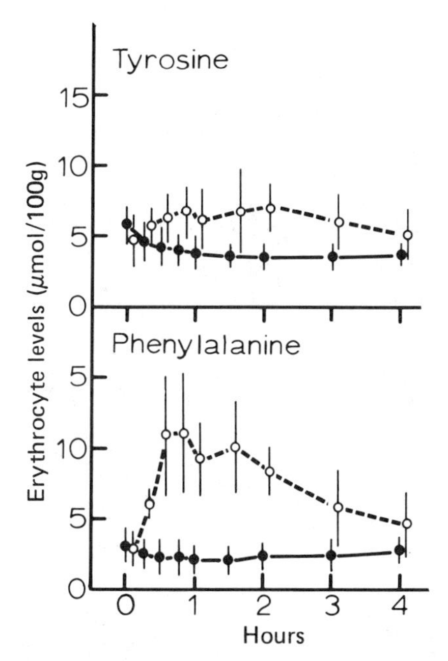

Figure 7 Mean (±SD) erythrocyte phenylalanine and tyrosine concentrations (μmol/100 g) in normal lactating females administered 50 mg/kg body weight of aspartame (○- - -○) or lactose (●——●). (From Ref. 46.)

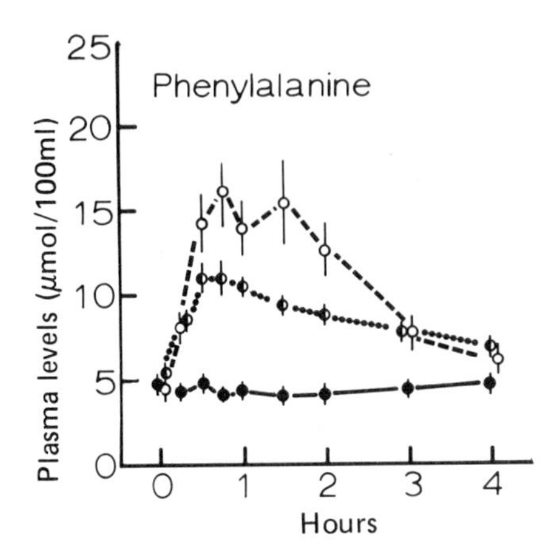

Figure 8 Comparison of plasma phenylalanine concentrations in normal adult subjects administered lactose (●——●), 34 mg/kg body weight aspartame (◑- - -◑), or 50 mg/kg body weight aspartame (○- - -○) (data shown as the mean ± SD).

In these studies, aspartame doses of 100, 150, and 200 mg/kg body weight were administered. These large doses represent intakes that only would be ingested accidentally. Table 3 compares dose size to expected intake of aspartame under abnormal conditions of use. The 100 mg/kg body weight dose is equivalent to drinking 10 liters of aspartame-sweetened beverage in a single setting [assuming a 10% sucrose content (wt/vol) and a 50-kg subject]. This dose is approximately 3 times the 99th percentile of projected aspartame ingestion and is 10-12 times the quantity of aspartame ingested if aspartame replaced all dietary sucrose on a sweetness basis. The highest dose studied (200 mg/kg body weight) was calculated to be the potential dose received by a 1-year-old child (10 kg) accidentally ingest-

Table 3 Comparison of Aspartame Doses Administered to Potential Abuse Situations

100 mg/kg body weight of aspartame is equivalent to

1. Ingestion of 10 liters of aspartame-sweetened beverage
2. Three times the total daily ingestion of aspartame estimated for the 99th percentile (34 mg/kg body weight)
3. Approximately 10 times the mean daily ingestion expected if aspartame replaces sucrose sweetness at current levels of sucrose ingestion

200 mg/kg aspartame dose is equivalent to

1. The dose ingested by a 1-year-old child accidentally ingesting the entire contents of aspartame coffee sweetener (100 tablets of 20 mg each)
2. Ingestion of 20 liters of aspartame-sweetened beverage by a 50-kg soldier in the tropics (equivalent to maximal daily water ingestion)

ing the entire contents of a container of aspartame coffee sweetener (100 tablets of 20 mg each). A similar level of ingestion was calculated as being possible for a 50-kg soldier in the tropics ingesting his entire water intake for the day (20 liters maximum) as aspartame-sweetened beverage in a single setting.

In this study, normal fasted adults were administered aspartame doses of either 100, 150, or 200 mg/kg body weight dissolved in 500 ml of orange juice. Six subjects were studied at each dose level, with plasma and erythrocyte amino acid levels measured with time. The studies were carried out sequentially. All data at the lowest dose were evaluated before proceeding to the next higher dose.

Figure 9 shows mean plasma phenylalanine levels in these subjects. Plasma phenylalanine levels increased rapidly over base-line levels ($p \leqslant 0.001$) 15 min to

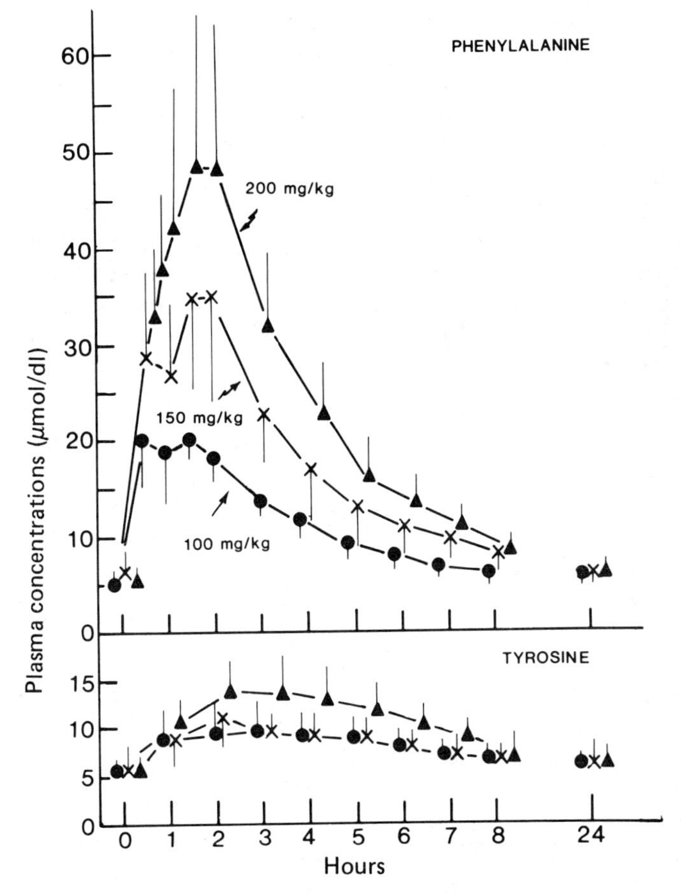

Figure 9 Mean (±SD) plasma phenylalanine concentrations (μmol/dl) in normal adults administered aspartame at 100 (●), 150 (X), or 200 mg/kg body weight (▲).

6 hr after ingestion, the increase being proportional to the dose. Similarly, the area under the plasma phenylalanine concentration-time curve increased with increasing aspartame load.

Mean (±SD) peak plasma phenylalanine levels were 20.3 ± 2.05, 35.1 ± 11.3, and 48.7 ± 15.1 μmol/dl after administration of aspartame at 100, 150, and 200 mg/kg body weight, respectively. The maximum plasma phenylalanine level in one subject was 74.4 μmol/dl after administration of aspartame at 200 mg/kg body weight, suggesting that this subject may be a phenylketonuric heterozygote. The mean peak plasma phenylalanine level noted at each dose in our studies was lower than the phenylalanine levels noted in either normal or phenylketonuric heterozygotes tested for heterozygosity for phenylketonuria by administration of 100 mg/kg body weight phenylalanine (47,49-53). Under these conditions, mean (±SD) plasma phenylalanine levels reach 78 ± 45 μmol/dl. No adverse effects have been reported as the result of such testing.

Table 4 compares the plasma phenylalanine levels observed in these patients with plasma phenylalanine levels in certain clinical situations. Normal fasting plasma phenylalanine concentrations are about 6 μmol/dl, with postprandial levels being 12 ± 3 μmol/dl (38-40). Children with classic phenylketonuria have plasma phenylalanine levels ranging from 120 to 600 μmol/dl continually. These children are mentally retarded. There are children with a benign variant form of phenylalaninemia who are not mentally retarded despite plasma phenylalanine levels ranging from 24 to 48 μmol/dl continuously (24). Peak plasma phenylalanine levels in normal children tested for the heterozygous state of phenylketonuria by phenylalanine loading (47) are also higher than values noted in these subjects administered abuse doses of aspartame. Children with classic phenyl-

Table 4 Plasma Phenylalanine Levels Under Various Conditions

	Dose (μmol/dl)
Normal subjects	
Fasting	6 ± 3
Postprandial	12 ± 3
Phenylalaninemia	
Classic phenylketonuria	120-600
Questionable variants	60-120
Benign variants	24-48
After 34 mg/kg body weight aspartame	
Normal subjects	11 ± 3
Phenylketonuric heterozygotes	16 ± 3
After 100 mg/kg body weight aspartame	
Normal subjects	20 ± 7
Phenylketonuric heterozygotes	42 ± 3

ketonuria treated with diets low in phenylalanine are permitted plasma phenylalanine levels of 24-48 μmol/dl. Although the plasma phenylalanine levels noted after administration of abuse doses of aspartame are outside the postprandial range for this amino acid, maximal plasma phenylalanine levels are well below those concentrations associated with toxic effects. Thus we feel that short-term elevation of mean plasma phenylalanine levels to 49 μmol/dl (approximately 8 mg/dl) after an abuse dose of aspartame of 200 mg/kg body weight does not represent substantial risk of toxicity.

Studies in Normal Infants

To evaluate the question of whether young children metabolize aspartame as well as adults, we carried out studies in 1-year-old infants (54). These studies, summarized by Filer et al. in Chap. 29 (55) demonstrate that 1-year-old infants metabolize and clear the phenylalanine portion of aspartame as well as adults.

Studies in Phenylketonuric Heterozygotes

Individuals heterozygous for phenylketonuria represent a population that might metabolize the phenylalanine portion of aspartame less well than normal.

The heterozygous state for phenylketonuria is estimated to occur in 1 person in every 50-70 persons. Such individuals will obviously ingest aspartame. Thus it was important to determine whether they metabolized aspartame normally.

In the earlier studies of Koch and colleagues, aspartame was administered to phenylketonuric heterozygotes for several weeks in a total daily dose approximating that used in the present study (56); however, the aspartame was given in divided doses in capsules with meals. Plasma phenylalanine levels were measured following an overnight fast. These investigators reported no difference in blood phenylalanine levels between subjects given aspartame and those given placebo; however, no measurements of maximal plasma phenylalanine levels were made. As a result, we reevaluated whether heterozygous subjects could adequately handle the phenylalanine present in an aspartame load.

A total of 8 subjects heterozygous for phenylketonuria and 12 normal subjects were studied after aspartame administration at a dose of 34 mg/kg body weight (57). Figure 10 shows plasma phenylalanine and tyrosine levels for these subjects. Plasma phenylalanine levels increased significantly after aspartame loading in both normal and phenylketonuric heterozygotes ($p \leqslant 0.001$ for both groups at 60 min). However, plasma phenylalanine levels in heterozygotes were significantly higher ($p \leqslant 0.001$ at 60-180 min) and the plasma concentration-time curve broader than that noted in normals. This was expected in view of the decreased level of liver phenylalanine hydroxylase present in the heterozygote. Maximal plasma phenylalanine levels in normal subjects (11.1 ± 2.49 μmol/dl) were in the range noted postprandially in healthy normal adults and infants (38-

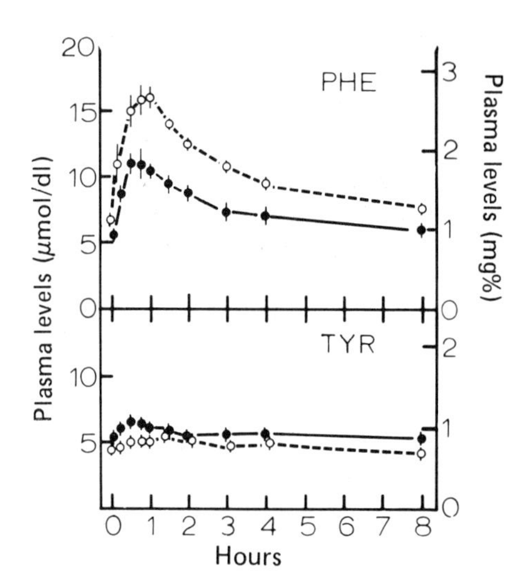

Figure 10 Mean (±SEM) plasma phenylalanine and tyrosine concentrations in normal adults (●——●) and PKU heterozygotes (○- - -○) administered aspartame at 34 mg/kg body weight. (From Ref. 57.)

40). Maximal plasma phenylalanine levels (mean ±SD) in phenylketonuric heterozygotes (16.0 ± 2.25 μmol/dl) were slightly above this postprandial range. Similarly, erythrocyte phenylalanine levels (Fig. 11) differed significantly between normal subjects and phenylketonuric heterozygotes. When the number of normal and heterozygous subjects was increased, similar results were noted (58).

These data clearly indicate that phenylketonuric heterozygotes metabolize the phenylalanine portion of aspartame slower than normal subjects. However, peak phenylalanine values in heterozygous subjects were well below those associated with toxic effects. Thus the 99th percentile of projected daily aspartame ingestion, given as a single dose to the phenylketonuria heterozygote, poses no significant risk.

Following completion of these studies, the effects of a potential abuse dose of aspartame were investigated in phenylketonuric heterozygotes. Five phenylketonuric heterozygotes and six normal subjects were administered 100 mg/kg body weight loads of aspartame dissolved in orange juice, and plasma and erythrocyte amino acid concentrations were measured with time (59).

Figure 12 shows plasma phenylalanine and tyrosine levels in these subjects. Plasma phenylalanine levels increased rapidly after aspartame loading in both normal and heterozygous subjects. Plasma phenylalanine levels in heterozygous subjects were slightly but significantly higher ($p \leqslant 0.02$) at zero time (fasting) than levels in normal subjects. Metabolism of the phenylalanine portion of

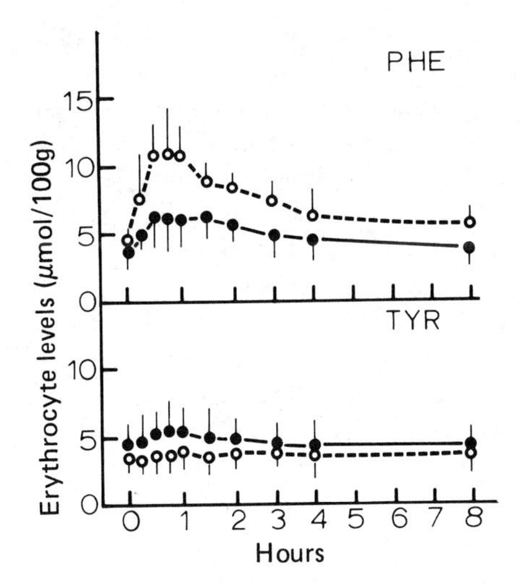

Figure 11 Mean (±SD) erythrocyte phenylalanine and tyrosine concentrations (μmol/100 g) in normal adults (●——●) and PKU heterozygotes (○- - -○) administered aspartame at 34 mg/kg body weight. (From Ref. 57.)

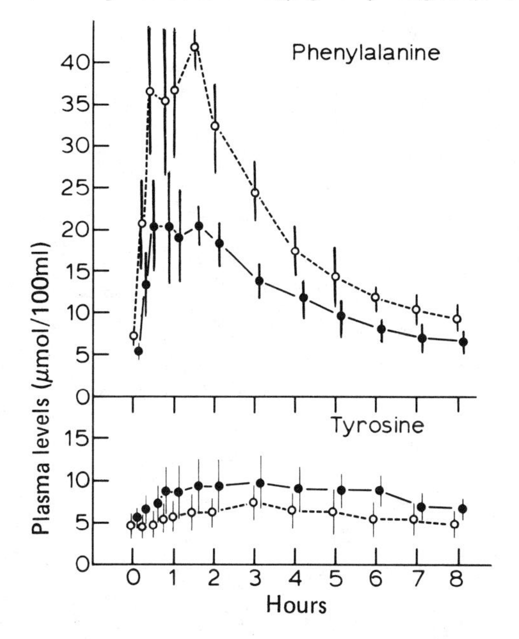

Figure 12 Mean (±SD) plasma phenylalanine and tyrosine concentrations in normal adults (●——●) and PKU heterozygotes (○- - -○) administered aspartame at 100 mg/kg body weight. (From Ref. 59.)

aspartame was slower in heterozygous subjects than in normal subjects. Plasma phenylalanine levels after aspartame loading in heterozygotes were significantly higher (p values ranging from 0.02 to 0.001) from 30 to 180 min than values in normal subjects, and the area under the plasma concentration-time curve greater. This was expected in view of the decreased liver phenylalanine hydroxylase activity levels reported in heterozygotes (60). The maximal mean (±SD) plasma phenylalanine level in phenylketonuric heterozygotes after aspartame loading at 100 mg/kg body weight (41.7 ± 2.33 μmol/dl) was approximately two times greater than the mean peak value (20.2 ± 6.77 μmol/dl) noted in normal subjects.

Plasma tyrosine levels in normal subjects were higher 60 min after aspartame loading in normal subjects than in heterozygotes. In normal subjects, plasma tyrosine levels remained significantly elevated over base-line levels from 15 min to 7 hr. The increase in plasma tyrosine levels in heterozygous subjects was slow, reaching maximal values 3 hr after dosing. Plasma tyrosine levels were significantly higher in normal subjects than in heterozygotes from 15 to 45 min after loading, with differences losing significance after that time.

Figure 13 shows changes in erythrocyte phenylalanine and tyrosine levels in these subjects. As expected, erythrocyte phenylalanine and tyrosine values follow those noted in plasma.

Based upon the observation that the plasma concentration-time curve of heterozygotes administered aspartame at 34 mg/kg body weight is similar to that of normal subjects administered aspartame at 50 mg/kg body weight, we con-

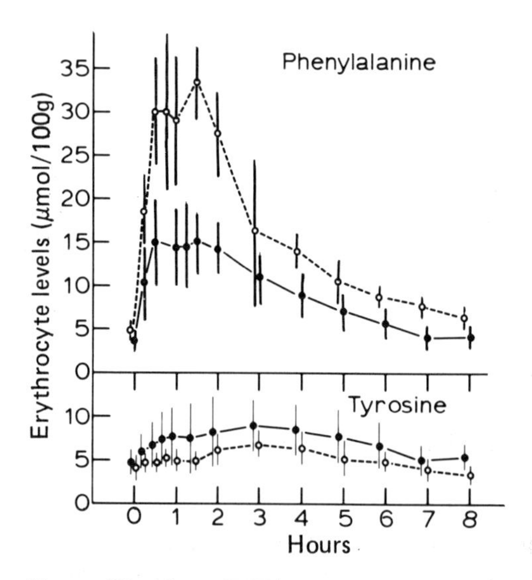

Figure 13 Mean (±SD) erythrocyte phenylalanine and tyrosine concentrations (μmol/100 g) in normal adults (●——●) and PKU heterozygotes (○- - -○) administered aspartame at 100 mg/kg body weight. (From Ref. 59.)

cluded that heterozygotes metabolized the phenylalanine portion of aspartame approximately one-half as well as normal subjects. The possibility exists, however, than an aspartame dose of 34 mg/kg body weight represents the maximal quantity readily handled by the heterozygote. If this were true, the ingestion of large quantities of aspartame would exceed metabolic capability and lead to a dramatic increase in plasma phenylalanine level. However, peak plasma phenylalanine concentrations and the area under the plasma phenylalanine concentration-time curves indicate that phenylketonuric heterozygotes ingesting aspartame at 100 mg/kg body weight metabolize the phenylalanine portion of aspartame at least as rapidly as normal adults metabolize and clear aspartame doses of 200 mg/kg body weight (Fig. 14). Thus it does not appear as though an abuse dose of aspartame would overwhelm the metabolism-clearance mechanisms for phenylalanine in the phenylketonuric heterozygote.

The observation that phenylketonuric heterozygotes metabolize and clear phenylalanine approximately one-half as well as normal subjects is consistent with the intravenous phenylalanine loading studies of Bremer and Neumann (61) and Woolf et al. (62). These investigators demonstrated that the disappearance rate of intravenously administered phenylalanine (80 mg/kg body weight) was twice as high in normals as in individuals heterozygous for classic phenylketonuria. Although it has been suggested that the heterozygote for classic phenylketonuria may have less than 50% of normal phenylalanine hydroxylase activity (60), intravenous infusion studies indicate that metabolism and clearance of phenylalanine in the heterozygote occur at approximately one-half the normal rate (61-63). Such studies suggest that phenylketonuric heterozygotes are not at a significantly increased risk from abuse doses of aspartame (Table 4).

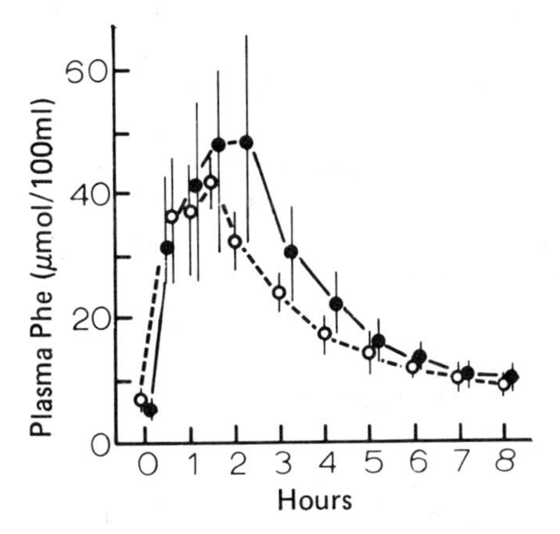

Figure 14 Mean (±SD) plasma phenylalanine concentrations in normal adults (●——●) administered aspartame at 200 mg/kg body weight and PKU heterozygotes (○- - -○) administered aspartame at 100 mg/kg body weight. (From Ref. 59.)

Plasma phenylalanine levels in heterozygous subjects after aspartame loading were below levels noted in normal subjects and phenylketonuric heterozygotes tested for heterozygosity by phenylalanine loading (47,49-53). Under these conditions, mean (±SD) peak plasma phenylalanine levels reach 78 ± 45 μmol/dl (47). No ill effects have been reported as the result of such testing. Children with classic phenylketonuria treated with diets low in phenylalanine are permitted plasma phenylalanine levels of 24-48 μmol/dl. Although the plasma phenylalanine levels noted in this study are outside the postprandial range for this amino acid (38-40), the maximal plasma phenylalanine levels observed are much below those associated with toxic effects. Thus even abuse doses of aspartame taken in a single dose are unlikely to have serious effects in either normal subjects or those heterozygous for phenylketonuria.

Kang and Paine (64) have shown that serum phenylalanine levels in the pregnant heterozygote are higher than those in either the nonpregnant heterzygote or normal pregnant subjects. This observation raises the question of whether the slower metabolism of aspartame by the pregnant heterozygote under abuse conditions of aspartame ingestion could lead to maternal plasma phenylalanine levels that would be detrimental to the fetus. The placenta normally maintains a gradient for most amino acids, including phenylalanine, of about 2:1 toward the fetal circulation (65). This active transport is more marked early in pregnancy (66,67). In the pregnant female who is homozygous for phenylketonuria, the elevated maternal plasma phenylalanine levels will be concentrated toward the fetal circulation. Maternal plasma phenylalanine levels of 120 μmol/dl, as occur continually in the phenylketonuric homozygote, will result in fetal plasma phenylalanine levels ranging from 240 to 350 μmol/dl. In view of this concentrating effect, it is not surprising that children born to homozygous mothers are often mentally retarded (68) even if they are only heterozygous for the genetic defect and would normally not be affected.

This information can be used to evaluate potential hazard to the pregnant phenylketonuric heterozygote from aspartame ingestion. When the 99th percentile of projected aspartame ingestion for the entire day (34 mg/kg body weight) is given to a phenylketonuric heterozygote in a single dose, mean peak plasma phenylalanine levels are less than 18 μmol/dl (3 mg %), with peak values ranging from 12 to 24 μmol/dl over the course of a few hours (57). Since these values are only slightly above normal postprandial plasma phenylalanine values, the data suggest little risk from normal use levels of aspartame.

The final question to be considered is whether aspartame poses undue risk to individuals with metabolic defects in phenylalanine metabolism who are not detected by screening programs. The incidence rate for the various forms of phenylketonuria and hyperphenylalaninemia is approximately 1 in 10,000-20,000 births. Most of these subjects are detected by neonatal screening programs and given dietary therapy. However, some infants may be missed. Olney (69) and Turner (70) have suggested that aspartame's phenylalanine content would greatly increase the risk to such undiagnosed individuals. This position fails to consider the size of normal protein intake. As pointed out earlier, the human infant normally ingests large quantities of protein as a source of the amino acids needed for growth. These proteins contain phenylalanine. For example, we have estimated that the 3.5-kg breast-fed infant ingests 79 mg of phenylalanine per kilogram body weight per day (46). Similarly, dietary protein provides older children large quantities of phenylalanine. Typical levels of phenylalanine ingestion from dietary protein range from 52 to 229 mg/kg body weight per day (Table 2). Projections of expected aspartame ingestion by infants under 2 years of age indicate that aspartame will add 3-4 mg/kg body weight phenylalanine to this pool (33). This quantity of phenylalanine will have little effect when compared to the quantity of phenylalanine resulting from normal protein ingestion. As shown in Table 2, an 8-kg infant normally ingests 229 mg of phenylalanine per kilogram body weight from his or her diet. The addition of 3-4 mg/kg body weight phenylalanine from projected levels of aspartame ingestion (33) will not significantly affect phenylalanine intake.

Available data indicate that dietary therapy of children with phenylketonuria must take place during the first month of life (preferably within the first 2-3 weeks of life) if therapy is to be effective. Thus the undetected phenylketonuric infant will sustain brain damage from phenylalanine derived from dietary protein long before the infant is exposed to aspartame-sweetened beverage. Furthermore, the quantity of phenylalanine added to the diet by such aspartame ingestion represents only a small fraction of total phenylalanine intake. Even later in life, the quantity of phenylalanine expected to be contributed to a normal diet by projected aspartame ingestion levels is relatively small compared to total phenylalanine intake.

The data suggest that the amount of phenylalanine that would result from aspartame ingestion under normal and abuse conditions would have no significant effect in normal subjects or phenylketonuric heterozygotes. This conclusion is also supported by chronic and acute animal studies evaluating the effect of enormous doses of aspartame, as well as by the human studies where normal and abuse doses of aspartame failed to increase blood phenylalanine to levels associated with any toxic findings.

Repeated Ingestion of Aspartame-Sweetened Beverage by Normal Subjects

The previous studies described single acute dosing studies in which large doses of aspartame were administered to normal human subjects. These studies did not evaluate the effect of successive smaller doses on plasma phenylalanine levels.

Aspartame is used to sweeten beverages, and individuals are likely to drink several aspartame-sweetened beverages between meals. If the phenylalanine load produced by aspartame loading is not completely cleared prior to ingestion of the next beverage serving, a cumulative increase in plasma concentrations might be noted. The population group of greatest concern is children. A 20-kg 4-year-old child drinking 12 oz of aspartame-sweetened beverage would ingest about 10 mg/kg body weight of aspartame. Morgan and Stults (71) reported data on the beverage consumption habits of children. They reported that 5-year-old children ingest an average of 11.4 oz of noncarbonated sweetened beverage and 10.8 oz of carbonated soft drinks daily. Thus it is possible that some infants will ingest three 12-oz beverage servings between meals. Accordingly, we elected to study the effects of the successive ingestion of three beverage servings each providing aspartame at 10 mg/kg body weight upon plasma amino acid concentration (72). This would seem to be an extreme test of whether normal use of aspartame would increase plasma phenylalanine concentration. Although these studies were carried out in adults, we assume they hold for children as well, since 1-year-old infants absorb and metabolize aspartame as well as adults (54,55).

Eight healthy, normal adults were studied (four male and four female). The study was conducted in two stages in a crossover design. Subjects ingested three successive 12-oz servings of beverage, with a 2-hr interval between servings. In one stage of the study, subjects drank three servings of unsweetened beverage, while in the other stage the three servings of beverage each provided aspartame at 10 mg/kg body weight.

The mean plasma phenylalanine levels in these subjects are plotted in Figure 15. The observed increase in plasma phenylalanine concentration after individual aspartame doses ranged from 1.64 to 2.05 μmol/dl when compared to base-line values observed just prior to dosing. "Base-line" plasma phenylalanine concentra-

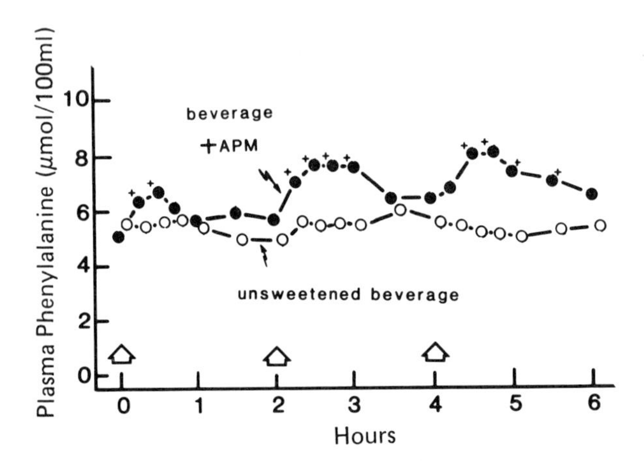

Figure 15 Mean plasma phenylalanine concentrations in normal adults ingesting repeated servings of either unsweetened beverage (○——○) or beverage providing 10 mg/kg body weight aspartame (●——●). Values designated with a + differ significantly ($p < 0.05$) from base-line values.

tions noted at 0, 2, 4, and 6 hr (just prior to beverage ingestion) increased slowly during the course of the study when the three successive beverage servings contained aspartame. Little change in plasma phenylalanine concentration was noted after ingestion of unsweetened beverage.

The slight increase in base-line plasma phenylalanine levels noted between zero time and 6-hr values after ingestion of aspartame-sweetened beverage suggests that phenylalanine was not completely cleared and metabolized during the 2-hr time period between servings. However, this small increase in plasma phenylalanine was not statistically significant and should have no physiological significance. These data indicate that a fasting 4-year-old child could ingest three successive 12-oz servings of aspartame-sweetened beverage without appreciably increasing plasma phenylalanine levels.

The observed increase in plasma phenylalanine concentrations after aspartame ingestion closely approximate values calculated theoretically for this dose. Figure 16 summarizes the response of peak plasma phenylalanine concentrations to increasing doses of aspartame in normal adults based on data from our earlier studies (35,46,48). Peak plasma phenylalanine levels increase in proportion to aspartame dose in normal subjects. This curve predicts that peak plasma phenylalanine levels in normal subjects would increase approximately 1 μmol/dl over base-line levels at an aspartame dose of 4 mg/kg body weight, and about 2-3 μmol/dl over base line at a dose of 10 mg/kg body weight. The observed increase

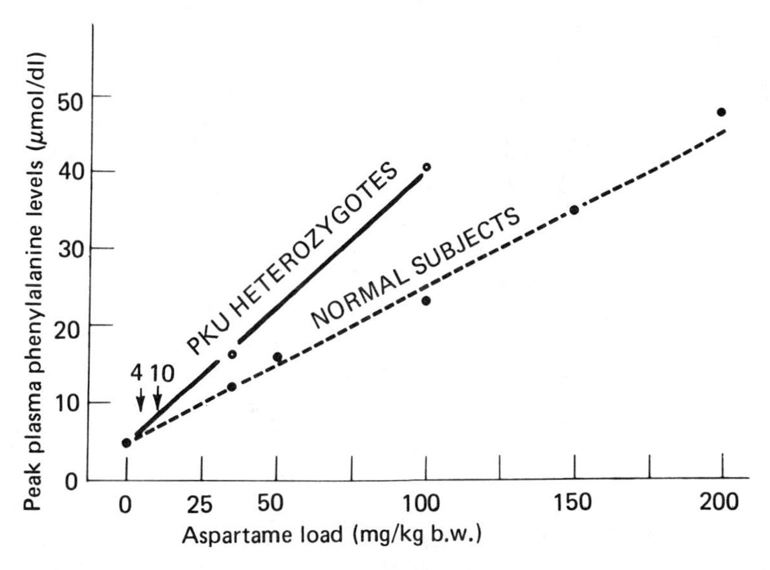

Figure 16 Correlation of mean peak plasma phenylalanine concentration with aspartame dose in normal subjects (●- - -●) and PKU heterozygotes (●——●). (Data obtained from Refs. 35, 46, 48, 57, and 59.)

in plasma phenylalanine concentration in subjects administered aspartame at 10 mg/kg body weight ranged from 1.64 to 2.05 μmol/dl.

ASPARTATE

Like other dicarboxylic amino acids, aspartate may exert toxic effects when administered at very high dose levels, although species and age susceptibility vary considerably. Neonatal rodents administered large doses of glutamate or aspartate, either orally or by injection, develop hypothalamic neuronal necrosis (14-18,73-79). However, sensitivity in the rodent is age and strain dependent. Adult mice are susceptible to dicarboxylic amino acid-induced neuronal necrosis, but much higher dose levels are required (74). In *both* neonatal and adult animals, blood levels of the dicarboxylic amino acids must be grossly elevated for lesion formation to occur (14-16,76-79).

There is disagreement over the ability of the dicarboxylic amino acids to produce neuronal necrosis in the infant primate. Although a glutamate-induced lesion in infant nonhuman primates has been reported (80,81), no one has demonstrated hypothalamic neuronal necrosis in infant nonhuman primates resulting from oral aspartame or aspartate administration. In fact, Reynolds et al. (20,21) reported no detectable lesions in infant monkeys administered large doses of aspartame with and without added monosodium L-glutamate (MSG).

Olney and colleagues (80,81) originally reported that large doses of glutamate given to neonatal nonhuman primates produced hypothalamic neuronal necrosis. However, four other laboratories, in a series of some 11 papers reporting studies in a large number of monkeys, were not able to find lesions in infant monkeys given equally large doses of MSG (20,21,82-90). In general, these groups had little difficulty in producing neuronal necrosis in rodents. Olney et al. (81) have suggested that the failure of research groups other than their own to produce a lesion in the nonhuman primate reflects a failure to elevate plasma glutamate and aspartate levels. This is not the case. A number of the monkeys studied by Reynolds et al. (20,21,82) had blood samples taken for plasma amino acid analyses. Plasma glutamate and aspartate levels in these animals were grossly elevated (Fig. 17) (21,86). Despite elevated plasma glutamate and aspartate levels, no evidence of hypothalamic neuronal necrosis was found (20,21,82,86).

Even in the sensitive neonatal mouse, gross elevations of plasma glutamate and/or aspartate concentrations must occur prior to lesion formation. Although both glutamate and aspartate appear to pass into the immature rodent brain and produce a lesion when plasma levels are grossly elevated, even the sensitive neonatal mouse tolerates substantial elevations of plasma glutamate plus aspartate levels without developing neuronal necrosis (14-16,76-79).

The results of animal studies with aspartame can be used to interpret data obtained in humans. There is no doubt that the administration of large aspartame doses to neonatal rodents elevates plasma aspartate levels and results in neuronal

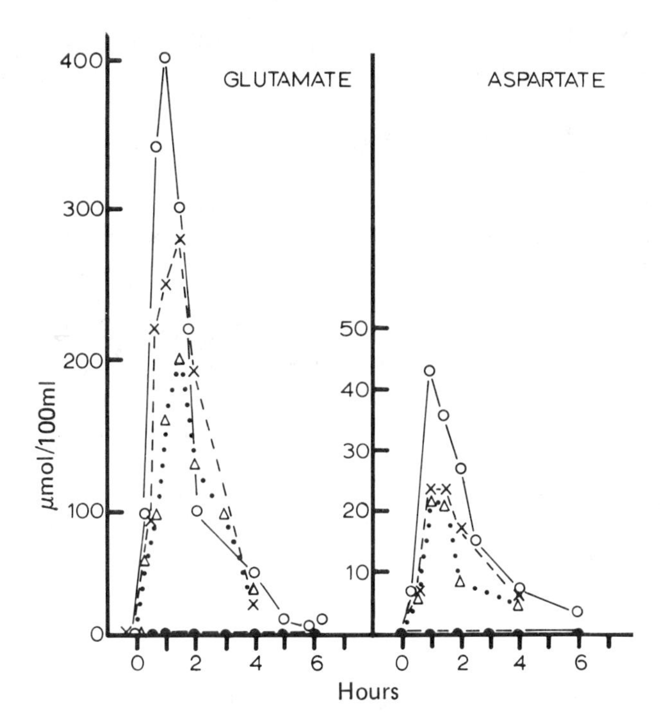

Figure 17 Mean plasma glutamate and aspartate concentrations in infant monkeys administered monosodium L-glutamate at 1 (△· · ·△), 2 (X- - -X), or 4 g/kg body weight (○——○). (From Ref. 86.)

damage (19,20). On the other hand, the administration of aspartame at 2 g/kg body weight to infant nonhuman primates, with and without glutamate (1 g/kg body weight), did not produce neuronal necrosis, despite grossly elevated plasma aspartate and glutamate levels (20,21).

The critical question, however, is how do these data relate to man. The neonatal human probably resembles the neonatal monkey to a greater extent than the infant mouse. Thus the negative findings noted in monkeys after administration of large doses of either aspartame alone or aspartame plus MSG (20,21) indicate that aspartame is safe. However, the controversy over the sensitivity of the nonhuman primate to glutamate and aspartate led us to explore additional ways of evaluating the potential neurotoxicity of compounds containing dicarboxylic amino acids.

The neonatal mouse has been shown to tolerate substantial doses of dicarboxylic amino acids without developing neuronal necrosis. Doses of aspartate or glutamate that produce plasma glutamate plus aspartate concentrations lower than 50 µmol/dl (four to six times normal) do not produce lesions in the infant rodent (14-16,76-79). Lesions are noted only when very large doses of these amino acids are administered and plasma aspartate plus glutamate concentrations exceed 60-100 μmol/dl. When dicarboxylic amino acids are given in water, the lowest dose producing neuronal necrosis in infant rodents is 500 mg/kg body weight (14,15,17,74,75,79). Intakes greater than 500 mg/kg body weight fail to produce neuronal necrosis when ingested with food by normally hydrated animals (91,92), although dehydrated animals have been shown to be sensitive (93). Thus dicarboxylic amino acid-induced toxicity appears to require at least two factors: a susceptible species and grossly elevated plasma levels of glutamate and/or aspartate. These data suggest that plasma glutamate plus aspartate concentrations below 50-60 μmol/dl are not toxic to infant nonhuman primates, since even the infant mouse tolerates this blood level.

It must be emphasized that the absence of neuronal necrosis in neonatal nonhuman primates given large doses of glutamate, aspartame, or aspartame plus glutamate does not reflect a failure of the doses to elevate plasma glutamate plus aspartate levels. Plasma glutamate plus aspartate levels in these neonatal nonhuman primates were significantly elevated (21,86) above those levels that produced lesions in the infant mouse.

Normal Subjects Given Aspartame at 34 mg/kg Body Weight

A total of 12 normal subjects (6 male and 6 female) were administered aspartame at 34 mg/kg body weight or aspartate at 13 mg/kg body weight, in a randomized crossover design (35). Plasma aspartate levels in these subjects, as well as levels of other amino acids that might be derived metabolically from aspartate, are shown in Figure 18. Particular attention was given to plasma values at early time points after dosing. No significant differences were noted in plasma aspartate, asparagine, glutamate or glutamine values from baseline values, indicating rapid metabolism of the aspartate portion of aspartame.

Erythrocyte amino acid concentrations were also determined. Levels of glutamate and aspartate are maintained at much higher levels in the erythrocyte than in plasma (Table 5). The plasma-to-erythrocyte gradient for glutamate and aspartate differs from the ratio noted for most other amino acids, such as phenylalanine, where levels are similar between the two compartments even when plasma phenylalanine concentrations increase (Fig. 5). Figure 19 shows that no significant changes occurred in erythrocyte glutamate or aspartate levels.

Administration of the 99th percentile of projected aspartame intake for an entire day, when given as a single dose to a fasting subject, increased plasma and erythrocyte phenylalanine levels to normal postprandial limits, but did not change plasma or erythrocyte aspartate or glutamate values. These data demonstrate that aspartame intake at this level does not represent a risk to the consumer on the basis of its aspartate or phenylalanine content.

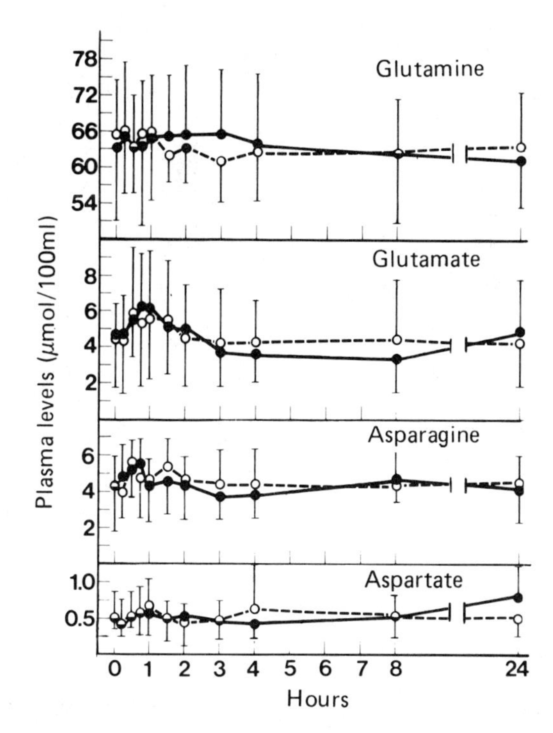

Figure 18 Mean (±SD) plasma glutamine, glutamate, asparagine, and aspartate concentrations in normal adults administered 34 mg/kg body weight aspartame (●——●) or 13 mg/kg body weight aspartate (○- - -○). (From Ref. 35.)

Table 5 Comparison of Mean (±SD) Plasma and Erythrocyte Levels of Selected Amino Acids in Normal Adults

	Plasma (μmol/dl)	Erythrocyte (μmol/100 g)
Glutamate	6 ± 3	20 ± 8
Aspartate	0.4 ± 0.2	23 ± 6
Phenylalanine	6 ± 3	5 ± 2
Tyrosine	5.5 ± 1	4.5 ± 1.5

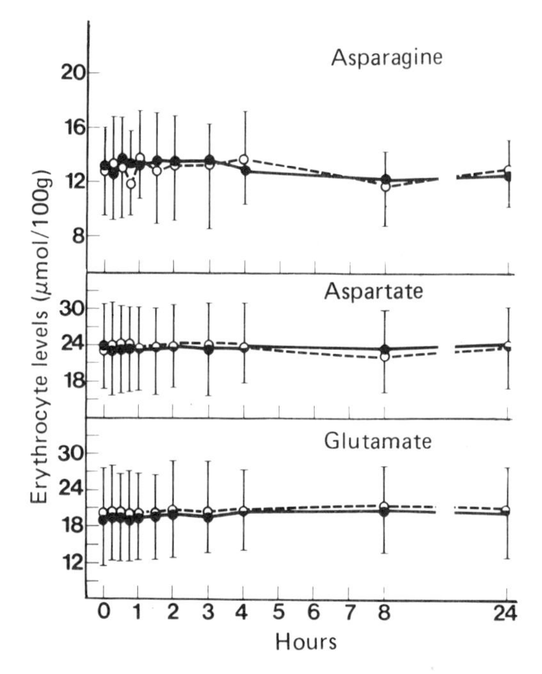

Figure 19 Mean (±SD) erythrocyte asparagine, aspartate, and glutamate concentrations in normal adults administered 34 mg/kg body weight aspartame (●——●) or 13 mg/kg body weight aspartate (○- - -○). (From Ref. 35.)

Normal Adults Given Aspartame at 50 mg/kg Body Weight

Six female subjects were administered 50 mg/kg body weight of aspartame or lactose in a randomized crossover design (46). Figure 20 shows that no changes occurred in plasma aspartate or glutamate levels. Concentrations of aspartate and glutamate in erythrocytes also remained unchanged (Fig. 21). These data demonstrate that aspartame ingestion at nearly twice the level projected for the 99th percentile of intake fails to increase plasma or erythrocyte aspartate or glutamate levels.

Normal Adults Administered Abuse Doses of Aspartame

Normal fasted adults were administered aspartame dissolved in cold orange juice at doses of either 100, 150, or 200 mg/kg body weight. Six subjects were studied at each level. The studies were carried out sequentially with all of the data at the lowest dose evaluated before proceeding to the next higher dose (48).

Table 6 shows that plasma aspartate levels do not increase proportionally to the dose. Plasma aspartate levels are no higher after the 200 mg/kg body weight dose than after the 150 mg/kg body weight dose. All plasma aspartate values

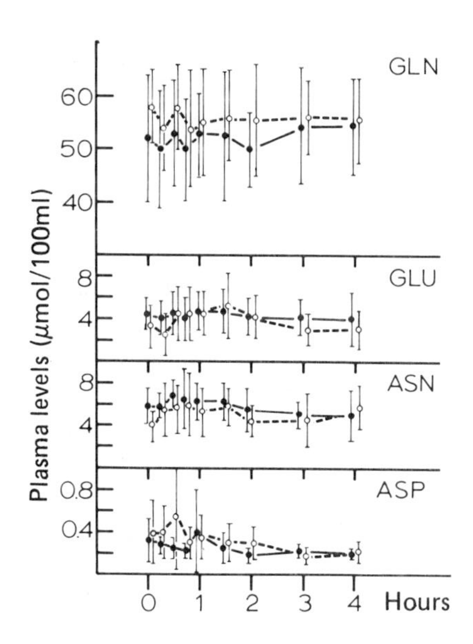

Figure 20 Mean (±SD) plasma glutamine (GLN), glutamate (GLU), asparagine (ASN), and aspartate (ASP) concentrations in lactating women administered 50 mg/kg body weight of aspartame (○- - -○) and lactose (●——●). (From Ref. 46.)

Table 6 Plasma Aspartate Concentrations (μmol/dl) in Normal Adults Administered Aspartame at 100, 150, or 200 mg/kg Body Weight

	Aspartame dose		
Time (hr)	100 mg/kg body weight	150 mg/kg body weight	200 mg/kg body weight
0	0.16 ± 0.05[a]	0.27 ± 0.10	0.22 ± 0.08
0.25	0.38 ± 0.15[b]	0.65 ± 0.50[c]	0.50 ± 0.40
0.5	0.43 ± 0.23[d]	1.00 ± 0.70	0.76 ± 0.57
0.75	0.29 ± 0.15	0.65 ± 0.42[b]	0.65 ± 0.47[c]
1.0	0.20 ± 0.09	0.47 ± 0.17[d]	0.57 ± 0.23[c]
1.5	0.29 ± 0.14[c]	0.79 ± 0.47[c]	0.71 ± 0.38[c]
2.0	0.27 ± 0.19	0.57 ± 0.36	0.58 ± 0.28[c]
3.0	0.14 ± 0.08	0.24 ± 0.11	0.27 ± 0.08
4.0	0.11 ± 0.04[c]	0.19 ± 0.15	0.23 ± 0.08
5.0	0.09 ± 0.02[d]	0.17 ± 0.13	0.20 ± 0.05
6.0	0.09 ± 0.02[d]	0.17 ± 0.15	0.13 ± 0.03
7.0	0.08 ± 0.06[d]	0.20 ± 0.13	0.14 ± 0.02
8.0	0.08 ± 0.05[d]	0.29 ± 0.19	0.25 ± 0.16
24.0	0.14 ± 0.13	0.15 ± 0.04[c]	0.28 ± 0.14

[a]Values listed as the mean ± SD for the six subjects studied at each dose.
[b]Differ significantly from zero time values; $p < 0.01$.
[c]Differ significantly from zero time values; $p < 0.005$.
[d]Differ significantly from zero time values; $p < 0.02$.

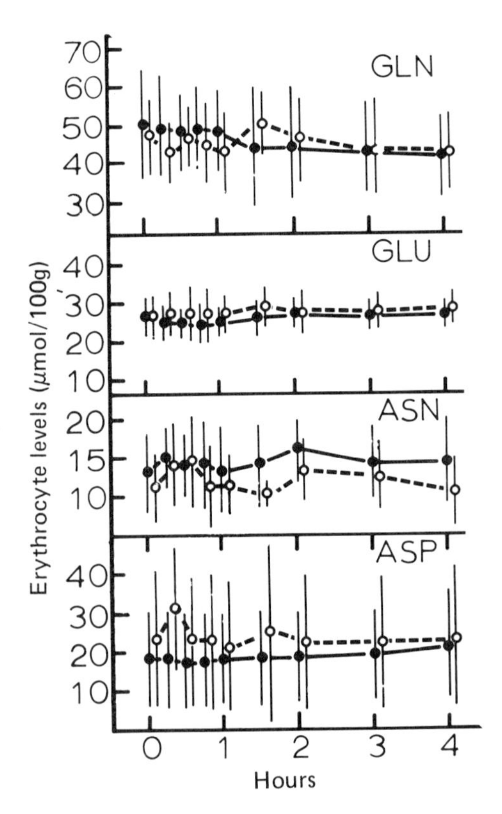

Figure 21 Mean (±SD) erythrocyte glutamine (GLN), glutamate (GLU), asparagine (ASN), and aspartate (ASP) concentrations in lactating women administered 50 mg/kg body weight of aspartame (○- - -○) and lactose (●——●). (From Ref. 46.)

noted are well below normal postprandial plasma aspartate levels observed in orally fed human infants and adults (38-40). Despite the large aspartame loads, no significant changes were noted in erythrocyte aspartate levels (Table 7). Plasma glutamate levels increased slightly after aspartame loading; however, the mean increase was not proportional to the dose (Table 8). Plasma levels were well within the normal postprandial limits observed for orally fed infants and adults (38-40) at all times during this study.

These data demonstrate the rapid metabolism and clearance of the aspartate portion of aspartame, even when the compound is administered at abuse levels. Since plasma aspartate levels are not elevated by such abuse doses, it is difficult to conceive of danger from aspartate-induced neuronal damage.

Table 7 Erythrocyte Aspartate and Glutamate Concentrations (μmol/100 g) in Normal Adults Administered Aspartame at 100, 150, or 200 mg/kg Body Weight[a]

	Aspartate levels observed at an aspartame dose of			Glutamate levels observed at an aspartame dose of		
Time (hr)	100 mg/kg body weight	150 mg/kg body weight	200 mg/kg body weight	100 mg/kg body weight	150 mg/kg body weight	200 mg/kg body weight
0	23.6 ± 3.82	21.1 ± 3.94	13.9 ± 5.57	23.2 ± 4.83	23.5 ± 5.94	22.1 ± 5.82
0.25	24.2 ± 5.34	22.5 ± 4.50	13.7 ± 5.84	23.9 ± 4.77	23.5 ± 7.17	21.2 ± 5.11
0.5	24.9 ± 5.00	21.7 ± 4.01	13.5 ± 5.77	24.9 ± 5.60	22.5 ± 7.17	22.6 ± 6.16
0.75	23.4 ± 3.77	22.5 ± 4.18	13.7 ± 6.05	25.1 ± 6.03	25.5 ± 8.74	23.1 ± 6.48
1.0	24.8 ± 4.60	21.9 ± 5.28	13.9 ± 5.80	24.7 ± 5.31	24.6 ± 8.32	23.1 ± 5.82
1.5	25.9 ± 5.91	22.5 ± 4.52	13.0 ± 5.97	24.1 ± 6.91	23.4 ± 7.97	22.9 ± 6.45
2.0	23.8 ± 4.18	22.8 ± 4.06	13.5 ± 6.17	23.8 ± 6.02	23.3 ± 6.61	23.0 ± 6.24
3.0	24.7 ± 6.07	22.7 ± 5.50	14.2 ± 6.65	24.7 ± 6.10	24.2 ± 7.92	22.9 ± 5.51
4.0	24.4 ± 5.04	22.8 ± 4.43	14.0 ± 5.99	24.8 ± 5.86	25.2 ± 7.51	23.3 ± 7.04
5.0	23.9 ± 4.89	21.6 ± 4.72	14.7 ± 6.61	24.2 ± 5.42	24.3 ± 8.04	23.1 ± 7.40
6.0	24.2 ± 4.93	23.9 ± 5.74	12.8 ± 4.87	23.3 ± 4.51	27.1 ± 6.81	22.7 ± 7.45
7.0	24.0 ± 4.11	21.5 ± 4.00	14.2 ± 6.80	23.7 ± 2.49	25.8 ± 7.24	23.5 ± 7.12
8.0	23.4 ± 4.45	21.3 ± 3.80	14.0 ± 5.63	25.1 ± 3.53	26.2 ± 7.62	23.9 ± 6.82
24.0	25.4 ± 5.47	22.1 ± 4.14	14.3 ± 6.46	24.5 ± 4.09	23.5 ± 5.70	22.8 ± 6.79

[a]Values listed as the mean ± SD for the six subjects studied at each dose.

Table 8 Plasma Glutamate Concentrations (μmol/dl) in Normal Adults Administered Aspartame at 100, 150, or 200 mg/kg Body Weight[a]

	Aspartame dose		
Time (hr)	100 mg/kg body weight	150 mg/kg body weight	200 mg/kg body weight
0	3.93 ± 2.65	3.43 ± 1.40	2.80 ± 1.50
0.25	4.34 ± 1.19	4.92 ± 3.05	3.40 ± 2.62
0.5	5.10 ± 1.50	6.77 ± 2.94[b]	4.50 ± 3.38
0.75	5.08 ± 1.54	7.08 ± 2.75[b]	5.52 ± 2.95[c]
1.0	3.83 ± 1.66	6.49 ± 1.47[c]	5.49 ± 2.54[d]
1.5	4.51 ± 2.50	7.23 ± 2.93[c]	5.96 ± 2.42[d]
2.0	4.98 ± 2.13	6.53 ± 3.01[c]	5.60 ± 2.50[c]
3.0	3.69 ± 1.99	4.44 ± 2.28	3.25 ± 1.63
4.0	3.37 ± 1.83	5.02 ± 2.40	3.19 ± 1.94
5.0	2.96 ± 1.44	4.33 ± 2.78	3.40 ± 2.52
6.0	2.63 ± 0.99	3.76 ± 3.24	2.43 ± 1.13
7.0	3.14 ± 1.59	3.06 ± 2.52	2.66 ± 1.75
8.0	2.51 ± 1.25	3.27 ± 2.26	2.86 ± 1.24
24.0	2.41 ± 0.95	2.81 ± 1.46	2.92 ± 1.03

[a]Values listed as the mean ± SD for the six subjects studied at each dose.
[b]Differs significantly from zero time values; $p < 0.05$.
[c]Differs significantly from zero time values; $p < 0.01$.
[d]Differs significantly from zero time values; $p < 0.005$.

Populations with Potential for Increased Risk

Special population groups that may have increased sensitivity to aspartame ingestion (on the basis of an altered physiologic state or genetic defect) were also studied. These conditions included infancy, pregnancy, lactation, and heterozygosity for the genetic defect of phenylketonuria (PKU). Studies of pregnancy were limited to an investigation of the placental transfer of aspartate in the nonhuman primate.

Infants

Olney suggested that the infant is at greater risk from aspartame ingestion than adults (6-9) because of a potential delay in the expression of genetic material coding for the synthesis of enzymes metabolizing glutamate and aspartate. This question is addressed by Filer et al. later in this book (see Chap. 29, Ref. 55). The data indicate that the aspartate portion of aspartame is well metabolized by normal infants at both use and abuse levels.

Pregnancy/Lactation

The results of studies in lactating women have been summarized by Baker (94). These data indicate that aspartame administration at 50 mg/kg body weight has only a small effect upon milk aspartate levels when compared to values noted after lactose loading. The biological importance of these small differences in the milk content of free amino acids is insignificant.

In pregnancy, the placenta normally maintains a concentration gradient of most amino acids in favor of the fetal circulation. Fetal plasma levels of most amino acids are approximately twice maternal levels. We addressed the question of whether maternal ingestion of abuse doses of aspartame would increase maternal plasma aspartate levels, and consequently fetal plasma aspartate concentration.

In earlier studies, we reported that glutamate was not concentrated to the fetal circulation of the pregnant monkey, even in the presence of elevated maternal plasma glutamate levels (95,96). These observations should also apply to aspartate, the other plasma dicarboxylic amino acid.

The results of these studies (97) are summarized by Pitkin (see Chap. 27) in this volume (98). The data indicate that aspartate, like glutamate, but unlike most other amino acids, is not concentrated toward the fetal circulation. Such observations indicate that it is virtually impossible for humans to ingest aspartame in quantities sufficient to increase maternal plasma aspartate levels to values that would allow transfer of significant quantities of aspartate to the fetal circulation.

Phenylketonuric Heterozygotes

As described in a previous section of this chapter, phenylketonuric heterozygotes metabolize the phenylalanine portion of aspartame less rapidly than normal subjects; however, such persons should metabolize the aspartate portion of the aspartame normally. Plasma aspartate levels in phenylketonuric heterozygotes administered aspartame (34 mg/kg body weight) are compared to values noted in normal subjects after ingestion of a similar dose in Figure 22. No differences were noted in aspartate handling. Similarly, no differences in erythrocyte aspartate levels were noted (Fig. 23).

Plasma and erythrocyte aspartate levels in the phenylketonuric heterozygotes administered aspartame at 100 mg/kg body weight are compared with values noted in normal subjects in Figure 24 and 25. Again, no differences in plasma or erythrocyte aspartate levels were noted. These data indicate that phenylketonuric heterozygotes handle the aspartate portion of aspartame as well as normal individuals, and that only the metabolism and clearance of the phenylalanine portion are slower in the heterozygote.

Repeated Ingestion of Aspartame-Sweetened Beverage

Since aspartame will be used as a sweetener for beverages, individuals are likely to ingest several aspartame-sweetened beverages between meals. If the aspartate

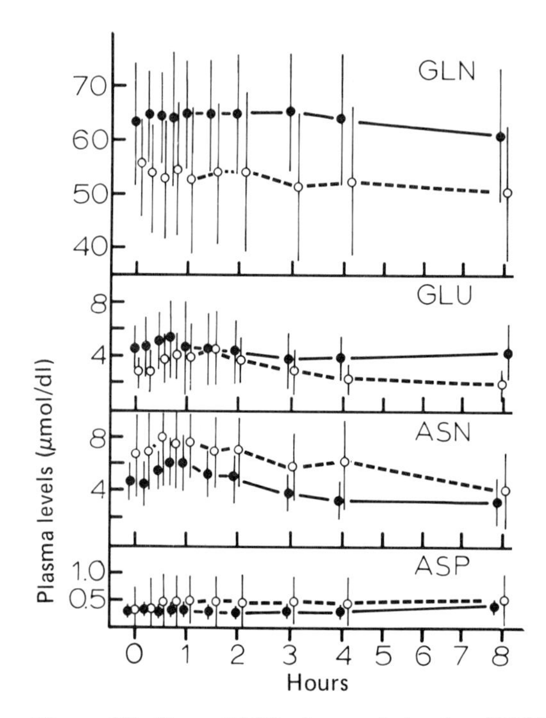

Figure 22 Mean (±SD) plasma glutamine (GLN), glutamate (GLU), asparagine (ASN), and aspartate (ASP) in normal adults (●——●) and PKU heterozygotes (○- - -○) administered aspartame at 34 mg/kg body weight. (From Ref. 57.)

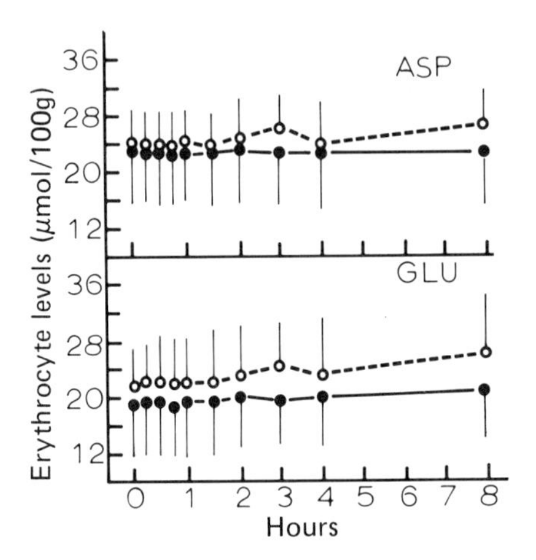

Figure 23 Mean (±SD) erythrocyte glutamate (GLU) and aspartate (ASP) concentrations in normal adults (●——●) and PKU heterozygotes (○- - -○) administered aspartame at 34 mg/kg body weight. (From Ref. 57.)

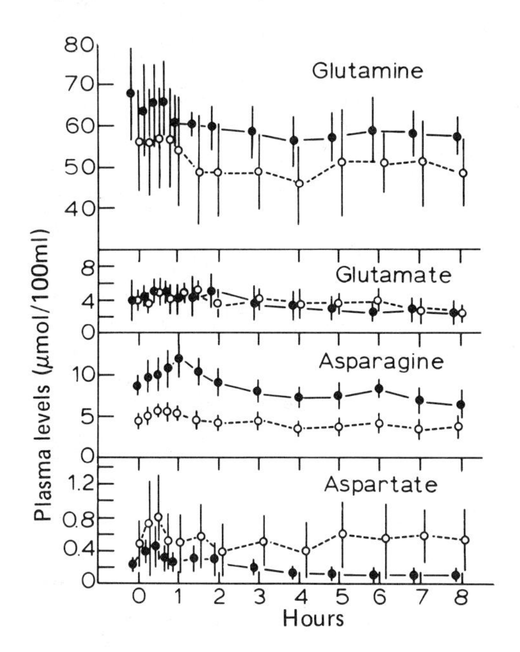

Figure 24 Mean (±SD) plasma glutamine, glutamate, asparagine, and aspartate concentrations in normal adults (●——●) and PKU heterozygotes (○- - -○) administered aspartame at 100 mg/kg body weight. (From Ref. 59.)

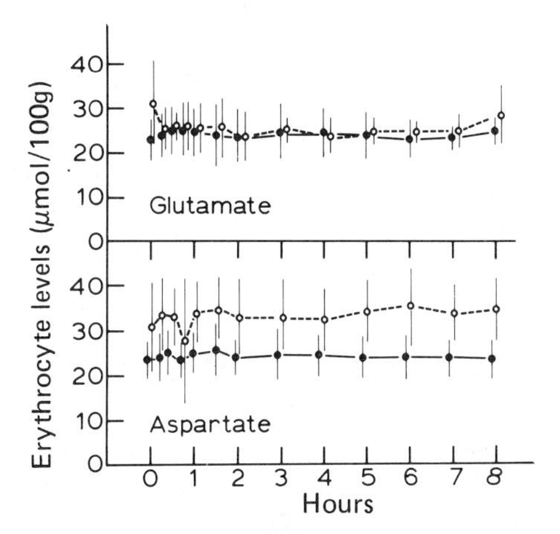

Figure 25 Mean (±SD) erythrocyte glutamate and aspartate concentrations in six normal subjects (●——●) and five subjects heterozygous for phenylketonuria (○- - -○) administered 100 mg aspartame per kilogram body weight. (From Ref. 59).

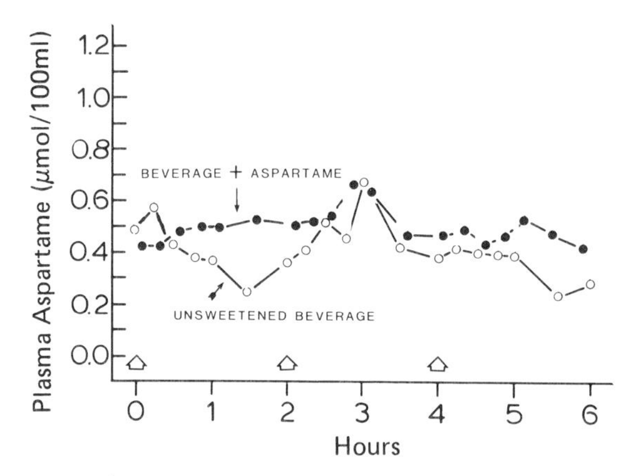

Figure 26 Mean plasma aspartate concentrations in normal adults ingesting repeated servings of either unsweetened beverage (○- - -○) or beverage providing aspartame at 10 mg/kg body weight (●——●).

load produced by aspartame loading is not completely cleared and/or metabolized prior to ingestion of a second dose of aspartame, a cumulative increase in plasma aspartate concentration might occur with the ingestion of successive aspartame doses. To test this, we studied the effects of successive ingestion of three servings of aspartame-sweetened beverage upon plasma aspartate concentration.

Eight normal healthy adults were studied, four male and four female (72). The study was conducted in two stages in a standard crossover design. Subjects ingested three successive 12-oz servings of beverage, with a 2-hr interval between servings. In one stage of the study, subjects drank three servings of unsweetened beverage, while in the other stage the three servings of beverage each provided aspartame at 10 mg/kg body weight.

As shown in Figure 26, the addition of aspartame to the beverage had no significant effect on plasma aspartate concentration. Similarly, plasma glutamate concentrations were not affected by the addition of aspartame to the beverage.

METHANOL

Aspartame is a dipeptide methyl ester, and during absorption and metabolism of the peptide, methanol is released. The ingestion of large quantities of methanol is known to result in elevated blood methanol and formate concentrations and leads to a variety of adverse effects, including metabolic acidosis and blindness (27-30). Blood methanol concentrations were measured in normal subjects administered aspartame at 34, 50, 100, 150, and 200 mg/kg body weight (99).

In the first study, blood methanol levels were measured in 12 subjects ingesting aspartame at 34 mg/kg body weight in a randomized crossover design with aspartic acid at a dose of 13 mg/kg body weight. Blood methanol concentrations were below the limits for detection (0.35-0.4 mg/dl).

Blood methanol concentrations were also measured in the six female subjects administered aspartame at 50 mg/kg body weight in a randomized crossover design with lactose at a dose of 50 mg/kg body weight. Blood methanol concentrations were below the detection limit when lactose was ingested; however, blood methanol was detected in normal adults administered 50 mg/kg body weight aspartame. The mean (±SEM) peak blood methanol level was 0.34 ± 0.12 mg/dl.

Blood methanol concentrations in subjects administered aspartame at 100, 150, and 200 mg/kg body weight are shown in Figure 27. Blood methanol concentrations were increased significantly over preloading values ($p < 0.05$) from 15 min to 7 hr after aspartame ingestion, with the increase in blood methanol proportional to the aspartame dose. Mean (±SD) peak methanol concentrations were 1.27 ± 0.48 mg/dl at the 100 mg/kg body weight dose, 2.14 ± 0.35 mg/dl at the 150 mg/kg body weight dose, and 2.58 ± 0.78 mg/dl at the 200 mg/kg body weight dose of aspartame. Similarly, the area under the blood methanol concentration-time curve increased in proportion to the increase in aspartame load. Blood methanol concentrations returned to preloading values 8 hr after administration of the 100 mg/kg body weight dose. Blood methanol was still detected 8 hr after subjects received aspartame at 150 or 200 mg/kg body weight; however, no methanol was detected 24 hr after administration of aspartame.

Recent studies (30,31) indicate that many of the toxic effects of methanol in the nonhuman primate are due to formate accumulation rather than to formaldehyde or methanol. Accordingly, blood and urine samples from subjects administered the highest aspartame dose were assayed for formate content by the method

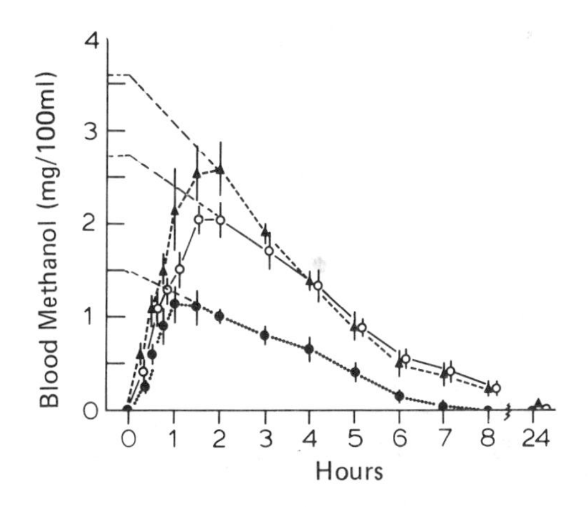

Figure 27 Mean (±SD) blood methanol concentrations in normal adults administered aspartame at 100 (●· · ·●), 150 (○- - -○), or 200 mg/kg body weight (▲- - -▲). (From Ref. 99.)

of Makar et al. (100). No significant changes in blood formate concentration were noted after aspartame administration at 200 mg/kg body weight (Fig. 28); however, urinary formate excretion (Table 9) was significantly increased over preloading values in urine samples collected 0-4 and 4-8 hr after aspartame loading. Urinary formate excretion returned to preloading values in those samples obtained 8-24 hr after loading. The urinary excretion data indicate conversion of methanol to formate. Since the rate of formate synthesis apparently did not exceed the rate of formate excretion, blood formate levels were not detectably elevated. Thus there appears to be little risk from aspartame's methanol content at the doses studied.

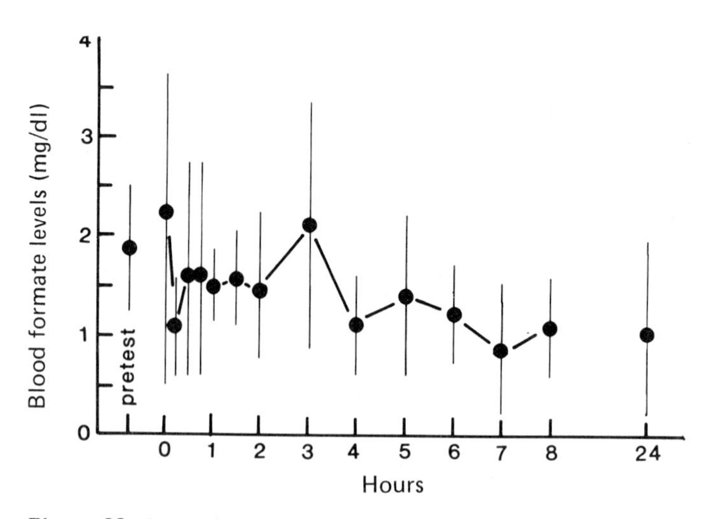

Figure 28 Mean (±SD) blood formate concentrations in normal adults administered aspartame at 200 mg/kg body weight.

Table 9 Urinary Formate Excretion in Normal Adults Ingesting 200 mg/kg Body Weight Aspartame

Urine sample (hr)[a]	Formate excreted (μg/mg creatinine)
-8 to 0	34 ± 22[b]
0 to 4	101 ± 30[c]
4 to 8	81 ± 22[c]
8 to 24	38 ± 12

[a]Negative values are preloading times.
[b]Data are expressed as the mean ± SD.
[c]Differs statistically from base-line levels; $p < 0.01$.

Methanol is a normal constituent of saliva and expired air and can be detected in blood (101,102). Because there is considerable variability in breath methanol concentration, Eriksen and Kulkarni (101) concluded that dietary sources are only partial contributors to the total body methanol pool. Indeed, Axelrod and Daly (103) reported the presence of a methyltransferase enzyme in pituitary extracts catalyzing the conversion of S-adenosylmethionine to methanol and S-adenosylhomocysteine. Later work (104,105) demonstrated that this methanol-forming enzyme was protein carboxylmethylase, an enzyme that methylates free carboxyl groups of proteins. Methanol is formed as the end product of this protein methylation through the action of protein methylesterases (106,107), as illustrated in Figure 29.

Earlier, we reported a method for predicting maximal blood methanol concentrations from a given aspartame dose in adults (99). These calculations suggest that a mean peak blood methanol concentration of 0.42 mg/dl might be expected to result at an aspartame intake of 34 mg/kg body weight. Although this concentration is at the outside limits of the assay sensitivity, it should be possible to detect. However, the equation developed for calculating the theoretical peak blood methanol concentration was developed using data obtained after loading with high doses of aspartame. Methanol may be cleared from the blood more rapidly at the levels produced by smaller doses of aspartame than was noted after ingestion of larger doses. This would lead to a lower blood methanol level than that predicted by our method. This factor probably accounts for our failure to detect methanol at an aspartame intake of 34 mg/kg body weight.

The projected methanol load resulting from the consumption of an aspartame-sweetened beverage might be less than that resulting from the consumption of an equivalent quantity of fruit juice. For example, an aspartame-sweetened beverage would have an aspartame content of about 555 mg/liter if sweetened at a typical soft drink level (10% sucrose wt/vol; aspartame sweetness 180 times that of

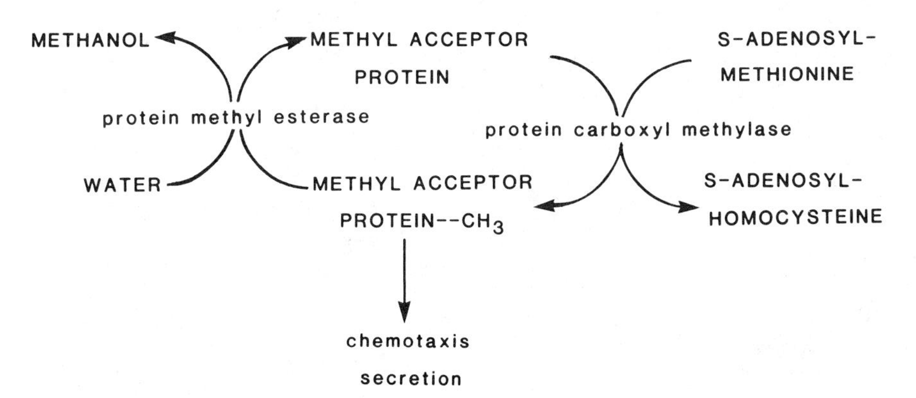

Figure 29 Synthesis of methanol in mammilian tissues. (From Refs. 107, 113.)

sucrose). In terms of aspartame's methanol content, this would be equivalent to 60 mg of methanol per liter—considerably less than the amount listed by Francot and Geoffroy (108) as the average methanol content of fruit juice (140 mg/liter). Other authors have reported similar levels of methanol in fruit juices (109-112).

Studies in Infants

Olney suggested that human infants metabolize the amino acids of aspartame less well than adults (6-9). While testing that hypothesis in infants and adults, blood methanol levels were also measured (113). Those data, summarized by Filer et al. in Chap. 29 (55), indicate similar blood methanol concentrations in infants and adults administered equivalent doses of aspartame.

REFERENCES

1. Oppermann, J. A., Muldoon, E., and Ranney, R. E. (1973). Metabolism of aspartame in monkeys. *J. Nutr.* 103, 1454-1459.
2. Ranney, R. E., Oppermann, J. A., Muldoon, E., and McMahon, F. G. (1976). The comparative metabolism of aspartame in experimental animals and man. *J. Toxicol. Environ. Health* 2, 441-451.
3. Ranney, R. E., and Oppermann, J. A. (1979). A review of metabolism of the aspartyl moiety of aspartame in experimental animals and man. *J. Environ. Pathol. Toxicol.* 2, 979-985.
4. Addison, J. M., Burston, D., Dalrymple, J. A., Matthews, D. M., Payne, J. W., Sleisenger, M. H., and Wilkinson, S. (1975). A common mechanism for transport of di- and tri-peptides by hamster jejunum in vitro. *Clin. Sci. Mol. Med.* 49, 313-322.
5. Matthews, D. M. (1984). Absorption of peptides, amino acids and their methylated derivatives. In *Aspartame: Physiology and Biochemistry*. (Stegink, L. D., and Filer, L. J., Jr., eds.), Marcel Dekker, New York, Chap. 3.
6. Olney, J. W. (1975). Another view of aspartame. In *Sweeteners, Issues and Uncertainties*, National Academy of Sciences, Washington, D.C. pp. 189-195.
7. Olney, J. W. (1976). L-Glutamic and L-aspartic acids—A question of hazard? *Food Cosmet. Toxicol.* 13, 595-596.
8. Olney, J. W. (1980). Testimony. In *Aspartame: Public Board of Inquiry, Transcript of Proceedings*, Vol. 3, Department of Health, Education and Welfare, Food & Drug Administration, Rockville, Md., pp. 117-118.
9. Olney, J. W. (1981). Excitatory neurotoxins as food additives: An evaluation of risk. *Neurotoxicology* 2, 163-192.
10. Reif-Lehrer, L. (1976). Possible significance of adverse reactions to glutamate in humans. *Fed. Proc.* 35, 2205-2211.
11. Stegink, L. D., Shepherd, J. A., Brummel, M. C., and Murray, L. M. (1974). Toxicity of protein hydrolysate solutions: Correlation of glutamate dose and neuronal necrosis to plasma amino acid levels in young mice. *Toxicology* 2, 285-299.

12. Itoh, H., Kishi, T., Iwasawa, Y., Kawashima, K., and Chibata, I. (1979). Plasma aspartate levels in rats following administration of monopotassium aspartate via three routes. *J. Toxicol. Sci.* 4, 377-388.
13. Stegink, L. D., Reynolds, W. A., Filer, L. J., Jr., Baker, G. L., Daabees, T. T., and Pitkin, R. M. (1979). Comparative metabolism of glutamate in the mouse, monkey, and man. In *Glutamic Acid; Advances in Biochemistry and Physiology* (Filer, L. J., Jr., Garattini, S., Kare, M. R., Reynolds, W. A., and Wurtman, R. J., eds.), Raven Press, New York, pp. 85-102.
14. Daabees, T. T., Finkelstein, M., Applebaum, A. E., and Stegink, L. D. (1982). Protective effect of carbohydrates or insulin on aspartate-induced neuronal necrosis in infant mice. *Fed. Proc.* 41, 396.
15. Finkelstein, M. W., Daabees, T. T., Stegink, L. D., and Applebaum, A. E. (1983). Correlation of aspartate dose, plasma dicarboxylic amino acid concentration and neuronal necrosis in infant mice. *Toxicology* 29, 109-119.
16. Applebaum, A. E., Finkelstein, M. W., Daabees, T. T., and Stegink, L. D. (1984). Aspartate-induced neurotoxicity in infant mice. In *Aspartame: Physiology and Biochemistry* (Stegink, L. D., Filer, L. J., Jr., eds.), Marcel Dekker, New York, Chap. 17.
17. Olney, J. W., and Ho, O.-L. (1970). Brain damage in infant mice following oral intake of glutamate, aspartate, or cysteine. *Nature* 227, 609-611.
18. Okaniwa, A., Hori, M., Masuda, M., Takeshita, M., Hayashi, N., Wada, I., Doi, K., and Ohara, Y. (1979). Histopathological study on effects of potassium aspartate on the hypothalamus of rats. *J. Toxicol. Sci.* 4, 31-46.
19. Olney, J. W. (1976). Brain damage and oral intake of certain amino acids. In *Transport Phenomena in the Nervous System. Physiological and Pathological Aspects* (Levi, G., Battistin, L., and Lajtha, A., eds.), Plenum, New York, pp. 497-506.
20. Reynolds, W. A., Bulter, V., and Lemkey-Johnston, N. (1976). Hypothalamic morphology following ingestion of aspartame or MSG in the neonatal rodent and primate: A preliminary report. *J. Toxicol. Environ. Health* 2, 471-480.
21. Reynolds, W. A., Stegink, L. D., Filer, L. J., Jr., and Renn, E. (1980). Aspartame administration to the infant monkey: Hypothalamic morphology and plasma amino acid levels. *Anat. Rec.* 198, 73-85.
22. Turner, J. S. (1979). Quoted in Aspartame: Ruling on objections and notice of hearing before a public board of inquiry. *Fed. Reg.* 44, 31716-31718.
23. Knox, W. E. (1972). Phenylketonuria. In *The Metabolic Basis of Inherited Disease* (Stanbury, J. B., Wyngaarden, J. B., and Fredrickson, D. S., eds.), McGraw-Hill, New York, pp. 266-295.
24. Blaskovics, M. E., Schaeffler, G.E., and Hack, S. (1974). Phenylalaninaemias. Differential diagnosis. *Arch. Dis. Child.* 49, 835-843.
25. Blaskovics, M. E. (1974). Phenylketonuria and phenylalaninaemias. *Clin. Endocrinol. Metab.* 3, 87-105.
26. Koch, R., Blaskovics, M., Wenz, E., Fishler, K., and Schaeffler, G. (1974). Phenylalaninemia and phenylketonuria. In *Heritable Disorders of Amino Acid Metabolism* (Nyhan, W. L., ed.), Wiley, New York, pp. 109-140.
27. Röe, O. (1955). The metabolism and toxicity of methanol. *Pharmacol. Rev.* 7, 399-412.

28. Tephly, T. R., Watkins, W. D., and Goodman, J. E. (1974). The biochemical toxicity of methanol. *Essays Toxicol.* 5, 149-177.
29. Posner, H. S. (1975). Biohazards of methanol in proposed new uses. *J. Toxicol. Environ. Health* 1, 153-171.
30. Tephly, T. R., and McMartin, K. E. (1984). Methanol metabolism and toxicity. In *Aspartame: Physiology and Biochemistry* (Stegink, L. D., and Filer, L. J., Jr., eds.), Marcel Dekker, New York, Chap. 6.
31. McMartin, K. E., Martin-Amat, G., Makar, A. B., and Tephly, T. R. (1977). Methanol poisoning. V. Role of formate metabolism in the monkey. *J. Pharmacol. Exp. Ther.* 201, 564-572.
32. Title 21–Food and Drugs (1974). Chapter 1–Food and Drug Administration, Department of Health, Education, and Welfare. Subchapter B–Food and Food Products. Part 121–Food Additives. Subpart D–Food Additives Permitted in Food for Human Consumption. Aspartame. *Fed. Reg.* 39, 27317-27320.
33. *Potential Aspartame Consumption Estimation* (1976). Research Summary. Prepared by the Market Research Department of General Foods, March 1976, Food and Drug Administration, Hearing Clerk File, Administrative Record, Aspartame 75F-0355. File Volume 103.
34. *Consumption of Sweeteners in the United States and Projected Consumption of Aspartame* (1976). A Report to General Foods by the Market Research Corporation. Food and Drug Administration, Hearing Clerk File, Administrative Record, Aspartame 75F-0355, File Volume 103.
35. Stegink, L. D., Filer, L. J., Jr., and Baker, G. L. (1977). Effect of aspartame and aspartate loading upon plasma and erythrocyte free amino acid levels in normal adult volunteers. *J. Nutr.* 107, 1837-1845.
36. *Food Intake and Nutritive Values of U.S. Diets* (1965). U.S.D.A. Study, Spring 1965.
37. Roak-Foltz, R., and Leveille, G. A. (1984). Projected aspartame intake: daily ingestion of aspartic acid, phenylalanine, and methanol. In *Aspartame: Physiology and Biochemistry* (Stegink, L. D., and Filer, L. J., Jr., eds.), Marcel Dekker, New York, Chap. 9.
38. Stegink, L. D., and Baker, G. L. (1971). Infusion of protein hydrolysates in the newborn infant: Plasma amino acid concentrations. *J. Pediatr.* 78, 595-603.
39. Stegink, L. D., Schmitt, J. L., Meyer, P. D., and Kain, P. H. (1971). Effect of diets fortified with DL-methionine on urinary and plasma methionine levels in young infants. *J. Pediatr.* 79, 648-659.
40. Vaughan, D. A., Womack, M., and McClain, P. E. (1977). Plasma free amino acid levels in human subjects after meals containing lactalbumin, heated lactalbumin or no protein. *Am. J. Clin. Nutr.* 30, 1709-1712.
41. Aoki, T. T., Brennan, M. F., Muller, W. A., Moore, F. D., and Cahill, G. F., Jr. (1972). Effect of insulin on muscle glutamate uptake Whole blood versus plasma glutamate analysis. *J. Clin. Invest.* 51, 2889-2894.
42. Elwyn, D. H., Launder, W. J., Parikh, H. C., and Wise, E. M., Jr. Roles of plasma and erythrocytes in interorgan transport of amino acids in dogs. *Am. J. Physiol.* 222, 1333-1342.

43. Elwyn, D. H. (1966). Distribution of amino acids between plasma and red blood cells in the dog. *Fed. Proc.* 25, 854-861.
44. Felig, P., Wahren, J., and Räf, L. (1973). Evidence of inter-organ amino-acid transport by blood cells in humans. *Proc. Nat. Acad. Sci. U.S.A.* 70, 1775-1779.
45. Drews, L. R., Conway, W. P., and Gilboe, D. D. (1977). Net amino acid transport between plasma and erythrocytes and perfused dog brain. *Am. J. Physiol.* 233, E320-E325.
46. Stegink, L. D., Filer, L. J., Jr., and Baker, G. L. (1979). Plasma, erythrocyte and human milk levels of free amino acids in lactating women administered aspartame or lactose. *J. Nutr.* 109, 2173-2181.
47. Tocci, P. M., and Berber, B. (1973). Anomalous phenylalanine loading response in relation to cleft lip and cleft palate. *Pediatrics* 52, 109-113.
48. Stegink, L. D., Filer, L. J., Jr., and Baker, G. L. (1981). Plasma and erythrocyte concentrations of free amino acids in adult humans administered abuse doses of aspartame. *J. Toxicol. Environ. Health* 7, 291-305.
49. Blaskovics, M. E., Schaeffler, G. E., and Hack, S. (1974). Phenylalaninemia: Differential diagnosis. *Arch. Dis. Child.* 49, 835-843.
50. Blau, L., Summer, G. K., Newsome, H. C., and Edwards, C. H. (1973). Phenylalanine loading and aromatic amino acid excretion in normal subjects and heterozygotes for phenylketonuria. *Clin. Chim. Acta* 45, 197-205.
51. Guttler, F., and Wamberg, E. (1972). Persistent hyperphenylalaninemia. *Acta Paediatr. Scand.* 61, 321-328.
52. Anderson, J. A., Fisch, R., Miller, E., and Doeden, D. (1966). Atypical phenylketonuria heterozygote. *J. Pediatr.* 68, 351-360.
53. Cunningham, G. C., Day, R. W., Berman, J. L., and Hsia, D. Y.-Y. (1969). Phenylalanine tolerance tests in families with phenylketonuria and hyperphenylalaninemia. *Am. J. Dis. Child.* 117, 626-635.
54. Filer, L. J., Jr., Baker, G. L., and Stegink, L. D. (1983). Effect of aspartame loading upon plasma and erythrocyte free amino acid concentrations in one-year-old infants. *J. Nutr.* 113, 1591-1599.
55. Filer, L. J., Jr., Baker, G. L., and Stegink, L. D. (1984). Aspartame ingestion by human infants. In *Aspartame: Physiology and Biochemistry* (Stegink, L. D., and Filer, L. J., Jr., eds.), Marcel Dekker, New York, Chap. 29.
56. Koch, R., Shaw, K. N. F., Williamson, M., and Haber, M. (1976). Use of aspartame in phenylketonuric heterozygous adults. *J. Toxicol. Environ. Health* 2, 453-457.
57. Stegink, L. D., Filer, L. J., Jr., Baker, G. L., and McDonnell, J. E. (1979). Effect of aspartame loading upon plasma and erythrocyte amino acid levels in phenylketonuric heterozygotes and normal adult subjects. *J. Nutr.* 109, 708-717.
58. Stegink, L. D., Koch, R., Blaskovics, M. E., Filer, L. J., Jr., Baker, G. L., and McDonnell, J. E. (1981). Plasma phenylalanine levels in phenylketonuric heterozygous and normal adults administered aspartame at 34 mg/kg body weight. *Toxicology* 20, 81-90.
59. Stegink, L. D., Filer, L. J., Jr., Baker, G. L., and McDonnell, J. E. (1980).

Effect of an abuse dose of aspartame upon plasma and erythrocyte levels of amino acids in phenylketonuric heterozygous and normal adults. *J. Nutr.* 110, 2216-2224.

60. Kaufman, S., Max, E. E., and Kang, E. S. (1975). Phenylalanine hydroxylase activity in liver biopsies from hyperphenylalaninemia heterozygotes: Deviation from proportionality with gene dosage. *Pediatr. Res.* 9, 632-634.
61. Bremer, H. J., and Neumann, W. (1966). Tolerance of phenylalanine after intravenous administration in phenylketonuria, heterozygous carriers and normal adults. *Nature* 209, 1148.
62. Woolf, L. I., Cranston, W. I., and Goodwin, B. L. (1967). Genetics of phenylketonuria. *Nature* 213, 882-885.
63. Rey, F., Blandin-Savoja, F., and Rey, J. (1979). Kinetics of phenylalanine disappearance after intravenous loads in phenylketonuria and its genetic variants. *Pediatr. Res.* 13, 21-25.
64. Kang, E., and Paine, R. S. (1963). Elevation of plasma phenylalanine levels during pregnancy of women heterozygous for phenylketonuria. *J. Pediatr.* 63, 283-289.
65. Kerr, G. R., and Waisman, H. A. (1966). Phenylalanine: Transplacental concentrations in rhesus monkeys. *Science* 151, 824-825.
66. Ghadimi, H., and Pecora, P. (1963). Free amino acids of cord plasma as compared with maternal plasma during pregnancy. *Pediatrics* 33, 500-511.
67. Schain, R. J., Carver, M. J., and Copenhaver, J. H. (1967). Protein metabolism in the developing brain. Influence of birth and gestational age. *Science* 156, 984-986.
68. Pueschel, S. M., Hum, C., and Andrews, M. (1977). Nutritional management of the female with phenylketonuria during pregnancy. *Am. J. Clin. Nutr.* 30, 1153-1161.
69. Olney, J. W. (1980). Testimony. In *Aspartame: Public Board of Inquiry, Transcript of Proceedings*, Vol. 1, Department of Health, Education, and Welfare, Food and Drug Administration, Rockville, Md., pp. 36-44.
70. Turner, J. S. (1979). Quoted in Aspartame: Ruling on objections and notice of hearing before a public board of inquiry. *Fed. Reg.* 44, 31716-31718.
71. Morgan, K. J., and Stults, V. J. (1981). Beverage consumption habits of children 5-18 years of age. *Fed. Proc.* 40, 857.
72. Stegink, L. D., Filer, L. J., Jr., and Baker, G. L. (1983). Effect of repeated ingestion of aspartame-sweetened beverages upon plasma aminograms in normal adults. *Am. J. Clin. Nutr.* 36, 704.
73. Olney, J. W. (1969). Brain lesions, obesity, and other disturbances in mice treated with monosodium glutamate. *Science* 164, 719-721.
74. Lemkey-Johnston, N., and Reynolds, W. A. (1974). Nature and extent of brain lesions in mice related to ingestion of monosodium glutamate: A light and electron microscope study. *J. Neuropathol. Exp. Neurol.* 33, 74-97.
75. Takasaki, Y. (1978). Studies on brain lesions by administration of monosodium L-glutamate to mice. I. Brain lesions in infant mice caused by administration of monosodium L-glutamate. *Toxicology* 9, 293-305.
76. Stegink, L. D., Shepherd, J. A., Brummel, M. C., and Murray, L. M. (1974). Toxicity of protein hydrolysate solutions: Correlation of glutamate dose

and neuronal necrosis to plasma amino acid levels in young mice. *Toxicology* 2, 285-299.

77. Finkelstein, M. W. (1982). Aspartate-induced neuronal necrosis in infant mice. Effect of dose, insulin and carbohydrate. M.S. thesis in Oral Pathology, University of Iowa, Iowa City, Iowa.
78. O'Hara, Y., and Takasaki, Y. (1979). Relationship between plasma glutamate levels and hypothalamic lesions in rodents. *Toxicol. Lett.* 4, 499-505.
79. Takasaki, Y., Matsuzawa, Y., Iwata, S., O'Hara, Y., Yonetani, S., and Ichimura, M. (1979). Toxicological studies of monosodium L-glutamate in rodents: Relationship between routes of administration and neurotoxicity. In *Glutamic Acid: Advances in Biochemistry and Physiology* (Filer, L. J., Jr., Garattini, S., Kare, M. R., Reynolds, W. A., and Wurtman, R. J., eds.) Raven Press, New York, pp. 255-275.
80. Olney, J. W., and Sharpe, L. C. (1969). Brain lesions in an infant rhesus monkey treated with monosodium glutamate. *Science* 166, 386-388.
81. Olney, J. W., Sharpe, L. G., and Feigin, R. D. (1972). Glutamate-induced brain damage in infant primates. *J. Neuropathol. Exp. Neurol.* 31, 464-488.
82. Reynolds, W. A., Lemkey-Johnston, N., Filer, L. J., Jr., and Pitkin, R. M. (1971). Monosodium glutamate: Absence of hypothalamic lesions after ingestion by newborn primates. *Science* 172, 1342-1344.
83. Abraham, R. W., Dougherty, W., Golberg, L., and Coulston, F. (1971). The response of the hypothalamus to high doses of monosodium glutamate in mice and monkeys. Cytochemistry and ultrastructural study of lysosomal changes. *Exp. Mol. Pathol.* 15, 43-60.
84. Newman, A. J., Heywood, R., Palmer, A. K., Barry, D. H., Edwards, F. P., and Worden, A. N. (1973). The administration of monosodium L-glutamate to neonatal and pregnant primates. *Toxicology* 1, 197-204.
85. Wen, C., Hayes, K. C., and Gershoff, S. M. (1973). Effects of dietary supplementation of monosodium glutamate on infant monkeys, weanling rats and suckling mice. *Am. J. Clin. Nutr.* 26, 803-813.
86. Stegink, L. D., Reynolds, W. A., Filer, L. J., Jr., Pitkin, R. M., Boaz, D. P., and Brummel, M. C. (1975). Monosodium glutamate metabolism in the neonatal monkey. *Am. J. Physiol.* 229, 246-250.
87. Abraham, R., Swart, J., Golberg, L., and Coulston, F. (1975). Electron microscopic observations of hypothalami in neonatal rhesus monkeys (*Macaca mulatta*) after administration of monosodium-L-glutamate. *Exp. Mol. Pathol.* 23, 203-213.
88. Reynolds, W. A., Lemkey-Johnston, N., and Stegink, L. D. (1979). Morphology of the fetal monkey hypothalamus after *in utero* exposure to monosodium glutamate. In *Glutamic Acid: Advances in Biochemistry and Physiology* (Filer, L. J., Jr., Garattini, S., Kare, M. R., Reynolds, W. A., and Wurtman, R. J., eds.), Raven Press, New York, pp. 217-229.
89. Heywood, R., and Worden, A. N. (1979). Glutamate toxicity in laboratory animals. In *Glutamic Acid: Advances in Biochemistry and Physiology* (Filer, L. J., Jr., Garattini, S., Kare, M. R., Reynolds, W. A., and Wurtman, R. J., eds.), Raven Press, New York, pp. 203-215.
90. Heywood, R., and James, R. W. (1979). An attempt to induce neurotoxicity

in an infant rhesus monkey with monosodium glutamate. *Toxicol. Lett.* 4, 285-286.

91. Takasaki, Y. (1978). Studies on brain lesions after administration of monosodium L-glutamate. II. Absence of brain damage following administration of monosodium L-glutamate in the diet. *Toxicology* 9, 307-318.
92. Anantharaman, K. (1979). In utero and dietary administration of monosodium L-glutamate to mice: Reproductive performance and development in a multigeneration study. In *Glutamic Acid: Advances in Biochemistry and Physiology* (Filer, L. J., Jr., Garattini, S., Kare, M. R., Reynolds, W. A., and Wurtman, R. J., eds.), Raven Press, New York, pp. 231-253.
93. Olney, J. W., Labruyere, J., & de Gubareff, T. (1980). Brain damage in mice from voluntary ingestion of glutamate and aspartate. *Neurobehav. Toxicol.* 2, 125-129.
94. Baker, G. L. (1984). Aspartame ingestion during lactation. In *Aspartame: Physiology and Biochemistry* (Stegink, L. D., and Filer, L. J., Jr., eds.), Marcel Dekker, New York, Chap. 28.
95. Stegink, L. D., Pitkin, R. M., Reynolds, W. A., Filer, L. J., Jr., Boaz, D. P., and Brummel, M. C. (1975). Placental transfer of glutamate and its metabolites in the primate. *Am. J. Obstet. Gynecol.* 122, 70-78.
96. Pitkin, R. M., Reynolds, W. A., Stegink, L. D., and Filer, L. J., Jr. (1979). Glutamate metabolism and placental transfer in pregnancy. In *Glutamic Acid: Advances in Biochemistry and Physiology* (Filer, L. J., Jr., Garattini, S., Kare, M. R., Reynolds, W. A., and Wurtman, R. J., eds.), Raven Press, New York, pp. 103-110.
97. Stegink, L. D., Pitkin, R. M., Reynolds, W. A., Brummel, M. C., and Filer, L. J., Jr. (1979). Placental transfer of aspartate and its metabolites in the primate. *Metabolism* 28, 669-676.
98. Pitkin, R. M. (1984). Aspartame ingestion during pregnancy. In *Aspartame: Physiology and Biochemistry* (Stegink, L. D., and Filer, L. J., Jr., eds.), Marcel Dekker, New York, Chap. 27.
99. Stegink, L. D., Brummel, M. C., McMartin, K., Martin-Amat, G., Filer, L. J., Jr., Baker, G. L., and Tephly, T. R. (1981). Blood methanol concentrations in normal adult subjects administered abuse doses of aspartame. *J. Toxicol. Environ. Health* 7, 218-290.
100. Makar, A. B., McMartin, K. E., Palese, M., and Tephly, T. R. (1975). Formate assay in body fluids: Application in methanol poisoning. *Biochem. Med.* 13, 117-126.
101. Eriksen, S. P., and Kulkarni, A. B. (1963). Methanol in normal human breath. *Science* 141, 639-640.
102. Larsson, B. T. (1965). Gas chromatography of organic volatiles in human breath and saliva. *Acta Chem. Scand.* 19, 159-164.
103. Axelrod, J., and Daly, J. (1965). Pituitary gland: Enzymic formation of methanol from S-adenosylmethionine. *Science* 150, 892-893.
104. Kim, S. (1973). Purification and properties of protein methylase II. *Arch. Biochem. Biophys.* 157, 476-484.
105. Morin, A. M., and Lis, M. (1973). Evidence for a methylated protein intermediate in pituitary methanol formation. *Biochem. Biophys. Res. Commun.* 52, 373-378.

106. Gagnon, C. (1979). Presence of a protein methylesterase in mammalian tissues. *Biochem. Biophys. Res. Commun.* 88, 847-853.
107. Gagnon, C., and Heisler, S. (1979). Protein carboxyl-methylation: Role in exocytosis and chemotaxis. *Life Sci.* 25, 993-1000.
108. Francot, P., and Geoffroy, P. (1956). Le methanol dans les jus de fruits, les boissons, fermentées, les alcools et spiritueux. *Rev. Ferment. Ind. Aliment.* 2, 279-287.
109. Lund, E. D., Kirkland, C. L., and Shaw, P. E. (1981). Methanol, ethanol, and acetaldehyde contents of citrus products. *J. Agric. Food Chem.* 29, 361-366.
110. Kazeniac, S. J., and Hall, R. M. (1970). Flavor chemistry of tomato volatiles. *J. Food Sci.* 35, 519-530.
111. Kirchner, J. G., and Miller, J. M. (1957). Volatile water-soluble and oil constituents of Valencia orange juice. *J. Agric. Food Chem.* 5, 283-291.
112. Heatherbell, D. A., Wrolstad, R. E., and Libbey, L. M. (1971). Carrot volatiles: Characterization and effects of canning and freeze drying. *J. Food. Sci.* 36, 219-224.
113. Stegink, L. D., Brummel, M. C., Filer, L. J., Jr., and Baker, G. L. (1983). Blood methanol concentrations in one-year-old infants administered graded doses of aspartame. *J. Nutr.* 113, 1600-1606.

27

Aspartame Ingestion During Pregnancy

Roy M. Pitkin
University of Iowa College of Medicine, Iowa City, Iowa

INTRODUCTION

During pregnancy, the maternal organism undergoes a number of remarkable physiological adjustments, many of them involving the absorption, distribution, and metabolism of ingested compounds. Gastrointestinal motility is generally slowed, presumably reflecting an inhibitory action of progesterone on smooth muscle, and the extent of absorption tends to increase. The extracellular fluid space expands, resulting in a greater apparent volume of distribution, and urinary excretion increases concomitantly. While these physiological adjustments have not been studied with particular reference to aspartame, there is every reason to expect that they apply. Nevertheless, it seems likely that any such differences between pregnant and nonpregnant animals are relatively minor with respect to maternal metabolism per se, leaving the fetoplacental unit as the principal focus of interest and concern. Accordingly, the emphasis of this chapter will be on the placental transfer and metabolism of the individual components of aspartame, aspartate, phenylalanine, and methanol.

GENERAL ASPECTS OF PLACENTAL TRANSFER OF AMINO ACIDS

In all mammalian species plasma levels of most free amino acids in the fetus exceed those of the mother. A number of different studies have established five general characteristics of placental transfer of amino acids: (a) transport against a concentration gradient; (b) stereospecific discrimination, with L forms transported substantially more rapidly than D forms; (c) insufficient binding protein characteristics to account for concentration differences; (d) competitive inhibition of transfer by similar amino acids; and (e) saturation by increasing amino acid levels. Taken together, these characteristics suggest strongly that an active transport process is involved in this aspect of maternofetal transfer.

The transplacental gradient for most amino acids is maintained such that the fetal plasma concentration is typically one and one-half to two times the maternal concentration. Teleologically, this concentrating effect offers certain advantages in ensuring adequate provision of amino acids for the developing fetus. It is not, however, an unmixed blessing, for it carries some potential risk if the amino acid is toxic at high levels and the transplacental gradient is maintained as maternal levels rise. The best-known example of this phenomenon is maternal phenylketonuria in which infants born to women with high plasma phenylalanine levels typically demonstrate evidence of prenatal brain damage as a result of placental "amplification" of already elevated phenylalanine loads (1). It appears that dietary therapy to normalize blood phenylalanine levels in phenylketonuric women, if instituted early enough in gestation, can lessen the risk of fetal damage (2).

ASPARTATE

The placental transfer of aspartate has received considerable study because it, along with glutamate, the other dicarboxylic acid normally present in plasma, can cause neurotoxic effects in infant mice (3), leading to suggestions that maternal ingestion might lead to neuronal necrosis in the developing fetal brain (4). Evidence from a number of sources indicates that aspartate and glutamate represent exceptions to the general rule of fetal concentration and, furthermore, that neither crosses the placenta to any appreciable degree.

Lemons and associates (5) measured whole blood concentrations of 22 amino acids over the last third of gestation in a chronically catheterized unstressed fetal lamb preparation. No umbilical uptake of glutamate or aspartate was demonstrated, and, in fact, there was a net flux out of the fetus, indicating that these amino acids are produced within the fetus. Similar conclusions were reached by Smith et al. (6) in another study in pregnant sheep.

In vitro studies of the human placenta have also yielded results consistent with a lack of placental transfer of the dicarboxylic amino acids. Schneider and Dancis (7) incubated placental slices with various amino acids and found that glutamate and aspartate were avidly taken up intracellularly but not released by placental tissues. In an extension of this work involving the perfusion of isolated portions of the placenta, Schneider et al. (8) observed a failure of glutamate transfer from the maternal to the fetal side.

The most detailed studies to date are those of Stegink and colleagues (9) of the metabolism and placental transfer of infused aspartate in pregnant rhesus monkeys near term. The dosages administered by maternal intravenous infusion were 100, 200, and 400 mg/kg body weight per hour, all with added tracer [U-^{14}C] aspartate. In all cases maternal plasma aspartate levels increased proportionally with the dose. At the 100 mg/kg body weight per hour dose fetal levels increased only slightly, albeit to a statistically significant degree (0.42 ± 0.31 to 0.98 ± 0.24 μmol/dl), as illustrated in Figure 1. Animals given 200 mg/kg body weight per hour demonstrated a greater degree of placental transfer (Fig. 2), and in those receiving 400 mg/kg body weight per hour an even larger amount reached the fetus. Thus, unlike most amino acids, aspartate is not concentrated to the fetal circulation. These data demonstrate that some placental transfer of aspartate occurs when maternal plasma aspartate levels are 100-500 times normal plasma values. This level of plasma aspartate is never approached under expected intake levels of aspartame (see Ref. 10).

The addition of radioactive tracer in the studies by Stegink et al. (9) permitted determination of the metabolic fate of infused aspartate. As illustrated in Figure 3, the major metabolite was glucose, which diffused the placenta freely. The second aspartate-derived metabolite, lactate, was actually in higher concentration on the fetal side of the placenta than on the maternal side, perhaps reflecting the relatively anaerobic condition of the fetus favoring conversion of labeled glucose to lactate.

From these studies it seems clear that the placenta of the sheep, monkey, and human is essentially impermeable to aspartate. Acutely infused aspartate is

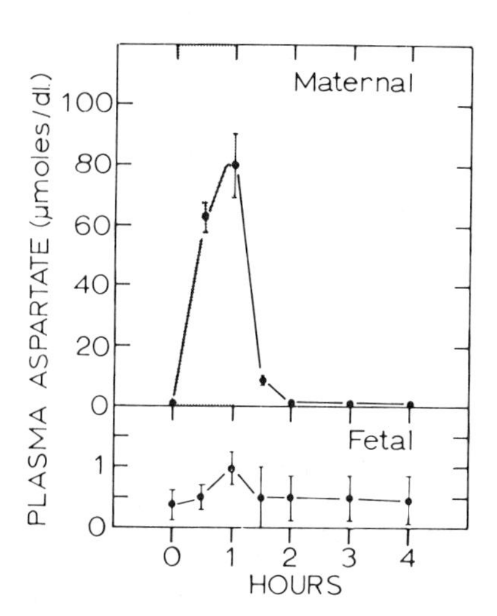

Figure 1 Maternal and fetal plasma aspartate levels (mean ± SD of five experiments) with maternal infusion of sodium aspartate at 100 mg/kg body weight per hour. The stippled area represents the time of infusion. (From Ref. 9.)

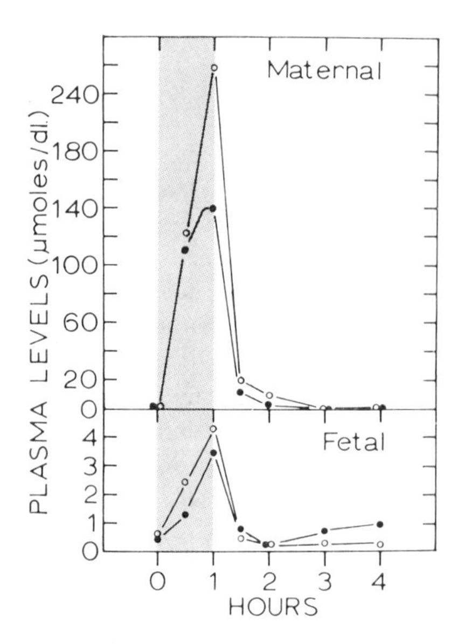

Figure 2 Maternal and fetal plasma aspartate levels in two monkeys infused with sodium aspartate at 200 mg/kg body weight per hour. The stippled area represents the time of maternal infusion. (From Ref. 9.)

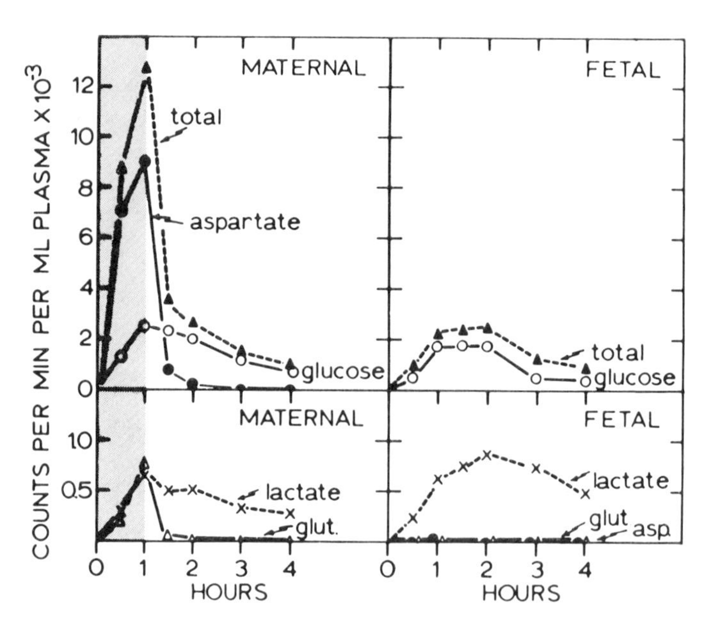

Figure 3 Radioactivity profile of aspartate-derived metabolites in maternal and fetal plasma during and following maternal infusion of aspartate (100 mg/kg body weight per hour, 50 μCi [^{14}C]-L-aspartate). (From Ref. 9.)

partially converted to glucose and lactate, which traverse the placenta freely, although the dicarboxylic acid itself is not appreciably transferred from mother to fetus, at least at maternal plasma levels of up to 100 times normal. Under conditions of enormous and artificial maternal loads of a type produced by intravenous infusion of an enormous dose, a limited transfer occurs. These same findings apparently apply in the case of glutamate, the other dicarboxylic acid normally present in plasma (11).

PHENYLALANINE

The fasting plasma concentration of phenylalanine in normal pregnant women at term averages 3-5 μmol/dl (0.5-0.8 mg/dl), and fetal plasma levels are approximately twice maternal ones (12). Evidence from several sources indicates that this transplacental gradient tends to be maintained in the face of elevation of maternal levels.

Pueschel and colleagues (13) have recently reported acute studies of materno-fetal phenylalanine transfer in nonhuman primates. Maternal intravenous administration of [^{14}C]phenylalanine led to a significant accumulation of radioactivity in the fetal circulation, compared with only minimal transfer to the maternal circulation when the compound was given to the fetus, findings interpreted as indicating an active placental transport mechanism from mother to fetus. In experiments involving chronic feeding of up to 6% aspartame to pregnant rabbits, Ranney and associates (14) found both phenylalanine and tyrosine to increase in concentration in fetal blood and amniotic fluid without affecting the normal transplacental gradient of either.

As noted earlier, infants born to women with untreated phenylketonuria (PKU) regularly exhibit neurotoxic and other adverse effects attributable to prolonged intrauterine exposure to high phenylalanine levels. This circumstantial evidence, coupled with findings that cord blood phenylalanine levels exceed maternal in such instances (15), indicates that the transplacental gradient for phenylalanine is maintained in the face of elevated maternal levels. Further support for this concept comes from investigations aimed at developing an experimental model of maternal PKU. Kerr and associates (16) fed rhesus monkeys a diet high in phenylalanine throughout gestation and found elevated plasma phenylalanine levels at birth. Other models, utilizing chemical agents to inhibit phenylalanine metabolism, such as α-methylphenylalanine (17) or p-chlorophenylalanine (18), along with high phenylalanine intake, have resulted in marked increases of phenylalanine in fetal blood and tissues.

In normal pregnant humans fasting plasma phenylalanine levels of 3-5 μmol/dl might be expected to double following intake of phenylalanine-containing foodstuffs such as protein. Assuming that the customary fetomaternal gradient of 2:1 is maintained under such circumstances, this would result in exposure of the fetus to a maximum fetal plasma phenylalanine concentration of 20 μmol (3.2 mg)/dl. This plasma level of phenylalanine cannot be considered toxic, since similar concentrations occur postprandially after ingestion of protein-containing meals.

The maximum maternal plasma phenylalanine concentration occurring after ingestion of aspartame-sweetened foods (assuming normal use levels) should be less than that occurring after ingestion of protein meals. For example, Stegink et al. (19) administered normal adults a dose of aspartame estimated to be the 99th percentile of projected use (34 mg/kg body weight) as a single dose. Maternal plasma phenylalanine concentrations increased by only 5.5 μmol/dl despite the fact that the entire aspartame load was given as a single bolus. Thus the maximal fetal plasma phenylalanine concentration resulting from administration of aspartame at this high level should be similar to those occurring after ingestion of a protein meal.

When aspartame is given to normal adults at potential abuse concentrations (100-200 mg/kg body weight), the increase in plasma phenylalanine concentration is considerably higher (10,20). Thus fetal plasma phenylalanine concentrations would be approximately twice maternal levels. However, the exposure would be intermittent and therefore almost surely less dangerous than the constant high levels characteristic of maternal PKU. Some degree of caution in this regard is indicated, however, by studies in the rat (21) indicating that the fetus is less effective than the mother in metabolizing phenylalanine, so that elevated fetal blood levels may persist for unexpectedly long times following a large maternal load of phenylalanine. Nevertheless, there seems to be little reason for concern about fetal damage in the normal situation at currently projected levels of aspartame ingestion.

The question of whether individuals heterozygous for PKU, a genetic trait occurring in approximately 2% of the population, should use aspartame when pregnant is less clear. Stegink and colleagues (22) found that aspartame doses of 34 mg/kg body weight (estimated as the 99th percentile of ingestion if aspartame replaced all dietary sucrose sources of sweetness) administered to nonpregnant adult PKU heterozygotes produced peak plasma phenylalanine levels averaging 16 μmol/dl. Thus this dose would be expected to produce fetal plasma values, again on an intermittent basis, of 32 μmol (5.3 mg)/dl. The maternal plasma phenylalanine value resulting from aspartame ingestion at this level is somewhat greater than the postprandial plasma phenylalanine level produced by protein ingestion. Fetal plasma phenylalanine levels would also be higher than values noted after maternal ingestion of a protein meal. However, the dose of aspartame administered in this study was eight times greater than normal projected single use levels of the sweetener. For example, a 12-oz bottle of carbonated beverage might provide 200 mg of aspartame, a 4-mg/kg body weight dose for a 50-kg woman.

METHANOL

Little is known of placental transfer and fetal effects of methanol. Ethanol, by contrast, has been studied extensively because of its well-known association with the fetal alcohol syndrome. In all animal species investigated, ethanol diffuses rapidly across the placenta and is evenly distributed throughout fetal body water. Maternal and fetal blood levels equilibrate quickly following maternal administration of ethanol; however, fetal clearance is lower than adult clearance, apparently because of lower levels of alcohol dehydrogenase in fetal liver and a greater proportional extracellular volume. With high levels of ethanol administration, there exists some tendency toward metabolic acidosis in the fetus.

Whether these fetal effects of ethanol apply to methanol as well is problematical. A suggestion that placental transfer characteristics may be similar comes from findings that both alcohols exhibit a placental transfer index (the ratio of tracer in fetal blood to that injected) greater than 1 (23).

The projected methanol load resulting from the consumption of aspartame-sweetened products is difficult to compare with normal levels of methanol ingestion (24-26). However, Stegink et al. (25,26) have pointed out that the amount of methanol ingested from an aspartame-sweetened beverage might be less than that from ingesting an equivalent volume of fruit juice. They noted that 12 oz of aspartame-sweetened beverage provided approximately 20 mg of methanol. Francot and Geoffroy (27) list the methanol content of a 12-oz serving of fruit juice as 50 mg.

SUMMARY

Of the three principal components of aspartame, there seems to be little concern for safety in pregnancy with aspartate, since this amino acid does not traverse the primate placenta at anything less than enormous maternal loads. Phenylalanine, on the other hand, does cross the placenta, and the normal transplacental gradient seems to be maintained when maternal levels increase. However, it seems highly unlikely that customary intakes in normal pregnancy could raise fetal levels close to a neurotoxic range. Theoretically, abuse levels of intake in a PKU heterozygote might approach a zone where fetal toxicity would be possible. Virtually nothing is known of fetal toxicity of methanol.

REFERENCES

1. Lenke, R. R., and Levy, H. L. (1980). Maternal phenylketonuria and hyperphenylalaninemia. *N. Engl. J. Med.* 303, 1202-1208.
2. Lenke, R. R., and Levy, H. L. (1982). Maternal phenylketonuria—Results of dietary therapy. *Am. J. Obstet. Gynecol.* 142, 548-553.

3. Olney, J. W. (1974). Toxic effects of glutamate and related amino acids in the developing central nervous system. In *Heritable Disorders of Amino Acid Metabolism* (Nyhan, W. L., ed.), Wiley, New York, pp. 501-512.
4. Olney, J. W. (1975). L-Glutamic and L-aspartic acids—A question of hazard? *Food Cosmet. Toxicol.* 13, 595-596.
5. Lemons, J. A., Adcock, E. W., Jones, M. D., Jr., Naughton, M. A., Meschia, G., and Battaglia, F. C. (1976). Umbilical uptake of amino acids in the unstressed fetal lamb. *J. Clin. Invest.* 58, 1428-1434.
6. Smith, R. M., Jarrett, I. G., King, R. A., and Russell, G. R. (1977). Amino acid nutrition of the fetal lamb. *Biol. Neonat.* 31, 305-310.
7. Schneider, H., and Dancis, J. (1974). Amino acid transport in human placental slices. *Am. J. Obstet. Gynecol.* 120, 1092-1098.
8. Schneider, H., Mohlen, K.-H., Challier, J.-C., and Dancis, J. (1979). Transfer of glutamic acid across the human placenta perfused *in vitro*. *Br. J. Obstet. Gynaecol.* 86, 299-306.
9. Stegink, L. D., Pitkin, R. M., Reynolds, W. A., Brummel, M. C., and Filer, L. J., Jr. (1979). Placental transfer of aspartate and its metabolites in the primate. *Metabolism* 28, 669-676.
10. Stegink, L. D. (1984). Aspartame metabolism in human subjects: Acute dosing studies. In *Aspartame: Physiology and Biochemistry* (Stegink, L. D., and Filer, L. J., Jr., eds.), Marcel Dekker, New York, Chap. 26.
11. Stegink, L. D., Pitkin, R. M., Reynolds, W. A., Filer, L. J., Jr., Boaz, D. P., and Brummel, M. C. (1975). Placental transfer of glutamate and its metabolites in the primate. *Am. J. Obstet. Gynecol.* 122, 70-78.
12. Young, M. (1976). The accumulation of protein by the fetus. In *Fetal Physiology and Medicine* (Beard, R. W., and Nathanielsz, P. W., eds.), W. B. Saunders, London, pp. 59-79.
13. Pueschel, S. M., Boylan, J. M., Jackson, B. T., and Piaseck, G. J. (1982). A study of placental transport mechanisms in nonhuman primates using (^{14}C) phenylalanine. *Obstet. Gynecol.* 59, 182-188.
14. Ranney, R. E., Mares, S. E., Shroeder, R. E., Hutsell, T. C., and Raczialowski, F. M. (1975). The phenylalanine and tyrosine content of maternal and fetal body fluids from rabbits fed aspartame. *Toxicol. Appl. Pharmacol.* 32, 339-346.
15. Cockburn, F., Farqular, J. W., Forfar, J. O., Giles, M., and Robins, S. P. (1972). Maternal hyperphenylalaninemia in the normal and phenylketonuric mother and its influence on maternal and fetal fluid amino acid concentrations. *J. Obstet. Gynaecol. Br. Commonw.* 79, 698-707.
16. Kerr, G. R., Chamove, A. S., Harlow, H. F., and Waisman, H. A. (1968). Fetal PKU: The effect of maternal hyperphenylalaninemia during pregnancy in the rhesus monkey (*Macaca mulatta*). *Pediatrics* 42, 27-36.
17. Brass, C. A., Isaacs, C. E., McChuney, R., and Greengard, O. (1982). The effects of hyperphenylalaninemia on fetal development: A new animal model of maternal phenylketoneuria. *Pediatr. Res.* 16, 388-394.
18. Andersen, A. (1976). Maternal hyperphenylalaninemia: An experimental model in rats. *Ber. Psychobiol.* 9, 157-166.

19. Stegink, L. D., Filer, L. J., Jr., and Baker, G. L. (1977). Effect of aspartame and aspartate loading upon plasma and erythrocyte free amino acid levels in normal adult volunteers. *J. Nutr.* 107, 1837-1845.
20. Stegink, L. D., Filer, L. J., Jr., and Baker, G. L. (1981). Plasma and erythrocyte concentrations of free amino acids in adult humans administered abuse doses of aspartame. *J. Toxicol. Environ. Health* 7, 291-305.
21. Lines, D. R., and Waisman, H. A. (1971). Placental transport of phenylalanine in the rat: Maternal and fetal metabolism. *Proc. Soc. Exp. Biol. Med.* 136, 790-793.
22. Stegink, L. D., Filer, L. J., Jr., Baker, G. L., and McDonnell, J. E. (1979). Effect of aspartame loading upon plasma and erythrocyte levels of amino acids in phenylketonuric heterozygotes and normal adult subjects. *J. Nutr.* 109, 708-717.
23. Bissonette, J. M., Cronan, J. Z., Richards, L. L., and Wickham, W. K. (1979). Placental transfer of water and nonelectrolytes during a single circulatory passage. *Am. J. Physiol.* 236, 47-52.
24. Roak-Foltz, R., and Leveille, G. A. (1984). Projected aspartame intake: Daily ingestion of aspartic acid, phenylalanine, and methanol. In *Aspartame: Physiology and Biochemistry* (Stegink, L. D., and Filer, L. J., Jr., eds.), Marcel Dekker, New York, Chap. 9.
25. Stegink, L. D., Brummel, M. C., McMartin, K., Martin-Amat, G., Filer, L. J., Jr., Baker, G. L., and Tephly, T. R. (1981). Blood methanol concentrations in normal adult subjects administered abuse doses of aspartame. *J. Toxicol. Environ. Health* 7, 281-290.
26. Stegink, L. D., Brummel, M. C., Filer, L. J., Jr., and Baker, G. L. (1983). Blood methanol concentrations in 1-year-old infants administered graded doses of aspartame. *J. Nutr.* 113, 1600-1606.
27. Francot, P., and Geoffroy, P. (1956). Le méthanol dans les jus de fruits, les boissons fermentées, les alcools et spiritueux. *Rev. Ferment. Ind. Aliment.* 11, 279-287.

28

Aspartame Ingestion During Lactation

George L. Baker*
University of Iowa College of Medicine, Iowa City, Iowa

INTRODUCTION

Approximately 60% of infants born in the United States receive human milk as their primary source of food for the first few months of life (1). At age 6 months, 27% of infants are still receiving significant amounts of human milk. It is thus important to examine the effect of any potential food additive on human milk. Aspartame (L-aspartyl-L-phenylalanine methyl ester), a new dipeptide sweetener, has recently been approved as a food additive in various countries. Questions have been raised about the safety of aspartame because of the potential toxic effects of its component amino acids, aspartate and phenylalanine. It should be noted that both aspartate and phenylalanine are present naturally in human milk (2).

Our group has studied the effect of another amino acid used as a food additive, glutamate, on human milk (3). Six grams of monosodium L-glutamate (MSG) administered orally to lactating women elevated mean plasma glutamate from 4.4 μmol/dl to a mean peak level of 31 μmol/dl. Despite this sevenfold increase in plasma levels, milk glutamate levels were not significantly increased (4).

We have previously reported the effect of aspartame loading on plasma amino acid concentrations in normal adult subjects given aspartame at 34 mg/kg body weight (99th percentile of projected daily ingestion). Aspartame ingestion increased plasma and erythrocyte phenylalanine concentrations to values seen in the postprandial state. There was no effect upon plasma and erythrocyte aspartate levels (5).

*Present affiliation: Mead Johnson & Company, Evansville, Indiana

In this chapter we will review data previously reported on the effect of aspartame or lactose administered to lactating women (6). Because aspartame administration at 34 mg/kg body weight did not elevate plasma and erythrocyte aspartate or phenylalanine concentrations above normal postprandial values, a level of 50 mg/kg body weight was selected for these studies.

MATERIALS AND METHODS

Six healthy women with well-established lactation were studied. The length of lactation ranged from 42 to 159 days, with a mean (±SD) of 98 ± 39 days. The women ranged in age from 20 to 29 years with a mean (±SD) age of 26 ± 3 years. Their weights ranged from 45 to 63.6 kg, with a mean (±SD) of 55.2 ± 6.42 kg. The protocol for the study was reviewed and approved by the Committee on Research Involving Human Subjects of the University of Iowa. The study was fully explained to each subject and informed, written consent was obtained.

The subjects were administered aspartame or lactose (50 mg/kg body weight) dissolved in 300 ml of cold orange juice at 0800 in a randomized crossover design. Each segment of the study was separated by an interval of at least 2 weeks. Subjects were fasted for an additional 4 hr following administration of the test dose, after which they were allowed a normal diet. Milk samples were collected at 0, 1, 2, 3, 4, 8, 12, and 24 hr following dosing. Samples were obtained by manual expression from both breasts with a total of 10-15 ml collected. All women suckled their infant one time during the first 4 hr of the study.

Blood samples for amino acid analyses were obtained at 0, 0.25, 0.5, 0.75, 1, 1.5, 2, 3, and 4 hr through an indwelling catheter with a heparin lock placed in an arm vein.

RESULTS

No significant differences were noted in mean plasma levels of aspartate asparagine, glutamate, or glutamine following administration of aspartame or lactose (Fig. 1).

As expected, plasma phenylalanine levels increased significantly over base-line levels ($p \leqslant 0.001$), reaching peak values of 16.2 ± 4.9 μmol/dl after aspartame ingestion. Peak plasma phenylalanine levels observed at this dose were higher than values found after aspartame ingestion at 34 mg/kg body weight (5). Plasma tyrosine levels also increased significantly ($p \leqslant 0.001$) after aspartame loading, reflecting conversion of phenylalanine to tyrosine. Plasma phenylalanine levels were not affected by lactose loading (Fig. 2). Tyrosine levels decreased slightly ($p \leqslant 0.01$) after lactose loading compared with zero-time levels following the pattern reported after aspartate loading (5) (Fig. 2).

Figure 3 indicates that plasma levels of alanine and proline increased significantly over zero-time levels ($p \leqslant 0.02$) after both aspartame and lactose loads.

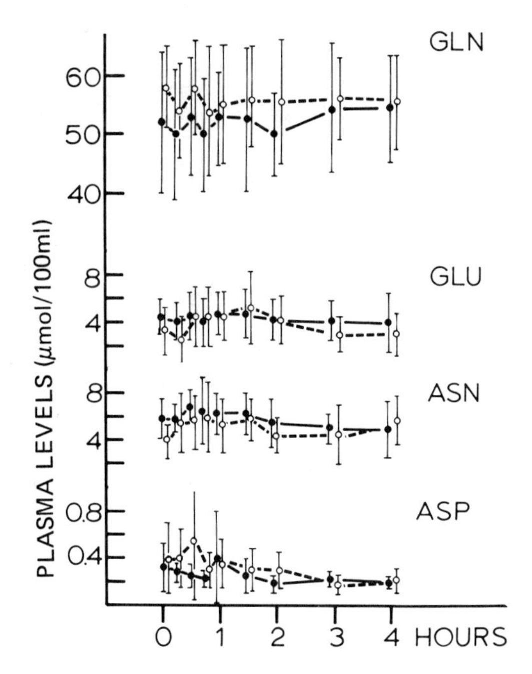

Figure 1 Mean (±SD) plasma glutamine (GLN), glutamate (GLU), asparagine (ASN), and aspartate (ASP) levels in six lactating women administered aspartame (○- - -○) or lactose (●——●) at 50 mg/kg body weight (From Ref. 6.)

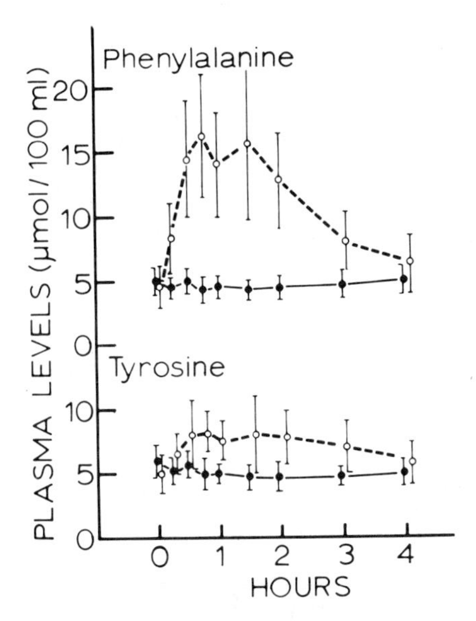

Figure 2 Mean (±SD) plasma phenylalanine and tyrosine levels in six lactating women administered aspartame (○- - -○) or lactose (●——●) at 50 mg/kg body weight. (From Ref. 6.)

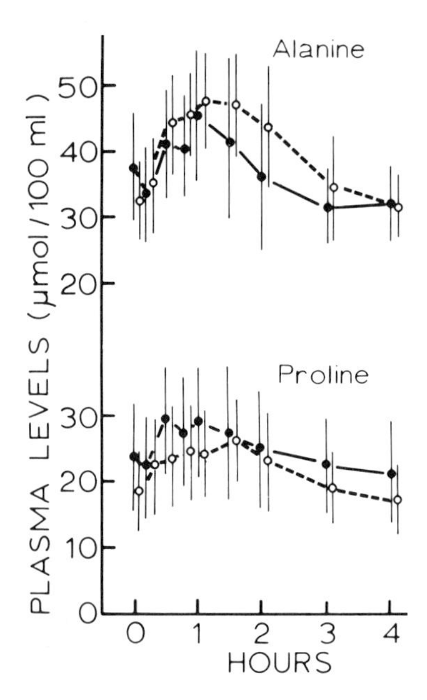

Figure 3 Mean (±SD) plasma alanine and proline levels in six lactating women administered aspartame (○- - -○) or lactose (●——●) at 50 mg/kg body weight. (From Ref. 6.)

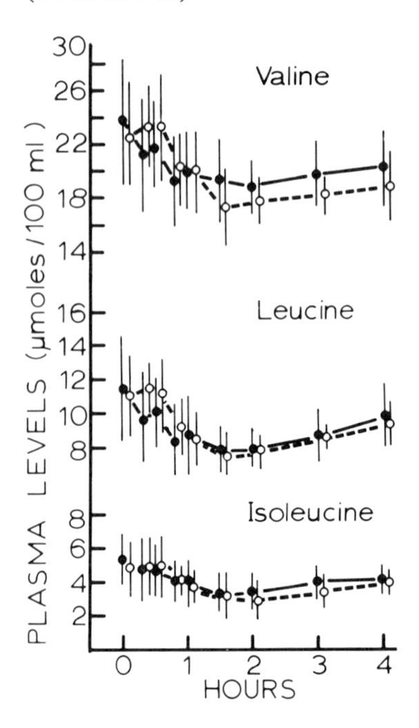

Figure 4 Mean (±SD) plasma valine, leucine, and isoleucine values in six lactating women administered aspartame (○- - -○) or lactose (●——●) at 50 mg/kg body weight. (From Ref. 6.)

Similar effects were seen after both aspartame (34 mg/kg body weight) and aspartate (13 mg/kg body weight) loading in studies which we reported earlier (5). This may be the result of either the orange juice carrier or the stress of blood sampling (7).

In both groups plasma levels of branched chain amino acids decreased significantly (Fig. 4: leucine, $p \leqslant 0.001$; isoleucine, $p \leqslant 0.01$; valine $p \leqslant 0.002$). Plasma levels of each of these amino acids returned to fasting values 3-4 hr later. This decrease presumably reflects insulin release secondary to the carbohydrate content of the orange juice (8). No differences were noted between the treatment groups for any other amino acids.

Several studies have suggested that under some circumstances amino acids are transported in the erythrocyte to a greater extent than plasma (9-14). Erythrocyte levels of free amino acids were measured following aspartame administration to assure that aspartate was not carried by the red cell. Erythrocyte glutamate, glutamine, aspartate, and asparagine levels were unchanged after either aspartame or lactose administration (Fig. 5). Erythrocyte phenylalanine but not tyrosine levels increased significantly ($p = 0.001$) after aspartame loading (Fig. 6). The increase in erythrocyte phenylalanine levels was expected in view of the increased plasma levels noted. Erythrocyte phenylalanine and tyrosine levels decreased significantly after lactose loading ($p \leqslant 0.01$).

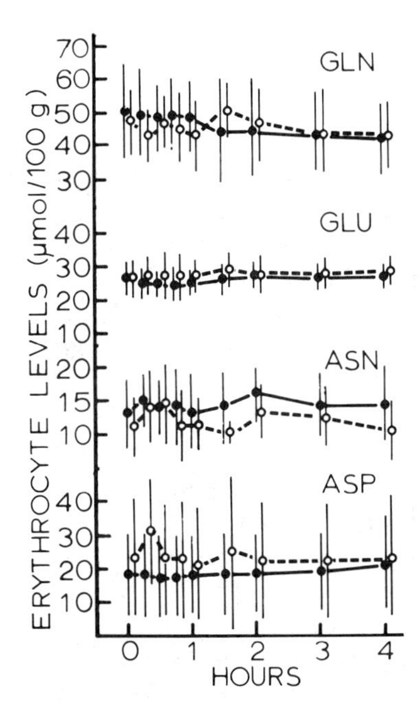

Figure 5 Mean (±SD) erythrocyte glutamine (GLN), glutamate (GLU), asparagine (ASN), and aspartate (ASP) levels in six lactating women administered aspartame (○- - -○) or lactose (●——●) at 50 mg/kg body weight. (From Ref. 6.)

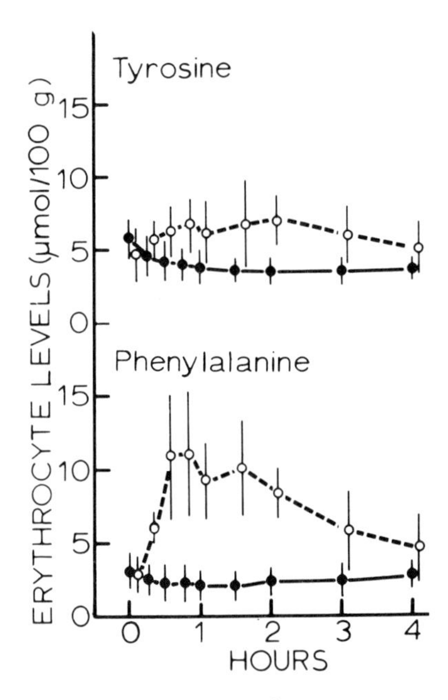

Figure 6 Mean (±SD) erythrocyte tyrosine and phenylalanine levels in six lactating women administered aspartame (○- - -○) or lactose (●——●) at 50 mg/kg body weight. (From Ref. 6.)

Following aspartame loading human milk glutamate levels increased from 1.09 to 1.20 μmol/100 ml, while milk aspartate levels increased from 2.3 to 4.8 μmol/100 ml (Fig. 7). These changes were not statistically significant using the paired t test. When the data were subjected to curve analysis using orthgonal polynomials, a small but statistically significant difference in milk aspartate time effect scores ($p \leqslant 0.04$, linear only) was recorded over the immediate 4-hr postabsorptive period. The mean cubic and quadratic time effect scores using orthogonal polynomials did not differ significantly. When the data on milk for the entire 24-hr study period were analyzed by analysis of variance using orthogonal polynomials, no statistically significant differences were noted. No significant differences in milk glutamate levels were noted using any statistical analysis.

The differences in milk phenylalanine and tyrosine levels between test substances were not significant using the paired t-test analysis (Fig. 8). However, when data were subjected to curve analysis using orthogonal polynomials, milk phenylalanine levels were significantly higher during both the immediate 4-hr postdosing period ($p = 0.034$, mean only) and during the entire 24-hr period ($p = 0.02$, mean only). However, linear, quadratic, and cubic time effect scores showed no significant effects. Milk tyrosine levels increased significantly ($p = 0.018$) in the immediate 4-hr period after aspartame loading compared to lactose

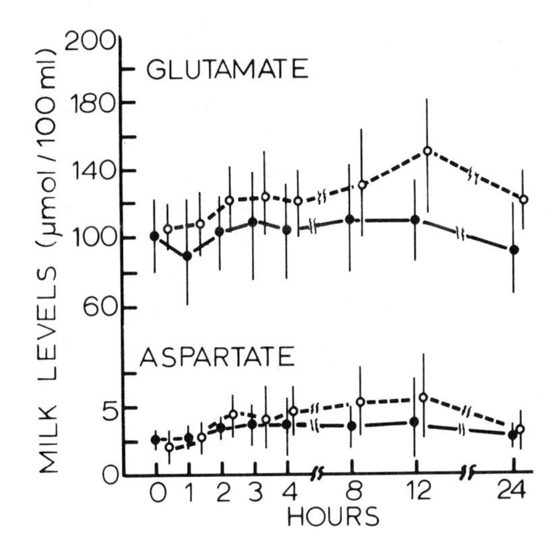

Figure 7 Mean (±SD) milk free glutamate and aspartate levels in 6 lactating women administered aspartame (○- - -○) or lactose (●——●) at 50 mg/kg body weight. (From Ref. 6.)

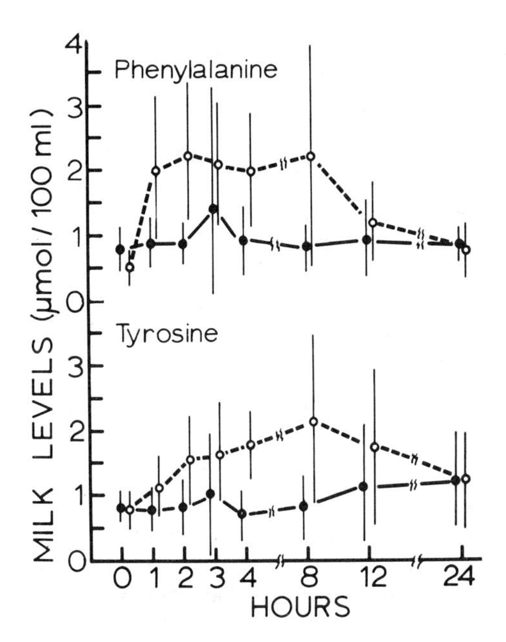

Figure 8 Mean (±SD) milk free phenylalanine and tyrosine levels in six lactating women administered aspartame (○- - -○) or lactose (●——●) at 50 mg/kg body weight. (From Ref. 6.)

(orthogonal polynomials, linear only), but did not differ significantly when data for the entire 24-hr period were analyzed.

Glutamate, aspartate, phenylalanine, and tyrosine levels in milk samples collected postprandially (8 and 12 hr) were similar to those noted 4 hr after aspartame loading. The 8-hr milk samples were obtained approximately 3 hr after the subjects' noon meal, while 12-hr samples were obtained approximately 2 hr after the evening meal. No free aspartame or aspartylphenylalanine was detected in the plasma, milk, or erythrocytes of the subjects at any time. The limit of detection was 0.5 μmol/100 ml. These data are consistent with animal studies indicating hydrolysis of aspartame either in the intestinal lumen (15,16) or intestinal mucosal cells after uptake (17) with subsequent release of aspartame components (aspartate, phenylalanine, and methanol) to the portal blood.

DISCUSSION

The amount of aspartame given in this study is considerably greater than the total daily projected intake of this compound under any but abusive conditions. We have previously documented (5) that daily intake of aspartame would approximate 7.5-8.5 mg/kg body weight per day if all usual sucrose intake (17% of energy) was replaced by aspartame. An aspartame intake of 34 mg/kg body weight represents the 99th percentile of projected daily intake. The aspartame load given in the present study is approximately 1.5 times this higher level.

The results of the present study demonstrate rapid metabolism of the aspartate portion of aspartame. Neither plasma nor erythrocyte aspartate levels increased after aspartame loading. There was no evidence of accumulation of amino acids derived from aspartate (asparagine, glutamine, glutamate).

Plasma levels of phenylalanine and tyrosine increased after the administration of aspartame. Levels of these amino acids remained constant or decreased slightly after lactose loading. In subjects given aspartame at 50 mg/kg body weight, mean peak plasma phenylalanine values and the area under the plasma phenylalanine concentration-time curve were greater than those noted in subjects given 34 mg/kg body weight (Fig. 9) (5). These plasma phenylalanine levels were only slightly above those found postprandially (12 ± 2 μmol/100 ml) in young infants or adults fed formula or protein meals (18,19). The phenylalanine levels observed are far below peak levels routinely noted (78 ± 45 μmol/100 ml) in subjects tested for the heterozygous condition of phenylketonuria (20). In children with classic phenylketonuria elevated plasma phenylalanine levels are associated with mental retardation. However, in this condition plasma phenylalanine levels are much higher, ranging between 120 and 300 μmol/100 ml (20-50 mg/100 ml) (21-23).

Aspartame administration at 50 ml/kg body weight has little effect upon milk aspartate levels when compared to values noted after lactose loading. No significant difference was noted in the aspartate time effect scores noted by the linear orthogonal polynomial method for the entire 24-hr period. There was a small

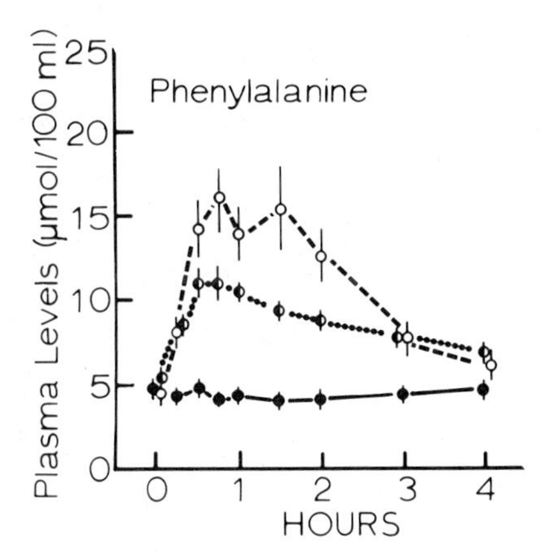

Figure 9 Mean (±SEM) plasma phenylalanine levels in adult subjects administered lactose (●——●), aspartame at 34 mg/kg body weight (◑· · ·◑) and aspartame at 50 mg/kg body weight (○- - -○). (From Ref. 6.)

statistically significant increase in aspartate time effect scores noted for the immediate 4-hr postloading period. Aspartame loading also had a small but significant effect on milk phenylalanine and tyrosine levels when the data were analyzed by the orthogonal polynomial method. The biological importance of these small differences is minimal.

In these studies the subjects fasted until after the 4-hr milk sample was obtained. After that time they were allowed their usual meal pattern. The levels of phenylalanine and tyrosine after aspartame loading were no greater than the values noted in postprandial milk samples collected after ingestion of meals. In addition, the phenylalanine and tyrosine levels noted in the milk samples obtained after aspartame loading were similar to those observed, but not published in a previous study of lactating women (3).

The slightly increased quantities of phenylalanine and tyrosine present in human milk after aspartame loading would have little impact on the total level of these amino acids ingested by the infant. This can be demonstrated by calculation of the daily intake of selected amino acids by a 3.5-kg breast-fed infant (Table 1). Such infants fed ad libitum have a mean milk intake of 164 ml/kg per day (24). This volume provides approximately 110 kcal/kg body weight per day. The levels of various free amino acids in pooled human milk have been reported by Stegink (2), and the total (free plus protein-bound) glutamate, aspartate, phenylalanine, and tyrosine have been reported by Macy et al. (25). The total glutamate value (230 mg/100 ml) reported by Macy et al. (25) includes both glutamate and glutamine and the total aspartate value (116 mg/100 ml) includes both aspartate and asparagine, since the values were obtained after acid hydrolysis. When these data are corrected for the quantities of free glutamate (20.7 mg/100 ml), glutamine (6.5 mg/100 ml), asparagine (1.5 mg/100 ml), and aspartate (0.5 mg/100 ml) in human milk (2), the total protein-bound glutamate plus glutamine

Table 1 Free and Total Aspartate, Glutamate, Phenylalanine, and Tyrosine Levels Ingested by a Normal 3.5-kg Breast-Fed Infant

Amino acids	Content of human milk (mg/100 ml)	Intake per day[a] (mg/kg body weight)
Free[b]		
Aspartate	0.53	0.87 ± 0.13
Glutamate	20.7	34.0 ± 5.0
Phenylalanine	0.21	0.34 ± 0.05
Tyrosine	0.26	0.43 ± 0.06
Protein bound + free		
Aspartate[c]	63	103 ± 15.1
Glutamate[c]	139	228 ± 33.3
Phenylalanine[d]	48	79 ± 11.5
Tyrosine[d]	61	100 ± 14.6

[a]Assuming a mean intake of 164 ± 24 ml/kg body weight (mean ± SD), with human milk having an energy density of 67 kcal/100 ml. Data are shown as the mean ± SD.
[b]Data from Stegink (2).
[c]Levels for glutamate and aspartate shown were obtained from correcting the data of Macy et al. (25) for free glutamine and asparagine, as well as for the approximate glutamine and asparagine content of the milk protein, using the factors published for typical proteins by Jukes et al. (26).
[d]Data from Macy et al. (25).

content of human milk is 203 mg/100 ml, and the total protein-bound aspartate plus asparagine content is 114 mg/100 ml.

The approximate quantities of glutamine and asparagine which contribute to the "total protein-bound glutamate" and the "total protein-bound aspartate" can be calculated using the data of Jukes et al. (26). These data indicate that 55% of total aspartate released from a typical protein upon acid hydrolysis arises from aspartate, while 45% arises from asparagine. Similarly, 58% of total glutamate resulting from acid hydrolysis of a protein arises from glutamate, while 42% comes from glutamine. Using these correction factors, the protein-bound glutamate content of human milk is 118 mg/100 ml, and the protein-bound aspartate content is 63 mg/100 ml.

From these data we can calculate the quantity of free and protein-bound amino acids ingested daily by the breast-fed infant (Table 1). These calculations in turn allow us to evaluate the potential effect of the small changes noted in human milk composition after aspartame loading. If we assume that the lactating mother ingests sufficient aspartame to increase her milk phenylalanine levels during the entire 24-hr period to the extent reported here for the 4-hr sampling period, milk phenylalanine levels would increase by about 1.8 μmol/100 ml. This increase would provide the breast-fed infant with an additional 3.1 μmol of phenylalanine per kilogram body weight per day (0.51 mg/kg body weight per day). This may be compared to the infant's normal intake of this essential amino acid of 79 mg/kg body weight per day (Table 1). Thus even high doses of aspartame would not meaningfully affect the infant's phenylalanine intake.

Similar calculations can be made for aspartate. The small increase in aspartate levels noted in human milk after aspartame loading, although not statistically significant, result in the ingestion of an additional 4.6 μmol of aspartate per kilogram body weight per day (0.77 mg/kg body weight per day). This is a trace quantity compared to the approximately 105 mg/kg body weight per day of ingested aspartate.

Similar calculations for tyrosine and glutamate indicate that tyrosine intake might increase by 1.71 μmol/kg body weight per day (0.31), and glutamate intake by 11.0 μmol/kg body weight per day (1.6 mg). Neither increase would meaningfully affect the total daily intake of those respective amino acids.

In conclusion, aspartate loading of lactating women at a level of 50 mg/kg body weight resulted in (a) no change in blood aspartate levels, (b) increased plasma phenylalanine levels to high postprandial levels, and (c) a small but statistically significant increase in human milk aspartate, phenylalanine, and tyrosine levels, increasing their levels from the fasting range to the postprandial range.

REFERENCES

1. Martinez, G. A., and Dodd, D. D. (1983). 1981 Milk patterns in the United States during the first 12 months of life. *Pediatrics* 71, 166-170.
2. Stegink, L. D. (1976). Absorption, utilization and safety of aspartic acid. *J. Toxicol Environ. Health* 2, 215-242.
3. Stegink, L. D., Filer, L. J., Jr., and Baker, G. L. (1972). Monosodium L-glutamate: Effect on plasma and breast milk amino acid levels in lactating women. *Proc. Soc. Exp. Biol. Med.* 140, 836-841.
4. Baker, G. L., Filer, L. J., Jr., and Stegink, L. D. (1979). Factors influencing dicarboxylic amino acid content of human milk. In *Glutamic Acid: Advances in Biochemistry and Physiology* (Filer, L. J. Jr., Garattini, S., Reynolds, W. A., Kare, M. R., and Wurtman, W. J., eds.), Raven Press, New York, pp. 111-123.
5. Stegink, L. D., Filer, L. J., Jr., and Baker, G. L. (1977). Effect of aspartame or aspartate loading upon plasma and erythrocyte free amino acid levels in normal adult volunteers. *J. Nutr.* 107, 1837-1845.
6. Stegink, L. D., Filer, L. J., Jr., and Baker, G. L. (1979). Plasma erythrocyte and human milk levels of free amino acids in lactating women administered aspartame or lactose. *J. Nutr.* 109, 2173-2181.
7. Heath, D. F., George, D. R., and Rose, J. G. (1971). The effects of the stress caused by experimental procedures on alanine, aspartate, glutamate, glutamine in rat liver. *Biochem. J.* 125, 765-771.

8. De Montis, M. G., Olianas, M. C., Haber, B., and Tagliamonte, A. (1978). Increase in large neutral amino acid transport into brain by insulin. *J. Neurochem.* 30, 121-124.
9. Elwyn, D. H. (1966). Distribution of amino acids between plasma and red blood cells in the dog. *Fed. Proc.* 25, 854-861.
10. Felig, P., Wahren, J., and Räf, L. (1973). Evidence of interorgan amino acid transport by blood cells in humans. *Proc. Nat. Acad. Sci. U.S.A.* 70, 1775-1779.
11. Aoki, T. T., Brennen, M. F., Müller, W. A., Moore, F. D., and Cahill, G. F., Jr. (1972). Effect of insulin on muscle glutamate uptake. Whole blood versus plasma glutamate analysis. *J. Clin. Invest.* 51, 2889-2894.
12. Elwyn, D. H., Launder, W. J., Parikh, H. C., and Wise, E. M., Jr. (1972). Roles of plasma and erythrocytes in interorgan transport of amino acids in dog. *Am. J. Physiol.* 222, 1333-1342.
13. Young, J. D., and Ellory, J. C. (1977). Red cell amino acid transport. In *Membrane Transport in Red Cells* (Ellory, J. C., and Lew, V. L., eds.), Academic Press, New York, pp. 301-326.
14. Aoki, T. T., Muller, W. A., Brennan, M. F., and Cahill, G. F., Jr. (1973). Blood cell and plasma amino acid levels across forearm muscle during a protein meal. *Diabetes* 22, 768-775.
15. Oppermann, J. A., Muldoon, E., and Ranney, R. E. (1973). Metabolism of aspartame in monkeys. *J. Nutr.* 103, 1454-1459.
16. Ranney, R. E., Oppermann, J. A., Muldoon, E., and McMahon, F. G. (1976). Comparative metabolism of aspartame in experimental animals and humans. *J. Toxicol. Environ. Health* 2, 441-451.
17. Addison, J. M., Burston, D., Dalrymple, J. A., Matthews, D. M., Payne, J. W., Sleisenger, M. H., and Wilkinson, S. (1975). A common mechanism for transport of di- and tri-peptides by hamster jejunum in vitro. *Clin. Sci. Mol. Med.* 49, 313-322.
18. Stegink, L. D., Schmitt, J. L., Meyer, P. D., and Kain, P. H. (1971). Effects of diets fortified with DL-methionine on urinary and plasma methionine levels in young infants. *J. Pediatr.* 79, 648-659.
19. Vaughan, D. A., Womack, M., and McClain, P. E. (1977). Plasma free amino acid levels in human subjects after meals containing lactalbumin, heated lactallbumin or no protein. *Am. J. Clin. Nutr.* 30, 1709-1712.
20. Tocci, P. M., and Beber, B. (1973). Anomalous phenylalanine loading response in relation to cleft lip and cleft palate. *Pediatrics* 52, 109-113.
21. Blaskovics, M. E. (1974). Phenylketonuria and other phenylalaninemias. *Clin. Endocrinol. Metabol.* 3, 87-105.
22. Koch, R., Blaskovics, M. E., Wenz, E., Fishler, K., and Schaeffler, G. (1974). Phenylalaninemia and phenylketonuria. In *Heritable Disorders of Amino Acid Metabolism* (Nyhan, W. L., ed.), Wiley, New York, pp. 109-140.
23. Blaskovics, M. E., Schaeffler, G. E., and Hack, S. (1974). Phenylalaninemia: Differential diagnosis. *Arch. Dis. Child.* 49, 835-843.
24. Fomon, S. J. (1974). Voluntary food intake and its regulation (Table 2-3). In *Infant Nutrition*, 2nd ed, W. B. Saunders, Philadelphia, pp. 20-33.

25. Macy, K. G., Kelly, H. J., and Sloan, R. E. (1953). *The Composition of Milks*, A compilation of the comparative composition and properties of human, cow and goat milk, colostrum and transitional milk, Publication 254, National Academy of Sciences–National Research Council, Washington, D.C., p. 40.
26. Jukes, T. H., Holmquist, R., and Moise, H. (1975). Amino acid composition of proteins: Selection against the genetic code. *Science* 189, 50-51.

29

Aspartame Ingestion by Human Infants

L. J. Filer, Jr., George L. Baker, and Lewis D. Stegink*
University of Iowa College of Medicine, Iowa City, Iowa

INTRODUCTION

The safety of aspartame has been questioned on the basis of its constituent amino acids, aspartate and phenylalanine (1-5). Others have expressed concern that the metabolism of aspartame releases methanol. High blood levels of each of these compounds are associated with toxicity. Administration of high doses of aspartate to neonatal mice and rats results in elevated plasma aspartate concentrations and hypothalamic neuronal necrosis (6-14). Infant mice given large doses of aspartame develop hypothalamic neuronal necrosis (15,16), presumably from elevated blood aspartate concentrations. However, administration of aspartame at 2 g/kg body weight to infant nonhuman primates with and without added monosodium L-glutamate (1 g/kg) does not result in neuronal necrosis, even though plasma aspartate and glutamate levels are elevated (15,17). Grossly elevated blood phenylalanine concentrations, such as those found in children with classic phenylketonuria, are associated with mental retardation (18-20). This led to the suggestion that aspartame ingestion by children might increase dietary intake of phenylalanine enough to produce elevated plasma phenylalanine concentrations. The ingestion of large quantities of methanol results in elevated blood methanol and formate concentrations and leads to a variety of adverse effects, including metabolic acidosis and blindness (21-24).

Infant mice and monkeys metabolize glutamate, a dicarboxylic amino acid similar to aspartate, more slowly than adult animals (25). On the basis of these

*Present affiliation: Mead Johnson & Company, Evansville, Indiana

studies, Olney (3) suggested that human infants metabolize glutamate and aspartate less rapidly than adults. He predicted that a fixed aspartame dose would produce higher plasma aspartate concentrations in infants as compared to adults and associated this hypothetically increased plasma dicarboxylic amino acid concentration with increased risk. Similarly, Turner (5) suggested that aspartame's phenylalanine content might greatly increase plasma phenylalanine concentrations with potentially deleterious effects.

Because of these concerns, we investigated the ability of 1-year-old infants to metabolize aspartame (26,27). Previously we had studied the effects of aspartame loading at 34, 50, 100, 150, and 200 mg/kg body weight upon plasma amino acid concentrations in adult subjects (28-30). In a follow-up study we investigated the effect of aspartame loading at 34, 50, and 100 mg/kg body weight upon the plasma amino acid concentrations and blood methanol concentrations in 1-year-old infants to determine whether infants metabolize aspartame as well as adults.

MATERIALS AND METHODS

A total of 24 infants aged 8-14 months were studied following an overnight fast. The proposed study was fully explained to at least one of the child's parents, and informed, written consent was obtained. The protocol of the study was reviewed and approved by the Committee on Research Involving Human Subjects of the University of Iowa.

A physical examination, urinalysis, and complete blood count were carried out on each infant prior to entry into the study. Values for all infants were within normal limits for the laboratory.

This investigation was carried out in three stages. In the first study 10 infants were administered 34 mg/kg body weight aspartame dissolved in 180 ml of cherry-flavored beverage mix. In the second study six infants were administered 50 mg/kg body weight aspartame dissolved in the flavored beverage mix. In the third study eight infants were administered 100 mg/kg body weight aspartame dissolved in the flavored beverage mix. The first study was completed and results evaluated prior to starting the second study. Similarly, the second study was completed and results evaluated prior to starting the final study. The details of these studies have been published elsewhere (26,27).

A carbohydrate-free cherry-flavored beverage was used as the vehicle for aspartame administration. The beverage was prepared from dry powder according to package instructions. Aspartame was dissolved in 180 ml (6 oz) of beverage and fed to the fasted infant from a bottle or cup.

RESULTS

Amino Acid Studies

Aspartame Ingestion at 34 mg/kg Body Weight

Plasma aspartate and phenylalanine concentrations in infants administered 34 mg/kg body weight aspartame are shown in Figure 1. These data are compared

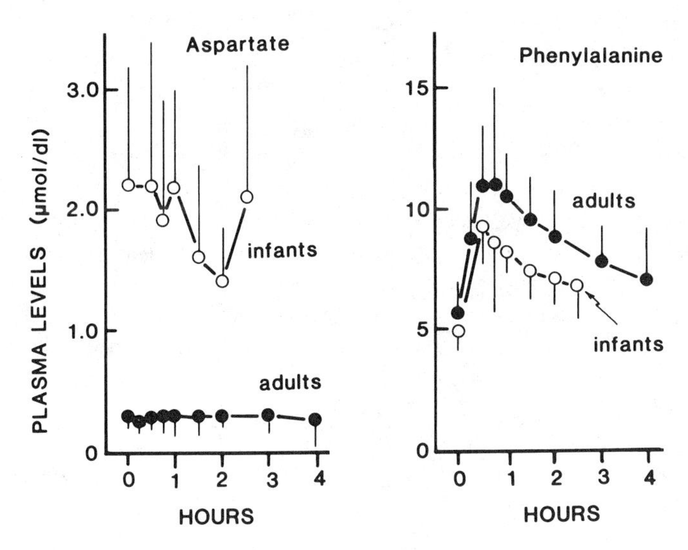

Figure 1 Mean (±SD) plasma aspartate and phenylalanine concentrations in normal adults (●) and 1-year-old infants (○) administered aspartame at 34 mg/kg body weight. Data from Refs. 26 and 28.

to concentrations observed in 12 adult subjects administered an equivalent amount of aspartame in orange juice (28). No significant change in plasma aspartate concentration was observed after dosing in either infants or adults. Similarly, the areas under the plasma aspartate concentration-time curve (AUC) did not differ significantly between infants and adults. Base-line levels of plasma aspartate were significantly higher in infants than adults. While we do not have an explanation for these differences, we have observed them in other studies (31).

In infants, mean (±SD) plasma phenylalanine concentration increased significantly from a base-line value of 4.92 ± 0.77 μmol/100 ml to 9.37 ± 1.44 μmol/100 ml 30 min after dosing. Values remained significantly elevated over base line until 120 min. Similarly, aspartame loading in adults increased mean (±SD) plasma phenylalanine concentration from a base line of 5.66 ± 1.21 μmol/100 ml to 11.1 ± 2.49 μmol/100 ml 30 min after loading. The increase over base line was similar in infants and adults (+4.45 and +5.44 μmol/100 ml, respectively). The plasma phenylalanine AUC values did not differ significantly between infants (398 ± 80.5 units) and adults (437 ± 106 units) 2.5 hr postdosing.

Since certain amino acids are transported in the erythrocyte to a greater extent than in plasma (32-35), erythrocyte concentrations of free amino acids were measured. No significant changes in erythrocyte aspartate or glutamate levels were noted in either age group. In both groups changes in erythrocyte phenylalanine concentration paralleled those occurring in plasma after aspartame loading.

Aspartame Ingestion at 50 mg/kg Body Weight

Plasma aspartate and phenylalanine concentrations in infants administered aspartame at 50 mg/kg body weight are shown in Figure 2. These data are compared to data previously reported for six lactating women ingesting an equivalent amount of aspartame in orange juice (29). No significant change in plasma aspartate concentration was observed after dosing in either infants or adults. Baseline levels for plasma aspartate were significantly higher in infants than adults.

Plasma phenylalanine concentration in infants increased significantly over base line (5.67 ± 0.53 μmol/100 ml) to a mean peak value of 11.62 ± 4.44 μmol/100 ml 60 min after aspartame dosing. Values remained significantly elevated over base-line levels at 150 min. In adults, aspartame loading increased plasma phenylalanine concentration from a mean (±SD) base-line value of 4.61 ± 1.72 μmol/100 ml to a mean peak value of 16.21 ± 4.86 μmol/100 ml 45 min after loading. The increase over base line was approximately twice as large in adults (+11.6 μmol/100 ml) as in infants (+6.14 μmol/100 ml). Similarly, the plasma phenylalanine AUC value was significantly smaller ($p = 0.01$) in infants (389 ± 225 units) than in adults (1065 ± 315 units) over a 2.5-hr postdosing time.

No significant changes in erythrocyte glutamate or aspartate concentrations were noted in either group. Erythrocyte phenylalanine and tyrosine concentrations showed a response similar to that noted in plasma after aspartame loading in both subject groups.

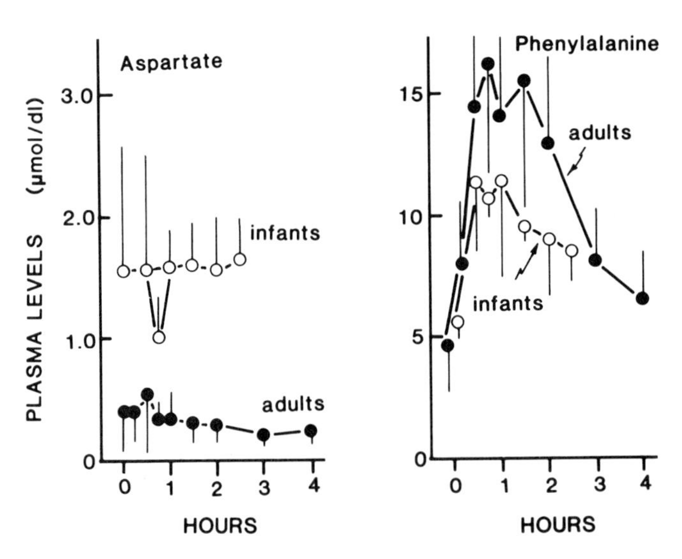

Figure 2 Mean (±SD) plasma aspartate and phenylalanine concentrations in adults (●) and 1-year-old infants (○) administered aspartame at 50 mg/kg body weight. Data from Refs. 26 and 29.

Aspartame Ingestion at 100 mg/kg Body Weight

The data obtained following aspartame administration at low doses indicated that infants metabolized aspartame as well as adults. Accordingly, we proceeded to the final stage of the study.

Plasma aspartate and phenylalanine concentrations in infants administered aspartame at 100 mg/kg body weight are shown in Figure 3 compared with values noted in six normal adults administered an equivalent dose in orange juice (30). Plasma aspartate concentrations were increased significantly over values observed at zero time in both infants and adults. However, the increase was small in each case, and maximal levels observed were below those noted postprandially (2.2 ± 2.1 μmol/100 ml) in formula-fed infants (36,37). In each case plasma aspartate values rapidly returned to base line. These data indicate rapid metabolism of the aspartate present in aspartame both by infant and adult subjects. As noted in studies in lower doses, plasma aspartate levels were higher for infants than adults.

In infants, plasma phenylalanine levels increased from a mean (±SD) base-line value of 4.59 ± 0.77 μmol/100 ml to a mean peak value of 22.3 ± 11.5 μmol/100 ml at 30 min. Aspartame loading in adults increased plasma phenylalanine concentrations from a mean (±SD) base-line value of 5.40 ± 1.05 μmol/100 ml to a mean peak value of 20.2 ± 6.77 μmol/100 ml 45 min after loading. The mean increase over base-line was approximately equal in infants (+16.7 μmol/100 ml) and adults (+14.8 μmol/100 ml). Similarly, the mean (±SD) plasma phenylalanine AUC value did not differ significantly between infants (1524 ± 741 units) and adults (1578 ± 339 units).

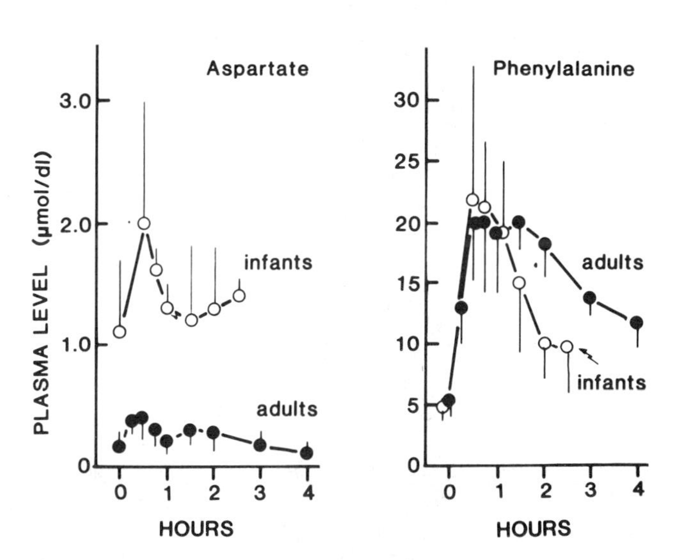

Figure 3 Mean (±SD) plasma aspartate and phenylalanine concentrations in normal adults (●) and 1-year-old infants (○) administered aspartame at 100 mg/kg body weight. Data from Refs. 26 and 30.

Erythrocyte aspartate and phenylalanine concentrations are shown in Figure 4. No significant changes in erythrocyte aspartate concentration over baseline were noted. Erythrocyte phenylalanine concentration showed a similar but reduced rise to that noted in plasma after aspartame loading in both groups.

Blood Methanol Response

Blood methanol concentrations were at the limits of detection in the 10 infants administered aspartame at 34 mg/kg body weight. Analysis of blood for low concentrations of methanol is complicated by the presence of small quantities of material that elutes at the methanol position. This material is also present in baseline samples. This made it difficult to detect low concentrations of methanol. The "apparent" blood methanol concentrations in the infants, corrected for this interfering substance, were below the limits of detection for methanol (0.35 mg/dl). The infant blood methanol response was similar to that noted in 12 normal adults administered an equivalent dose of aspartame.

Similar problems with methanol analyses were noted in infants and adults administered aspartame at 50 mg/kg body weight. In some subjects, at some time points, the blood methanol level did not rise to detectable levels, while in other subjects at other times blood methanol was detectable. In infants the mean (±SEM) peak blood methanol level was 0.30 ± 0.10 mg/dl 30-90 min after loading. In normal adults mean peak blood methanol level was 0.34 ± 0.12 mg/dl.

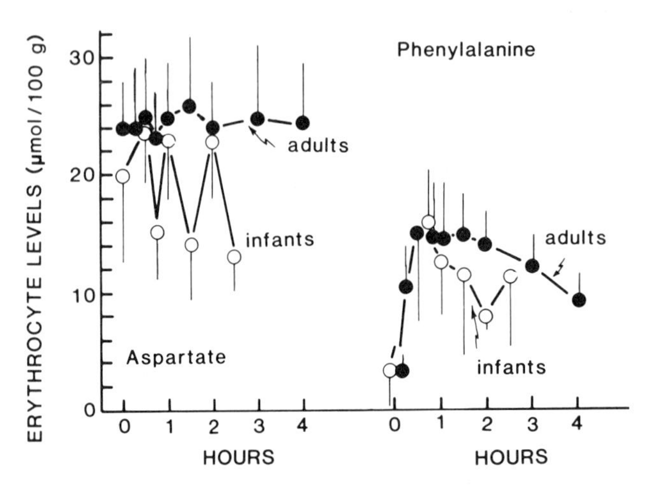

Figure 4 Mean (±SD) erythrocyte aspartate and phenylalanine concentrations in 6 normal adults (●—●) and eight 1-year-old infants (○—○) administered aspartame at 100 mg/kg body weight. Variation in infant erythrocyte aspartate concentration at alternative time points reflects the division of infants into two groups for blood sampling. One group was sampled at 0, 30, 60, and 120 min, the other at 0, 45, 90, and 150 min. No significant change in erythrocyte aspartate concentration occurred relative to base-line levels in either group, base-line erythrocyte aspartate levels differed between groups. This leads to the up-and-down appearance of the graph. From Ref. 26.

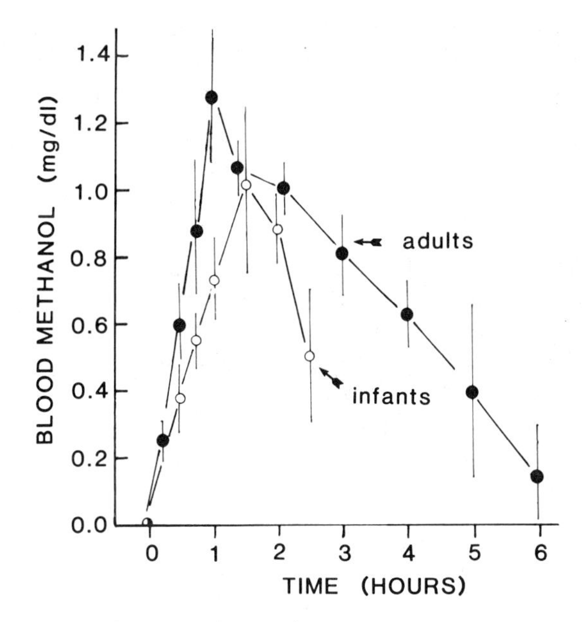

Figure 5 Mean (± SEM) blood methanol concentrations in normal adults (●) and 1-year-old infants (○) administered aspartame at 100 mg/kg body weight. From Ref. 27.

Blood methanol levels in the eight infants administered aspartame at 100 mg/kg body weight are shown in Figure 5. Blood methanol concentrations increased from base-line values at zero time to a mean (±SEM) peak value of 1.02 ± 0.28 mg/dl 90 min postdosing. This response was similar to that observed in six normal adults administered aspartame at 100 mg/kg body weight (38).

These data demonstrate that the infant metabolizes aspartame as well as a normal adult at these levels of ingestion.

DISCUSSION

In earlier studies (25) we demonstrated that infant mice and monkeys metabolize glutamate less rapidly than adults, particularly when very large doses are given. Olney (3) extended these data to suggest that the human infant metabolizes glutamate and aspartate less rapidly than an adult at levels likely to be provided by diets containing added monosodium L-glutamate and aspartame. He associated this hypothetical decrease in metabolic capacity with potential for increased plasma glutamate and/or aspartate concentrations and neuronal damage if compounds containing these amino acids are ingested.

Despite Olney's suggestion, we had little reason to expect that infants would metabolize the aspartate present in aspartame poorly at expected use levels of the product. Available data indicate adequate metabolism of ingested or infused aspartate by 1-year-old infants, term infants, and low birth weight infants at levels of intake comparable to those present in foods or parenteral solutions. These data are summarized in the following paragraphs.

One-Year-Old Infants

Although we had not tested aspartate metabolism in infants per se, we had compared protein digestion and metabolism in 6 normal adults, 16 normal 1-year-olds, and 8 normal 2-year-olds fed a milk-egg custard meal providing 1 g of protein per kilogram body weight (31,39). Infants fed custard consumed 66 mg of phenylalanine and 80 mg of aspartate per kilogram body weight. This intake compares with the 50 mg/kg body weight phenylalanine and 40 mg/kg body weight aspartate provided by a 100-mg/kg body weight dose of aspartame. The results of the custard study indicated that infants metabolized peptide-bound amino acids, including aspartate and phenylalanine, as well as adults (37,39).

Orally Fed Term and Low Birth Weight Infants

Postprandial plasma amino acid levels were measured in term and low birth weight infants fed either conventional formulas with little free glutamate and aspartate, or a casein hydrolysate formula containing large quantities of these amino acids (31,37,40,41). Plasma glutamate and aspartate levels were within normal ranges for both groups, indicating adequate handling of free glutamate and aspartate at these levels of intake.

Parenterally Fed Term and Premature Infants

Stegink and Baker (36) measured plasma amino acid levels in infants fed totally by vein using casein or fibrin hydrolysate-based preparations. The casein hydrolysate provided 1.2 mmol glutamate and 0.3 mmol aspartate per kilogram per day, while the fibrin hydrolysate provided 0.2 mmol glutamate and 0.4 mmol aspartate per kilogram per day. Plasma glutamate and aspartate levels in these infants were normal. Filer et al. (31) reported plasma amino acid levels in two low birth weight infants (1.38 and 1.51 kg body weight) administered a casein hydrolysate-based parenteral solution. These infants received 1.1-1.4 mmol glutamate and 0.3-0.4 mmol aspartate per kilogram body weight per day. Plasma glutamate and aspartate levels were normal in each case, indicating adequate metabolism of these amino acids. Recently Bell et al. (42) infused young infants with an amino acid-based parenteral regimen providing 1.5 mmol/kg body weight glutamate and 1.0 mmol/kg body weight aspartate per day. Normal plasma glutamate and aspartate concentrations were observed.

These summarized data suggest little risk to normal infants fed moderate quantities of glutamate or aspartate. Since the studies of parenterally and enterally fed infants outlined above did not support Olney's hypothesis at the dose levels studied, Olney's theory (3) was directly tested at aspartame concentrations likely to be ingested by infants. An aspartame-sweetened beverage mix sufficient to prepare 2 qt of beverage could contain 1 g of aspartame; thus, 8 oz of this beverage would contain approximately 125 mg. A 10-kg 1-year-old would ingest 12.5 aspartame per kilogram body weight from such a serving. The possibility exists that a curious 1-year-old might try to ingest the dry contents of an entire aspartame-sweetened instant soft drink packet. Under these circumstances the aspartame intake of a 10-kg infant would be 100 mg/kg body weight.

The data from this study indicate no obvious differences between the ability of infants and adults to metabolize the aspartate, phenylalanine, and methanol provided by aspartame at the doses studied.

Only a small change in plasma aspartate concentration was noted at the highest aspartame dose studied. While we cannot speculate on the potential effects on plasma aspartate concentrations of higher intakes of aspartate or aspartame, plasma aspartate concentrations were within the normal postprandial range seen in formula-fed infants (36,37).

Peak plasma phenylalanine concentrations and plasma phenylalanine AUC values increased with aspartame dosing in infants and adults. Although significant differences between infants and adults were noted at the 50-mg/kg dose level, no significant difference was noted at the 34- or 100-mg/kg body weight dose level. Thus it is assumed that infants metabolize phenylalanine as well as adults. Similarly, blood methanol concentrations were similar in infants and adults administered aspartame at 100 mg/kg body weight, suggesting that infants metabolize methanol as well as adults.

Since marked elevations of plasma aspartate, phenylalanine, or methanol are required for these compounds to exert any toxic effect, the failure to markedly elevate plasma concentrations of these amino acids at the doses studied indicates little hazard to the infant from a single acute dose of aspartame. While the infant mouse is acutely sensitive to glutamate- and/or aspartate-induced neuronal necrosis (6-14), plasma glutamate plus aspartate concentrations must exceed 50 μmol/dl for lesion formation to occur (9,10,12-14,43). In the infant nonhuman primate, which is insensitive to dicarboxylic amino acid-induced neuronal necrosis, plasma glutamate plus aspartate concentrations as high as 460 μmol/dl were not associated with neurotoxicity (17,44). The small changes noted in plasma glutamate or aspartate concentrations of infants administered aspartame are far below these critical levels.

Children with untreated classic phenylketonuria have plasma phenylalanine concentrations ranging from 120 to 300 μmol/dl (18-20), and children treated with diets low in phenylalanine are permitted plasma phenylalanine concentrations of 24-48 μmol/100 ml (19,20,45). Thus, while the plasma phenylalanine concentrations noted after administration of abuse doses of aspartame are outside

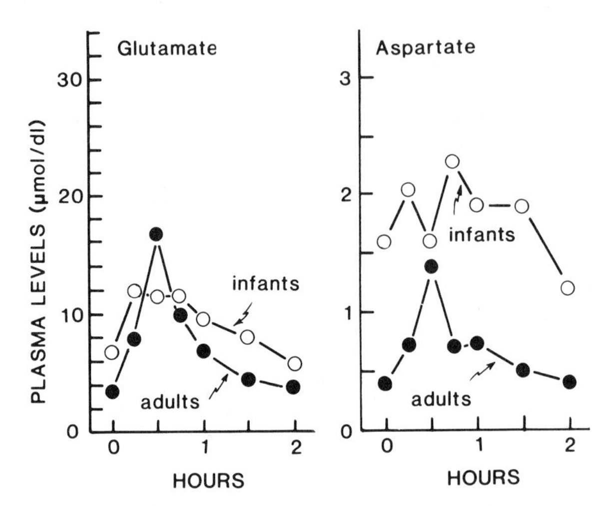

Figure 6 Mean plasma glutamate and aspartate concentrations in one-year-old infants (○) and normal adults (●) administered 50 mg/kg body weight monosodium L-glutamate in consommé.

the usual postprandial range (12 ± 3 μmol/100 ml) (36,37,46), maximal concentrations are well below those associated with toxic effects.

Since the data in infants administered aspartame did not support Olney's hypothesis (3), we recently examined whether human infants metabolize monosodium L-glutamate less well than adults (47). Olney's hypothesis was made on the basis of studies on glutamate metabolism, and it was possible that an age-related effect would be noted with glutamate, but not aspartame. Normal 1-year-old infants and adults were administered beef consommé providing monosodium L-glutamate at 0, 25, and 50 mg/kg body weight. Consommé was chosen as the vehicle because it contains little carbohydrate. The monosodium L-glutamate doses selected were those likely to be ingested by infants fed commercially prepared soups. Mean plasma glutamate and aspartate concentrations in infants and adults administered the highest dose of monosodium L-glutamate studied (50 mg/kg body weight) are shown in Figure 6. The increase in plasma aspartate concentration over baseline levels postdosing was similar in both infants and adults. These data indicate that infants metabolize glutamate as well as adults at levels tested. However, differences in metabolism might still be noted at very high dose levels, as has been observed for glutamate in experimental animals (25).

REFERENCES

1. Olney, J. W. (1975). Another view of aspartame. In *Sweeteners, Issues and Uncertainties*, National Academy of Sciences, Washington, D.C., pp. 189-195.

2. Olney, J. W. (1975). L-Glutamic and L-aspartic acids–A question of hazard? *Food Cosmet. Toxicol.* 13, 595-596.
3. Olney, J. W. (1981). Excitatory neurotoxins as food additives: An evaluation of risk. *Neurotoxicology* 2, 163-192.
4. Reif-Lehrer, L. (1976). Possible significance of adverse reactions to glutamate in humans. *Fed. Proc.* 35, 2205-2211.
5. Turner, J. S. (1979). Quoted in Aspartame: Ruling on objections and notice of hearing before a public board of inquiry. *Fed. Reg.* 44, 31716-31718.
6. Olney, J. W., and Ho, O.-L. (1970). Brain damage in infant mice following oral intake of glutamate, aspartate and cysteine. *Nature* 227, 609-610.
7. Fujiwara, T., Hirai, K., Nakamura, T., Kubo, M., Imai, T., Hasegawa, S., Arai, Y., and Matsumoto, A. (1976). Fundamental study on the toxicity of amino acids given intravenously to infant rats. Part 1. *Jpn. J. Surg. Metabolism Nutr.* 10, 385-389.
8. Fujiwara, T., Hirai, K., Nakamura, T., Kubo, M., Imai, T., Hasegawa, S., Aria, Y., and Matsumoto, A. (1977). Fundamental study on the toxicity of amino acids given intravenously to infant rats. Part 2. *Jpn. J. Surg. Metabolism Nutr.* 11, 426-430.
9. Stegink, L. D., Shepherd, J. A., Brummel, M. C., and Murray, L. M. (1974). Toxicity of protein hydrolysate solutions: Correlation of glutamate dose and neuronal necrosis to plasma amino acid levels in young mice. *Toxicology* 2, 285-299.
10. Okaniwa, A., Hori, M., Masuda, M., Takeshita, M., Hayashi, N., Wada, I., Doi, K., and Ohara, Y. (1979). Histopathological study of effects of potassium aspartate on the hypothalamus of rats. *J. Toxicol. Sci.* 4, 31-46.
11. Oppermann, J. A., and Ranney, R. E. (1979). The metabolism of aspartate in infant and adult mice. *J. Environ. Pathol. Toxicol.* 2, 987-998.
12. Daabees, T. T., Finkelstein, M., Applebaum, A. E., and Stegink, L. D. (1982). Protective effect of carbohydrate or insulin on aspartate-induced neuronal necrosis in infant mice. *Fed. Proc.* 41, 396.
13. Applebaum, A. E. (1984). Aspartate-induced toxicity in infant mice. In *Aspartame: Physiology and Biochemistry* (Stegink, L. D., and Filer, L. J., Jr., eds.), Marcel Dekker, New York, Chap. 17.
14. Finkelstein, M. W., Daabees, T. T., Stegink, L. D., and Applebaum, A. E. (1983). Correlation of aspartate dose, plasma dicarboxylic amino acid concentration, and neuronal necrosis in infant mice. *Toxicology* 29, 109-119.
15. Reynolds, W. A., Butler, V., and Lemkey-Johnson, N. (1976). Hypothalamic morphology following ingestion of aspartame or MSG in the neonatal rodent and primate: A preliminary report. *J. Toxicol. Environ. Health* 2, 471-480.
16. Olney, J. W. (1976). Brain damage and oral intake of certain amino acids. In *Transport Phenomena in the Nervous System. Physiological and Pathological Aspects* (Levi, G., Battistin, L., and Lajtha, A., eds.), Plenum, New York, pp. 497-506.
17. Reynolds, W. A., Stegink, L. D., Filer, L. J., Jr., and Renn, E. (1980). Aspartame administration to the infant monkey: Hypothalamic morphology and plasma amino acid levels. *Anat. Rec.* 198, 73-85.

18. Knox, W. E. (1972). Phenylketonuria. In *The Metabolic Basis of Inherited Disease* (Stanbury, J. B., Wyngaarden, J. B., and Fredrickson, D. S. eds.), McGraw-Hill, New York, pp. 266-295.
19. Blaskovics, M. E., Schaeffler, G. E., and Hack, S. (1974). Phenylalaninemia: Differential diagnosis. *Arch. Dis. Child.* 49, 835-843.
20. Koch, R., Blaskovics, M., Wenz, E., Fishler, K., and Schaeffler, G. (1974). Phenylalaninemia and phenylketonuria. In *Heritable Disorders of Amino Acid Metabolism* (Nyhan, W. L., ed.), Wiley, New York, pp. 109-140.
21. Roe, O. (1955). The metabolism and toxicity of methanol. *Pharmacol. Rev.* 7, 399-412.
22. Tephly, T. R., Watkins, W. D., and Goodman, J. I. (1974). The biochemical toxicology of methanol. *Essays Toxicol.* 5, 149-177.
23. Posner, H. S. (1975). Biohazards of methanol in proposed new uses. *J. Toxicol. Environ. Health* 1, 153-171.
24. Tephly, T. R., and McMartin, K. E. (1984). Methanol metabolism and toxicity. In *Aspartame: Physiology and Biochemistry* (Stegink, L. D. and Filer, L. J., Jr., eds.), Marcel Dekker, New York, Chap. 6.
25. Stegink, L. D., Reynolds, W. A., Filer, L. J., Jr., Baker, G. L., Daabees, T. T., and Pitkin, R. M. (1979). Comparative metabolism of glutamate in the mouse, monkey, and man. In *Glutamic Acid: Physiology and Biochemistry* (Filer, L. J. Jr., Garattini, S., Kare, M. R., Reynolds, W. A., and Wurtman, R. J., eds.), Raven Press, New York, pp. 85-110.
26. Filer, L. J., Jr., Baker, G. L., and Stegink, L. D. (1983). Effect of aspartame loading on plasma and erythrocyte free amino acid concentrations in one-year-old infants. *J. Nutr.* 113, 1591-1599.
27. Stegink, L. D., Brummel, M. C., Filer, L. J., Jr., and Baker, G. L. (1983). Blood methanol concentrations in one-year-old infants administered graded doses of aspartame. *J. Nutr.* 113, 1600-1606.
28. Stegink, L. D., Filer, L. J., Jr., and Baker, G. L. (1977). Effect of aspartame and aspartate loading upon plasma and erythrocyte free amino acid levels in normal adult volunteers. *J. Nutr.* 107, 1837-1845.
29. Stegink, L. D., Filer, L. J., Jr., and Baker, G. L. (1979). Plasma erythrocyte and human milk levels of free amino acids in lactating women administered aspartame or lactose. *J. Nutr.* 109, 2173-2181.
30. Stegink, L. D., Filer, L. J., Jr., and Baker, G. L. (1981). Plasma and erythrocyte concentrations of free amino acids in adult humans administered abuse doses of aspartame. *J. Toxicol. Environ. Health* 7, 291-305.
31. Filer, L. J., Jr., Baker, G. L., and Stegink, L. D. (1979). Metabolism of free glutamate in clinical products fed infants. In *Glutamic Acid: Advances in Biochemistry and Physiology* (Filer, L. J., Jr., Garattini, S., Kare, M. R., Reynolds, W. A., and Wurtman, R. J., eds.), Raven Press, New York, pp. 353-362.
32. Elwyn, D. H. (1966). Distribution of amino acids between plasma and red blood cells in the dog. *Fed. Proc.* 25, 854-861.
33. Felig, P., Wahren, J., and Raf, L. (1973). Evidence of inter-organ amino-acid transport by blood cells in humans. *Proc. Nat. Acad. Sci. U.S.A.* 70, 1775-1779.

34. Aoki, T. T., Brennen, M. F., Muller, W. A., Moore, F. D., and Cahill, G. F., Jr. (1972). Effect of insulin on muscle glutamate uptake. Whole blood versus plasma glutamate analysis. *J. Clin. Invest.* 51, 2889-2894.
35. Elwyn, D. H., Launder, W. J., Parikh, H. C., and Wise, E. M., Jr. (1972). Roles of plasma and erythrocytes in interorgan transport of amino acids in dogs. *Am. J. Physiol.* 222, 1333-1342.
36. Stegink, L. D., and Baker, G. L. (1971). Infusion of protein hydrolysates in the newborn infant: Plasma amino acid concentrations. *J. Pediatr.* 78, 595-602.
37. Stegink, L. D., Schmitt, J. L., Meyer, P. D., and Kain, P. H. (1971). Effect of diets fortified with DL-methionine on urinary and plasma methionine levels in young infants. *J. Pediatr.* 79, 648-655.
38. Stegink, L. D., Brummel, M. C., McMartin, K., Martin-Amat, G., Filer, L. J., Jr., Baker, G. L., and Tephly, T. R. (1981). Blood methanol concentrations in normal adult subjects administered abuse doses of aspartame. *J. Toxicol. Environ. Health* 7, 281-290.
39. Filer, L. J., Jr., Baker, G. L., and Stegink, L. D. (1977). Plasma aminograms in infants and adults fed an identical high protein meal. *Fed. Proc.* 36, 1181.
40. Stegink, L. D., and Lampy Schmitt, J. (1971). Postprandial serum amino acid levels in young infants fed casein hydrolysate-based formulas. *Nutr. Rep. Int.* 3, 93-99.
41. Filer, L. J., Jr., Stegink, L. D., and Chandramouli, B. (1977). Effect of diet on plasma aminograms of low-birth-weight infants. *Am. J. Clin. Nutr.* 30, 1036-1043.
42. Bell, E. F., Filer, L. J., Jr., Wong, A. P., and Stegink, L. D. (1983). Effects of a parenteral nutrition regimen containing dicarboxylic amino acids on plasma, erythrocyte, and urinary amino acid concentrations of young infants. *Am. J. Clin. Nutr.* 37, 99-107.
43. O'hara, Y., and Takasaki, Y. (1979). Relationship between plasma glutamate levels and hypothalamic lesions in rodents. *Toxicol. Lett.* 4, 499-505.
44. Stegink, L. D., Reynolds, W. A., Filer, L. J., Jr., Pitkin, R. M., Boaz, D. P., and Brummel, M. C. (1975). Monosodium glutamate metabolism in the neonatal monkey. *Am. J. Physiol.* 229, 246-250.
45. Dobson, J. C., Williamson, M. L., Azen, C., and Koch, R. (1977). Intellectual assessment of 111 four-year-old children with phenylketonuria. *Pediatrics* 60, 822-827.
46. Vaughan, D. A., Womack, M., and McClain, P. E. (1977). Plasma free amino acid levels in human subjects after meals containing lactalbumin, heated lactalbumin or no protein. *Am. J. Clin. Nutr.* 30, 1709-1712.
47. Baker, G. L., Bell, E. F., Filer, L. J., Jr., and Stegink, L. D. (1983). Metabolism of monosodium L-glutamate by infants and adults. *Fed. Proc.* 43, 1311.

30

Aspartame Ingestion by Phenylketonuric Heterozygous and Homozygous Individuals

Richard Koch
Childrens Hospital of Los Angeles, and University of Southern California School of Medicine, Los Angeles, California

Elizabeth J. Wenz
Childrens Hospital of Los Angeles, Los Angeles, California

INTRODUCTION

Aspartame is approximately 50% phenylalanine by weight. Persons with phenylketonuria (PKU) are unable to convert phenylalanine into tyrosine because of a defect in or a deficiency of the enzyme phenylalanine hydroxylase. In normal individuals this enzyme facilitates the conversion of about 90% of ingested phenylalanine to tyrosine (1), which in turn is catabolized. In individuals with PKU the enzyme is inactive. This enzyme deficiency results in an accumulation of excessive quantities of phenylalanine in fluid and tissues of persons with PKU. In some way this accumulation causes brain damage and severe mental retardation. The usual treatment for newly diagnosed PKU infants and children is a phenylalanine-restricted diet administered from infancy through childhood. Such dietary treatment, if initiated soon enough, prevents mental retardation. Dietary treatment is usually terminated prior to adolescence; however, some physicians recommend use of a low-phenylalanine diet until adulthood.

Since phenylketonuric persons may be on a diet restricted in phenylalanine, it is important to establish what effect aspartame ingestion might have upon daily dietary intake of phenylalanine. Searle Laboratories estimated that a dose of 34 mg/kg body weight per day would be the 99th percentile of projected intake of aspartame. Subsequent calculations by other groups (2-5) support that estimate as reasonable.

Aspartame is a dipeptide methyl ester (6). As expected, the body metabolizes this compound just as it does dietary protein, hydrolyzing it to aspartate, phenylalanine, and methanol (7). Normal individuals ingest approximately 3-5 g of phenylalanine daily as dietary protein. Ingestion of 34 mg/kg body weight aspartame by a 50-kg person would provide 1.7 g of aspartame, or 0.85 g of phenylalanine. Thus aspartame ingested at this level could increase the daily phenylalanine intake by 20-30%. The child with PKU fed a phenylalanine-restricted diet normally has a phenylalanine intake of 0.3-0.8 g. An aspartame intake of 34 mg/kg body weight for a 20-kg child would provide 0.68 g of aspartame, or 0.34 g of phenylalanine. This represents a significant increase in phenylalanine intake and would affect dietary control. The actual intake of phenylalanine from aspartame-sweetened foods may, however, be substantially less than this theoretical maximum projected intake.

Because aspartame is approximately 200 times sweeter than sugar, only a small amount of aspartame is needed to provide a sweetness equivalent to sucrose. For example, a child might ingest aspartame- rather than sucrose-sweetened Kool-Aid; 8 oz of this beverage would provide about 140 mg of aspartame, or 70 mg of phenylalanine. Families with PKU children will need to know that beverage drinks usually classified as phenylalanine free now need to be scrutinized for their phenylalanine content.

ASPARTAME METABOLISM BY PHENYLKETONURIC HETEROZYGOTES AND HOMOZYGOTES SUBJECTS HETEROZYGOUS FOR PHENYLKETONURIA

To evaluate the use of aspartame in PKU heterozygotes, 45 parents of known PKU children were studied over a period of 27 weeks (8). Such parents are obligate heterozygotes for PKU.

A double-blind study design was utilized to administer the test substances. Participants were divided by gender and assigned random study numbers. Identical capsules containing either aspartame or placebo were supplied. The containers were numbered to correspond with the randomly assigned study number of the participant. Table 1 describes the schedule of capsule intake and the daily intake of aspartame and phenylalanine consumed in the capsules.

The study was divided into two phases. During the first 6 weeks (phase A), aspartame ingestion was increased at weekly intervals from 3 to 27 capsules per day. For the next 21 weeks (phase B), six capsules were ingested daily for a total of 1.8 g of aspartame per day.

Before starting the study, each participant received a medical and ophthalmic examination to determine general health status. During each week of the study, blood pressures and weights were obtained. Medications were recorded, and participants' observations and complaints were noted and reviewed weekly. A daily diary of food intake was recorded by the participants and reviewed weekly by a consulting nutritionist. Medical and ophthalmic examinations were repeated upon completion of the study.

Table 1 Dosing Schedule

	Phase A						Phase B
	Week 1	Week 2	Week 3	Week 4	Week 5	Week 6	(weeks 7-21)
Number of capsules taken with each meal (three times per day)	1	2	4	5	7	9	2
Amount of aspartame per capsule (g)	0.2	0.2	0.2	0.3	0.3	0.3	0.3
Total amount of aspartame per day (g)	0.6	1.2	2.4	4.5	6.3	8.1	1.8
Total daily phenylalanine from aspartame (g)	0.34	0.67	1.35	2.53	3.54	4.55	1.01

Results

Of the 45 PKU heterozygotes who agreed to participate in phase A, 20 were male and 25 were female. Their ages ranged from 28 to 50 years, averaging 35 years. All were in apparent good health.

The initial physical examinations revealed two persons (one male and one female) who were overweight and thus excluded. Initial laboratory results for one female suggested hypothyroidism. Clinically, however, she was not sluggish, did not exhibit dry hair or skin, and was not constipated; therefore she was continued in the study. Total participants numbered 43 in phase A and 41 in phase B.

Ophthalmic examinations, including slit-lamp studies, were performed. All results were normal, and no significant ophthalmic changes were observed during the entire study.

During the second week one woman developed an acute dental abscess that was treated with antibiotics and dental extraction. No changes occurred in her chemistry values.

Two women, one in late menopause and the other aged 36 years, developed vaginal spotting and were referred to an obstetrician for further evaluation. In both cases the spotting subsided spontaneously.

Numerous laboratory tests were performed during the study. No significant changes in serum phenylalanine, tyrosine, or urinary metabolite levels were noted following an overnight fast. All follow-up medical examinations were unchanged from initial evaluations. Other examinations included complete blood count, prothrombin time, urinalysis, blood urea nitrogen, T_4, bilirubin, serum glutamic oxalacetic transaminase, alkaline phosphatase, uric acid, creatinine, cholesterol, triglycerides, glucose tolerance test, urine chromatography for phenylketones, and pregnancy test. No significant medical or biochemical changes were noted during the 28-week study of PKU heterozygotes receiving aspartame. Fasting serum phenylalanine and tyrosine levels remained within normal limits throughout the study period.

Stegink et al. (9-11) have demonstrated that PKU heterozygotes metabolize aspartame-derived phenylalanine well, even when abusive doses of aspartame are administered. When normal subjects were administered 100 mg/kg body weight of aspartame, plasma phenylalanine levels increased from fasting values (5.40 μmol/dl) to a mean (±SD) peak concentration of 20.2 ± 6.77 μmol/dl. In heterozygous subjects the mean (±SD) peak plasma phenylalanine concentration was approximately twice as high, 41.7 ± 2.33 μmol/dl. These data indicate a slower but adequate metabolism and clearance of this dose of aspartame.

SUBJECTS HOMOZYGOUS FOR PHENYLKETONURIA

The phenylalanine content of aspartame is of little concern in normal individuals and individuals heterozygous for PKU because of its rapid rate of metabolism.

However, the amount of phenylalanine supplied by aspartame must be considered carefully in homozygous patients with classic PKU (12). These subjects have a genetic defect that results in the absence of or a defect in the activity of the enzyme phenylalanine hydroxylase. Therefore dietary phenylalanine in excess of that needed for protein synthesis is not converted to tyrosine and catabolized, and phenylalanine accumulates in body tissues and fluids. In affected persons, abnormal metabolites of phenylalanine (o-hydroxyphenylacetic acid, phenylpyruvate, phenylacetylglutamine) appear in large amounts in the urine.

Phenylketonuric homozygotes represent about 1 birth per 15,000. Such children are born to heterozygous parents (i.e., to matings in which mother and father each carry one PKU allele). Since a homozygous child may suffer brain damage and mental retardation unless placed on a special low-phenylalanine diet within the first few weeks of life, phenylalanine blood tests are now routinely conducted on newborn infants at most hospitals.

Individuals responsible for the diets of PKU children must be made aware of the phenylalanine content of aspartame, since this sweetening agent should be excluded from the diet of PKU children on a low-phenylalanine diet.

Table 2 compares the daily phenylalanine intake from all sources for normal adults (about 2.5-10 g), with the range of phenylalanine intake allowed in PKU diets (20-30 mg/kg per day or about 1-1.5 g per day for a 50-kg adult). In such individuals the ordinary use of aspartame as a sweetening agent might provide as much as 340 mg/day of phenylalanine, a significant contribution to the intake of PKU homozygotes on a low-phenylalanine diet.

However, dietary phenylalanine levels are less of a problem for older children with PKU allowed a liberalized diet. We have studied two boys homozygous for

Table 2 Total Dietary Intake of Phenylalanine

Type of individual	Range of daily intake of phenylalanine (mg/kg body weight)	Daily intake of phenylalanine (g)
		For a 50-kg person
Normal	50-200	2.5-10
PKU homozygous adult on phenylalanine-restricted diet	20-30	1.0-1.5
PKU homozygous adult (liberalized diet)	50-100	2.5-5.0
		For a 55-kg woman
PKU homozygous pregnant woman	10-15	0.6-0.8
		For a 20-kg child
PKU homozygous child	20-30	0.4-0.6

phenylketonuria who were given aspartame (12). One of the boys was on a liberalized Lofenalac* diet supplemented with an allowable phenylalanine dietary intake of 70 mg/kg body weight per day. The other boy was on a well-controlled diet (phenylalanine intake, 17 mg/kg body weight per day), utilizing Lofenalac as the base formula.

The two subjects were given a single loading dose of 34 mg/kg body weight aspartame in 180 ml of unsweetened orange juice. Two weeks later they were given an equimolar quantity of L-phenylalanine (19 mg/kg body weight) in orange juice.

A standardized diet consistent with each subject's clinical state was prescribed for each of three 24-hr periods before the aspartame and the phenylalanine loads, as well as for each of three 24-hr periods following each load. The mother or nurse was instructed to record carefully the items and amounts actually consumed.

All urine was collected from each patient for three successive 8-hr periods before each load and for three successive 8-hr periods after each load. Additional early morning urines were collected at 48 and 72 hr after each load. Urine samples were analyzed for amino acids, phenolic acids, phenylpyruvic acid, and phenylacetylglutamine.

Serum levels of phenylalanine and tyrosine were determined in the fasting state before the loading doses and at 1, 2, 4, 8, 24, 48, and 72 hr after each load. Urine and serum samples levels were also assayed for methanol.

Two normal adolescents (male aged 12 years and female aged 15 years) served as controls and were given aspartame (34 mg/kg body weight) in 180 ml of unsweetened orange juice. Blood and urine samples were collected using the same schedule as for the two PKU homozygous adolescents. Two weeks later the control subjects were given an equimolar dose of L-phenylalanine (19 mg/kg body weight), and similar laboratory data were obtained.

Results

Table 3 shows the amounts of phenylalanine ingested by the two PKU subjects on the test days. Comparison of the data presented in Tables 2 and 3 reveals that

Table 3 Phenylalanine Intake of Adolescent PKU Subjects

Subject	Weight (kg)	Phenylalanine dietary intake (mg)	Phenylalanine added as aspartame (mg)	Total phenylalanine	
				mg	mg/kg body weight
1	36.3	2539	689	3228	90
2	56.4	965	1072	2037	36

*Lofenalac (Mead Johnson, Evansville, Indiana) is a low-phenylalanine preparation used as a protein source for phenylketonuric children.

subject 1 fed the liberalized diet remained well within the permitted range of phenylalanine intake during the loading study. However, the phenylalanine intake of subject 2 exceeded dietary limitations for individuals on a low-phenylalanine diet.

The urinary excretion of phenylalanine, o-hydroxyphenylacetic acid, phenylpyruvic acid, and phenylacetylglutamine for each subject, measured before and after loading doses of aspartame and L-phenylalanine, were not significantly altered (13,14). Patient 1, on the liberalized diet, was already excreting large quantities of these phenylalanine metabolites. This may have obscured small changes in excretion that might result from ingestion of the test compounds. The PKU patient under close dietary restriction (patient 2) did not excrete detectable quantities of phenylpyruvic acid after either dose.

Aspartame (Table 4) and L-phenylalanine dosing (Table 5) increased the serum phenylalanine concentration by 4-6 mg/dl over base-line values, but did not affect tyrosine levels. Each blood and urine sample was also tested for methanol. None was detected.

Serum concentrations of phenylalanine and tyrosine, urine chromatography for phenylalanine metabolites, and urine methanol concentrations were also determined in the two control subjects (Tables 6 and 7). All values were unchanged throughout the period of observation. The subjects remained well, without any signs or symptoms.

These data show that a single loading dose of aspartame or its L-phenylalanine equivalent did not provoke a significant metabolic change in either PKU heterozygotes, two PKU homozygous patients, or two normal adolescents. It would appear that aspartame, in the dose utilized, is well tolerated. Our data are consistent with those reported by Stegink, et al. (15).

In young homozygous PKU children on a phenylalanine-restricted diet the amount of phenylalanine contributed to the diet by aspartame ingestion is sufficiently large, so that it must be calculated as part of the total phenylalanine allowed in each child's diet (Table 8). However, aspartame's effect in adolescents would appear to be small, since those individuals are usually permitted diets providing increased quantities of phenylalanine.

DISCUSSION

The amount of phenylalanine provided by a normal diet is much larger than that likely to be ingested as aspartame. For example, an individual ingesting an amount of aspartame equivalent in sweetness to one teaspoon of sugar would consume about 10 mg of phenylalanine. By comparison, 50 times this amount of phenylalanine is contained in an 8-oz glass of milk (Table 8).

It is not expected that usual use of the sweetener will significantly affect plasma phenylalanine concentrations in the heterozygous PKU person, since the body retains its ability to metabolize phenylalanine, albeit somewhat more slowly.

Table 4 Aspartame Loading Studies in Two PKU Adolescents

Subject 1[a]					Subject 2[b]				
Time	Activity	Intake	Phenylalanine[c] (mg/dl)	Tyrosine[c] (mg/dl)	Time	Activity	Intake	Phenylalanine[c] (mg/dl)	Tyrosine[c] (mg/dl)
7/10/72					7/10/72				
0800	Base line		18.7	4.8	0810	Base line		3.0	5.0
0805	Loading dose, 34 mg/kg	240 cm^3 of orange juice			0815	Loading dose, 34 mg/kg	240 cm^3 of orange juice		
0902			22.9	5.9	0915			4.5	5.8
1005					1020				
1130		Lunch			1130		Lunch		
1205					1215				
1600					1610				
1800		Dinner			1800		Dinner		
7/11/72 0700	24-hr postload	Usual meals	20.2	4.5	7/11/72 0800	24-hr postload	Usual meals	5.0	5.5
7/12/72 0645	48-hr postload	Usual meals	22.0	4.9	7/12/72 0815	48-hr postload	Usual meals	4.9	5.5
7/13/72 0705	72-hr postload	Usual meals	18.5	6.0	7/13/72 0815	72-hr postload	Usual meals	4.5	5.7

[a]Age, 14 1/12 years; height, 57 3/4 in.; weight, 79 3/4 lb; blood pressure, 90/62; diet recipe: phenylalanine, 2539 mg; protein, 101 g; calories, 4155; loading dose, 1.232 g aspartame.
[b]Age 13 11/12 years; height, 65 1/4 in.; weight, 124 lb; blood pressure, 96/94; diet recipe: phenylalanine, 965 mg; protein, 61.2 g; calories, 2662; loading dose, 1.917 g aspartame.
[c]Phenylalanine and tyrosine concentrations are expressed as milligrams per deciliter, since those are the units used by most PKU patients. A phenylalanine concentration of 1 mg/dl is equivalent to 6.06 μmol/dl. A tyrosine concentration of 1 mg/dl is equivalent to 5.52 μmol/dl.
Source: Ref. 14.

Table 5 Phenylalanine Loading Studies in Two PKU Adolescents

Subject 1[a]					Subject 2[b]				
Time	Activity	Intake	Phenylalanine[c] (mg/dl)	Tyrosine[c] (mg/dl)	Time	Activity	Intake	Phenylalanine[c] (mg/dl)	Tyrosine[c] (mg/dl)
7/24/72					7/24/72				
0810	Preload		17.49	4.62	0805	Preload		5.22	6.88
0812	Loading dose, 19 mg/kg	180 cm^3 of orange juice			0808	Loading dose 19 mg/kg	180 cm^3 of orange juice		
0915			18.58	4.85	0910			6.30	4.24
1015		240 cm^3 of Coke			1010		240 cm^3 of Coke	6.62	5.44
1200		Lunch			1200		Lunch		
1215			23.8	6.80	1210			6.15	5.45
1615			21.37	5.90	1610			8.95	4.77
1800		Dinner			1800		Dinner		
7/25/72	24-hr	Usual meals			7/25/72	24-hr	Usual meals		
0710	postload		17.7	3.4	0815	postload		6.5	4.1
7/26/72	48-hr	Usual meals			7/26/72	48-hr	Usual meals		
0700	postload		18.4	4.1	0800	postload		7.3	4.6
7/27/72	72-hr	Usual meals			7/27/72	72-hr	Usual meals		
0645	postload		17.7	4.3	0730	postload		6.1	4.8

[a]Age, 14 1/12 years; height, 57 3/4 in.; weight, 79 3/4 lb; blood pressure, 94/62; diet recipe: phenylalanine, 2539 mg; protein, 101 g; calories, 4155; loading dose, 689 g of L-phenylalanine.
[b]Age, 13 11/12 years; height, 65 1/4 in.; weight 124 lb; blood pressure, 94/60; diet recipe: phenylalanine, 965 mg; protein, 61.2 g; calories, 2662; loading dose, 1.072 g of L-phenylalanine.
[c]Phenylalanine and tyrosine concentrations are expressed as milligrams per deciliter, since those are the units used by most PKU patients. A phenylalanine concentration of 1 mg/dl is equivalent to 6.06 μmol/dl. A tyrosine concentration of 1 mg/dl is equivalent to 5.52 μmol/dl.
Source: Ref. 14.

Table 6 Aspartame Loading Studies in Two Normal Adolescents

Subject 1[a]					Subject 2[b]				
Time	Activity	Intake	Phenylalanine[c] (mg/dl)	Tyrosine[c] (mg/dl)	Time	Activity	Intake	Phenylalanine[c] (mg/dl)	Tyrosine[c] (mg/dl)
10/29/72					10/29/72				
0805	Preload		1.7	4.1	0815	Preload		1.5	3.8
0808	Loading dose, 34 mg/kg	180 cm^3 of orange juice			0817	Loading dose 34 mg/kg	180 cm^3 of orange juice		
0905			2.0	4.7	0912			1.8	4.0
1003			2.2	4.8	1010			2.0	4.1
1200		Lunch	1.5	4.7	1208		Lunch	2.0	4.1
1603			1.8	4.3	1610			2.2	4.5
1800		Dinner			1800		Dinner		
10/30/72	24-hr	Usual meals			10/30/72	24-hr	Usual meals		
0800	postload		2.5	5.0	0700	postload		2.3	4.6
10/31/72	48-hr	Usual meals			10/31/72	48-hr	Usual meals		
0800	postload		2.0	5.3	0800	postload		2.1	4.0
11/01/72	72-hr	Usual meals			11/01/72	72-hr	Usual meals		
0630	postload		1.9	4.9	0715	postload		1.4	4.6

[a]Age, 15 1/6 years; height 66 in.; weight, 135 lb; blood pressure, 98/66; diet, regular; loading dose, 2.084 g of aspartame.
[b]Age, 12 1/2 years; height, 62 1/2 in.; weight, 80 1/2 lb; blood pressure 96/64; diet regular; loading dose, 1.245 g of aspartame.
[c]Phenylalanine and tyrosine concentrations are expressed in milligrams per deciliter, since those are the units used by most PKU patients. A phenylalanine concentration of 1 mg/dl is equivalent to 6.06 μmol/dl. A tyrosine concentration of 1 mg/dl is equivalent to 5.52 μmol/dl.

Table 7 Phenylalanine Loading Studies in Two Normal Adolescents

Subject 1[a]					Subject 2[b]				
Time	Activity	Intake	Phenylalanine[c] (mg/dl)	Tyrosine[c] (mg/dl)	Time	Activity	Intake	Phenylalanine[c] (mg/dl)	Tyrosine[c] (mg/dl)
11/12/72					11/12/72				
0808	Preload		1.8	5.9	0830	Preload		1.5	4.6
0810	Loading dose, 19 mg/kg	180 cm^3 of orange juice			0832	Loading dose, 19 mg/kg	180 cm^3 of orange juice		
0910			2.5	6.9	0930			2.3	5.3
1010			2.2	5.9	1030			2.2	5.6
1210			1.8	5.3	1230			2.2	4.7
		Lunch					Lunch		
1610			1.3	3.8	1630			1.7	3.9
1800		Dinner			1800		Dinner		
11/13/72	24-hr	Usual meals			11/13/72	24-hr	Usual meals		
0700	postload		1.8	5.7	0700	postload		1.6	5.2
11/14/72	48-hr	Usual meals			11/14/72	48-hr	Usual meals		
0715	postload		1.7	qns	0700	postload		1.6	3.8
11/15/72	72-hr	Usual meals			11/15/72	72-hr	Usual meals		
0715	postload		2.1	5.4	0715	postload		2.1	4.2

[a]Age, 15 1/6 years; height, 66 in.; weight, 135 lb; blood pressure, 98/66; diet, regular; loading dose, 1.165 g of L-phenylalanine.
[b]Age, 12 1/12 years; height, 62 1/2 in.; weight, 80 1/2 lb; blood pressure, 94/62; diet, regular; loading dose, 0.684 g of L-phenylalanine.
[c]Phenylalanine and tyrosine concentrations are expressed in milligrams per deciliter, since those are the units used by most PKU patients. A phenylalanine concentration of 1 mg/dl is equivalent to 6.06 μmol/dl. A tyrosine concentration of 1 mg/dl is equivalent to 5.52 μmol/dl.

Table 8 Some Natural Sources of Phenylalanine Compared to Aspartame

Typical serving	Phenylalanine (mg)
Sweetener	
Aspartame, 20 mg (equivalent to 1 teaspoon of sugar)	10
Meat-dairy	
Hamburger, 4 oz	882
Chicken, 4 oz	907
Egg, 1	323
Milk, 8 oz	542
Vegetables	
Spinach, 4 oz	150
Lima beans, 4 oz dry	1355
Soybeans, 4 oz dry	2330
Tomato juice, 8 oz	45
Fruit	
Banana, 4 oz	49
Cherries, 4 oz	16
Pear, 4 oz.	13

However, homozygous PKU persons on a phenylalanine-restricted diet *must* take the phenylalanine content of aspartame into their dietary calculations. This would be especially important to the infant or young child. For the adolescent PKU homozygote on a restricted phenylalanine intake, aspartame poses less of a risk. It is still not clear whether elevated plasma phenylalanine levels that remain between 10 to 15 mg/dl are damaging to the adolescent. There is general agreement that plasma phenylalanine levels continuously greater than 20 mg/dl are damaging to the brain, but the Collaborative Study of Children Treated for Phenylketonuria has not been able to document a difference in IQ outcome in children treated with diets that maintained blood levels of phenylalanine at two different levels (1-5.4 mg/dl compared to 5.5-9.9 mg/dl) (16).

No data on aspartame ingestion in pregnant PKU women are available. However, the gradient maintained from maternal to fetal blood for phenylalanine levels (17) suggests that such individuals must consider this source of dietary phenylalanine.

SUMMARY

Our data indicate that aspartame is well tolerated by normal and PKU heterozygote persons at the dose studied and does not cause a clinically significant rise in plasma phenylalanine content.

In the young homozygous PKU child on a phenylalanine-restricted diet, aspartame ingestion at 34 mg/kg body weight would contribute significantly to the phenylalanine content of the diet, and its use should therefore be avoided. In older PKU persons on a more normal diet, aspartame usage does not greatly increase blood phenylalanine levels over usual concentrations unless doses greater than normal projected use levels are ingested. Such individuals should be made aware of the phenylalanine content of this sweetener, and its use discouraged.

Aspartame ingestion for the pregnant PKU homozygous women is not recommended.

REFERENCES

1. Udenfriend, S., and Bessman, S. P. (1953). The hydroxylation of phenylalanine and antipyrine in phenylpyruvic oligophrenia. *J. Biol. Chem.* 203, 961-966.
2. Stegink, L. D., Filer, L. J., Jr., and Baker, G. L. (1977). Effect of aspartame and aspartate loading upon plasma and erythrocyte free amino acid levels in normal adult volunteers. *J. Nutr.* 107, 1837-1845.
3. Anonymous (1974). Aspartame. *Fed. Reg.* 39, 27317-27319.
4. Hattan, D. (1980) FDA presentation. In *Aspartame: Public Board of Inquiry, Transcript of Proceedings*, Vol. 1, U.S. Department of Health, Education, and Welfare, Food and Drug Administration, Rockville, Md., pp. 58-76.
5. Anonymous (1976). General Foods–Market Research Corporation of America Food Consumption Study, Food and Drug Administration, Hearing Clerk File, Administrative Record–Aspartame 75F-0355, File Vol. 103.
6. Mazur, R. H. (1976). Aspartame–A sweet surprise. *J. Toxicol. Environ. Health*, 2, 245-249.
7. Matthews, D. M. (1984). Absorption of peptides, amino acids, and their methylated derivatives. In *Aspartame: Physiology and Biochemistry* (Stegink, L. D., and Filer, L. J., Jr., eds.), Marcel Dekker, New York, Chap. 3.
8. Koch, R., Shaw, K. N. F., Williamson, M., and Haber, M. (1976). Use of aspartame in phenylketonuric heterozygous adults. *J. Toxicol. Environ. Health* 2, 453-457.
9. Stegink, L. D., Filer, L. J., Jr., Baker, G. L., and McDonnell, J. E. (1979). Effect of aspartame loading upon plasma and erythrocyte amino acid levels in phenylketonuric heterozygotes and normal adult subjects. *J. Nutr.* 109, 708-717.
10. Stegink, L. D., Filer, L. J., Jr., Baker, G. L., and McDonnell, J. E. (1980). Effect of an abuse dose of aspartame upon plasma and erythrocyte levels of amino acids in phenylketonuric heterozygous and normal adults. *J. Nutr.* 110, 2216-2224.
11. Stegink, L. D. (1984). Aspartame metabolism in humans:Acute dosing studies. In *Aspartame: Physiology and Biochemistrv* (Stegink, L. D., and Filer, L. J., Jr., eds.), Marcel Dekker, New York, Chap. 26.
12. Koch, R., Shaw, K. N. F., Acosta, P. B., Fishler, K., Schaeffler, G., Wenz, E., and Wohlers, A. (1970). Clinical aspects of phenylketonuria. *J. Pediatr.* 76, 815.

13. Shaw, K. N. F., and Gortatowski, M. (1971). Analytical methods used in phenylketonuria. In *Phenylketonuria* (Bickel, H., Hudson, F. P., and Woolf, L. I., eds.), George Thieme Verlag, Stuttgart, pp. 47-56.
14. Koch, R., Schaeffler, G., and Shaw, K. N. F. (1976). Results of loading doses of aspartame by two phenylketonuric (PKU) children compared with two normal children. *J. Toxicol. Environ. Health* 2, 459-469.
15. Stegink, L. D., Koch, R., Blaskovics, M. E., Filer, L. J., Baker, G., and McDonnell, J. E. (1981). Plasma phenylalanine levels in phenylketonuric heterozygous and normal adults administered aspartame at 34 mg/kg body weight. *Toxicology* 20, 81-90.
16. Williamson, M. L., Koch, R., Azen, C., and Chang, C. (1981). Correlates of intelligence test results in treated phenylketonuric children. *Pediatrics* 68, 161-167.
17. Pitkin, R. M. (1984). Aspartame ingestion during pregnancy. In *Aspartame: Physiology and Biochemistry* (Stegink, L. D., and Filer, L. J., Jr., eds.), Marcel Dekker, New York, Chap. 27.

31

Interactions of Aspartame and Glutamate Metabolism

Lewis D. Stegink
University of Iowa College of Medicine, Iowa City, Iowa

INTRODUCTION

Aspartame is a dipeptide (L-aspartyl-L-phenylalanine methyl ester) sweetener that has recently been approved for use in the United States, Canada, and Europe. When ingested orally it is hydrolyzed in both the intestinal lumen and mucosa to its component amino acids and methanol, which in turn are handled in a manner similar to aspartate, phenylalanine, and methanol arising from the digestion of dietary protein and methylated polysaccharides (Fig. 1).

Olney (1-3) and Reif-Lehrer (4) have suggested that aspartame not be added to the food supply because of interactions between the aspartate contained in aspartame and the glutamate contained in monosodium L-glutamate (MSG). Their concerns were based on findings that the administration of large amounts of glutamate and aspartate to infant rodents produced hypothalamic neuronal necrosis.

There is no doubt that the administration of large amounts of glutamate and/or aspartate to neonatal rodents produces elevated plasma glutamate and aspartate concentrations (5-10) and hypothalamic neuronal necrosis (8-17). However, the ability of these amino acids to produce neuronal necrosis in infant nonhuman primates is questionable. Although Olney and colleagues report neuronal necrosis in infant nonhuman primates given large doses of glutamate (18,19), four other research groups have been unable to produce a lesion (20-30), even when plasma glutamate plus aspartate concentrations are grossly elevated (24,29,30). These

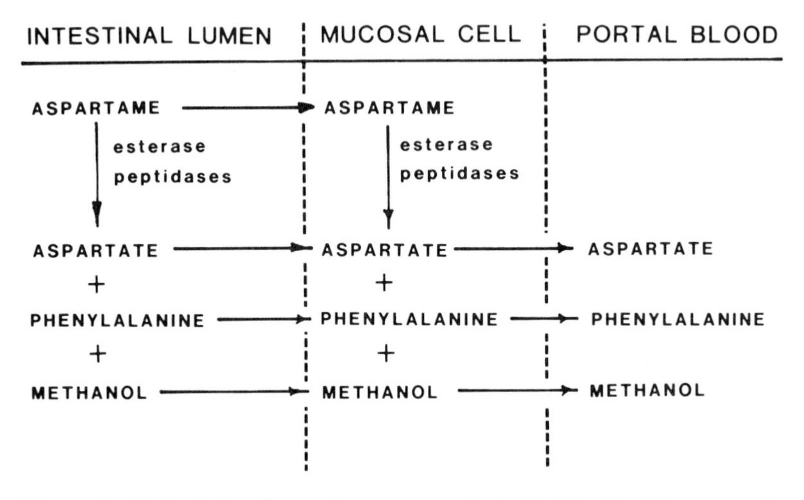

Figure 1 Hydrolysis of aspartame occurs in both intestinal lumen and mucosal cells, ultimately releasing methanol, aspartate, and phenylalanine to the portal blood.

other groups generally had no difficulty in producing the glutamate-induced lesion in infant rodents. Aspartate- or aspartame-induced lesions have not been reported in primates. In fact, Reynolds et al. (26,29) were unable to detect lesions in infant nonhuman primates given large doses of aspartame with large doses of MSG, although these doses produced grossly elevated plasma glutamate and aspartate levels.

The controversy over nonhuman primate sensitivity to dicarboxylic amino acids led us to explore additional ways of evaluating the potential of these compounds to produce neurotoxicity. Even the highly sensitive neonatal mouse tolerates substantial doses of glutamate and aspartate without developing neuronal necrosis. For example, our data (5,10) and those of Takasaki and colleagues (8,9) indicate that plasma glutamate plus aspartate concentrations must reach at least 60-100 μmol/dl (5-10 times normal) in infant mice before neuronal necrosis is noted. When MSG is given in water, the lowest dose producing lesions in infant mice is 500 mg/kg body weight (2.68 mmol/kg body weight) (12,13). Interestingly, MSG doses greater than 500 mg/kg body weight fail to produce neuronal necrosis when the MSG is ingested with food by normally hydrated animals (31, 32), although dehydrated animals are apparently more sensitive (33). When aspartate is given in water to infant mice, the lowest dose producing lesions is 650 mg/kg body weight (4.87 mmol/kg body weight) (10,34,35). Thus dicarboxylic amino acid toxicity appears to require at least two factors: a sensitive animal species and grossly elevated plasma levels of glutamate and/or aspartate.

The metabolism and toxicity of glutamate and aspartate are interrelated. For example, Olney and Ho (13) report that glutamate and aspartate are additive in their ability to produce neuronal necrosis in mice. Furthermore, each amino acid is rapidly converted to the other. Thus the administration of large glutamate loads to animals or humans increases plasma concentrations of both glutamate and aspartate (6,9,24,36,37). Similarly, aspartate loads increase both plasma aspartate and glutamate concentrations (7,10,34,35). These considerations led Olney (1-3) and Reif-Lehrer (4) to suggest that plasma concentrations of the individual dicarboxylic amino acids would increase greatly when both aspartame and MSG were ingested as a part of a meal, increasing the potential for neurotoxicity.

This possibility was tested by measuring plasma and erythrocyte amino acid concentrations in normal adult volunteers fed high-protein meals with and without aspartame plus MSG. Three different meal studies were carried out to test this point. The first study investigated possible interactions of MSG and aspartame when each was ingested at 34 mg/kg body weight (38,39). The second study (40) investigated the interaction of aspartame and MSG when each was present at levels estimated at the time to approximate the acceptable daily intake (ADI) level expected to be set by the Food and Agriculture Organization of the United Nations and the World Health Organization (FAO/WHO). The third study investigated the interactions of aspartame and MSG when ingested together in a soup-beverage meal providing little carbohydrate (41).

STUDY 1

The first study investigated the interactions of MSG and aspartame when each was added to a high protein meal at 34 mg/kg body weight. This study addressed two of the concerns raised by Olney about MSG and aspartame (1-3). Olney's first concern was that free glutamate (MSG) added to the food would be metabolized differently than protein-bound glutamate. Olney concedes that the large quantities of peptide-bound glutamate and aspartate contained in dietary protein are readily metabolized. However, he suggested that free glutamate added to meals would be absorbed and metabolized differently than protein-bound glutamate (1-3). Olney's suggestion that free glutamate was absorbed more rapidly than peptide-bound glutamate, producing a rapid early rise in blood glutamate levels and an increased potential for neuronal damage, had some scientific basis. Protein-bound glutamate and aspartate are absorbed by peptide absorption sites in the gut, whereas free glutamate and aspartate are absorbed by transport sites specific for these amino acids in free form (for a review see Refs. 42 and 43). Olney's second concern was that the addition of aspartame to meals already containing MSG would produce even higher plasma glutamate and aspartate concentrations than would result from the addition of MSG alone.

Olney's first hypotheses was tested by evaluating plasma and erythrocyte amino acid concentrations in normal adult volunteers fed a high-protein meal with and without added MSG. The level of MSG selected for study (34 mg/kg body weight) approximates the level estimated to be the 90th percentile for the total daily ingestion of added MSG by the age group in the United States having the highest ingestion level (44).

A total of 12 normal, healthy adult subjects were studied, 6 male and 6 female. Subjects, fasted 8-10 hr, ingested one of the two test meals at 0800. Six subjects (three male, three female) ingested the test meal without added MSG. The other six subjects ingested the test meal to which MSG had been added at a level providing 34 mg/kg body weight.

The test meal (composition shown in Table 1) consisted of lean ground beef served with a bun, and 200 ml of a vanilla-flavored milk shake. The test meals were fed at a level providing 1 g of protein per kilogram body weight. The quantities of total free and peptide-bound glutamate and aspartate provided by the two meals are shown in Table 2.

Figure 2 shows plasma glutamate and aspartate concentrations in the two groups of individuals ingesting the test meal with and without added MSG. Plasma glutamate concentration increased significantly ($p \leqslant 0.05$) after ingestion of the meal without added MSG, rising from a mean (±SD) base-line value of 4.02 ± 1.84 μmol/dl to a mean peak value of 9.23 ± 3.77 μmol/dl at 3 hr. This increase presumably reflects the protein-bound glutamate in the meal. The addition of 34 mg/kg body weight MSG to the meal did not significantly increase plasma glutamate concentration above values noted in individuals ingesting the meal without added MSG. Particular attention was given to early time points after meal ingestion, since Olney had proposed (1,2) that the addition of MSG to a meal would result in a rapid, early elevation of plasma glutamate. Similarly, the mean areas under the plasma glutamate concentration-time curve did not differ significantly between groups.

Plasma aspartate concentration was also considered. Aspartate is a major glutamate metabolite (24,36,37), and Olney and Ho (13) reported that aspartate and glutamate are additive in producing hypothalamic neuronal necrosis in the

Table 1 Composition of the Test Meal for a 70-kg Adult[a]

Component	Quantity (g)	Protein (g)	Fat (g)	CHO (g)	Energy (kcal)
Hamburger	214	60	24.6	0	461
Bun	50	4.5	1.5	25.5	133
Milk	100	3.5	3.5	5	65
Ice cream	50	2.0	5.0	11	97
Total	414	70	34.6	41.5	757

[a]The hamburger content of the meal was varied by subject to provide protein at 1 g/kg body weight. The hamburger assayed at 28 g protein/100 g cooked meat. Protein provided 37% of total dietary energy, fat 41%, and carbohydrate 22%.

Table 2 Estimated Intake of Protein, Aspartate, Glutamate, and Phenylalanine in Hamburger-Milk Shake Meal Studies

Aspartame added (mg/kg) body weight	MSG added (mg/kg) body weight	Protein (g/kg) body weight	Total aspartate[a,b] (mg/kg) body weight	Total glutamate[a,b] (mg/kg) body weight	Phenylalanine[a] (mg/kg) body weight
0	0	1.0	89	169	42
0	34	1.0	89	196[c]	42
34	34	1.0	102	196[c]	59
0	150	1.0	89	286[c]	42
23	150	1.0	98	286[c]	54

[a]Total glutamate, aspartate, and phenylalanine values calculated from the data of Orr and Watt (69).

[b]Since the values were obtained after acid hydrolysis, the total glutamate values reported by Orr and Watt (69) include both glutamate and glutamine, while total aspartate values include both aspartate and asparagine. The approximate quantities of glutamate and aspartate can be calculated using the data of Jukes et al. (70). These data indicate that 55% of the total glutamate resulting from acid hydrolysis of a typical protein arises from glutamate. The precise quantity of each dicarboxylic amino acid or its amide presented for mucosal metabolism will vary considerably because of glutamine and asparagine's tendency to undergo deamination.

[c]Corrected for the sodium content and water of hydration of MSG (78% of MSG is glutamate).

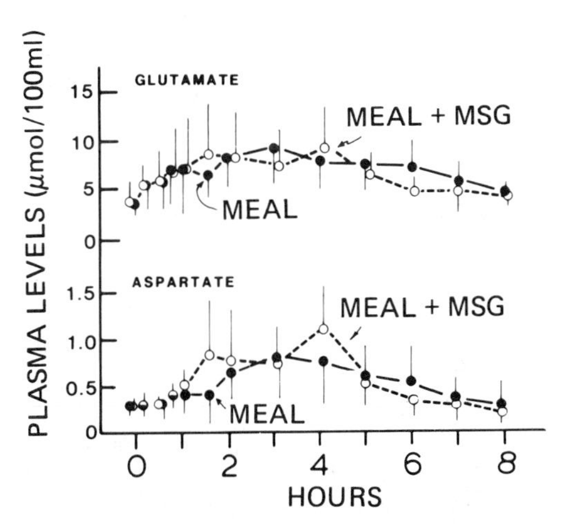

Figure 2 Mean (±SD) plasma glutamate and aspartate concentrations in normal adults ingesting a hamburger-milk shake meal with (○- - -○) and without (●——●) added monosodium L-glutamate (34 mg/kg body weight).

neonatal mouse. Ingestion of the meal's protein-bound aspartate significantly increased plasma aspartate concentrations. The mean (±SD) plasma aspartate concentration increased from a base-line value of 0.32 ± 0.15 μmol/dl to a mean peak value of 0.83 ± 0.33 μmol/dl 3 hr after meal ingestion. However, the addition of MSG to the test meal had no significant effect on plasma aspartate concentration beyond the effect of the meal alone. Similarly, the area under the plasma aspartate concentration-time curves did not differ significantly between groups.

It has been suggested that certain amino acids, including glutamate and aspartate, are transported via the erythrocyte to a greater extent than plasma (45-48). Accordingly, erythrocyte-free amino acid concentrations were measured after ingestion of the test meals to determine if the failure of added MSG to increase plasma glutamate concentration reflected glutamate transfer by erythrocytes rather than by plasma. The effect of the test meals on erythrocyte aspartate and glutamate concentrations is shown in Figure 3. Neither meal had a significant effect on erythrocyte glutamate or aspartate concentration.

Thus our data did not support Olney's suggestion (1-3) that free glutamate added to a meal is absorbed much more rapidly than protein-bound glutamate, producing higher plasma glutamate levels. The failure of added free glutamate

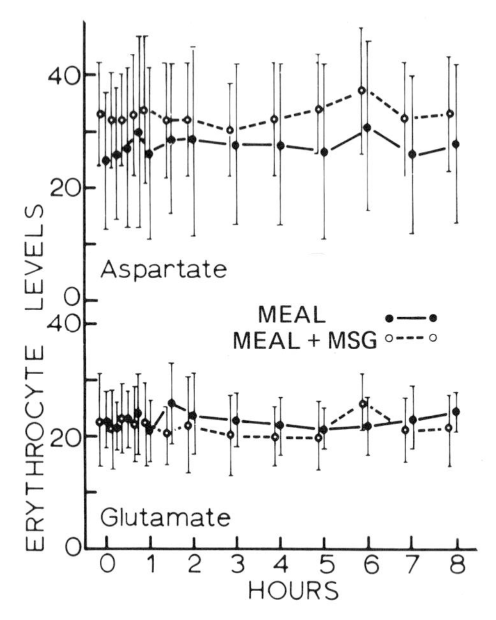

Figure 3 Mean (±SD) erythrocyte glutamate and aspartate concentrations (μmol/100 g) in normal adults ingesting hamburger-milk shake meals with (○- - -○) or without (●——●) added monosodium L-glutamate (34 mg/kg body weight). (Redrawn from Ref. 53.)

to increase plasma glutamate concentration above values produced by the meal alone probably reflects the rapid metabolism of glutamate by intestinal mucosal cells. The rapid metabolism of glutamate by mucosal cells appears to be independent of whether glutamate is absorbed in peptide-bound or free form. Meal studies in humans (49,50) indicate that glutamate present in dietary protein is absorbed as small peptides rather than as the free amino acid. However, the gut also has specific transport sites for free glutamate (51,52). Thus glutamate is probably absorbed differently, depending on whether present in the diet in protein-bound or free form.

The ingestion of large quantities of protein-bound glutamate alone (169 mg/kg body weight) had only a small effect on plasma glutamate concentration. This indicates the considerable metabolic capacity available for handling dietary protein-bound glutamate. If a similar quantity of free glutamate (150 mg/kg body weight) is ingested in water, mean plasma glutamate concentration increases rapidly, peaking at 72 μmol/dl (53). The relatively small increase in plasma glutamate concentration noted after ingestion of the meal's protein-bound glutamate probably reflects rapid metabolism due to the presence of carbohydrate in the meal (53) and lower availability of glutamate for absorption due to slow release of peptide-bound glutamate from intact protein by proteolysis. The usual slow release of protein-bound amino acids to the absorptive process after ingestion of the meal is illustrated by the plasma values for the branched-chain amino acids (Fig. 4). Peak values of these amino acids are not reached until 4-6 hr after ingestion of the meal.

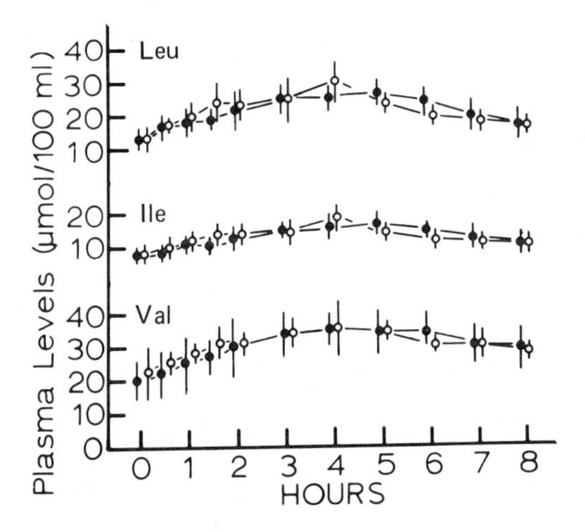

Figure 4 Mean (±SD) plasma concentrations of leucine (Leu), isoleucine (Ile), and valine (Val) in normal adults ingesting hamburger-milk shake meals with (○- - -○) and without (●——●) added monosodium L-glutamate (34 mg/kg body weight). (From Ref. 38.)

Having shown that the first of Olney's assumptions was not correct, we turned our attention to the second. Olney (1-3) and Reif-Lehrer (4) had suggested that the metabolism of glutamate and aspartate might individually decrease when both aspartame and MSG were ingested as part of a meal. Such conditions were postulated to produce elevated plasma glutamate and aspartate concentrations and to increase the potential for neurotoxicity. This interaction should be greatest when the meal also provides large quantities of protein-bound glutamate and aspartate.

This hypothesis was tested by measuring plasma and erythrocyte amino acid concentrations in normal adult volunteers fed high-protein meals with and without added aspartame plus MSG (39).

Six normal healthy adult subjects, three male and three female, were fed two meals. No additions were made to one test meal, while aspartame (34 mg/kg body weight) and MSG (34 mg/kg body weight) were both added to the other. Subjects, fasted 8-10 hr, ingested the two test meals in a randomized cross-over design. The test meal was the same hamburger-milk shake meal used in the previous study (Tables 1 and 2) and was fed at a level providing 1 g of protein per kilogram body weight.

Figure 5 shows plasma glutamate and aspartate concentrations in these subjects. Particular attention was given to early time points after meal ingestion, since Olney (1-3) had proposed that the addition of MSG and/or aspartame (APM) to meals would result in rapid early elevation of plasma glutamate concentration.

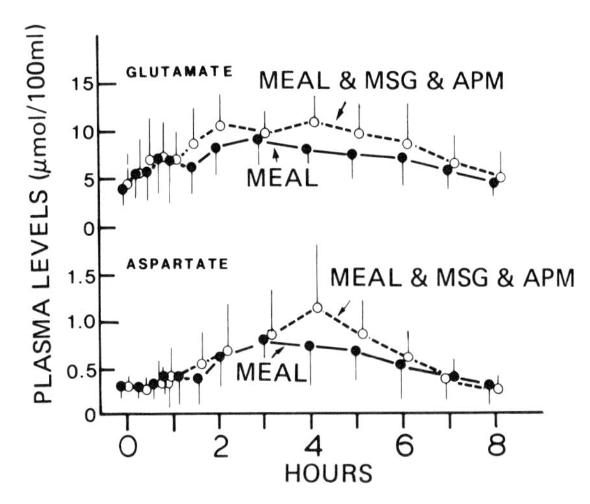

Figure 5 Mean (±SD) plasma glutamate and aspartate concentrations in normal adult subjects ingesting hamburger-milk shake meals with no addition (●——●) or with added monosodium L-glutamate (34 mg/kg body weight) and aspartame (APM, 34 mg/kg body weight) (○- - -○).

Contrary to Olney's prediction, the addition of MSG plus aspartame to the meal did not significantly increase plasma glutamate concentration above values noted after ingestion of the meal. The only significant difference in plasma glutamate response was noted at 4 hr, when a small but statistically significant increase was noted when MSG plus aspartame were added to the test meal. Furthermore, the area under the plasma glutamate concentration-time curve did not differ significantly between groups. The addition of aspartame plus MSG to the test meal had no significant effect on plasma aspartate concentration beyond the effect of the meal alone. Similarly, the area under the plasma aspartate concentration-time curve did not differ significantly between meals.

Erythrocyte levels of glutamate and aspartate were measured to determine if the failure of added MSG and aspartame to increase plasma glutamate and aspartate concentration reflected transfer by erythrocytes. Erythrocyte glutamate and aspartate concentrations are shown in Figure 6. Neither meal had a significant effect on erythrocyte glutamate and aspartate concentration.

Plasma phenylalanine and tyrosine concentrations are shown in Figure 7. Plasma phenylalanine concentration increased significantly after ingestion of the meal alone, rising from a mean (±SD) fasting level of 4.72 ± 0.52 μmol/dl to a mean peak value of 7.14 ± 0.82 μmol/dl at 4 hr, reflecting the meal's protein-bound phenylalanine content. Plasma phenylalanine levels were significantly

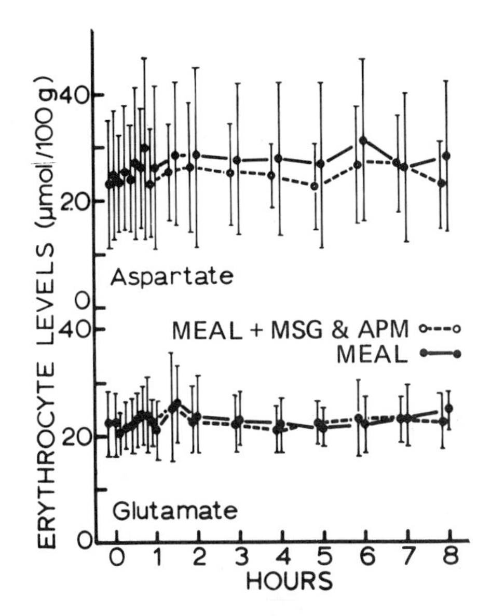

Figure 6 Mean (±SD) erythrocyte glutamate and aspartate concentrations in normal adults ingesting hamburger-milk shake meals with no additions (●——●) or with added monosodium L-glutamate (34 mg/kg body weight) and aspartame (34 mg/kg body weight (○- - -○).

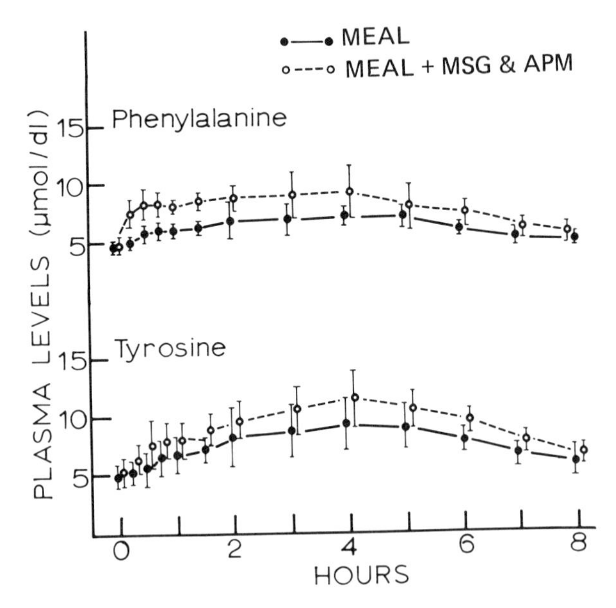

Figure 7 Mean (±SD) plasma phenylalanine and tyrosine concentrations in normal adult subjects ingesting hamburger-milk shake meals with no additions (●——●) or with added monosodium L-glutamate (34 mg/kg body weight) and aspartame (34 mg/kg body weight) (○- - -○).

higher after ingestion of the meal with added aspartame. The mean plasma phenylalanine concentration was significantly higher ($p < 0.05$) from 0.25 to 4 hr after ingestion of the meal. Similarly, the mean (±SD) area under the plasma phenylalanine concentration-time curve was significantly higher ($p < 0.05$) after ingestion of the meal with added MSG and aspartame (1440 ± 817 units) than after ingestion of meal alone (740 ± 333 units). These differences presumably reflect aspartame's phenylalanine content.

The mean plasma tyrosine concentration also increased significantly after meal ingestion. When the meal alone was ingested, the mean (±SD) plasma tyrosine concentration increased from a fasting value of 4.96 ± 1.09 μmol/dl to a mean peak value of 9.31 ± 2.26 μmol/dl at 4 hr. The addition of aspartame and MSG to the meal resulted in a small increase in the plasma tyrosine concentration with a mean (±SD) peak value of 11.5 ± 2.30 μmol/dl noted at 4 hr, presumably reflecting the conversion of phenylalanine provided by aspartame to tyrosine. These differences were statistically significant at 0.75, 3, 4, and 6 hr after meal ingestion. However, the mean (±SD) area under the plasma tyrosine concentration-time curve did not differ significantly between ingestion of the meal alone (1331 ± 675 units) and the meal with added MSG and aspartame (1915 ± 653 units).

These data do not support the suggestion (1-4) that addition of aspartame and MSG to meals will substantially elevate plasma concentrations of the dicarboxylic amino acids. The failure of added aspartame and MSG to increase plasma aspartate or glutamate concentrations above values produced by ingestion of the meal alone at any time point, except at 4 hr, undoubtedly results from the rapid metabolism of aspartate and glutamate by intestinal mucosal cells (42). The present study clearly demonstrates that plasma glutamate and aspartate concentrations are not elevated beyond usual postprandial values at this level of aspartame and MSG addition. Thus the available data indicate little hazard from simultaneous ingestion of aspartame plus MSG.

Aspartame addition to the meal increased dietary intake of phenylalanine from 42 to 63 mg/kg body weight (Table 2). As a result, plasma phenylalanine and tyrosine concentrations were slightly higher when aspartame was present. Phenylalanine derived from aspartame is already in peptide form and readily accessible to the peptide absorption sites in the intestinal mucosa (54,55). Thus aspartame ingestion would be expected to increase plasma phenylalanine levels early in the time curve, a point where release of free amino acids and small peptides by proteolysis is limited.

The plasma phenylalanine levels noted after ingestion of the meal with added aspartame were well within the normal postprandial range for term infants given formula feedings (56) and adults ingesting a variety of meals (57). These phenylalanine levels are markedly below those associated with toxicity in the young infant (see the discussion in Ref. 58).

The doses of MSG and aspartame studied were relatively large. Data taken from the Committee on GRAS List Survey–Phase III (44) indicate a mean expected daily intake of 6.8 mg MSG per kilogram body weight in the age group with the highest ingestion level (Table 3). In this age group 30 mg/kg body weight

Table 3 Expected Daily Intake of Monosodium Glutamate Based on Person-Days (Means and Percentiles by Age)

	Total sample intake (mg/kg per day)			
Age	Mean	90th percentile	99th percentile	99.9th percentile
0-5 months	0.3	0	11	25
6-11 months	1.9	1.9	36	46
12-23 months	6.8	30	43	61
2-5 years	5.5	23	37	56
6-17 years	2.7	10	35	40
18+ years	1.5	7	12	19

Source: From Appendix E of Ref. 62.

MSG represents the 90th percentile for total expected daily ingestion of added MSG. The 34 mg/kg body weight dose of MSG studied is slightly above the 90th percentile of total daily ingestion and was administered in a single dose.

Similarly, the aspartame dose administered was considerable and should represent a reasonable test of whether aspartame addition to a meal already containing high levels of MSG would significantly increase plasma concentrations of glutamate and aspartate. If aspartame sweetness were to replace all sucrose sweetness in the typical American diet, daily aspartame ingestion would be 7.5-8.5 mg/kg body weight, with the 99th percentile of projected daily intake being 34 mg/kg body weight (59). Similar projected intake values have been calculated by the Food and Drug Administration (60) and the Market Research Corporation of America (61). The aspartame dose given closely approximates the acceptable daily intake value (40 mg/kg body weight) recently established by the FAO/WHO (62).

STUDY 2

The second study extended our initial one and tested the interaction between MSG and aspartame when MSG was ingested in large amounts (40). Plasma amino acid levels were measured to determine whether the addition of aspartame to the meal increased plasma glutamate plus aspartate levels beyond those produced by the meal providing only MSG.

Six normal, healthy adult subjects (three male and three female) were studied. Subjects were studied in a Latin square design (63) using three different meal systems, each providing 1 g of protein per kilogram body weight. The meals included (a) the hamburger-milk shake meal alone, (b) the hamburger-milk shake meal with MSG added to provide 150 mg/kg body weight, and (c) the hamburger-milk shake meal with MSG added to provide 150 mg/kg body weight and aspartame added to provide 23 mg/kg body weight. The composition of the test meal is shown in Table 1. Estimated intakes of protein, aspartate, glutamate, and phenylalanine are shown in Table 2.

Figure 8 shows plasma glutamate concentrations. After ingestion of the meal alone, the mean (±SD) plasma glutamate concentration increased from a fasting value of 5.47 ± 2.16 μmol/dl to a peak value of 12.3 ± 3.47 μmol/dl and then declined slowly. The mean (±SD) area under the plasma glutamate concentration-time curve for the 8 hr period was 1259 ± 259 units. The addition of MSG at 150 mg/kg body weight significantly ($p = 0.001$) increased peak plasma glutamate and the area under the curve above values noted after ingestion of the meal alone. The mean peak plasma glutamate concentration was 21.1 ± 8.6 μmol/dl 45 min after meal ingestion, and the area under the curve (AUC) was 2837 ± 191 units. The addition of aspartame (23 mg/kg) and MSG (150 mg/kg) to the meal did not significantly increase peak plasma glutamate concentration or the AUC above

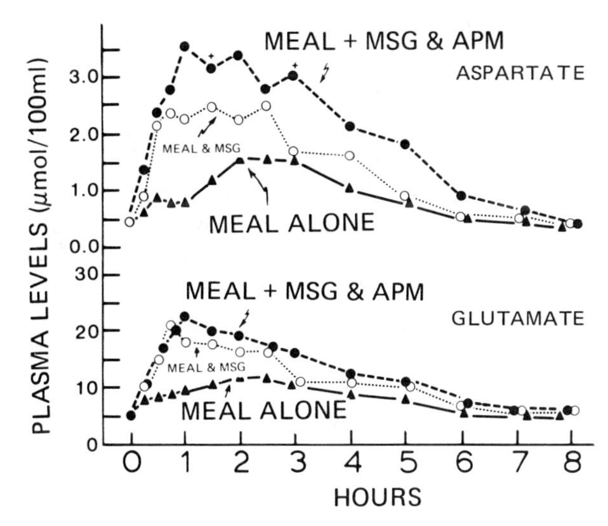

Figure 8 Mean plasma glutamate and aspartate concentrations in normal adults ingesting hamburger-milk shake meals with and without added monosodium L-glutamate and aspartame: no additions (▲—▲), with MSG added at 150 mg/kg body weight (○- - -○), with MSG added at 150 mg/kg body weight and aspartame added at 23 mg/kg body weight (●- - -●). Values designated with a + symbol differ significantly from values observed after ingestion of the meal providing MSG ($p < 0.05$).

values noted after ingestion of the meal plus MSG. A mean peak plasma glutamate level of 23.0 ± 9.29 μmol/dl was noted 60 min after this meal; the area under the curve was 3647 ± 1339 units.

Similar changes were noted in plasma aspartate concentrations (Figure 8). The mean (±SD) plasma aspartate concentration increased significantly from a fasting value of 0.38 ± 0.24 μmol/dl to a peak value of 1.58 ± 0.67 μmol/dl 2.5 hr after the meal alone, with an AUC of 287 ± 111 units. Addition of MSG to the meal significantly ($p < 0.05$) increased the mean peak plasma aspartate level above that noted following the meal alone, with a mean peak value of 2.52 ± 0.66 μmol/dl noted 2.5 hr after meal ingestion. However, the AUC (414 ± 120 units) did not differ significantly. The addition of aspartame and MSG to the meal resulted in a small but statistically significant ($p < 0.05$) increase in plasma aspartate level above that noted after the meal with MSG. A mean peak plasma aspartate value of 3.65 ± 1.86 μmol/dl was noted 60 min after meal ingestion. However, the area under the plasma aspartate concentration-time curve (685 ± 352) did not differ significantly ($p > 0.05$) from the value noted after the meal with MSG alone.

The combined values for plasma glutamate plus aspartate concentrations are shown in Figure 9. Plasma glutamate plus aspartate levels increased from a mean (±SD) fasting level of 5.86 ± 1.99 μmol/dl to a peak value of 13.9 ± 3.45 μmol/dl 2.5 hr after the meal alone. The area under the plasma glutamate plus aspartate concentration-time curve is 1521 ± 402 units. Addition of MSG to the meal resulted in a significantly higher mean plasma glutamate plus aspartate concentration, with the peak value reaching 23.5 ± 9.44 μmol/dl 45 min after loading. Similarly, the AUC increased significantly (3276 ± 267 units). However, the addition of aspartame plus MSG to the meal did not significantly increase the mean peak plasma glutamate plus aspartate level or AUC beyond those values resulting from ingestion of the meal with MSG. The mean (±SD) peak plasma glutamate plus aspartate concentration was 26.4 ± 10.5 μmol/dl 60 min after ingestion of the meal containing aspartame and MSG; the AUC was 4190 ± 1790 units.

These studies demonstrate that the addition of aspartame to meal systems containing large quantities of both free and protein-bound glutamate does not produce higher plasma glutamate plus aspartate concentrations than ingestion of a meal containing MSG alone.

The mean (±SD) plasma phenylalanine concentration (Fig. 10) increased from a fasting level of 5.09 ± 0.44 μmol/dl to a peak value of 9.75 ± 2.55 μmol/dl 2.5 hr after ingestion of the meal alone. The AUC was 1106 ± 515 units. Addition of MSG to the meal had no significant effect upon either peak plasma phenylalanine

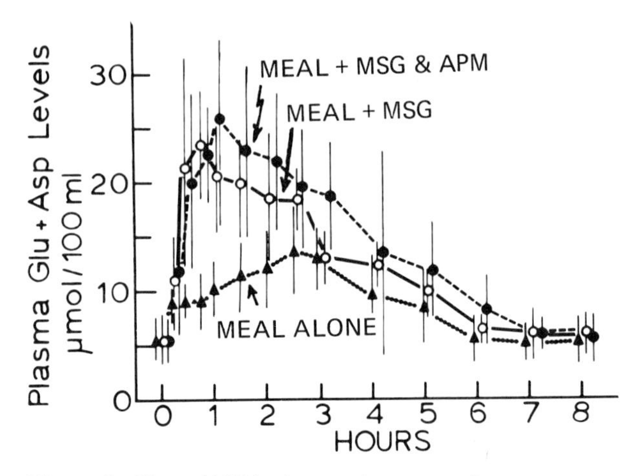

Figure 9 Mean (±SD) plasma glutamate plus aspartate concentrations in normal adults ingesting hamburger-milk shake meals with and without added monosodium L-glutamate and aspartame: no additions (▲- - -▲), with MSG added at 150 mg/kg body weight (○——○), with MSG and aspartame added at 150 and 23 mg/kg body weight, respectively (●- - -●).

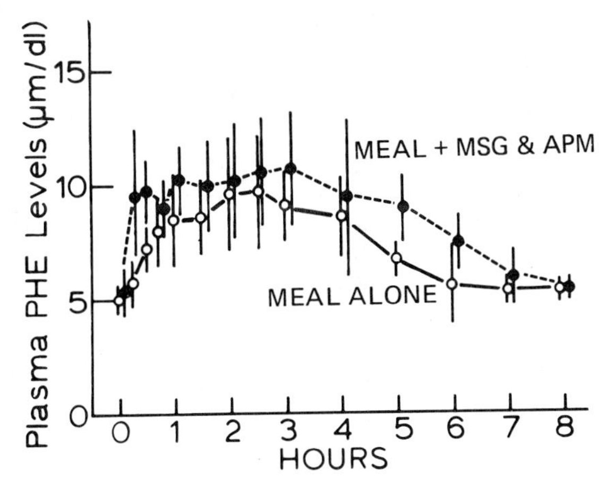

Figure 10 Mean (±SD) plasma phenylalanine concentrations in normal adults ingesting hamburger-milk shake meals with no additions (○——○) or with monosodium L-glutamate and aspartame added at 150 and 23 mg/kg body weight, respectively (●- - -●).

levels (8.95 ± 1.12 μmol/dl) or AUC (881 ± 282 units). However, the addition of MSG and aspartame significantly increased both the mean peak plasma phenylalanine level and AUC. The mean peak plasma phenylalanine value was 10.2 ± 1.62 μmol/dl 2.5 hr after eating the meal, and the AUC was 1617 ± 474 units.

The quantities of MSG and aspartame administered in this study are large. In the present study the level of MSG added (150 mg/kg body weight) closely approximated the acceptable daily intake set by the FAO/WHO (153 mg/kg body weight). In addition to this quantity of free glutamate, the meal's protein content provided approximately 169 mg/kg body weight protein-bound glutamate (includes glutamine) and 89 mg/kg body weight protein-bound aspartate (included asparagine). The total amount of ingested glutamate and aspartate is shown in Table 2.

At the time this study was carried out, the FAO/WHO had not set an acceptable daily intake value for aspartame (the current value is 40 mg/kg body weight; see Ref. 62). We estimated that the acceptable daily intake might approximate 23 mg/kg body weight, an assumption based on calculations of the total carbohydrate intake of normal adults (59). If the total carbohydrate intake of a 70-kg adult is assumed to be 50% of the total energy intake, 313 g of carbohydrate would be required. If all 313 g of carbohydrate were consumed as sucrose, this is equivalent to 4.47 g/kg body weight sucrose. On an equal sweetness basis, this amount of sucrose is equivalent to 23 mg/kg body weight aspartame.

Ingestion of a high-protein meal with added MSG (150 mg/kg body weight) increased the mean (±SD) plasma glutamate plus aspartate concentration from a normal fasting value of 5.45 ± 2.73 μmol/dl to a peak value of 23.5 ± 9.44 μmol/

dl. The maximum value for plasma glutamate plus aspartate in one subject following a meal with added MSG was 35.3 μmol/dl. The addition of 23 mg/kg body weight aspartame to the meal containing MSG did not significantly increase peak glutamate plus aspartate levels (23.5 ± 9.44 μmol/dl). The maximum plasma glutamate plus aspartate level following the meal with MSG plus aspartame was 35.4 μmol/dl. These observations on mean and maximum plasma glutamate plus aspartate concentrations agree well with those in the previous studies (38,39).

The aspartame dose administered provides approximately 12.5 mg phenylalanine per kilogram body weight, in addition to the 42 mg/kg body weight provided by the protein. As shown in Figure 10, aspartame addition significantly increased plasma phenylalanine levels and the plasma phenylalanine AUC above values noted after ingestion of the meal with MSG alone. The peak plasma phenylalanine levels observed (10.7 ± 2.28 μmol/dl) are well within the normal postprandial range for normal, orally fed human infant and adult subjects (56,57) and represent no hazard.

STUDY 3

The final study represented in our minds the ultimate test of Olney's (1-3) and Reif-Lehrer's (4) hypotheses on the interactions between glutamate and aspartame. To understand the rationale for this experiment it is necessary to briefly review data on the interaction between carbohydrate ingestion and glutamate metabolism (53). Our studies have shown that a marked difference in plasma glutamate concentration occurs depending on whether glutamate ingestion is in water, with a meal containing carbohydrate, or in water with carbohydrate. When glutamate is ingested with water, plasma glutamate concentrations increase proportionally to dose (53). However, when carbohydrate is present, plasma glutamate levels are much lower and may barely increase above fasting levels (Fig. 11) (53,64-66). When MSG (150 mg/kg body weight) is ingested with a high-carbohydrate meal (64), the mean plasma glutamate concentration does not exceed average postprandial values. However, when an equivalent dose of MSG is administered in water, the plasma glutamate concentration increases dramatically, peaking at 75 μmol/dl (65).

It is not clear how ingestion of carbohydrate with glutamate modulates the absorption of glutamate and/or its metabolism. A slower rate of absorption would permit greater catabolism of glutamate by the intestinal mucosa, resulting in a decreased release of glutamate to portal blood. Alternatively, carbohydrate may directly stimulate glutamate metabolism in intestinal mucosal cells by the metabolic pathway shown in Figure 12. Our studies suggest that carbohydrate is absorbed into the intestinal mucosa and converted to glucose and pyruvate, facilitating the transamination and metabolism of glutamate by mucosal cells (66).

Thus if we were to pick a meal system in which the additive effect of aspartame upon plasma glutamate and aspartate concentration should be greatest, it

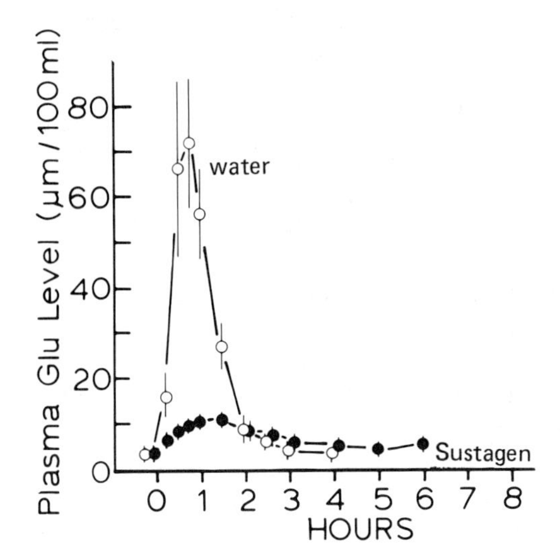

Figure 11 Mean (±SD) plasma glutamate concentration in normal adult subjects ingesting 150 mg/kg body weight of monosodium L-glutamate in either water (○—○) or Sustagen (Mead Johnson, Evansville, IN) (●—●). (From Ref. 66.)

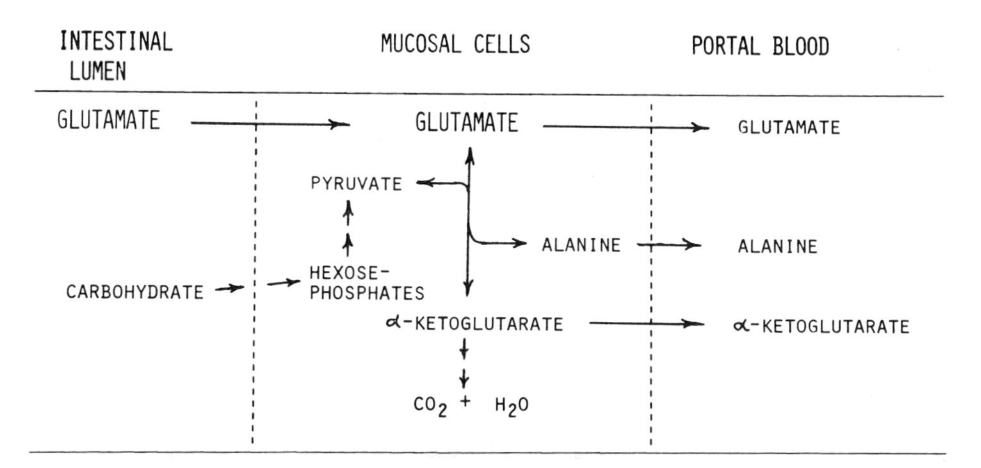

Figure 12 Suggested role of carbohydrate in facilitating glutamate metabolism in intestinal mucosa. Dietary carbohydrate serves as a pyruvate source facilitating glutamate transamination. (From Ref. 65.)

would be a meal with little carbohydrate, such as a soup-beverage meal. Our final study (41) investigated the interaction of MSG and aspartame in such a "dieter's lunch."

A recent review in *Consumer Reports* (67) listed the MSG content of 42 commercially available soups. These data indicate a mean level of 735 mg of MSG per serving, with serving sizes ranging from 6 to 15 oz. A 70-kg adult ingesting an average serving could receive a dose of 10.5 mg/kg body weight of MSG, a 26-kg child a dose of 28 mg/kg body weight, and a 15-kg child a dose of 49 mg/kg body weight. For this study we chose a soup essentially free of MSG that contained as little carbohydrate as possible. The composition of this soup is shown in Table 4. Consommé usually contains MSG at levels varying from 0.15 to 0.30%. When 50 mg/kg body weight MSG was added to this consommé, a 70-kg person received 3.50 g of MSG in 300 ml of soup, approximately a 1.16% solution.

The quantity of aspartame added to the beverage (34 mg/kg body weight) was considerably greater than the expected use level per meal. The usual soft drink contains about 10% sucrose (wt/vol). A 12-oz bottle of such a beverage would provide about 36 g of sucrose, equivalent to 200 mg of aspartame (using a sweetening factor of aspartame relative to sucrose of 180:1). A 70-kg adult ingesting 12 oz of this beverage would ingest 2.9 mg/kg body weight aspartame, a 26-kg child 7.7 mg/kg body weight, and a 15-kg child 13.3 mg/kg body weight. The aspartame dose used in our studies (34 mg/kg body weight) represents the 99th percentile of projected aspartame ingestion (59).

A total of 12 normal adult volunteers (6 males, 6 females) ingested soup-beverage meals with and without added MSG, with and without added aspartame in a Latin square design (63). Three soup-beverage meal systems were studied: (a) soup (no added MSG) with unsweetened beverage, (b) soup providing 50 mg/kg body weight MSG with unsweetened beverage, and (c) soup providing 50 mg/

Table 4 Composition of Beef Consommé Soup Used[a]

Component	Percent dry weight
Crude protein	6.4
Fat	11.3
Carbohydrate	22.5
Ash	56.0
Water	1.6
Free glutamate	1.2

[a]Soup base reconstituted at a level of 1.94 g base/100 ml hot water. Adult subjects ingested 4.3 ml of reconstituted soup per kilogram body weight. This level provides approximately 1 mg MSG per kilogram body weight and 18.8 mg carbohydrate per kilogram body weight if ingested without any additions.

kg body weight and sweetened beverage providing 34 mg aspartame/kg body weight.

Figure 13 shows plasma phenylalanine levels in these subjects. As expected, plasma phenylalanine levels did not increase after ingestion of the soup with or without MSG. When aspartame was added at 34 mg/kg body weight, plasma phenylalanine levels increased into the normal postprandial range. Plasma phenylalanine concentration reached a mean (±SD) peak level of 14.5 ± 4.53 μmol/dl 30 min following ingestion of the meal, returning to near base-line concentrations by 4 hr. Values noted in these subjects were similar to those noted in normal subjects ingesting the same aspartame dose dissolved in orange juice (59).

Figure 14 shows plasma glutamate plus aspartate concentrations in these subjects. The mean plasma glutamate plus aspartate concentration did not increase significantly after ingestion of the soup alone. However, the addition of MSG to the soup significantly increased the plasma glutamate plus aspartate concentration. Values increased from a mean (±SD) base-line value of 5.64 ± 2.62 μmol/dl to a mean peak concentration of 23.1 ± 7.29 μmol/dl at 30 min. The addition of aspartame to the glutamate-containing meal had no significant effect upon the plasma glutamate plus aspartate concentration beyond those caused by MSG addition alone at time points where maximal plasma glutamate and aspartate concentration were observed. The mean (±SD) peak plasma glutamate plus aspartate concentration was 26.8 ± 9.74 μmol/dl 30 min after aspartame plus MSG addition to the meal. When MSG and aspartame were present, a small but statis-

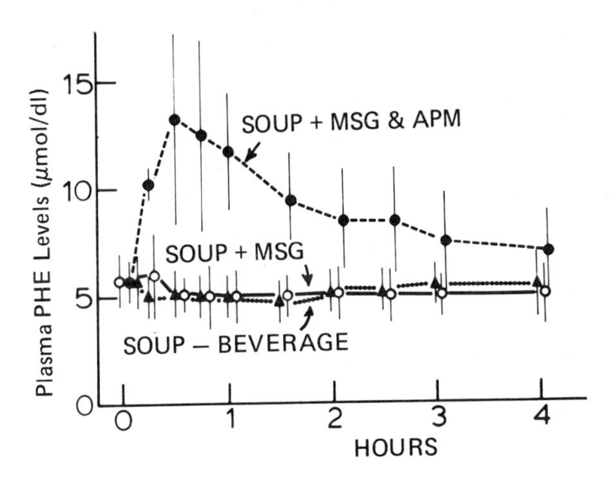

Figure 13 Mean (±SD) plasma phenylalanine concentrations in normal adult subjects ingesting soup-beverage meals with no additions (▲- - -▲), monosodium L-glutamate added at 50 mg/kg body weight (○—○), and monosodium L-glutamate and aspartame added at 50 and 34 mg/kg body weight respectively, (●- - -●).

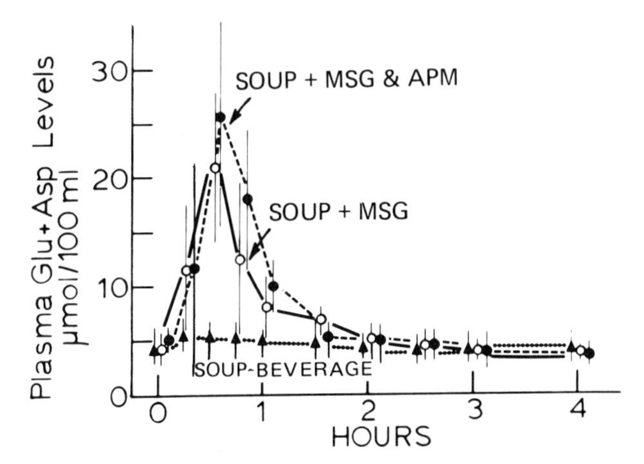

Figure 14 Mean (±SD) plasma glutamate + aspartate concentrations in normal adults ingesting soup beverage meals with no additions (▲- - -▲), monosodium L-glutamate added at 50 mg/kg body weight (○——○), and monosodium L-glutamate and aspartame added at 50 and 34 mg/kg body weight, respectively (●- - -●).

tically significant difference was noted in the plasma glutamate plus aspartate concentration 60 min after dosing. The mean (±SD) peak plasma glutamate plus aspartate concentration was 8.72 ± 3.22 μmol/dl after the addition of MSG along, and 10.5 ± 3.19 μmol/dl after addition of MSG plus aspartame.

Thus the addition of a large quantity of aspartame to the soup-beverage meal resulted in only a small increase in the plasma glutamate plus aspartate concentration, with no significant difference noted between mean peak concentrations.

The data were also examined by the method used by Olney in his review (3) of our preliminary report of data from nine subjects (68). He examined the peak plasma glutamate plus aspartate concentration for each subject after ingestion of the various meals, ignoring the time at which the value was noted. This method compares peak concentrations regardless of time. A similar analysis for all 12 subjects studied is shown in Table 5. Ingestion of the meal alone significantly

Table 5 Peak Plasma Glutamate Plus Aspartate Concentrations in Normal Adults Ingesting Soup-Beverage Meals With and Without MSG and Aspartame

Meal system fed	Peak plasma concentration[a,b] (μmol/dl)
Meal alone	7.29 ± 2.77
Meal plus MSG	23.0 ± 6.61
Meal plus MSG and aspartame	27.3 ± 9.07

[a]Ignoring time.
[b]Mean ± SD.

increased the peak plasma glutamate plus aspartate concentration above zero-time values, with concentrations peaking at 7.29 ± 2.77 μmol/dl. Mean peak plasma glutamate plus aspartate concentrations were significantly increased over baseline values after ingestion of soup with added MSG, reaching a value of 23.0 ± 6.61 μmol/dl when time factors were ignored. The mean peak value after glutamate plus aspartame ingestion was slightly but not significantly greater at 27.3 ± 9.07 μmol/dl than values observed after ingestion of soup with MSG addition alone. This indicates no significant effect of aspartame upon plasma glutamate plus aspartate beyond the effects of MSG alone.

Thus our data indicate that the addition of large quantities of aspartame to food containing large quantities of MSG has only a minimal effect upon the plasma glutamate and aspartate concentration.

SUMMARY

The addition of normal and high use levels of aspartame to meals containing large amounts of monosodium L-glutamate does not significantly affect plasma dicarboxylic amino acid levels beyond those noted from MSG alone. This is true whether aspartame is ingested with a high-protein meal providing substantial quantities of protein-bound glutamate and aspartate, or a soup-beverage meal providing little carbohydrate.

REFERENCES

1. Olney, J. W. (1975). L-Glutamic and L-aspartic acids—A question of hazard? *Food Cosmet. Toxicol.* 13, 595-596.
2. Olney, J. W. (1975). Another view of aspartame. In *Sweeteners, Issues and Uncertainties*, Academy Forum, National Academy of Sciences, Washington, D.C., pp. 189-195.
3. Olney, J. W. (1981). Excitatory neurotoxins as food additives: An evaluation of risk. *Neurotoxicology* 2, 163-192.
4. Reif-Lehrer, L. (1976). Possible significance of adverse reactions to glutamate in humans. *Fed. Proc.* 35, 2205-2211.
5. Stegink, L. D., Shepherd, J. A., Brummel, M. C., and Murray, L. M. (1974). Toxicity of protein hydrolysate solutions: Correlation of glutamate dose and neuronal necrosis to plasma amino acid levels in young mice. *Toxicology* 2, 285-299.
6. Stegink, L. D., Reynolds, W. A., Filer, L. J., Jr., Baker, G. L., Daabees, T. T., and Pitkin, R. M. (1979). Comparative metabolism of glutamate in the mouse, monkey, and man. In *Glutamic Acid: Advances in Biochemistry and Physiology* (Filer, L. J., Jr., Garattini, S., Kare, M. R., Reynolds, W. A., and Wurtman, R. J., eds.), Raven Press, New York, pp. 85-102.
7. Itoh, H., Kishi, T., Iwasawa, Y., Kawashima, K., and Chibata, I. (1979). Plasma aspartate levels in rats following administration of monopotassium aspartate via three routes. *J. Toxicol. Sci.* 4, 377-388.

8. Takasaki, Y., Matsuzawa, Y., Iwata, S., O'Hara, Y., Yonetani, S., and Ichimura, M. (1979). Toxicological studies of monosodium L-glutamate in rodents: Relationship between routes of administration and neurotoxicity. In *Glutamic Acid: Advances in Biochemistry and Physiology* (Filer, L. J. Jr., Garattini, S., Kare, M. R., Reynolds, W. A., and Wurtman, R. J., eds.), Raven Press, New York, pp. 225-275.
9. O'Hara, Y., and Takasaki, Y. (1979). Relationship between plasma glutamate levels and hypothalamic lesions in rodents. *Toxicol. Lett.* 4, 499-505.
10. Daabees, T. T., Finkelstein, M., Applebaum, A. E., and Stegink, L. D. (1982). Protective effect of carbohydrate or insulin on aspartate-induced neuronal necrosis in infant mice. *Fed. Proc.* 41, 396.
11. Olney, J. W. (1969). Brain lesions, obesity, and other disturbances in mice treated with monosodium glutamate. *Science* 164, 719-721.
12. Lemkey-Johnston, N., and Reynolds, W. A. (1974). Nature and extent of brain lesions in mice related to ingestion of monosodium glutamate. A light and electron microscope study. *J. Neuropathol. Exp. Neurol.* 33, 74-97.
13. Olney, J. W., and Ho, O.-L. (1970). Brain damage in infant mice following oral intake of glutamate, aspartate or cysteine. *Nature* 227, 609-610.
14. Okaniwa, A., Hori, M., Masuda, M., Takeshita, M., Hayashi, N., Wada, I., Doi, K., and Ohara, Y. (1979). Histopathological study on effects of potassium aspartate on the hypothalamus of rats. *J. Toxicol. Sci.* 4, 31-46.
15. Olney, J. W., Rhee, V., and de Gubareff, T. (1973). Neurotoxic effects of glutamate. *N. Engl. J. Med.* 289, 1374-1375.
16. Tafelski, T. J., and Lamperti, A. A. (1977). The effects of a single injection of monosodium glutamate on the reproductive neuroendocrine axis of the female hamster. *Biol. Reprod.* 17, 404-411.
17. Lamperti, A., and Blaha, G. (1976). The effects of neonatally-administered monosodium glutamate on the reproductive system of adult hamsters. *Biol. Reprod.* 14, 362-369.
18. Olney, J. W., and Sharpe, L. G. (1969). Brain lesions in an infant rhesus monkey treated with monosodium glutamate. *Science* 166, 386-388.
19. Olney, J. W., Sharpe, L. G., and Feigin, R. D. (1972). Glutamate-induced brain damage in infant primates. *J. Neuropathol. Exp. Neurol.* 31, 464-488.
20. Reynolds, W. A., Lemkey-Johnston, N., Filer, L. J., Jr., and Pitkin, R. M. (1971). Monosodium glutamate: Absence of hypothalamic lesions after ingestion by newborn primates. *Science* 172, 1342-1344.
21. Abraham, R., Dougherty, W., Golberg, L., and Coulston, F. (1971). The response of the hypothalamus to high doses of monosodium glutamate in mice and monkeys. Cytochemistry and ultrastructural study of lysosomal changes. *Exp. Mol. Pathol.* 15, 43-60.
22. Newman, A. J., Heywood, R., Palmer, A. K., Barry, D. H., Edwards, F. P., and Worden, A. N. (1973). The administration of monosodium-L-glutamate to neonatal and pregnant rhesus monkeys. *Toxicology* 1, 197-204.
23. Wen, C.-P., Hayes, K. C., and Gershoff, S. N. (1973). Effects of dietary supplementation of monosodium glutamate on infant monkeys, weanling rats and suckling mice. *Am. J. Clin. Nutr.* 26, 803-813.

24. Stegink, L. D., Reynolds, W. A., Filer, L. J., Jr., Pitkin, R. M., Boaz, D. P., and Brummel, M. C. (1975). Monosodium glutamate metabolism in the neonatal monkey. *Am. J. Physiol.* 229, 246-250.
25. Abraham, R., Swart, J., Golberg, L., and Coulston, F. (1975). Electron microscopic observations of hypothalami in neonatal rhesus monkeys (*Macaca mulatta*) after administration of monosodium-L-glutamate. *Exp. Mol. Pathol.* 23, 203-213.
26. Reynolds, W. A., Butler, V., and Lemkey-Johnston, N. (1976). Hypothalamic morphology following ingestion of aspartame or MSG in the neonatal rodent and primate: a preliminary report. *J. Toxicol. Environ. Health* 2, 471-480.
27. Reynolds, W. A., Lemkey-Johnston, N., and Stegink, L. D. (1979). Morphology of the fetal monkey hypothalamus after *in utero* exposure to monosodium glutamate. In *Glutamic Acid: Advances in Biochemistry and Physiology* (Filer, L. J., Jr., Garattini, S., Kare, M. R., Reynolds, W. A., and Wurtman, R. J., eds.), Raven Press, New York, pp. 217-229.
28. Heywood, R., and Worden, A. N. (1979). Glutamate toxicity in laboratory animals. In *Glutamic Acid: Advances in Biochemistry and Physiology* (Filer, L. J., Jr., Garattini, S., Kare, M. R., Reynolds, W. A., and Wurtman, R. J., eds.), Raven Press, New York, pp. 203-215.
29. Reynolds, W. A., Stegink, L. D., Filer, L. J., Jr., and Renn, E. (1980). Aspartame administration to the infant monkey: Hypothalamic morphology and plasma amino acid levels. *Anat. Rec.* 198, 73-85.
30. Heywood, R., and James, R. W. (1979). An attempt to induce neurotoxicity in an infant rhesus monkey with monosodium glutamate. *Toxicol. Lett.* 4, 285-286.
31. Takasaki, Y. (1978). Studies on brain lesions after administration of monosodium L-glutamate to mice. II. Absence of brain damage following administration of monosodium L-glutamate in the diet. *Toxicology* 9, 307-318.
32. Anantharaman, K. (1979). *In utero* and dietary administration of monosodium L-glutamate to mice: Reproductive performance and development in a multigeneration study. In *Glutamic Acid: Advances in Biochemistry and Physiology* (Filer, L. J., Jr., Garattini, S., Kare, M. R., Reynolds, W. A., and Wurtman, R. J., eds.), Raven Press, New York, pp. 231-253.
33. Olney, J. W., Labruyere, J., and de Gubareff, T. (1980). Brain damage in mice from voluntary ingestion of glutamate and aspartate. *Neurobehav. Toxicol.* 2, 125-129.
34. Finkelstein, M. W., Daabees, T. T., Stegink, L. D., and Applebaum, A. E. (1983). Correlation of aspartate dose, plasma dicarboxylic amino acid concentration and neuronal necrosis in infant mice. *Toxicology* 29, 109-119.
35. Applebaum, A. E. (1984). Aspartate-induced neurotoxicity in infant rodents. in *Aspartame: Physiology and Biochemistry* (Stegink, L. D., and Filer, L. J., Jr., eds.). Marcel Dekker, New York, Chap. 17.
36. Stegink, L. D., Filer, L. J., Jr., and Baker, G. L. (1972). Monosodium glutamate: Effect on plasma and breast milk amino acid levels in lactating women. *Proc. Soc. Exp. Biol. Med.* 140, 836-841.
37. Stegink, L. D., Brummel, M. C., Boaz, D. P., and Filer, L. J., Jr. (1973).

Monosodium glutamate metabolism in the neonatal pig: Conversion of administered glutamate into other metabolites in vivo. *J. Nutr.* 103, 1146-1154.
38. Stegink, L. D., Filer, L. J., Jr., and Baker, G. L. (1982). Plasma and erythrocyte amino acid levels in normal adult subjects fed a high protein meal with and without added monosodium glutamate. *J. Nutr.* 112, 1953-1960.
39. Stegink, L. D., Filer, L. J., Jr., and Baker, G. L. (1982). Effect of aspartame plus monosodium L-glutamate ingestion on plasma and erythrocyte amino acid levels in normal adult subjects fed a high protein meal. *Am. J. Clin. Nutr.* 36, 1145-1152.
40. Stegink, L. D., Filer, L. J., Jr., and Baker, G. L. (1983). Plasma amino acid concentrations in normal adults fed meals with added monosodium L-glutamate and aspartame. *J. Nutr.* 113, 1851-1860.
41. Stegink, L. D., Filer, L. J., Jr., and Baker, G. L. (1983). Interaction of aspartame and monosodium L-glutamate when ingested together as part of a low carbohydrate meal. *Fed. Proc.* 42, 544.
42. Stegink, L. D. (1976). Absorption, utilization and safety of aspartic acid. *J. Toxicol. Environ. Health* 2, 215-242.
43. Stegink, L. D. (1984). Metabolism of aspartate and glutamate. In *Aspartame: Physiology and Biochemistry* (Stegink, L. D., and Filer, L. J., Jr., eds.), Marcel Dekker, New York, Chap. 4.
44. Appendix E, Estimating Distribution of Daily Intakes of Monosodium Glutamate (MSG) (1976). In *Estimating Distribution of Daily Intakes of Certain GRAS Substances*, Committee on GRAS List Survey–Phase III, Food and Nutrition Board, Division of Biological Sciences, Assembly of Sciences, National Research Council, National Academy of Sciences, December 1976, Washington, D.C.
45. Elwyn, D. H. (1966). Distribution of amino acids between plasma and red blood cells in the dog. *Fed. Proc.* 25, 854-861.
46. Felig, P., Wahren, J., and Räf, L. (1973). Evidence of inter-organ amino-acid transport by blood cells in humans. *Proc. Nat. Acad. Sci. U.S.A.* 70, 1775-1779.
47. Aoki, T. T., Brennen, M. F., Müller, W. A., Moore, F. D., and Cahill, G. F., Jr. (1972). Effect of insulin on muscle glutamate uptake. Whole blood versus plasma glutamate analysis. *J. Clin. Invest.* 51, 2889-2894.
48. Elwyn, D. H., Launder, W. J., Parikh, H. C., and Wise, E. M., Jr. (1972). Roles of plasma and erythrocytes in interorgan transport of amino acids in dogs. *Am. J. Physiol.* 222, 1333-1342.
49. Nixon, S. E., and Mawer, G. E. (1970). The digestion and absorption of protein in man. I. The site of absorption. *Br. J. Nutr.* 24, 227-240.
50. Nixon, S. E., and Mawer, G. E. (1970). The digestion and absorption of protein in man. 2. The form in which digested protein is absorbed. *Br. J. Nutr.* 24, 241-258.
51. Schultz, S. G., Yu-Tu, L., Alvarez, O. O., and Curran, P. E. (1970). Dicarboxylic amino acid influx across brush border of rabbit ileum. *J. Gen. Physiol.* 56, 621-639.
52. Ramaswamy, K., and Radkakrishnan, A. N. (1966). Patterns of intestinal uptake and transport of amino acids in the rat. *Indian J. Biochem.* 3, 138-143.

53. Stegink, L. D., Filer, L. J., Jr., Baker, G. L., Mueller, S. M., and Wu-Rideout, M. Y.-C. (1979). Factors affecting plasma glutamate levels in normal adult subjects. In *Glutamic Acid: Advances in Biochemistry and Physiology* (Filer, L. J., Jr., Garattini, S., Kare, M. R., Reynolds, W. A., and Wurtman, R. J., eds.), Raven Press, New York, pp. 333-351.
54. Addison, J. M., Burston, D., Dalrymple, J. A., Matthews, D. M., Payne, J. W., Sleisenger, M. H., and Wilkinson, S. (1975). A common mechanism for transport of di- and tri-peptides by hamster jejunum *in vitro*. *Clin. Sci. Mol. Med.* 49, 313-322.
55. Matthews, D. M. (1984). Absorption of peptides, amino acids, and their methylated derivatives. In *Aspartame: Physiology and Biochemistry* (Stegink, L. D., and Filer, L. J., Jr., eds.), Marcel Dekker, New York, Chap. 3.
56. Stegink, L. D., Schmitt, J. L., Meyer, P. D., and Kain, P. H. (1971). Effect of diets fortified with DL-methionine on urinary and plasma methionine levels in young infants. *J. Pediatr.* 79, 648-655.
57. Vaughan, D. A., Womack, M., and McClain, P. E. (1977). Plasma free amino acid levels in human subjects after meals containing lactalbumin, heated lactalbumin, or no protein. *Am. J. Clin. Nutr.* 30, 1709-1712.
58. Stegink, L. D., Filer, L. J., Jr., and Baker, G. L. (1981). Plasma and erythrocyte concentrations of free amino acids in adult humans administered abuse doses of aspartame. *J. Toxicol. Environ. Health* 7, 291-305.
59. Stegink, L. D., Filer, L. J., Jr., and Baker, G. L. (1977). Effect of aspartame and aspartate loading upon plasma and erythrocyte free amino acid levels in normal adult volunteers. *J. Nutr.* 107, 1837-1845.
60. Hatten, D. (1980). FDA presentation. In *Aspartame: Public Board of Inquiry, Transcript of Proceedings*, Vol. 1, U.S. Department of Health, Education, and Welfare, Food and Drug Administration, pp. 58-76.
61. Anonymous (1976). *General Foods–Market Research Corporation of America, Food Consumption Study*, Food and Drug Administration, Hearing Clerk File, Administrative Record–Aspartame 75F-0355, File Vol. 103.
62. Anonymous (1980). *Evaluation of Certain Food Additives*, Twenty-fourth Report of the Joint FAO/WHO Expert Committee on Food Additives, World Health Organization Technical Report Series 653, World Health Organization, Geneva, pp. 20-21.
63. Cochran, W. G., and Cox, G. M. (1950). *Experimental Design*, Wiley, New York, p. 86.
64. Stegink, L. D., Baker, G. L., and Filer, L. J., Jr. (1983). Modulating effect of Sustagen upon plasma glutamate concentration in humans ingesting monosodium L-glutamate. *Am. J. Clin. Nutr.* 37, 194-200.
65. Stegink, L. D., Filer, L. J., Jr., and Baker, G. L. (1983). Effect of carbohydrate on plasma and erythrocyte glutamate levels in humans ingesting large doses of monosodium L-glutamate in water. *Am. J. Clin. Nutr.* 37, 961-968.
66. Daabees, T. T., Andersen, D. W., Zike, W. L., Filer, L. J., Jr., and Stegink, L. D. (1984). Effect of meal components on peripheral and portal plasma glutamate levels in young pigs administered large doses of monosodium L-glutamate. *Metabolism* 33, 58-67.

67. Anonymous (1978). Dried soup mixes (this is soup?). *Consumer Reports*, November, 615-619.
68. Stegink, L. D., Filer, L. J., Jr., and Baker, G. L. (1981). Effect of aspartame and monosodium glutamate (MSG) upon plasma amino acid levels when ingested as part of a soup-beverage meal. *Am. J. Clin. Nutr.* 34, 649.
69. Orr, M. L., and Watt, B. K. (1957). *Amino Acid Content of Foods*, Home Economics Research Report No. 4, Household Economics Research Division, Institute of Home Economics, U.S. Department of Agriculture, Agricultural Research Service, Superintendent of Documents, U.S. Government Printing Office, Washington, D.C., pp. 1-45.
70. Jukes, T. H., Holmquist, R., and Moise, H. (1975). Amino acid composition of proteins: Selection against the genetic code. *Science* 189, 50-51.

32

Aspartame Use by Persons with Diabetes

David L. Horwitz
University of Illinois Health Science Center, Chicago, Illinois

INTRODUCTION

In the United States, 2.4% of women and 1.6% of men have diabetes (1). Treatment modalities for insulin-dependent and non-insulin-dependent diabetes consist of diet, medication, and exercise. In many cases diet therapy alone is used to treat a non-insulin-dependent diabetic patient.

The most frequently restricted item in the diet of both non-insulin-dependent and insulin-dependent diabetic individuals is refined carbohydrate, or mono- and disaccharides. It is generally taught that persons with diabetes cannot consume glucose and glucose-containing disaccharides and still maintain good blood glucose control (2). Because of an innate or acquired craving for sweetness, restricting the intake of disaccharides by a person with diabetes may be difficult, and alternative sweeteners have been suggested. Mehnert questioned 500 diabetic persons and only 84 were willing to do without sweet-tasting food. Of the persons questioned, 57 used sorbitol, 8 used fructose, and 351 used cyclamate or saccharin as sweeteners (3). Court questioned 100 mothers of diabetic children and found that 72% of them considered artificially sweetened foods helpful in increasing dietary compliance (4). Achieving compliance to any dietary restriction is difficult, however. Dunbar and Stunkard estimate less than 50% adhere to their prescribed regimen (5). In order to increase adherence and satisfy the craving for sweetness while avoiding glucose and glucose-containing disaccharides, a safe and acceptable sweetener may be useful for the person with diabetes.

In reviewing the current status of saccharin and nonglucose carbohydrates (fructose, sorbitol, and xylitol) for use by individuals with diabetes, Talbot and Fisher concluded (6), "leading authorities on the dietary management of diabetes mellitus affirm that the main emphasis in obese patients should be on control of calorie intake to achieve and maintain ideal body weight." Some diabetologists believe there may be a need for a selection of foods sweetened with nonglucose carbohydrates; for instance, if nonnutritive sweeteners such as saccharin are available, then soft drinks containing a nonglucose carbohydrate sweetener might be useful "in terms of palatability and acceptability provided that their long-term influence on blood sugar levels was innocuous." This statement was written prior to the general availability of aspartame, but the same conclusion applies to all nonglucose sweeteners. Aspartame contributes far fewer calories for a given amount of sweetness than does fructose, xylitol, or sorbitol. As a result, it is likely to be even more useful, in the above context, than nonglucose carbohydrates. Before general use by persons with diabetes can be recommended, however, its effects (if any) on metabolism in diabetic individuals must be defined.

EFFECTS OF ASPARTAME ON BLOOD GLUCOSE

Although persons with diabetes are likely to be frequent users of aspartame-sweetened foods and beverages, there are few published data on any blood glucose changes in this situation. Stern et al. (7) studied 43 adults with non-insulin-dependent diabetes who were given either placebo or 1.8 g of aspartame per day, divided into capsules given with each meal, for 90 days (13 weeks). Diabetic control was classified as excellent, with a majority of fasting blood sugar (FBS) values less than 110 mg/dl; good, with a majority of FBS values over 110 but less than 130 mg/dl; or poor, with a majority of FBS values over 130 mg/dl. Prior to starting the study, seven aspartame-taking subjects showed excellent control, seven had good control, and nine had poor control. For the placebo group, the numbers of subjects showing each degree of control were 11, 3, and 6, respectively. The investigators reported that all subjects showed the same degree of control while taking capsules as during the before and after periods; however, no specific glucose values were reported for these subjects, nor were other measures of diabetic control specified.

The same investigators reported data for 26 additional subjects, given aspartame or placebo according to the same schedule, where specific glucose values were stated (7). These values are shown in Table 1. The specific method of glucose determination was not stated, nor was it mentioned whether serum, plasma, or whole blood glucose was determined. Base-line glucose values prior to aspartame or placebo administration were not stated. It appears that week-14 values are those noted on completion of the study, but the length of the aspartame-free interval is not given. Thus we do not have any true base-line values for either group. It appears that the persistently higher values in the aspartame group are

Table 1 Fasting Glucose in Subjects with Non-Insulin-Dependent Diabetes Taking Aspartame or Placebo

	Fasting glucose (mg/dl)			
Group	Week 1	Week 5	Week 9	Week 14[a]
Aspartame[b]	116 ± 14 (10)[c]	116 ± 11 (10)	136 ± 17 (9)	135 ± 21 (10)
Placebo	89 ± 9 (11)	107 ± 11 (11)	118 ± 14 (10)	96 ± 11 (11)
t	1.64	0.58	0.87	1.67
p	>0.10	>0.50	>0.40	>0.10

[a]Week 14 is presumed to be after completion of experimental period, but this point is not explicitly stated in the original paper.
[b]1.8 g/day.
[c]The number in parentheses is the number of subjects. The original paper does not state whether the value after the ± sign is the standard deviation or the standard error.
Source: From Ref. 7.

not significantly different than those in the control group, and the difference may be due to higher base-line values in the former group. Stern et al. (7) do not state whether the week-9 values are significantly different from week-1 values, and insufficient data are reported to allow this calculation to be made. In any case, the change from week 1 to week 9 is, if anything, less in the aspartame group than in the placebo group.

Both groups of subjects in the study of Stern et al. (7) appeared to tolerate aspartame well, with no effect of aspartame seen on blood pressure, pulse, weight, blood counts, liver function tests, cholesterol, triglycerides, ophthalmic examination, or electrocardiograms in the 36 subjects taking aspartame (33 of whom completed the study). Plasma phenylalanine and tyrosine levels in the fasting state did not change in either aspartame-taking or placebo-consuming subjects with non-insulin-dependent diabetes.

Paccalin et al. have reported their experiences in 177 subjects using a mean of 5 (range 2-15) aspartame tablets per day, each tablet containing 20 mg of aspartame, for 3-15 months (mean of 6 months) (8). An unspecified number of their subjects had diabetes, and the authors reported the harmlessness ("l innocuité") of aspartame on blood sugar, triglycerides, and total lipids. Specific data are not given, however. No interactions with antidiabetic treatment were noted.

Reynolds and Bauman (personal communication, 1983) have studied the effects of an acute dose of aspartame on serum glucose and insulin levels in subhuman primates. The results are shown in Table 2. It should be noted that the monkeys used in these studies were acutely stressed by restraint and sedation by phencyclidine; these stimuli can also potentially affect glucose levels. In spite of this, no important changes were seen in serum glucose or insulin levels after acute administration of aspartame.

Table 2 Acute Effects of Aspartame on Serum Glucose and Insulin in Monkeys[a,b]

	Time (min) after aspartame dosing					
	0	30	60	90	120	180
Serum glucose (mg/gl)	69 ± 11	79 ± 13	80 ± 4	68 ± 4	60 ± 4	54 ± 4
Serum insulin (μunits/ml)	11 ± 3	8 ± 1	15 ± 1	12 ± 4	8 ± 2	6 ± 2

[a]All values, mean ± S.E. (N = 3).
[b]The aspartame dose was 54 mg/kg body weight.
Source: Unpublished data were kindly furnished by W. A. Reynolds and A. Bauman.

ASPARTAME AND HORMONES OF GLUCOREGULATION

Ingestion of protein and intravenous infusion of amino acids are known to stimulate secretion of insulin, growth hormone, and glucagon. Each of these hormones can, in turn, affect the blood glucose level. Because aspartame is a dipeptide that is hydrolyzed to its amino acids in the gastrointestinal tract, it can potentially alter secretion of these hormones. The oral threshold level of each amino acid that must be reached to affect insulin secretion is not known. Most studies are based on intravenous administration of amino acids. In general, 500 mg/kg of most amino acids, given intravenously, causes substantial secretion of insulin, glucagon, and growth hormone. The expected 99% consumption level of aspartame is 34 mg/kg body weight, which would provide about 14 mg/kg body weight of aspartic acid and 17 mg/kg body weight of phenylalanine. As these amounts would probably be taken in over an 8- to 14-hour period, it is unlikely that hormone secretion would be appreciably altered.

Guttler et al. gave an oral dose of 0.6 mmol/kg body weight phenylalanine (99 mg/kg body weight) to healthy (nondiabetic) men with no family history of phenylketonuria (9). This dose raised serum phenylalanine levels to peak values of 100 μmol/dl. This compares to peak phenylalanine levels of 12 μmol/dl reported after ingestion of 34 mg/kg body weight aspartame (10). At the time of peak phenylalanine levels in the Guttler study, insulin increased from 9 to 20 μunits/ml, and glucagon from 100 to 200 pg/ml. The rise in insulin levels is less than that seen after a normal protein-containing meal, and that in glucagon similar to that seen after such a meal. Thus the smaller change in phenylalanine levels seen after typical aspartame ingestion is unlikely to affect insulin or glucagon secretion in a meaningful way. Similar data are not available for aspartic acid, or for either amino acid's effect on growth hormone secretion. Such data are also not available for diabetic individuals, where abnormal regulation of insulin, glucagon, and growth hormone secretion has been reported.

The previously cited data of Reynolds and Bauman (Table 2) also suggest no effect of aspartame on insulin secretion. These data suggest in particular that ingestion of a sweet substance, in the absence of significant calories, does not initiate centrally mediated insulin secretion.

POTENTIAL ASPARTAME TOXICITY IN PERSONS WITH DIABETES

Since its discovery in 1965, aspartame has been subject to a more extensive safety evaluation than probably any other sweetening agent, and has been found to be remarkably free of known or potential toxicity. As aspartame use by persons with diabetes is likely, it is important to determine whether this wealth of toxicity data is relevant to individuals with altered carbohydrate tolerance.

In infant rodents, parenteral or forced enteral administration of large amounts of monosodium L-glutamate or aspartate causes focal necrosis of neurons. Initially it was postulated that the aspartic acid component of aspartame could cause similar consequences. Although massive amounts of aspartame can cause lesions in infant mice, infant subhuman primates are not affected. For normal man the aspartame Public Board of Inquiry concluded that the risk of neuronal toxicity is negligible; evidence for this is reviewed elsewhere in this volume.

Takasaki and Yugari (11) demonstrated that injection of large amounts of insulin in mice (0.02 units per 4-6 g mouse, or 4 units/kg body weight) suppressed the monosodium L-glutamate-induced neuronal lesions otherwise seen in this species. The authors suggest that insulin has this protective effect because it stimulates muscle uptake of glutamate. Daabees et al. reported similar findings of a protective effect of insulin on aspartate-induced neuronal necrosis in infant mice (12). After 1.0 g/kg body weight aspartate, peak plasma aspartate levels were 333 μmol/dl, and 80 necrotic neurons per section were seen. Pretreatment with insulin (0.02 units per animal) reduced the number of necrotic neurons per section to 32, although the peak plasma aspartate level was unchanged (326 μmol/dl). Thus the mechanism for the protective effect of insulin is unclear.

The above studies (11,12) do not report the insulin levels attained, but the dose given (4 units/kg body weight) is likely to produce unphysiologically high levels. For comparison, a dose of 0.1 unit/kg body weight of insulin in man generally produces severe hypoglycemia and insulin levels similar to those seen after a meal. The significance of these findings in humans is not clear. Even modest degrees of diabetic control generally permit adequate muscle uptake of amino acids, and treated insulin-independent diabetic subjects generally have free insulin levels comparable to those seen in nondiabetic individuals (13). Non-insulin-dependent diabetic patients often have increased insulin levels. Thus it is unlikely that either diabetic or normal individuals are at risk of neuronal necrosis from aspartame.

Insulin may also increase phenylalanine hydroxylase activity, as has been reported in serum-deprived hepatoma cells (14). Therefore aspartame ingestion by insulin-deficient individuals could potentially produce higher plasma phenylalanine levels and lower tyrosine levels than are seen in normal individuals. For the same reasons as stated above for aspartate and glutamate, this potential problem, even if it were demonstrated to occur, is unlikely to be of clinical significance, and in any case its magnitude would be no greater than that seen after a typical protein-containing meal.

CONCLUSIONS

Although it is likely that many persons with diabetes will elect to use aspartame as a sweetener, few data are available on aspartame metabolism in such individuals. The few published studies related to diabetes suggest that aspartame does not alter glycemic control of diabetes. However, only crude measures of control were used and the published results do not permit critical analysis. Preliminary results from the author's laboratory (15) indicate that, following ingestion of 2.7 g of aspartame per day for 18 weeks, fasting plasma glucose and glucosylated hemoglobin levels in subjects with insulin-dependent or non-insulin-dependent diabetes did not differ from baseline values, or from values in a control group given a placebo instead of aspartame. The existing literature indicates that the amino acid content of aspartame is unlikely to affect secretion of insulin, glucagon, or growth hormone, but, again, there are no direct data on this point. Finally, while insulin has been shown to have a protective effect on glutamate- or aspartate-induced neuronal necrosis in mice, and to stimulate phenylalanine hydroxylase activity in vitro, these findings are unlikely to be of critical relevance to the potentially insulin-deficient diabetic individual. At present, studies are underway in the author's laboratory which will provide data on all of these matters, for both insulin-dependent and non-insulin-dependent diabetes.

It is our present practice to recommend limited consumption of saccharin by young persons and by pregnant women because of potential long-term carcinogenicity (16). Furthermore, the taste of saccharin is unacceptable to many persons. We also limit the use of nonglucose carbohydrate sweeteners, especially fructose, in most non-insulin-dependent diabetic patients because of their caloric value (17). The availability of aspartame may enable us to substantially liberalize the diabetic diet.

REFERENCES

1. U.S. Department of Health, Education and Welfare (1978). Diabetes data, Public Health Services, National Institutes of Health, DHEW Publication No. (NIH) 78-1468, Washington, D.C.
2. American Diabetes Association (1979). Principles of nutrition and dietary recommendations for individuals with diabetes mellitus: 1979. *Diabetes* 28: 1027-1030.
3. Mehnert, H. (1971). Über den relativen Wert von Zuckeraustauschstoffen und Susstoffen in der Diabetesdiat. *Wiss. Veroeff. Dtsch. Ges Ernaehr.* 20, 80-84.

4. Court, J. M. (1976). Diet in the management of diabetes: Why have special diet foods? *Med. J. Aust.* 1:841-843.
5. Dunbar, J. M., and Stunkard, A. J. (1979). Adherence to diet and drug regimen. In *Lipids and Coronary Heart Disease* (Levy, R., Rifkind, B., and Dennis, B., eds.), Raven Press, New York, pp. 391-423.
6. Talbot, J. M., and Fisher, K. D. (1978). The need for special foods and sugar substitutes by individuals with diabetes mellitus. *Diabetes Care* 1:231-240.
7. Stern, S. B., Bleicher, S. J., Flores, A., Gambos, G., Recitas, D., and Shu, J. (1976). Administration of aspartame in noninsulin-dependent diabetics. *J. Toxicol. Environ. Health* 2:429-439.
8. Paccalin, J., Lambert, J., and La Comere, R. (1980). L'aspartam: Une nouvelle génération d'édulcorants. *Cah. Nutr. Diet.* 15:41-48.
9. Güttler, F., Kuhl, C., Pedersen, L., and Paby, P. (1978). Effects of oral phenylalanine load on plasma glucagon, insulin, amino acid and glucose concentrations in man. *Scand. J. Clin. Lab. Invest.* 38:255-260.
10. Stegink, L. D., Koch, R., Blaskovics, M. E., Filer, L. J., Baker, G. L., and McDonnell, J. E. (1981). Plasma phenylalanine levels in phenylketonuric heterozygous and normal adults administered aspartame at 34 mg/kg body weight. *Toxicology* 20:81-90.
11. Takasaki, Y., and Yugari, Y. (1980). Protective effect of arginine, leucine and preinjection of insulin on glutamate neurotoxicity in mice. *Toxicol. Lett.* 5:39-44.
12. Daabees, T. T., Finkelstein, M., Applebaum, A. E., and Stegink, L. D. (1982). Protective effect of carbohydrate or insulin on aspartate-induced neuronal necrosis in infant mice. *Fed. Proc.* 41:396.
13. Kuzuya, H., Blix, P. M., Horwitz, D. L., Steiner, D. F., and Rubenstein, A. H. (1977). Determination of free and total insulin and C-peptide in insulin-treated diabetics. *Diabetes* 26:22-29.
14. Tourian, A. (1975). Control of phenylanine hydroxylase synthesis in tissue culture by serum and insulin. *J. Cell Physiol.* 87:15-23.
15. Bauer-Nehrling, J. K., McLane, M. P., Kobe, P. J., Kamath, S. K., and Horwitz, D. L. (1983). Evaluation of aspartame in diabetes control. *Diabetes* 32 (suppl. 1):64A (abstract).
16. Horwitz, D. L., and Bauer, J. K. (1980). How do we feed the person with diabetes? *Conn. Med.* 44:707-712.
17. Bauer, J. K., and Horwitz, D. L. (1979). New trends in diabetic diets. *Compr. Ther.* 5:12-18.

33

Effects of Acute Aspartame Ingestion on Large Neutral Amino Acids and Monoamines in Rat Brain

John D. Fernstrom
University of Pittsburgh School of Medicine, Western Psychiatric Institute and Clinic, Pittsburgh, Pennsylvania

Aspartame–aspartylphenylalanine methyl ester (APM)–was studied in vivo for its ability to influence brain levels of the large neutral amino acids (LNAAs). the rates of hydroxylation of tryptophan and tyrosine, and the ability of α-methyl dopa, an LNAA drug, to reduce blood pressure. The administration by gavage of APM (200 mg/kg body weight) caused large increments in blood and brain levels of phenylalanine and tyrosine by 60 min. Brain tryptophan level was occasionally reduced significantly, but the brain levels of the branched-chain amino acids were always unaffected. Smaller APM doses (50 and 100 mg/kg body weight) also raised blood and brain tyrosine and phenylalanine concentrations, but did not reduce brain tryptophan levels. At the highest dose (200 mg/kg body weight), APM gavage caused an insignificant increase in tyrosine hydroxylation rate, and a modest reduction in the tryptophan hydroxylation rate (indices of catecholamine and serotonin synthesis rates, respectively). No changes in the brain levels of serotonin, 5-hydroxyindoleacetic acid, dopamine, dihydroxyphenylacetic acid, homovanillic acid, or norepinephrine were produced by APM administration (200 mg/kg body weight). When hypertensive rats received α-methyldopa soon after APM gavage, no significant effect was noted on the drug's ability to reduce blood pressure. Thus these results indicate that APM, even when administered to animals in amounts that cause large increments in brain tyrosine and phenylalanine concentrations, produces minimal effects on the brain levels of other large neutral amino acids, on the rates of formation of the monoamine transmitters, and on the pharmacological potency of centrally acting LNAA drugs.

INTRODUCTION

The artificial sweetener aspartame (APM) is a phenylalanine-containing dipeptide (aspartylphenylalanine methyl ester). Although only milligram quantities of the compound are normally to be consumed with foods, the possibility exists that excessive or abuse amounts might acutely be ingested by some individuals. If so, such a "treatment" would be analogous to receiving a large dose of phenylalanine, and might therefore elicit predictable biochemical abnormalities. Of the biochemical changes that might be postulated to follow from the ingestion of substantial amounts of phenylalanine, one involves the alterations in the uptake into brain of the large neutral amino acids (LNAAs) (of which phenylalanine is one). These amino acids are transported into brain by a competitive carrier shared among all LNAAs (1,2). Consequently, if blood phenylalanine levels became very high, one might expect that the brain uptake of this amino acid would increase, while the uptakes of the other LNAAs would decrease (owing to competition). In fact, this logic has been used to explain some of the LNAA changes seen in the brains of animals with experimental phenylketonuria, or injected with phenylalanine (e.g., see Ref. 3).

A second type of change that might be seen follows from these postulated changes in brain levels of the LNAAs. Specifically, if APM administration raised brain phenylalanine (and perhaps also tyrosine; see below) levels while lowering brain tryptophan levels, the rates at which these amino acids are converted to their respective monoamine neurotransmitters might be influenced (since the rates of formation of these compounds are to different extents dependent on precursor concentrations; see Ref. 4). The formation of serotonin (5-HT) from tryptophan might be diminished if brain tryptophan levels were reduced, and the formation of the catecholamines from tyrosine and phenylalanine might be elevated or reduced, depending on whether more of the substrate stimulated (5-7) or inhibited (7) tyrosine hydroxylation.

Yet a third effect, which might be anticipated to follow from increased LNAA competition for brain uptake (secondary to the increments in blood tyrosine and phenylalanine), relates to the efficacy of centrally acting drugs that are also LNAAs. For example, α-methyldopa is an LNAA drug that must gain access to brain in order to lower blood pressure. Might the presence of high circulating levels of tyrosine and phenylalanine following APM ingestion inhibit α-methyldopa uptake into brain, and thereby compromise the drug's antihypertensive actions? Data in the literature indicate that other treatments that elevate blood levels of the natural LNAAs do have these effects on α-methyldopa action (8,9).

Because such effects of APM, if present, might ultimately influence brain function, and APM is to have wide use in the general population, it seemed important to determine whether this compound had any of these postulated effects on the brain.

MATERIALS AND METHODS

Adult male Sprague-Dawley rats (COBS, 150 g) or spontaneously hypertensive rats of the Okamoto strain were obtained from Charles River Breeding Laboratories (Wilmington, Mass.). They were housed in our animal quarters and received food (Charles River Rat, Mouse, and Hamster Maintenance Formula) and tap water ad libitum. The room was lighted 12 hr/day, and the ambient temperature was maintained at 22°C. Animals were deprived of food (but not water) overnight before an experiment the next morning. Rats were killed by decapitation, and brains were immediately removed, dissected, frozen on dry ice, and then stored at -70°C until assayed. Blood was collected from the cervical wound, and sera were stored at -20°C until assayed.

Aspartame (aspartylphenylalanine methyl ester; G. D. Searle and Company, Chicago, Ill.), phenylalanine (Sigma Chemical Company, St. Louis, Mo.), and aspartate (Sigma) were administered by gavage, as slurries in distilled water. In some experiments m-hydroxybenzylhydrazine dihydrochloride (NSD 1015, Sigma Chemical Co., St. Louis, Mo.) was injected intraperitoneally at a dose of 100 mg/kg body weight; animals were killed 30 min after NSD 1015 injection. In other experiments, α-methyldopa was injected intraperitoneally at 50 mg/kg body weight. Blood pressures were estimated via tail vein plethysmography just before and then at various intervals following drug injection.

Tryptophan (10-12), tyrosine (13), and phenylalanine (14) levels were determined fluorometrically in half brain or cerebellum samples and in blood. In some experiments the brain and serum concentrations of tyrosine, phenylalanine, and the branched-chain amino acids were determined using a Beckman model 119C Amino Acid Analyzer (Table 1). (When the analyzer was employed, it yielded values for tyrosine and phenylalanine that were lower than those obtained using the fluorometric method. However, this difference did not affect the outcome of the experiments or conclusions drawn from them. Within a given experiment, the same methodology was always employed; i.e., the analyzer or the fluorometric methods). Tissues (sera, half brains, cerebellums) were prepared for amino acid analysis as described by Fernstrom and Faller (15). Tissue dihydroxyphenylalanine, dopamine, dihydroxyphenylacetic acid, and homovanillic acid levels were measured using a Bioanalytical Systems model LC-304 high performance liquid chromatography-electrochemical detection system (Bioanalytical Systems, West Lafayette, Ind.) and a Waters μBondapak C-18 column (6). Norepinephrine levels were measured using the same methodology as that for dopamine (see Ref. 6), except that the buffer contained 0.3 mM octane sulfonate (norepinephrine, which previously eluted with the solvent front, separated and eluted at about 4.5 min). Serotonin 5-HT and 5-hydroxyindoleacetic acid (5-HIAA) levels were determined by the high-performance liquid chromatography method of Reinhard et al. (16); 5-hydroxytryptophan (5-HTP) levels were also

Table 1 Effect of Aspartame Administration on Blood and Brain Levels of the Branched-Chain Amino Acids[a]

Treatment	Tyrosine	Phenylalanine	Leucine	Isoleucine	Valine
Brain amino acid (nmol/g)					
Vehicle	41 ± 2	32 ± 2	55 ± 3	26 ± 3	61 ± 3
Aspartate	54 ± 6	40 ± 1	57 ± 3	27 ± 1	63 ± 6
Phenylalanine	102 ± 6[b]	74 ± 3[b]	57 ± 2	28 ± 1	65 ± 3
Aspartame	117 ± 6[b]	77 ± 5[b]	56 ± 2	27 ± 2	63 ± 4
Blood amino acid (nmol/ml)					
Vehicle	73 ± 6	70 ± 3	167 ± 6	110 ± 8	198 ± 12
Aspartate	70 ± 3	69 ± 4	145 ± 7	97 ± 4	187 ± 1
Phenylalanine	161 ± 8[b]	126 ± 6[b]	151 ± 4	107 ± 2	182 ± 5
Aspartame	171 ± 6[b]	120 ± 8[b]	145 ± 9	106 ± 3	187 ± 12

[a]Groups of five overnight-fasted male rats received APM (200 mg/kg body weight), vehicle, aspartate, or phenylalanine (doses equimolar with that of APM) per os, and were killed 60 min later. Data are presented as the means ± SE.
[b]Compared to vehicle values, $p < 0.01$ (analysis of variance/Newman-Keuls test).

determined using this methodology (16), except that methanol was deleted from the mobile phase.

Data were analyzed by analysis of variance, followed by the Newman-Keuls test (17), or in some cases simply by Student's t test.

RESULTS

In initial experiments, we studied the time course and dose-response effects of APM on serum and brain levels of the aromatic amino acids. Using the highest dose (200 mg/kg body weight), we noted that the peak effects on these amino acids occurred 30-60 min after intubation (Table 2). The treatment elevated blood and brain levels of both tyrosine and phenylalanine by similar amounts. Brain tryptophan was significantly reduced 30 min (Table 2) and occasionally 60 min (Table 3) postintubation.

In dose-response studies (Table 4), an APM dose of 100 mg/kg body weight also elevated blood and brain levels of both tyrosine and phenylalanine; the 50-mg/kg body weight dose caused significant increments in blood phenylalanine and brain tyrosine levels only (Newman-Keuls test). In this particular experiment, APM had no dose-related effects on blood or brain levels of tryptophan (Table 4).

The 200-mg/kg body weight dose of APM was also tested for its effects on the blood and brain levels of the branched-chain amino acids (Table 1). Although aspartame administration raised brain (and blood) levels of both tyrosine and phenylalanine in these experiments, it had no effect on brain (or blood) levels of

Table 2 Time Course of the Effects of Aspartame on Blood and Brain Levels of the Aromatic Amino Acids[a]

	Time after aspartame ingestion		
Parameter	30 min	60 min	120 min
Phenylalanine			
Serum (nmol/ml)			
Control	116 ± 2	–	127 ± 7
Aspartame	194 ± 8[b]	176 ± 5[c]	151 ± 2[b]
Brain (nmol/g)			
Control	94 ± 8	–	89 ± 2
Aspartame	116 ± 4[b]	147 ± 7[c]	96 ± 4
Tyrosine			
Serum (nmol/ml)			
Control	91 ± 6	–	101 ± 4
Aspartame	176 ± 7[b]	180 ± 9[c]	150 ± 8[b]
Brain (nmol/g)			
Control	169 ± 8	–	153 ± 5
Aspartame	224 ± 7[b]	234 ± 8[c]	214 ± 8[b]
Tryptophan			
Serum (nmol/ml)			
Control	94 ± 3	–	128 ± 5
Aspartame	110 ± 5	96 ± 5	114 ± 3
Brain (nmol/g)			
Control	19 ± 1	–	15 ± 1
Aspartame	14 ± 1[b]	16 ± 1[c]	15 ± 1

[a]Groups of seven male rats, fasted overnight, received APM (200 mg/kg body weight, per os) or its vehicle the next morning. They were killed at the indicated times thereafter. Data are presented as the means ± SE.
[b]Significantly different from respective control, $p < 0.05$.
[c]Significantly different from 30-min control value, $p < 0.05$ (Student's t test).

Table 3 Effect of Aspartame on the Brain Levels of 5-Hydroxyindoles[a]

Parameter	Vehicle	Aspartame
Phenylalanine (nmol/g)	137 ± 10	179 ± 12[b]
Tryptophan (nmol/g)	32 ± 1	25 ± 2[b]
Serotonin (ng/g)	514 ± 22	536 ± 24
5-HIAA (ng/g)	349 ± 27	316 ± 17

[a]Groups of six male rats, fasted overnight, received APM (200 mg/kg, per os), and were killed 60 min later. Data are presented as means ± SE.
[b]$p < 0.05$ differs from vehicle group (Student's t test).

Table 4 Dose Response of the Effects of Aspartame on Blood and Brain Levels of the Aromatic Amino Acids[a]

Parameter	Dose of aspartame (mg/kg body weight)			
	0	50	100	200
Phenylalanine				
Serum[b] (nmol/ml)	56 ± 3	72 ± 6	78 ± 4	92 ± 2
Brain[b] (nmol/g)	129 ± 6	132 ± 4	157 ± 5	159 ± 7
Tyrosine				
Serum[b] (nmol/ml)	74 ± 4	92 ± 6	119 ± 9	177 ± 16
Brain[b] (nmol/g)	106 ± 4	133 ± 4	159 ± 4	190 ± 10
Tryptophan				
Serum (nmol/ml)	68 ± 5	92 ± 6	83 ± 6	70 ± 6
Brain (nmol/g)	23 ± 1	24 ± 1	25 ± 1	24 ± 1

[a]Groups of six male rats, fasted overnight, received APM (at the indicated dose, per os) or its vehicle the next morning. They were killed 60 min thereafter. Data are presented as the means ± SE.
[b]A significant ($p < 0.01$) effect of dose was noted for this variable (analysis of variance).

leucine, isoleucine, or valine. This result is identical to that obtained when an equimolar dose of phenylalanine itself was administered by gavage (Table 1); that is, blood and brain levels of both tyrosine and phenylalanine rose, but those of the branched-chain amino acids were unchanged. Table 1 also shows that the administration of aspartate, the other amino acid moiety of APM, had no effects on brain or blood levels of any of the aromatic or branched-chain amino acids.

Since APM affected brain levels of phenylalanine, tyrosine, and occasionally tryptophan, we next studied whether these changes influenced the rate of tyrosine or tryptophan hydroxylation in brain. Data are presented in Table 5 for a dose-response experiment. As before, APM elevated brain tyrosine, but in this experiment the only significant rise occurred at the highest dose tested (200 mg/kg body weight). At this high dose the increment in brain tyrosine level was associated with only an insignificant increment in tyrosine hydroxylation rate (as measured here via dopa accumulation). Administration of APM also had minimal effect on tryptophan hydroxylation (as measured here by 5-HTP accumulation). No significant reductions in 5-HTP accumulation were noted at any dose of APM tested (though a trend was evident). However, in this study no significant reduction in brain tryptophan occurred either. In some other, similar studies in which brain tryptophan level *was* significantly reduced (at the 200-mg/kg body weight dose), a significant reduction in 5-HTP accumulation was noted (data not shown). However, such effects were not a consistent finding, even at this high dose.

In related studies we also examined the effect of aspartame administration on brain levels of serotonin, 5-HIAA, dopamine, the major brain metabolites of

Table 5 Dose Response of the Effects of Aspartame on Brain Levels of Aromatic Amino Acids and Their Hydroxylation Rates[a]

	Dose of aspartame (mg/kg body weight)			
Parameter	0	50	100	200
Tyrosine (nmol/g)	140 ± 9	164 ± 4	176 ± 8	229 ± 14[b]
Dopa (ng/g)	677 ± 35	627 ± 110	639 ± 62	763 ± 67
Tryptophan (nmol/g)	33 ± 1	36 ± 4	30 ± 2	30 ± 1
5-HTP (ng/g)	161 ± 8	165 ± 10	146 ± 6	138 ± 13

[a]Groups of six male rats, fasted overnight, received APM (at the indicated dose, per os) or its vehicle the next morning. Thirty minutes later, they were given an intraperitoneal injection of NSD 1015 (100 mg/kg body weight) and were killed 30 min thereafter. Dopa was determined in the corpora striata, and 5-HTP in whole brain less these structures and the cerebellums. Tryptophan and tyrosine were measured in cerebellums. Data are presented as means ± SE.
[b]Compared to vehicle values, $p < 0.05$ (analysis of variance Newman-Keuls test).

Table 6 Effect of Aspartame on Brain Catecholamines[a]

Parameter	Vehicle	Aspartame
Experiment A		
Phenylalanine (nmol/g)	98 ± 3	135 ± 3[b]
Tyrosine (nmol/g)	102 ± 6	224 ± 10[b]
Dopamine (ng/g)	920 ± 27	976 ± 21
Norepinephrine (ng/g)	393 ± 28	349 ± 17
Experiment B		
Phenylalanine (nmol/g)	98 ± 5	125 ± 5[b]
Tyrosine (nmol/g)	118 ± 3	178 ± 3[b]
Dopamine (ng/g)	1142 ± 19	1129 ± 43
Dihydroxyphenylacetic acid (ng/g)	87 ± 15	74 ± 14
Homovanillic acid (ng/g)	82 ± 10	80 ± 4

[a]Groups of six male rats, fasted overnight, received APM (200 mg/kg body weight, per os), and were killed 60 min later. Data are presented as means ± SE.
[b]$p < 0.05$, differs from vehicle group (Student's t test).

dopamine, and norepinephrine. As indicated in Table 3, although APM gavage elevated brain phenylalanine and lowered brain tryptophan, it did not alter the levels of 5-HT and 5-HIAA. In other experiments this same treatment was noted also to have no effects on the brain levels of dopamine or its major metabolites dihydroxyphenylacetic acid and homovanillic acid (Table 6). Similarly, brain levels of norepinephrine were unaffected by APM administration (Table 6).

Table 7 Time Course of the Effects of Aspartame on Blood Pressure in Hypertensive Rats Injected with α-Methyldopa[a]

Group	Change in blood pressure (mmHg) after		
	45 min	75 min	90 min
Experiment 1			
Control	33 ± 7	–	36 ± 4
Aspartame	21 ± 2	–	26 ± 4
Experiment II			
Control	–	30 ± 9	–
Aspartame	–	42 ± 9	–

[a]Groups of five to six male, spontaneously hypertensive rats, fasted overnight, received APM (200 mg/kg body weight, per os) or its vehicle (water). Fifteen minutes later all animals received an intraperitoneal injection of α-methyldopa (50 mg/kg body weight). Blood pressures were then measured at the indicated times after methyldopa injection. Data are presented as the means ± SE; pretreatment blood pressures ranged between 170 and 235 mmHg. By analysis of variance, no effect of treatment (APM versus vehicle) was observed.

Finally, experiments were conducted in spontaneously hypertensive rats to determine if APM administration would inhibit the ability of a centrally acting LNAA drug, α-methyldopa, to lower blood pressure. Data from two studies are presented in Table 7 and show that when APM (200 mg/kg body weight) was administered just prior to α-methyldopa, aspartame caused no significant alterations in the drug's antihypertensive efficacy.

DISCUSSION

These data show that administration of the artificial sweetener aspartame, while producing substantial elevations in both blood and brain levels of tyrosine and phenylalanine, causes only modest changes in the brain level of tryptophan, and no significant alterations in the hydroxylation rates of tryptophan or tyrosine, or in the levels of 5-HT, dopamine, norepinephrine, or a number of their metabolites. In addition, APM was without significant effects on the brain levels of the branched-chain amino acids, and appeared not to alter the uptake into brain of an LNAA drug, methyldopa, as evidenced by APM causing no change in methyldopa's antihypertensive efficacy. The dominant effects on blood and brain levels of tyrosine and phenylalanine appear attributable to the phenylalanine moiety of the APM molecule, inasmuch as injection of phenylalanine at a dose equimolar to that of APM produced effects indistinguishable from those seen with the sweetener (Table 1). Tyrosine levels in blood probably rose in response to APM (or phenylalanine), because a significant amount of the phenylalanine was hydroxylated in the liver (18). Brain levels of tyrosine may have increased for two reasons. First, the substantial rise in its blood levels may have allowed it to compete more effectively for transport into brain via the competitive uptake mechanism (even though blood phenylalanine levels were also increased). Second, phenylalanine is a substrate for tyrosine hydroxylase in brain (7). Consequently, the large increments in brain levels of phenylalanine may have enhanced this amino acid's conversion to tyrosine. Regardless of mechanism, the effects noted here on brain tyrosine and phenylalanine levels are largely consistent with other published data on the effects of administering large doses of phenylalanine (1000 mg/kg body weight) on these amino acids in brain (see, e.g., Refs. 3 and 19).

The occasional decrease in brain tryptophan level probably followed from reduced uptake of this amino acid into brain, secondary to the large increments in blood levels of tyrosine and phenylalanine. These increases would be expected to inhibit tryptophan transport via the competitive uptake carrier into brain (1-3). Since the reduction was only occasional, and always small, the competitive inhibition offered by the increased blood levels of phenylalanine and tyrosine was probably not substantial at the doses of APM studied. Compatible with this view is the finding that APM treatment produced only very small decrements in a good predictor of brain tryptophan uptake, the "serum tryptophan ratio" (a ratio of the blood concentration of tryptophan to the sum of the serum levels of the other aromatic and branched-chain amino acids—APM ingestion lowers this ratio by increasing the denominator; data not shown). The serum ratio has been shown in other, similar studies to predict accurately a change in brain tryptophan level (e.g., see Refs. 4 and 15). In the present experiments the changes in the tryptophan ratio were only a fraction of those found in other studies to be associated with large changes in brain tryptophan level (15). This finding would thus predict that the change in brain tryptophan produced by APM should have been modest at best, which was the observed result. Similarly, the uptake of each of the other aromatic and branched-chain amino acids can be predicted by a similar serum ratio (15). For example, to calculate this value for valine, the serum level of valine would be divided by the sum of the serum concentrations of tryptophan, the aromatic amino acids, and the other branched-chain amino acids. When the ratio for each branched-chain amino acid was calculated in the present studies, only very small reductions were noted (data not shown). Compared with data from other acute studies, in which large reductions in the brain levels of each branched-chain amino acid were produced (15), the reductions in the serum ratios that occurred in the present experiments should not have been large enough to elicit reductions in the brain level of each branched-chain amino acid. This finding is thus also consistent with the absence of significant changes in the brain levels of the branched-chain amino acids. Hence, overall, it seems that the increments in the blood levels of tyrosine and phenylalanine produced by APM ingestion were sufficient to enhance their own uptake into brain, but not to diminish consistently the uptakes of the other large neutral amino acids.

Aspartame administration, despite eliciting increments in brain levels of both tyrosine and phenylalanine, did not stimulate tyrosine hydroxylation rate significantly (Table 5). This lack of effect may simply reflect the finding that brain tyrosine levels must increase by at least twofold before stimulation of hydroxylation rate becomes apparent in normal rats (e.g., see Ref. 5); such did not occur at any APM dose tested (Table 5). In the absence of any significant alteration in tyrosine hydroxylation rate produced by aspartame, it is not surprising that no alterations were noted in the brain levels of dopamine or norepinephrine, or in the striatal levels of dihydroxyphenylacetic acid and homovanillic acid, two metabolites whose levels have been noted to rise under conditions in which a tyrosine-induced stimulation of dopamine synthesis was suspected (e.g., see Refs. 20 and 21).

The tryptophan hydroxylation rate also did not decline significantly following APM administration, though a trend was evident (Table 5). Perhaps a consistent effect would ultimately have been observed at a dose much higher than 200 mg/kg body weight, if one assumes that the brain tryptophan level would have shown a larger decrement. Such would indeed be expected based on other work involving the use of LNAAs to reduce brain levels of tryptophan and its rate of hydroxylation (e.g., Ref. 22). The fact that the rise in brain phenylalanine level itself did not inhibit tryptophan hydroxylation rate is also not surprising, given the increments in brain phenylalanine produced by APM gavage. The K_i of tryptophan hydroxylase for phenylalanine is reported to be about 210 nmol/g (3×10^{-4}M) (23). Brain phenylalanine did not approach this level, even in rats receiving 200 mg/kg body weight APM. Hence, overall, even at an APM dose of 200 mg/kg body weight, anticipated effects on the rate of conversion of tryptophan to 5-HT were not realized, most likely because this dose did not elevate blood tyrosine and phenylalanine sufficiently to (a) cause serious inhibition of tryptophan uptake into brain or (b) increase brain phenylalanine level enough to inhibit directly tryptophan hydroxylase.

In formulating possible effects of APM ingestion on LNAA uptake into brain, we posited that the sweetener-induced increments in blood phenylalanine and tyrosine should inhibit the brain uptakes of not just the natural LNAAs, but also of unnatural LNAAs which might be employed as centrally acting drugs. One important example is α-methyldopa (Aldomet). α-Methyldopa is a major antihypertensive agent; its uptake into brain appears to occur via the same transport mechanism shared among the natural LNAAs (8). Consequently, treatments that elevate blood levels of the natural LNAAs, should (and do) inhibit α-methyldopa access to brain (8,9). Since α-methyldopa must gain access to brain to lower blood pressure (24), these same treatments should (and do) also diminish the antihypertensive response to α-methyldopa (8,9). We therefore thought that perhaps, since APM elevates blood levels of two of the natural LNAAs, its administration would also diminish α-methyldopa efficacy. However, as indicated in Table 7, such appears *not* to be the case: α-Methyldopa administration was found to be equally efficacious in lowering blood pressure in hypertensive rats receiving either APM or its vehicle. In retrospect, perhaps if a much smaller dose of α-methyldopa (or a larger dose of APM) had been employed, a small effect of the sweetener might have been discerned. However, since 50 mg/kg body weight of α-methyldopa is within the normal antihypertensive dose range in rats, a modest effect of APM seen with a much smaller dose of α-methyldopa would probably import very little therapeutic significance. A similar conclusion could be drawn were a small effect to be seen at a much higher APM dose.

The present studies were undertaken to determine if the ingestion of APM, by elevating blood levels of phenylalanine, would elicit any of the changes in blood and brain known to result from phenylalanine loading. Some changes were noted, but, by and large, they were pharmacologically modest. No doubt greater effects could have been elicited had a dose 10 times larger been employed. The effects of phenylalanine in rats have most often been reported for doses in the range of 1000 mg/kg body weight (see Refs. 3,19, and 22); to achieve this phenylalanine dosage using APM, the dose of the sweetener would have had to be 2000 mg/kg body weight. However, the highest dose employed, 200 mg/kg body weight, produced blood levels of phenylalanine in rats comparable to those achieved in humans receiving a dose projected to represent the 99th percentile of intake when APM replaces sucrose in the diet (34 mg/kg body weight, a very high dose) (25). Hence it seemed that no additional information of relevance to human consumption would be obtained by testing extremely high doses in rats. Overall, the data from the present studies suggest that the phenylalanine contributed to the body following APM ingestion (50-200 mg/kg body weight) does not elicit profound chemical effects on the brain. It therefore seems likely that in the dose range projected to be consumed in the human population, negligible effects can be anticipated.

ACKNOWLEDGMENTS

These studies were supported by a grant from G. D. Searle and Company. The author is the recipient of a Research Scientist Development Award from the National Institute of Mental Health (MHOO254).

REFERENCES

1. Pardridge, W. M. (1977). Regulation of amino acid availability to the brain. In *Nutrition and the Brain*, Vol. 1 (Wurtman, R. J., and Wurtman, J. J., eds.), Raven Press, New York, pp. 141-204.
2. Oldendorf, W. H. (1971). Brain uptake of radiolabelled amino acids, amines, and hexoses after arterial injection. *Am. J. Physiol.* 221, 1629-1639.
3. McKean, C. M., Boggs, D. E., and Peterson, N. A. (1968). The influence of high phenylalanine and tyrosine on the concentrations of essential amino acids in brain. *J. Neurochem.* 15, 235-241.
4. Fernstrom, J. D. (1983). Role of precursor availability in control of monoamine biosynthesis in brain. *Physiol. Rev.* 63, 484-546.

5. Wurtman, R. J., Larin, F., Mostafapour, S., and Fernstrom, J. D. (1974). Brain catechol synthesis: Control by brain tyrosine concentration. *Science* 185, 183-184.
6. Sved, A. F., and Fernstrom, J. D. (1981). Tyrosine availability and dopamine synthesis in the striatum: Studies with gammabutyrolactone. *Life Sci.* 29, 743-748.
7. Kaufman, S. (1974). Properties of pterin-dependent aromatic amino acid hydroxylases. In *Aromatic Amino Acids in the Brain* (Wolstenholme, G. E. W., and FitzSimons, D. W., eds.) Elsevier-North Holland, Amsterdam, pp. 85-108.
8. Markovitz, D. C., and Fernstrom, J. D. (1977). Diet and uptake of Aldomet by brain: Competition with natural, large neutral amino acids. *Science* 197, 1014-1015.
9. Sved, A. F., Goldberg, I. M., and Fernstrom, J. D. (1980). Dietary protein intake influences the antihypertensive potency of methyldopa in spontaneously hypertensive rats. *J. Pharmacol. Exp. Ther.* 214, 147-151.
10. Denckla, W. D., and Dewey, H. K. (1967). The determination of tryptophan in plasma, liver, and urine. *J. Lab. Clin. Med.* 69, 160-169.
11. Bloxam, D. L., and Warren, W. H. (1974). Error in the determination of tryptophan by the method of Denckla and Dewey, a revised procedure. *Anal. Biochem.* 60, 621-625.
12. Lehmann, J. (1971). Light–A source of error in the fluorimetric determination of tryptophan. *Scand. J. Lab. Clin. Invest.* 28, 49-55.
13. Waalkes, T. P., and Udenfriend, S. (1957). A fluorimetric method for the estimation of tyrosine in plasma and tissues. *J. Lab. Clin. Med.* 50, 733-736.
14. McCaman, M. W., and Robins, E. (1962). Fluorimetric method for the determination of phenylalanine in serum. *J. Lab. Clin. Med.* 59, 885-890.
15. Fernstrom, J. D., and Faller, D. V. (1978). Neutral amino acids in the brain: Changes in response to food ingestion. *J. Neurochem.* 30, 1531-1538.
16. Reinhard, J. F., Moskowitz, M. A., Sved, A. F., and Fernstrom, J. D. (1980). A simple, sensitive, and reliable assay for serotonin and 5-HIAA in brain tissue using HPLC with electrochemical detection. *Life Sci.* 27, 905-911.
17. Zivin, J. A., and Bartko, J. J. (1976). Statistics for disinterested scientists. *Life Sci.* 18, 15-26.
18. White, A., Handler, P., Smith, E. L., Hill, R. L., and Lehman, I. R. (1978). *Principles of Biochemistry*, 6th ed., McGraw-Hill, New York, p. 740.
19. Carver, M. J. (1965). Influence of phenylalanine administration on the free amino acids of brain and liver in the rat. *J. Neurochem.* 12, 45-50.
20. Sved, A. F., Fernstrom, J. D., and Wurtman, R. J. (1979). Tyrosine administration decreases serum prolactin levels in chronically-reserpinized rats. *Life Sci.* 25, 1293-1300.
21. Fuller, R. W., and Snoddy, H. D. (1982). L-Tyrosine enhancement of the elevation of 3,4-dihydroxyphenylacetic acid concentrations in rat brain by spiperone and amfonelic acid. *J. Pharm. Pharmacol.* 34, 117-118.
22. Carlsson, A., and Lindqvist, M. (1978). Dependence of 5-HT and catecholamine synthesis on concentrations of precursor amino acids in rat brain. *Naunyn-Schmiedebergs Arch. Pharmacol.* 303, 157-164.

23. Lovenberg, W., Jequier, E., and Sjoerdsma, A. (1968). Tryptophan hydroxylation in mammalian systems. *Adv. Pharmacol.* 6A, 21-36.
24. VanZwieten, P. A. (1975). Antihypertensive drugs with a central action. *Prog. Pharmacol.* 1, 1-63.
25. Stegink, L. D., Filer, L. J., Jr., and Baker, G. L. (1977). Effect of aspartame and aspartate loading upon plasma and erythrocyte free amino acid levels in normal adult volunteers. *J. Nutr.* 107, 1837-1845.

Index